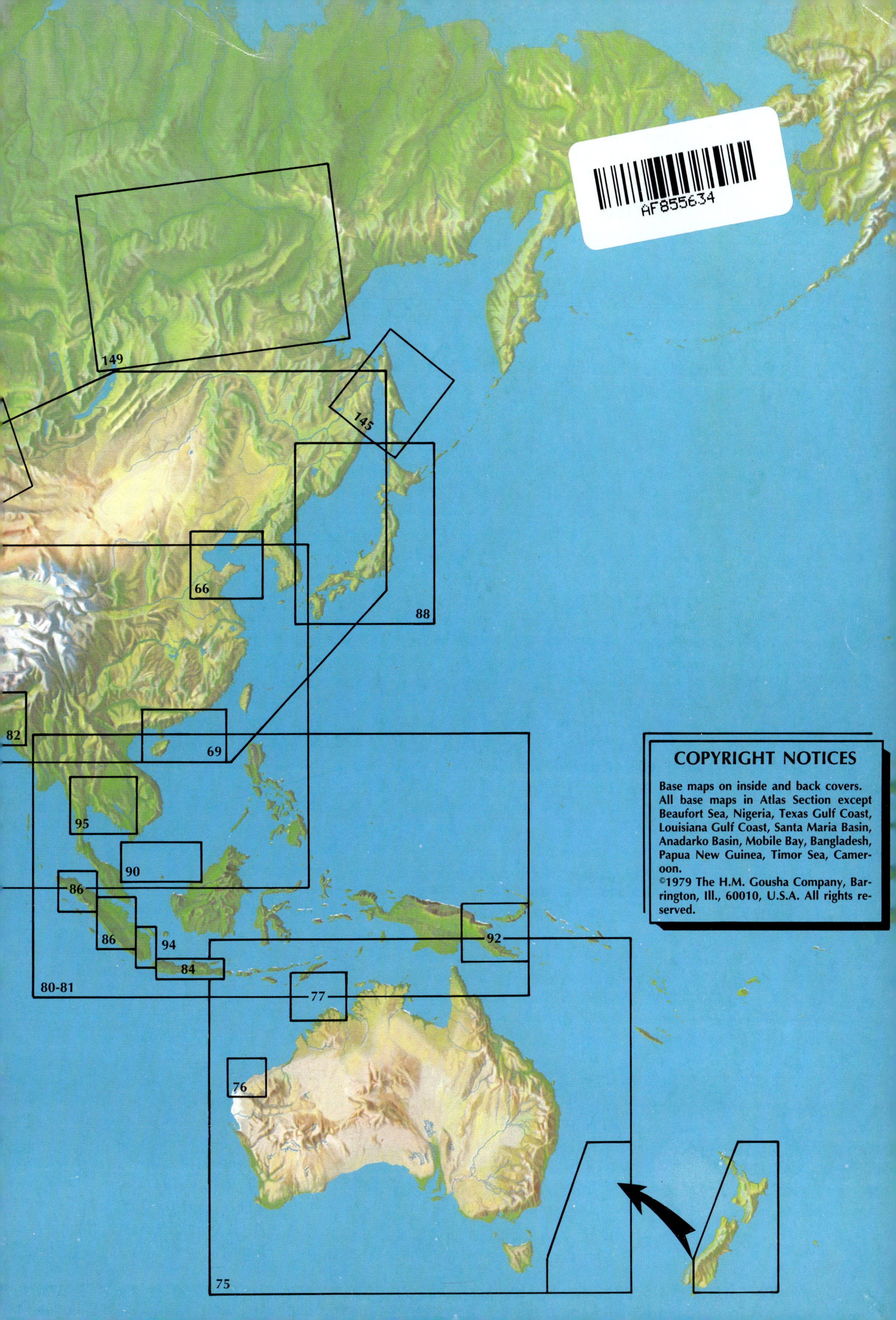
149
145
88
66
82
69
95
90
86
86
94
84
92
80-81
77
76
75

International Petroleum Encyclopedia

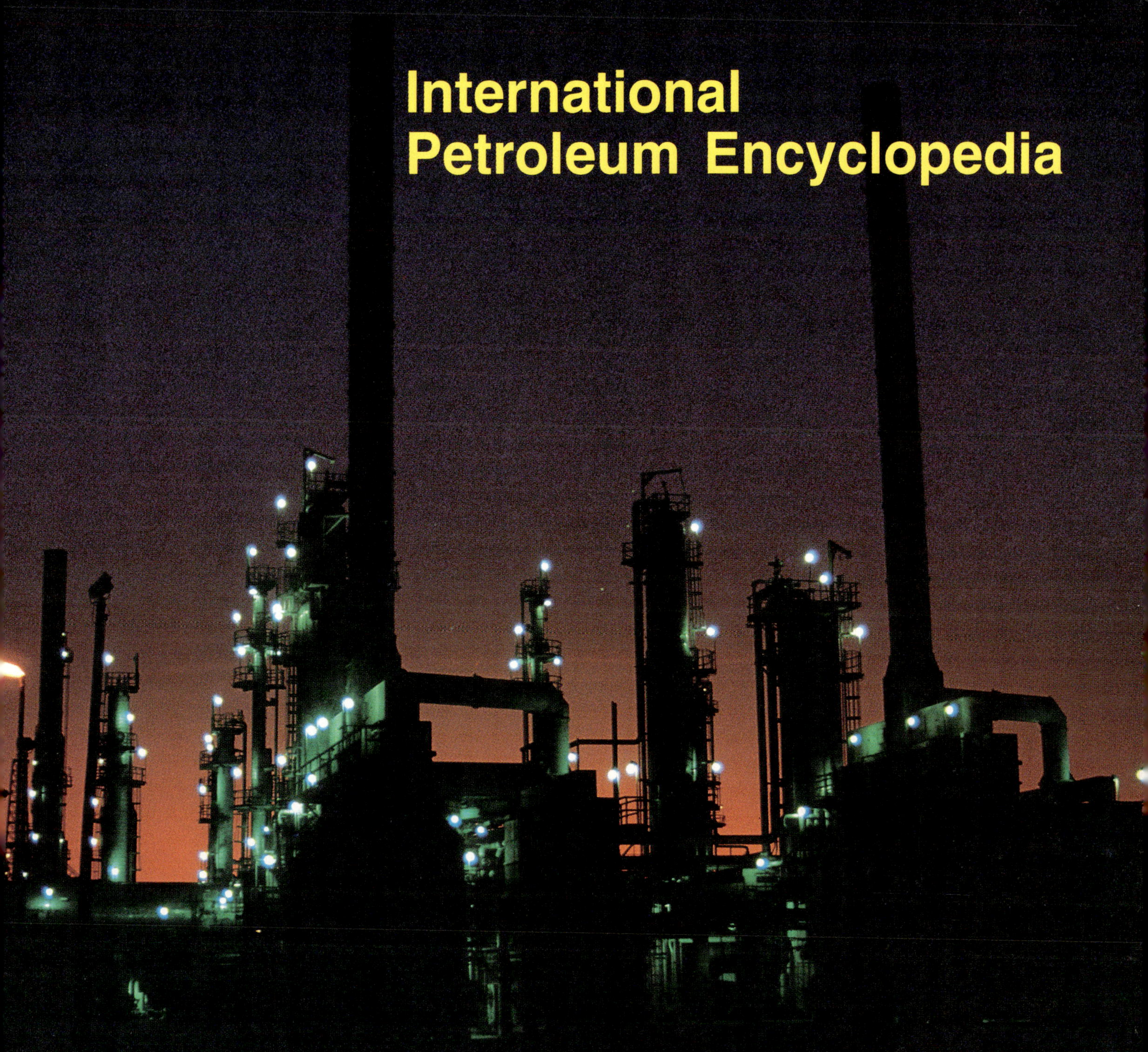

Library of Congress
CIP 77-76966
ISBN 0-87814-420-X
Printed in U.S.A.

TABLE OF CONTENTS

ATIN AMERICA

FRICA

EUROPE

SPECIAL FEATURES

International Petroleum Encyclopedia

International Petroleum Encyclopedia is published annually by PennWell Publishing Co., 1421 S. Sheridan Rd., Box 1260, Tulsa, Okla. 74101. Phone: (918) 835-3161.
Telex: 20-3604.
Fax: (918) 831-9555.

Asia-Pacific demand to fuel petroleum industry growth

The booming Asia-Pacific region's seemingly unslakeable thirst for energy may well prove to be the key to the future health of the world petroleum industry.

Under all forecasts, the surging economies of this region will fuel whatever significant growth in oil and gas demand the world will see heading into the 21st century.

The rapidly industrializing Asia-Pacific region has become an economic powerhouse to be reckoned with on its own. No longer is its economy tied to the fortunes of the more mature economies of Europe and the U.S.

During the most recent recession in the industrialized West, the Asia-Pacific region showed it can grow on its own. While Europe and the U.S. slumped, major oil consuming countries east of Suez expanded their economies by 5-6%/year, more than 1½ times average growth worldwide.

Nor is Japan still the focus of the region's projected economic growth. Most analysts see the Asia-Pacific region outside Japan as the world's primary growth market for oil demand.

East-West Center, Honolulu, in 1993 projected Asia-Pacific average oil demand growth at 3.6%/year through 1999, almost double the volume most analysts expect for oil demand growth worldwide. Analysts point out that a moderation in regional oil demand that had been anticipated in 1992 did not occur. The region's demand jumped by 862,000 b/d to 14.5 million b/d in 1992.

Despite projections that oil's share of Asia-Pacific's energy mix will continue to decline, East-West Center estimates the region's oil consumption will climb to 16.4 million b/d in 1995 and 19.1 million b/d by 2000.

That view is not universal, however. Energy Security Analysis Inc. (ESAI), Washington, D.C., in 1993 estimated demand in Southeast Asia will grow at an average 500,000 b/d/year, about half the growth rate predicted by other analysts. ESAI's pessimism hinges on doubts about the potential for contintued skyrocketing oil consumption in China.

According to the U.S. unit of Australia's BHP Petroleum Pty. Ltd., Asia's total energy market will double in size the next 20 years, led by a boom in natural gas trade.

Yukong Ltd's 490,000 b/d Ulsan, South Korea, refinery helps feed voracious consumer demand for petroleum products in one of the pacesetters of the Asia-Pacific region.

How Far East gas pipeline grid* would look

*Asian Energy Community Pipeline System in 2020
Source: BHP Petroleum (Americas) Inc.

How East Asian economic growth compares

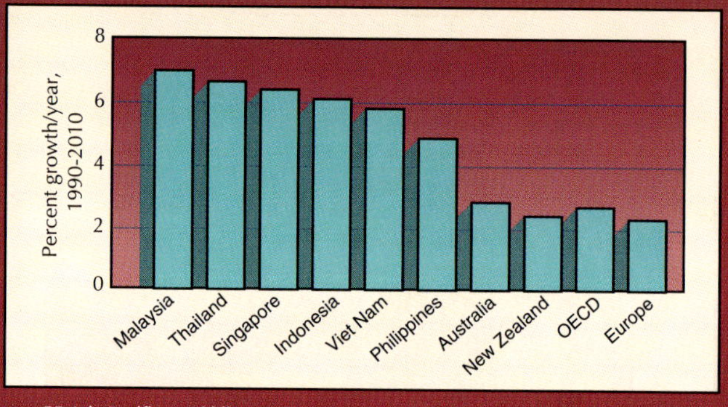

Source: BP Asia-Pacific & Middle East

Oil's share of that market will shrink to 42% from 51% as environmental concerns focus on increasing the role of natural gas in Asia's energy mix.

At the same time, any discussion of the world trade in natural gas, especially in liquefied natural gas, must begin with Asia-Pacific, from a supply as well as a demand standpoint.

Economic strength in Asia-Pacific

Joint ventures between oil producing countries and oil marketers are boosting the flow of investment into the Asia-Pacific region's oil consuming sector, Caltex Petroleum Corp. Chairman Patrick J. Ward told a Dubai petroleum conference in 1993.

Malaysia, Thailand, Indonesia, Singapore, and other countries in Southeast and South Asia have received large amounts of capital from the flow of trade into Japan and newly industrialized nations of the region. High domestic savings rates further help capital formation.

The growth stimulates oil demand throughout the region, even in Japan, where energy use efficiency gains and fuel switching cut oil use in the early 1980s. Since 1987, Japanese oil demand has risen at the rate of 5%/year, adding 900,000 b/d to the country's requirements. Oil accounts for 55-60% of total energy supply in Japan.

In South Korea, the economy in recent years has grown 8%/year, while oil demand has soared 20%/year. Ward expects South Korean oil demand growth to steady at about 7%/year during the next 3 years. Vehicle registrations in South Korea during 1987-92 have increased 22%/year, while gasoline demand has jumped 29%/year.

Perhaps the fastest growing economy in the region is Thailand's, averaging more than 10%/year since 19986.

"Thailand has created a very favorable climate for foreign investment, and the capital influx has been driving economic growth at a very high rate," Ward said.

He expects a new Thai government to maintain the high growth policies of its predecessor. Thai oil consumption has risen at the rate of 14%/year since 1987, gasoline use more than 10%/year. Lack of roads and highways is a major constraint in Thailand, but the government is expanding the transportation system.

East Asia focus of growth

East Asia in particular is the locomotive pulling this growth juggernaut. Looking at those 10 countries—China, Hong Kong, Indonesia, Malaysia, Philippines, Singapore, South Korea, Taiwan, Thailand, and Viet Nam—it's not difficult to see why.

Those nations hold 1.55 billion people, or about 30% of the world's population. Several of them—Taiwan, Singapore, Hong Kong, and South Korea—are well along the path of burgeoning industrialization and soaring economic growth fed by feverish consumer demand. Their growth has been so startling in recent years that geopolitical analysts and economists have had to devise a new category to distinguish them from industrialized and developing nations: "newly industrialized countries."

Yet those four account for less than 5% of the region's population. A much bigger population contingent is fast on the heels of the so-called "Four Tigers" as the economies of Thailand, Malaysia, and Indonesia continue to show growth spurts and a shift to heavy industry and exports. Since 1987, gross domestic product (GDP) has risen 11%/year in Thailand and 9% in Malaysia and Indonesia.

Waiting in the wings, however, are the still heavily agrarian giants: Philippines, Viet Nam, and, biggest of them all, China.

China's GDP has jumped 6%/year since 1987.

One intriguing development overlooked by many in 1993 is that the widely accepted economic triumvirate of the U.S., Japan, and Germany is no more. China has overtaken

Asia-Pacific energy demand*

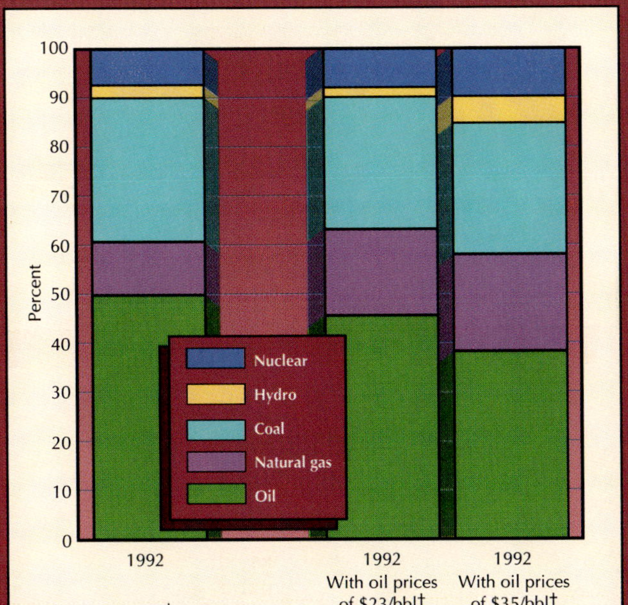

*Excluding China. †1992 dollars
Source: East-West Center

Asia-Pacific oil demand

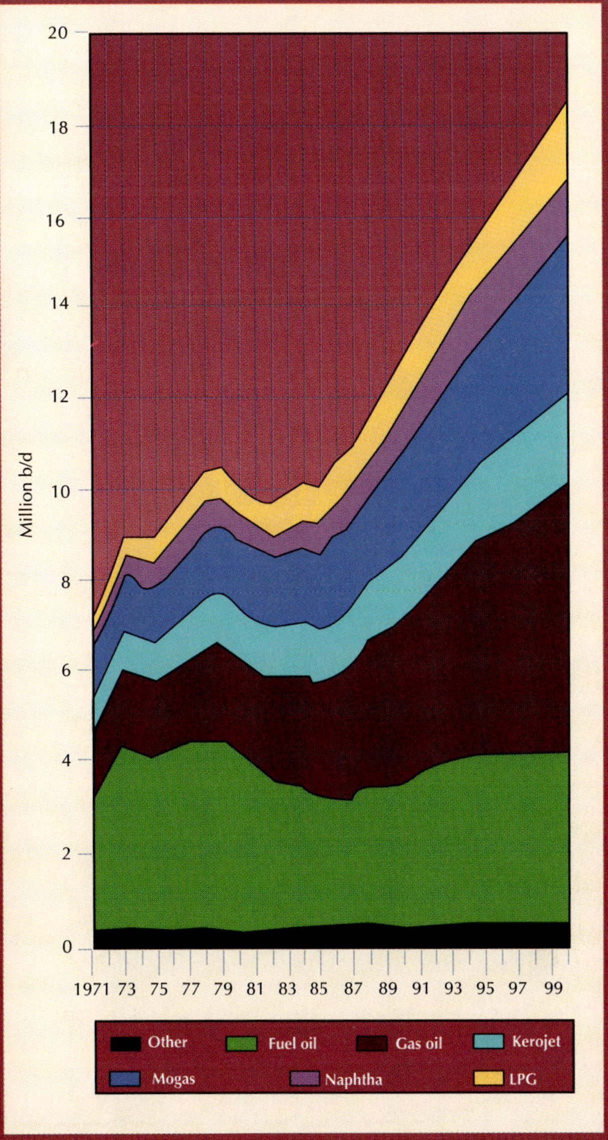

*Base case Source: East-West Center

Germany as the world's third largest economy in terms of GDP and is nipping at Japan's heels for the No. 2 slot, which many predict could occur by the turn of the century. The International Energy Agency's outlook for world energy to 2010 assumes economic growth in China at an average 7.6%/year through the forecast period.

As more Asia-Pacific developing nations look to tiny dynamo Singapore as an economic example, perhaps even the most sanguine forecasts for economic growth in this region will come to pass. And that poses heady potential for explorationists and producers and exporters of oil, natural gas, and their products.

Pacific Century

East Asia has been the world's fastest growing economic region the past 3 decades. Since 1960, economic growth has averaged 6.4%/year. That's about double that of western Europe and the U.S.

International Monetary Fund (IMF) projections place Asia's economic growth in 1994 at a sizzling 7%. That represents something of a modest slowdown from 1993 but still outstrips the combined average projected growth of only 2.7% in Japan, Europe, and the U.S.

Specifically, IMF estimates GDP will grow in 1993 and 1994, respectively, by these percentages: in China 12% and 8%, Thailand 7% and 8%, Malaysia 7% and 8%, Indonesia 6.5% and 8%, South Korea 4.5% and 6%, Singapore 7.5% and 6%, Taiwan 6% and 6.5%, Hong Kong 5.5% and 5%, India 4.5% and 6%, Philippines 7% and 7.5%, Australia 2.75% and 3%, and Japan 2% and 2.5%.

That growth is not just fueling ballooning consumer demand in those countries. It also points to an export boom. China's exports to the U.S. and the European Community in first half 1993 grew by about one fourth from the prior year period. Indonesia, Malaysia, and Singapore also showed double digit increases in exports in a comparison of the same periods.

"Many economists believe East Asia will continue to achieve growth well above the world average for the next 20 years," said Hugh Norton, chairman of BP Asia-Pacific and Middle East, at a 1991 Singapore petroleum conference. "Even if the phrase 'Pacific Century' is not on as many lips now as 5 years ago, few observers from outside this region—let alone inside it—doubt its capacity to grow impressively well into the next century.

"Yet while it supports a third of the world's population and accounts for 20% of the world's gross domestic product, a proportion that clearly is going to increase, it has only some 4% and 5% of world proved oil and gas reserves, respectively."

Consultant A.W. Jessup, writing in Geopolitics of Energy, concurred, noting the combined per capita GDP growth of the East Asian countries grew at a pace of 5.2%/year the past 25 years. That is more than double the pace of other regions: U.S. 1.6%, Organisation for Economic Cooperation and Development 2.5%, and Latin America 1.8%.

"The economic dynamism of East Asia promises to continue through the rest of this decade and into the 21st Century," Jessup wrote. Citing World Bank projections that the region's real GDP per capita will increase an average 5.3%/year, effectively doubling the regional economy in 10

Asia-Pacific's falling crude export capacity*

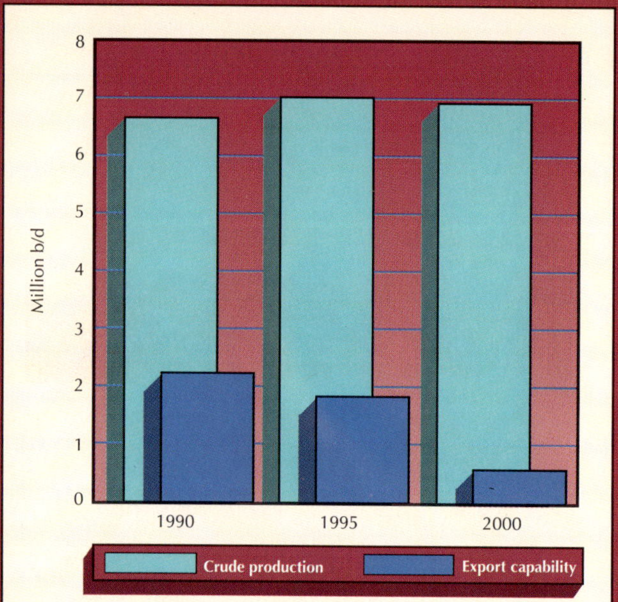

*Export capacity is the difference between crude production and demand for domestic crudes Source: East-West Center

Southeast China's oil demand

	Oil demand 1,000 b/d		
Province	1990	1995	2000
Guangxi	25.5	41.3	N/A
Yunnan	18.2	23.3	N/A
Guizhou	10.7	13.3	N/A
Hainan	N/A	6.0	7.0
Fujian	N/A	50.9	78.8
Guangdong	155.3	227.2	337.6

Southeast China's coastal refining boom

Refinery location	Capacity (1,000 b/d)	Sponsor(s)	aCompletion date
Fujian	50	Sinopec	1992
Chongqing	50	N/A	N/A
Zhanjiang	30	N/A	N/A
Nanning	50	N/A	N/A
Dalian	100	Sinopec, Total, Hong Kong	1993
Shenzhen	60	Sinopec, Hong Kong	1995
Huizhou	100	Sinopec, Royal Dutch/Shell	1996
Haikou	120	N/A	1996

years, Jessup notes the World Bank's downside case is still a robust 4.2%/year gain.

"No other region will come close to matching this performance."

Much of that projected growth comes from the expected transition of these countries' economies to developed/industrialized from developing. Those still in the latter category are expected to have the highest rate of economic growth and with it the biggest surge in energy demand during the next few decades.

The surge represents a structural change as well as growth in absolute terms. In 1965, agriculture accounted for 42% of East Asia's GDP, industry 35%. Currently, agriculture accounts for less than 25% of the region's GDP, while industry's share has soared to 44% and reflects a much higher degree of technological sophistication.

In 1965, East Asian trade involved mostly primary commodities, with fuels, metals, ores, and other raw materials accounting for about 90% of exports and manufactured goods—including textiles—the remainder. That has shifted dramatically.

Currently, primary commodities account for about 30% of East Asia's exports and manufactured goods about 70%, with a growing volume of products such as electronics, telecommunications, ships, and automobiles.

Correspondingly, imports have shifted in nature as well. Food accounts for only 7% of the region's import bill today vs. 16% in 1965, while imports of machinery and transportation equipment have jumped 9%.

The limited development of the region points to that sort of growth and fundamental economic change as only the beginning, for there is still much more progress that can be made.

Speaking at a Singapore petroleum conference, S. Samy Vellu, Malaysia's minister of Energy, Telecommunications, and Posts, provided this portrait of the region:

"The per capita income of developing countries in this region averages about 1/10th that of the rich industrialized countries. Average life expectancy is about 50 years compared with more than 70 years in industrialized countries. One in every five persons in developing countries suffers from malnutrition. One in two has little chance of becoming literate.

"Per capita commercial energy consumption of developing countries in this region ranks among the lowest in the world. It averages less than 1 metric ton of oil equivalent (TOE)/per capita/year, compared with 8 TOE in North America and 4 TOE in Europe."

Energy use characteristics

Malaysia's Vellu noted large variations in energy intensities among the countries of the Asia-Pacific region in terms of final energy consumption per unit of GDP.

"This is reflective of the dominance of traditional energies that have very poor use efficiencies," he said. "Dependence on biomass fuel, especially fuel wood, in (some) developing countries continues to be high, at about 70% of total energy consumption."

Vellu also pointed out that the use of hydrocarbons in energy systems in the region continues to dominate, although there is evidence of declining dependence on oil, mainly through substitution by coal, natural gas, and nuclear power. Notable exceptions are China and India, where coal accounts for 60% and 35%, respectively, of primary energy consumption.

Vellu acknowledged frequent regional disparities in energy supply and demand within many Asian developing countries, where the balance between centralized and decentralized distribution often is not achieved.

"Developing countries are largely dependent on foreign resources, and domination by foreign corporations in energy production, transport, and distribution is very evident. Even with subsequent takeovers by national governments financially and legally, dependence still exists in various forms, such as in the choice of technology, scale of projects, choice of production and use of equipment...Generally, energy systems have been constructed by copying industrialized countries without due regard to appropriateness to

Asia-Pacific net imports/exports by subregion

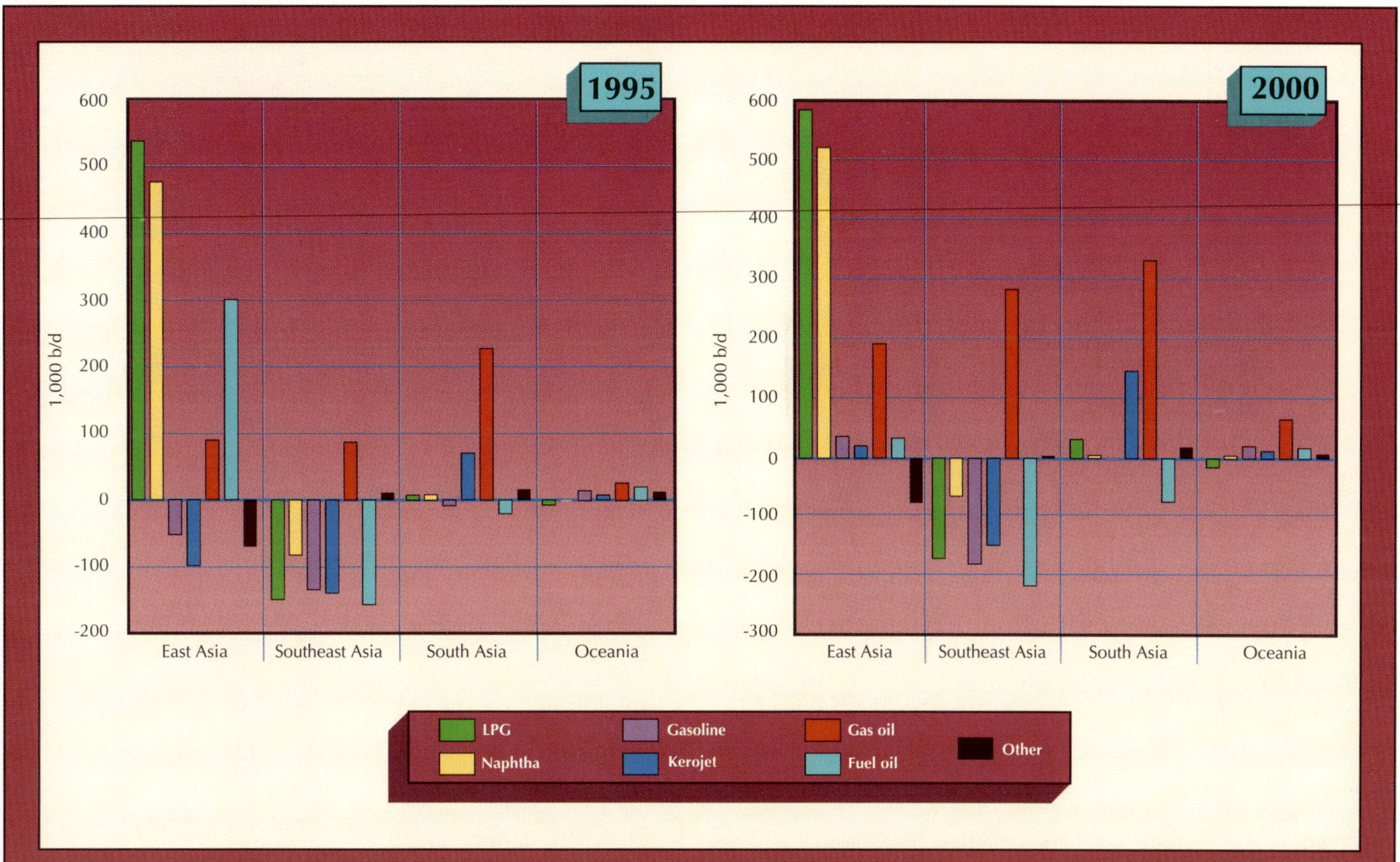

Source: East-West Center

local conditions, end use, or supply.

"Many developing countries have based their energy systems on petroleum products, although they do not possess natural hydrocarbon resources. At the same time, their own resources remain unexploited. Electrification of urban and rural areas were carried out in imitation of industrialized countries' experiences. Large scale plants, oil refineries, and power stations entailing costly investments are constructed on too large a scale, inappropriate to the needs of the country.

"Past energy policies of most developing countries have been merely supply policies, ignoring the importance of demand management through a rational use of energy. Various measures such as interfuel substitution, combined heat and electricity production, and byproduct recovery have been largely neglected."

Urgency of changing energy use patterns

Vellu sees an urgent task ahead in meeting the energy needs of a region that contains half the world's population, specifically in changing the societal patterns of energy use, the kinds of energy using equipment deployed, and the kinds and amounts produced.

"For example," he said, "the use of fuel efficient cooking stoves should be promoted to ease the pressure on fuel wood use. There is increasing pressure on wood resources by a rapidly growing population of poor people who rely on wood to heat and light their homes and cook their food.

"This has serious environmental repercussions. At present consumption rates, the annual fuel wood deficit will more than double between 1980 and 2000, from 400 million cu m to 900 million cu m, leading to serious consequences of deforestation."

Vellu contends many Asian developing countries face another energy crisis, hitting rural areas hardest because of their heavy dependence on traditional, noncommercial sources of energy, such as firewood.

Because some estimates put more than 1 billion people in Asia at risk of energy shortfall, Vellu wants to see a high priority placed on harnessing renewable energy technologies, especially for rural areas. He recommends that reforestation and efforts to generate income and promote development complement and be integrated with efforts to develop energy supplies.

At the same time, Vellu claims, developing nations face the tradeoff between economic development and environmental preservation.

"Developing countries are being called upon to undertake measures that could increase the economic gap between North and South," he said.

"The challenge before us is to ensure an equitable sharing of the responsibility for the global environment between developed and developing countries, ensuring that the economic growth of developing countries is not stunted, and ensuring an equitable place for developing countries in the world economic order."

Soaring energy demand growth

Energy consumption in East Asia has jumped by almost 8%/year since 1965, consultant Jessup wrote.

Consumption of commercial energy has risen almost fourfold to 3.8 bbl of oil equivalent (BOE)/person/year. That compares with a world average of about 8.9 BOE/person-/year.

Oil accounts for about 48% of East Asia's energy, coal about 30%, natural gas 10%, and hydro and nuclear com-

Asia-Pacific refining capacity additions

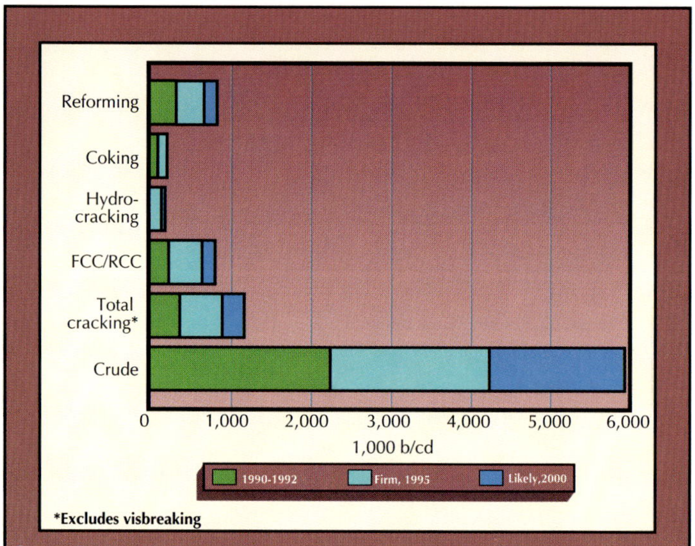

Source: East-West Center

bined about 12%. East Asian oil demand is expected to jump to about 11.7 million b/d in 1995 and 14 million b/d in 2000 from about 9.7 million b/d in 1991. Adding in Japan's oil demand, the region's oil demand will rise to 16.8 million b/d in 1995 and 19.25 million b/d in 2000, Jessup predicted.

The sea change in personal transportation will account for much of that oil demand growth. The automobile fleet in East Asia expands by almost 3 million passenger vehicles, trucks, and buses every year, Jessup noted. In South Korea alone there are 4.1 million cars on the road, with about 2,000 added each day. Accordingly, South Korean oil demand has rocketed by 16%/year since 1987.

Even with the region's abundant energy resources, there is a shortfall of about 40% of total energy requirements, Jessup wrote. The difference is covered by imports, mainly crude oil and petroleum products from the Middle East.

Under every scenario, that oil consumption is projected to grow in absolute terms even if its share of the region's energy mix declines. Jessup puts that growth at 5-6 million b/d by 2000, or a 50% jump from 1991 levels, that he contends can be filled only by imports from the Middle East.

Natural gas use will increase sharply in the same period, making inroads into oil's dominance of the region, albeit only marginally. Jessup estimates the expected doubling of gas demand in East Asia from the 1990 level of 4.4 tcf/year during the next 20 years would increase the gas share of energy use in the region to 20% only if energy consumption remained at current levels. However, that's unlikely, he contended, because energy growth is almost 1:1 with GDP growth.

Japan hopes to trim the share of oil in its energy mix to 51.3% by 2000. Even if it succeeds, Jessup noted, Japan's volume of imports still will be 5.25 million b/d. And a greater volume of oil imports than that might be needed if Japan's ambitious nuclear power expansion plans slip.

Japan is hardly alone in its limited success in reining oil demand growth, Jessup pointed out.

"In East Asia, pricing and taxation policies to promote oil conservation have generally failed. In South Korea, gasoline consumption rose even when the price was the equivalent of $5/gal. Price increases in Indonesia have had no noticeable impact on growing domestic demand.

"Rapid economic development appears to be accompanied by a rapid rise in energy consumption and reliance on petroleum produts as the easiest and cheapest to use. As a result, East Asia plus Japan will be importing 10 million b/d of Middle East oil in 1995 and over 12 million b/d in 2000 while making major investments in gas production and distribution projects."

Regional oil demand growth

Oil demand in the region of the Suez increased almost 6%/year during 1985-90, accounting for about half the world's demand growth in the period, said Caltex's Ward.

He projected growth in demand for refined products in the region to more than 19.5 million b/d by 2000 from slightly more than 14.5 million b/d in 1992.

Transportation fuels will account for most of the consumption gain, rising to 12 million b/d in 2000 from almost 8 million b/d in 1991. Motor vehicle ownership will grow at an average rate of 4%/year.

"If China could expand auto ownership to match the U.S.," Ward said, "nearly 700 million vehicles would be added—equal to the number in Europe, Japan, and the U.S. combined.

"We know that's not going to happen, but it does illustrate the potential for continued growth in Asia and the impact on transportation fuel demand."

Air travel is growing rapidly east of Suez as well. In the consuming countries of the region, international air passenger traffic grew by 12%/year during 1970-90 and is expected to increase by 8.5%/year through 2000—twice the rate of growth of the rest of the world. Air traffic in the region's oil producing countries will increase at an average rate of slightly less than 5%/year.

In 1992, in the Asia-Pacific region alone, product demand increased by about 6.3%, according to East-West Center. Diesel fuel demand increased by 7.6% and naphtha by 13.6%, the latter due to a 44% jump in South Korea.

Product demand in 1992 increased by 28% in Viet Nam, 21% in Philippines, 20% in South Korea, 11% in Thailand, and 9.3% in China.

"Despite a 20% increase in the price of oil products instituted in 1991 in Indonesia, the country's oil demand grew by 8.9% in 1991 and 6.2% in 1992," East-West Center said.

The center sees oil demand growth through 1999 in the Asia-Pacific region averaging 4.4%/year during 1993-95, then slowing to 3.1% through the rest of the decade. Diesel fuel demand will rise the fastest at 5.3%/year during 1993-99, with gasoline second fastest at 4.6%/year. Demand for fuel oil, mostly low sulfur, will increase at only 0.5%/year.

East Asia's' energy security

BP's Norton contended that a shift of priorities for the oil industry in the Asia-Pacific region has taken place in recent years, noting that discussions about the state of the industry centered on the question of security of crude oil supply.

Because many East Asian countries depend heavily on imported crude and the region overall has a net supply deficit, including a low share of the world's proved and probable reserves in tandem with underexplored frontiers, much of the debate a few years ago focused on finding and developing indigenous hydrocarbon reserves, Norton pointed out. For hydrocarbon-poor Japan, South Korea, and Taiwan, there has been a focus on ensuring a choice between imported oil and imported gas, a choice recently accelerated by environmental pressures favoring gas.

East Asia still imports more than 12 million b/d of oil, compared with indigenous production of only 5 million b/d. While the outcome of the Persian Gulf war showed that heavy reliance on Middle East crude supplies is not the onus it once was, "It would be a mistake to think that experience removes the risk for the East of overreliance on imports from outside the region in the long run.

S.E. Asia distillation capacity additions

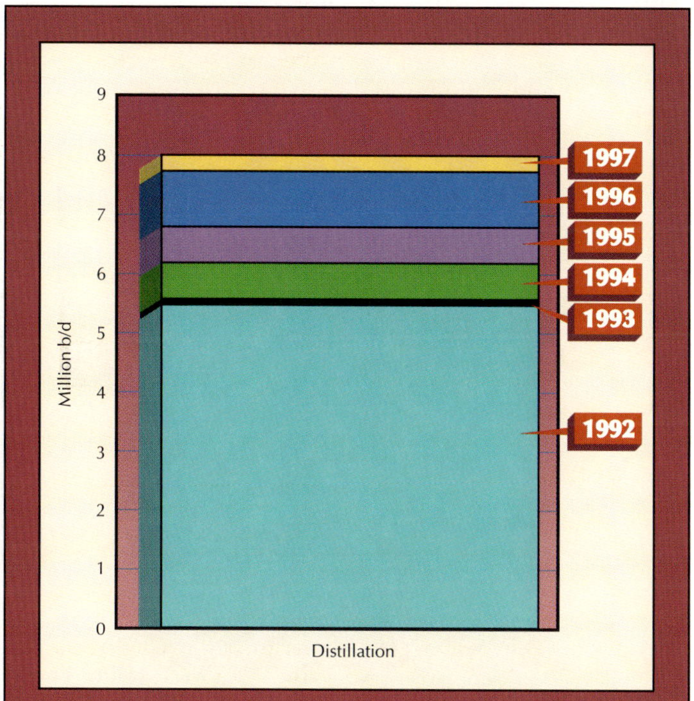

Source: Energy Security Analysis Inc.

"Even given a more benign climate of political willingness all around, global oil supply and demand forecasts point to a tightening of supply, with resulting pressure on prices, no later than the second half of this decade."

Norton cited a number of variables in assessing the speed with which that tightening of supply occurs, including the status of oil exports from the former Soviet Union (FSU) and how quickly and smoothly needed investments are made to maintain—let alone increase—productive capacity from the aging fields of the Persian Gulf region.

Resuscitating the FSU's oil sector and boosting crude productive capacities in the Middle East will require perhaps a combined $250 billion, according to a variety of forecasts.

With that in mind, Asian Development Bank estimated developing countries in Asia will need to invest $500 billion in energy production in the next decade to offset consumption that has jumped by 50% the last 10 years.

And that doesn't include massive outlays needed to upgrade and expand world refining capacities to meet increasing overall demand for refined products as well as for environmentally cleaner fuels.

Pressures on Asia-Pacific refining

Strong oil demand growth will overwhelm supplies of products refined in the consuming countries east of Suez, said Caltex's Ward.

"By the year 2000, we could see a deficit of some 2.7 million b/d," he told the conference in Dubai. "And this could happen despite an expected addition of almost 5.5 million b/d of new refining capacity in this area."

Growing product supplies from Japan and South Korea, as well as imports from outside the region, have kept up with rising demand. But Japan and South Korea are approaching capacity, and Ward expects economics to deteriorate for imports from distant sources.

The result will be rising prices of the products in highest demand—transportation fuels—which will improve refining margins enough to stimulate refinery investment. Already, Ward said, companies are debottlenecking plants and adding distillation and upgrading capacity in Japan, South Korea, Singapore, Philippines, and Australia.

In the east of Suez producing countries, product supply exceed demand by almost 2 million b/d during 1990, and major refinery modernization projects are under study. The work could boost the producing region's product surplus to 2.3 million b/d by 2000, Ward said.

Refiners in the consuming countries east of Suez face a challenge in addition to that of keeping up with rising product demand. Ward said environmental regulations are reducing allowable sulfur levels of diesel and fuel oil and lead content of gasoline.

Japan will phase down diesel sulfur content from 0.5 wt % to 0.05 wt % in 1997. Ward said many countries in the region now have unleaded gasoline, and several plan to ban lead altogether in the mid-1990s.

Asia-Pacific shortfall seen

In just the Asia-Pacific countries, crude production last year averaged 6.8 million b/d, of which 2.2 million b/d left the region, according to East-West Center.

The center expects Asia-Pacific crude output to rise to 7 million b/d by 1995 and remain at that level through 2000 largely on the strength of new production from Widuri, Duri, and Belida fields in Indonesia and Dulang field in Malaysia.

With regional demand rising, Asia-Pacific exports will drop to 1.8 million b/d in 1995 and 632,000 b/d in 2000.

China and Indonesia will be net oil importers by 2000, contends East-West Center. Exporters at that time will include Malaysia, Brunei, and Papua New Guinea.

Prospective loss of Indonesia's ability to produce more oil than it needs has scuttled plans by the government to build several export refineries.

Exports from the Asia-Pacific region of low sulfur, waxy crudes will decline. Among Asia-Pacific crudes, light sweet crude supply will lose ground to heavy sweet crudes.

"The changing crude availability will have a major impact in the refining investments geared toward lower sulfur fuel oil production to meet environmental requirements," East-West Center said.

Asia-Pacific's dependence on imported oil will increase to 57% by 1995 and 64% by 2000 from 50% in 1992, according to the center.

In line with Ward's projection of a growing surplus in the producing countries east of Suez, East-West Center sees much of Asia-Pacific's growing imports coming from Persian Gulf producers.

By 2000, more than 90% of that the Asia-Pacific countries import from outside the region will come from the Persian Gulf, compared with about 70% at present.

"In terms of security of supply, the Asia-Pacific region is in a more precarious situation than the U.S. or Europe," East-West Center said.

ESAI's slant on Asia-Pacific

To ESAI, the question is not whether refining capacity expansions will cover the lower rate of demand growth that it expects by whether product yields will match consumption patterns.

During 1993-97, distillation capacity in Southeast Asia will increase at the rate of about 600,000 b/d/year, more than enough to cover the 500,000 b/d/year demand growth ESAI expects.

"Viewed with North Asia and the Middle East, capacity expansion is more than adequate," the analyst said.

Distillate demand in ESAI's projection for Southeast Asia will increase at the rate of 250,000 b/d/year, gasoline at 125,000 b/d/year, and fuel oil and feedstocks at 100,000 b/d/year.

Refiners in Southeast Asia during 1993-97 will add about 290,000 b/d of fluid catalytic cracking capacity, 40,000 b/d of hydrocracking capacity, 30,000 b/d of thermal capacity, and 180,000 b/d of reforming capacity.

The upgrading capacity additions will enable refiners in the region to meet demand for gasoline but not for distillate, ESAI said.

Asia-Pacific refining construction activity

In line with its development as a growth market for oil, Asia-Pacific has become a center of refining construction activity.

Singapore is in the midst of a construction push, much of its focused on refinery upgrading capacity.

Singapore Refining Co. Pte. Ltd., for example, in 1993 announced an $840 million upgrade of its 220,000 b/d Pulau Marlimau Island refinery. It will enable the company, a joint venture of British Petroleum Co. plc, Singapore Petroleum Co., and Caltex (Asia) Ltd. to produce more transportation fuel and less fuel oil.

All the other big refiners in Singapore have construction projects under way, most involving upgrading capacity and some adding to distillation capacity.

In Japan, construction centers on upgrading capacity, some distillation capacity expansion, and significant addition of methyl tertiary butyl ether capacity.

Along with upgrading and modernization work planned or in progress elsewhere in the Asia-Pacific region are several grassroots projects.

Construction is booming in Thailand. Rayong Refinery Co., a joint venture of Shell International BV and Petroleum Authority of Thailand (PTT), plans a 145,000 b/d grassroots refinery in Mab Ta Phud, Rayong Province, on the Thai eastern seaboard.

Nearby, Star Petroleum, a joint venture of Caltex and PTT, plans a 130,000 b/d grassroots facility with a 37,000 b/d residual fluid catalytic cracker. Also in Rayong Province, Esso Standard Thailand Ltd. is adding 82,000 b/d of distillation capacity to its 75,000 b/d refinery at Sri Racha.

In Malaysia, BHP Petroleum Pty., in a joint venture with Taiwanese state oil company Chinese Petroleum Corp. (CPC) and Malaysia's state owned Petroliam Nasional Bhd., plans a 150,000 b/d refinery at Bintulu, Sarawak. Petronas plans a 100,000 b/d sweet crude refinery at Malacca as part of integrated complex that eventually is to have 100,000-130,000 b/d of sour crude capacity as well.

At least two new refineries are in prospect in Indonesia. BP Asia Pacific, in a joint venture with C. Itoh, plans a 120,000 b/d refinery oin Bintan Island.

Also, state owned Pertamina has plans for a 125,000 b/d refinery at Balongan, West Java.

Pilipinas Shell Petroleum Corp. plans to build a 110,000 b/d refinery at Tobangao, Philippines, next to a 65,8000 b/d refinery it operates there. Caltex (Philippines) Inc. is adding 11,400 b/d of crude capacity to its 65,400 b/d at Batangas and upgrading and expanding downstream processing capacity as well.

In Taiwan, CPC is adding crude distillation and downstream processing capacity at its Kaohsiung and Tao-yuan refineries.

Viet Nam's state owned Petrovietnam wants to build its first modern grassroots refinery at Vung Tau with distillation capacity of 130,000 b/d. Taiwan's CPC and France's Total are conducting a feasibility study of the project.

China's refined products demand

Nowhere in the Asia-Pacific region—likely in the world—is there a more ambitious refining capacity expansion program than in China.

S.E. Asia upgrading capacity expansions

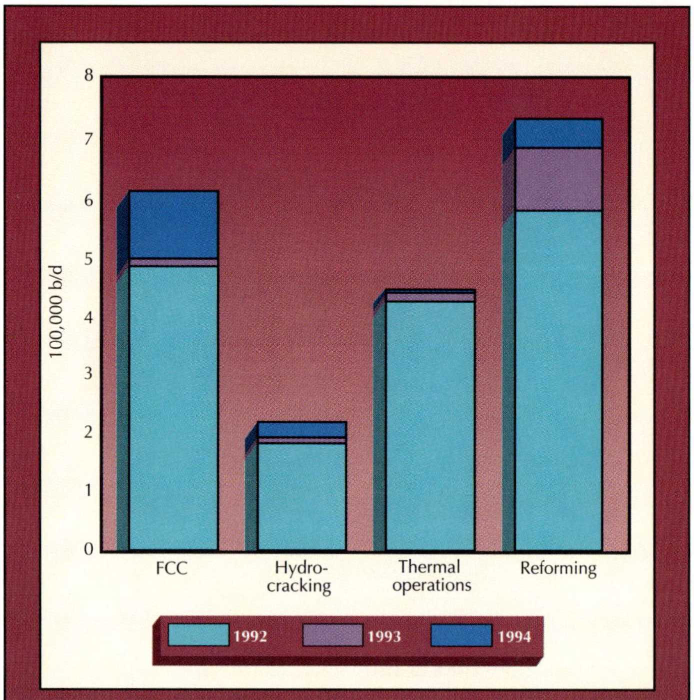

Source: Energy Security Analysis Inc.

China's refining sector is in the midst of a major expansion and reorganization in response to booming domestic demand for petroleum products.

Plans call for hiking crude processing capacity to 3.9 million b/d in 1995 from the current 3.085 million b/d. Much of that 26% increase will come where the products demand growth is the strongest: China's coastal provinces, notably those in the Southeast.

Despite the demand surge, China's refineries operated at only 74% of capacity in 1991, and projections for 1992 weren't much better. Domestic crude supply is limited because of Beijing's insistence on maintaining crude export levels, a major source of hard currency foreign exchange.

While the government scales back crude exports to feed voracious domestic consumption, it is restructuring the refining sector to reflect regional disparities and foreign participation in Chinese refinery projects. Those steps are part of overall reforms Beijing is implementing throughout the nation's petroleum industry.

China's superheated economy is fueling a surge in oil demand. China consumes more than 2 million b/d of oil currently, a level most observers expect to climb by 120,000-140,000 b/d/year. In first half 1992 alone, China's refined products consumption jumped by 18.2% from the previous year's period vs. a planned increase of 12.2%.

In the southeastern coastal provinces—which if taken separately would constitute the world's fastest growing economy at perhaps 20%/year or more —oil demand is projected to more than double by 2000 from 1990 levels.

Domestic demand for oil is outpacing Chinese crude production. In 1991, China's crude production reached a record 2.792 million b/d, but the need to export crude for hard currency reined processing of domestic crude to less than 2.3 million b/d.

China's current refinery crude throughput capacity of 3,085,200 b/d makes it potentially the world's fourth biggest oil processor after the U.S., the former Soviet Union, and Japan.

In 1992, China processed 2.38 million b/d of crude oil, up about 4% from the 1991 level. Of the total, 2.127 million b/d

Asia-Pacific gas demand

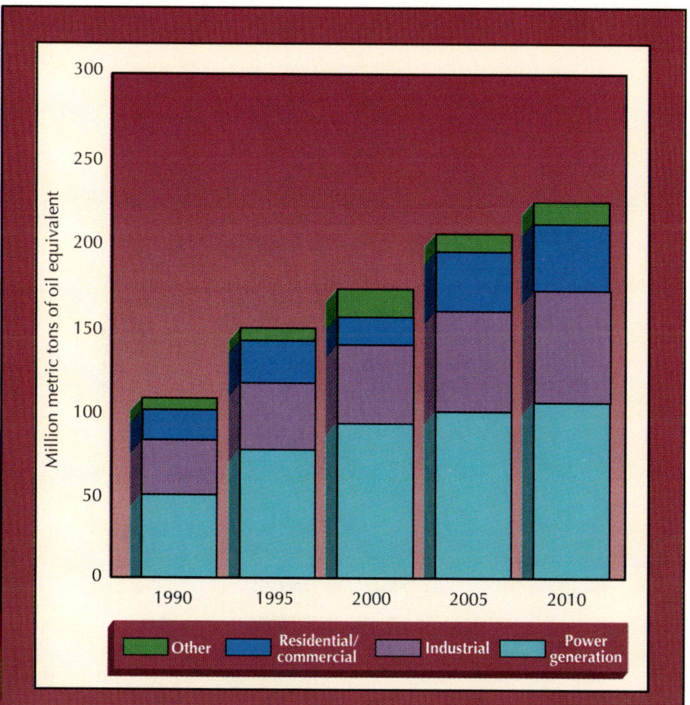

Source: Purvin & Gertz Inc.

was processed by state owned Sinopec. The rest was processed by 25 small refineries operated by China National Petroleum Corp. (CNPC), the state owned company that oversees exploration and development operations.

The combined output of gasoline, diesel, oil, kerosine, and base lubes totaled 1.209 million b/d in 1992, up 8.5% from 1991. In addition to domestic production, refineries along the coastal areas processed 300,000 b/d of imported crude, or 14.1% of the total processed volume.

In 1991, Sinopec refineries accounted for about 90% of the 2,272,600 b/d of crude oil processed in China. Of those Sinopec refineries, 30 have crude processing capacity exceeding 50,000 b/d.

Sinopec plans to shift its attention to areas along the coast and the Yangtze River to turn them into export-oriented oil processing bases, according to Sinopec Pres. Sheng Huaren. Sinopec officials point to refineries under construction with domestic and foreign funding in coastal areas.

China's refining construction boom

Chinese officials look to the refinery under construction at Shenzhen as a prototype for sophisticated Chinese refineries of the future.

A joint venture of Sinopec, the city of Shenzhen, and Hong Kong, the $500 million project will have crude processing capacity of 60,000 b/d when complete in 1994. The crude will be imported from the Middle East, and 60% of its products will be exported to Hong Kong.

The project will have eight processing units, including an atmospheric resid desulfurization/hydrogenation complex that is a first for China. At present, most of China's 100 refineries can process only low sulfur crude. The atmospheric resid desulfurization/hydrogenation unit and Shenzhen's several heavy oil hydrotreating units, catatlytic cracking units, and continuous catalytic reformer can process a range of high sulfur imported crudes.

Shenzhen's sulfur unit can recover 60,000 metric tons/year of sulfur, compared with an average of less than 10,000 tons/year at other Chinese refineries with sulfur units.

Shenzhen also will yield a much whiter barrel than other domestic refineries, about 81.9% gasoline and middle distillates. In 1990, the average recovery yield of gasoline and middle distillates at Chinese refineries was 51.5%.

The continuous catalytic reformer at Shenzhen can produce 100 RON gasoline, a significant advance for China. Even with the addition of tetraethyl lead, very few Chinese refineries can produce 90 RON gasoline.

Shenzhen is one of eight refineries China plans or is building along its coastal areas. The others are in Fujian, Chongqing, Zhanjiang, Nanning, Dalian, Huizhou, and Haikou. The eight will have a combined capacity of 560,000 b/d.

The Fujian refinery is the first to be completed. The 50,000 b/d plant started up in September 1992 in Fuzhou, capital of Fujian province. Three of the refineries in Southeast China were still under construction at IPE presstime. The combined crude throughput of old and new refineries in the area is expected to reach 362,000 b/d in 1995. Of that, only 137,000 b/d is earmarked for Guangdong, 89,2000 b/d short of projected demand. That shortfall won't be made up by 2000.

Although throughput is expected to reach 532,000 b/d in 2000, Guangdong will account for only 199,000 b/d—138,600 b/d short of projected demand.

Foreign participants are becoming increasingly common in China's refining sector.

Construction began in May 1992 of a 100,000 b/d export refinery at Dalian in Northeast China's Liaoning Province. It is a $437 million joint venture of four Chinese companies, Hong Kong's Sinochem International (H.K.) Co. Ltd., and Total. The foreign partners have a combined 42.5% stake in the venture. It is expected to produce 38,000 b/d of gasoline, 12,000 b/d of naphtha, and 2,600 b/d of liquid petroleum gas. Crude oil will be imported from the Middle East, and processed with technology and equipment provided by Total. Total also will help the Chinese partners purchase crude and sell products on the international market.

Royal Dutch/Shell Group is considering participation in a 100,000 b/d refinery near the coastal town of Aoutou, Guangdong Province. Pending results of a feasibility study, Shell could take a 50% stake in the joint venture, with the rest held by a Chinese group led by the Guangdong provincial government and including China National Offshore Oil Corp. The study is to be complete this year. If it receives timely approval, the project could be complete in 1997-99.

In addition to the refineries to be built along the coast, Sinopec plans to build a 160,000 b/d refinery at Shanghai in cooperation with France's Ste. Nationale Elf Aquitaine in 1994. The planned refinery has been given a green light by Beijing's State Council.

Meanwhile, China Chemicals Import & Export Corp. (Sinochem) is negotiating with Saudi Arabia to build another 120,000 b/d refinery at Quingdao, a port city in East China's Shandong Province.

Asia-Pacific refining bottleneck

BP's Norton contends that the most recent Persian Gulf crisis underscores a new concern for the Asia-Pacific region: refining capacity bottlenecks.

The elimination of a large volume of high grade export refining capacity in Kuwait, and to a less extent Iraq, resulted in a global squeeze on refining capacity, he noted.

"While crude supplies remained generally plentiful, certain refined products, mainly naphtha and middle distillates, became scarce, and that tightness in capacity began to look like a lasting phenomenon...The world was used to regrading crude oil supply capacity as the crucial bottleneck in the pipeline to the consumer. A new bottleneck now appeared, in secondary and tertiary refining capacity."

So the shift in priorities for the Asia-Pacific region's oil sector is how to best plan ahead for the provision of new refining capacity in East Asia in order to avoid capacity pressures while still avoiding the mistakes of overbuilding that occurred in Europe and the U.S. during the 1970s.

With the prospect that incremental oil demand in East Asia by 2005 will exceed that of the U.S. and western Europe combined and the continuing global emphasis to develop oil and gas supplies outside the Middle East, it behooves East Asian refiners to expand refining capacity, thereby expanding products supply flexibility as well as adding to value of indigenous oil.

Norton contends there are a number of considerations a country must take into account before committing large scale funds to new refinery projects.

"If it is a producer country, the driving consideration may be the wish to add value to what would otherwise be exported as crude oil—in which case a new kind of customer must be identified.

"If if is a consumer country, (the driving consideration may be) a degree of import substitution, for either: How much imported crude should be planned for and in what grades; (or)what environmental pressures on product quality need to be met?

"If predominantly export oriented, even for part of its lifespan, is a location available that is topographically competitive? Is it wise to depend totally on imported crude if the investor has no equity crude of his own? Is there scope for commercial optimization by integrating the yield structures of existing refineries or supply systems?

"In some instances, it makes more sense to upgrade existing refineries to take advantage of the built-in economic advantage or at least to integrate new development with existing refining infrastructures. In other cases, a phased developmental approach may be the most suitable."

Above all, said Norton, three questions must be answered:

- Is the regional supply/demand equation in refining capacity correctly balanced to justify the decision to invest?
- Will a new refinery have to compete with debottlenecking or expansion options in existing sites?
- Where will the financing come from?

"The management of the refinery building program in the region over the coming years must succeed in avoiding a series of poor choices, leading to an unbalanced regional refining industry full of stresses and conflicts. There is no single recipe or mechanism to avoid this, nor should there be.

"Individual, commercially motivated investors or consortia of investors will continue to be the driving force. But I must conclude that creating timely and fundable new capacity across the region as a whole, not just country by country, is the new challenge to governments, state oil companies, and the private sector oil industry. This calls for collaboration between these three sectors. The winners will be those who can best seize the opportunities to create strategic collaborations."

Asia-Pacific's natural gas boom

As strong as oil demand growth is expected to be in the Asia-Pacific region, the surge in demand for natural gas will far outstrip it, most industry analysts concur.

"The infrastructure to bring gas to market will be an enormous opportunity in the coming decades," John Froning, vice-president of exploration for BHP Petroleum (Americas) Inc., Houston, told a National Oilfield Manufacturers & Delegates Society luncheon during the American Petroleum Institute annual meeting in New York in late 1992.

Gas consumption grew by 130% in Asia during the 1980s, while average gas use worldwide grew by less than 40%.

Asia's gas consumption growth is likely to slow to 50% during the 1990s, but that will still be more than double the growth in the rest of the world, BHP contends.

Pacific Rim LNG supply, demand

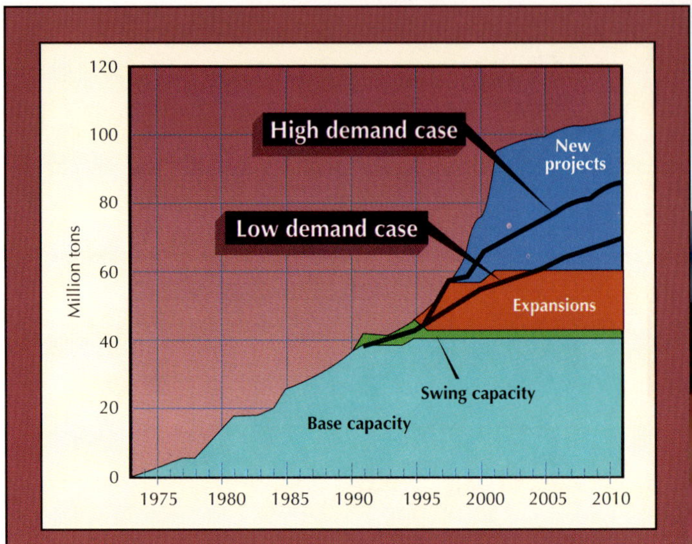

Source: BHP Petroleum (Americas) Inc.

"The existing price structure won't justify many gas development projects," Froning said, so nations will have to pay more to reap the environmental benefits of natural gas.

Though there is a large supply of natural gas in the region, a greatly expanded gas pipeline network is needed to link supplies with consuming areas. A high pressure gas pipeline network in the Far East that has been proposed by a group of Japanese companies would require construction of about 27,000 km of new pipelines by 2020 at an estimated cost of $42 billion (Australian). According to Froning, 10-1[illegible] new trunk line projects are currently planned or under way.

The network would include a line from Russia's Sakhalin Island to Japan and connections with other fields in Russia and China.

Pacific Rim LNG trade also will continue to expand at a healthy pace until 2020. Froning said the cost of submarine gas transmission lines is competitive with LNG transportation up to a distance of 1,500 km, and a land transmission line can compete with LNG shipment over distances of as much as 3,000 km.

Asia-Pacific underpins LNG trade growth

World trade in liquefied natural gas could double or triple in the next 2 decades, under a range of industry forecasts.

Estimates are that industry may have to spend as much as $40-60 billion in new facilities and expanded fleets to meet projected LNG demand by 2010.

Fueled by soaring demand in the Asia-Pacific region, the surge in LNG trade may overwhelm current projections of the industry's ability to accommodate that growth. Some industry analysts see shortfalls on the horizon in terms of LNG plant capacity and LNG tanker fleets.

Underlying those possible shortfalls are current and projected prices for LNG, which are tied to a consensus view that oil prices will remain relatively low for the foreseeable future.

Developers are unlikely to press costly grassroots LNG projects without the prospect of substantial price increases. And prospective LNG customers balk at signing contracts for speculative supplies at prices needed to justify those investments while energy prices are expected

East Asian LNG demand

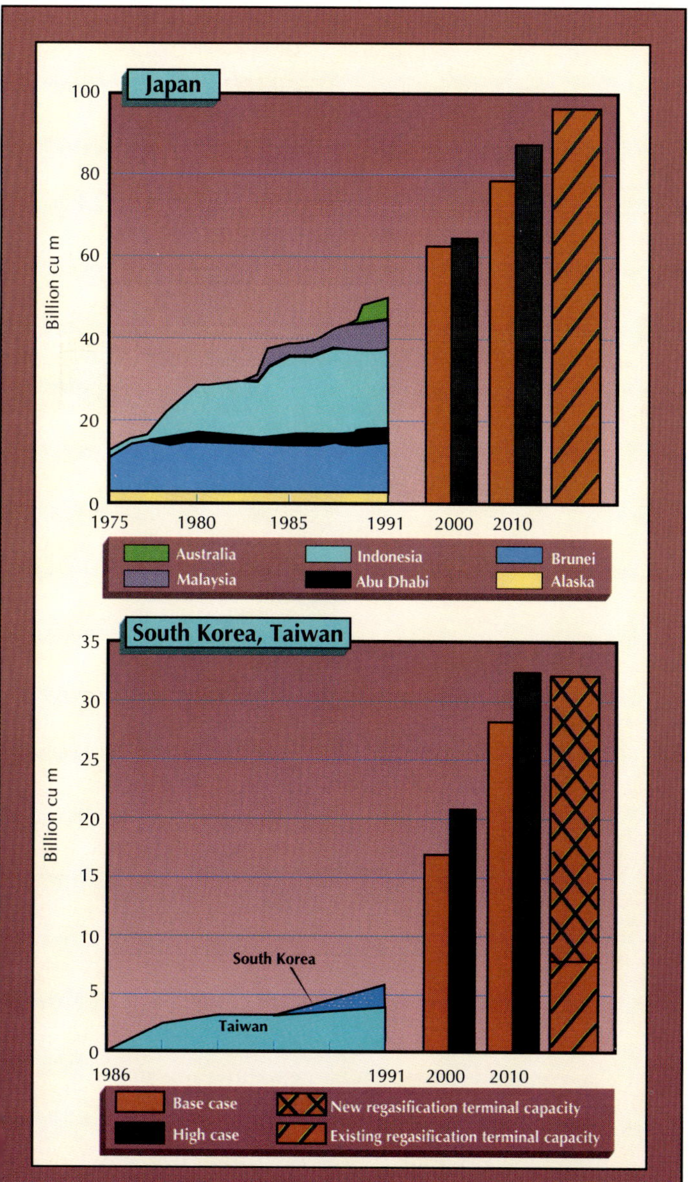

Source: Cedigaz

to remain relatively flat.

The result likely will be a shakeout of the many projects emerging in recent years. That could LNG far short of fulfilling its market potential absent some preferential treatment by governments on boths sides of the supply/demand equation.

Asia-Pacific's LNG dominance

The Asia-Pacific region's LNG producers and consumers will continue to dominate world trade in the fuel.

The Pacific Rim accounts for almost two thirds of the LNG transported and three fourths of the LNG delivered in the world.

World LNG trade rose 3.5% in 1992 to 59.5 million metric tons (79.2 billion cu m gasified), of which the Far East accounted for 73.4%, according to the Groupe Internationale des Importateurs de Gas Natural Liquefie (Giignl). Indonesia accounted for 39% of all LNG exports in 1992, followed by Algeria 24%, Malaysia 12%, Brunei 9%, and Australia almost 8%, Giignl estimated.

Japan accounted for almost 65% of all LNG imports in 1992, followed by France with 11.5%. Along with Japan, South Korea and Taiwan are expected to be the major new sources of LNG import growth, and with Japan imported almsot 42 million tons of LNG in 1991.

The three East Asian countries could absorb LNG imports of as much as 63-66 million tons by 2000 and as much as 91-100 million tons by 2010, said Ichiro Yokose, managing director, fuels, Mitsubishi Corp.

Yokose expects incremental demand to 2000 in the three nations to be covered by expansion of existing supply sources in Australia, Malaysia, and Abu Dhabi, and by new projects in Qatar, Oman, and Sakhalin Island.

The big question is what new supply sources will come to the fore to meet the roughly 50% further increase to 2010, Yokose said.

He contends the big driver in the LNG market, Japanese demand, will be affected by:

- The outlook for electrical power demand.
- Delays in construction of nuclear power plants.
- Increases in town gas use.
- The Japanese government's policy to diversify energy sources among coal, oil, gas, and alternate fuels such as methanol.
- The current gloomy overall economic condition, which is likely to continue in the years to come.

Asia-Pacific LNG trade outlook

Key LNG exporters such as Brunei, Malaysia, and Indonesia are scrambling to implement projects to meet the expected boom in Asia-Pacific LNG demand.

Uncertainties cloud the outlook for Far East LNG trade, however. Australia, for one, is more cautious in pressing expansion of its LNG export capacity as more competing LNG expansions spring up around the world, notably in the Middle East and Africa. Potential LNG exporters such as Papua New Guinea and Venezuela also loom on the horizon.

At the same time, new LNG import markets may be opening up in Asia. India is considering several pipeline and LNG import projects to accommodate an expected spurt in natural gas demand for environmental reasons—high sulfur coal provides more than half of India's primary commercial energy consumption.

One such project under consideration is being pushed by an Enron Corp. group to build a 2 million kw gas fired power plant on India's western coast that will be fueled by LNG. Enron in 1993 signed a letter of intent to buuy 2.5 million tons/year of LNG from Qatar's Ras Laffan LNG project beginning in 1997-98.

Complicating the picture further are widely varying industry estimates of LNG market growth. Environmental concerns have led many analysts to jump sharply their estimates of projected LNG demand growth from levels predicted just a few years ago.

Projections of LNG trade

France's Cedigaz contends world LNG trade could leap to 120-130 billion cu m/year in 2000 and more than 200 billion cu m/year in 2010, following an average 20%/year increase in demand for LNG since the first liquefaction plant started up in 1964.

Cedigaz cites the Far East as the linchpin in projections of spectacular growth in LNG supply and demand, noting Japanese LNG demand alone could soar to 79-88 billion cu m/year in 2010 from 51 billion cu m in 1991.

Most analysts contend LNG demand in the Asia-Pacific region will exceed supply before the end of the decade.

Involved in the booming LNG trade are grassroots projects as well as expansions that will call for investment of tens of billions of dollars. Sources of capital needed for those projects are tightening because of competing investments in the formerly Communist eastern Europe and U.S.S.R. and in efforts to ramp up productive capacity in OPEC nations to meet the expected call on OPEC oil after the mid-1990s.

Adding to the uncertainty for Asian LNG trade are unresolved political and economic problems among the industrializing nations of the region.

However, it certain that the most intensive development of LNG export capacity will occur in Indonesia and Malaysia, the biggest and third biggest LNG exporters, respectively. Algeria holds the No. 2 spot.

Malaysian LNG projects

Malaysia expects to double its LNG export capacity to almost 16 million metric tons/year by 1997, which would net it projected earnings of $4.5 billion/year. By 1995, Malaysian revenues from LNG exports are expected to outstrip those from oil exports, $2.92 billion vs. $2.76 billion.

State owned Petronas plans to retain a major interest in a new LNG plant at Bintulu on the Sarawak coast if it proceeds with the project. Plans call for construction of another three trains at Bintulu that would boost capacity to 15.8 million tons/year by 1996 from the current 7 million tons/year. Cost estimates vary, but if the plant could share infrastructure with the existing Bintulu plant, it could be built at a cost of $1.6 billion.

Petronas is seeking customers, financing, and equity partners in the project. Among partners under consideration is the group led by Occidental Petroleum Corp. that in 1992 discovered a giant gas field in the Luconia basin off Sarawak.

Malaysia LNG Sdn. Bhd. (MLNG) in 1992 let contract to Coprimo (Malaysia) for construction of a second LNG tanker loading berth at Bintulu. MLNG is owned by Petronas, Shell, Mitsubishi, and the Sarawak government. The project , to install a second berth capable of accommodating LNG carriers of 130,000 cu m capacity, is estimated to cost $400 million.

Petronas let contract in 1991 to French shipbuilder Chantiers de L'Atlantique for five LNG carriers, each with capacity of 62,500 tons of LNG, to handle deliveries ot Japan. Plans call for delivery of the first vessel in July 1994 and the fifth in July 1997.

Indonesian LNG

LNG exported from Indonesia's Arun and Bontang complexes covers half of Japan's consumption and all of South Korea's and Taiwan's.

Indonesia in 1991 exported 18.5 million tons of LNG to Japan, 2.8 million tons to South Korea, and 1.5 million tons to Taiwan.

Indonesia's current export capacity is 24 million tons, divided 50-50 between Arun's six trains and Bontang's five. A sixth train is under construction at Bontang to provide supplies covered under a recent contract signed with Japanese buyers, with first deliveries planned for January 1994.

In addition, a third LNG complex is under consideration based on reserves to be developed in giant Natuna gas field in the Natuna Island area of the South China Sea. Negotiations have been under way between Pertamina and Exxon Corp., operator of the field, for a number of years, stymied by technical (high carbon dioxide content) and economic concerns. Those negotiations were still under way at IPE presstime.

Brunei, Australia LNG work

Brunei Cold Gas, a joint venture of Brunei's government, Shell, and Mitsubishi, have supplied 5.14 million tons/year of LNG to Japanese utilities Tokyo Electric, Tokyo Gas, and Osaka Gas since 1972 under a contract that was to have expired in March 1993.

Northwest Shelf LNG project on Australia's Burrup Peninsula is one of the key supply sources for Pacific Rim LNG, driver in the world LNG demand outlook. Photo courtesy M.W. Kellogg Co.

A new contract, signed in early 1993, allows for imports to be increased 10% and extends deliveries another 20 years.

Off Australia's western coast, a group led by Woodside Petroleum Ltd. is pursuing expansion of gas productive capacity in the Northwest Shelf project to sustain LNG exports covered by existing Japanese supply contracts in the second half of the 1990s.

The Woodside group at IPE presstime was in the midst of installing Goodwyn platform to produce condensate as well as gas. The Northwest Shelf group in 1992 signed agreements to supply another 1 million tons/year of LNG to its Japanese buyers beginning in 1995. The increase will boost deliveries to almost 7 million tons/year. A third LNG train at the Northwest Shelf LNG complex on Western Australia's Burrup Peninsula started up in December 1992.

Taiwan's LNG demand

Taiwan plans a huge increase in its consumption of LNG especially to fuel electrical power plants.

State owned Taiwan Power Co. expects installed capacity for power generated by LNG to increase to 8.59 million kw by 2001 from 751,000 kw in 1991.

LNG's share of Taiwan's power market in that period is expected to jump to 22.9% from the current 4.1%. At the same time, nuclear's share of the Taiwanese power market will fall to 19% from the current 30%, Taiwan's Economic Ministry predicted.

CPC hopes to begin importing 2.25 million tons/year from Malaysia in 1995. That would make Malaysia the biggest supplier of LNG to Taiwan, surpassing Indonesia with its current level of 1.6 million tons/year.

CPC wants to further diversify its sources of natural gas supplies, citing concerns that principal LNG supplier Indonesia might not be able to meet Taiwan's gas needs in a few years.

North America

UNITED STATES

CAPITAL: Washington, D.C.
MONETARY UNIT: Dollar
REFINING CAPACITY: 15,131,556 b/cd
PRODUCTION: 6.855 million b/d
RESERVES: 23.745 billion bbl

The U.S. petroleum industry entered 1994 with its refining segment under increasing pressure from government to produce cleaner burning motor fuels.

Nowhere was the pressure more intense than in pollution-plagued California, America's No. 1 gasoline consuming state. Motorists there burn about 12% of all gasoline consumed in the U.S. State authorities are determined to achieve a sharp reduction in tailpipe emissions from burning of gasoline and diesel fuel.

U.S. refiners also were under pressure from company management to improve profits despite lackluster demand for products.

Figures compiled by the American Petroleum Institute pegged U.S. demand for all products at 17.241 million b/d through the first 9 months of 1993, up a bare 1.8% from the same period of 1992. Demand for motor gasoline, the industry's big ticket item, averaged 7.422 million b/d during the first 9 months of 1993, up only 2.1% from the same period of 1992.

Refiners were running virtually flat out with a 90.3% utilization rate of operable capacity, API figures showed. The rate was up slightly from 87.8% in the first 9 months of 1992.

A rising share of refinery crude runs was coming from imports as domestic crude oil production continued its decline and petroleum products consumption kept inching up.

Refiners' environmental cost

The National Petroleum Council (NPC) in September 1993 reported to Energy Sec. Hazel O'Leary the U.S. refining industry will be able to meet current stringent environmental rules on refineries and products. But the costs, already high, will climb much higher.

If several of the study scenarios prove correct, the U.S. refining industry, the public, and Department of Energy (DOE) will be paying a great deal more attention to the NPC report as the environmental bills fall due in the balance of the 1990s, Oil & Gas Journal (OGJ) reported.

O'Leary said DOE will use the NPC report in discussions with the Environmental Protection Agency (EPA) on ways to identify cost effective means to meet the government's environmental goals.

Deputy Energy Sec. Bill White and his task group working on the administration's domestic oil and gas initiative were scheduled to review the NPC study and consider its findings when developing their report.

The NPC study, which had been under way for 2½ years and was billed as the most comprehensive ever done of the U.S. refining industry, predicted that during 1991-2000 the U.S. refining industry will face capital outlays of $36.5 billion in 1990 dollars to comply with environmental rules.

About two thirds of that sum will be spent in 1991-95. About $22.8 billion will go for stationary sources facilities such as refinery emission controls, while $13.7 will be required to meet product quality—units and equipment to make reformulated gasoline, for example.

Total forecast outlays were $5.5 billion more than the present book value of the entire U.S. refining industry.

It was estimated that about half the $13.7 million for product quality had already been spent vs. about 15% of the projected $22.8 billion for stationary source facilities.

Longer term, the report said, outlays of $106 billion are projected during 1991-2010 for new facilities and programs required for compliance with existing and anticipated stationary sources rules related to air, water, and waste and to safety and health requirements in U.S. refineries. Those are capital expenditures, one-time expenses, and operating and maintenance costs.

In addition, $46 billion will be spent to operate and maintain similar facilties and programs already in place. That will bring total stationary facilities environmental health and safety expenditures to $152 billion.

Those costs are in addition to outlays for manufacturing reformulated gasoline and ultralow sulfur diesel fuel.

A result of the required outlays will be an average cost increase for light products of 6¢/gal in 1995 over 1989 and another 10¢/gal in 2000 (Fig. 1). Reformulated fuel taken alone will cost in 1995 about 8¢/gal more than 1989 conventional gasoline did and 12¢ and 14¢/gal more in 2000 and 2010, respectively.

On a positive note, the study saw an adequate supply of oxygenates worldwide in 1995 for at least the legislated mininum requirement in oxygenated gasoline. By 2000 the potential supply is expected to cover esentially any situation permitted by the U.S. Clean Air Act (CAA) amendments of 1990.

Refinery shutdowns

Heavy capital and operating costs for environmental protection programs will create further pressure to shut down U.S. refineries.

Kenneth T. Derr, chairman of the NPC study group and chairman of the board and chief executive officer of Chevron Corp., pointed out there are major environmental disincentives to shutting down even a poorly performing refinery. However, the NPC report said there will be pressure to shut down refineries in the U.S. if products demand declines or lower cost refined products enter the country.

The study compared world regions and found that each had lower standards than the U.S. refining industry in 1990. Refiners on the Pacific Rim and in the Middle East and Latin America must cope with far fewer regulations than U.S. refiners and could be sources of low cost product. Many must beef up their refineries if they wish to supply advanced fuels to the U.S.

However, the biggest threat to the U.S. refining system, according to comments by the study group, is a February 1993 proposal by the EPA that would require stringent separation of reformulated gasoline in the country's pipeline system. Derr said if such a regulation is implemented, the game is over, and "we can all go home."

A member of the study group at a Houston press conference corrected him with: "We can all walk home."

API voices concern

API said the NPC study showed the U.S. refining industry faced an imminent, staggering financial challenge to meet environmental regulations.

API Pres. Charles DiBona said, "We are concerned that this challenge comes at a critical point for the nation's petroleum refining industry, which is vital to the nation's economic health."

He too reminded that U.S. investments in the 1990s to meet environmental requirements on refineries and petroleum products are predicted to be billions more than the current book value of all refineries in the country.

He said that follows the 1980s, when the industry earned a profit of only 2.5¢/gal and realized an average return on investment of 8.8%, and 120 refineries were forced to close. Profits in 1991 and 1992 were much lower than in the previous decade, and refineries continued to shut down.

DiBona said, "In effect, the consumer pays an indirect tax for mandated environmental improvements and deserves to be assured of commensurate benefits. That's why it is critical, as the NPC study recommends, that policymakers carefully and realistically evaluate the impacts of policy before legislation is passed or policies established.

Fig. 1

Cost increase for reformulated gasoline*

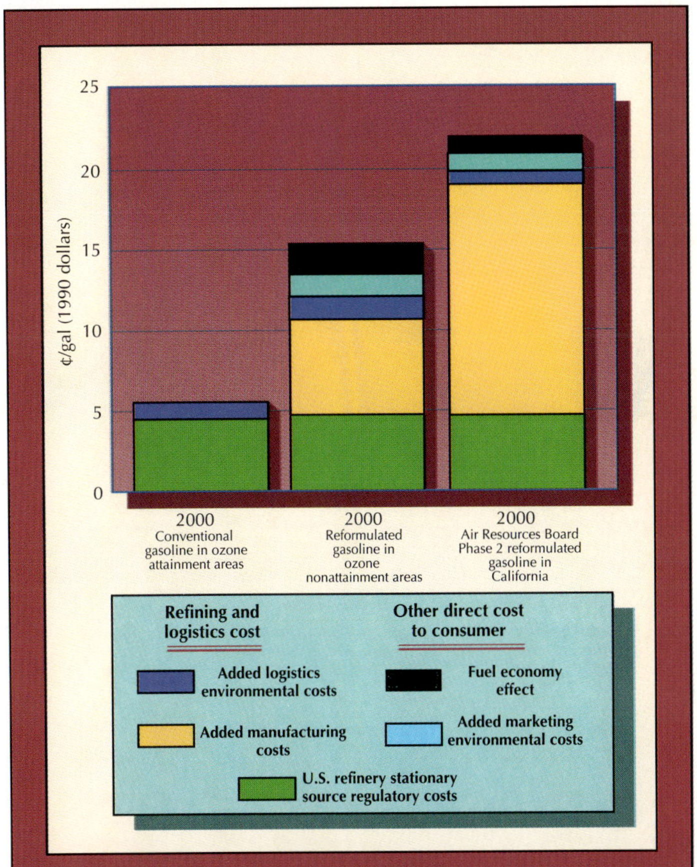

Source: National Petroleum Council

Table 1

What's ahead for U.S. oil and gas supply/demand

	1993	1995	2000	1993-95 % change	1995-2000 % change
	1,000 b/d				
PETROLEUM					
Demand	18,082	18,543	19,458	+1.3	+1.0
Production	8,652	8,286	7,673	−2.1	−1.5
Imports	8,210	8,841	10,148	+3.8	+2.8
Total supply*	**18,075**	**18,297**	**19,136**	**+0.6**	**+0.9**
Imports as % of demand	47.8	50.2	54.9	–	–
NATURAL GAS	bcf				
Demand	23,129	23,892	25,527	+1.6	+1.3
Production†	18,276	18,814	19,739	+1.5	+1.0
Imports	2,211	2,536	3,124	+7.1	+4.3
Withdrawals from storage	2,524	2,659	2,747	+2.6	+0.7
Supplemental gas fuels	117	125	148	+3.4	+3.4
Total supply	**23,128**	**23,279**	**25,758**	**+0.3**	**+2.0**

*Includes refinery processing gain, unaccounted for crude oil, and other hydrocarbons. Excludes Strategic Petroleum Reserve. †Dry.

Source: Supply/demand committee, Independent Petroleum Association of America

"Policies that unnecessarily reduce demand for petroleum products also result in lost jobs. And, over the long term, refining expenditures will be reflected in the marketplace."

UNITED STATES

State	No. Refineries
Alabama	3
Alaska	6
Arizona	1
Arkansas	3
California	25
Colorado	3
Delaware	1
Georgia	2
Hawaii	2
Illinois	7
Indiana	4
Kansas	4
Kentucky	2
Louisiana	18
Michigan	4
Minnesota	2
Mississippi	5
Montana	4
Nevada	1
New Jersey	6
New Mexico	3
New York	1
North Dakota	1
Ohio	4
Oklahoma	7
Oregon	1
Pennsylvania	7
Tennessee	1
Texas	29
Utah	6
Virginia	1
Washington	7
West Virginia	2
Wisconsin	1
Wyoming	4
TOTAL	**178**

CANADA

Province	No. Refineries
Alberta	5
British Columbia	3
New Brunswick	1
Northwest Territories	1
Nova Scotia	2
Ontario	7
Quebec	3
Saskatchewan	2
Newfoundland	1
TOTAL	**25**

For individual locations and capacities, see detailed maps on following pages.

Spending by major refiners

Derr's Chevron in autumn 1993 unveiled an example of what refiners must spend to comply with product environmental standards. The company plans outlays of nearly $1 billion at two big California refineries to produce reformulated gasoline.

The new fuels will meet environmental standards required by federal CAA amendments and the California Air Resources Board (CARB). Federal law requires oxygenated motor gasoline in all 39 urban areas exceeding the carbon monoxide standard during winter months, generally from November through February, when carbon monoxide levels are higher.

About $370 million will be spent at Chevron's 229,000 b/cd Richmond refinery, near San Francisco, and $580 million at the 244,000 b/cd El Segundo refinery, near Los Angeles. Construction was scheduled during the ensuing 3 years.

The projects show that preparing to produce environmentally improved gasoline will be very costly, said Bruce Frolich, vice-president of manufacturing for Chevron U.S.A. Products Co.

"We want consumers to recognize that these kinds of mandated invest

LEGEND

- Oil field
- Oil shale
- Oil sand
- Gas field
- Crude oil pipeline
- Natural gas pipeline
- Products pipeline
- Coal slurry pipeline
- Pipeline planned or under construction
- Represents two or more pipelines
- Refinery in operation
- 600 Refinery capacity in 10,000 b/d.
- Tanker terminal
- Cities
- Capital
- International boundaries
- State and provincial boundaries
- Water depth 0 to 200 meters
- 200 meters and deeper

HALLIBURTON CUSTOMIZED SERVICES PROVIDE BIGGER PAYBACKS ON WELLS THE WORLD OVER.

Sources and uses of U.S. oil supply

Fig.

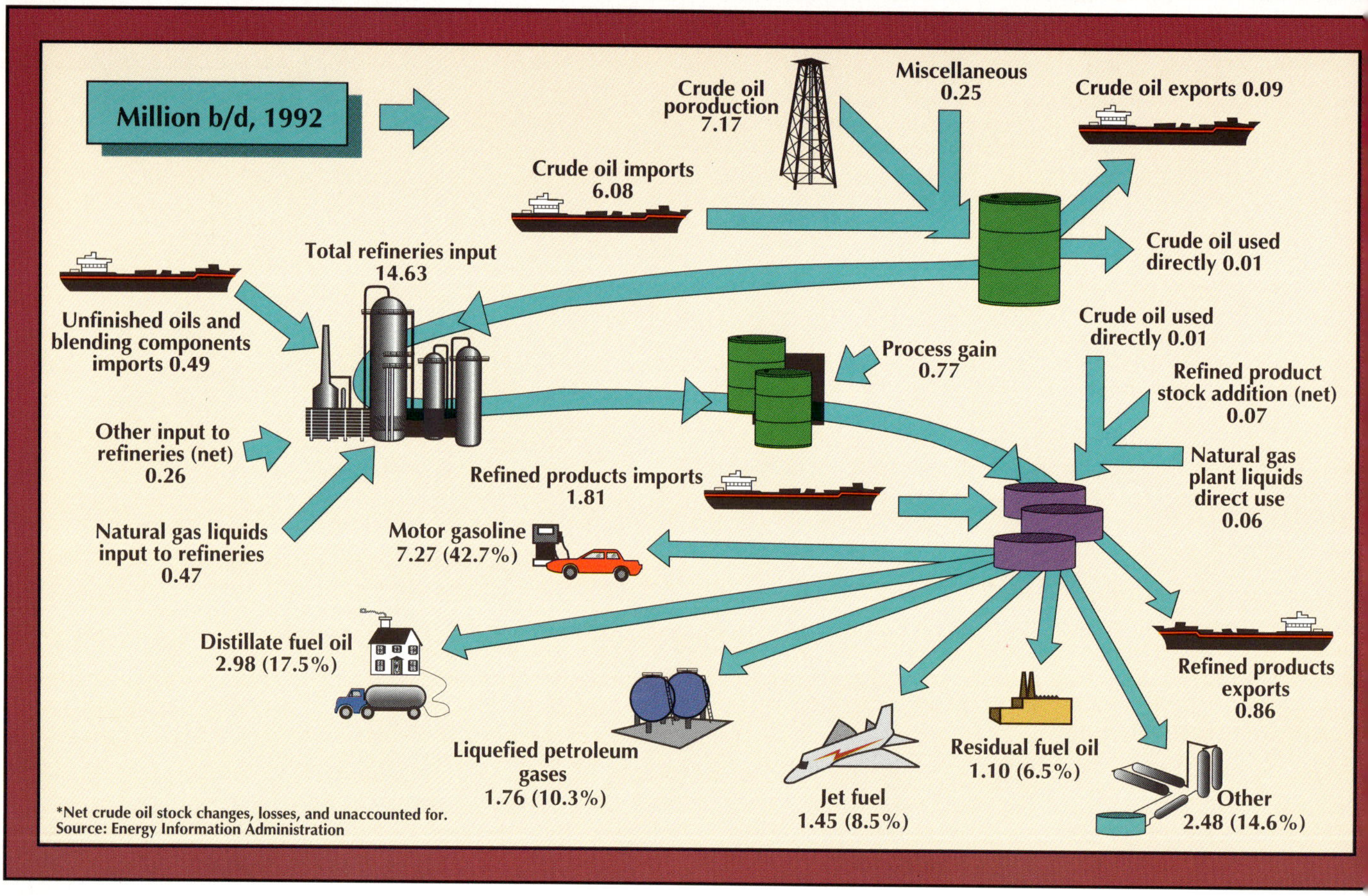

ments—and the higher costs of making reformulated gasolines—will likely mean higher fuel prices."

Frolich said consumer cost increases in California could amount to 8¢/gal beyond the added cost stemming from CAA amendment standards. That's because the state's standards are tougher than national ones.

In addition to the two big environmental projects, Chevron approved a $290 million upgrade of the Richmond refinery's fluid catalytic cracking unit. The upgrade was scheduled to employ new instrumentation and computer technology to boost yields of gasoline and jet fuel.

Among other large refiners, a joint venture of Shell Oil Co. and Mexico's state owned Petroleos Mexicanos in July 1993 began a major refinery upgrade at Shell's Deer Park, Tex., manufacturing complex.

Shell and Pemex said ground breaking for the 2 year long construction project marks the beginning of a new era for their Deer Park Refining Ltd. Partnership (Dprlp), formed in February 1993.

Upon completion in mid-1995, the $1 billion upgrading program will enable the 215,900 b/cd refinery to process as much as 100,000 b/d of heavy Mayan crude oil to be supplied by Pemex, while at the same time reducing air emissions. The plant's crude charge capacity will not change.

Elsewhere, Lyondell Petrochemical Co., Houston, and Citgo Petroleum Co., Tulsa, signed a final agreement calling for an $800 million upgrade of Lyondell's 265,000 b/d Houston refinery.

The accord resulted in creation of a new company, Lyondell-Citgo Refining Co. Ltd., jointly owned by affiliates of Lyondell and Citgo.

The latter is an indirect, wholly owned unit of Venezuela's state owned Petroleos de Venezuela SA. Also involved in the agreement is a long term, firm supply of heavy Venezuelan crude. Volume was not disclosed.

IPAA forecasts record oil imports

A midyear report by the Independent Petroleum Associ tion of America predicted that a modest increase in U. petroleum product demand will combine with a decline i U.S. crude production to yield a record level of crude oil an petroleum products imports when all figures are in for 19 (Table 1). A breakout by the Energy Information Administr tion (EIA) of U.S. oil supply and consumption showed th crude oil imports amounted to 6.08 million b/d, or 45% of th country's 1992 crude supply of 13.5 million b/d. Unfinished c and blending components added 49,000 b/d to supply th year, while refined products imports amounted to 1.8 million b/d (Fig. 2).

EIA said the volume of total U.S. petroleum impor climbed to a record of more than 8 million b/d in 1977, fell t a little more than 4 million b/d in the first half of the 1980s i response to a flurry of U.S. drilling and well completions, the began a rebound to more than 6 million b/d in the late 198 and early 1990s (Fig. 3). The merchandise trade value of net o imports peaked at more than $9 billion/year in 1980.

IPAA's supply/demand committee forecast a 1.1% increas in U.S. demand for petroleum products in 1993, coupled wit a 2.5% drop in crude production. That opened a wideni gap to be met by petroleum imports, forecast to increas 4.7%.

Petroleum product demand was expected to continu rising while crude production declines through 2000. Tot imports will move up to fill the growing gap, increasing t 54.9% of U.S. oil supply by 2000, the committee reported.

Consumption, production, and imports of natural gas als will increase steadily through 2000, with gas expected t capture a growing share of the energy market.

The committee predicted the second straight year of risin U.S. petroleum product demand following 2 years of decline

Demand will increase 1.1% in 1993 to 17.189 million b/d. That compares with an increase of 1.7% in 1992 to 17.006 million b/d. The recession and sluggish recovery led to a decline in demand of 1.9% in 1990 and a decline of 1.6% in 1991 to 16.714 million b/d.

Motor gasoline demand will increase for the second straight year in 1993 after 3 years of decline. Motor gasoline demand for 1993, was forecast at 7.33 million b/d, up 0.8%, while demand in 1992 moved up 1.2% to 7.271 million b/d. The record high was 7.412 million b/d in 1978.

Fuel efficiency gains will be more than offset by additional miles driven, the IPAA committee said. In recent years motor vehicle fleet efficiency gains were slowed by sluggish auto sales. Demand for aviation fuel will resume its growth in 1993 after a 2 year slide, increasing 1.7% to 1.475 million b/d. Demand fell 1.4% in 1992. The air traffic industry forecast an increase in revenue from passenger miles and freight ton miles flown.

Demand for distillate fuel oil will rise 2.8% in 1993 to 3.064 million b/d following an increase of 2% in 1992. Gains were expected in all end use sectors except residential and commercial heating.

The rate of decline in residual fuel consumption was predicted to slow to 1.8% in 1993 to 1.072 million b/d. Demand fell 5.7% in 1992.

Demand for all other petroleum products will increase 0.9% in 1993 to 4.248 million b/d. Demand for other products increased 5.2% in 1992. The growth in 1993 was expected mainly from higher demand for liquefied petroleum gases, asphalt, and petrochemicals.

The IPAA committee's 1993 report warned the slide in U.S. crude oil production that started in 1985 will continue for the ensuing few years.

Production, which fell 3.6% to 7.153 million b/d in 1992, will slip another 2.5% to 6.974 million b/d in 1993, the lowest level since 1958.

The decline in number of wells drilled, number of rotary rigs working, and other industry indicators pointed to continued erosion in crude oil production and reserves.

Natural gas liquids production also will slip in 1993, down 1.1% at 1.678 million b/d. As a result, total liquids output will be down 2.2% at 8.652 million b/d, the lowest since 1963.

Slumping production and rising demand will lead to another increase in imports in 1993, the committee said.

Total crude oil and petroleum product imports will advance 4.7% in 1993 to 8.212 million b/d, after a 2.9% increase in 1992. The record level of petroleum imports was 8.876 million b/d in 1977.

Final figures were destined to show U.S. import dependency reached a record high in 1993 at 47.8% of demand—up from 46.1% in 1992 and 45.6% in 1991. The current record was 47.7% of supply in 1977.

...oducers seek ...lief from imports

IPAA early in December 1993 asked the Clinton administration to grant U.S. oil producers relief from low priced oil imports.

IPAA and 20 allied oil associations asked the Commerce Department for expedited relief under Section 232 of the 1962 Trade Expansion Act.

They did not seek a specific type of relief, such as an oil import fee, leaving options open for the administration. IPAA said the administration could declare an import fee, quotas, or a floor price, offer tax incentives for marginal production, or take a variety of diplomatic actions.

The request was filed just a few days before the Department of Energy released an oil and gas initiative designed to aid the U.S. oil industry. The initiative did contain substantive relief for producers, but IPAA officials noted it was premised on $18-19/bbl oil.

George Alcorn, IPAA chairman, in a letter to Clinton said,

How U.S. oil imports have varied Fig. 3

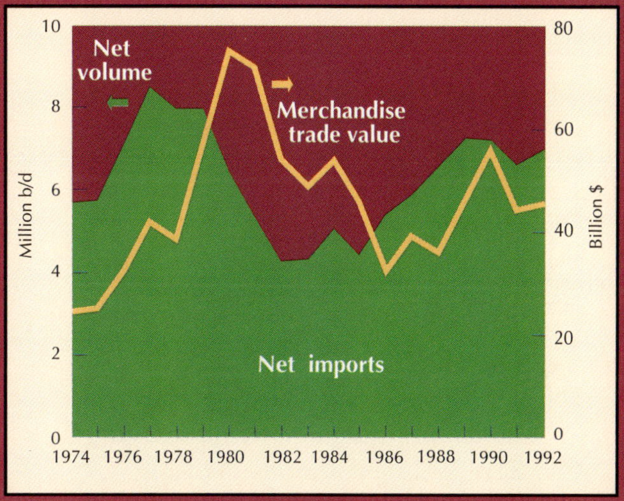

Source:Energy Information Administration

"Since the beginning of this year, oil prices have fallen from $19/bbl to $14.95 on Dec. 3. In the field, producers are receiving around $13. In California, where Alaskan oil skews the market, producers are receiving $8.25.

"In constant dollars, this price is more than $1 lower than in 1986, when the oil market crashed, bringing down businesses and financial institutions across the country and sending the Southwest into a depression.

"Over 20% of our production in the U.S.—about 1.3 million b/d—is marginally economic and labor intensive, and current prices are well below the cost of producing this oil. If this situation persists, the U.S. could easily lose an additional 100,000 oil industry jobs on top of the 450,000 jobs we have lost in the last decade.

"With this growing level of cheap imports, energy conservation measures are hopeless, and work on alternative fuels will grind to a halt. Already, utilities and industrial users that can switch to cheap imported fuel oil are turning away from natural gas. Imported fuel oil is now cheaper than coal."

IPAA was the lead association filing the petition, although it had declined to join some of its member companies in a similar petition after the 1986 price collapse. IPAA officials said the Clinton administration's receptive attitude made the situation different.

After the earlier petition was filed, Commerce conducted a lengthy study that concluded oil imports threaten U.S. national security. But the Reagan and Bush administrations declined to take action to restrict oil imports. IPAA maintained that study is still valid and another review, which could take a year or longer to prepare, is not necessary.

Alcorn quoted Clinton as saying during his campaign in 1992, "Our reliance on foreign oil is a genuine threat to our national and economic security. (In 1988) foreign oil made up a third of our trade deficit...Now we import nearly half of our oil, which accounts for two thirds of our trade deficit. That kind of dependence makes us vulnerable, and we must change the situation."

Sens. David Boren (D-Okla.), a finance committee member, and Bennett Johnston (D-La.), energy committee chairman, urged the administration to seriously consider the petition.

Boren said, "If prices were to stay at this artificially low level for 6 months to a year, Oklahoma could lose over 10,000 jobs and see its economy shrink by $1 billion."

Johnston said, "The collapse of oil prices will mean further curtailments in exploration and drilling, more layoffs in the

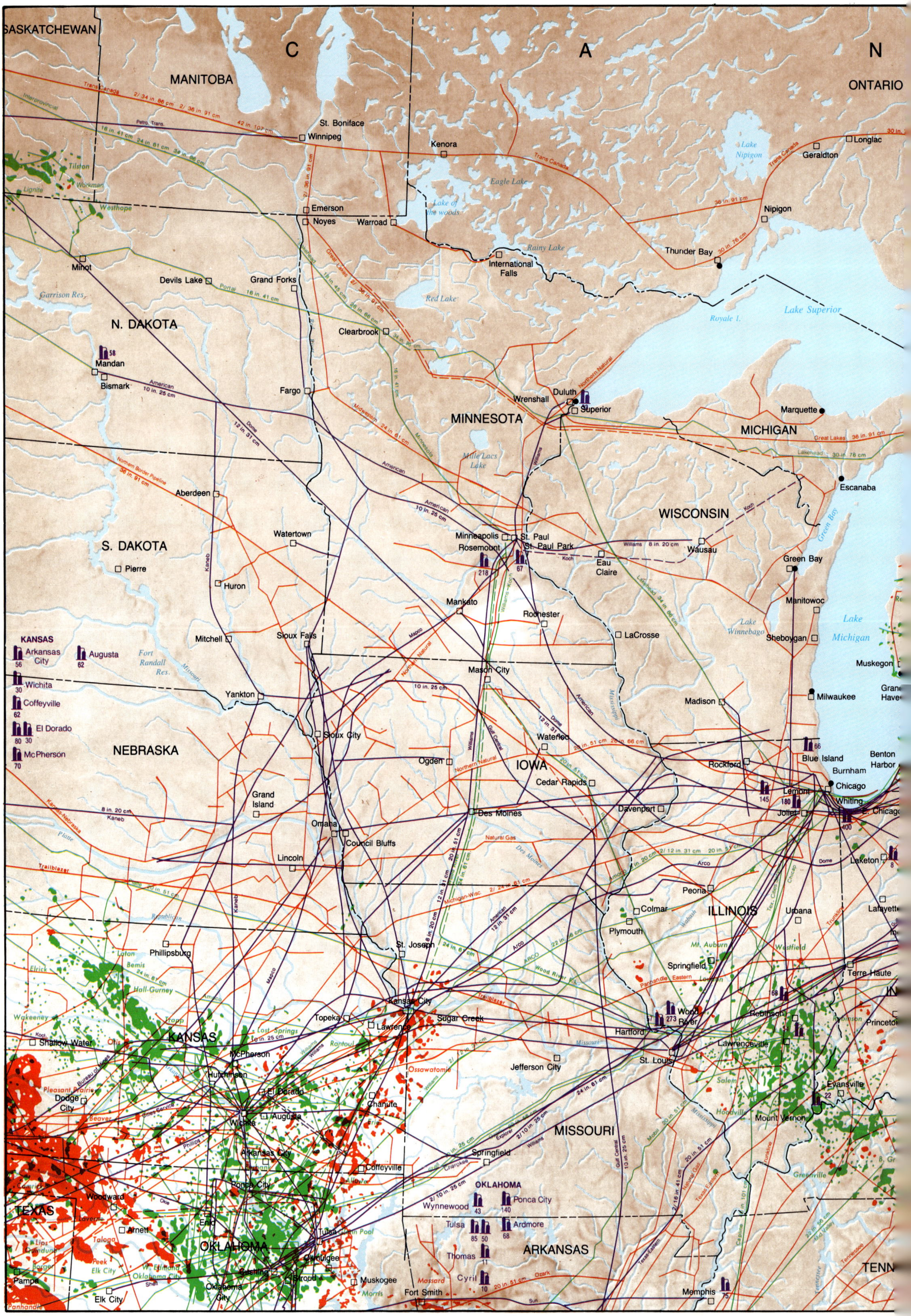

SASKATCHEWAN
MANITOBA
C
A
N
ONTARIO
St. Boniface
Winnipeg
Kenora
Longlac
Geraldton
Lake Nipigon
Eagle Lake
Lake of the woods
Emerson
Noyes
Warroad
Nipigon
Rainy Lake
International Falls
Thunder Bay
Minot
Devils Lake
Grand Forks
Garrison Res.
Red Lake
Lake Superior
Royale I.
N. DAKOTA
Clearbrook
Mandan
Bismark
Fargo
Duluth
Wrenshall
Superior
Marquette
MINNESOTA
MICHIGAN
Mille Lacs Lake
Aberdeen
Escanaba
WISCONSIN
Watertown
Minneapolis
St. Paul
Rosemount
St. Paul Park
Wausau
S. DAKOTA
Pierre
Eau Claire
Green Bay
Huron
Mankato
Rochester
Manitowoc
KANSAS
Arkansas City
Augusta
Mitchell
Sioux Falls
LaCrosse
Lake Winnebago
Sheboygan
Lake Michigan
Fort Randall Res.
Muskegon
Wichita
Coffeyville
El Dorado
McPherson
Yankton
Mason City
Madison
Milwaukee
Sioux City
Waterloo
Blue Island
Benton Harbor
Burnham
NEBRASKA
Ogden
IOWA
Rockford
Chicago
Lemont
Joliet
Whiting
Cedar Rapids
Grand Island
Des Moines
Davenport
Omaha
Council Bluffs
Lincoln
Laketon
Peoria
Colmar
Plymouth
ILLINOIS
Urbana
Phillipsburg
St. Joseph
Springfield
Terre Haute
Kansas City
Topeka
Lawrence
Sugar Creek
Wood River
Robinson
Shallow Water
KANSAS
McPherson
Hartford
Lawrenceville
Hutchinson
Jefferson City
St. Louis
El Dorado
Dodge City
Chanute
Augusta
Wichita
Evansville
Mount Vernon
MISSOURI
Springfield
Arkansas City
Coffeyville
OKLAHOMA
Ponca City
Wynnewood
Ponca City
TEXAS
Woodward
Tulsa
Ardmore
Arnett
Enid
OKLAHOMA
Thomas
ARKANSAS
Cushing
Okmulgee
Cyril
Muskogee
Pampa
Oklahoma City
Elk City
Fort Smith
Memphis
TENN
Trans Canada
Interprovincial
Northern Natural
Great Lakes
Williams
American
Dome
Kaneb
Northern Border Pipeline
Koch
Lakehead
Mapco
Gulf Central
Trailblazer
Natural Gas
Michigan-Wisc
Arco
Texas Eastern
Panhandle Eastern

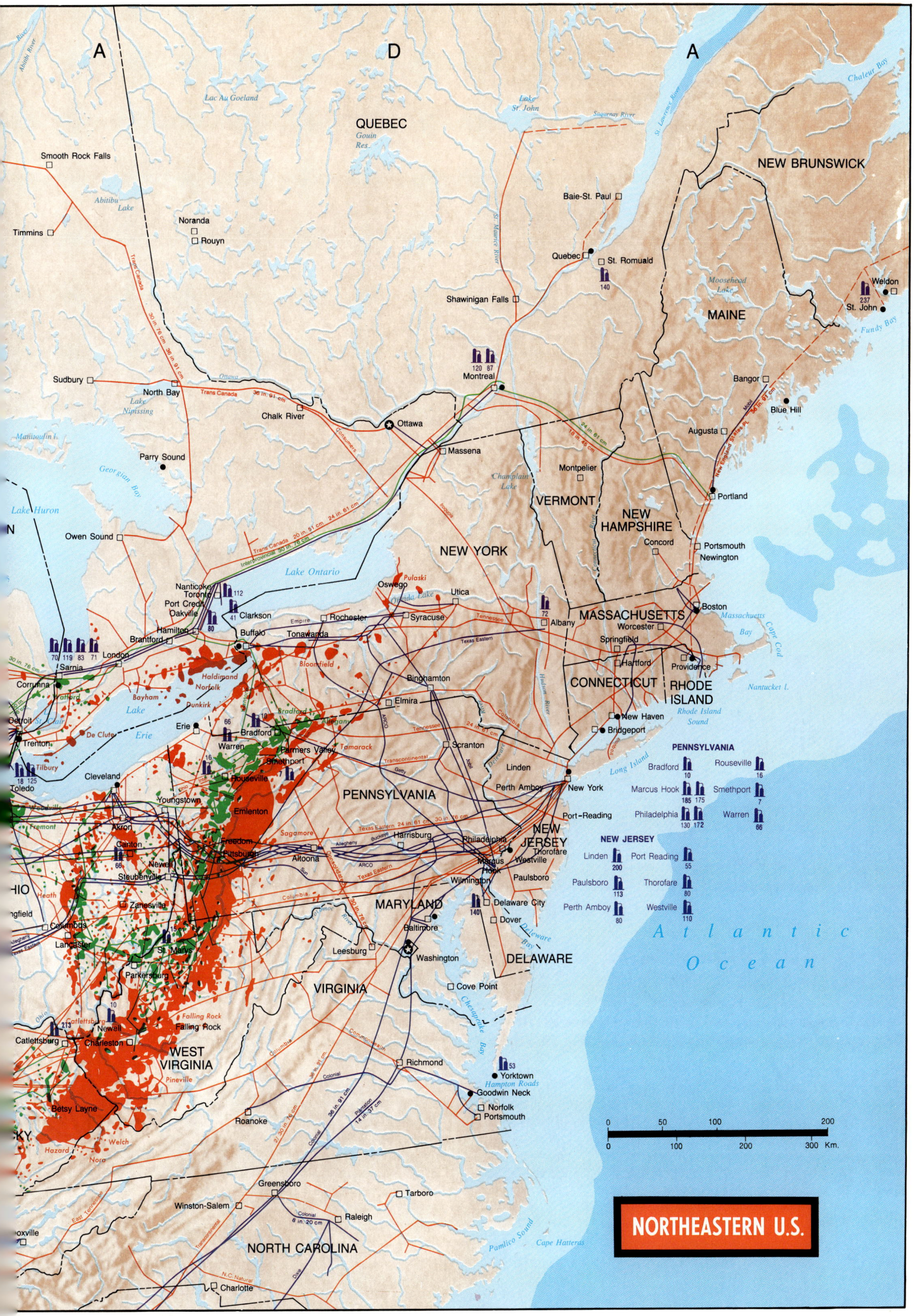
NORTHEASTERN U.S.
QUEBEC
NEW BRUNSWICK
MAINE
VERMONT
NEW HAMPSHIRE
NEW YORK
MASSACHUSETTS
CONNECTICUT
RHODE ISLAND
PENNSYLVANIA
NEW JERSEY
MARYLAND
DELAWARE
VIRGINIA
WEST VIRGINIA
NORTH CAROLINA
Atlantic Ocean
Lake Huron
Lake Ontario
Lake Erie
Georgian Bay
Smooth Rock Falls
Timmins
Noranda
Rouyn
Sudbury
North Bay
Chalk River
Ottawa
Parry Sound
Owen Sound
Montreal
Quebec
St. Romuald
Baie-St. Paul
Shawinigan Falls
Massena
Montpelier
Portland
Augusta
Bangor
Blue Hill
St. John
Weldon
Concord
Portsmouth
Newington
Boston
Worcester
Springfield
Hartford
Providence
New Haven
Bridgeport
Albany
Utica
Syracuse
Oswego
Rochester
Buffalo
Tonawanda
Clarkson
Toronto
Nanticoke
Port Credit
Oakville
Hamilton
Brantford
London
Sarnia
Corunna
Detroit
Trenton
Toledo
Cleveland
Erie
Warren
Bradford
Farmers Valley
Smethport
Rouseville
Emlenton
Youngstown
Akron
Canton
Newell
Steubenville
Freedom
Pittsburgh
Zanesville
Columbus
Lancaster
Parkersburg
St. Marys
Catlettsburg
Charleston
Falling Rock
Betsy Layne
Binghamton
Elmira
Scranton
Linden
Perth Amboy
New York
Port-Reading
Harrisburg
Altoona
Philadelphia
Marcus Hook
Thorofare
Westville
Paulsboro
Wilmington
Delaware City
Dover
Baltimore
Washington
Leesburg
Cove Point
Richmond
Yorktown
Goodwin Neck
Norfolk
Portsmouth
Roanoke
Greensboro
Winston-Salem
Raleigh
Tarboro
Charlotte
Knoxville
PENNSYLVANIA
Bradford 10
Rouseville 16
Marcus Hook 185 175
Smethport 7
Philadelphia 130 172
Warren 66
NEW JERSEY
Linden 200
Port Reading 55
Paulsboro 113
Thorofare 80
Perth Amboy 80
Westville 110
0 50 100 200
0 100 200 300 Km.

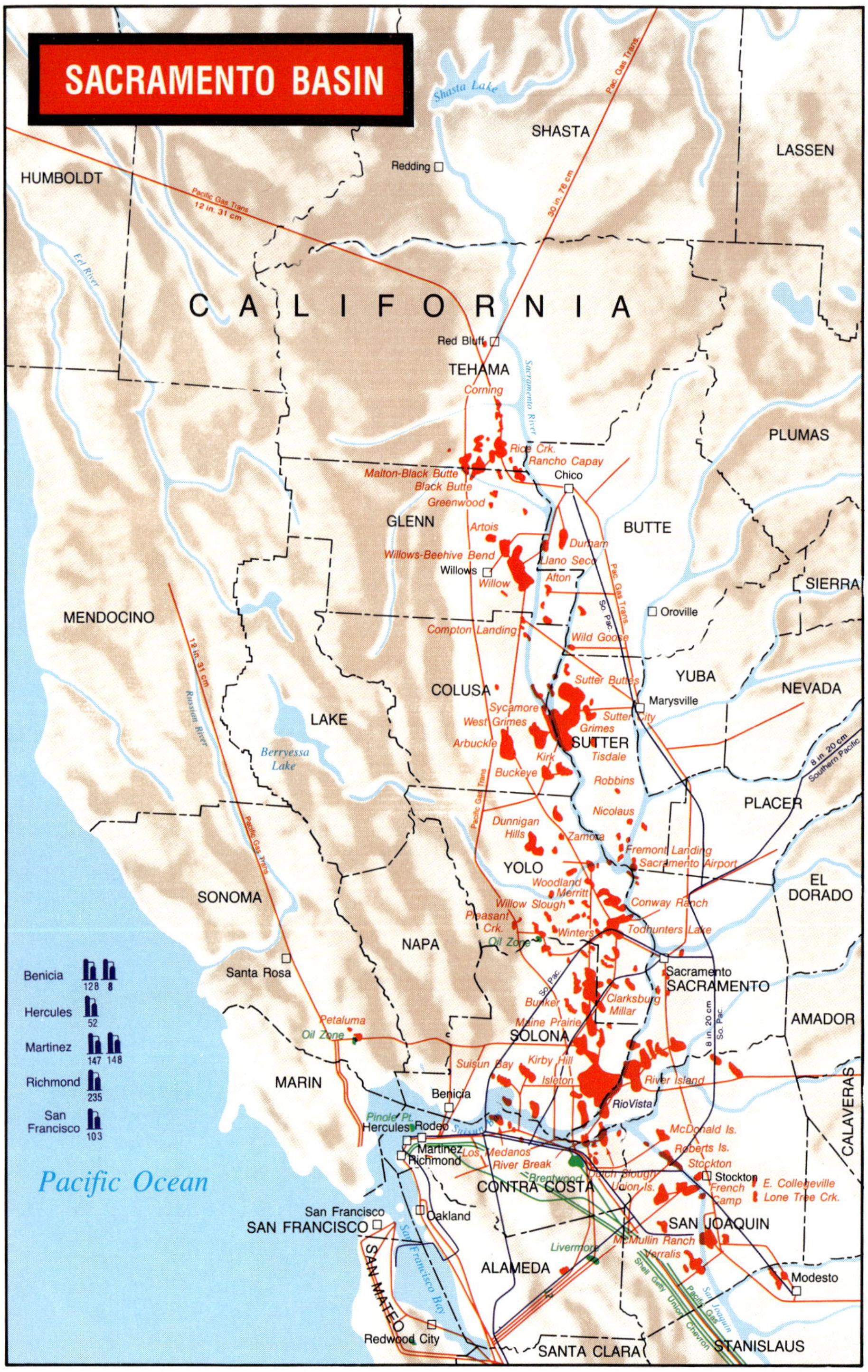

industry, and even greater long term dependence on imported oil."

American Gas Association Pres. Mike Baly said low oil prices could result in flat U.S. gas consumption in 1994, rather than the 2.4% increase AGA predicted.

DOE's O'Leary early in December 1993 unveiled the outline of a presidential initiative aimed at boosting markets for U.S. oil and gas while developing a long term strategy to reduce dependence on imported oil. Largely involving only plans for studies, the administration planned to:

- Review tax incentives to attract capital into the petroleum industry.
- Increase oil and gas production and environmental protection through development and application of technology.
- Stimulate markets for natural gas, including natural gas derived products.
- Cut the cost of regulation where applicable while maintaining environmental protection.

Natural gas supply/demand

IPAA's supply/demand committee's midyear 1993 report said the year's increases in natural gas consumption and production will be closely aligned at 2.1% and 2.5%, respectively, while imports of natural gas will increase 7.2%.

Increased economic activity and

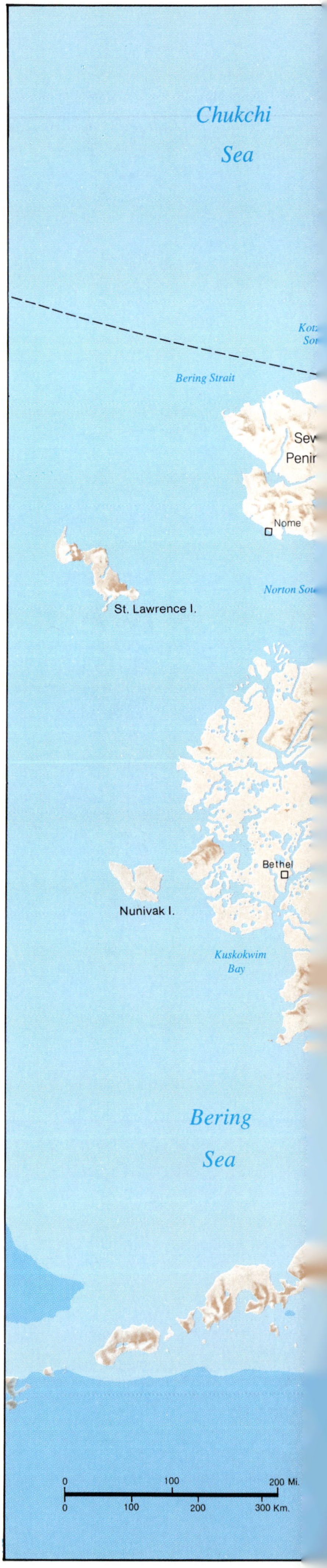

Point Barrow
Barrow
Barrow
Smith Bay
Simpson
Beaufort Sea
Arctic Plains
Harrison Bay
Prudhoe Bay field unit
Prudhoe Bay
Kuparuk
Prudhoe Bay
Endicott
Point Thomson
Flaxman Is.
Camden Bay
Umiat
Gubik
E. Umiat
Kavik
Petroleum Reserve No. 4
Arctic Wildlife Reserve
rooks Range
10 in. 25 cm
Tuktoyaktuk Pen.
Koakoak
Kopanoar
Amauligak
Issungnak
Atkinson
Pitsiulak
Tarsiut
Itiyok
Richard I.
Ivik
Mayogiak
Adgo
Mallik
Tuk
Niglintgak
Taglu
Kumak
Ya Ya
Parsons Lake
Titalik
Reindeer
Mackenzie Bay
Inuvik
Polar Gas
Mackenzie River
Fort Good Hope
Norman Wells
Arctic Circle
Fort Yukon
Porcupine
Circle
Dempster lateral
YUKON
NORTHWEST TERRITORIES
Tanana
Yukon River
Yukon River
Fairbanks
North Pole
LASKA
Delta Jct.
Dawson
48 in. 122 cm
Northwest Alaskan Pipeline
Tanana River
Alaska Range
Susitna River
CANADA
Willow
Palmer
Chugach Mountains
Anchorage
Ivan River
Beluga River
Granite Point
North Cook Inlet
Trading Bay
Birch Hill
McArthur River
Swanson River
MGS
Nikiski
Drift River
Kenai
Kenai
Kenai Pen.
Valdez
Gravina Point
Cordova
Montague I.
Whitehorse
Haines Jct.
Foothills
56 in. 142 cm
Watson Lake
Yakataga
Yakutat
Skagway
Haines
Juneau
Cook Inlet
Kodiak
Kodiak I.
Gulf of Alaska
Sitka
Wrangell
Ketchikan
ASKA

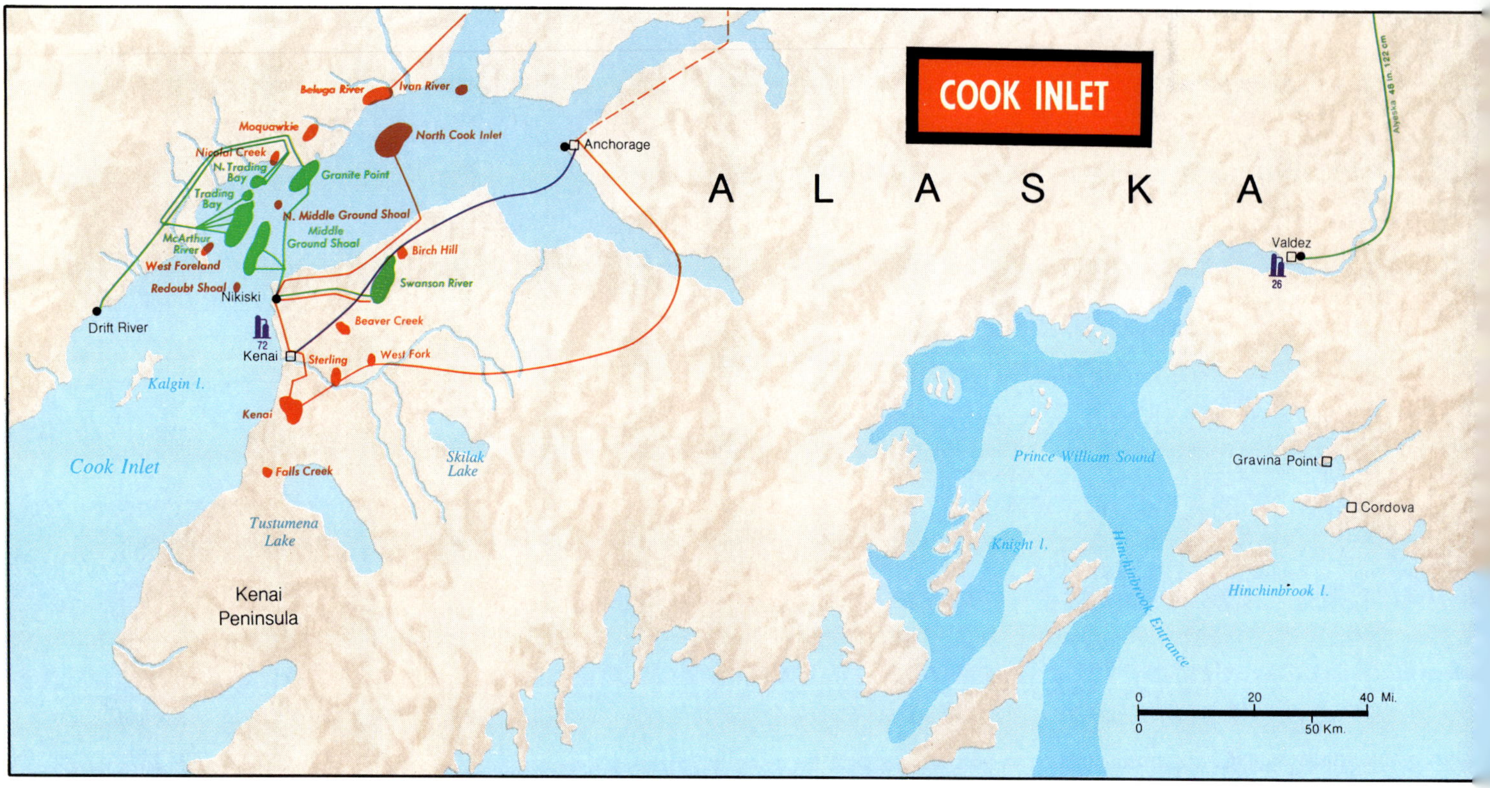

colder weather in first quarter 1993 was predicted to boost the year's gas consumption 2.1% to 20.148 tcf following a 3.5% increase in 1992, the IPAA committee said.

EIA's breakout of that year's supply and disposition showed that 17.9 tcf went to U.S. consumers, with industrial users accounting for nearly 42% of sales volumes (Fig. 4).

The industrial sector posted the strongest gain in 1992, mostly due to continued growth of gas fired cogeneration, IPAA's analysis showed. Growth in industrial demand was expected to lead the way again in 1993, rising 2.7% to 7.846 tcf.

Consumption increased in the residential and commercial sectors during 1992 as many customers began to take advantage of new pipeline capacity, particularly in the Northeast.

Growth of customer numbers in the residential and commercial sectors has been strong. Since 1987 natural gas has captured a dominant share of the new housing market. Conversion to natural gas from other fuels also has increased in recent years. And cogeneration in the commercial sector is increasing. Residential consumption will move up 2.6% in 1993 to 4.825 tcf, and commercial demand will be up 1.3% at 2.809 tcf, committee figures showed.

In 1992 the electric utility sector posted a slight decrease in consumption due to mild summer weather in many regions, especially the Southwest. Utility demand was expected to grow

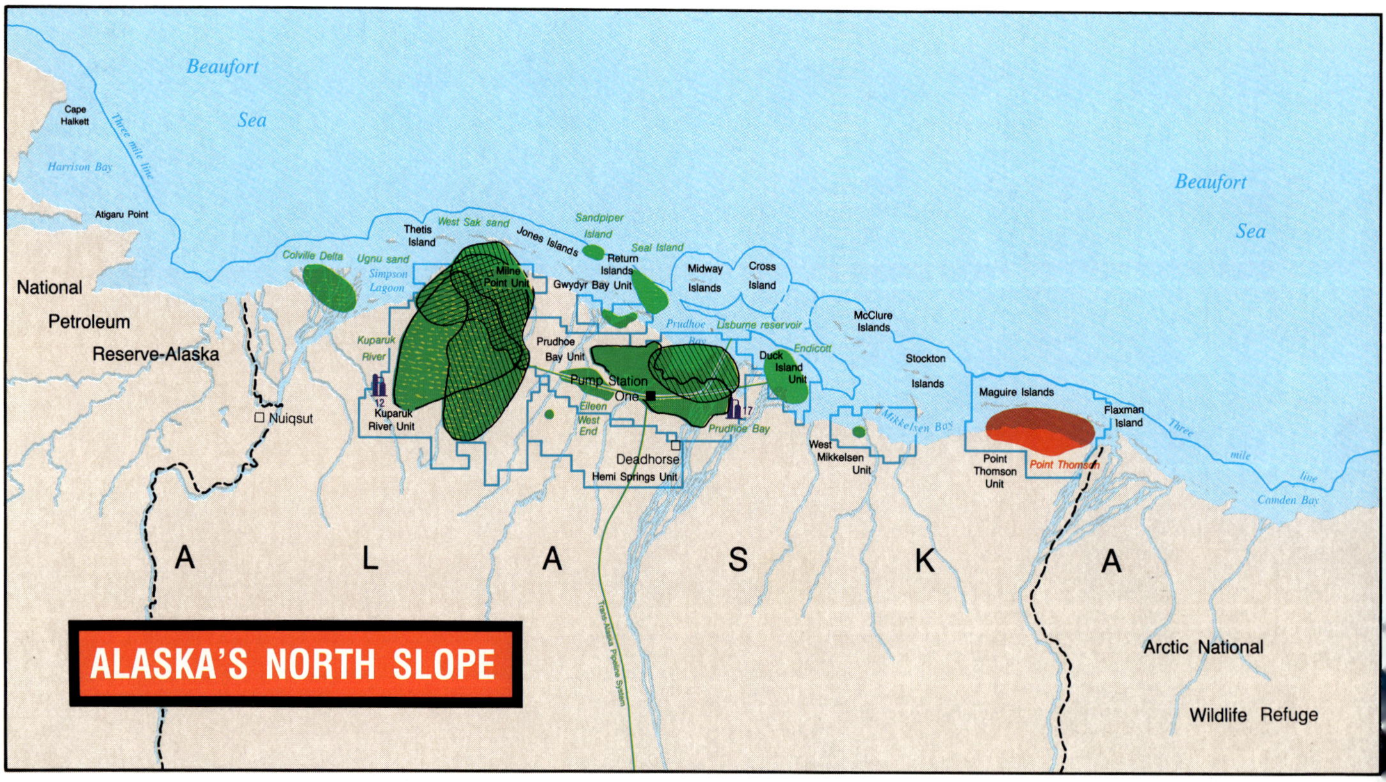

CALIFORNIA
OREGON
IDAHO
NEVADA
CALIFORNIA
ARIZONA
MEXICO
Pacific Ocean
Grants Pass
Medford
Talent
Ashland
Klamath Falls
Rome
Triangle
Fields
Alturas
Eureka
Tompkins Hill
Grizzly Bluff
Petrolia
Susanville
Humboldt
Elko
Sparks
Reno
Virginia City
Carson City
Santa Rosa
Sacramento
Hercules
Benicia
Martinez
Richmond
San Francisco
Oakland
Hayward
Redwood City
San Jose
Santa Cruz
Salinas
Modesto
Fresno
Hanford
Bishop
Tonopah
Warm Springs
Oildale
Bakersfield
San Luis Obispo
Santa Maria
Santa Barbara
Ventura
Oxnard
Newhall
Los Angeles
Torrance
Long Beach
Santa Fe Springs
San Bernardino
Barstow
Carlsbad
San Diego
Tijuana
Centro
Mexicali
Yuma
Indian Springs
Overton
Las Vegas
Henderson
Boulder City
Hoover Dam
Nelson
Davis Dam
Kingman
Needles
Yucca
No. Refineries
State
CALIFORNIA
Bakersfield....4
Benicia....2
Carson....2
El Segundo....1
Hercules....1
Long Beach....1
Los Angeles....1
Martinez....2
Oildale....1
Oxnard....1
Paramount....1
Richmond....1
Rodeo....1
Santa Fe Springs....1
Santa Maria....1
Signal Hill....1
South Gate....1
Torrance....1
Wilmington....3
TOTAL....27
Carson
El Segundo
Long Beach
Los Angeles
Oxnard
Paramount
Santa Fe Springs
Signal Hill
South Gate
Torrance
Wilmington
50
100 Mi.
50
100 Km.

Sources and uses of U.S. gas supply

Fig. 4

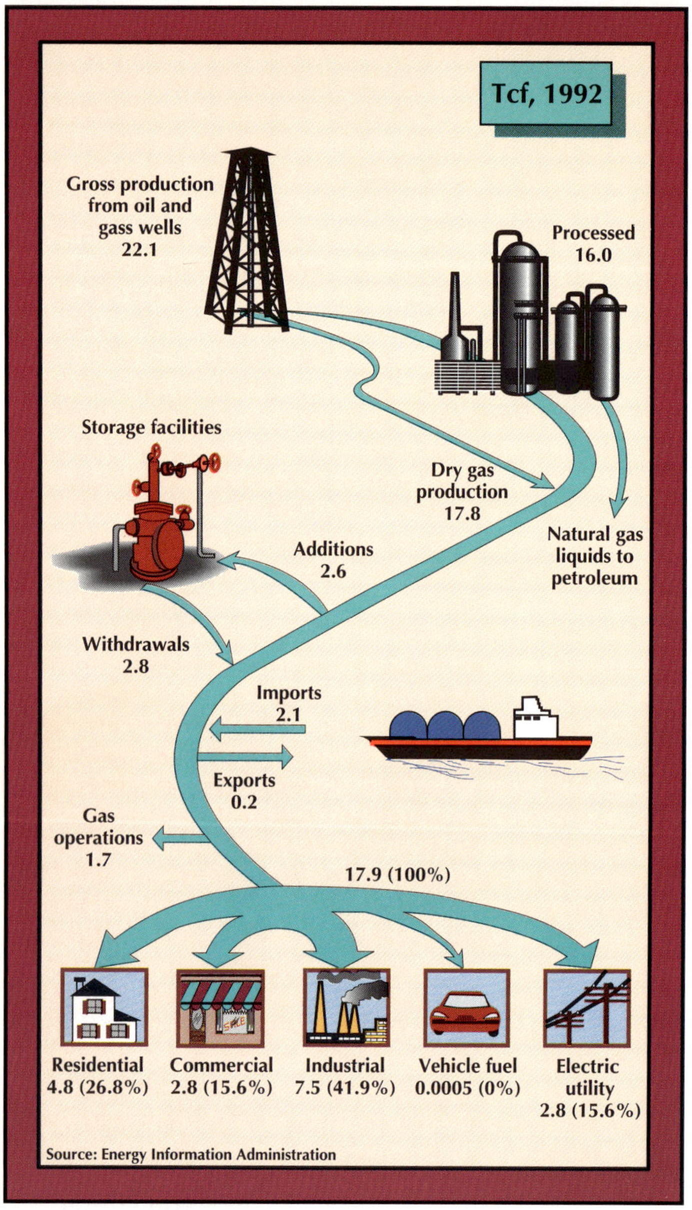

Source: Energy Information Administration

only modestly in 1993, increasing 0.7% to 2.785 tcf. Additions of coal and nuclear power plants in the Southwest and the expected strong increase in hydropower during 1993 will slow utility demand for natural gas.

Natural gas exports, particularly to Mexico, will continue to increase, IPAA said. For 1993 total U.S. exports will move up 4.9% to 257 bcf, including about 55 bcf of liquefied natural gas exports from Alaska to Japan.

The committee said it expects U.S. natural gas flow to advamce 3% in 1993 to 18.276 tcf following 3 years of relatively steady production of 17.7-17.8 tcf.

Production from New Mexico, Colorado, and Alabama will continue to increase, mostly due to the success of coalbed methane development.

In other areas, natural gas productive capacity declined for lack of drilling. Texas, Louisiana, Oklahoma, and Kansas will continue to log production declines.

The increase in gas consumption in recent years has been met mainly by higher imports. U.S. natural gas imports, almost entirely from Canada, have set records each year since 1988.

Imports were predicted to advance 7.2% in 1993 to 2.211 tcf. Imports increased 16.3% in 1992 to 2.062 tcf. Imports provided 10% of U.S. consumption in 1992 and were predicted to be abou 11% in 1993.

The 1993 estimate of Canadian pipeline export capacity wa 2.4 tcf. This is destined to increase as pipeline expansio projects go on stream.

Outlook for 1993-2000

U.S. petroleum product demand will keep growing at abou 1%/year throughout the rest of the 1990s, the IPAA committe said in its midyear 1993 report. Demand will reach 18. million b/d in 2000 as all major products post increases.

Growth in motor gasoline consumption will be slower tha in other products, with increased fuel efficiency and conse vation offsetting gains in miles traveled. More stringe corporate average fuel economy standards or added taxes o gasoline could dampen gasoline demand even further.

Long term, competition among fuels and efficiency gai are likely to limit growth in demand for all petroleu products.

The committee said the downward trend in crude c production will continue. Production will slide to an averag 5.964 million b/d in 2000, the lowest level since 1950.

Oil imports will continue to move up during 1993-2000 fill the gap between demand and production. The grow rate in oil imports will amount to 4%/year during 1993-9 slowing to 2.8%/year in 1995-2000.

Imports will amount to 10.148 million b/d in 2000, with import dependency rising to 50.2% of supply in 1995 and 54.9% in 2000.

Natural gas consumption will reach 22.7 tcf in 2000, reflecting growing demand for electrical power generation and strong gains in industrial demand. Demand will rise 2.5%/year in 1993-95 and 1.5%/year in 1995-2000.

Natural gas is expected to gain market share in power generation due to improved efficiency of new combined cycle plants and, to a lesser extent, environmentally motivated fuel switching. In addition, some coal fueled plants and nuclear plants will be taken out of service.

Some growth will come from compressed natural gas vehicles, which will continue to penetrate the fleet vehicle market. Growth in the commercial and residential sectors will be constrained by conservation, demographics, and high saturation rates.

Natural gas production will rise to 19.739 tcf in 2000, moving up 1.5%/year in 1993-95 and 1%/year in 1995-2000.

Natural gas trade is expected to increase dramatically. Imports will increase to 2.536 tcf in 1995 and 3.124 tcf in 2000. Exports to northern Mexico will grow substantially as total U.S. exports advance to 335 bcf in 1995 and 380 bcf in 2000.

The committee noted that for this relatively optimistic forecast to play out, current regulatory uncertainties will have to be resolved favorably and there must be a rebound in drilling.

A pinpoints eserve decline

A tally by the EIA underscored the increasingly grim situation in U.S. hydrocarbon reserves (Table 2).

The agency's latest figures, disclosed in September 1993, show U.S. proved reserves of oil, gas, and gas liquids declined in 1992 due to low oil and gas prices that reined drilling.

Oil reserves were down 937 million bbl, or 3.8%, to 23.745 billion bbl, largely because low prices pushed 1992 oil well completions down 28% to 8,640, a 20 year low. The 3.8% drop in reserves was more than twice the average decline of 1.7%/year for the previous decade.

While most areas had declines, two areas accounted for 73% of the U.S. decline. They were Texas (356 million bbl) and California (324 million bbl).

EIA said the 20 largest oil and gas producing companies in 1992 owned 76% of U.S. proved reserves of crude oil. The top 20 producing companies had a 6% decline in their U.S proved reserves of crude oil during 1992. The rest of the producers had a 3% increase.

Table 2

U.S. reserves at a glance

CRUDE OIL (billion bbl)	
Yearend 1991	24.682
Yearend 1992	23.745
Decrease	0.937
NATURAL GAS (tcf)	
Yearend 1991	167.062
Yearend 1992	165.015
Decrease	2.047
NATURAL GAS LIQUIDS (billion bbl)	
Yearend 1991	7.464
Yearend 1992	7.451
Decrease	0.013

Source: Energy Information Administration

Revisions and adjustments for crude reserves totaled 1.025 billion bbl, only 65% of the average of 1.574 billion bbl for the prior decade. This component of reserves change is usually the largest and has consistently helped sustain U.S. crude oil proved reserves. Three areas had 71% of the positive net of revisions and adjustments: Alaska with 451 million bbl, the federal offshore with 147 million, and Texas with 125 million bbl. California adjustments were a negative 45 million bbl, the largest percentage drop in proved reserves, due to depressed heavy oil recovery and a rapid decline in offshore reserves. Total discoveries of crude oil were 484 million bbl in 1992, down from 1991 and 38% lower than the 783 million bbl average during the previous 10 years. Texas, Alaska, and the Gulf of Mexico federal offshore accounted for 74% of total discoveries. For the previous 7 years, total discoveries were relatively low, reflecting a similar trend in exploratory drilling that followed the crude oil price collapse of 1986. In 1985 there were more than four times as many successful exploratory oil wells drilled as in 1992. However, total discoveries

WESTERN U.S.
For detail on California refineries, see California maps
BRITISH COLUMBIA
ALBERTA
WASHINGTON
OREGON
IDAHO
MONTANA
NEVADA
UTAH
CALIFORNIA
ARIZONA
Pacific
Ocean
Vancouver
Cherry Point
Ferndale
Anacortes
Port Angeles
Puget Sound
Seattle
Aberdeen
Olympia
Tacoma
Longview
Vancouver
Portland
The Dalles
Salem
Corvallis
Eugene
Coos Bay
Roseburg
Medford
Klamath Falls
Upper Klamath Lake
Nelson
Trail
Spokane
Coeur d'Alene
Moses Lake
Yakima
Richland
Pasco
Walla Walla
Pendleton
Moscow
Lewiston
Missoula
Helena
Anaconda
Butte
Great Falls
Medicine Hat
Babb
Cut Bank
Kevin
Gallup City
Billings
Laurel
Dry Creek
Elk Basin
Frannie
Cody
Oregon Basin
Thermopolis
Yellowstone Lake
Dillon
Baker
Bend
Ontario
Caldwell
Boise
Mountain Home
Malheur Lake
Twin Falls
Pocatello
American Falls Res.
Fort Washakie
Beaver Creek
Piney
Pinedale
La Barge
Bear Lake
Logan
Brigham City
Ogden
Great Salt Lake
Woods Cross
Salt Lake City
Rock Springs
Baxter
Church Buttes
Hiawatha
Asphalt Ridge
Roosevelt
Altamont
Rangely
Provo
Utah Lake
Elko
Tomera Ranch
N. Willow Creek
Blackburn
Reno
Fallon
Carson City
Yerington
Ely
Currant
Trap Spring
Eagle Springs
Kate Spring
Bacon Flat
Grant Canyon
Tonopah
Sevier Lake
Bitter Creek
Grand Junction
Cedar City
Glen Canyon Res.
Aneth
Dove Creek
Bondad
Fredonia
Kirtland
Chaco
Las Vegas
Boulder City
Hoover Dam
Mead Res.
Kingman
Flagstaff
Winslow
Ciniza
Gallup
Prescott
Phoenix
Ajo
Tucson
Nogales
Douglas
Yuma
El Centro
Mexicali
Salton Sea
San Diego
Tijuana
Rosarito
Carlsbad
Eureka
Shasta Lake
Goose Lake
Pyramid Lake
Lake Tahoe
Mono Lake
Rice Creek
Rancho Capay
Willows-Beehive Bend
Sutter Butte
Arbuckle
Santa Rosa
Sacramento
Conway Ranch
Rio Vista
Hercules
Richmond
San Francisco
Oakland
Stockton
Hayward
Redwood City
San Jose
Santa Clara
Modesto
Santa Cruz
Monterey Bay
Salinas
Chowchilla
Gill Ranch
Fresno
Hanford
Coalinga
Kettleman Hills
Trico
San Ardo
Estero Bay
San Luis Obispo
Santa Maria
Mc Kittrick
Oildale
Bakersfield
Mount Poso
Kern River
Fruitvale
Mountain View
Lompoc
Santa Barbara
Ventura
Oxnard
Newhall
Los Angeles
Long Beach
Santa Fe Springs
San Bernardino
San Nicolas I.
San Clemente I.
San Miguel I.
Santa Rosa I.
Donner Pass
Extremeno
Opata
0
100
200 Mi.
0
100
200 Km.

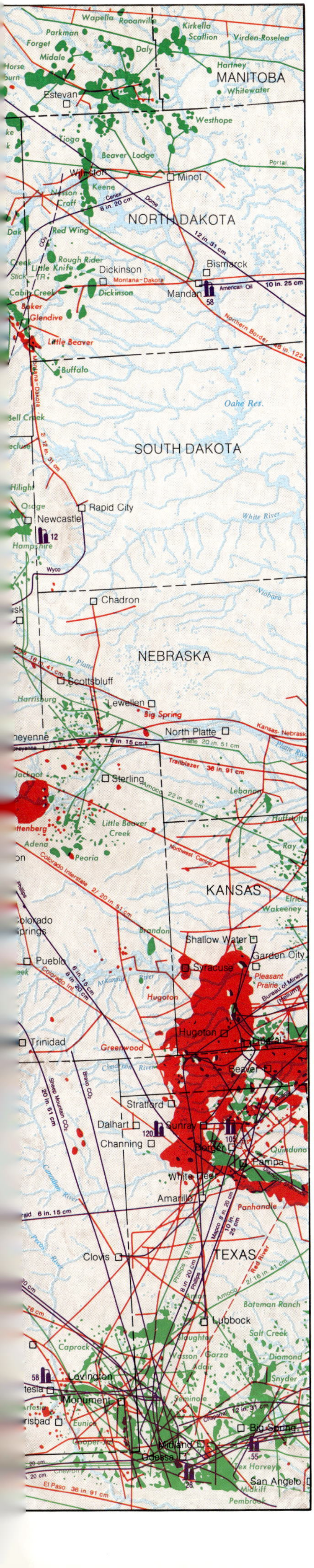

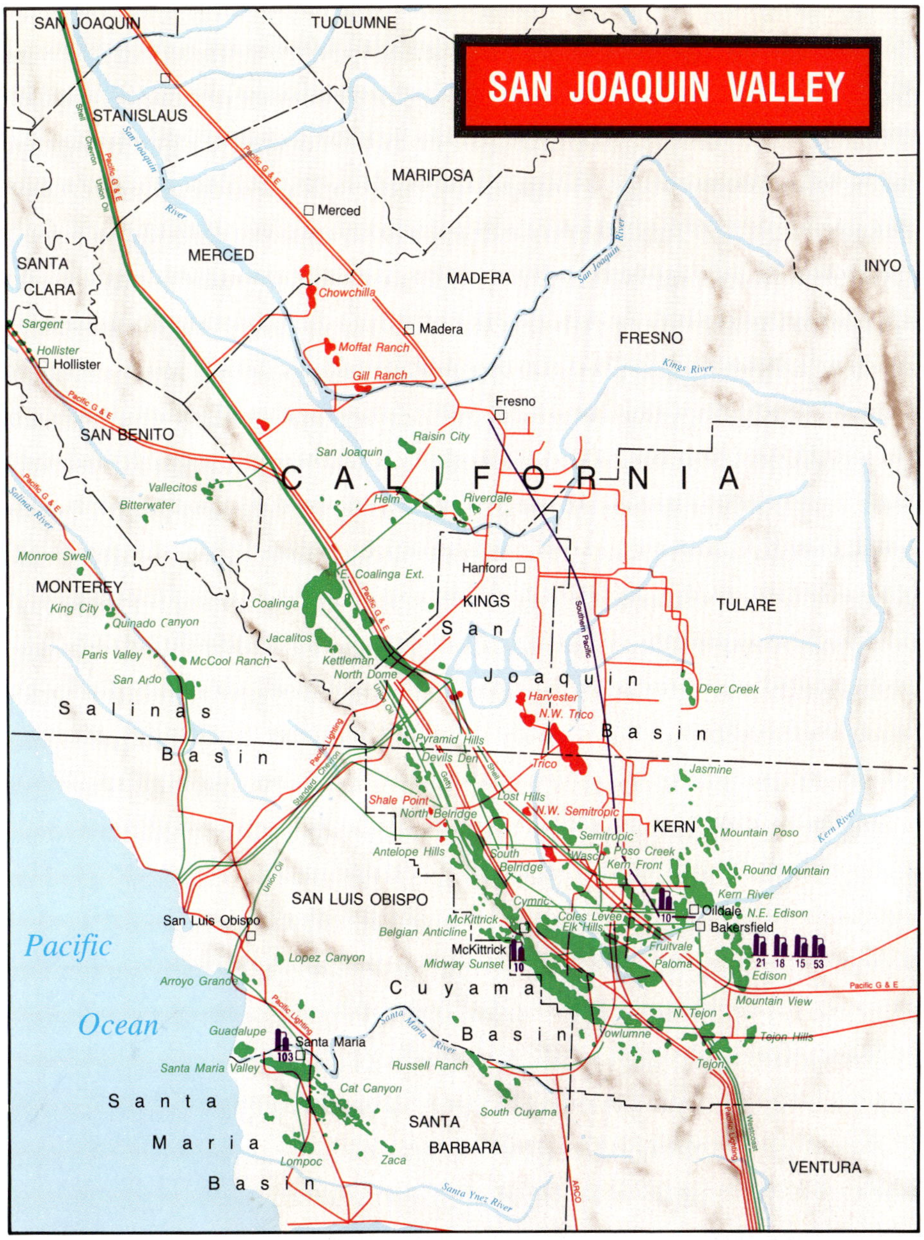

added per exploratory well have roughly doubled, easing the effect of the drop in drilling.

Proved reserves added by extensions (391 million bbl) were up from 1991 but well below the prior 10 year average of 533 million bbl.

Horizontal drilling in the Austin chalk formation was the major factor behind the 118 million bbl of extensions in Texas, which accounted for 30% of U.S. crude oil extensions.

New field discoveries of crude oil have been depressed since 1985 and were "an exceptionally low" 8 million bbl in 1992. The prior 10 year average for new field discoveries was 15 times as high.

EIA said, "New discoveries, the Sunfish prospect in the Cook Inlet of Alaska and the Kuvlum prospect 15 miles offshore from the Arctic National Wildlife Refuge, that are still being evaluated and are not yet proved, may turn this trend around in 1993."

Proved reserves of new reservoirs discovered in old fields were 85 million bbl, down from 1991 and well below the 135 million bbl average for the prior 10 years. There were proved reserves of 2.638 billion bbl of crude oil in nonproducing reservoirs. They are included in the total proved reserves figures and represent 11% of the total.

Indicated additional crude reserves were 3.782 billion bbl, down 11% from 1991.

"These reserves are crude oil volumes that may become economically recoverable from known reservoirs through application of improved recovery techniques using current technology," EIA said. "The presence of large indicated additional reserves in the Alaskan North Slope, California, West Texas, and New Mexico implies that significant upward revisions to crude oil proved reserves could occur in the future."

EIA said the scheduled end of federal tax credits for new unconventional gas wells in December 1992 spurred drilling for coalbed methane and tight gas to high levels in the last half of that year.

Gas completions dropped only

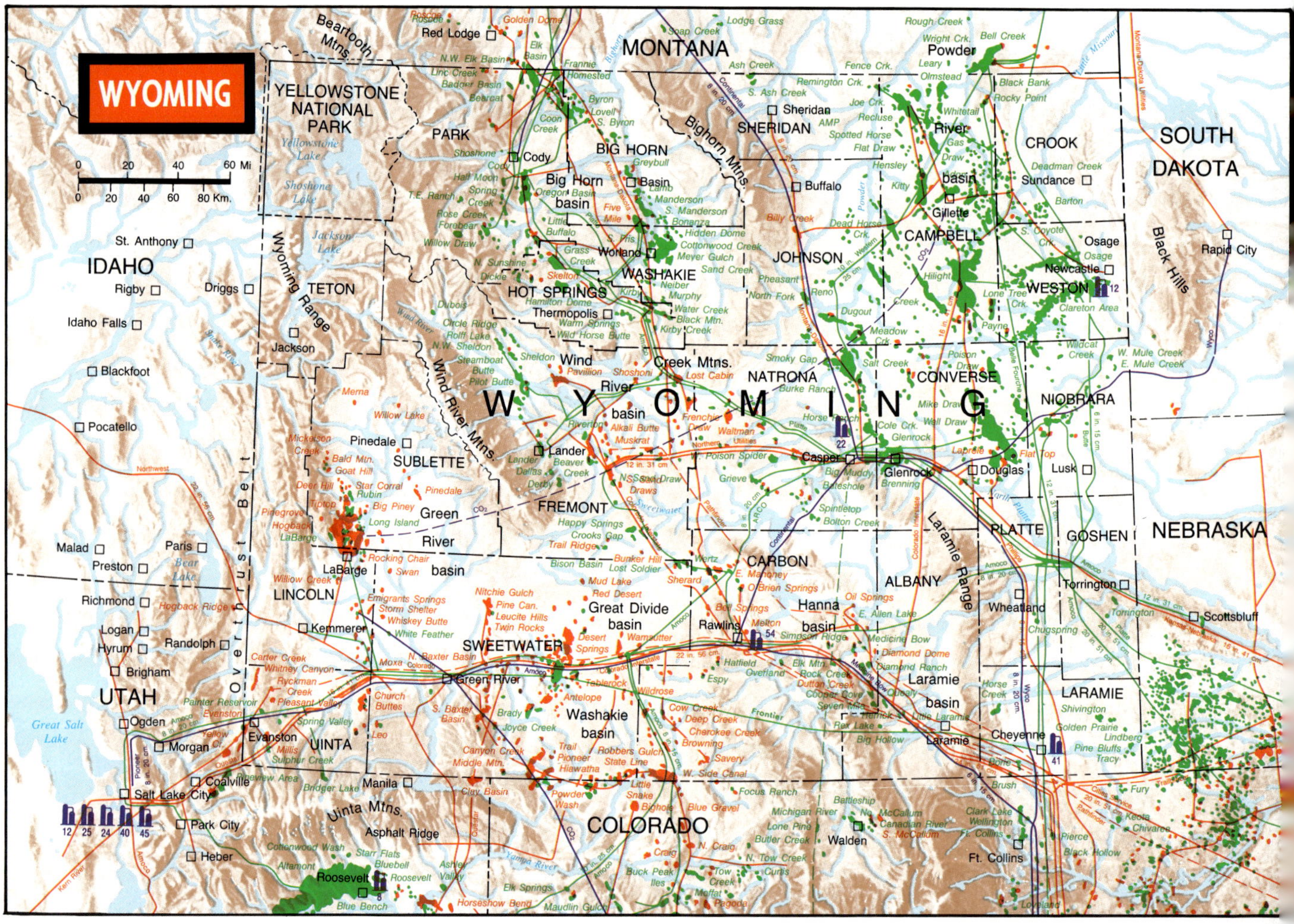

16% to 7,640, but exploratory gas well completions reached a 20 year low after dropping 27% during 1992.

Proved reserves of dry natural gas declined a little more than 1% or 2.047 tcf in 1992. All four leading gas producing areas—Texas, the Gulf of Mexico, Oklahoma, and Louisiana—had large proved reserves declines totaling 4.649 tcf.

Partially offsetting those declines were large increases in coalbed methane reserves totaling 1.957 tcf in Virginia, New Mexico, Colorado, and Alabama.

Coalbed methane reserves increased 23% to 10.034 tcf to account for 6% of total U.S. gas reserves.

Coalbed methane production increased almost sixfold in 3 years to account for 3% of U.S. gas production.

Exploitation of the coalbed methane resource has expanded rapidly because of a large tax credit incentive and improved understanding of the underlying production technology. The tax credit in 1992 was about 90¢/Mcf, or about half of the average U.S. wellhead price.

Total gas discoveries in 1992 were 7.048 tcf, down from 1991 and 34% lower than the average during the prior 10 years. More than half of them were in Texas and the Gulf of Mexico federal offshore.

New field discoveries were 649 billion cu ft, down substantially from the 1991 level and 59% lower than the average of 1.592 tcf during the previous 10 years. Extensions were 4.675 tcf, 31% less than the prior 10 year average. Gas reserve additions from new reservoirs increased slightly to 1.724 bcf, but were well below the average of the prior decade.

EIA said the net of revisions and adjustments for natural gas in 1992 was 8.328 tcf.

In recent years, the net of revisions and adjustments played a growing role in sustaining Lower 48 states natural gas proved reserves. In 1992, 7.883 tcf of these were added, 74% more than the Lower 48 states average for the past 10 years.

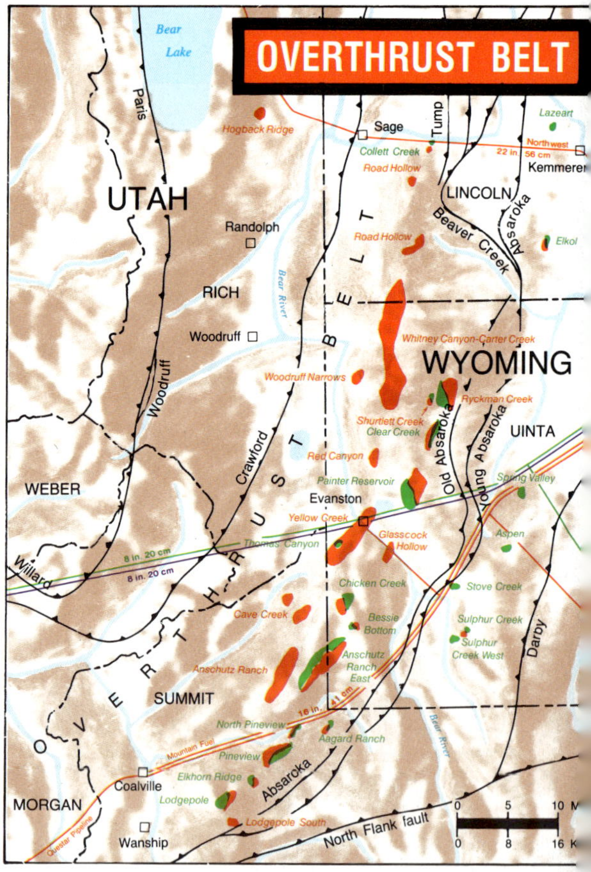

Table 3

Lower 48 resource base*

	Tcf
Proved reserves	160
Conventional resources	
Reserve appreciation	203
New fields	413
Subtotal	**616**
Nonconventional resources	
Coalbed methane	98
Shales	57
Tight sands	349
Other	15
Subtotal	**519**
Total	**1,295**

*As of Jan. 1, 1991, assuming offshore drilling moratoriums expire as scheduled and incorporating technology advancement through 2010. Assuming no advancement, the total is 1,065 tcf.

Source: National Petroleum Council

Table 4

How reserve additions, replacement costs compare

	1988	1989	1990	1991	1992	3 year total 1990-92	5 year total 1988-92
	Million bbl of oil equivalent						
RESERVE ADDITIONS*							
U.S.	4,988	3,743	4,057	3,210	3,088	10,355	19,086
Canada	1,538	1,397	731	462	611	1,804	4,739
North America	6,526	5,140	4,789	3,671	3,699	12,159	23,825
Europe/North Sea	2,617	1,307	2,923	1,677	1,664	6,264	10,188
Other overseas	2,482	3,287	4,059	3,376	2,669	10,104	15,873
Worldwide	11,625	9,734	11,771	8,724	8,032	28,527	49,886
	$/bbl of oil equivalent						
RESERVE REPLACEMENT COSTS*							
U.S.	5.00	4.86	4.84	5.90	4.82	5.17	5.06
Canada	4.31	4.68	5.23	8.89	4.54	5.93	5.04
North America	4.84	4.81	4.90	6.28	4.78	5.28	5.06
Europe/North Sea	5.31	5.45	4.13	9.49	8.05	6.60	6.12
Other overseas	4.01	3.64	2.77	3.68	4.64	3.57	3.65
Worldwide	4.77	4.50	3.97	5.89	5.41	4.96	4.83

*Discoveries, extensions, revisions, and acquisitions.

Source: John S. Herold Inc.

"This increasing recovery of gas from the resource base of old fields is enhanced by application of new technologies like 3D seismology, horizontal drilling, and better fracturing treatments and well completions, " EIA said.

Total discoveries for the Lower 48 states were only 6.994 tcf in 1992, the second time since 1977 they were lower than the net of revisions and adjustments.

There were proved reserves of 34.118 tcf of natural gas, wet after lease separation, in nonproducing reservoirs. They were included in the total proved reserves data and represent 20% of the total.

EIA said U.S. natural gas liquids proved reserves, the sum of natural gas processing plant liquids and lease condensate reserves, dropped slightly to 7.451 billion bbl in 1992.

Natural gas potential

About 1 year before IEA issued its reserves report, NPC unveiled a study showing that natural gas can make a greater contribution to U.S. energy needs and environmental goals. That's because gas can meet demand at competitive prices, it is clean burning, and it is a secure, mainly domestic energy source.

The study, which took 2 years to prepare, involved contributions from 200 persons in industry, government, trade groups, academia, and private enterprise.

NPC said estimates of gas resources should rise at least through 2010 from the then-present volume of 1.295 quadrillion cu ft technically recoverable in the Lower 48 states alone. NPC's study covered proved reserves in current fields, probable reserves in current fields, possible reserves in new fields, speculative volumes in frontier areas, and noncovential resources such as coalbed methane (Table 3).

Much of the groundwork needed to develop a more competitive, consumer oriented gas industry has been laid.

But NPC said, "Perceptions of natural gas that arise from its heavily regulated past represent the greatest challenge to be overcome by the industry. In particular, the industry must pay more attention to meeting customer needs through greater efficiency and more competitive services."

The report also said, "The natural gas resource base is abundant and can be produced and delivered at prices that

Table 5

Gas pipeline construction in the U.S.

1993	1994	1995	1996	1997
Miles of line				
2,343	2,478	2,355	1,703	1,421

Source: Energy and Environmental Analysis Inc.

allow expansion of the market and continued developmen of the resource."

Sufficient new supplies can be delivered at competitiv prices, even though it will require sustained growth i prices to bring on the new production.

"The continued evolution of technology and energ efficiency improvements combine to reduce delivere prices below what they would have been without thes advances and tend to mitigate the cost of developing mor expensive supplies. With the proper emphasis on researc and development, technology impacts in the future will b even greater than in the past."

NPC found the U.S. gas market to be increasingly divers with new challenges and growth opportunities in all ma ket segments and several new technologies.

"New regulatory policies are encouraging competition NPC said. "The natural gas industry is responding b providing additional value added services, flexible contrac ing options, and increased attention to customer needs.

"Sound management, operational mechanisms, contra diversity, and use of financial markets can work together manage risk and reduce the short term volatility that is like to occur as the industry moves through transitional phases

But the report also said potential customers are co cerned about industry's ability to deliver gas when an where it is needed, particularly during peak demand.

"Additionally, the industry does not have a good publ image and has not been sufficiently effective at marketir natural gas, in spite of its inherent beneficial qualities, an

2½" —63MM

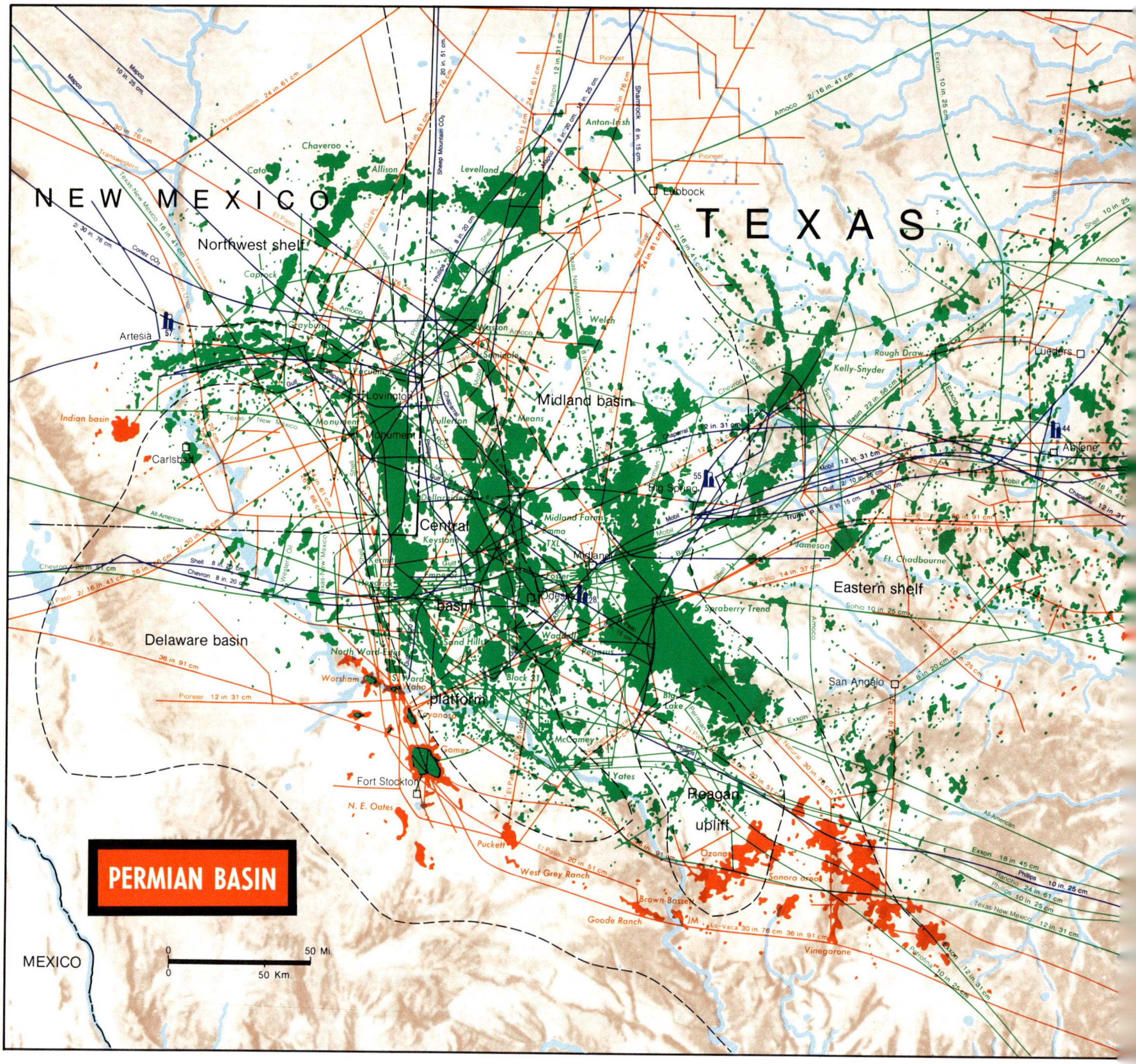

has been overdependent on regulation for setting direction and resolving conflicts."

The study recommended federal, state, and local officials allow competitive market forces to continue to develop and work. And governments should exercise restraint during periods of price and supply volatility as the industry adjusts to the changing marketplace.

It also said industry needs to help the market work by accelerating technologies and procedures that reduce gas prices, promoting and commercializing new end use technologies, developing strategies for environmental issues, improving supply and delivery systems, and increasing customer focus in marketing.

The study said the industry's exploration and production outlays have averaged $35-40 billion/year in 1990 dollars during the previous few years, equal to the level of the mid-1970s but only about half the peak spending years of the 1980s.

If U.S. production increases to about 21 tcf/year by 2010, industry would need to invest about $60 billion/year during 2000-2010.

NPC warned increases in environmental regulation cou rein production, and if incremental environmental cos exceed $30 billion, its estimates of production would reduced nearly 20 tcf over the period.

It noted technology advancement has had a significa effect on supplies.

"During the past two decades, technology advanceme has reduced drilling costs by almost 3%/year below what th would have been in the absence of the advancing technolo The rate of technology advance appears to be acceleratir with more effect in the 1980s than in the 1970s."

That trend should continue.

NPC predicted Canadian gas imports will increase a may reach 3 tcf/year or more, but LNG imports will rema low and "price and demand levels appear to be inadequa for developing the Alaskan North Slope gas resources or t northern frontier gas in Canada for U.S. consumption pri to 2010."

The gas transmission and storage network can supp the increased demand.

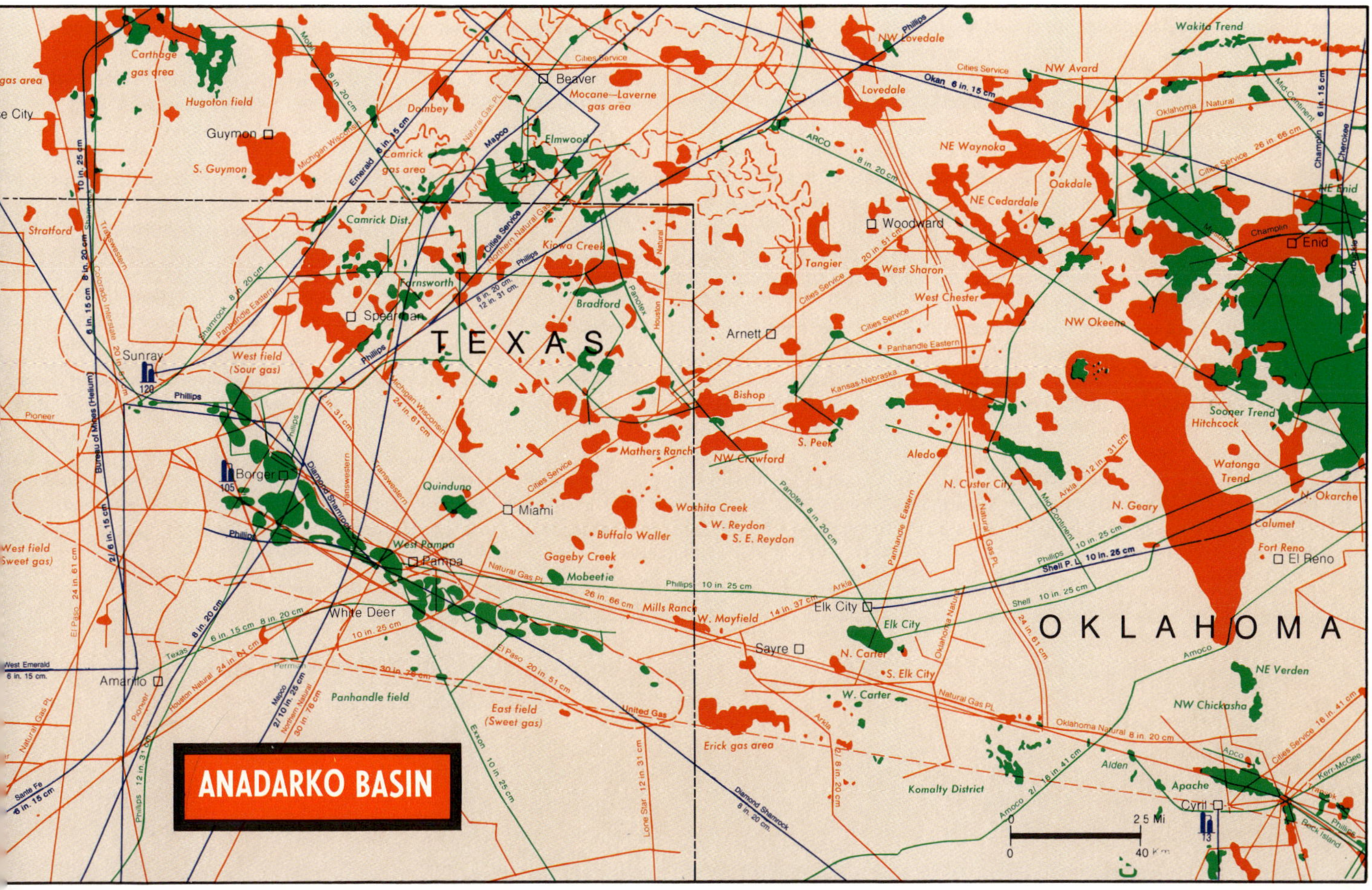

Although U.S. gas consumption peaked in 1972 at 22.1 tcf, the transmission and storage system has continued to expand due to geographic shifts in supply and demand. In 1992, the report said, there were more than 280,000 miles of gas transmission pipeline and about 8 tcf of storage capacity.

The system had a capacity of 24 tcf/year and a peak capacity of 120 bcfd in 1991, well above consumption of 19.2 tcf/year and firm peak demand of 102 bcfd in 1991.

NPC said it expects a significant shift in gas supply and consumption patterns by 2010, which creates a need for construction of new transportation and storage facilities.

eplacement costs
or reserves

Contrarians take notice, John S. Herold Inc. told the petroleum industry in a May 1993 report.

The Greenwich, Conn., analyst's study of 1992 results for 128 oil companies found that U.S. and Canadian reserve replacement costs (RRCs) were among the world's most competitive (Table 4). The studied firms accounted for more than 95% of capital spending and reserve additions among all public oil companies.

"Apparently the recent draconian cost reduction efforts implemented by the oil industry for the North American upstream sector are paying off," Herold said.

"After years of hearing the mantra that oil company spending must emigrate to more hospitable international climes, it is refreshing to observe that the oil industry's renewed commitment to cost containment and efficiency and focus of operations is paying dividends."

What's more, Herold said, 1992 could well represent the low water mark of oil industry activity in North America for the rest of the 1990s.

"With oil industry fundamentals on the mend and a bull market in oil and gas shares under way in 1993, we expect a solid recovery in upstream capital commitments to become increasingly evident."

Herold's figures show U.S. and Canadian RRCs in 1992 averaged \$4.82/bbl of oil equivalent (BOE) and \$4.54/BOE added, respectively, vs. \$5.95/BOE for the rest of the world.

Worldwide RRCs for 1992 came in at \$5.41/BOE of reserve additions, a significant improvement over the dismal industry performance of 1991 when U.S. RRCs surged 20% to \$6/BOE and international RRCs jumped "an astonishing" 54% to \$6.12/BOE.

Reserves added exclusively from exploration and development outlays showed similar trends. Those finding and development costs for the U.S. and Canada amounted to \$5.29/BOE and \$7.37/BOE, respectively, against year earlier levels of \$6.88/BOE and \$15.65/BOE.

Oil companies slashed upstream capital spending by more than 15.4% in 1992, Herold's figures showed.

Responding to tough industry operating fundamentals and low natural gas price realizations that prevailed during first quarter 1992, total outlays fell to \$43.4 billion from \$51.4 billion in 1991.

Not surprisingly, Herold said, most of the 1992 spending cuts took place in North America. U.S. spending shrank \$4.1 billion, or 21.4%, to \$14.9 billion. Canadian outlays plunged by an even more dramatic \$1.3 billion, or 32.5%, to \$2.8 billion.

The dollar exodus from the U.S. continued to gain speed. For the sixth straight year, spending outside the U.S. at \$28.6 billion outpaced U.S. outlays of \$14.9 billion. That means 66% of the total upstream dollar was spent outside the U.S.

Herold said, "It is significant to note that 10 years ago the geographic spending weights were the exact opposite, with two thirds of 1982's \$48.4 billion capital program occurring in the U.S.

"It is noteworthy that 1992 RRC expenditures in North America of \$17.7 billion were roughly half the \$35.7 billion average expenditure level of the first half of the 1980s.

Moreover, last year's outlays were below the previous low point—$18.7 billion—that occurred in 1987 in reaction to the world oil price crash of 1986."

Companies covered by Herold's study booked 1992 reserve additions of 6.6 billion BOE excluding acquired reserves and 8 billion BOE including acquired reserves—by either measure the lowest level of net additions in 5 years.

Reserve additions from exploration and development efforts alone did not replace 1992 production. The replacement rate of 85% from finding efforts climbed to a "barely acceptable" 104% when reserves added from acquisitions were included.

The 1992 replacement ratios compared poorly with the 5 year record of 102% replacement from finding and development efforts and 132% including acquisitions.

The dollar volume of reserves changing hands via merger and acquisition (M&A) and the implied cost of proved reserve acquisitions declined sharply in 1992. Worldwide, about 1.5 billion BOE changed hands—essentially level with 1991—but the dollar volume slumped from $7.1 billion in 1991 to $4.5 billion. Acquisition values fell from $4.53/BOE in 1991 to a 5 year low of $3.12/BOE.

By comparison, Herold's M&A Transaction Review Yearbook tallied 1.3 billion BOE of reserve transactions in 1992. The universe of transactions in the M&A research report were those in which there were adequate disclosures to permit analysis. The yearbook counted $5.4 billion in oil and gas reserve trades value at an implied $4.15/BOE.

Canada, where reserve acquisitions roughly doubled in 1992, afforded the world's lowest cost M&A theater. Canadian acquisitions came in at a rock bottom $2.12/BOE, less than half the 1991 level of $4.45/BOE.

Natural gas leads pipeline construction

Yearly figures compiled by OGJ showed the usual accent on construction of natural gas systems by the U.S. pipeline industry during 1993.

OGJ's annual survey of industry activity revealed plans for construction of 3,751 miles of gas pipeline during 1993 in the U.S. Sizes ranged from 4 through 30+ in. About two-thirds of construction involved 22-30+ in. lines.

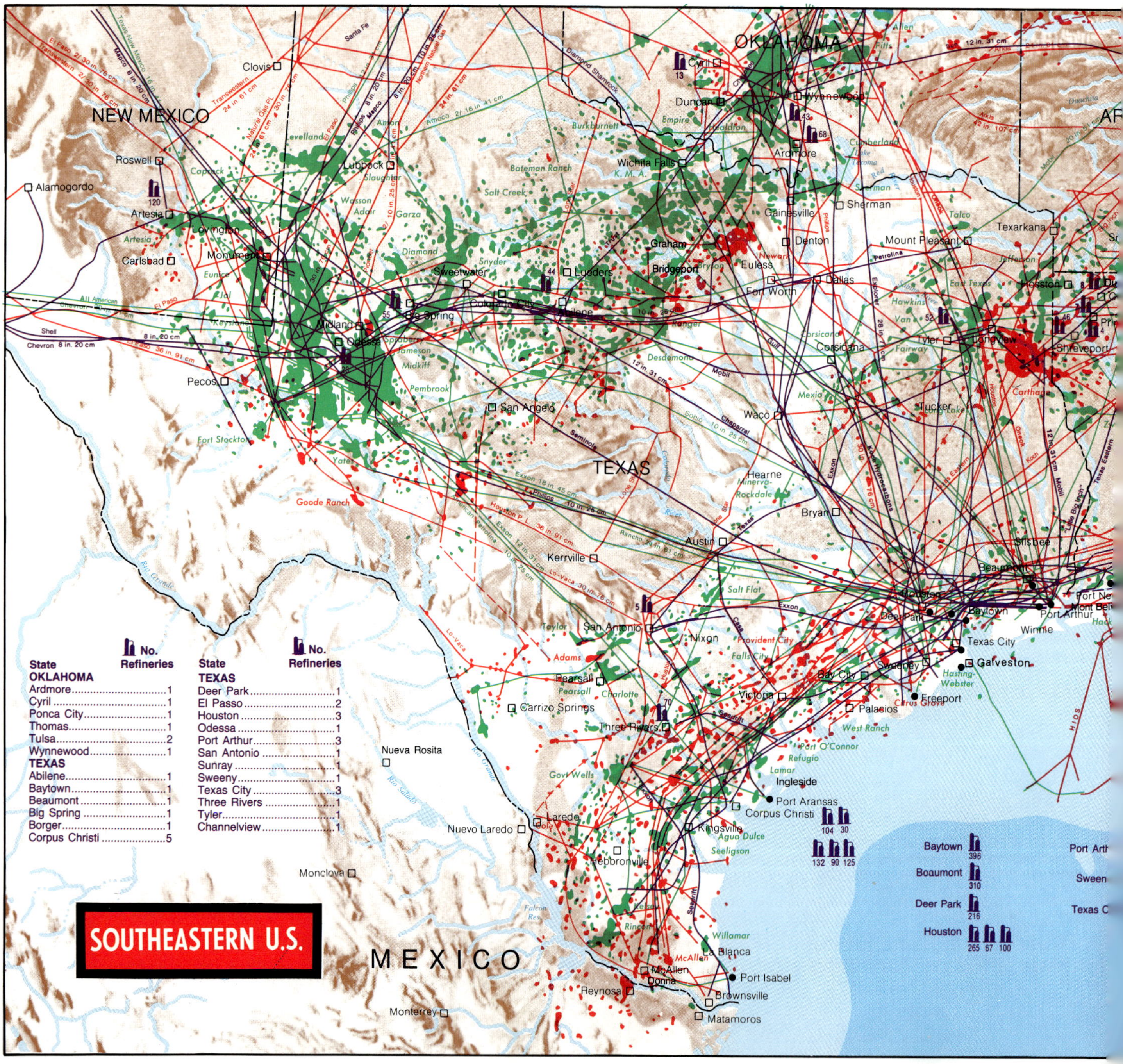

Industry plans also called for construction of 2,558 miles of product pipeline and 446 miles of crude oil line during the year.

Much of the gas pipeline activity centered on the West Coast and in the Northeast and Southeast, where markets were growing rapidly.

Construction projects also were afoot in far South Texas.

Increased environmental awareness and low cost U.S. natural gas spot market supplies were driving up U.S. gas exports to Mexico and spawning a series of proposals aimed at increasing pipeline capacity across the U.S.-Mexican border. Mexico's Petroleos Mexicanos predicted U.S. gas exports to Mexico could increase to more than 900 MMcfd from only 43 MMcfd in 1990.

U.S.-Mexico gas pipeline capacity was estimated at about 549 MMcfd in spring 1992. Since then, plans were disclosed for system expansions or new gas lines across the international border with combined capacity of more than 1.4 bcfd.

Meantime, another pipeline industry study predicted 10,300 miles of gas transmission line will be laid in the U.S. during 1993-97 (Table 5).

Energy & Environmental Analysis Inc., in a study commissioned by the Ingaa Foundation, Washington, D.C., surveyed North American gas pipeline and storage companies.

It figured the construction projects will cost $12.06 billion during the study period, compared with $16.05 billion in 1988-92 when 13,411 miles of gas pipeline were laid.

The study said, "The decline in expenditures can be attributed to the fact that pipeline capacity is coming more in line with current and expected future natural gas demand nationally and regionally, necessitating less capacity expansion."

Canadian construction in 1993-97 will total 1,570 miles, costing $3 billion (Canadian), while Mexico will add 350 miles of pipeline during the same period. That will boost North American construction to 12,200 miles of transmission line during the study period.

U.S. companies will spend $2.22 billion during 1993-97 to increase underground storage, adding working gas capacity of 500 bcf.

"The expected growth in underground storage is due to the changing structure of U.S. gas markets and growing gas demand, particularly for electric generation, which tends to

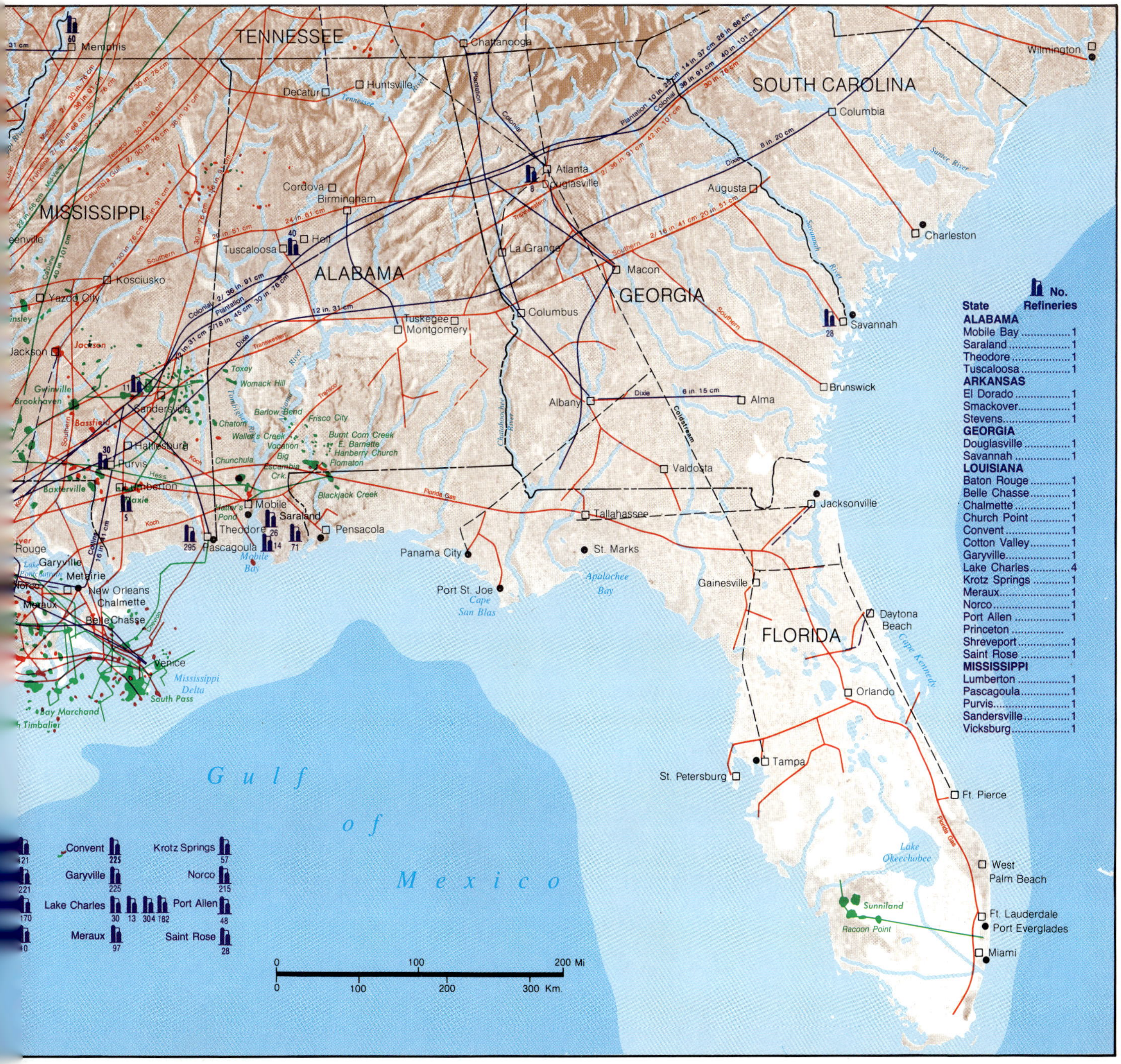

TEXAS
BEAUREGARD
ALLEN
ST. LANDRY
LOUISIANA
CALCASIEU
JEFFERSON DAVIS
ACADIA
LAFAYETTE
ST. MARTIN
IBERVILLE
CAMERON
VERMILION
IBERIA
ST. MARTIN
ST. MARY
Lake Charles
Jennings
Lafayette
New Iberia
Abbeville
Cameron
Port Arthur
Sabine Lake
Calcasieu Lake
Grand Lake
White Lake
Vermilion Bay
WEST CAMERON
EAST CAMERON
VERMILION
SOUTH MARSH ISLAND NORTH ADDITION
SOUTH MARSH ISLAND
EUGENE ISLAND
WEST CAMERON WEST ADDITION
HIGH ISLAND EAST ADDITION
WEST CAMERON SOUTH ADDITION
EAST CAMERON SOUTH ADDITION
VERMILION SOUTH ADDITION
SOUTH MARSH ISLAND SOUTH ADDITION
EUGENE ISLAND SOUTH ADDITION
HIGH ISLAND EAST ADDITION SOUTH EXTENSION
SABINE PASS
THREE LINE
GARDEN BANKS NG 15-2

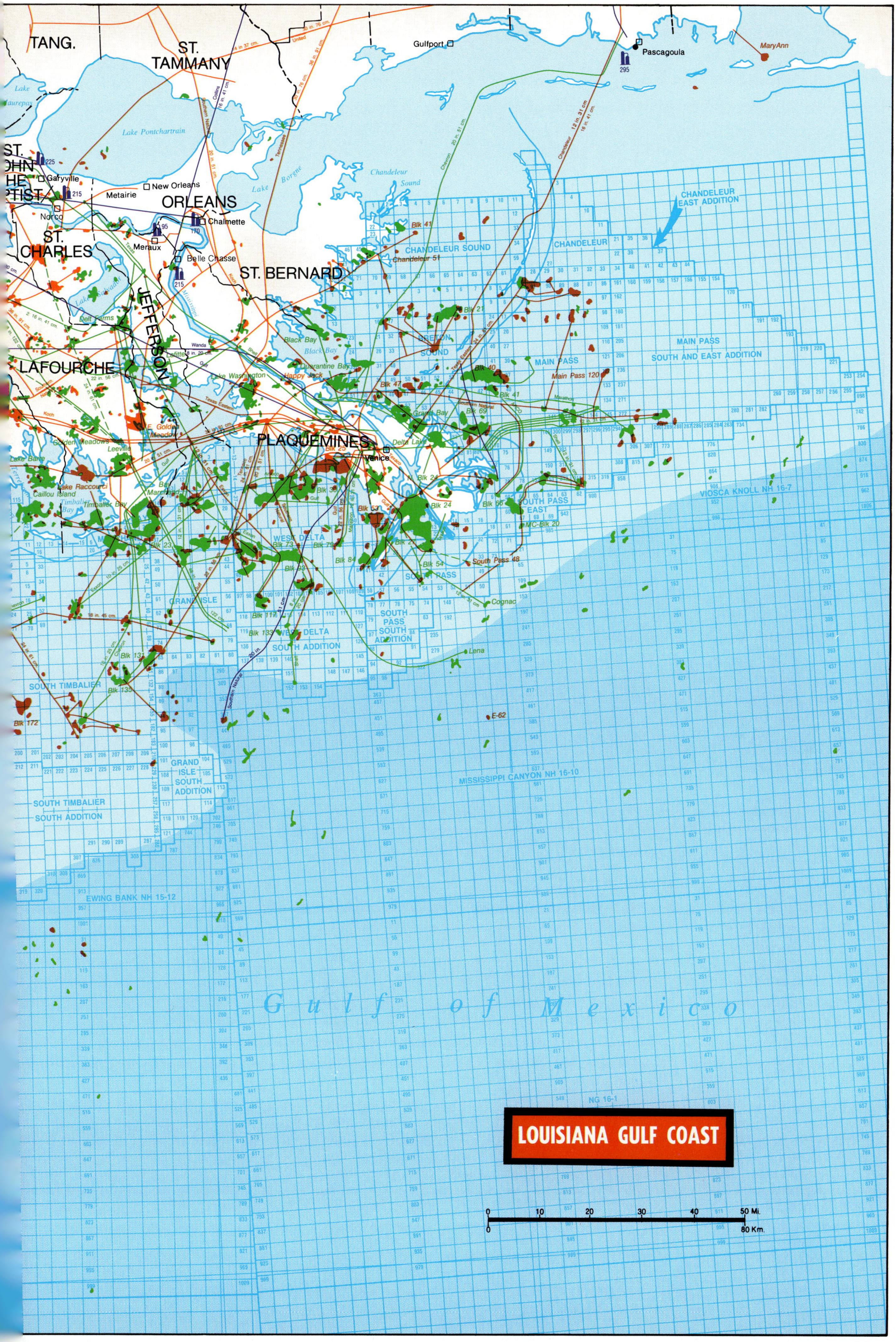
LOUISIANA GULF COAST
Gulf of Mexico
TANG.
ST. TAMMANY
ORLEANS
ST. BERNARD
ST. CHARLES
JEFFERSON
LAFOURCHE
PLAQUEMINES
Lake Pontchartrain
Lake Borgne
Chandeleur Sound
Gulfport
Pascagoula
MaryAnn
New Orleans
Metairie
Garyville
Norco
Chalmette
Meraux
Belle Chasse
Venice
CHANDELEUR
CHANDELEUR EAST ADDITION
CHANDELEUR SOUND
BRETON SOUND
MAIN PASS
MAIN PASS SOUTH AND EAST ADDITION
VIOSCA KNOLL NH 16-7
SOUTH PASS EAST
SOUTH PASS
SOUTH PASS SOUTH ADDITION
WEST DELTA
WEST DELTA SOUTH ADDITION
GRAND ISLE
GRAND ISLE SOUTH ADDITION
SOUTH TIMBALIER
SOUTH TIMBALIER SOUTH ADDITION
EWING BANK NH 15-12
MISSISSIPPI CANYON NH 16-10
NG 16-1
Cognac
Lena
E-62
0 10 20 30 40 50 Mi.
0 80 Km.

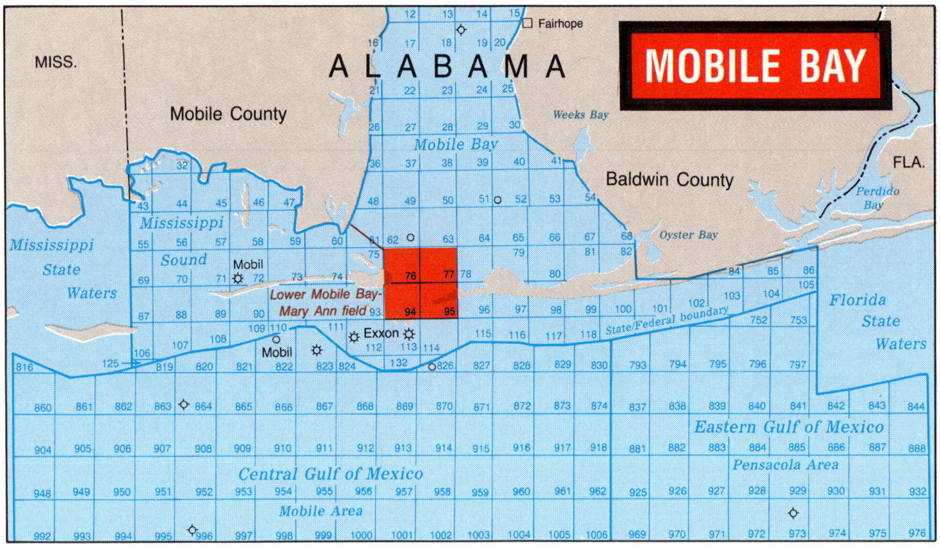

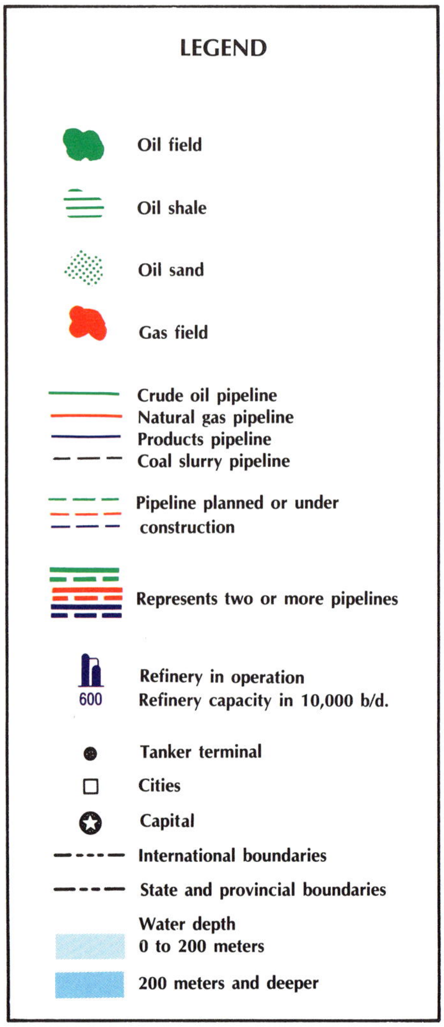

require more gas storage due to uncertain, fluctuating loads," the study said.

Canada is expected to add 13 bcf of storage in 1993-97.

The study noted two major uncertainties facing the gas market.

"A faster than expected growth in gas demand would lead to greater need for transmission and storage capacity than reflected here, while a lower than expected growth in demand would have the opposite effect."

There is regulatory uncertainty due to Federal Energy Regulatory Commission orders to unbundle interstate transmission services and allow risk sharing for new construction.

The Interstate Natural Gas Association of America formed the Ingaa Foundation in 1990 to sponsor research on ways to advance the use of natural gas in North America.

Free trade accord in place

The U.S. began 1994 with a North American Free Trade Agreement (Nafta) in place with trading partners Canada and Mexico. The U.S. and Canada had long enjoyed virtually free trade with each other, but the petroleum industry's dealings with Mexico was quite a different matter.

Beginning Jan. 1, 1994, Nafta will phase out all tariffs among the three agreement signatories during a 15 year period.

Congress approved the accord in November 1993, and President Clinton signed the agreement a short time later.

Clinton had to make many concessions to individual congressmen to snatch a come-from-behind victory on Nafta, which created the world's largest free trade zone.

It will make the sale of U.S. and Canadian oil services and supplies in Mexico easier but not allow foreign ownership of Mexican oil and gas reserves.

To win votes for the agrement, Clinton agreed to restrain U.S. imports of orange juice, winter vegetables, sugar, peanut butter, and wheat.

U.S. labor unions, assisted by some environmental groups, fought hard to defeat Nafta.

Despite two side agreements that gave the U.S. added assurances on labor and environmental issues, those groups believed Nafta will cost the U.S. too many jobs. IPE

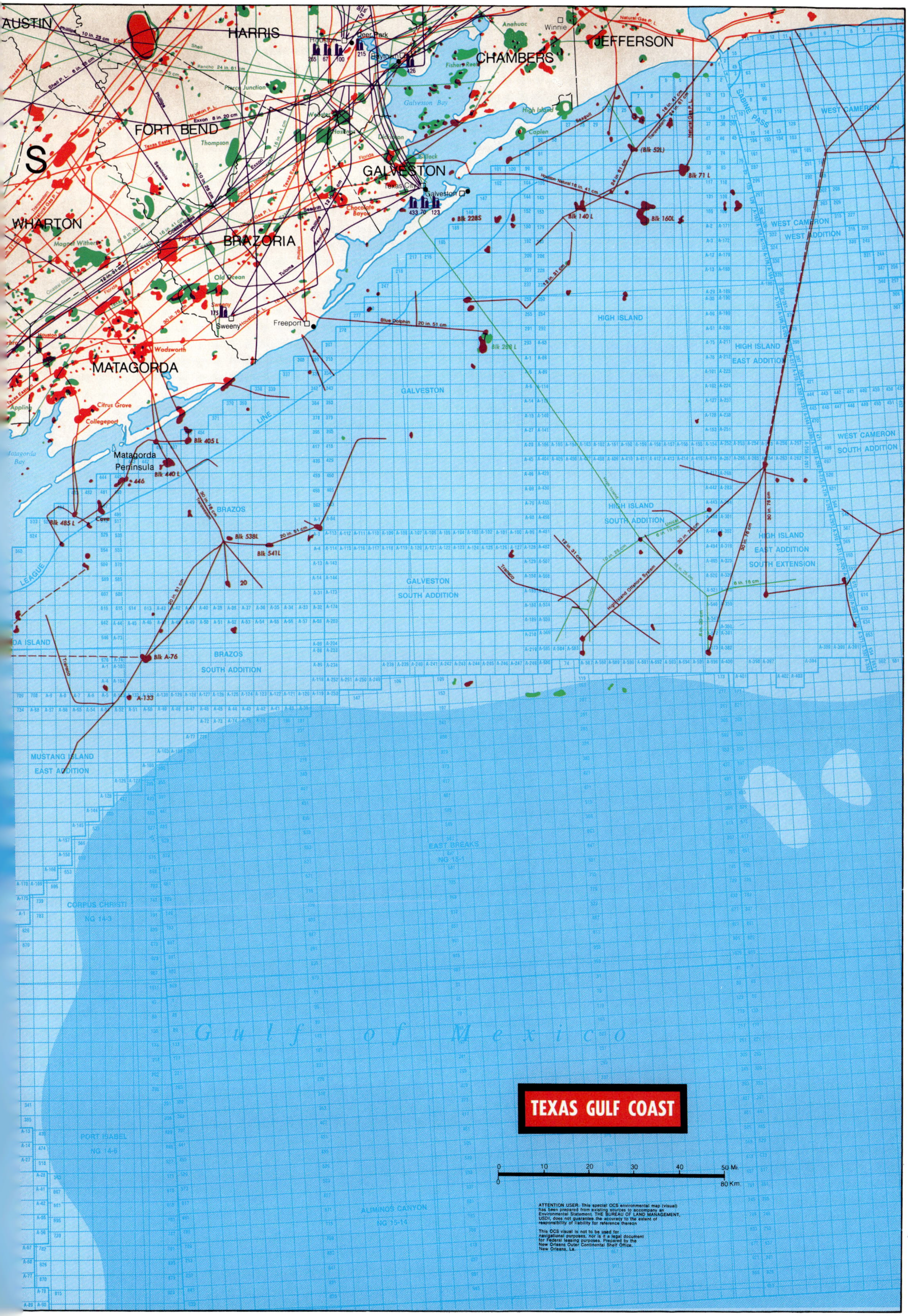
AUSTIN
HARRIS
JEFFERSON
CHAMBERS
FORT BEND
GALVESTON
WHARTON
BRAZORIA
MATAGORDA
S
Deer Park
Baytown
Winnie
Anahuac
Galveston Bay
High Island
Texas City
Galveston
Chocolate Bayou
Old Ocean
Sweeny
Freeport
Wadsworth
Citrus Grove
Collegeport
Matagorda Peninsula
Matagorda Bay
Blk 405 L
Blk 440 L
Blk 485 L
Blk 538L
Blk 541L
Blk A-76
A-133
Blk 228S
Blk 140 L
Blk 160L
Blk 71 L
(Blk 52L)
Blk 288 L
Blue Dolphin 20 in. 51 cm
SABINE PASS
WEST CAMERON
WEST CAMERON WEST ADDITION
WEST CAMERON SOUTH ADDITION
HIGH ISLAND
HIGH ISLAND EAST ADDITION
HIGH ISLAND SOUTH ADDITION
HIGH ISLAND EAST ADDITION SOUTH EXTENSION
GALVESTON
GALVESTON SOUTH ADDITION
BRAZOS
BRAZOS SOUTH ADDITION
MUSTANG ISLAND EAST ADDITION
LINE
LEAGUE
EAST BREAKS NG 15-1
CORPUS CHRISTI NG 14-3
PORT ISABEL NG 14-6
ALAMINOS CANYON NG 15-14
Gulf of Mexico
TEXAS GULF COAST
0 10 20 30 40 50 Mi
0 80 Km
ATTENTION USER: This special OCS environmental map (visual) has been prepared from existing sources to accompany an Environmental Statement. THE BUREAU OF LAND MANAGEMENT, USDI, does not guarantee the accuracy to the extent of responsibility of liability for reference thereon
This OCS visual is not to be used for navigational purposes, nor is it a legal document for Federal leasing purposes. Prepared by the New Orleans Outer Continental Shelf Office, New Orleans, La.

Phillips subsalt strike, independents help spark Gulf of Mexico

Reports in third quarter 1993 of the Gulf of Mexico's first commercial subsalt discovery highlighted that year's activity but weren't the only signs of renewal in the most important U.S. offshore oil and gas province.

Drilling rebounded gulf-wide at a modest rate throughout the year, sustaining a recovery that began about mid-1992 in response to higher gas prices.

Major companies and independents alike stepped up exploration and development in the gulf, mostly pursuing strategies based on technological advances. By many measures, independents were driving much of the drilling and development rebound.

Among new technical capabilities leading the activity were 3D seismic data collection and processing and horizontal drilling. Deepwater development in the gulf also made important strides in 1993, with majors and independents claiming regional and world water depth records.

Evolving technology and savvy logistical tactics helped reduce upfront E&D costs and shorten development timetables. Growing know-how and better economics gave more companies the confidence to operate offshore and spawned a few sizable acquisitions.

Competition for good prospects caused gulf operators to increase participation in both area-wide lease sales conducted in 1993 by the Minerals Management Service (MMS). MMS at yearend began seeking comments on its leasing policies and procedures in the central and western gulf.

Meanwhile, a deepwater oil port to be built on Corpus Christi Bay began seeking environmental permits, while plans were unveiled for a deepwater oil port off Texas.

Gulf data trail

Public and commercial sources of Gulf of Mexico oil and gas data gauged the gulf's modest activity rebound.

MMS data showed gulf operators in 1993 through June spudded 167 exploratory wells and 230 development wells. By comparison, offshore operators through first half 1992 spudded 83 exploratory wells and 144 development wells in the entire U.S. federal Outer Continental Shelf (OSC).

Gulf production in 1993 through July totaled 2.642 tcf of gas and 167.87 million bbl of oil, MMS figures show.

James K. Dodson Co., Grapevine, Tex., reported gulf operators in 1993 through October drilled 245 exploratory wells, 42% more than the 10 month total in 1992. Operators through October spudded 375 development wells, 82% more than in the same period a year earlier. Dodson expected well starts in the gulf by yearend to reach nearly 750, almost 300 more than 1992's yearly total.

Offshore Data Services Inc. (ODS), Houston, reported 82.7% of the gulf's 162 available mobile offshore drilling rigs under contract in November 1993, including 124 drilling wells, nine under contract but not working, and one under contract for nondrilling service. That near yearend 1993 count compared to a nadir of 60 active units in May 1992 and 105 active units in November 1992. ODS's Gulf of Mexico newsletter reported that gulf operators in 1993 through November had filed 599 permits outlining plans to drill 1,222 wells. Gulf oil and gas producers in 1992 through November filed 412 permits to drill 945 wells on the way to yearly totals of 439 permits and 989 wells.

ODS counted 279 exploratory locations in the gulf through November 1993 and 386 exploratory sites, 48.4% and 63.6%, respectively, more than year-long 1992 totals. ODS also counted 50 oil and gas discoveries in the gulf during 1993 through November, including 43 in less than 600 ft of water, five in 600-1,000 ft of water, and two in more than 1,000 ft of water.

Flood of independents

Dodson's data showed that eight of the gulf's 10 most active drillers were major companies.

Shell Offshore Inc. was the gulf's most active exploratory driller in 1993 and Chevron U.S.A. Inc. the most active development driller.

Overall, the two companies were the region's most active drillers during the year.

Through the first three quarters of 1993, Shell, Chevron, Amoco Production Co., and Arco Oil & Gas Co. accounted for nearly one third of all wells drilled in the gulf.

However, better economics combined with technical capabilities to broaden the range of exploratory prospects independent operators were willing to undertake. As a result, more independents than ever in 1993 were developing or planning to develop deepwater gulf prospects, in the past high dollar projects reserved for deep pocketed majors. Also, some independents applied horizontal drilling techniques at selected sites.

By many statistical measures, independents began dominating Gulf of Mexico oil and gas activity as never before. Data collected by Oceandril Inc., Houston, showed that 70% of all rigs working in the gulf from first quarter 1991 through August 1993 were hired by independents. Independents in first quarter 1993 contracted 73 of 103 active rigs and 77 of 114 units in the second quarter. In August 1993, 86—71.7%—of the gulf's 120 active offshore rigs worked for independent clients, Oceandril data showed.

Independents' leasing activity

One of the main reasons independents ramped up drilling activity in the gulf likely was that independent companies sought and acquired more acreage in the region.

MMS data on recent federal lease sales showed independents in most years since federal area-wide leasing began in 1983 spent more money and acquired more acreage than major companies, regardless of the size of the sale.

Independents offered 83% of bonuses at Gulf of Mexico federal lease sales in 1993, up from 46.5% in 1986. Similarly, independents in 1992 federal offshore lease sales won 76% of all acreage acquired, up from 44.8% in 1988 sales.

J. Connor Consulting Inc., Houston, calculated that independents won 51.5% of 3,551 offshore federal leases issued in the Gulf of Mexico from 1988 through the March 1993 federal sale for the central OCS. Perhaps more telling, Connor's data showed how quickly independent companies began developing gulf acreage after acquiring it.

Connor found that independents had drilled 85.1% of all wells on tracts leased since 1988, recorded 120 of 12 discoveries, and installed 48 of 52 offshore structures. Of 7 recently leased tracts with production, independents operated 66. Independents held more active leases in the gulf as well as more leases with production, in primary term and with at least one well drilled or completed.

In 1993 through August independent operators filed 14 of the 205 plans of exploration to test gulf prospects, 71 o 125 development plans, and 241 of 397 drilling permits Connor's statistics showed.

Mahogany prospect

Phillips Petroleum Co.'s Mahogany subsalt prospect spudded in mid-May off Louisiana in 370 ft of water on Shi Shoal South Addition Block 349, was the gulf's most cele brated well during 1993.

On preliminary tests, the 16,500 ft wildcat flowed 3,70

b/d of oil and 559 Mcfd of gas through a 14⁄64 in. choke with 6,800 psi flowing tubing pressure. On a later test of a larger interval in the same subsalt pay, the Mahogany strike flowed 7,256 b/d of oil and 9.944 MMcfd of gas through a ½ in. choke with 7,063 psi flowing tubing pressure.

Operator Phillips and Anadarko Petroleum Corp. each own 37.5% interest in Mahogany and Amoco Production Co. 25%. An appraisal well was to spud in December 1993.

When Phillips and partners in September announced the Mahogany find beneath a 3,000 ft thick salt sill, many said it marked the rebirth of oil and gas exploration and development in the gulf.

Time will tell whether that is so, but it was hard to argue with such suggestions. By some estimates, large mostly horizontal allochthonous salt sheets conceal oil and gas bearing strata under as much as 60% of the gulf in water less than 5,000 ft.

Before Mahogany, several companies had tested subsalt strata without success. But mostly because of improved subsalt imaging resulting from 3D seismic acquisition and processing, interest in the gulf's subsalt potential was increasing.

A month after Phillips spudded Mahogany, Amoco and partners Oryx Energy Co., Amerada Hess Corp., and Petroleo Brasileiro SA spudded another subsalt test, the Mattaponi prospect on South Marsh Island Block 169. Mattaponi was the first subsalt wildcat drilled in the gulf on the basis of 3D seismic data.

Other subsalt tests were planned in the gulf before yearend 1993. In addition to its Mahogany appraisal well, Phillips planned a subsalt test on South Timbalier Block 260, again with Anadarko as a partner. Anadarko was operator of another subsalt wildcat expected to spud by yearend.

Oryx was studying subsalt prospects at two other locations in the gulf.

All told, gulf operators had identified at least another 15-20 subsalt prospects in the gulf.

The Gulf of Mexico's subsalt setting

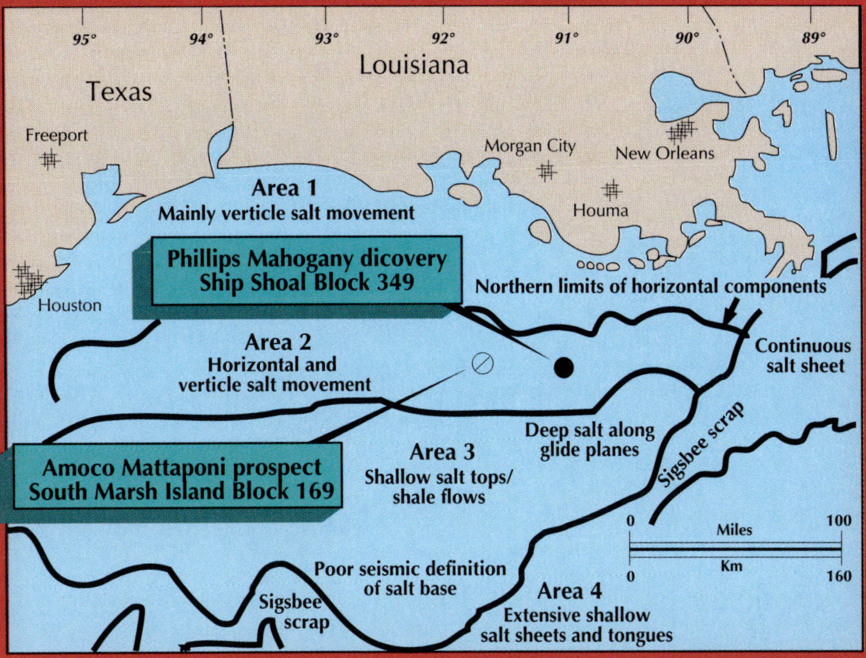

Source: TGS/GECO

Focusing the play

Due mostly to improved 3D seismic data, some operators concluded that salt in the gulf occurs in a variety of structural styles. Similarly, subsalt hydrocarbons likely are trapped by many mechanisms, depending on local geology.

Early concepts envisioned reservoir rock trapped against a salt face with salt as the seal. But in the greater play, any geologic structure obscured by salt could contain hydrocarbons, regardless of the trapping mechanism.

To help focus the study of salt sheets in the gulf and subsalt exploration, TGS-Calibre Geophysical Co., Houston, grouped the region's allochthonous salt sheets into eight structural styles and divided the gulf into four areas based on plots of each structural style's distribution across the gulf's shelf and slope off Louisiana and Texas. In TGS-Calibre's system:

- Area 1 includes roughly the northern tier of federal planning areas closest to shore.
- Area 2 covers the second tier of planning areas, including 10-15 miles beyond the southern edge of the shelf.
- Area 3 includes most of East Breaks, Garden Banks, and Green Canyon planning areas and northern parts of Alaminos Canyon, Keathley Canyon, and Walker Ridge areas.
- Area 4 is made up of a narrow belt in deep water north of and parallel to Sigsbee scarp.

Debate at yearend 1993 was continuing over many salt concepts, including the timing of salt deposition and differential loading that influenced salt formation. But many gulf operators were believed to be queing subsalt acreage higher in their portfolios of exploratory prospects.

Deepwater projects

Deepwater development in the Gulf of Mexico gathered momentum in 1993.

Exxon Co. U.S.A. early in the year completed installation of the gulf's largest multiwell satellite subsea template over Zinc field in 1,460 ft of water on Mississippi Canyon Block 354. The 445 ton, 118 ft by 75 ft by 28 ft drilling and production template and 285 ton, 10 well manifold rest on four 54 in. piles driven 225 ft into the seabed.

Zinc gas production at yearend was flowing 6½ miles from the first three of six wells on the $150 million Zinc template to the Alabaster platform in 468 ft of water on Mississippi Canyon Block 397. Exxon expected Zinc production to peak in 1995 at 125 MMCfd of gas and 5,000 b/d of oil and condensate.

Zinc field covers parts of Mississippi Canyon Blocks 354, 355, 398, and 399. Exxon drilled the discovery well in 1977 on Block 354.

Late in 1993, Shell began towing its Auger tension leg platform (TLP) into the Gulf of Mexico for installation in 2,860 ft of water on Garden Banks Block 426. When installed about yearend, Auger TLP claimed the gulf's water depth record for a permanent drilling and production platform and the world water depth record for TLPs. Auger's wells bettered by nearly 800 ft the gulf's water depth record for producing wells.

From seabed to the top of the flare tower, Auger TLP stood 3,280 ft tall.

Some common Gulf of Mexico salt related traps

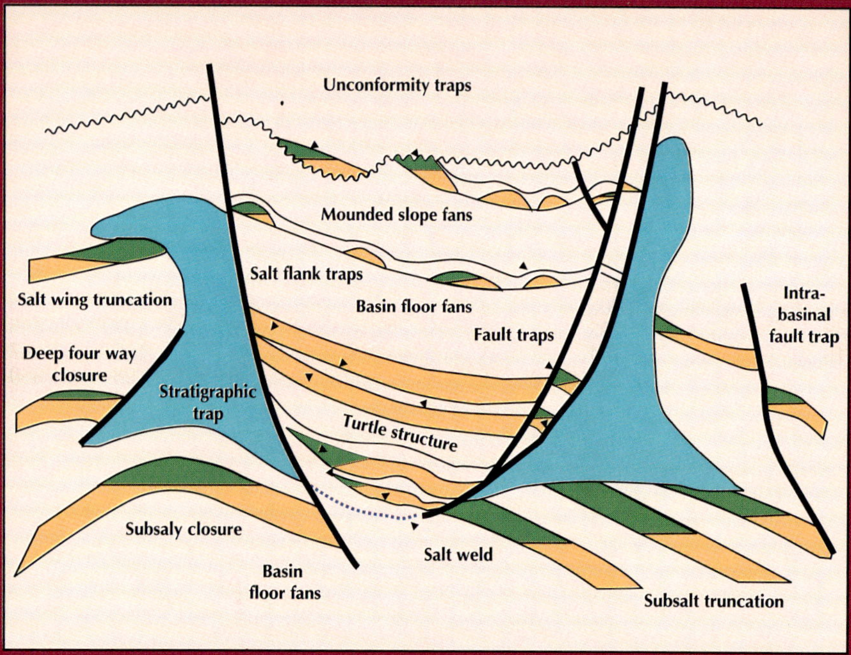

Source: Oryx Energy Co.

Shell and BP plan to predrill about 10 of the 26 Mars development wells. The George Richardson semisubmersible, owned by Sonat Offshore Drilling Inc., Houston, was to begin predrilling in fourth quarter 1993. The remaining wells are to be drilled by a rig mounted on the TLP after installation.

Mars TLP details

Drilling and production topsides will have slots for 24 wells in the Mars development project.

The Mars TLP deck is to include five modules—drilling, processing, power, wellbay, and quarters—fabricated by McDermott Inc. at Morgan City, La. The deck is to be an open truss framing design with dimensions of 245 ft by 245 ft by 45 ft high and is to weigh about 7,200 tons.

Production and processing equipment installed on the TLP will include complete oil and gas separation, dehydration, and treatment facilities capable of handling 100,000 b/d of oil, 110 MMcfd of gas, and 25,000 b/d of produced water. The TLP accomodation module will include a control room, emergency response center, and housing for 94 persons.

Mars oil will move through a 116 mile, 18 in. pipeline to a shore based oil terminal near Clovelly, La. Mars gas is to flow about 55 miles through a 14 in. pipeline to a pipeline interconnect in the South Pass area.

The 15,560 ton Mars TLP hull is to consist of four circular 66.5 ft by 162 ft steel columns, connected at the bottoms by four 27 ft by 24 ft pontoons. Belleli SpA, Taranto, Italy, is to fabricate the hull.

The Mars TLP is to be anchored to piles on the seafloor by 12 tendons—3 at each corner—each about 2,852 ft long with a diameter of 28 in. and 1.2 in. WT. Combined weight of the 12 tendons is 6,150 tons. Each of the 12 piles is to be 84 in. in diameter, 375 ft long, and weigh 260 tons.

The Auger prospect covers parts of Garden Banks Blocks 426, 427, 470, and 471, all 100% Shell owned tracts about 255 miles southeast of Houston and 214 miles southwest of New Orleans.

Gross oil production on the $1.2 billion development is expected to peak in 2001 at about 46,000 b/d and gas production at 125 MMcfd. Auger oil is to move by pipeline 72 miles to a platform in 243 ft of water on Eugene Island Block 331. Gas will flow 38 miles to a platform in 398 ft of water on Vermilion Block 397.

Auger's gross ultimate yield is expected to total about 220 million bbl of oil equivalent (BOE) with a 2:1 oil-gas ratio.

Shell completed its Auger discovery well in May 1987 in 2,862 ft of water on Garden Banks 426. The well was drilled by the Zane Barnes semisubmersible rig. Since 1990, Shell predrilled 10 wells on the unit with the George Richardson semisubmersible.

Mars development announced

About the end of third quarter 1993, Shell and BP Exploration Inc. disclosed plans to use a TLP in first phase development of giant Mars field about 130 miles off Louisiana.

The 3,250 ft tall Mars TLP is to be installed in 1996 in 2,933 ft of water on Mississippi Canyon Block 807. Upon installation, it will eclipse water depth records claimed by the Auger TLP for a permanent drilling and production platform.

Mars oil and gas is to begin flowing in fourth quarter 1996, increasing steadily to a peak in 2000 of about 100,000 b/d of oil and 110 MMcfd of gas. Shell's production in the Gulf of Mexico in late 1993 amounted to 132,000 b/d of oil and 753 MMcfd of gas.

Shell pegged ultimate recovery of the $1.2 billion first phase Mars development at about 500 million BOE and recovery of more than 200 million BOE during the next development phase. Recoverable reserves of 700 million BOE would rank Mars as the Gulf of Mexico's largest oil accumulation.

Shell in deep water

In addition to Mississippi Canyon 807, the Mars unit includes all or parts of five other blocks in the Mississippi Canyon area: 762, 763, 806, 850, and 851.

Shell acquired the leases encompassing Mars field at federal offshore lease sales in 1985 and 1988. Operator Shell holds a 71.5% interest in Mars and BP Exploration Inc. 28.5%.

Shell and BP drilled the Mars discovery well in 1989 on Mississippi Canyon 763, with the Discoverer Seven Seas drillship. The companies by late 1993 had drilled another four wells and six sidetracks and shot two 3D seismic surveys to delineate the prospect.

Mars and Auger TLPs will surpass water depth records set in 1988 by Shell's Bullwinkle platform in 1,350 ft of water on Green Canyon Block 65 as the world's tallest fixed platform. For 10 years before Bullwinkle was installed, Shell's Cognac platform in 1,025 ft of water on Mississippi Canyon Block 194 held the gulf's water depth record among fixed platforms.

CNG Producing Co., New Orleans, in January 1993 acquired half of Shell Offshore's 75% interest in the Popeye gas reservoir in 2,000 ft of water on Green Canyon Block 116 Unit.

Oil and gas completions and footage for gulf areas

	Oil well completions					Gas well completions				
	Alabama offshore	Gulf of Mexico Northern Federal Waters	Louisiana offshore	Texas offshore	Total Gulf of Mexico	Alabama offshore	Gulf of Mexico Northern Federal waters	Louisiana offshore	Texas offshore	Total Gulf of Mexico
1982	NA	17	113	17	147	NA	10	196	94	300
1983	NA	13	231	21	265	NA	8	124	129	261
1984	0	6	310	2	318	1	1	278	75	355
1985	0	7	139	22	168	3	3	150	90	246
1986	0	16	159	10	185	1	2	117	90	210
1987	0	24	163	9	196	1	2	106	35	144
1988	0	13	141	1	155	1	12	113	57	183
1989	0	47	205	13	265	0	28	293	154	475
1990	0	16	243	33	292	3	10	305	137	455
1991	0	22	110	9	141	4	10	244	109	367
1992	0	70	145	14	229	1	25	226	98	350
1993*	0	29	134	6	169	2	16	112	57	187
	Oil well footage					Gas well footage				
1982	NA	183,832	1,117,360	181,734	1,482,926	NA	123,817	1,991,511	824,041	2,939,369
1983	NA	158,276	2,264,523	187,287	2,610,086	NA	91,322	1,430,347	1,189,822	2,711,491
1984	0	56,310	2,952,490	15,953	3,024,753	21,000	10,275	2,658,371	707,151	3,396,797
1985	0	75,900	1,402,400	195,600	1,673,900	64,400	27,400	1,540,400	907,800	2,540,000
1986	0	176,320	1,457,400	96,602	1,730,322	22,390	15,481	1,298,367	998,752	2,334,990
1987	0	260,845	1,534,846	100,091	1,895,782	21,710	21,128	1,168,353	376,961	1,588,152
1988	0	131,995	1,306,788	10,200	1,448,983	22,950	92,953	1,171,555	593,957	1,881,415
1989	0	504,356	1,861,900	126,060	2,492,316	0	254,838	2,915,083	1,499,214	4,669,135
1990	0	169,057	2,405,389	327,070	2,901,516	27,272	103,962	2,875,241	1,183,752	4,190,227
1991	0	252,154	1,003,045	81,675	1,336,874	13,679	107,763	2,518,805	1,004,246	3,644,493
1992	0	633,306	1,160,221	113,455	1,906,982	21,842	182,678	2,030,695	880,752	3,115,967
1993*	0	308,080	1,081,846	69,654	1,459,580	7,307	141,641	1,033,587	524,602	1,707,137

*1993 well completions and footage through October.

Source: American Petroleum Institute

Data provided by Oil & Gas Journal Energy Database

Shell at yearend 1993 owned 32% of leased Gulf of Mexico acreage in water depths of more than 1,500 ft. In 1987, Shell set a drilling water depth record with a wildcat in 7,520 ft of water on the gulf's Mississippi Canyon Block 657.

Independents in deep water

Several independent operators claimed water depth records in the Gulf during 1993.

Oryx Energy Co., Dallas, set several global and regional water depth records for subsea wells and flexible gas flow lines on its Mississippi Canyon 400.

Three subsea wells on the six tract unit in 1,350-2,100 ft of water about 30 miles off Louisiana were producing 60 MMcfd of gas, making it Oryx's largest U.S. gas field. Gas flowed more than 7 miles through flexible subsea flow lines to a platform on West Delta Block 152.

With a fast track development plan, Oryx achieved production on Mississippi Canyon 400 unit in summer 1993, more than 1 year ahead of expectations.

According to Subsea Data Services, Wimberley, Tex., Oryx's development of the 400 unit's 1 Mississippi Canyon Block 445 well in 2,088 ft of water set world water depth records for deepest water producing subsea gas well and deepest water and longest all flexible, gas well flow line. The well is the deepest producing subsea well to date in the Gulf of Mexico.

Another well in the unit, 2 Mississippi Canyon Block 401 in 1,696 ft of water, is the gulf's second deepest water producing well.

Enserch Exploration Partners Ltd., a unit of Enserch Corp., Dallas, also started production in the gulf in unprecedented water depths. All six wells—three each on two subsea templates in 1,410-1,520 ft of water on Enserch's Mississippi Canyon Block 441—were among the gulf's 12 deepest water producing subsea wells.

Gas from Block 441 was being shipped 6 miles through subsea flow lines to a fixed platform in 380 ft of water on adjacent Ewing Bank Block 482. Block 441 subsea wells in mid-September were producing about 72 MMcfd of gas.

Also during 1993, Enserch drilled and cased four wells in 2,100 ft of water on its Garden Banks Block 388 unit. On that project Enserch was to install a floating production system (FPS) and 24 slot deepwater drilling and production template that will allow simultaneous drilling and production. Enserch planned to position the FPS—Enserch Garden Banks, formerly the Glomar Biscay I semisubmersible drilling rig—in 2,190 ft of water above 388 unit's subsea template.

Production from the 388 unit is to start in mid-1995. Enserch will ship the flow to a processing platform in about 250 ft of water on Eugene Island Block 315. The unit's production system was designed with capacity of 40,000 b/d of oil or condensate and 120 MMcfd of gas.

Tatham FPS prospects

At the same time Enserch was converting the Glomar Biscay I into Enserch Garden Banks, Deepwater Production Systems Inc, a unit of Deeptech International Inc., Houston, was converting the former Penrod 75 semisubmersible rig to FPS Lafitt Pincay. The two units were berthed side by side at Pascagoula, Miss.

Another Deeptech unit, Tatham Offshore Inc., plans to use the FPS Lafitt Pincay at Seattle Slew on Ewing Bank Block 915 unit, a six block unit in 980 ft of water about 50 miles off Louisiana, or Sunday Silence, a four block unit in 1,400-1,500 ft of water on Ewing Bank Blocks 958, 959, 1002, and 1003.

In a late August 1993 test, Tatham's 2 Ewing Bank Block 914 well flowed 12.5 MMcfd of gas and 2,300 b/d of condensate through a 25/64 in. choke with 4,400 psi flowing tubing pressure.

The company used a refurbished tree, fabricated a guide base to set over the template, and laid and tied back two 7 mile pipelines to a platform on Ewing Bank Block 826. Engineering and prefabrication took about 9 months before installation.

Tatham estimated both Seattle Slew and Sunday Silence reserves at more than 100 million bbl of oil equivvalent. An early 1994 exploratory well was being planned for Sunday Silence.

Baker Hughes active rig count

	Central gulf	Western gulf
1982	137	78
1983	124	48
1984	132	55
1985	129	48
1986	69	19
1987	65	19
1988	88	22
1989	72	22
1990	76	22
1991	57	15
1992	38	6
1993*	58	12

*1993 is average rigs through November.

Source: Baker Hughes Inc.
Data provided by Oil & Gas Journal Energy Database

Deepwater pipelay activity

For each regional or global deepwater production record claimed during 1993 in the gulf, a pipelining mark of equal note was set by companies designing and installing flow lines.

Among the gulf's most significant pipelay projects during 1993:

- Exxon used a diverless lateral deflection tie-in to connect Zinc's three 6 mile long flow lines, including two 8 in. and one 4 in., to the project's subsea template on Mississippi Canyon 354. Exxon laid the Zinc lines in July, August, and December 1992. The 60 mile, 20 in. Southern Natural Gas Co. pipeline through which Zinc gas flows to shore from the Alabaster platform is the gulf's deepest water, large diameter, subsea line.
- McDermott International Inc. in April used a J-lay technique to lay the deepwater portions of 12 in. oil and gas flow lines for Shell's Auger TLP. McDermott's Derrick Barge 50 (DB-50), following installation of the Auger TLP, in early 1994 was to install the last 3 miles of both lines with a diverless lateral deflection tie-in.
- The flow line on Oryx's Mississippi Canyon 400 Unit connecting the 1 Mississippi Canyon Block 445 well to the West Delta platform claimed the world record for deepest water and longest all flexible gas flow line. Oryx about mid-1993 began transporting 60 MMcfd of gas through the unit's flexible subsea flow lines.
- On the Mississippi Canyon 441 flow lines, Enserch assembled the project's integrated oil and gas flow line and control umbilical bundles on Matagorda Island near Bay City, Tex., and bottom towed the bundles 500 miles to Block 441.

There were technological capabilities in the gulf during 1993 to allow installations of pipelines in water 3,000 ft deep and more. But whether deepwater oil and gas transportation capacity in the gulf will expand far enough and fast enough to allow timely development of marginal deepwater projects likely will come down to how willing operators are to cooperate in regional, phased development schemes.

Horizontal drilling activity

While becoming relatively well known on shore, horizontal drilling in the gulf was in its infancy in 1993. However, several independents were among companies helping to nurture it.

McMoRan Oil & Gas Inc., New Orleans, in 1993 drilled three horizontal wells in about 250 ft of water at Freeport McMoRan Inc.'s sulfur and oil development on Main Pass Block 299 on the heels of a redrilling and workover project in late 1992 and early 1993.

Horizontal sections of the new wells ranged from 500 to 1,000 ft, while displacement from surface location averaged about 2,000 ft. Block 299's reservoir was producing from 1,400 to 1,500 ft subsea.

McMoRan early in September was producing oil from 11 wells on Main Pass, including eight horizontal and three at 70° angles. All told, the company had drilled about 24 wells at Main Pass, including horizontal and deviated wellbores. Some of the wells are injectors and some producers.

The company was injecting more than 238,000 b/d of hot water into the flanks of the field to maintain reservoir pressure and drive oil and sulfur to producing wells.

Main Pass flush oil production in early 1992 was almost 50,000 b/d. After the remediation program that ended in first quarter 1993, the facility's production through 1993 held at about 25,000 b/d. McMoRan's cumulative Main Pass oil production at the end of August had surpassed 15 million bbl.

Aran's horizontal drilling

Also in 1993, the first horizontal well in the gulf by Aran Energy Corp., Houston, led to a three well horizontal drilling program.

Aran drilled all three wells on South Pass 37 from Platform A on the southern half of the block. Horizontal sections in the wells averaged 1,100 ft, and the targeted reservoir was about about 2,600 ft vertical depth.

Because Aran wasn't drilling into an established reservoir and had little well control and only over-the-counter 2D seismic data to go on, the company first drilled pilot holes "toward the point where the seismic lines intersected," Aran said.

Aran gained a foothole on South Pass 37 in 1990 by taking a small nonoperating interest from a major company that had identified more than 30 reservoirs at 2,600-16,000 ft. At the time Aran became operator of the tract, the original development was producing 5-6 MMcfd of gas. The company expected 17-20 MMcfd of combined production from the three new horizontal wells.

Aran in November 1993 planned to remobilize on the platform in the northern half of the field and begin another horizontal drilling program at the same depth. Horizontal wells on the northern half of the block were expected to add 15-20 MMcfd of gas production, lifting total tract production by the end of first quarter 1994 to 40-50 MMcfd.

During horizontal completion operations in spring 1993, Aran had trouble inserting a 7 in. slotted liner with a prepacked screen assembly into one of the holes, so two of the wells were completed with 7 in. liners and the third with a 5 in. Also, in the last well—A-16—Aran had to cut a window to sidetrack out of the horizontal section to assure it would pass through the heart of the reservoir, possibly the first such operation in the gulf on a horizontal well.

Pogo Producing Co., Houston, in November 1993 announced results of a five well horizontal development program on Eugene Island Block 295 in the central gulf.

The five wells, drilled to true vertical depths of 1,203-2,482 ft with horizontal intervals ranging from 840 to 1,801 ft, flowed at a combined calculated rate of 100.1 MMcfd of gas.

Exxon with an 80% interest in deeper horizons had been producing from deeper wells on the Eugene Island tract since 1973. Pogo owned a 100% interest in the shallow gas reservoirs and the remaining 20% interest in deeper production.

Pogo at yearend had a platform in position and expected to begin production before the end of January 1994.

Property acquisitions

Higher gas prices in 1993 prompted acquisitions in the gulf and spurred participation in gulf federal offshore lease sales.

In one of the most noteworthy acquisitions, Apache Corp., Houston, in July 1993 closed a $125.5 million two step deal for most of the developed leases in the gulf operated by Hall-Houston Oil Co., also of Houston. The deal included proved and probable reserves of about 142 bcf of gas equivalent (bcfe) and increased Apache's gulf reserves to 275 bcfe.

Apache said the acquisition about doubled its presence in the gulf at a comparable cost and much faster than would have been possible by acquiring leases at federal sales. Also, Apache was able to get a better idea of what it was acquiring and could see upside potential in developing nonproducing properties and in reducing operating costs on wells already on line, all without accruing upfront evaluation, leasing, exploration, and development costs.

Hall-Houston benefited in selling the portfolio of leases by gaining financial flexibility.

The perceived advantages of strategic acquisitions did not divert attention from federal lease sales in the gulf.

At OCS Sale 142 in March 1933 covering acreage in the central gulf, 60 companies offered 262 bids totaling more than $86 million for 200 tracts. At Sale 143 in September for the western gulf, 48 companies offered 197 bids totaling more than $80 million for 157 tracts. Both sales reversed 3 years of declining participation at federal lease sales in the gulf.

At Sale 139 in March 1992, companies offered just $65.3 million for 151 tracts in the central gulf, and high bids amounted to $56.2 million. At MMS Sale 141 for the western gulf in August 1992, 38 companies offered 81 bids totaling $36.7 million for 61 tracts, and high bids amounted to $30.6 million.

Apparent winning bids at Sale 142 totaled more than $69 million and at Sale 143 more than $64 million.

Leasing policies studied

Meanwhile, concerns were raised that federal area-wide lease sale policies and procedures might make planning and assessing the effects of leasing difficult for states to address and could exaggerate the negative effects of price volatility on communities.

At yearend 1993, MMS Director Tom Fry announced that the agency was seeking comments on gulf leasing. Fry said changes affecting the OCS program in the central and western gulf—including low oil prices, maturity of the region, advances in 3D seismic techniques, and the influx of new companies participating in federal gulf lease sales—indicated a review was needed.

"We need to assure that we're managing the public's resources in the most efficient way possible, encouraging competition, providing all necessary protection for the environment, and obtaining fair market value for taxpayers," Fry said.

However, he said, no major policy changes would be considered until MMS had heard from all concerned parties.

"We have already heard comments on these complex issues, and we will consider them thoroughly," Fry said.

Changes MMS was considering included whether to restrict the number of tracts offered or the number of tracts leased at each sale. While either response could answer claims that too much acreage was being leased each year in the gulf, area-wide leasing was encouraging companies to apply new technologies offshore and to use a variety of exploration and development strategies.

Deepwater ports proposed

Not all news in the Gulf of Mexico in 1993 concerned upstream oil and gas activity. Proponents of the proposed Safeharbor onshore port on Corpus Christi Bay for deepdraft oil tankers in August announced start up of the project's formal environmental permitting process. Then in September, Petroport Inc., Corpus Christi, proposed building a deepwater oil port in the gulf using salt dome storage caverns about 35 miles off Freeport, Tex.

Port of Corpus Christi Authority (PCCA) said the 750,000 b/d Safeharbor port was the best alternative to lightering for handling oil imported across the gulf with very large crude carriers (VLCCs). PCCA in May 1993 had sponsored a series of workshops in and near Corpus Christi to air preliminary plans for and stimulate public discussion about Safeharbor. PCCA in June announced in the Federal Register plans to begin drafting Safeharbor's environmental impact statement (EIS). The draft EIS was expected to be available by August 1994 for public review.

According to preliminary estimates, both Safeharbor and Petroport would cost about $450 million to build. With fast track regulatory approval, both could be ready to begin operating by mid to late 1997.

As conceived, Safeharbor would include dredging a channel about 10 miles long, 80 ft deep, and 500-700 ft wide, constructing two or more berthing docks for VLCCs on Harbor Island across from Mustang Island on the Corpus Christi Ship Channel, building a tank farm and laying a pipeline to connect offloading and storage facilities, and laying an oil pipeline from Safeharbor's site to Texas City, Tex., so the facility could be used by refiners in the Houston area. Safeharbor proponents estimated about 50% of the deepdraft port's 750,000 b/d capacity would be used by refiners in and near Corpus Christi. The balance of oil imported through the facility could be shipped to other refiners along the Texas Gulf Coast.

Preliminary Petroport plans described an offloading platform with three single point moorings to be built in about 110 ft of water, deep enough to allow the world's largest VLCCs and ultralarge crude carriers to discharge as much as 100,000 bbl/hr of imported oil bound for U.S. refineries.

Oil offloaded at the facility would be injected into eight subsea salt dome caverns with combined capacity of 30 million bbl for either long or short term storage. Oil from the storage complex would be shipped to shore through a single 36 in. pipeline to Freeport for redelivery to refining centers onshore along the Texas Gulf Coast.

Smaller tankers capable of safely entering coastal Texas oil terminals also could be loaded at the facility, a direct replacement of lightering.

Petroport officials estimated market areas served by the port in 7 years would be importing about 2.3 million b/d of oil into the U.S., up from about 2 million b/d at present.

Several deepwater oil port projects have been proposed in the past two or three decades for installation off Texas in the gulf. But the Louisiana Offshore Oil Port, 18 miles off Louisiana in 155 ft of water, at yearend 1993 remained the gulf's only operating deepwater port. IPE

CANADA

CAPITAL: Ottawa
MONETARY UNIT: Dollar
REFINING CAPACITY: 1,810,100 b/cd
PRODUCTION: 1.64 million b/d
RESERVES: 5.9 billion bbl

The Canadian petroleum industry entered 1994 with its eyes on a rising level of drilling activity at home and increasing demand for its oil and gas shipments to the U.S.

Buoyed largely by firm prices for gas, Canadian drilling activity was setting a faster pace in 1993 than it did in 1992. The weekly active rig count by Baker & Hughes Inc. stood at 198 as 1993 drew to a close, up 82 rigs from the same time in 1992.

The Canadian Association of Oilwell Drilling Contractors (Caodc) forecast 9,000 wells will be drilled in western Canada in 1994, up from an estimated 8,800 wells for 1993.

Caodc Pres. Jim Strain said the 1994 forecast took into account a possible drop in crude oil prices. But the effect of such a drop would not be too pronounced on drilling activity because the Canadian industry's focus would continue to be on natural gas exploration and development. Wellhead gas prices were expected to remain firm.

Richard Wyman of Peters & Co., Calgary, said Alberta wellhead prices were forecast to stay in the $1.65-2 (Canadian)/Mcf range in 1994.

The Petroleum Services Association of Canada (PSAC) predicted oil field activity and employment in 1994 will match 1993 levels. PSAC said increased activity in 1993 lifted oil field service jobs by 3,000 to 15,000, up 20% from 1992. But employment was still only half of the peak work force in boom years of the late 1970s and early 1980s.

1992 crude oil and gas exports

Figures compiled by Canada's National Energy Board (NEB) showed a rising level of crude oil and natural gas shipments to the U.S. during 1992, the last year for which full data were available.

Total crude oil exports, including synthetic crude, were 134,000 cu m/day (844,000 b/d), up 10% from the 1991 level of nearly 122,000 cu m/day (768,000 b/d). The 1992 total was made up of about 74,000 cu m/day (466,000 b/d) of light crude and equivalent and nearly 60,000 cu m/day (378,000 b/d) of blended heavy crude.

NEB reckoned the value of 1992 crude oil exports at $6 billion, compared with $5.2 billion in 1991. The increase flowed from higher volumes and prices for heavy crude. Prices for light crude exports were about the same as in 1991.

The U.S. Midwest continued to be Canada's biggest market. Smaller volumes were shipped to the U.S. East, West, and Gulf coasts.

The largest buyers of light Canadian crude oil exports in 1992 were Amoco Oil Co., Clark Oil & Refining Corp., Koch Refining Co., Mobil Oil Corp., Sun Co. Inc., Ashland Oil Inc., and United Refining Co. For heavy Canadian crude oil, the largest foreign buyers were Koch, Mobil, Amoco, and Conoco Inc.

Gas shipments to the U.S., Canada's only export customer for that fuel, rose by about 17% in 1992 to 56.1 billion cu m (2 tcf) from 47.8 billion cu m (1.7 tcf) in 1991. Total gas export revenues also rose, by about 22%,

Canadian oil and gas supply and disposition*

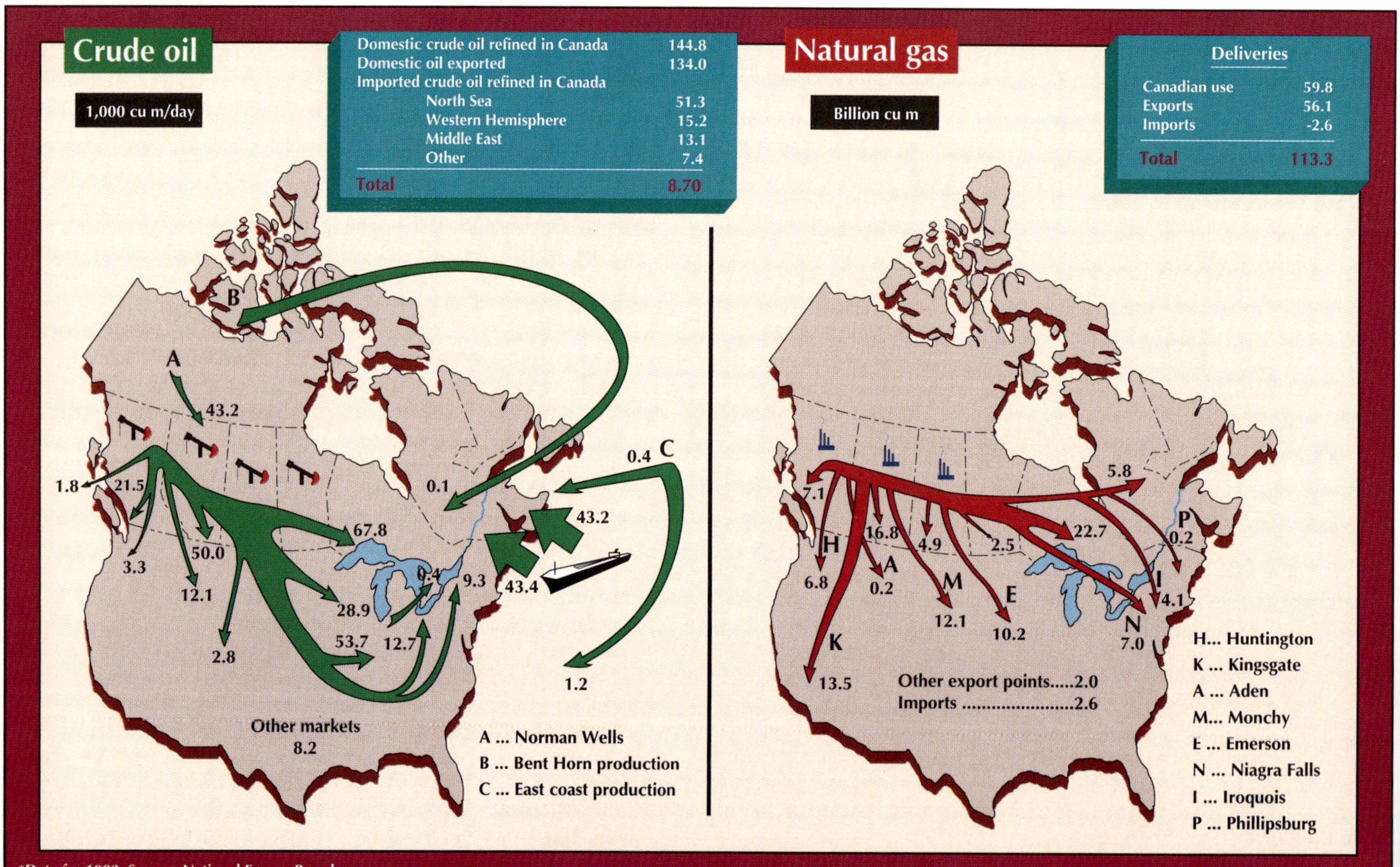

*Data for 1992. Source: National Energy Board

to $4.4 billion from $3.6 billion in 1991.

The central U.S. accounted for 42% of Canada's 1992 gas export sales, California 23%, the Northeast 21%, Pacific Northwest 12%, and the Rocky Mountain region 2%. Canada supplied 10% of gas consumption in the U.S. during 1992.

Average export price rose to $1.69 (U.S.)/Mcf in 1992 from $1.64 (U.S.)/Mcf in 1991.

Pipeline construction in 1992 gave Canada's four largest interprovincial gas pipelines—TransCanada Pipe Lines, Westcoast Energy, Foothills Pipe Line, and Alberta Natural Gas—a combined delivery capacity of 310 million cu m/day (10.9 bcfd) by yearend. Of that volume, 185 million cu m/day (6.5 bcfd) was available for exports. And exports amounted to 90% of available export capacity.

In total, domestic and export volumes averaged 90% of pipeline capacity.

A flurry of gas pipeline construction projects was under way or proposed as 1994 drew near.

A measure of Canada's increased gas shipments came from Nova Corp., Calgary. It reported an 11.4% increase for delivery volumes in its Alberta gas pipeline system in first half 1993. The company said volume in the system that transmits Alberta gas to Canadian markets and export points to the U.S. handled 1.86 tcf compared with 1.67 tcf in first half 1992.

Nova's deliveries were up to all markets served by its system in 1993. Deliveries averaged 10.3 bcfd.

Gas transported within Alberta in first half 1993 was up 15.3% from the same period in 1992. Exports were up 14.4% at the Empress border point, 6% at the McNeill point, and 2.8% at the Alberta-British Columbia border point.

Nova ships gas to Alberta border points for markets in Central Canada, as well as the U.S. Midwest, Northeast, and California.

NEB estimated Canada's 1992 production of natural gas liquids at 75,400 cu m/day (475,000 b/d). NGL exports amounted to 18,000 cu m/day (113,000 b/d) in 1992.

Where Canada's established reserves are

	Initial	Remaining*
	Million cu m	
CONVENTIONAL CRUDE OIL		
British Columbia	93.2	18.5
Alberta	2,221.6	424.5
Saskatchewan	511.4	116.4
Manitoba	36.3	7.7
Ontario	12.0	1.6
Northwest Territories and Yukon		
Mackenzie Delta and Beaufort Sea	0.0	0.0
Arctic Islands, eastern Arctic offshore†	1.0	0.8
Mainland territories§	35.2	19.3
Nova Scotia¶	7.6	7.6
Newfoundland▪	80.0	80.0
Subtotal	**2,998.3**	**676.4**
CRUDE BITUMEN√		
Surface mineable	644.0	451.0
In situ	102.8	50.7
Subtotal	**746.8**	**501.7**
Total	**3,745.1**	**1,178.1**
MARKETABLE NATURAL GAS		
	Billion cu m	
British Columbia	475	238
Alberta	3,348	1,630
Saskatchewan	151	87
Ontario	40	9
Northwest Territories and Yukon		
Mackenzie Delta and Beaufort Sea	309	309
Arctic Islands, eastern Arctic offshore	0	0
Mainland Territories	15	7
Total	**4,338**	**2,280**

*As of Dec. 31, 1991. †Bent Horn field only. §Norman Wells field only. ¶Cohasset and Panuke fields only. ▪Hibernia field only. √Alberta Energy Resources Conservation Board data based on developed projects only.

Source: National Energy Board

Oil and gas export outlook

All signs point to rising U.S. demand for Canadian exports of crude oil and natural gas.

In fact, forecast demand for crude oil in the U.S. supports at least two new projects to deliver Alberta oil to U.S. markets.

Unveiled in October 1993, a study by the Canadian Energy Research Institute (CERI), Calgary, said the U.S. will need added shipments of 500,000-600,000 b/d of crude oil by 2000 as established fields deplete.

CERI, funded by Canadian governments and industry, said the U.S. demand outlook supports two oil pipeline projects in the regulatory mill.

NEB scheduled hearings for late November 1993 on export pipeline plans by Alberta Energy Ltd., Calgary, and Interprovincial Pipe Lines Ltd. (IPL), Edmonton.

Express Pipeline Inc., a unit of Alberta Energy, sought approval for a $470 million, 777 mile, 180,000 b/d crude oil line from the Edmonton area to a pipeline hub at Casper, Wyo.

IPL applied for a $258 million expansion of its crude oil system line to Chicago. It would increase capacity by 125,000 b/d with an option of a further expansion to 190,000 b/d.

CERI study said as much as 267,000 b/d of incremental oil could be sold to refinery and distribution markets at Wood River and Patoka, Ill, and an additional 98,000 b/d into Chicago.

It said the Alberta Energy project had potential for as much as 113,000 b/d in new supplies to markets in Wyoming, Colorado, Montana, and the Pacific Coast. A link to Casper could serve markets in Illinois.

CERI said the Alberta Energy proposal also could improve the position of Alberta exporters who at the time lost as much as $1 (U.S.)/bbl to discounters in the Chicago market. The discounts were based on limited delivery capacity for Canadian crude into Chicago.

IPL generally operated at capacity on the Edmonton-Sarnia segment of its crude oil pipeline system during 1992.

Procedures developed by IPL and the industry to reduce apportionment were put in place on a trial basis early that year. While those procedures were partially successful, volumes nominated for shipment on the system exceeded capacity in all months except April, resulting in monthly apportionment.

In July 1991, IPL Line 9 from Sarnia to Monreal was shut down. However, in July 1992 the Alberta Petroleum Marketing Commission began injecting crude oil into Line 9 for delivery to Montreal, following approval of an integrated toll for west-east shipments.

Among other pipelines, Trans Mountain Pipe Line Co. Ltd. had spare capacity on its system during most of 1992, although throughput increased.

Most forecasters said they expect Canada to retain its role as the No. 1 source of U.S. gas supply during the next decade, aside from production in the Lower 48 states.

NCE Resources Group, Toronto, said in a 1993 report gas demand in North America is reaching record levels due to changing weather patterns, electrical power and transportation fuel conversions, and clean air concerns.

NCE Pres. John Driscoll forecast that U.S. productive capacity will slide sharply during the next 3 years and the U.S. likely will have to double its gas imports by 1996. Gas prices also may double in the next 3-5 years.

The NCE report said the energy policy of the Clinton adminstration is expected to encourage increased gas use to protect the U.S. environment, stimulate the economy, and reduce dependence on Middle East oil.

NCE predicted gas production in the U.S. will fall short of demand. The study said average gas well production in the U.S. has declined from 500 Mcfd in 1972 to 185 Mcfd at present.

"Demand for natural gas in the U.S. and the inability to produce it are on a collision course that inevitably will lead to much higher prices," Driscoll said.

The report noted U.S. gas imports from Canada have nearly tripled since 1986 and reached 2 tcf in 1992. It forecast Canadian imports will increase to 2.2 tcf in 1993.

Alberta's Energy Resources Conservation Board (ERCB) estimated Canada's main producing province has discoverable resources of 136 tcf, up 20% from earlier estimates and more than double present reserves of 59 tcf. But prices must rise to $3 (Canadian)/Mcf in 1991 dollars for recovery of 70% of its resource estimate.

ERCB predicted there will be gas shortages by 2002 even if prices do reach $3, and buyers will have to pay higher prices to ensure adequate supplies. It said a price of $5/Mcf would result in development of 87% of its resource estimate.

Canadian Gas Association (CGA) said in a January 1993 demand analysis Canadian gas consumption will jump 38% by 2005. It forecast a growth rate of 2.5%/year to 3 tcf by 1995.

CGA forecast a 46% hike in industrial demand during the next 13 years, fed by increased use of gas for electrical power generation and upstream industrial uses. Commercial demand in the same period will rise 34%, residential demand 25%.

Strongest demand growth was forecast for British Columbia, Alberta, Ontario, and Quebec, with more moderate growth in Manitoba and Saskatchewan. Oil and electricity are the main energy sources in the Atlantic provinces, and there were no current plans to extend gas pipelines to the East Coast.

Canada's federal energy department forecast Canadian gas exports to the U.S. will be 2.25 tcf by 1995 and worth about $4 billion (U.S.)/year, even at discounted prices. Ottawa said Canadian production will have to increase to 4.5 tcf/year from the current 3.6 tcf to meet export demand in 1995.

Irvine J. Koop, chairman of the Canadian Association of Petroleum Producers (CAPP), said producers see a bright future for gas in the North American energy mix. Koop cited a transition to a more balanced market and growing interest among buyers in long term contracts as the basis for their gas portfolios.

Products demand

Domestic demand for Canada's main petroleum products as well as refinery production were virtually unchanged in 1992.

Exports of main refined petroleum products—motor gasoline, middle distillates, and heavy fuel oil—and partially processed oil in 1992 fell 13% to 29,900 cu m/day (188,000 b/d) from the 1991 level. A 15% slide in shipments of middle distillates was the main reason, although all products declined.

On a regional basis, exports from Canada's Atlantic provinces and from Ontario showed declines. Atlantic provinces accounted for about 43% of total products exports.

Estimated value of main petroleum product exports, including partially processed oil, was $1.4 billion in 1992, down from $1.5 billion in 1991. This value excludes product exports from crude oil processing agreements for which prices were not assigned.

The U.S. continued to be Canada's largest buyer of petroleum products, receiving almost 91% of total exports vs. 87% in 1991. Exports also went to the Far East and Europe.

In the U.S., the East Coast continued to be the largest recipient of Canadian refined petroleum products, followed by the Midwest.

Largest exporters of main petroleum products were Newfoundland Processing Ltd., Imperial Oil Ltd. Irving Oil Ltd., Shell Canada Ltd., and Petro-Canada Products.

Crude oil imports represented 37% of total feedstock for Canadian refiners in 1992, compared with 38% in 1991. The Atlantic region depended 100% on imported crude. Quebec imported about 99% of its requirements vs. 94% in 1991. The increase stemmed from temporary closure of the Sarnia-Montreal pipeline in mid-1991. Ontario continued to import a small volume of crude oil, mainly from the U.S.

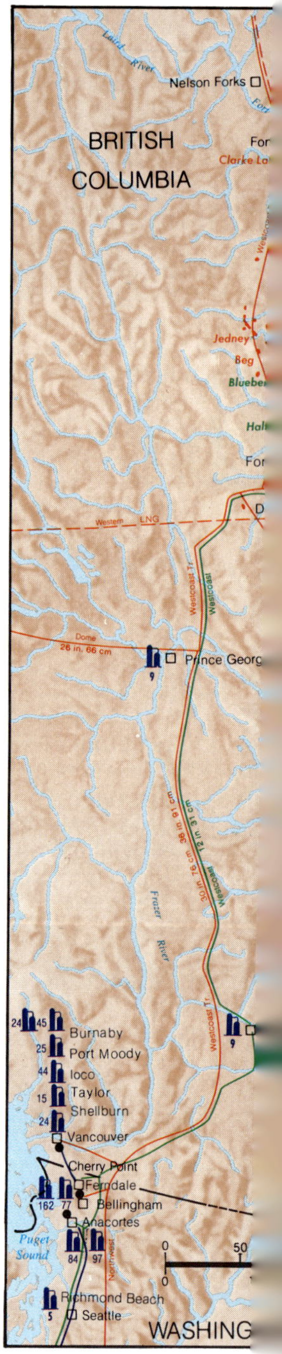

Crude oil purchased from members of the Organization of Petroleum Exporting Countries represented 29% of total imports in 1992, down from 32% in 1991. North Sea crude amounted to 59% of total crude imports, flat with the 1991 level. Crude oil imports from other sources accounted for 12%, up 3 percentage points from 1991.

Upstream activity

Figures compiled by Oil & Gas Journal (OGJ) showed Canadian crude oil production stood at 1.64 million b/d during the first 7 months of 1993, up 3.7% from the same period of 1992. Cumulative gas production for the first 7 months of 1993 was 3.168 tcf.

The Canadian industry achieved the increased volume of crude oil production despite a declining reserves base.

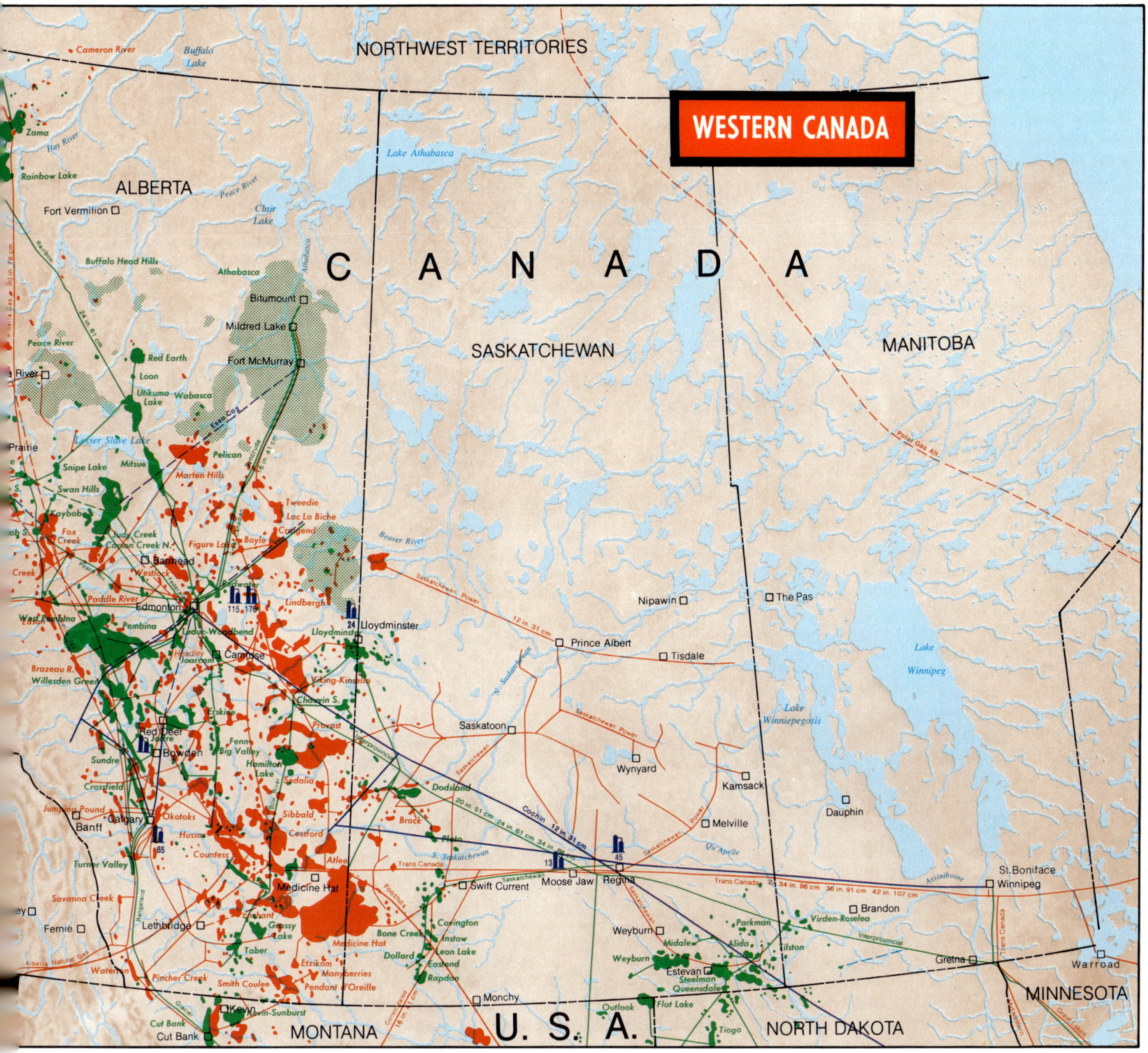

CAAP data show a low level of spending for exploration and development, along with low prices, caused a slide in conventional crude oil, natural gas, and NGL during 1992.

Industry spent $5.1 billion (Canadian) for E&D in conventional areas and $830 million (Canadian) in frontier regions in 1992. Corresponding figures for 1991 were $6.9 billion and $748 million.

At yearend 1992, conventional reserves had slipped by 198.65 million bbl to 5.1 billion bbl in the western sedimentary basin. Reserves in arctic and east coast frontiers were 800 million bbl, about flat with yearend 1991.

CAAP figures show light crude reserves slipped 7.2% to 3.31 billion bbl. Alberta, by far Canada's No. 1 producing province, replaced only 28% of its conventional production. Horizontal drilling added 141.8 million bbl to Saskatchewan's reserves, replacing 168% of the province's production.

Western Canada's natural gas reserves slipped 2% to 69.8 tcf. Revisions and additions replaced 67% of production.

With frontier areas included, Canadian marketable reserves slid by 1.4 tcf to 94.82 tcf.

Reserves of NGL fell 11.3% to 1.3 billion bbl.

Canada's total number of producing oil wells climbed to 42,070 at yearend 1992 from 40,580 at the same time a year earlier. Average production for an oil well in western Canada was 31 b/d in 1992, off a little from 32 b/d in 1991.

Total gas wells rose to 36,854 from 34,588. Average production was up at 381 Mcfd from 367 Mcfd.

Alberta oilsands play bigger role

Sweet synthetic crude from vast bitumen deposits in northern Alberta is playing an increasing part in Canadian oil production. That's largely because conventional crude reserves in Alberta are in steady decline.

Alberta oilsand deposits hold an estimated 1.7 trillion

bbl of oil in place. Athabasca sands, the largest deposit where two large commercial recovery and processing operations are sited, hold an estimated 870 billion bbl of bitumen. The bitumen, a thick mixture of oil-impregnated sand in sandstone and carbonate formations, must be processed to extract the oil.

As much as 307 billion bbl could be recovered from all oilsand deposits in the Athabasca, Cold Lake, and Peace River regions by in situ technology in use or under development in 1993.

Alberta's ERCB estimates remaining reserves of crude bitumen from deposits under development by surface mining as of December 1992 at 2.734 billion bbl. It estimates remaining reserves at in situ recovery programs at 304 million bbl.

Only a small fraction of oilsand assets is being worked. ERCB figures show the two commercial operations in the Athabasca sands produced a combined 88 million bbl of light crude oil in 1992.

NEB said combined production from oilsands and heavy oil operations could account for more than 50% of Canada's total crude oil production by 2010.

Improved technology and increased operating efficiency have steadily reduced the cost of bitumen production at Alberta's two major commercial oilsand operations in the Fort McMurray region, 300 miles north of Edmonton. Executives of both operators—Syncrude Canada Ltd. and Suncor Inc.—forecast that costs are close to and will reach levels competitive with new conventional crude production.

Eric P. Newell, Syncrude president and chief executive officer, said his company's unit operating cost of about $15/bbl (Canadian) is competitive with new sources of conventional oil. He said a survey by Powerwest Financial, an industry analyst, estimated finding costs for new conventional oil at $8-12/bbl.

"Our finding costs, of course, are zero because we know where the oil is," Newell said.

"When you add production costs to finding costs, new conventional oil in Canada costs $10-15/bbl to find and produce, which is very close to Syncrude's unit production costs."

Oilsand production from commercial plants in the Athabasca deposit accounted for more than 16% of Canadian oil production in 1992.

Surface mining and processing plants operated by Syncrude and Suncor produced a combined average of 240,000 b/d in 1992. That is forecast to rise to 256,000 b/d in 1993 when final figures are in.

Sun Co. Inc. is the majority shareholder in Suncor.

Syncrude, a combine of private companies and the Alberta government, is the world's largest synthetic

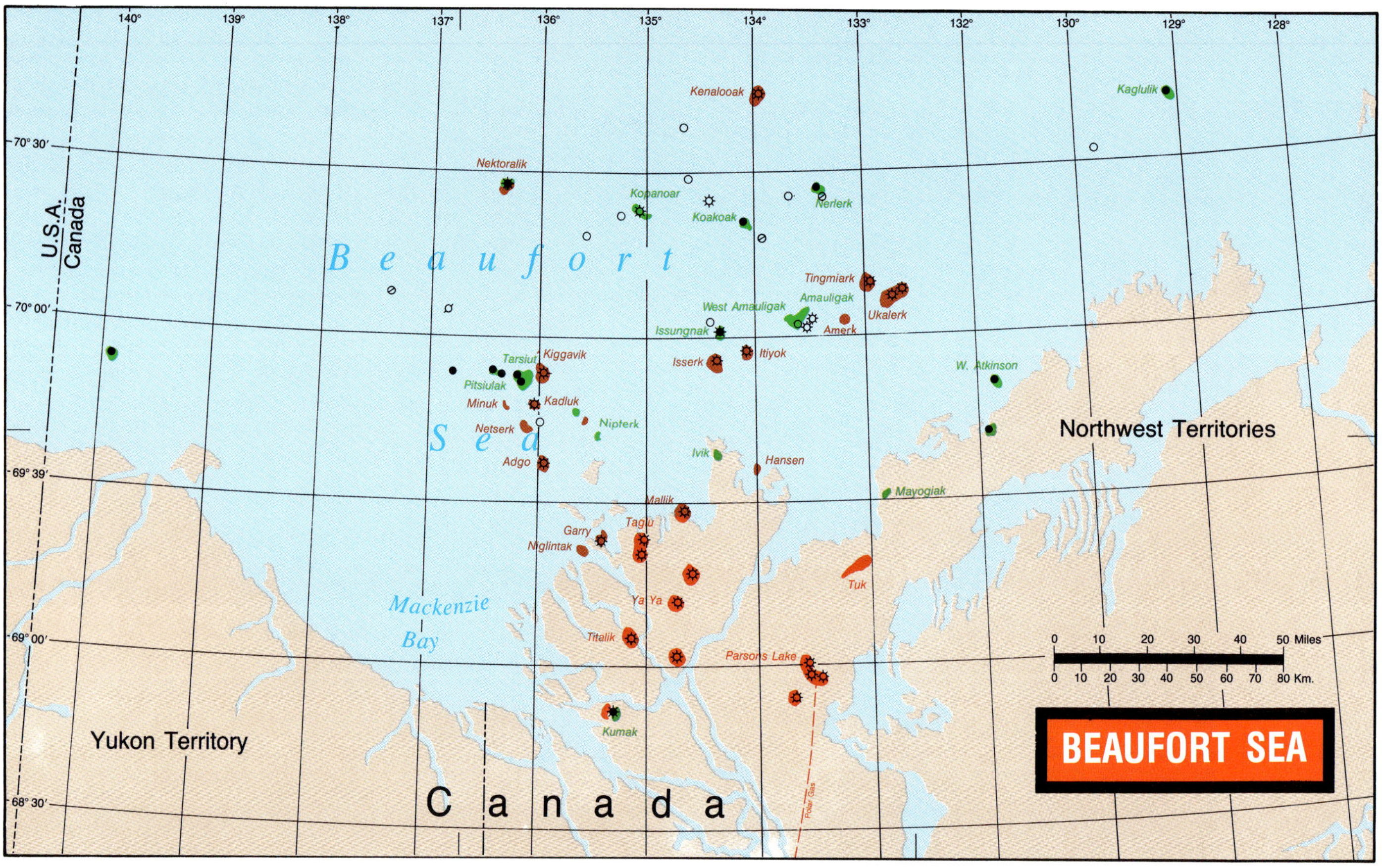

crude producer.

Group members include Alberta Energy Ltd., Alberta Oil Sands Equity (the Alberta government), Canadian Occidental Petroleum Ltd., Gulf Canada Resources Ltd., Imperial Oil Resources, Mocal Energy Ltd. (a unit of Japan's Mitsubishi), PanCanadian Petroleum Ltd., Petro-Canada, and HBOG-Oil Sands Ltd. (Talisman Energy Inc.).

The Syncrude joint venture group produced an average 180,000 b/d in 1992 and reached a production milestone of 600 million bbl in September 1992 since beginning operations in 1978. Production of 65.4 million bbl of light synthetic crude accounted for 12% of Canadian oil production in 1992.

Syncrude reckoned it was on track to reach a target of 68 million bbl in 1993, or an average 188,000 b/d. It processes about 325,000 metric tons/day of oil sand to produce 390,000 b/d of diluted bitumen, making it the largest mine operation in the world in terms of ore processed. The total is 600,000 metric tons/day if over burden and other materials handled by draglines ar included.

The average overburden at the Syncrude mine is 62 f and the oilsand formation averages a depth of 137 ft.

The company is Canada's third largest crude oi producer behind top ranked Imperial and Amoco Cana da Petroleum Ltd.

Syncrude has made more than $1.1 billion in pay ments to the Alberta government, and the government' 16.74% share has earned more than $530 million i profits on equity. Unlike many ventures involving in dustry and government, the Syncrude combine ha turned a profit every year since its inception. It curren ly pumps more than $1.1 billion/year in operational an capital funds into the Canadian economy.

Syncrude and Suncor set another landmark in au tumn 1992 when their combined cumulative produc tion topped the 1 billion bbl mark. That total make

Syncrude operations in Northeast Alberta include bucketwheel reclaimer (left) and hydrogen reformer (right).

Where Alberta's oilsand activity is under way

Source: Energy Resources Conservation Board

oilsand operations Canada's second largest cumulative oil producer, outranked only by Alberta's Pembina field with more than 1.2 billion bbl.

In addition to major operations in the Fort McMurray region, a number of companies are developing oilsand operations in other deposits. A large number of pilot production and experimental projects are under way.

Syncrude and Suncor are the only surface mining operations classified as commercial by ERCB, but most major companies operating in Alberta are involved in pilot and experimental programs for primary bitumen recovery programs or in situ recovery schemes.

Shell Canada, for example, operates a 10,000 b/d commercial in situ oilsand plant in the Peace River area of western Alberta. Shell acquired full ownership of the operation in 1992 with a buyout of the 50% interest of Pecten Canada Ltd., a unit of Shell Petroleum Inc.

The purchase included Pecten's 50% interest in 123,500 acres of oilsand leases in the Athabasca area and 197,600 acres in the Peace River area. Shell estimated its Peace River lease, about 311 miles northwest of Edmonton, has 14 billion bbl of bitumen in place. Syncrude and Suncor were poised in 1993 to spend large sums of money for equipment to step up production.

Shell's Caroline gas project

Shell Canada's $1 billion Caroline project began processing gas in March 1993 from Alberta's largest gas field discovery in 20 years.

Shell's plant near the village of Caroline, Alta., about 75 miles northwest of Calgary, was scheduled to process 8.5 million cu m/day (300 MMcfd) of gas from Caroline Devonian gas/condensate field.

Daily production was scheduled to amount to 3,000 cu m (18,800 bbl) of condensate, 4,200 cu m (26,400 bbl) of natural gas liquids, 4,000 metric tons (4,400 short tons) of sulfur, and 2.8 million cu m (100 MMcf) of sales gas. Inlet gas contains 35% hydrogen sulfide.

Shell said the plant's sulfur recovery facilities have the highest sulfur recovery rate of any major gas processing plant in North America. They are designed to achieve at least 99.8% sulfur recovery to meet new Alberta guidelines.

The Caroline project employs the world's longest underground liquid sulfur pipeline, which carries the product from the plant to a sulfur forming plant at Shantz, 25 miles south. Waste water and chemicals are not discharged. They are recycled and reused.

The project provided about 7.5 million man-hr of work during its 30 month construction period, with a peak work force of 2,500 in summer 1992.

More than 85% of the project's construction costs were spent in Canada.

About 150 persons are involved in daily operations at the plant, where operating costs amount to about $50 million/year. Caroline gas field was discovered in 1986. Plant construction began in October 1990.

Shell owns a 72% interest in Caroline field. Husky Oil Ltd. and Union Pacific Resources Inc. are among other working interest owners. IPE

China

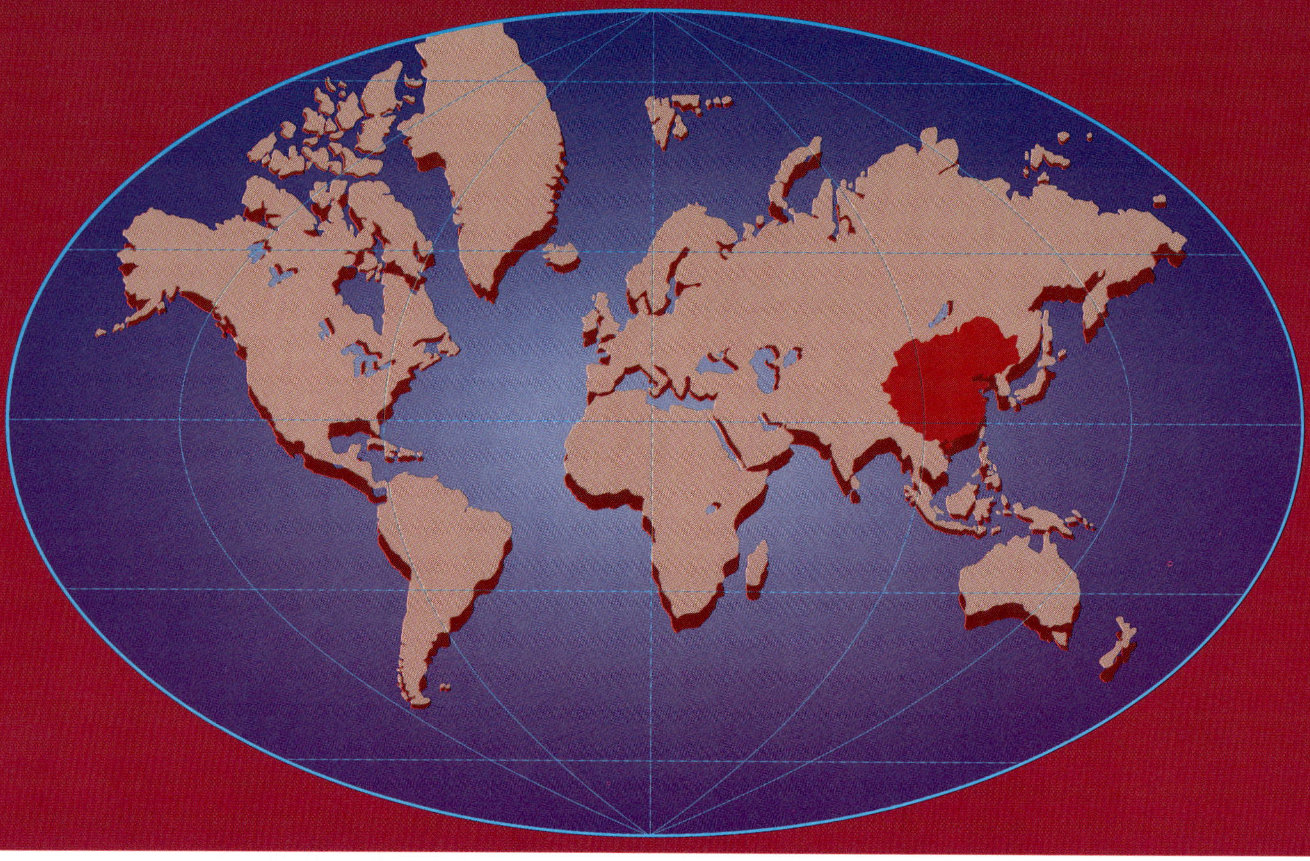

CHINA

CAPITAL: Beijing
MONETARY UNIT: Yuan Renminbi
REFINING CAPACITY: 2.2 million b/cd
PRODUCTION: 2.9 million b/d
RESERVES: 24 billion bbl

China in 1993 was stepping up its campaign to attract more foreign investment to boost onshore exploration.

After years of moribund drilling activity and lagging production growth, China has done an about-face on foreign exploration and development investment.

During the 1980s China focused its invitations to foreign companies to conduct E&D in its offshore frontiers. With rare exceptions, little onshore acreage was made available. And the most prospective area, the Tarim basin of China's remote, rugged northwestern region, was zealously guarded as the province solely of state petroleum agencies.

China's offshore has not yielded the spate of giant fields that had been hoped for, and foreign multinational interest in Chinese E&D dwindled in recent years.

Although Beijing opened the South China and Bohai seas to foreign exploration 14 years ago, the country's offshore has contributed little to production. Of China's total production of about 2.8 million b/d in 1992, only 80,000 b/d came from offshore fields.

The government in 1992 added the East China Sea to the list of foreign participation acreage.

Trying to cope with the world's most explosive economic and oil demand growth, China is projected to become a net oil importer by 1995—perhaps sooner.

That spurred Beijing to begin a campaign to attract E&D investment to onshore provinces.

Beijing opened onshore exploration acreage in 10 mostly northern and western provinces, including the prized Tarim basin, with prospects featuring a postulated combined potential resource of 40-60 billion bbl of oil and 87.5 tcf of natural gas. And joint venture opportunities in producing fields will be available.

New approaches

Chinese government officials disclosed plans to award additional acreage in the Tarim basin in a second bidding round, citing foreign companies' strong showing in the first round.

The enthusiasm shown for foreign investment in Tarim E&D was sparking plans for expansion of the region's petroleum infrastructure, including added refining capacity and a megaproject to transport Tarim crude to China's industrial heartland via pipeline to the central part of the country..

Beijing disclosed it will offer onshore acreage in 12 areas covering 420,000 sq km under a competitive tender for risk exploration contracts. Under the first bidding round, five blocks in the Tarim basin were up for bid.

Beijing also was opening 14 blocks in 10 major oil producing areas, including Daqing, Shengli, Liaohe, and Zhongyan, to foreign cooperation in an effort to boost oil recovery.

Apart from the open tender bidding rounds, a handful of joint venture exploration contracts signed in 1990-93 were expected to see some seismic work and drilling in 1993.

In addition to offering big chunks of acreage in large area-wide tenders, unlike the piecemeal approach in the past, the government sweetened terms with a new production sharing contract model.

In outlining its oil policy, state petroleum monopoly China National Petroleum Corp. (CNPC) now stresses the use of foreign as well as domestic resources and capital and targets foreign markets for investment. On average about $1.9 billion/year is spent on oil and gas exploration in China, with about half that coming from foreign sources.

At the same time, CNPC was considering bidding for

exploration rights in Peru, India, New Zealand, Indonesia, Australia, and Papua New Guinea. The company is seeking foreign sources of crude oil to make up for looming domestic supply shortfalls.

Tarim bidding

Beijing expected to award foreign oil companies exploration rights in the Tarim basin as early as yearend 1993, said Zeng Xingqiu, vice-president of CNPC subsidiary China National Oil & Gas Exploration & Development Corp. (Cnogedc).

CNPC authorized Cnogedc to handle all onshore bidding.

Deadline for submitting bids on first round Tarim acreage was Oct. 31, 1993. In late February 1994, Cnogedc was to inform bidders of results of bid evaluation, then contract negotiation was to begin Mar. 1, 1994.

The second bidding round, scheduled for first quarter 1994, was to offer more exploration blocks in the Tarim basin and elsewhere as well as 10 producing areas for improved oil recovery projects.

The Tarim first round acreage consists of five blocks totaling 72,730 sq km, about one eighth of the total basin area.

In all, more than 60 foreign oil companies from 17 countries submitted bids for exploration in the Tarim basin under the first round.

A number of companies disclosed plans to form bidding groups to assess the first round Tarim blocks.

They include:

- A group led by Exxon Corp. 35%, with other interests held by Mobil Corp. 35%, British Gas Exploration & Production Ltd. 15%, Sumitomo Corp. 7.5%, and Indonesia Petroleum Ltd. 7.5%.
- A group of Shell Exploration (China)/Pecten Thailand 36%, Amoco Orient Petroleum Co. 36%, Total China 18%, and Marubeni Corp. 10%.
- A group of British Petroleum Co. plc, Mitsubishi Corp., Itochu Corp., and Nippon Oil Co.
- A group led by Agip SpA and including Texaco Inc., Ste. Nationale Elf Aquitaine, and Nikko Kyodo Co.
- A group led by Phillips Petroleum Co. Phillips had not decided whether it will bid for Tarim acreage.

arim contract

Cnogedc was to have drafted and delivered to foreign applicants by May 31, 1993, a new model production sharing contract (PSC) said to include more favorable terms designed to attract risk exploration in the Tarim basin, China Features reported.

Those new terms included:

- An exploration term of 8-9 years vs. the usual 7 years because of the Tarim blocks' location in the remote Taklimakan Desert. As with the existing PSC, foreign companies must pay all exploration costs.
- Because of the region's inadequate infrastructure, foreign operators are allowed to delay a development decision for as many as 7 years after commercial hydrocarbons are found. Generally, companies will be allowed a 3 year waiting period on oil discoveries with the potential to produce more than 10,000 b/d. Beijing will set the term of the suspension period.
- Foreign companies will be allowed to market their shares of production domestically.
- Foreign companies involved in Tarim E&D also will be allowed to participate in downstream projects.

In the event of a commercial discovery, China will become an equity participant by paying 51% of development costs. Of resulting production, 62.5% is earmarked as "cost oil" and goes to the foreign partner. The foreign partner then earmarks sales of cost oil to cover production costs 20%, taxes 5%, and a progressive royalty. The rest of the 62.5% of production represents return on E&D investment.

Southeast China's booming oil demand

Province	Oil demand 1,000 b/d 1990	1995	2000
Guangxi	25.5	41.5	N/A
Yunnan	18.2	23.3	N/A
Guizhou	10.7	13.3	N/A
Hainan	N/A	6.0	7.0
Fujian	N/A	50.9	78.8
Guangdong	155.3	227.2	337.6

Currently, onshore royalties break out as 7% for production of 4,000-10,000 b/d, 10% for 10,001-15,000 b/d, and 12.5% for 15,001-20,000 b/d. Zeng Xingqiu said Beijing was considering a royalty exemption for production of less than 20,000 b/d.

Of the other 37.5% of production, called "profit oil," China will retain a certain percentage to be negotiated before a contract is signed. The remainder of the 37.5%, called "sharing oil," is split 51% China and 49% foreign partner.

Second Tarim round needed

Cnogedc said the five Tarim blocks up for bid were not enough to go around.

It plans to offer more acreage in the southeast arm of the Tarim basin outside the five first round blocks after the first round bidding closes.

Cnogedc expects to sign at least five contracts under the first round.

It has shot seismic surveys totaling 2,226 line km covering the five first round blocks.

Those five and their respective areas are Yatongguuzi 9,814.3 sq km, Xiaoertang 14,698.1 sq km, Qiemo and Tulabei each 14,475.4 sq km , and Washixia 19,267.2 sq km.

Prospects for both Tarim bidding rounds took on added luster earlier in 1993 when CNPC disclosed the biggest strike to date in the basin. Its 4 Tazhong discovery well initially flowed 1,824 b/d of oil and 1.86 MMcfd of gas. Later tests yielded flow rates of 3,705 b/d of oil and 2.28 MMcfd of gas at high wellhead pressure.

CNPC planned a 1,100 sq km 3D seismic survey of the central part of the basin and appraisal drilling to follow up the discovery. The Tazhong structure is the biggest identified to date in China, and CNPC's preliminary estimate of potential reserves is 730 million bbl.

The 4 Tazhong will see development starting in July 1994 and is slated to be on stream by 1996.

Tarim geology

Among the information CNPC was making available to foreign companies interested in Tarim E&D was an evaluation of the basin's hydrocarbon potential by Kenneth J. Hsu of Tarim Associates AG, Zurich.

The consulting firm has been involved in Chinese exploration studies for 15 years and Tarim petroleum geology studies for 7 years.

Hsu developed a buried euxenic basin model for generation of Tarim hydrocarbons that he contends explains the genesis of source beds containing postulated hydrocarbons totaling more than 350 million bbl of oil equivalent, the origin of structural and stratigraphic traps, including an oil bearing anticline 230 km long and 120 km wide with 2.4 km of closure, and the history of hydrocarbon maturation and migration.

There is evidence, Hsu says, that large volumes of trapped hydrocarbons have been prevented from being overly mature because of their migration into old traps prior to burial under Cenozoic sedimentary load. Tarim oil is being produced from Mesozoic and Paleozoic reservoirs at depths of more than 6,000 m. Paradoxically, Hsu said, gas and condensate are produced from 3,000 m deep Miocene reservoirs in the Tarim basin because of their late migration.

Contrary to current assumptions that the source beds of Tarim oil are Paleozoic platform carbonates, Hsu concludes the Paleozoic source beds are euxenic deposits of a deep marine basin. Oil migrated from lower Paleozoic source beds during the late Paleozoic or Mesozoic into karsts and other porous reservoirs on anticlinal structures and under seal beds above regional unconformities. Hsu's report also presents a new interpretation of the pre-Mesozoic structures on the Tanan uplift, the main structural elevation of the five first round blocks.

Most Tarim discoveries have been in the northern part of the basin. But drilling in the southeast portion of the basin, near the five first round blocks, turned up oil in Jurassic sands and as many as 10 pay zones with combined gross thickness of 150 m, the Ministry of Geology and Mineral Resources reported.

The first geological surveys in the Tarim basin started in 1952, but a major exploration and development effort did not begin there until 1985.

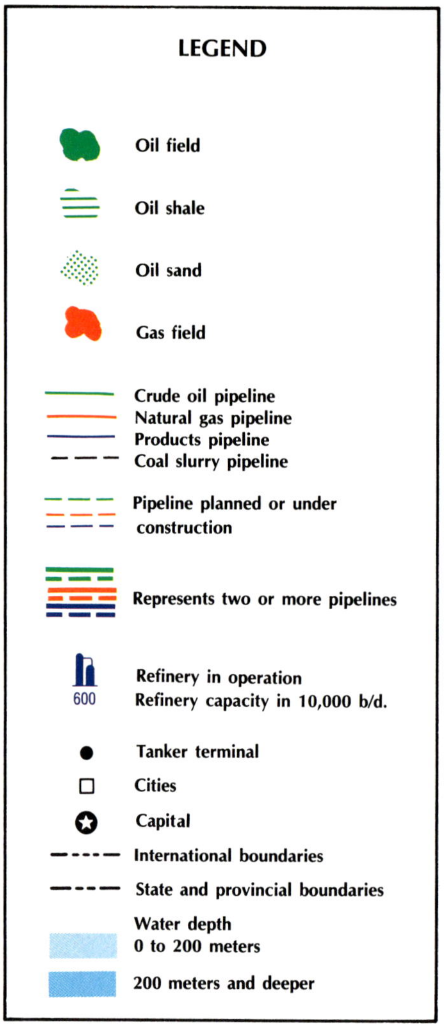

By yearend 1992, CNPC subsidiaries had acquired 107,000 line km of 2D seismic and 1,909 sq km of 3D seismic in the basin. They drilled 127 wildcats and appraisals, 58 of which flowed hydrocarbons at commercial rates.

Officials put the basin's postulated hydrocarbon resource at 134.7 billion bbl of oil equivalent, broken out as 74.1 billion bbl of oil and 290.5 tcf of natural gas. Proved and probable reserves are estimated at 1.46 billion bbl.

Tarim's mixed prospects

China's first Tarim basin oil and gas tender includes an unusual mix of prime and poor prospects, a U.S. geological and engineering company has found.

Of five Tarim basin blocks offered by China National Petroleum Co. (CNPC) in a bidding round that began last Mar. 1, 1993, Gustavson Associates Inc., Boulder, Colo., rates the geological characteristics of the most prospective tract as variable but generally positive.

Gustavson said another block is mediocre but probably will attract bidders, while the remaining three tracts should be considered very high risk.

Those findings appeared in a nonexclusive study of the basin by John Weiner, Gustavson's chief geologist, based mainly on public data.

"It is highly unusual for anyone to offer such widely differing concession blocks," said John B. Gustavson, president of the firm. "As large international companies learn about the drastic differences in source rocks, reservoirs, traps, seals, drilling depths, and discoveries, they will compete aggressively for the one or two favorable blocks while the others go begging."

That strategy by big companies or groups of majors could allow groups of independent oil companies a chance to win concessions on some higher risk acreage, he said.

Gustavson declined to disclose which tracts appear to be most promising.

Weiner said Gustavson rated each tract based on the presence or absence of lower Paleozoic source rock, suitable reservoir rocks of several geological ages, and extent of structuring or trap formation.

Weiner reached his conclusions by piecing together information from many sources. He cautioned reliability is suspect for much of the Chinese data on which the study is based.

For example, various Chinese geological reports place Tarim discoveries at different locales, and field names vary from source to source. Fields in some instances are known only from vague references or rumors, so some might not exist. Conversely, some real fields might not appear on available maps.

"I don't think there is a deliberate plan to mislead companies," Weiner said. "CNPC officials know where

everything is, but the literature is unclear."

But, he said, "Even Chinese officials might not yet be sure how big some Tarim fields are."

Information also is spotty about the first five Tarim basin blocks offered for concessions. For example, available information indicates blocks included in the first round were only sparsely explored by drilling, if at all.

"We're not sure, but some of the blocks might not have any wells on them whatsoever," Weiner said.

Gustavson's study estimated well costs in the area at $12-16 million for a 6,000 m Paleozoic test, partly because of the high costs of moving equipment into the remote region.

The company could not quantify current Tarim oil and gas production or determine which fields are on line. And it found scant evidence of oil and gas transportation infrastructure in the first round tracts.

One of the few certainties: The Qiemo block is among closest in the group to China's 1989 Tazhong oil and gas discovery on the Central Tarim uplift.

Despite the limitations of available data, Gustavson concluded that quality and maturity range widely among the basin's potential source rocks.

Based on discoveries to date, lower Paleozoic Cambro-Ordovician gray to black carbonaceous marine shales appear to have the best oil and gas potential in the basin. Carboniferous black shales and carbonates also might have potential but have not yet been linked to known oil and gas accumulations, Gustavson said.

Upper Triassic-middle Jurassic lacustrine mudstones and oil prone coals are the main sources of traps in the Kuqa and Southwest depressions but appear to be less important compared with Cambro-Ordovician strata. Cenozoic source rocks likely are of only localized importance.

Gustavson counted three main types of known or expected reservoir rocks in the basin:

- Clastic sandstone and conglomerate reservoirs of early Silurian, Devonian, Carboniferous, Permian, early and middle Jurassic, Cretaceous, and Tertiary ages.
- Carbonate, mostly marine sedimentary sequences of the upper Silurian, Cambrian, Ordovician, Carboniferous, lower Permian, Cretaceous, and Paleogene.
- Weathered layer reservoirs resulting from frequent uplifts and long exposures of thick zones, especially in the North Tarim and Central Tarim uplifts where extensive weathering and leaching has occurred to carbonates now lying beneath an unconformity.

Gustavson said Carbonate reservoirs, generally fractured, fossiliferous, and locally cavernous, karstified, or microcavernous, partially contain skeletal grains in algal mats and bioherm reef complexes. The most important weathered layer reservoirs associated with unconformities are beneath Devonian, Carboniferous, and Triassic sections and between Mesozoic and Cenozoic strata.

Gustavson found that the most important Tarim basin structural traps are anticlines. More than 200 anticlines have been mapped in the basin—137 on the surface and 68 in the subsurface.

The most obvious anticlines are compressional folds near mountain fronts bordering the basin.

Broad platform anticlines, including some with areas reported to exceed 1,000 sq km, are present in regions such as the Central uplift. Reverse fault related traps, which sometimes caused folding of overlying strata, are important in Paleozoic sections of the North Tarim uplift. Mesozoic anticlines draped over Paleozoic erosional and/or structural highs also are important in the North Tarim uplift.

Buried hill stratigraphic traps, partially controlled by an adjacent unconformity, have shown their importance in the North Tarim uplift. Also, truncated sandstone reservoirs in the area sealed by mudstones lying above the unconformity have oil and gas pays.

Gustavson said other types of stratigraphic traps in clastic rocks—submarine fans or river channels, for example—can be expected in varied marine and lacustrine deposits common to the basin. Reefs postulated in the Carboniferous-lower Permian also could bear hydrocarbons, but that is not confirmed.

The basin's Paleozoic seals include lower Cambrian black shales, middle Cambrian gypsum beds and clayey shales, lower Silurian clayey shales, and Carboniferous gypsum beds and clayey shales. Such Paleozoic seals in the Cambrian and Silurian could be somewhat regional, but Gustavson said there is a doubt about the regional nature of upper Paleozoic strata.

Gustavson also concluded adequate seals probably are not a problem in the basin's Mesozoic-Cenozoic formations, and shales and mudstones should be adequate seals in the Jurassic, upper Cretaceous, and Paloegene-Miocene.

Tarim drilling, production

Forty drilling rigs were working in the Tarim basin near yearend 1993.

Eleven Tarim oil and gas fields are productive from 14 pay zones. Five fields on stream in the northern and central portions of the basin—Lunnan, Donghetang, Sangtamu, Jilak, and Jiefangqudong—produced a combined average of 17,780 b/d in 1992, exceeding the basin's official target by about 2,000 b/d.

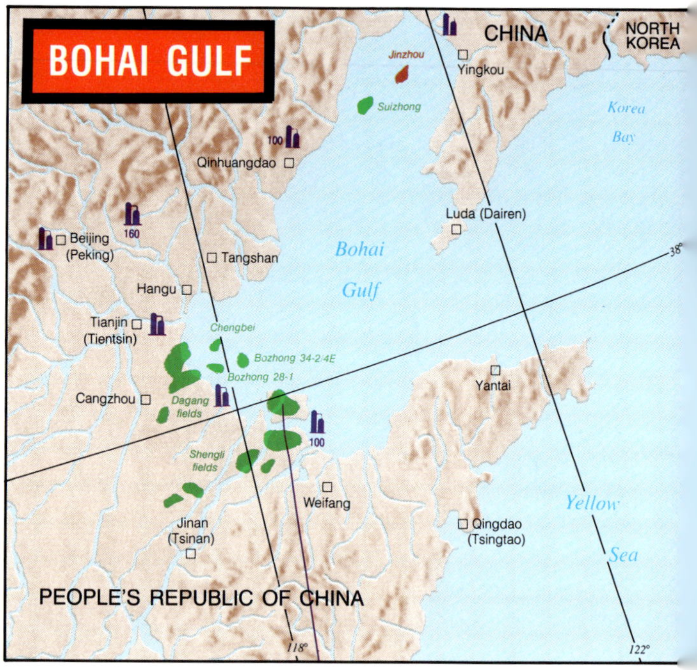

CNPC laid a 137 km pipeline to Lunnan in 1992. Th field's productive capacity from 53 wells has doubled t about 29,000 b/d while its water cut fell to 10%.

Some pay sands have thicknesses of almost 330 ft, an individual well flow rates average 630-3,711 b/d. In all, abou 77 Tarim development wells have been drilled, and anothe 55 are planned.

Average 1993 production was estimated at 24,000 b/ reaching 33,000 b/d by yearend and 48,000 b/d in 199 CNPC has targeted Tarim production of 100,000 b/d of o and 57.5 MMcfd of gas by 1995. The basin's cumulativ production is pegged at about 19.4 million bbl.

Accordingly, CNPC expected to invest $491 million i Tarim development in 1993, Beijing's China Daily reporte

Tarim infrastructure plans

CNPC was conducting a feasibility study of two refineri to process Tarim oil.

Plans call for construction of a 20,000 b/d refinery a Korla, the headquarters of Tarim E&D, with start-up i 1997.

CNPC also plans to build a 100,000 b/d refinery i Southwest China's Sichuan province. Southwest Chin where oil demand is soaring, does not have a refinery wit capacity greater than 4,000 b/d.

In addition, preliminary work was under way to co struct China's longest pipeline to transport Tarim oil. Th 3,500 km line from Korla in Xinjiang originally was to exten to Lanzhou, Sichuan, and Xi'an and terminate at Luoya in Henan province. From there, Tarim oil would have bee shipped through existing pipelines to the Jiangsu port ci of Lianyungang for export to world markets.

At a projected cost of $1.75 billion, the pipeline is to ha capacity of 400,000 b/d and is to be complete in 1997.

However, in light of soaring domestic demand, China State Planning Commission decided to earmark Tarim o for domestic consumption. Now the line is to terminate the proposed 100,000 b/d refinery at Mianyang, 100 k north of Sichuan's provincial capital of Chengdu.

It still must be approved by the commission, and Beiji says it can finance the project alone but will welco foreign investment.

Tarim oil currently moves via a 250 km pipeline from 30,000 b/d capacity, computer controlled oil productic gathering center to Korla, then via train to Lanzhou Gansu province.

The Tarim basin, in southern Xinjiang Ugyur Autonomous Region, covers 560,000 sq km. Two thirds of the basin is covered by the Taklimakan desert, the world's second largest.

Expected to expedite Tarim E&D and boost the regional economy is a 120 km highway, extending into the desert, under construction in South Tarim. At Tarim E&D headquarters in Korla, 90 operating teams drill and monitor wells, conduct geological studies, and provide technical services. As many as 15,000-20,000 Chinese oil workers are involved in Tarim E&D.

With its E&D focus weighted heavily on the Tarim basin, Beijing at first reserved the region for Chinese oil agencies, with only technology and consulting services coming from foreign companies.

China's about-face on self-reliance for Tarim E&D stems from Beijing's realization that work in the region requires huge infusions of capital and advanced technology if major discoveries are to be made. That means foreign multinational participation.

Southeast China's coastal refining boom

Refinery location	Capacity (1,000 b/d)	Sponsor(s)	Completion date
Fujian	50	Sinopec	1992
Chongqing	50	N/A	N/A
Zhanjiang	30	N/A	N/A
Nanning	50	N/A	N/A
Dalian	100	Sinopec, Total, Hong Kong	1993
Shenzhen	60	Sinopec, Hong Kong	1995
Huizhou	100	Sinopec, Royal Dutch/Shell	1996
Haikou	120	N/A	1996

Other onshore bidding

CNPC plans to make available exploration blocks in other onshore provinces under a second bidding round, expected to get under way early in 1994, China Features reported.

In addition to making more acreage available in the Tarim basin, CNPC was to offer blocks in the Sanjiang basin in Heilongjiang province, Baoding basin in Hebei, Zhoukou basin in Henan, Huahai basin in Gansu, Hailar basin in Inner Mongolia Autonomous Region, Bohai Sea tidelands including offshore acreage in less than 5 m of water, Dacheng basin in Hebei and the city of Tianjin, Qiuxian basin in Hebei and Henan, Chenhu Tuditang basin in Hubei, Hexi basin in Gansu and Inner Mongolia, and northern Qaidam basin in Qinghai.

All the second round acreage lies in sedimentary basins close to fields with well developed infrastructure.

At the same time, CNPC was to offer 14 blocks in 10 major producing areas to participation by foreign companies in an effort to boost recovery in China's aging fields beyond the current rate of 30% of original oil in place. Crude reserves under the 14 blocks are estimated at a combined 5.34 billion bbl. Cnogedc's Zeng told a Singapore conference earlier in 1993 talks were under way with Exxon and Texaco regarding methods to improve recovery.

Existing contracts

Work was beginning to pick up in the onshore exploration areas opened to foreign participation in 11 southern provinces beginning in 1984.

Beijing's piecemeal approach in this effort has resulted in very little exploration to date.

The first contract was signed in 1985 by Cnogedc forerunner China National Oil Development Corp. (Cnodc) and an Australian group led by Australia's CSR Orient Pte. Ltd. Under the 7 year contract, CSR was to explore an area in Fushan county on Hainan Island. CSR later became the first non-Chinese company to discover hydrocarbons onshore China, although CSR later declared its Jinfeng oil and gas find noncommercial.

Poor economics forced CSR to quit development there after discovering combined reserves estimated at 6.06 million bbl of oil equivalent, China Features reported.

A Tulsa, Okla., independent, Myung & Associates Inc., later became the first U.S. company to obtain an onshore PSC in China when it signed a joint venture E&D agreement covering about 20,000 acres of the original 600,000 acre CSR block that encompasses the Jinfeng strike.

At yearend 1990, a group made up of Fletcher Challenge Ltd., Santa Fe Energy Resources, and Nomeco Oil & Gas Co. signed a PSC with Cnodc covering a 15,900 sq km area in the Dongting basin of Hunan province. Plans called for the group to spud the first of two wildcats in August 1993.

Cnodc signed a third PSC with Energy Development Corp. (EDC) in October 1991 covering a 14,423 sq km block in the Boyang basin of Jiangxi province. EDC has acquired 660 line km of seismic data and identified a 70 sq km structural trap, China Features reported. Plans called for a wildcat to spud in the fourth quarter of 1993.

EDC's exploration program is expected to cost $30 million. In the event of a commercial discovery, development could begin in 1995.

The fourth onshore PSC was signed with Amoco Orient Petroleum Co. in May 1992. It calls for Amoco to explore 5,126 sq km in the Fuyang basin of Anhui province. The contract term is 30 years with the first 7 years for exploration. Amoco has identified a 40 sq km structure, Chinese industry sources say. Fuyang geology is comparable to that in Huabei oil field.

1993 PSCs

Three PSCs were signed in 1993, one involving a nearshore, shallow water prospect.

Shell Exploration (China) Ltd. and Pecten Orient Co. in February signed a PSC calling for exploration on an 8,930 sq km block encompassing the Yancheng, Baiju, and Hai'an areas of northern Jiangsu province. Just east of a number of producing oil and gas fields, the area is believed to be highly prospective. Seven structures have been identified on the Jiangsu block.

About the same time, Exploration Co. of Louisiana (XCL) signed a PSC with Cnodc covering acreage off the Bohai Sea coast. The Zhaodong block is a 197 sq km tract extending from the shoreline near Dagang oil field into coastal tidelands in about 5 m of water.

The 22 year PSC calls for XCL to drill at least seven wells during an initial 7 year exploration phase. Seismic and geological data show the presence of a 10,000 m deep sedimentary section that includes source rocks productive in Dagang oil field. Dagang is China's sixth biggest producer at 78,000 b/d. XCL had a seismic survey under way and was expected to spud the first well before yearend 1993.

The exploration joint venture agreement disclosed most recently is one covering the Nanpanjiang block in southern China's Hunan, Guangxi, and Guizhou provinces. Premier Consolidated Oilfields plc signed that agreement. Premier won exploration rights to the block under an initial 12 month term. It can then sign a PSC covering all or part of the area.

Lufeng field start-up

China brought on stream another oil field in the South China Sea, using a floating production, storage, and offloading (FPSO) unit.

JHN Oil Operating Co., a joint venture of Japan Hunan Oil Corp. and China National Offshore Oil Corp. unit Nanhai East Oil Corp. (NEOC), started up Lufeng 13-1 oil field.

The field holds proved reserves of 32.5 million bbl out of original oil in place of 140 million bbl.

China's ethylene capacity expansion boom

Plant location	Capacity (1,000 tons/year)	Comments
PROJECTS EXISTING OR UNDER WAY		
Daqing	300	Completed in 1986 for $900 million with equipment imported from Japan, U.S., U.K., and Germany
Qilu	300	Completed in 1987 for $1.2 billion with equipment imported from Japan, U.K., and Germany
Yangtze (Jinling)	300	Completed in 1990 for $1.7 billion with equipment imported from Germany and Japan
Shanghai	300	Completed in 1992 for $1.21 billion with equipment imported from Japan and U.S.
Maoming (Guangdong)	300	Draft contract signed in 1992 with 10 company group. Cost estimated at $1.9 billion. Completion targeted for 1995.

Plant location	Capacity (1,000 tons/year)
PROJECTS PLANNED BY 2000	
Haikou (Hainan)	450
Jilin	300
Fujian	300
Duzishan (Xinjiang)	140
Beijing	115
Tianjin	140
Zhongyuan (Henan)	115
Guangzhou	115
Sichuan	115
Hebei	115
Anhui	115
Balin (Hunan)	115
Dongyin (Shandong)	115
Wuxi (Jiangsu)	115

Production started at about 4,500 b/d and is expected to ramp to more than 20,000 b/d in 1995. Field life is projected at 8 years. Approved in 1990, Lufeng 13-1 development cost $300 million. NEOC contributed $75 million of that cost.

JHN in early August 1993 began installing an FPSO in 465 ft of water in Lufeng 13-1 field in the South China Sea about 150 miles off Hong Kong.

Nanhai Shengkai FPSO is equipped with a disconnectable turret mooring system designed and installed by Sofec Inc., Houston, a unit of FMC Corp. The mooring system was fabricated in Japan and installed on the Nanhai Shengkai in Singapore. Nanhai Shengkai FPSO formerly was the 128,000 dwt Sea Queen tanker.

Following conversion, the FPSO sailed July 22, 1993, from Singapore. JHN general contractor Clough Stena (Asia) installed Nanhai Shengkai at Lufeng in a period of several weeks, during which the vessel's mooring system was disconnected and reconnected several times for commissioning and ship crew training. Clough Stena subcontracted a unit of Sonsub Inc., Houston, to help set, hook up, and tension anchors and run a flexible flowline over a midwater arch to a previously installed jacket.

The FPSO is moored with a symmetric eight leg layout. The turret is attached by a collet connector to a buoyant anchor leg support structure, or spider buoy. In the Sofec design, the collet connector is in series with a hydraulic tensioner, which transfers connection forces through surfaces preloaded in compression at the outer diameter of the turret shaft. The FPSO's turret is mounted inboard near the FPSO's bow with the main roller bearing located on the top deck, far from direct exposure to seawater and easily available for inspection and maintenance.

Sofec's turret mooring design allows Nanhai Shengkai to disconnect quickly to avoid severe weather or for maintenance. The disconnect threshold is equal to a 100 year nontyphoon storm environment, and the system in mid-August 1993 during installation weathered Typhoon Tasha without disconnecting.

Liuhua development

Amoco Orient Petroleum Co. and China Offshore Oil Nanhai East, through general contractor Reading & Bates Development Co., hired John Brown, Houston, to provide engineering design as part of a $650 million development program for Liuhua 11-1 deepwater oil field in the South China Sea.

The field is 120 miles southwest of Hong Kong. John Brown will assist in design of the floating production system, a modified semisubmersible support vessel equipped for power generation, drilling, well completion, workover, and accommodations.

The partners also let a $19.3 million project services contract to the Brown & Root Marine unit of Brown & Root Inc., Houston. BRM is to design Liuhua 11-1 complex's safety, control, and communication systems, help monitor costs, scheduling, and procurement, coordinate subprojects, prepare tender documents for facilities installations, and help with permit applications, surveying, and soil boring.

Pipeline to Yacheng

ARCO China Inc. awarded a construction contract for a 440 mile, 28 in. gas pipeline from China's Pearl River estuary to Yacheng 13-1 gas field off Hainan Island in the South China Sea.

A joint venture of European Marine Contractors Ltd (EMC), London, and Saipem SpA of Italy will lay the line in EMC's first effort outside Europe.

ARCO China operates Yacheng 13-1 field in 300 ft of water. Other participants in the gas project are China National Offshore Oil Corp. and Kufpec (China) Inc.

The offshore project is valued at $1.2 billion. Construction contract value was not disclosed.

Beginning in January 1996, the line will move as much as 336 MMscfd of gas to Hong Kong for delivery to Castle Peak Power Co.

EMC's Semac 1 semisubmersible lay vessel was to move from the North Sea to begin work in December 1993. Project completion was scheduled for summer 1994.

EMC said the ARCO China line will be the area's first offshore infrastructure and the second longest offshore pipeline in the world. It will be exceeded only by the 510 mile, 40 in. Zeepipe installed by EMC in 1991-92 between Sleipner platform in the Norwegian North Sea to Zeebrugge, Belgium.

As part of the ARCO China project, Brown & Root Marine was carrying out engineering, procurement, and project support services for the offshore facilities and ancillary shore facilities at Hong Kong and Hainan Island.

An eight leg process platform, bridge-linked to a 16 slot wellhead platform, will be installed 60 miles off Hainan Island.

Refining expansion

China's refining sector is in the midst of a major expansion and reorganization in response to booming domestic

demand for petroleum products.

Plans call for hiking crude processing capacity to 3.9 million b/d in 1995 from 3.085 million b/d. Much of that 26% increase will come where the products demand growth is the strongest: China's coastal provinces, notably those in the Southeast.

Despite the demand surge, China's refineries operated at only 74% of capacity in 1991, and projections for 1992 weren't much better. Domestic crude supply is limited because of Beijing's insistence on maintaining crude export levels, a major source of hard currency foreign exchange.

While the government scales back crude exports to feed voracious domestic consumption, it's restructuring the refining sector to reflect regional disparities and foreign participation in Chinese refinery projects. Those steps are part of overall reforms Beijing is implementing throughout the nation's petroleum industry.

Superheated demand

Steady movement toward a market economy in China's southeastern coastal provinces has made that region the world's fastest growing economy. Some estimates place growth at 20%/year.

The expansion is heating oil demand. China consumes more than 2 million b/d of oil, a level expected by most observers to climb by 120,000-140,000 b/d/year. In first half 1992 alone, China's refined products consumption jumped by 18.2% from the previous year's period vs. a planned increase of 12.2%.

In the southeastern coastal provinces, notably Fujian and Guangdong, oil demand is projected to more than double by 2000 from 1990.

Domestic demand for oil is outpacing Chinese crude production. In 1991, China's crude production reached a record 2.792 million b/d, but the need to export crude for hard currency reined processing of domestic crude to less than 2.3 million b/d.

Crude production from onshore eastern China, which accounts for 80% of proved reserves and 90% of current output, is declining after 30 years of intense development.

Although production increased marginally in this region in 1992, annual growth has dropped from 115,600 b/d in 1986 to a paltry 8,000 b/d in 1990.

xports, imports

Because of flat production and the explosive growth in oil demand along the southeastern coast, Beijing has had to backpedal on efforts to promote crude oil exports.

In 1985, China exported 640,000 b/d of crude, which accounted for 20% of its foreign exchange revenue that year. In 1990, it exported 580,000 b/d of crude oil, 100,000 b/d above target because of favorable selling prices on the world market.

However, soaring domestic oil demand is reversing that trend. In 1991, China earned about U.S. $3.2 billion from oil exports, making up 8% of its foreign exchange earnings.

To maintain domestic refinery throughput, China National Chemicals Import & Export Corp. (Sinochem), the country's monopoly for crude imports and exports, cut crude exports to 420,000 b/d in 1992 from a previously planned 640,000 b/d. That garnered earnings of only $2.6 billion.

Sinochem stepped up imports of crude to feed refineries along the coastal provinces, notably Guangdong and Fujian. In 1991, Chinese refineries processed 100,000 b/d of imported crude. The figure was expected to double in 1992.

The regional oil shortages have even led some special economic zones such as Shenzhen to free themselves from the constraints of Sinochem to import their own crude.

The boom town Shenzhen, which borders Hong Kong, in 1991 consumed 23,600 b/d of oil, of which 21,400 b/d was imported, mainly from Singapore.

According to Jiang Yunlong, a Sinochem vice-president, inadequate oil storage capacity in China hampers Sinochem's efforts to import more crude. The present crude storage capacity in China totals only about 18.25 million bbl.

Although China's 450 oil terminals combined can handle as much as 1.8 million b/d, capacity of most tank farms is only about 146,000-365,000 bbl of crude and products each, with the largest at 1.095 million bbl at Zhenhai refinery in Zhejiang province.

It is not profitable to ship crude from the Middle East to China with single shipments of about 365,000 bbl, as would be required under Chinese storage constraints, because of high transportation costs, Jiang said.

Refining infrastructure

China's refinery crude throughput capacity of 3,085,200 b/d makes it potentially the world's biggest oil processor after the U.S., the former Soviet Union, and Japan.

In 1992, China processed 2.38 million b/d of crude oil, up about 4% from the 1991 level. Of the total, 2.127 million b/d was processed by 32 refineries operated by Sinopec. The rest was processed by 25 small refineries operated by China National Petroleum Corp. (CNPC), the state owned company that oversees exploration and development operations.

The combined output of gasoline, diesel oil, kerosine, and base lubes totaled 1.209 million b/d in 1992, up 8.5% from 1991.

In addition to domestic production, refineries along the coastal areas processed 300,000 b/d of imported crude, or 14.1% of the total processed volume.

In 1991, Sinopec refineries accounted for about 90% of the 2,272,600 b/d of crude oil processed in China. Of those Sinopec refineries, 30 have crude processing capacity exceeding 50,000 b/d.

Sinopec plans to shift its attention to areas along the coast and the Yangtze River to turn them into export-oriented oil processing bases, according to Sinopec Pres. Sheng Huaren. Sinopec officials point to refineries under construction with domestic and foreign funding in coastal areas.

While expanding its refining capacity, China has made progress in technology development.

China Research Institute of Petroleum Processing developed a deep catalytic cracking (DCC) process to produce gaseous olefins that feature cuts of propylene higher than

Floating production, storage, and offloading unit (left) Nanhai Shengkai installed in Lufeng 13-1 oil field off southern China features a disconnectable turret mooring system (right) designed and installed by Sofec Inc., a unit of FMC Corp. Photo courtesy FMC.

those of butylenes and ethylene. The process uses heavy hydrocarbon fractions such as vacuum gas oil as feedstock in tandem with special catalysts. With operating temperatures much lower than with steam cracking, the process treats liquid hydrocarbons with conventional alkali and water washing, thus separating out impurities without further refining.

In addition to the DCC process, China also has developed is own processes for heavy oil hydrotreating/cracking and atmospheric and vacuum distillation.

Shenzhen refinery

Chinese officials look to the refinery under construction at Shenzhen as a prototype for sophisticated Chinese refineries of the future.

A joint venture of Sinopec, the city of Shenzhen, and Hong Kong, the $500 million project will have crude processing capacity of 60,000 b/d when complete in 1994. The crude will be imported from the Middle East, and 60% of its products will be exported to Hong Kong.

The project will have eight processing units, including an atmospheric resid desulfurization/hydrogenation complex that is a first for China.

Most of China's 100 refineries can process only low sulfur crude. The atmospheric resid desulfurization/hydrogenation unit and Shenzhen's several heavy oil hydrotreating units, catalytic cracking units, and continuous catalytic reformer can process a range of high sulfur imported crudes. The sulfur unit can recover 60,000 metric tons/year of sulfur, compared with an average of less than 10,000 tons/year at other Chinese refineries with sulfur units.

Shenzhen also will yield a much whiter products barrel than other domestic refineries, about 81.92% gasoline and middle distillates. In 1990, the average recovery yield of gasoline and middle distillates at Chinese refineries was 51.53%.

The continous catalytic reformer at Shenzhen can produce 100 RON gasoline, a significant advance for China. Even with the addition of tetraethyl lead, very few Chinese refineries can produce 90 RON gasoline.

Shenzhen also will have a new deepwater terminal in Dapeng Gulf, several kilometers away from Hong Kong, that could handle oil cargoes of as much as 1.46 million bbl.

The transport cost of a single shipment of crude from the Middle East to the 365,000 bbl Sinopec Maoming Petrochemical Corp. terminal near Zhanjiang, a coastal city in Guangdong, runs as much as $2.74/bbl. But for the Shenzhen facility, transportation costs for the bigger cargo runs to only $1.03/bbl for the same distance.

Shenzhen, Hong Kong demand

Shenzhen is one of eight refineries China plans or is building along the coastal areas.

The eight refineries will have a combined capacity o 560,000 b/d.

Because most of China's refineries can produce only low octane gasoline, which must be further processed by foreign refineries when exported, China's gasoline expor business is a low margin operation. On the other hand finished gasoline is imported from Singapore to Shenzhen at a much higher price.

Hong Kong imported about 143,000 b/d of refined products in 1991, 70% from Singapore. Of the total, 25,000 b/d was exported to Shenzhen and Macao, leaving Hong Kong' net products consumption in 1991 at about 118,000 b/d The distance between Singapore and Shenzhen and Hon Kong is 2,685 km, putting the transportation cost at abou $4.10/bbl. When the Shenzhen refinery begins exportin products to Hong Kong, the transportation cost will drop t about 55¢/bbl.

The government estimates that the city of Shenzhen's o consumption will climb to about 38,000 b/d by 1995 an 58,400 b/d by 2000. In Hong Kong, net oil consumption i expected to climb at a rate of 4%/year by 1995 to reac 140,000 b/d and 3%/year thereafter to reach 160,000-164,00 b/d by 2000.

More than a dozen ammonia plants like this one have been built in China using Kellogg technology.

Southeast coast demand

The six provinces of Southeast China—Guangxi, Yunnan, Guizhou, Hainan, Fujian and Guangdong—rely mainly on Guangzhou Petrochemical Corp. and Sinopec Maoming Petrochemical Corp. for refined products supply.

Processing capacity of Guangzhou Petrochemical is 104,000 b/d and of Sinopec Maoming Petrochemical 170,000 b/d. In 1992, the two processed a total of 200,400 b/d of crude oil.

The region's booming economy is fueling a terrific thirst for refined products, triggering the refinery building boom.

The Fujian refinery is the first to be completed of the eight refineries Sinopec is building along the southeastern coast. The 50,000 b/d plant started up in September 1992 in Fuzhou, capital of Fujian province.

There are three refineries under construction in Southeast China. The combined crude throughput of old and new refineries in the area is expected to reach 362,000 b/d in 1995. Of that, only 137,600 b/d is earmarked for Guangdong, 89,200 b/d short of projected demand. That shortfall won't be made up by 2000.

Although throughput is expected to reach 532,000 b/d in 2000, Guangdong will account for only 199,000 b/d—138,600 b/d short of projected demand.

1993 plans

In 1993, China planned to process 2.54 million b/d of crude, with 2.24 million b/d processed by Sinopec refineries. Of the total crude processed, about 380,000 b/d will be imported by Sinopec.

In 1993, China's total production of gasoline, diesel, kerosine, and lubricants was expected to reach 1,306,200 b/d, with Sinopec accounting for 1,162,000 b/d. That compares with 1,209,000 in 1992, in turn a rise of 8.5% from 1991's level.

Sinopec has 51 catalytic cracking units with a total processing capacity of 840,000 b/d. In 1992, Sinopec cat crackers' throughput averaged 640,000 b/d. Sinopec now can produce 60,000 tons/year of seven kinds of catalysts.

Sinopec also has delegated authority to four refining and petrochemical corporations to import crude and conduct foreign trade.

In addition to Sinopec Maoming Petrochemical and Guangzhou Petrochemical, they are Zhenhai Petrochemical Corp., Zhejiang province, with refining capacity of 110,000 b/d, and Gaoqiao Petrochemical Corp., Shanghai, 104,000 b/d. In 1993, the four corporations were to import 80,000 b/d of crude.

Foreign investment

Foreign participants are becoming increasingly common in China's refining sector.

Construction began in May 1992 of a 100,000 b/d export refinery at Dalian in Northeast China's Liaoning province. It is a $437 million joint venture of four Chinese companies, Hong Kong's Sinochem International (H.K.) Co. Ltd., and Total. The foreign partners have a combined 42.5% stake in the venture.

It is expected to produce 38,000 b/d of gasoline, 12,000 b/d of kerosine, 3,200 b/d of naphtha, and 2,600 b/d of liquefied petroleum gas. Crude oil will be imported from the Middle East and processed with technology and equipment provided by Total. Total also will help the Chinese partners purchase crude and sell products on the international market.

Royal Dutch/Shell Group was considering participation in a 100,000 b/d refinery near the coastal town of Aoutou, Guangdong. Pending results of a feasibility study, Shell could take a 50% stake in the joint venture, with the rest held by a Chinese group led by the Guangdong provincial government and including China National Offshore Oil Corp. The study was to be completed in 1993. If it receives timely approval, the project could be complete in 1997-99.

In addition to the refineries to be built along the coast, Sinopec plans to build a 160,000 b/d refinery at Shanghai in cooperation with France's Ste. Nationale Elf Aquitaine in 1994. The planned refinery has been given a go ahead by Beijing's State Council.

Meanwhile, China Chemicals Import & Export Corp. (Sinochem) was negotiating with Saudi Arabia to build another 120,000 b/d refinery at Quingdao, a port city in East China's Shandong province.

Amoco, Sinopec venture

Amoco Corp. and China Petrochemical Corp. (Sinopec) signed a general protocol that will enable the two companies to explore potential projects, including joint venture refining and petrochemical operations, in China.

Under the 2 year accord signed by Amoco Vice Chairman Lawrason D. Thomas and Sinopec Vice President Wu Xie Gang in Beijing, the companies may cooperate on projects involving:

- Petroleum refining and natural gas conversion to produce and market motor fuels, petrochemicals, and other hydrocarbon products.
- Transportation, storage, and marketing of products, feedstocks, and crude oil.
- Use and marketing of natural gas, coal gas, methane, and other fossil fuels produced in China.

All of Amoco's operating companies are pursuing opportunities in China.

Amoco Oil Co., Amoco Corp.'s refining, marketing, and

transportation unit, coordinates its China business activities from Hong Kong.

Amoco Chemical Co., engaged in chemical manufacturing and marketing, has been doing business in China since the late 1970s, when it licensed technology there. Yizheng Amoco Fabrics Co. Ltd., a joint venture of Amoco Chemical and Yizheng Joint Corp. of Chemical Fiber Industry, began producing and marketing fabric products in May 1992. Amoco Chemical also is evaluating a purified terephthalic acid project in Zhuhai.

Amoco Chemical's Far East headquarters also is in Hong Kong. An office was opened in Beijing in 1992 to further strengthen and build business relationships in China.

Amoco Production Co., the exploration and production arm of Amoco Corp., has several projects under way in China. The company is developing Liuhua field in the South China Sea and plans to explore a mainland tract in Anhui province's Fuyang basin.

Under the direction of China's State Planning Commission, Sinopec is an oil and natural gas refiner and petrochemical manufacturer and marketer. It controls 90% of China's 2.2 million b/d refining capacity.

BP venture

British Petroleum Co. plc signed a joint venture agreement with Hong Leong, Singapore, and Fuzhou Gas Corp. (FGC) to develop a $15.5 million LPG plant at Fuzhou, China.

Official approval was expected, after which construction of a dockside storage terminal and bottling plant will take 18 months. Plant capacity will be 30,000 metric tons/year at start-up, rising to 60,000 metric tons/year after 12-18 months.

BP Hong Leong (Fujian) Enterprise Ltd. will build and operate the plant. BP and LPG bottle manufacturer Hong Leong will jointly hold 67% in the JV and FGC the remainder.

Golmud refinery

China expected to start up in October 1993, 1 year ahead of schedule, the 20,000 b/d Golmud refinery, a key state poverty relief project for the Tibet autonomous region and Qinghai province.

The plant is at the northern foot of the Kunlun Mountains in Qinghai province. Work started in August 1991.

Inner Mongolia plant

Inner Mongolia Chemical Fertiliser Works will use M.W. Kellogg Co.'s reduced energy ammonia synthesis technology in a plant it will build at Huhehot City, Inner Mongolia, China.

Kellogg will provide the technology to Toyo Engineering Corp. and Mitsui & Co., contractors for the 330,000 ton/year, oil based plant. Completion is scheduled for 1996.

It will be among more than a dozen plants in China using the Kellogg ammonia synthesis process introduced to that country 20 years ago. Kellogg said the technology is "significantly advanced" over that installed during the early 1970s.

Petrochemical expansion

China is keeping the fires stoked under its petrochemical expansion campaign.

Sparked by a continuing surge in domestic demand for petrochemical products and regional demand for petrochemical feedstock exports, Beijing has embarked on an ambitious program of expanding a range of upstream and downstream petrochemical capacities this decade.

Many are world scale projects. Some are proceeding with foreign technology and equity participation by foreign partners.

The implications could prove significant for China's neighbors in the Asia-Pacific region, where there is apprehension over possible lingering surplus petrochemical capacity during the 1990s. Although China's surging economy will absorb much of that expected future production, Beijing is moving aggressively to sustain growth in petrochemical exports as a key source of hard currency earnings.

Sheng Huaren, president of state owned China Petrochemical Corp. (Sinopec), estimated national petrochemical production has jumped at a rate of 8%/year the last 8 years, Xu said. In 1992, Sinopec remitted to Beijing about $3.3 billion in taxes, or about $270,000 more than its target.

Plans call for still more growth during the 1990s as a hefty slate of projects gets under way.

Background

In the past 10 years, Sinopec has sold 55 billion tons of refined and petrochemical products with sales totaling $11 billion. Tax remittances to Beijing have totaled more than 127 billion yuan ($21.9 billion).

China's petrochemical infrastructure largely consists of:

- Thirty oil processing complexes nationwide.
- Basic petrochemical complexes at Yanshan in Beijing province, Daqing in Heilongjiang, Qilu in Shandong, Jinshan in Shanghai, and Yangtze at Nanjing in Jiangsu province.
- Petrochemical fiber complexes at Liaoyang in Liaoning province, Tianjin, Shanghai, and Chendu in Sichuan province.
- Synthetic ammonia/urea plants at Urumqi in Xinjiang Autonomous Region, Xining in Ningxia Autonomous Region, Anqing in Anhui province, and Guangzhou in Guangdong province.

In recent years, Sinopec has cooperated with foreign companies in a bid to update its technology and facilities. The corporation has so far absorbed more than $6 billion in foreign investment and established cooperation with more than 1,000 companies in 50 countries and regions, Xu reported.

Exports

China exports 100 kinds of petrochemical products.

Petrochemical industry exports totaled $1.5 billion in 1992, including $254 million worth of products directly exported by Sinopec—mostly base oils, petroleum coke, paraffin wax, synthetic rubber, and synthetic resins.

In 1993, Sheng expected exports, including those by China National Chemicals Import & Export Corp. (Sinochem), to increase by 10%.

Sinopec has offices in the U.S., Japan, Hong Kong, Thailand, Germany, and Ecuador and plans to enter markets in Africa and South America via exports of paraffin wax and petroleum coke.

At a Sinopec conference in 1992, government officials proposed allowing Sinopec International, which oversees Sinopec's export and import business and accounts for about one fifth of China's petrochemical exports, to handle a bigger share of the petrochemical export market.

Recent performance

Sinopec in 1992 increased petrochemical production value by 8.3% from the previous year to 75.6 billion yuan ($13 billion). Sales totaled 97 billion yuan ($16.7 billion), up 20% from 1991's level.

The company's ethylene production jumped 10.2% from 1991's level to 1.72 million tons due to full operation of the Shanghai and Fushan complexes' expansions of 300,000 tons/year and 115,000 tons/year, respectively.

Investment in grassroots petrochemical projects in 1992 totaled $1.3 billion, with a further $480 million spent on revamps and modernization projects, notably key projects at Fushun, Fujian, and Urumqi. That compares with $1.[illegible] billion in capital outlays in 1991.

1993 outlook

Sinopec expected to produce 1.84 million tons of ethylene in 1993.

Under a conservative scenario, that level is expected to rise to 2.3 million tons in 1995 and 3 million tons in 2000.

Sinopec expected to remit levies to Beijing totaling 175 million yuan ($29.3 million) in 1993. Sales were expected to climb about 20% from 1992's level to more than 120 billion yuan ($20.6 billion).

Plans called for capital spending of more than 11 billion yuan ($2 billion) in 1993, with 7.82 billion yuan ($1.34 billion) going for grassroots construction and 4 billion yuan ($689 million) going for revamps and maintenance work.

Sinopec also will expand its petrochemical commodities trading capability in 1993. It established three petrochemical exchanges late in 1992: Pudong Petrochemical Exchange Center, opened in Shanghai; Northeast Petrochemical Exchange Center, opened in Shenyang, Liaoning province; and Tiajin Petrochemical Center.

Sales at the Pudong center by yearend 1992 totaled 800 million yuan ($138 million). Plans called for setting up three more exchange centers in 1993 at Guangzhou, Lanzhou in Gansu province, and Wuhan in Hubei province.

Ethylene capacity boost

The linchpin of China's petrochemical boom is ethylene.

With full commercial start-up of a 300,000 metric ton/year ethylene project at Shanghai in 1992, China is now the world's eighth largest ethylene producer with productive capacity of 2.1 million tons/year, officials claimed.

The Shanghai project, started in 1987, was the last to be completed among four 300,000 ton/year ethylene projects China implemented with imported technology and equipment. The other three were installed at Daqing, Qilu, and Jinling.

China in 1992 signed a draft contract with a group of 10 foreign companies to implement a fifth such 300,000 ton/year ethylene project at Maoming.

Maoming is one of 15 grassroots ethylene plants or major expansions China plans to complete by 2000.

Along with a number of minor expansions from debottlenecking and other steps, China's ethylene productive capacity could jump by at least another 3 million tons/year, Xu reported. By 2000, China's domestic demand for ethylene is expected to reach 4-5 million tons/year.

However, some Beijing officials figure China would be lucky if 60% of the planned projects materialize, based on the country's hard currency squeeze.

Maoming projects

Beijing in 1992 signed a draft contract with a group of 10 foreign companies to start what China calls its most ambitious petrochemical project in the current decade.

The ethylene project, to be built by Sinopec Maoming Petrochemical Corp. (SMPC) in South China's Guangdong province, involves a $1.9 billion investment. Of that amount, $735 million is to be foreign capital.

The draft was signed by Sinopec, Guangdong province, and 10 companies from Japan, Italy, and the U.S. led by JGC Corp. Companies in the group include Marubeni Corp., Mitsubishi Corp., Mitsui Shipbuilding & Engineering Co., Snamprogetti SpA, and Stone & Webster Engineering Corp.

The group will provide export credits valued at $200 million from the U.S. and Japan to install three trains to produce 300,000 tons/year of polymer grade ethylene, 100,000 tons/year of low density polyethylene, 140,000 tons/year of linear low density polyethylene (Lldpe), and 165,000 tons/year of propylene. Feedstock will be naphtha and heavier liquids.

Japan's JGC will oversee lump sum project responsibilities, including detailed engineering, procurement of foreign equipment, and construction supervision. Stone & Webster, Houston, will provide technology licensing and process design, detailed engineering of critical areas, training, and commissioning technical services.

Snamprogetti late in 1992 signed a $70 million contract to provide technology for the Maoming Lldpe unit. Sinopec also let a $38 million contract to Snamprogetti to build a 100,000 ton/year styrene monomer plant at Maoming.

Maoming complex construction was to start in October 1992, and production is expected to begin late in 1995. Delivery of first products is scheduled for early 1996 from the complex, which will serve China's booming coastal economic development zones as well as Hong Kong, Macau, and other export markets.

Meanwhile, Maoming also was mapping out a 300,000 ton/year synthetic ammonia project downstream of the ethylene project. The $275 million project was expected to get under way in early 1994.

Other ethylene projects

Sinopec also was implementing a 115,000 ton/year ethylene project at Guangzhou, the capital of Guangdong province in South China.

For the $715 million project, Sinopec will import ethylene units valued at $150 million from Italy's Tecnimont SpA and $120 million worth of polyethylene and polypropylene facilities from Tokyo Engineering Corp.

When complete in 1995, the complex will be able to produce 115,000 tons/year of ethylene, 100,000 tons/year of polyethylene, 70,000 tons/year of polypropylene, 80,000 tons/year of styrene, and 50,000 tons of polystyrene.

Construction began at midyear 1992 on a 140,000 ton/year ethylene plant at Duzishan Petrochemical Complex in Northwest China's Xinjiang Autonomous Region. The $660 million project will use eight production trains—four each imported from Japan and the U.S. The plant is to start up by yearend 1994.

In 1991, Beijing's Yanshan Petrochemical Corp. installed several new units, boosting ethylene capacity to 300,000 tons/year.

In mid-1992, China National Petroleum Corp. (CNPC) and Beijing's municipal government started work on a 115,000 ton/year ethylene project at Beijing. The $537 million project will have four units, all imported from Italy, Japan, and the U.S.

When production starts in September 1994, the project will have capacities of 115,000 tons/year of ethylene, 40,000 tons/year of ethyl vinyl acetate resins, 40,000 tons/year of ethylene glycol, and 70,000 tons/year of gasoline additives.

Other petrochemicals

Among other petrochemicals, China has capacity to produce 1.27 million tons/year of polyester.

That figure is expected to jump to 1.63 million tons in 1995 and 2.11 million tons in 2000.

Local officials have applied to implement 21 polyester projects with a total capacity of 1.7 million tons—860,000 tons more than the target level.

The polyester projects, scattered around the country, include a completed 60,000 ton/year unit at Foshan in Guangdong province, a 60,000 ton/year plant to be built with foreign capital at Guangzhou, and a cluster of four 60,000 ton plants in the Pearl River Mouth delta area.

Sinopec late in 1992 signed a letter of intent with Mitsui Engineering & Shipbuilding Co. and Marubeni Corp. to build a 100,000 ton/year polyester fiber plant in Guangdong province.

China also plans to build four petrochemical fiber plants in Tianjin, Liaoning, and Henan provinces with a combined capacity of 600,000 tons.

Completion of the projects will boost capacity of synthetic fiber monomers by 340,000 tons/year and product fibers by 200,000 tons/year. IPE

Asia-Pacific

AUSTRALIA

CAPITAL: Canberra
MONETARY UNIT: Dollar
REFINING CAPACITY: 705,000 b/cd
PRODUCTION: 512,800 b/d
RESERVES: 1.6 billion bbl

Australia's natural gas industry has undergone fundamental changes in the last few years.

The Australian Bureau of Agricultural and Resource Economics (Abare) predicts natural gas consumption in the country will grow at an average rate of 2.6%/year during 1991-2005. By then it will account for about 18.5% of Australia's total energy consumption.

Abare also projects that natural gas production from Australian fields will grow by an average 3.5%/year to 2005.

AGA study

A study by the Australian Gas Association (AGA) in 1992 found Australia could require a major trunk pipeline linking large untapped gas reserves off Western Australia with major consumer markets on the country's eastern seaboard as early as 2009.

In addition, other interstate gas pipeline connections are likely to be needed by 2030. Of special note is a probable pipeline to link Northern Territory with southern and eastern states.

AGA's study found Australia has enough gas reserves to meet domestic demand for the next 30 years in addition to servicing existing gas export contracts. It estimates current proved and probable gas reserves at 105.45 tcf, excluding coalbed methane reserves found mainly in New South Wales and Queensland.

Australian gas consumption in 1991 totaled about 630 bcf. Demand is projected to jump to about 1.52 tcf by 2030 under a base case that calls for growth of 2.3%/year.

Under a high demand case of 2.9%/year, Australian gas consumption could reach almost 2 tcf by 2030.

The study also found that if state and territorial governments persist with their current reluctance to remove interstate gas trade barriers, there will be a sharp jump in gas prices compared with a scenario that eliminates interstate gas trade barriers.

By 2030, gas prices in Sydney could be 32% higher than in an unrestricted market. In Melbourne, gas prices would be 24% higher, and in Brisbane and Adelaide 31% higher.

Interstate gas trade

Talk of change centers on the federal government's initiative, backed by the AGA, to create free trade of gas across state boundaries.

With one exception, the country's domestic gas industry has evolved as a series of separate entities, wherein a field's production is consumed only in the borders of the state in which the field lies. The exception is the supply of Cooper basin gas from South Australia under contract to Australian Capital Territory/New South Wales, where it is sold to Canberra, Sydney, Newcastle, and Wollongong markets.

The federal government's plan to put gas trade on a national basis is by no means a sure thing. There is considerable resistance to free trade from the government of South Australia in particular, which wants to ensure that its consumers have security of supply beyond the first few years of the next century, when existing contracts expire.

However, many observers believe increased cross border gas trade is inevitable and will be a feature of the domestic gas scene after 2000.

Pipeline prospects

Windows of opportunity for several more cross border gas pipeline projects are coming into view. Preliminary negotiations were under way for planning and construction for projects that have lead times of as much as 10 years.

One such project is the possibility of supplying the New

South Wales market with gas from Bass Strait fields off Victoria once the current contracts for supplies from South Australian Cooper basin fields begin to expire in 2001. That would involve construction of a new gas pipeline across Victoria to link with the existing trunk line into the Sydney-Canberra markets.

Intrastate proposals

Another feature of the changing gas scene in Australia is the move to expand pipeline networks within state boundaries.

A good example is the proposal to move gas by pipeline from Northwest Shelf fields to the gold and nickel fields region centered around Kalgoorlie and Kambalda in South Central Western Australia.

The scheme was originally put forward by a coalition of Western Australian state conservative parties just before an election early in 1993, and it has been confirmed since conservatives won control of the state government in February 1993.

What's more, the proposal has been enthusiastically embraced by private enterprise in Western Australia. As many as 16 groups have lodged expressions of interest, including a joint venture of Western Australia's three largest mining companies.

Considered to be the front-runner, the venture of Western Mining Corp., BHP Minerals, and Normandy/Poseidon has proposed a $400 million (Australian) project to lay and operate the gas line. The three are the main energy users along the proposed pipeline route and would take up as much as 90% of the initial throughput.

The proposal entails a pipeline designed to transport about 95 MMcfd of gas, which equals about 25% of the contract commitment for the domestic gas phase of the Northwest Shelf gas project. The venture estimates gas could be on stream 3 years after government approvals are

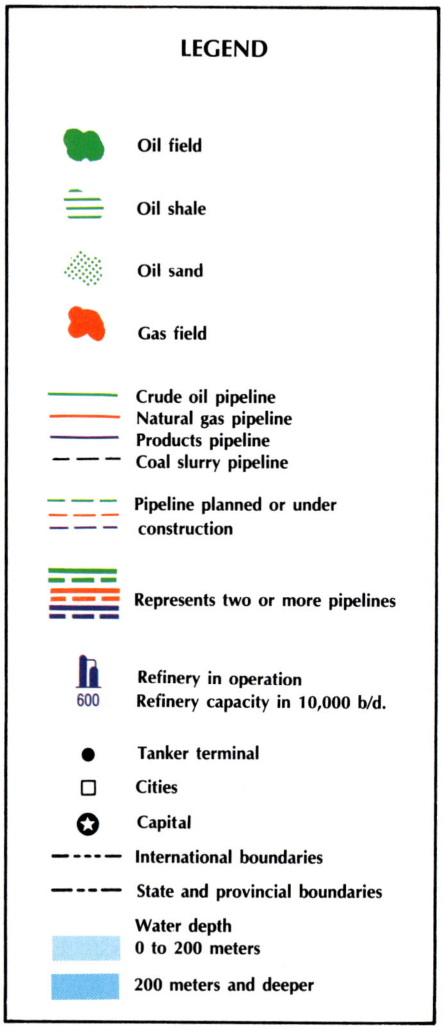

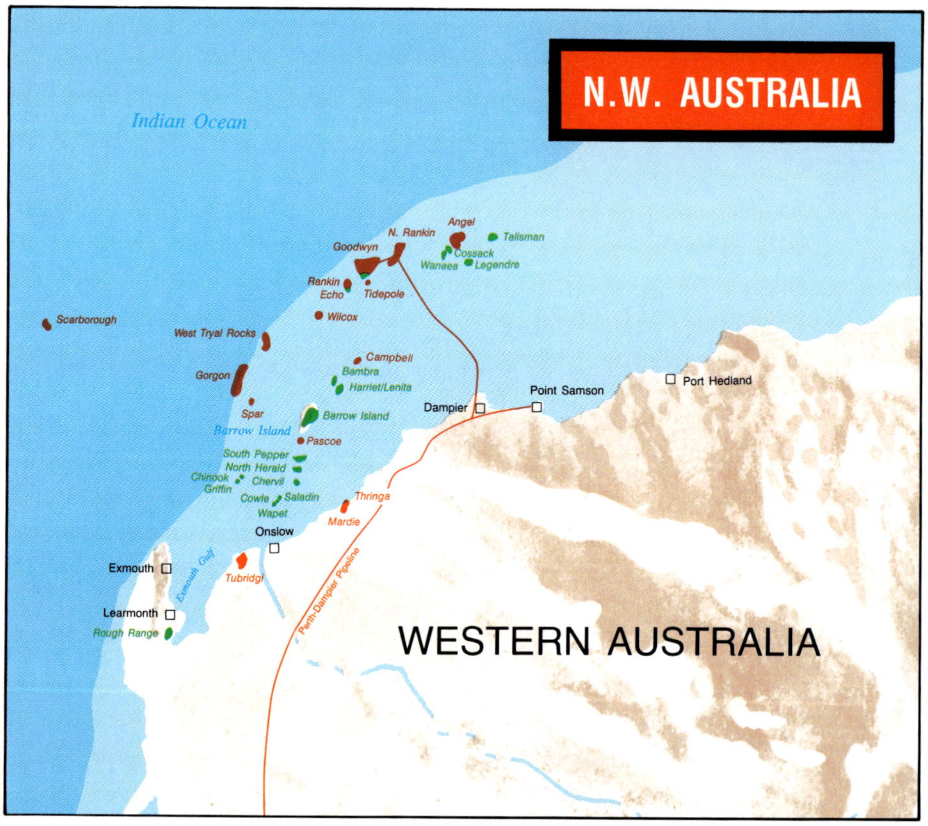

given.

Although not as advanced, there are similar sized pipeline proposals in other states. Notably these include an extension of the Palm Valley/Mereenie-Darwin gas line to include a spur connecting with the Nabalco aluminum refinery at Gove on the northeast tip of Arnhem Land in Northern Territory. Such a line also would extend beyond the proposed new lead/zinc mine at McArthur River near the border with Queensland. Currently the Gove plant is powered by fuel oil brought in by sea from Singapore, but a change to natural gas fuel is under consideration.

Another possibility is a major trunk line feeding Southwest Queensland gas to that state's existing eastern sector pipeline network via Roma, thus linking with Brisbane and the central coastal cities of Gladstone and Rockhampton.

There also is potential for a line running north from Southwest Queensland fields to the Mount Isa copper/lead/zinc mining province.

Field economics improve

The growing number of pipeline links has led to another aspect of changed attitudes toward the use of domestic gas in recent years.

With the expanded infrastructure, producers are finding that even relatively small gas reservoirs can be viable for development. That applies to fields such as Tubridgi near Onslow on the Western Australia coast, which has proved and probable gas reserves of 85 bcf, and to associated gas in offshore Griffin oil field.

In both cases the gas is economic because it can be connected at relatively small cost to the major Northwest Shelf domestic gas line that extends along the Western Australia coast to Perth and markets in the southwest corner of the state. In addition, new markets are being proposed in the form of gas fired power plants at various mining centers in the northwestern mining belt.

Another interesting proposal being looked at along these lines is the possibility of connecting gas reserves found in the Point Torment field in the Kimberley region of Western Australia to the towns of Derby and Broome on the far northwest coast.

Another example is the use of Palm Valley gas to produce a small volume of LNG in a plant at Alice Springs that is then trucked to the Ayers Rock complex 300 km away and used as fuel for power generation. And there is a new scheme that will involve piping gas from Gilmore field in central Queensland about 90 km to the town of Blackall, where it will be used to generate electricity for the region.

Domestic consumption

The industrial sector is the largest domestic market for gas in Australia.

According to a gas supply/demand study released late in 1992 this sector accounted for about 351.5 bcf/year in 1990-91, or 48% of total consumption.

One noteworthy new development in the sector is the green light BHP gave to construction of a $70 million methanol pilot plant just west of Melbourne in Victoria. It will use Bass Strait gas as feedstock in a demonstration of new methanol technology developed by U.K. company ICI Katalco.

Sponsors hope the plant's innovative natural gas conversion technique will contribute to development of a significant share of Australia's uncommitted gas reserves and perhaps lead to applications on floating production, storage, and offloading facilities.

Although the plant's primary emphasis will be on technology development, its production of about 160 metric tons/day of methanol will supply most of Victoria's requirements, which are currently imported from abroad. The plant was to go on stream in second half 1994.

Natural gas consumption for power generation in Australia stood at 95 bcf/year in 1990-91, accounting for about 15% of gas used in the country. South Australia, Victoria, and Western Australia are the largest users of gas for this purpose. Power generation in the Northern Territory also uses gas almost exclusively as its fuel.

With more than 2 million customers, the residential sector accounts for about 85 bcf/year, or about 13%, of Australia's gas consumption.

Commercial demand stood at 32.3 bcf/year in 1990-91, or 5% of gas use.

The sector with perhaps the best prospect for growth is transportation. Currently, the volume of natural gas—as opposed to liquefied petroleum gas—used in transport is only about 190 MMcf/year, but it is expected to rise to 28.5 bcf/year by 2005, if the gas industry's aim of achieving a 10% penetration of Australia's road transportation fleet within 10 years is successful.

Export gas

Despite renewed interest in the domestic scene, the main potential for Australia's natural gas reserves lies in exports.

Notwithstanding BHP's hopes for methanol production, the export of LNG offers the Australian gas industry its largest potential increase in dollar terms.

Abare points out that LNG shipments to Japan from the Northwest Shelf project began at 2 million tons/year in 1989-90 and reached the 5.4 million ton/year mark early in 1993 on their way to a contract volume of 7 million tons/year by 1995. Abare says LNG exports will plateau at this level until 2000, when a new phase of growth is expected to begin.

Shipments are expected to reach 9.3 million tons/year by 2002-03.

Northwest Shelf update

The most obvious candidate for further action is an extension of the Woodside Petroleum Ltd. group's fields and facilities.

Although gas reserves in North Rankin and Goodwyn fields are fully committed to current contracts, there is potential for additions using gas from nearby Angel field. This is a reservoir discovered in 1972 and kept as a backup to safeguard existing commitments to Japanese customers.

But the Woodside group has wider ambitions. Discoveries within the company's license area the past decade include Echo, Yodel, Wilcox, and Dixon fields. There also were plans to consolidate the reserves picture with a new drilling program beginning in 1993.

Gorgon/West Tryal

Rivaling the Woodside group's plans are those of West Australian Petroleum Pty. Ltd. (Wapet), Perth, which has an estimated 8 tcf of gas and 40 million bbl of condensate in Gorgon and West Tryal Rocks fields west of Barrow Island and south of the Woodside operations.

The fields were discovered in 1980 and 1972, respectively, but they have remained in the shadow of the Woodside Northwest Shelf project mainly for lack of a gas market.

During the last few years Wapet has taken a number of steps toward an LNG project. These include an extensive 3D survey over Gorgon and feasibility studies on development technology as well as market surveys, particularly of the Asian region.

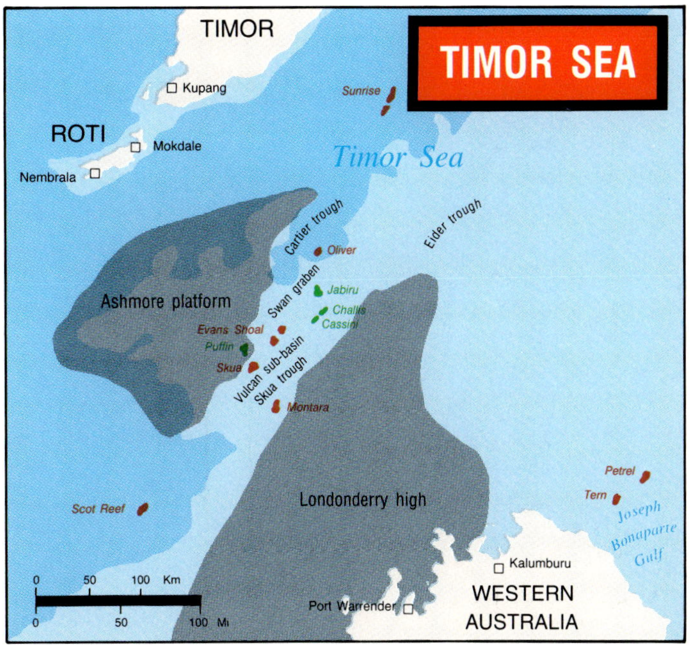

Development options being considered include building facilities next to the Woodside group's gas liquefaction plant on the Burrup Peninsula and building a grassroots plant on Barrow Island next to Wapet's oil facilities.

Decisions were expected in 1993 on an appraisal drilling program in Gorgon, larger of the two discoveries. The most optimistic timing for any development project is the late 1990s.

A new gas strike, East Spar, has been made close to Gorgon/West Tryal Rocks by Western Mining Corp. Although more appraisal was needed, first indications were of a substantial discovery that may enhance prospects for a regional development project based on LNG exports early next century.

Bonaparte Gulf, Browse basin

Australia's only other potential medium term LNG development prospect lies in the Bonaparte Gulf and embraces the Petrel and Tern fields, found in the late 1960s and early 1970s.

For many years various appraisal programs were operated by Ste. Nationale Elf Aquitaine, but Elf withdrew in the late 1980s, and operatorship is now in the hands of Santos.

Lying 250 km west of Darwin, Petrel has four wells. Tern, 50 km farther west, has only two wells.

Estimated combined gas reserves total about 3 tcf, but the reservoirs appear to be complex, and more appraisal work was planned.

Santos and Sumitomo Corp. were conducting feasibility studies for production of 2 million tons/year of LNG.

However, it may be found more economical to use gas from Tern and Petrel fields to supply gas to Darwin, with additional gas used for industry in Northern Territory. Another alternative is to reverse flow of the Palm Valley-Darwin pipeline and lay spurs to Moomba to allow supplies to reach Southeast Australian markets.

A wild card in the LNG stakes is combined development of the Woodside group's huge Scott Reef and Brecknock gas/condensate fields in the Browse basin about 350 km west of Broome on the northwest coast. They hold an estimated combined total of 17 tcf of gas and 200 million bbl of condensate.

BANGLADESH

CAPITAL: Dhaka
MONETARY UNIT: Taka
REFINING CAPACITY: 31,200 b/cd
PRODUCTION: 1,000 b/d
RESERVES: 4 million bbl

Bangladesh is pressing initiatives to boost foreign participation in its upstream and downstream petroleum sectors.

The impoverished country at midyear 1993 put into effec a new national petroleum policy intended to speed privat sector investment in petroleum E&D.

The government and state oil company Bangladesh Oi Gas, & Mineral Corp. (Petrobangla) opened almost th entire country's petroleum sector—downstream as well a upstream—to competitive bids from international petro leum companies. There was no bidding deadline, bu Petrobangla will negotiate terms and assign available acre

age on a first come, first served basis.

Bids were being accepted on 17 of 23 blocks created in 1989 under terms of a model production sharing contract (PSC) formulated in 1988. Two tracts—Blocks 13 and 21—were assigned, and Blocks 12, 14, 16, and 22 were under licensing negotiations. The 1988 model PSC will serve only as a guideline for new E&D proposals.

To meet current and future demand for petroleum products, Bangladesh must increase its oil and gas production and refining capacity.

Petrobangla's two production companies—Bangladesh Gas Fields Co. Ltd. and Sylhet Gas Fields Ltd.—produce 635 MMcfd gas, 1,000 b/d of condensate, and 200 b/d of oil. If gas output increases as expected to 950 MMcfd by 2000, net remaining reserves of 10.554 tcf will be exhausted by 2015.

The country's sole refinery produces about 28,000 b/d of petroleum products. However, Bangladeshi products consumption averages about 40,000 b/d, and demand is increasing at a rate of 2.5-3%/year.

Petroleum prospects

Most of the oil and gas discovered and produced under PSCs negotiated in Bangladesh's open ended tender is to be sold into the country's expanding domestic markets. Contractors also may arrange to export some output.

Because of gas and condensate discoveries in the country's eastern foldbelt, Bangladesh for the past 3 decades has been considered gas prone. However, three discoveries in the heart of the eastern foldbelt revealed a previously unknown system of oil bearing structures beneath the base of gas and above highly pressured zones, Petrobangla said.

Fifty-two exploratory wells drilled on Bangladeshi territory found 21.3 tcf of proven and probable gas reserves, including 12.43 tcf of recoverable gas, about 1.876 tcf of which has been produced. Recoverable oil and condensate is estimated at 64.69 million bbl, including 2.63 million bbl produced.

Most E&D attention has been directed at the eastern foldbelt, where Plio-Pleistocene folding dominates all other trapping mechanisms. However, seismic surveys also have detected undrilled anticlinal folds with significant stratigraphic variation.

Bangladesh's known gas and condensate reserves are distributed among 17 eastern foldbelt fields, often in stacked Miocene and Pliocene fluvial and deltaic sandstone reservoirs. Reservoir porosities of 20-22% are common at depths of 1,096-3,020 m.

About 85% of Bangladesh's condensate reserves are confined to four eastern foldbelt fields: Kailas Tila, Beani Bazar, Sylhet, and Jalabad.

New policy goals

Bangladesh's new petroleum policy is intended to boost exploitation of petroleum resources by encouraging systematic evaluation, exploration, and development in drilled and undrilled areas.

Government officials hope more attractive terms will help mobilize domestic and foreign financial and technical resources. Development of petroleum refining, import, export, storage, distribution, and marketing is integral to its plan.

A major goal involves replacing oil imports as much as possible with domestic gas production and augmenting energy supplies from other undeveloped sources with commercial potential, including coalbed methane, coal, peat, and liquid petroleum gas.

BRUNEI

CAPITAL: Bandar Seri Begawan
MONETARY UNIT: Dollar
REFINING CAPACITY: 8,500 b/cd
PRODUCTION: 158,800 b/d
RESERVES: 1.4 billion bbl

Brunei Shell Petroleum Co. Sdn. Bhd. was to install a $115.6 million compression unit to increase production from Champion field off Brunei to 80,000 b/d from 70,000 b/d.

Fabrication of the compressor module was under way, with work on the jacket and module support frame scheduled to start in March 1993.

Project completion is scheduled for September 1994.

Brunei Shell Petroleum Co. Sdn. Bhd.'s 1 Enggang wildcat, 22 miles off Seria in 40 ft of water, flowed at a rate of 2,250 b/d of oil and 18.4 MMcfd of gas. Four zones were tested between 5,000-12,000 ft.

Shell said the discovery could be developed early because it lies near the producing Seria/Tali fields. Further appraisal and development was planned.

MALAYSIA-INDONESIA
MYANMAR (BURMA)
LAO PEOPLE'S DEMOCRATIC REPUBLIC
SOCIALIST REPUBLIC OF VIET NAM
THAILAND
DEMOCRATIC CAMPUCHEA
HAINAN
MALAYSIA
MALAYA
SABAH
SARAWAK
BRUNEI
BORNEO (KALIMANTAN)
SUMATRA
JAVA
South China Sea
Andaman Sea
Gulf of Thailand
Sulu Sea
Java Sea
Indian Ocean
Strait of Malacca
Hanoi
Haik'ou
Vientiane
Fang
Hue
Da Nang
Bangkok
Sriracha
Phnom Penh
Sihanoukville
Ho Chi Minh City
Long Xuyen
Khanom
Si Thammarot
Kutaradja
Pinang
Kerteh
Pangkalanberandon
Belawan
Kuala Lumpur
Port Dickson
Sibolga
Dumai
SINGAPORE
Batam
Sungei Pakning
Bukittinggi
Padang
Djambi
Sungei Gerong
Plagu
Palembang
Telukbetung
Jakarta
Balongan
Bandung
Cilacap
Semarang
Jogjakarta
Tjepu
Rembang
Surabaja
Wonokromo
Singaradja
Raba
Kuching
Pontianak
Palangkaraya
Bandjarmasin
Balikpapan
Samarinda
Santan
Batuputih
Sangkulirang
Tawau
Sandakan
Kudat
Kota Kinabalu
Labuan
Bandar Seri Begawan
Lutong
Miri
Bintulu
Subic Bay
Limay
Quezon City
Manila
Batangas
Makassar
Arun
Perlak fields
North fields
Rantau fields
Duri fields
Minas fields
Lirik fields
Central fields
Djambi fields
Talang Akar fields
Limau fields
Radja fields
Muraraenim fields
South fields
Rembang fields
Bunju
Tandjung
0 100 200 300 400 Mi
0 200 400 600 Km.

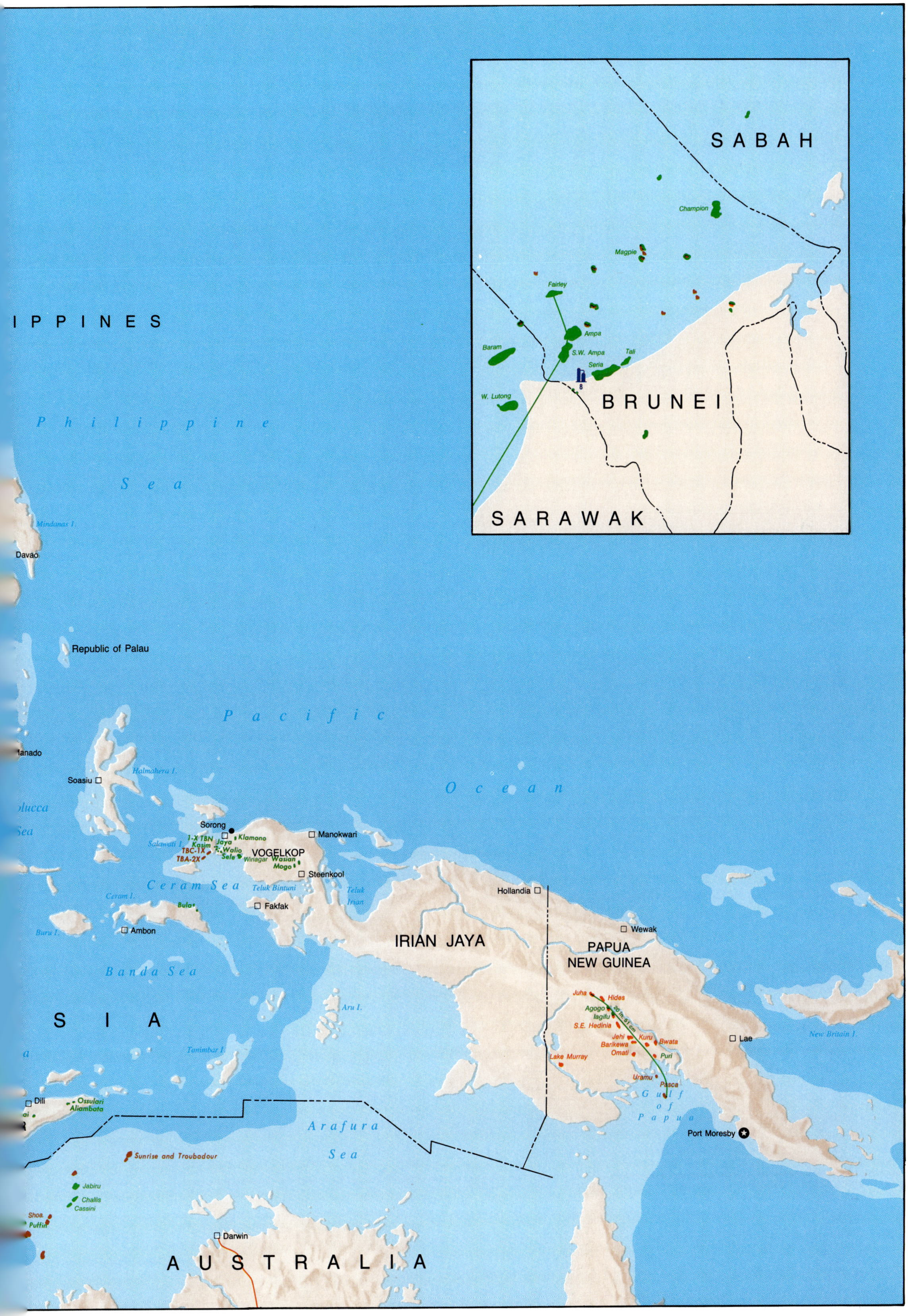

SABAH
Champion
Magpie
Fairley
Ampa
Baram
S.W. Ampa
Tali
Seria
W. Lutong
BRUNEI
SARAWAK
IPPINES
Philippine Sea
Mindanao I.
Davao
Republic of Palau
Pacific Ocean
Halmahera I.
Soasiu
Sorong
Klamono
Manokwari
VOGELKOP
Salawati I.
Ceram Sea
Teluk Bintuni
Steenkool
Teluk Irian
Fakfak
Ceram I.
Bula
Buru I.
Ambon
Banda Sea
Hollandia
IRIAN JAYA
PAPUA NEW GUINEA
Wewak
Juha
Hides
Agogo
Iagifu
S.E. Hedinia
Jehi
Barikewa
Kuru
Bwata
Omati
Puri
Lake Murray
Uramu
Pasca
Gulf of Papua
Lae
New Britain I.
Port Moresby
Aru I.
Tanimbar I.
S I A
Dili
Ossulari
Aliambata
Arafura Sea
Sunrise and Troubadour
Jabiru
Challis
Cassini
Puffin
Darwin
AUSTRALIA

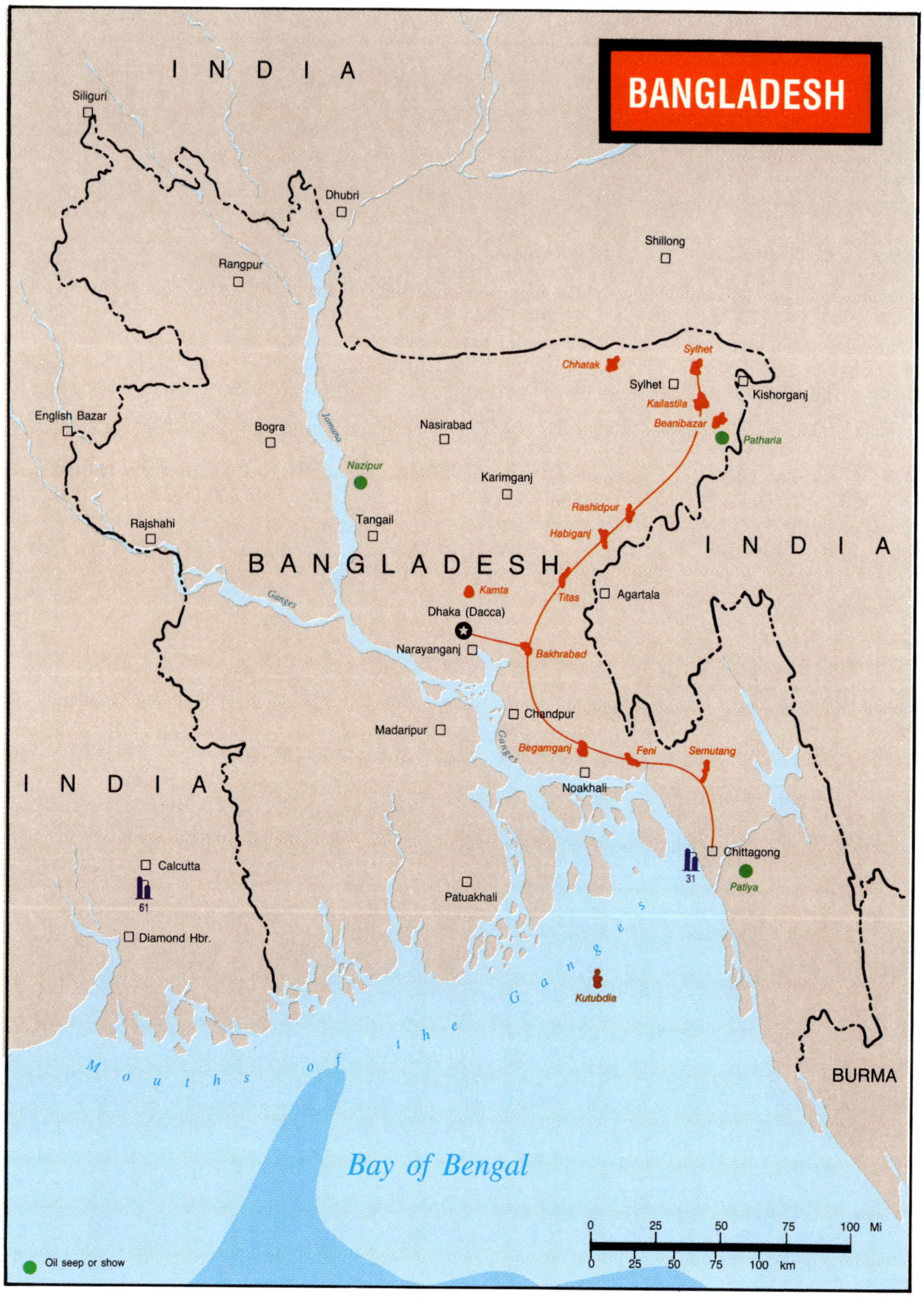

INDIA

CAPITAL: New Delhi
MONETARY UNIT: Rupee
REFINING CAPACITY: 1 million b/cd
PRODUCTION: 524,300 b/d
RESERVES: 5.9 billion bbl

India's government let contract to Czech Republic group Skoda-Strojoexport to lay a 1,300 km products pipeline from Kandla on the Gulf of Kutch in Gujarat on India's west coast to Bhatinda in Punjab in northern India.

Bids on the project were opened in October 1992, and the project originally was scheduled to be complete by May 1993. Civil unrest in Punjab state and concerns over boosting India's investment in the project to keep the foreign exchange component within World Bank credit limits helped cause the delay.

India secured almost half of its total oil import needs for fiscal 1993-94, projected to be flat with the previous year's level of 578,000 b/d.

State owned Indian Oil Corp. completed contracts for 290,000 b/d from five countries: Saudi Arabia 100,000 b/d, Kuwait 80,000 b/d, Iran 60,000 b/d, Abu Dhabi 40,000 b/d, and Malaysia 10,000 b/d.

Daewoo Shipbuilding & Heavy Machinery Ltd. (Korea), Seoul, let a $38 million contract to NPCC Singapore Pvt. Ltd., a unit of National Petroleum Construction Co., Abu Dhabi, U.A.E., to install a water injection platform and underwater pipeline off Bombay. NPCC will provide engineering, procurement, and construction of Oil & Natural Gas Commission Platform SHW and the pipeline in South Bombay High field.

Project completion is planned for December 1994.

India's government approved Indian Oil Corp.'s (IOC) proposed $65 million liquefied petroleum gas import terminal at Kandla, Gujarat.

It will allow IOC to import 600,000 metric tons/year of LPG vs. the current 200,000 tons/year through import sites at Bombay and Visakhapatnam.

The project was to be complete in 42 months.

In early 1993, India announced a fifth bidding round for private sector exploration, involving 45 onshore and offshore blocks.

Twenty-one of the blocks on offer were new, and the rest were reoffered from the fourth round. In that fourth round, held in late 1992, only 24 bids were submitted for 13 blocks out of 72 on offer. The government also has sweetened fiscal terms for bidders.

Indian Oil Corp. shortlisted three companies—Chevron Corp., UOP Inc., and Institut Francaise de Petrole—to provide technology for its proposed 120,000 b/d grassroots Karnal refinery.

The project involves capital outlay of 39 billion rupees ($1.19 billion), with the refinery accounting for 28 billion rupees ($853 million).

Plans also call for laying a pipeline from Chaksa, Gujarat, to Karnal, Haryana, at a cost of 5.5 billion rupees ($167.6 million).

INDONESIA

CAPITAL: Jakarta
MONETARY UNIT: Rupiah
REFINING CAPACITY: 860,200 b/cd
PRODUCTION: 1.3 million b/d
RESERVES: 5.8 billion bbl

In 1992, Indonesia produced an estimated 1.33 million bo/d, mostly through Pertamina's production sharing contracts with foreign companies.

Estimated 1992 exports were 0.8 million bo/d of both oil and condensate. The condensate is produced primarily at the Arun field with some coming from the gas fields in East Kalimantan. Exports go mostly to Japan.

Indonesia is also the world's largest exporter of liquefied

natural gas (LNG), providing about 39% of the world's total supply. Gas export revenues are expected to exceed oil export revenues in the next 10 years.

Oil and gas export revenue was $10.9 billion in 1991 and $10.7 billion in 1992. Forecasts for 1993 stood at $10.1 billion.

Although at this time a net exporter, Pertamina depends heavily on imports mainly because the heavy crudes it produces cannot be processed in local refineries. Pertamina has a refining capacity of about 865,000 bo/d in six refineries.

Demand in Indonesia is toward middle distillate products such as kerosine, light diesel oil and, to a lesser extent, gasoline.

This affects Indonesia's petroleum products because the refining of heavy Indonesian crudes produces a greater amount of heavier distillates such as fuel oils and low-sulfur waxy residue (LSWR) crude. These heavier fuels are mainly exported, while substantial quantities of middle distillate products are imported.

Pertamina imports about 150,000 b/d of light crude, with 110,000 b/d coming from the Middle East, 20,000 from Malaysia, and 20,000 b/d from Australia.

Proven reserves are estimated to be 5.8 billion bbl of oil and 64.4 tcf of gas. Estimates from Indonesia's Department of Mines and Energy indicate reserves, including probable, of 11 billion bbl of crude oil and 104 tcf of gas.

Foreign contractors

On the upstream side of the industry, Pertamina is very dependent on foreign contractors to explore, develop, and produce both gas and oil.

The Oil Law of 1960 permitted foreign companies to establish themselves under contract of work agreements (COWs) with the two state-owned oil companies then in operation.

Until 1967, foreign oil companies operated exclusively under COW agreements, which stipulated that 60% of profits from an exploration group's production went to the state-owned company. During 1993, the last remaining COWs will be converted to production-sharing contracts (PSCs).

Since 1967, Pertamina and foreign companies have operated primarily under PSCs. Under these, a contractor finances all exploration, development, and production costs within a specified area. Pertamina oversees the operations.

Pertamina signed 11 new exploration contracts in 1992 and expected 20 more to be signed in 1993.

Exploratory wells drilled under new and existing contracts numbered 131 in 1990, 162 in 1991, and 123 in 1992, with 207 projected for 1993.

Timor Gap

Signed by Indonesia and Australia in February 1991, the Timor Gap Oil Treaty governs management of oil and gas exploration and production in a 62,880 sq km zone of cooperation.

The zone is divided into three parts: Area A will be developed jointly by both countries and production will be shared on a 50/50 basis. Area B will be developed by Australia, and 10% gross resource rent tax will be paid to Indonesia. Area C will be developed by Indonesia, and Australia will receive 10% of the income tax collected by Indonesia from contractors.

New incentives

Indonesia's domestic petroleum product sales has steadily grown. In 1987, sales averaged 455,000 b/d, increasing to 661,000 b/d in 1992. The three main products were gas oil, 254,000 b/d; kerosine, 147,000 b/d; and gasoline, 124,000 b/d.

Because Indonesia has the world's fourth highest population (1992 estimate of 186 million), the domestic energy consumption is expected to grow. One scenario is that Indonesia will become a net importer in the next decade.

To temper the effects of this possibility, Indonesia is attempting to increase oil exports while reducing the rate of growth of domestic consumption.

Refining and marketing

Nearly all refining in Indonesia is carried out by Pertamina in six refineries with a combined capacity of 865,000 b/d. Overall refinery output totaled 300 million bbl in 1992 up from 286 million in 1991.

Improvements and expansions are under way at several refineries. Also, a new export-oriented facility at Balongan, West Java, with a capacity of 120,000 b/d is slated for completion in 1994.

Through the construction of three or four new refineries, Indonesia hopes to boost capacity by 600,000 b/d in the last half of this decade.

Pertamina has started to license a number of gasoline filling stations to private operators and may even sell some stations.

Natural gas

As with oil, foreign contractors produced most of Indonesia's natural gas. Gas production has more than doubled in the last 5 years from 1.7 tcf in 1987 to 2.5 tcf in 1991. In 1992, production was estimated to be 2.6 tcf, of which only 243.9 bcf was produced by Pertamina.

The two largest gas-producing areas are the Arun field in Sumatra and the Badak, Nilam, and adjacent fields in Kalimantan. Pertamina has set up two joint venture companies, PT Arun LNG Co. and PT Badak LNG Co., to liquefy the gas from these two areas.

Exports of LNG, which began in 1977 at $80 million, reached $4 billion in 1992.

Domestic demand for natural gas is expected to grow. Fertilizer, cement, and downstream petrochemical plants are the more significant consumers of gas.

To further facilitate Indonesia's domestic gas distribution, a new pipeline will link the Kangean block near Madura Island with a gas-fired combined cycle power plant under construction at Gresik near Surabaya in East Java.

Petrochemicals

About 56 chemical plants are in operation, producing $4 billion worth of products annually, of which over $550 million are exported.

Facilities include plants with installed annual capacity of 7 million tons of fertilizer, 1 million tons of adhesive resins, 330,000 tons of methanol, 270,000 tons of paraxylene, and 225,000 tons of purified terphthalic acid (PTA). Other products include synthetic resins, PVC polypropylene, polyvinyl chloride, polyester, and nylon fibers.

Two recent facilities are the $2 billion olefins complex at Cilagon, West Java, for producing ethylene, propylene, and related downstream products, and the $1.3 billion aromatics complex at Arun in North Sumatra. This facility will produce benzene, paraxylene, and fuels.

Java pipeline

North American Nippon Technologies Corp. (Nantco), Vancouver, B.C., filed an expression of interest with Indonesia's Pertamina to participate in the construction, ownership, and operation of a major petroleum products pipeline in Java.

The 1,200 km pipeline will serve eastern Java. Estimated cost is $600 million (Canadian).

It will replace eastern Java's existing products supply system and be designed to carry in batches a range of refined products, including gasoline, kerosine, diesel, avia-

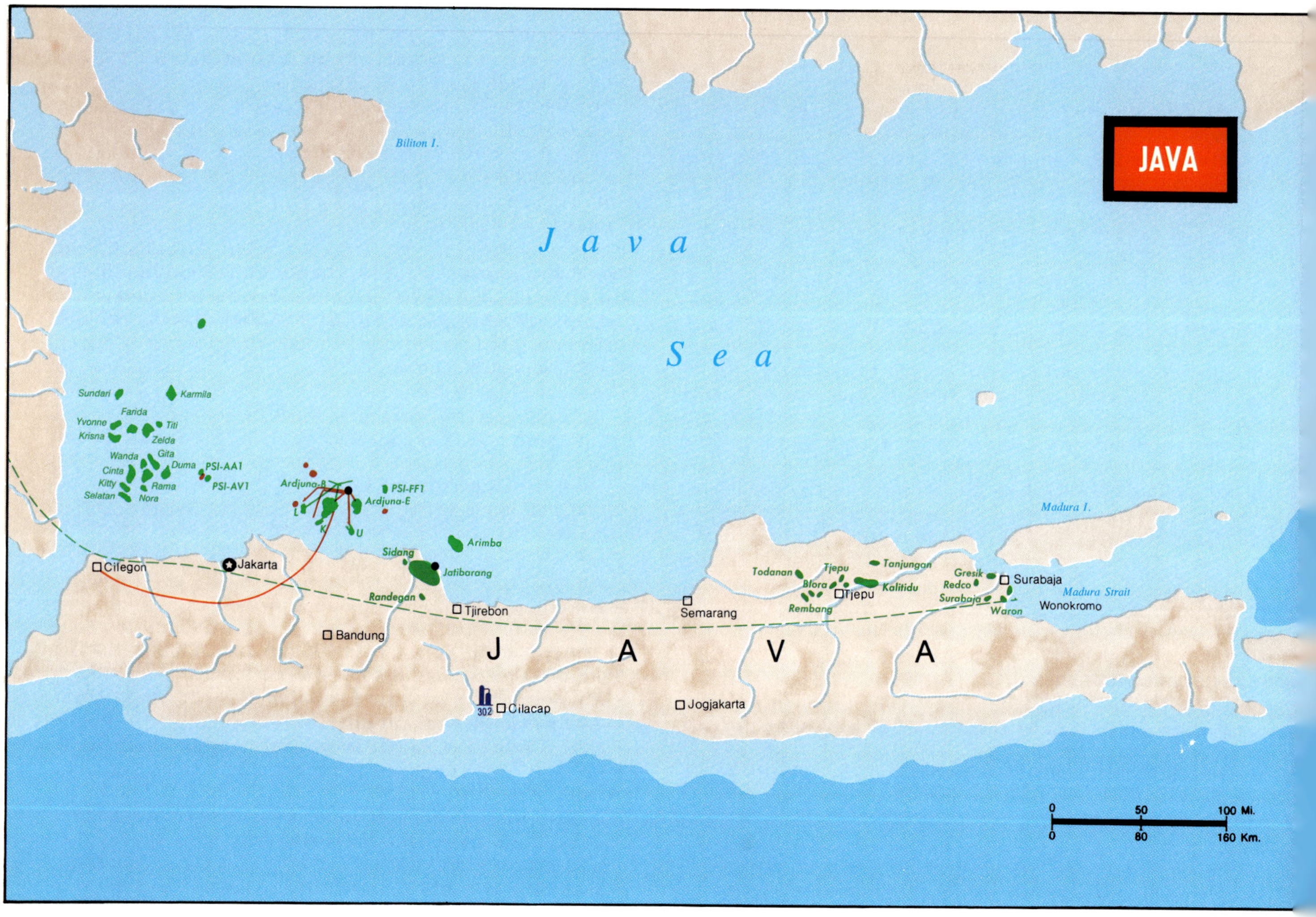

tion fuels, and fuel oil.

The pipeline will connect a new marine terminal with eight tank farms and include three main booster stations and several pumping facilities, Nantco said.

Pagerungan Besar production facilities

An ARCO unit scheduled completion of final commissioning in late 1993 for production facilities in Pagerungan Besar gas/condensate field in Indonesia.

Site of the Atlantic Richfield Bali North Inc. development project is on tiny Pagerungan Besar Island in the Bali Sea, 300 km east of Surabaya, East Java.

Clough Engineering Group, Perth, Western Australia, through its Indonesian subsidiary PT Petrosea, completed its $250 million (Australian) Pagerungan Besar field development contract with Atlantic Richfield. Atlantic Richfield holds the field under a production sharing contract with state owned Pertamina.

Initial gas processing involves removal of water and condensate from natural gas produced from two wellsites on the island. Dry gas will move by pipeline under a separate contract to markets in Java. Condensate will be shipped via tankers from the island by Atlantic Richfield.

Belida production

Conoco Indonesia Ltd. and partners have achieved phase two production in Belida oil field off Indonesia in the South Natuna Sea.

Phase two oil began flowing in late 1993 from the BB-01 directional development well. BB-01 is the first of as many as 20 directional development wells to be drilled and placed on production by April 1994 on Belida Platform B (WHP-B). Simultaneous drilling and production was to continue on WHP-B until then.

PT Gema SemBrown (GSB) completed WHP-B hookup and commissioning. Belida partners in August 1992 let a $200 million turnkey contract to GSB for detailed engineering, procurement, fabrication, and installation of Belida stage two facilities.

WHP-B, Belida's second platform and the first major phase two component installed, is a four pile, 20 slot platform designed to handle 65,000 b/d of produced fluid and 33 MMcfd of gas.

JAPAN

CAPITAL: Tokyo
MONETARY UNIT: Yen
REFINING CAPACITY: 4.8 million b/cd
PRODUCTION: 15,500 b/d
RESERVES: 54.2 million bbl

Here's a sample of Japanese processing activity in 1993:

Nippon Eurecat Co. Ltd., a new joint venture company at Niihama on Japan's Shikoku Island, was building a plant there that includes a 3,000 metric ton/year regeneration unit and an off site 10,000 metric ton/year presulfiding unit. The JV is owned 50-50 by Eurecat SA, La Voulte, France, and Sumitomo Metal Mining Co. Ltd. and related companies.

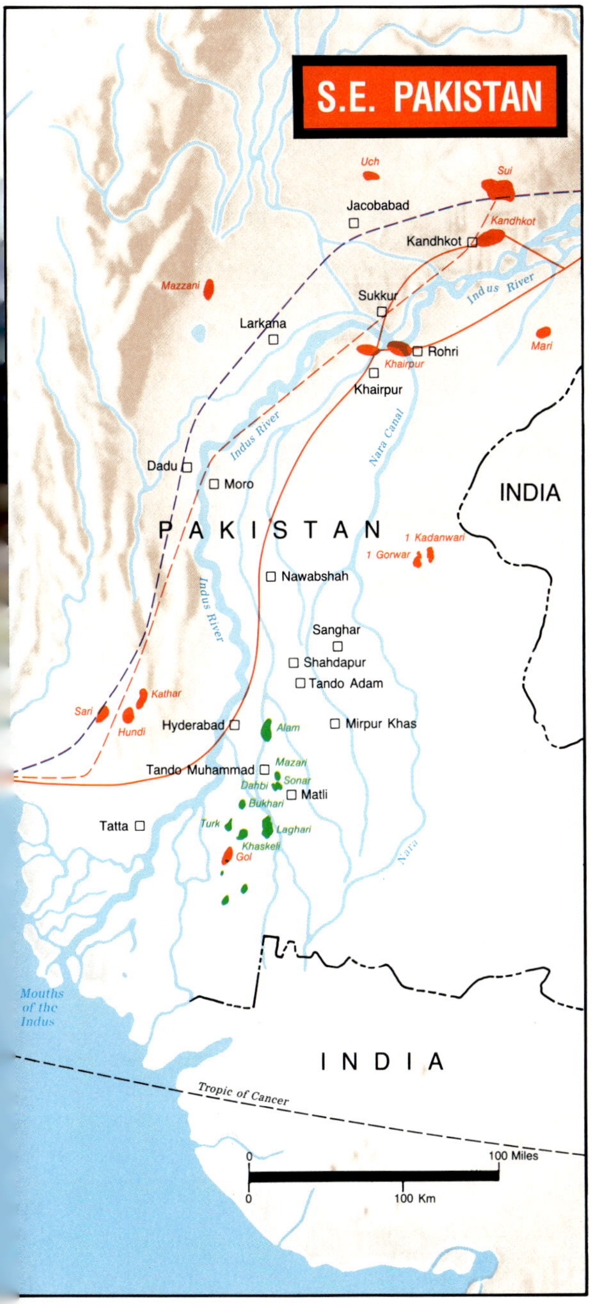

Plant start-up is scheduled for third quarter 1994.

Maruzen Petrochemical Co. and Asahi Glass Co. will form a joint venture to build a 200,000 metric ton/year vinyl chloride monomer plant at a Maruzen site in Ichihara, Chiba prefecture, Japan. Total investment is estimated at 13 billion yen ($117 million). Start-up is scheduled for April 1995, with ethylene feedstock supplied by a new plant Maruzen expected to be complete in late 1993.

Mitsubishi Gas Chemical Co. Inc. (MGC) and Mitsubishi Heavy Industries Ltd. (MHI) commissioned a methanol synthesis process that cuts overall plant energy consumption by 10%. Testing at MGC's Niigata, Japan, 520 metric ton/day methanol plant verified the new system's higher production efficiency. The double concentric tube method achieves higher conversion by cooling the catalyst and maximizing its activity. MHI said the new method's one pass conversion rate is more than double that of conventional converters.

Tosoh Corp. of Japan is to delay construction of an ethylene plant in Yokkaichi, Mie prefecture, 1 year. The plant will start up in spring 1996, with productive capacity of 400,000 tons/year. Nikkei Weekly reports Ube Industries Ltd., Mitsui Toatsu Chemicals Inc., and Nippon Petrochemicals Co. are considering delaying completion of a 500,000 ton/year ethylene plant at Ube until 1996.

Tohoku Oil Co. has started up a 25,000 b/d fluid catalytic cracker at its 83,900 b/d Sendai, Japan, refinery. The new unit incorporates the combined technology of M.W. Kellogg Co. and Mobil Oil Corp. Kellogg said it is the first unit of its kind in Japan and the first grassroots unit anywhere using the combined technology.

Nippon Petroleum Refining Co. started up a 30,000 b/d gas oil and residual catalytic cracker at its Negishi refinery at Yokohama City, Japan. The Stone & Webster Engineering Ltd. two stage resid unit was installed because of the increasing variety of imported crudes and reduced supplies of light, low sulfur crude.

MALAYSIA

CAPITAL: **Kuala Lumpur**
MONETARY UNIT: **Ringgit**
REFINING CAPACITY: **269,800 b/cd**
PRODUCTION: **640,000 b/d**
RESERVES: **4.3 billion bbl**

Private companies planned to spend $2.1 billion for oil and gas exploration, development, and production in Malaysia during the fiscal year ending Mar. 31, 1994, state owned Petronas predicted.

That sum was expected to be easily topped because it did not include upstream spending planned by Petronas Carigali Sdn. Bhd., a unit of Petronas billed as the third most active exploration and production company in Malaysia after the Royal Dutch/Shell and Esso groups.

A Petronas prospectus, issued when the company raised $500 million in the U.S., said Petronas Carigali expected to commit about $1.24 billion for exploration, development, and production.

The $2.1 billion in planned private investment exceeded the $1.8 billion spent in fiscal 1993. In fiscal 1992, private firms spent $1.8 billion, while in fiscal 1991 they spent $1.5 billion.

The prospectus also showed the firms drilled 32 wells and made 20 discoveries during the fiscal year. In fiscal 1992, they drilled 61 wells and were successful in 41. In fiscal 1991, they were successful in 21 wells out of 41 spudded.

The success rate for oil and gas discoveries was one in three exploration wells drilled in 1957-92, the prospectus said. For oil discovery it was one in six wells.

Present activities of nongovernment firms cover 210,200 sq km of the 523,100 sq km of land and seabed available for oil and gas exploration in the country.

The prospectus also disclosed that crude oil is the biggest contributor to Petronas revenue. However, Petronas' total gas revenue has been steadily increasing since 1990.

Sarawak contract

Occidental Petroleum Corp. unit Occidental Petroleum (Malaysia) Ltd. signed a production sharing contract with national company Petronas for the 3.2 million acre, 12,990 sq km SK-3 contract area off Sarawak, Malaysia.

SK-3 is adjacent to SK-8, where Oxy drilled four gas discoveries that included Serai and Jintan fields.

Processing activities

Royal Dutch/Shell Group started producing synthetic diesel from its $780 million natural gas to middle distillates synthesis (MDS) plant at Bintulu, Malaysia, after solving start-up problems caused by a fire at the plant.

It is the world's first commercial natural gas to middle

distillates plant.

Ethylene Malaysia Sdn. Bhd. let a turnkey contract to Toyo Engineering Corp. and ABB Lummus Crest Inc. for a 320,000 metric ton/year grassroots ethylene plant to be built at Kertih, Terengganu, Malaysia.

Conoco Inc. and Malaysia's Petroliam Nasional Bhd. (Petronas) signed a letter of intent to study the feasibility of adding 100,000 b/d of sour crude refining capacity to a site at Malacca.

The companies planned an 8-10 month engineering study that could lead to a joint venture refinery project.

The sour crude unit is the second phase of Petronas' planned $2.8 billion integrated refinery complex. Construction of the unit could begin as early as 1994 with start up sometime in 1997.

Petronas was building a 100,000 b/d sweet crude unit at the site, scheduled for completion in 1994.

Malaysia's Petronas, British Petroleum Chemicals Co., and Idemitsu Petrochemicals Co. agreed to build and develop Malaysia's first petrochemical complex. It will consist of a 320,000 metric ton/year ethylene plant and 200,000 ton/year polyethylene plant at Kertih, in Northeast Trengganu state. A $333 million contract for engineering, procurement, and construction (EPC) of the ethylene plant was let to a combine of Toyo Engineering Corp., Mitsui & Co. Ltd., ABB Lummus Crest, and Equator Engineering Sdn. Bhd. A $150 million EPC contract went to Technip, Paris, for the polyethylene plant. Start-up is set for mid-1995.

Mobil contracts

Production sharing contracts signed by Mobil Corp. and Malaysian state oil company Petroliam Nasional Bhd. (Petronas) called for an exploration campaign in the South China Sea off Sarawak.

The contracts for Blocks SK-A and SK-B were Malaysia's first deepwater acreage awards. They also were the first acreage awards in Malaysia since February 1991.

Wood Mackenzie Consultants Ltd., Edinburgh, said the awards contain the first significant revisions to Malaysia's fiscal regime since 1985.

But the analyst warned that Mobil will have to discover about 300 million bbl of oil on either block to justify field development on a rate of return basis.

The tracts lie about 280 km off the Malaysian state of Sarawak on the island of Borneo.

Exploration drilling is slated for late 1994 or early 1995

Borneo's Sandakan basin

Borneo's Sandakan basin was neglected for 15-20 years between the mid-1970s and 1990. In this time the oi industry underwent a revolution in all areas, from quantum leaps in seismic acquisition, processing, and interpretation technology to the great conceptual developments of plate tectonics and sequence stratigraphy.

Also during this time strong evidence emerged that oi could be generated from terrestrial source rocks.

Recent basin studies indicate that there is at least 8,000 n of quartz-rich deltaic sediment, with great potential fo previously overlooked low resistivity/low contrast pa zones. Previous lithostratigraphic correlations were incorrect, and modern bio/chronostratigraphic correlations indicate thick stratigraphic intervals that are virtually untested.

Coupled with recent developments in drilling technology, flexibility in field development, and booming Asian markets for gas and oil, these studies suggest that the Sandakan basin is poised to become Borneo's sixth hydrocarbon-productive Neogene deltaic basin.

MONGOLIA

CAPITAL: Ulaanbaatar
MONETARY UNIT: Tughrik
REFINING CAPACITY: N.A.
PRODUCTION: N.A.
RESERVES: N.A.

Mongolia's Mongol Petroleum Co. (MGT) and Nescc Energy Co., Austin, signed a joint venture agreement givir Nescor exploration and development rights in the East Go basin of Southeast Mongolia.

Nescor's area of interest, covering five blocks, includes Mongolia's only significant hydrocarbon discoveries, Zuunbayan and Tsagaan Els. Nescor and MGT will concentrate first efforts on starting production from Tsagaan Els.

Nescor was reprocessing several hundred line miles of old Soviet seismic data. Western Geophysical, Houston, completed a 1,500 line km seismic survey in Southeast Mongolia and was expected to conduct more survey work.

Meantime, Nescor was negotiating with Chinese petroleum officials to drill a delineation well in Tsagaan Els before yearend 1993, if weather allowed. Successful delineation of the reservoir could lead to a development involving 40-80 wells at a cost of $50-100 million, said Nescor Managing Director Neil Salsich III.

Nescor estimated original oil in place in Tsagaan Els at as much as 100 million bbl. The company expects to return to Mongolia to negotiate a production sharing contract (PSC) involving some of the area of interest acreage.

Other activity

Zuunbayan and Tsagaan Els were the centerpiece of an agreement with Amgol Inc., Houston, that expired.

Amgol is a partnership of SOCO International, an affiliate of Snyder Oil Corp., Fort Worth; Exploration Associates International of Texas Inc. (EAIT), Houston; and CP&G, a Houston oil field equipment supply firm. Snyder's SOCO Mongolia signed what it said was the first PSC in Mongolia, covering Contract Area XXII (Buir), a 2.8 million acre tract in the Tamtsag basin of far eastern Mongolia.

SOCO International Pres. Ed Story cited an encouraging sign from China concerning Mongolia's petroleum prospects. Officials of state oil company China National Petroleum Corp. said they were willing to buy any crude found in Mongolia, he said. That would eliminate the need for a costly export pipeline to the Chinese port of Dalian, Story said, thus removing a major roadblock to commerciality of Mongolian discoveries.

Mongolia's Tamtsag basin is much like China's Hailar basin, EAIT noted. The Chinese were expected to open an adjacent Hailar basin block for application in first quarter 1994, EAIT said. China was stepping up efforts to expand foreign participation in exploration and development of its vast, remote, interior basins, some of which border or extend into Mongolia.

Background

Mongolian and Soviet explorationists found Zuunbayan field in 1941. It produced about 4 million bbl of oil during 1950-69. A small refinery was built nearby, along with a rail terminal to ship products. Tsagaan Els is a 1953 discovery. Although 10 wells were completed as oil producers, none was placed on production due to lack of facilities.

Because of equipment failures, a fire that severely damaged the refinery, and a shift in Soviet exploration emphasis to western Siberia and the Caspian Sea in a low oil price climate, Mongolia abandoned efforts to boost oil production in favor of importing subsidized products from the former U.S.S.R. Oil development, production, and refining halted in Mongolia in 1969.

MYANMAR

CAPITAL: Rangoon
MONETARY UNIT: Kyat
REFINING CAPACITY: 32,000 b/cd
PRODUCTION: 14,000 b/d
RESERVES: 50 million bbl

Texaco Exploration Myanmar Inc.'s 1 East Yetagun discovery well flowed a combined 63 MMcfd and 1,922 b/d of 53.5° gravity condensate from three zones at 6,600-7,000 ft in 333 ft of water on Block M-13 off Myanmar.

The well, Texaco's third off Myanmar, is about 2 miles southeast of its 1 Yetagun gas discovery. Texaco's second well off Myanmar, 1 Zagawa, was a noncommercial discovery.

Interests in Block M-13 and adjoining Blocks M-12 and M-14 are operator Texaco 50%, Premier Petroleum Myanmar Ltd. 30%, and Nippon Oil (Myanmar) Exploration Ltd. 20%.

Myanmar began offshore gas production at its 1 Yadana, 112 km off Bogale township in the Andaman Sea.

The project was carried out under a production sharing contract between Myanma Oil & Gas Enterprise, Yangon, and an undisclosed French company.

PAKISTAN

CAPITAL: Islamabad
MONETARY UNIT: Rupee
REFINING CAPACITY: 132,900 b/cd
PRODUCTION: 62,500 b/d
RESERVES: 203 million bbl

Here's a roundup of industry events in Pakistan in 1993:

OMV (Pakistan) Exploration GmbH's 1 Miano wildcat on Block 20 in the Middle Indus Valley of Pakistan flowed dry gas at stabilized rates as much as 35.5 MMcfd from lower Cretaceous sandstone.

The well, drilled to 13,200 ft, was to be completed as a potential producer. More exploration and appraisal on Block 20 are planned for 1994.

A proposed marine oil terminal at Port Qasim, Pakistan, received a $32.8 million financing package from the Asian Development Bank and $19 million from Commonwealth Development Corp.

Fauji Foundation's Fauji Oil Terminal & Distribution Co., in collaboration with Canadian and U.S. participants, will build and operate the terminal. It will have 9 million metric tons/year capacity and berthing eventually to accommodate 75,000 dwt tankers.

At first, however, only 50,000 dwt vessels will be received because of channel restrictions.

Pakistan's State Engineering Corp. (SEC) will undertake turnkey engineering jobs in cooperation with foreign companies on the basis of technology transfer with emphasis on domestic content of oil and gas processing equipment.

SEC unit Heavy Mechanical Complex (HMC) had orders to supply a sulfur recovery unit for Meyal oil field, equipment for a 50 MMcfd gas dehydration plant for Pirkoh Gas Co., and facilities for state owned Oil & Gas Development Corp.'s Dakhni oil and gas fields.

In addition, the government ordered 20% of the work related to the hydrocracker project at Pakistan Refining Ltd.'s Karachi refinery to be done by HMC's Enar Petrotech division.

Lasmo Oil Pakistan Ltd. let contract to Foster Wheeler Petroleum Development Ltd., London, for engineering and construction supervision of Kadanwari gas field development in Sind province.

The 700 bcf field will start up before the end of 1994, moving 150 MMcfd to Sui Southern Gas Co., which is developing a transmission pipeline.

Occidental Petroleum Corp. started production of 2,100 b/d of oil and 22 MMcfd of gas from its Ratana discovery in the Potwar region of Pakistan.

State owned Oil & Gas Development Corp. started production of about 2,500 b/d of oil and 15 MMcfd of gas from Sadkal field in the Margalla Hills area south of Islamabad. In

Japan

Location	No. Refineries	Location	No. Refineries
Sakaide	1	Mizushima	2
Yokohoma	1	Okinawa	3
Chiba	4	Toyama	1
Sakai	2	Chita	2
Yokkaichi	2	Funakawa	1
Hyogo	1	Niigata	2
Kashima	1	Muroran	1
Marifu	1	Negishi	1
Osaka	1	Yamaguchi	2
Oita	1	Ehime	1
Tomakomai	1	Wakayama	1
Kawasaki	5	Owase	1
		Sendai	1

both instances, oil moves to the Attock refinery at Rawalpindi and gas to the Sui Northern gas system at Lahore.

North Coast Energy Inc. (NCE), Cleveland, signed a memorandum of understanding with Pakistan's Oil & Gas Development Corp. (OGDC) for an enhanced recovery project in OGDC's Toot oil field. Toot, discovered in 1988, has produced 11.2 million bbl of an original reserves estimate of 15.8 million bbl.

NCE will conduct a study and develop an EOR proposal under a joint venture with OGDC. NCE will bear project costs and share incremental production 50-50 with OGDC.

Pakistan's Oil & Gas Development Corp. (OGDC) was to start at midyear 1993 production of 3,000 b/d and 25 MMcfd from Sakdal oil and gas field in the Margalla Hills area, 45 km south of Islamabad.

Pay depth in the two well field is 4,500 m. Crude will move to the Attock refinery and gas via the Sui Northern Gas Pipeline after installation of a dehydration plant. The field will boost OGDC oil production to its target of 32,700 b/d and Pakistan's total to 70,000 b/d, compared with national oil consumption of 230,000 b/d.

Pakistan's OGDC let contract to Australia's Clough Engineering to design, build, and commission a gas/condensate processing plant in Dhodak gas field at a cost of $75 million.

It is to be located 20 km north of Taunsa, Dera Ghazi Khan district, Punjab. The plant, scheduled for completion in August 1994, will produce liquefied petroleum gas and other products.

Qatar was studying plans for a $3.5 billion, 1,600 km, mostly subsea pipeline to export 2 bcfd of natural gas to Pakistan.

Pakistan in 1992 signed a memorandum of understanding with Crescent Petroleum Co., Sharjah, to conduct a feasibility study of the project.

Crescent would lead an international group to lay, own, and operate the pipeline. The pipeline could be on stream by 1998, provided World Bank guarantees for the project are obtained. It would extend from Qatar across the Persian Gulf to the United Arab Emirates, overland to and then across the Gulf of Oman, then hug the shorelines of Iran and Pakistan before terminating at Gadani, near Karachi.

PAPUA NEW GUINEA

CAPITAL: Port Moresby
MONETARY UNIT: Kina
REFINING CAPACITY: N.A.
PRODUCTION: 40,400 b/d
RESERVES: 253.4 million bbl

A burst of exploration and development interests trades was under way in Papua New Guinea, newest addition to the world list of oil exporters.

Exxon Exploration Co. acquired a farmout on exploration and development permits in Papua New Guinea's highlands through a $59 million agreement with British Petroleum Co. plc.

It involves petroleum licenses near the country's sole commercial hydrocarbon production, Hides gas field and the Kutubu oil export project.

Under a separate agreement, BP agreed to transfer a 2.5% interest in Petroleum Prospecting License (PPL) 138 and Petroleum Development License (PDL) 1 to Australia's Oil Search Ltd.

In another sale of Papua New Guinea interests, Kerr-McGee Corp. acquired a 33.33% interest in the former PPL 126 and PPL 94 tracts, which have been combined into one exploration block operated by Phillips Petroleum Co. in the foreland region. The interests were obtained through sale of certain interests by Australian independents Mosaic Oil and Intrepid Oil.

Papua New Guinea's Department of Minerals & Petroleum approved a petroleum prospecting grant of 33⅓% interest in onshore Papua New Guinea License 163 to Kerr-McGee Corp. unit Kerr-McGee Papua New Guinea Ltd. The 1.9 million acre license area is 310 miles northwest of Port Moresby. Other interest holders are operator Phillips Petroleum Co. Niugini 56.2%, Mosaic Oil Niugini Pty. Ltd. 7.5%, and Pagini Oil & Gas NL 3%.

Exxon farmout

Exxon will pay BP's share of costs to drill the 3 Hides wildcat in exchange for an option to take a farmout on the three permits.

Operator BP, with a 95% interest, spudded 3 Hides early in 1993. Drilling was expected to require 80 days to a projected 3,690 m to assess the possibility of an oil pay below the Hides gas reservoir. Oil Search Ltd. holds the remaining 5% interest.

Upon completion of 3 Hides, Exxon will have 6 months to exercise its option to acquire a 47.5% interest in PPL 138 and PDL 1, which covers the 3 Hides area but excludes the two producing Hides wells. The 1 and 2 Hides wells produce gas to generate electricity for the nearby Porgera gold mine.

PHILIPPINES

CAPITAL: Manila
MONETARY UNIT: Peso
REFINING CAPACITY: 278,750 b/cd
PRODUCTION: 10,100 b/d
RESERVES: 279 million bbl

Philippines national company Energy Development Corp. let contract to California Energy Co. Inc., Omaha, and partners to build, own, and operate two geothermal plants with a combined capacity of 300,000 kw at a combined construction cost of more than $350 million.

One unit will be a 180,000 kw plant in Mahanagdong geothermal field, the other a 120,000 kw plant in Upper Mahiao geothermal field. CECI and Kiewit Diversified Group Inc. will share development costs and equity funding equally at the larger plant, scheduled for completion in June 1997.

At the smaller plant, CECI partner Ormat Inc. will be turnkey contractor and CECI the operator. It's scheduled for completion in July 1996.

Philippines National Oil Co. chose Magma Power Co., San Diego, to build and operate a 70 million kw geothermal power plant on Leyte Island. Under a proposed 15 year contract, the power generating plant will distribute electricity via underwater cable to Cebu Island.

The plant is expected to be in operation by mid-1996. Magma bought some of Unocal Corp.'s geothermal assets.

Pilipinas Shell Petroleum Corp., Manila, let contract to Parsons Corp. unit Parsons International Ltd. for engineering, procurement, and construction management services at its Tabangao refinery at Batangas, Philippines. Parsons will revamp the hydrocarbon solvents plant, allowing Shell to process new feedstocks of naphtha and kerosine.

The revamp is scheduled for completion in October 1994.

Subic Bay Metropolitan Authority leased to Coastal Corp. subsidiary Coastal Petroleum NV the POL Depot petroleum storage facilities at the former U.S. naval base at Subic Bay, Philippines.

The long term agreement includes 68 sites on 304 acres of land with 2.4 million bbl of mainly underground storage capacity and 40 miles of pipelines connecting the terminal with other facilities within Subic Bay Free Port.

Pilipinas Shell Petroleum Corp. completed financial ne-

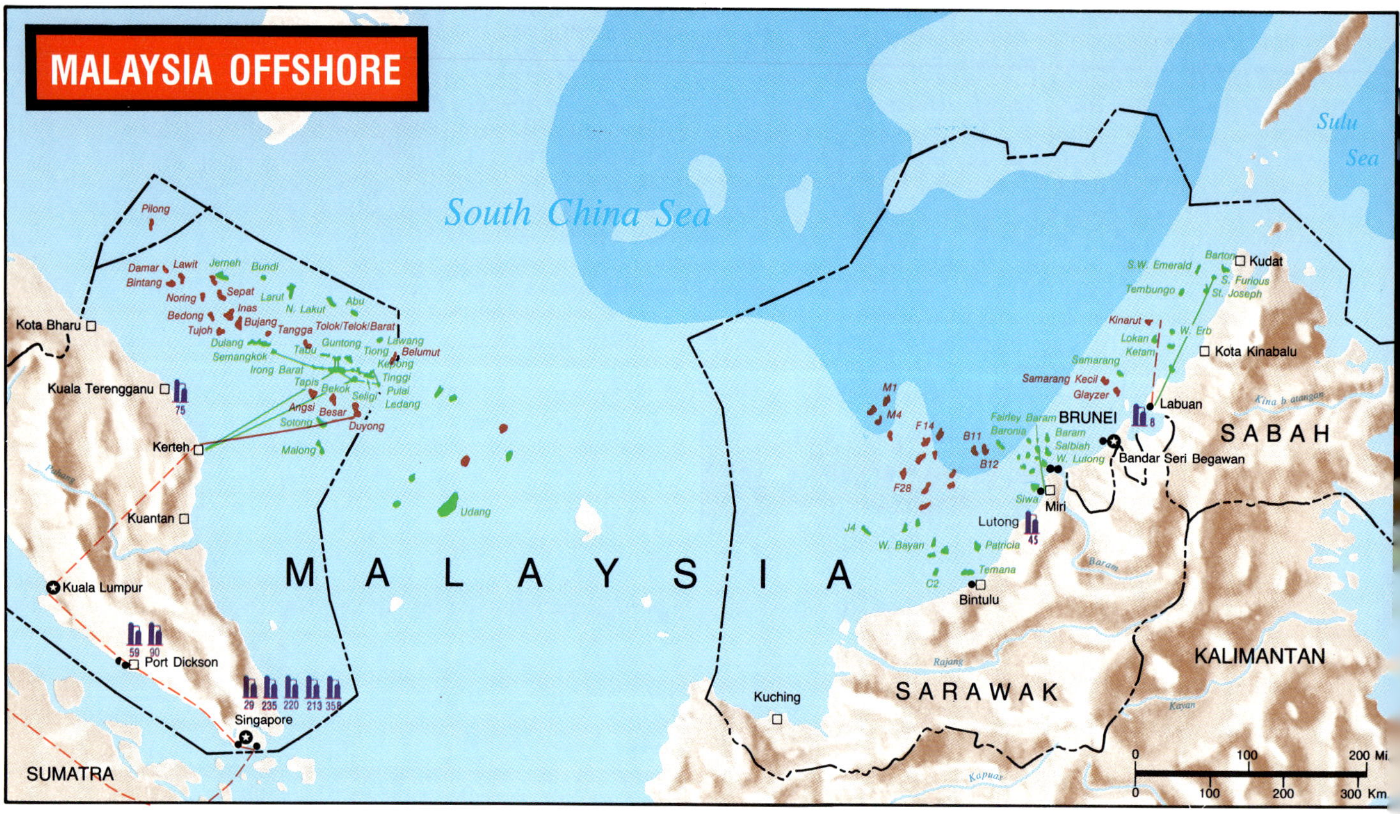

gotiations on a $667 million, 110,000 b/d refinery to be built next to its 65,800 b/d Tabangao, Philippines, refinery.

U.S. Export-Import Bank authorized a $106.6 million political risk guarantee that will facilitate the purchase of $120 million of U.S. equipment and services. Fluor Daniel will be the main supplier.

Occidental Petroleum Corp. acquired a 70% interest in the 213,600 acre Alva block off Palawan Island, Philippines, bringing its holdings to 1.3 million contiguous acres off the island.

The block is adjacent to the 405,500 acre West Culion block acquired by an Oxy group in October and is near Block SC-38 where Oxy and Royal Dutch/Shell made the Malampaya discovery. Oxy's program includes a seismic acquisition program and the option to drill two wildcats.

Privatization plan

Philippines President Fidel Ramos approved a plan to privatize the government's energy companies and deregulate the oil industry. The plan includes offering to domestic and foreign investors 65% of state owned Petron Corp., which had assets of $873 million as of December 1991 and net income of $47.3 million for the year.

Petron holds a 40.7% share of the domestic petroleum products market and has refining capacity of 155,000 b/d. The privatization plan is a reaction to the country's prolonged power shortages, which last 3-4 hr every day in Luzon and about 8 hr/day in Mindanao.

SINGAPORE

CAPITAL: Singapore
MONETARY UNIT: Dollar
REFINING CAPACITY: 1 million b/cd
PRODUCTION: N.A.
RESERVES: N.A.

Singapore was busy with processing activity in 1993.

Esso Singapore Pte. Ltd. let a $236 million contract tc Foster Wheeler Eastern Pte. Ltd., Singapore, a unit of Foster Wheeler Italiana SpA, Milan, for a 15,000 b/d hydroprocessing unit at its 220,000 b/cd Pulau Ayer Chawan refinery ir Singapore.

The unit will upgrade high sulfur vacuum gas oil into low sulfur diesel fuel, naphtha, and vacuum gas oil. Work was tc start early in 1994 with completion scheduled for late 1995.

Exxon Chemical Co., Amoco Chemical Co., and China American Petrochemical Co. Ltd. (Capco) of Taipei formed a joint venture to build a $1 billion (Singapore) ($617 millior U.S.) aromatics complex on Pulau Ayer Chawan Island of Singapore.

The complex, expected to be in operation by 1996, wi include a reformer and units to produce paraxylene anc benzene.

Singapore Refining Co. Pte. Ltd., a joint venture of Britisl Petroleum plc, Singapore Petroleum Co., and Caltex (Asia Ltd., will upgrade its 220,000 b/d refinery on Pulau Marli mau Island southwest of Singapore.

The $840 million project will enable the plant to produce more transportation fuel and less fuel oil by 1995. Work wi be carried out by Japanese Gasoline Corp., Tokyo.

Shell Singapore (Pte.) Ltd. let a $66 million contract t Chiyoda Corp. to design and build a unit to produce lo sulfur gas oil at its Bukom refinery complex in Singapore.

Upon completion in 1994, the new unit will have cos more than $97.4 million, enabling the plant to process 5,00 metric tons/day of high sulfur gas oil components. Th project is part of Shell's $365 million, 3 year plan ending i 1995 to upgrade Bukom.

Mobil Oil Singapore let contract to Chiyoda Singapore t expand crude oil and distillates storage at its Juron refinery in Singapore by 10%, or 1 million bbl, at a cost c $18.78 million.

The project involves adding a tank farm on Pulau Pesek, nearby island. The expansion will support increased crud runs at the refinery, where Mobil hiked crude capacity b 10,000 b/d to 245,000 b/d in 1992. Mobil also has und construction at Jurong a continuous catalytic reforme

aromatics complex and a catalytic hydrodesulfurization unit.

Shell Eastern Petroleum (Pte.) Ltd. will install a $38 million sulfur emission reduction plant at its Pulau Bukom refinery in Singapore.

To be on stream in 1995, the plant is designed to cut sulfur dioxide emissions by more than 20%, in line with the government's Green Plan.

TAIWAN

CAPITAL: Taipei
MONETARY UNIT: New Taiwan dollar
REFINING CAPACITY: 542,500 b/cd
PRODUCTION: 1,100 b/d
RESERVES: 4 million bbl

Taiwan's petrochemical sector continues to expand.

Construction of Formosa Plastics Group's (FPG) often delayed naphtha cracker in the Mailiao industrial zone was scheduled to get under way at midyear 1993. FPG and government agencies agreed on a series of issues that had threatened to delay the project again. Frequent delays had led FPG to consider a $6-7 billion petrochemical complex in China as an alternative, but the company later scrapped that project.

The $3.3 billion cracker would be Taiwan's sixth and the nation's first privately owned naphtha cracker. It will be the heart of a $7 billion complex that would produce 450,000 metric tons/year of ethylene and 225,000 tons/year of propylene to feed more than 20 downstream plants.

The giant naphtha cracker is expected to increase Taiwan's economic growth by 1%/year during construction and 1.5%/year when it goes on stream.

Meantime, Tuntex Distinct Corp. (Tainan) planned to spend $227 million to build plants at Mailiao to produce purified terephthalic acid (PTA) and paraxylene. Tuntex, one of Taiwan's biggest producers of polyester, currently produces 350,000 metric tons/year of PTA. Taiwan's three other producers of PTA, China American Petrochemical Corp., ICI, and FPG, planned to increase their PTA capacities by 100,000 tons/year each.

THAILAND

CAPITAL: Bangkok
MONETARY UNIT: Baht
REFINING CAPACITY: 313,750 b/cd
PRODUCTION: 51,400 b/d
RESERVES: 178.5 million bbl

Thailand's demand for gas will outstrip supplies in the late 1990s as rapid economic growth continues.

Gas will be a cornerstone for Thai energy policy throughout the growth, although sources in neighboring countries need development.

Thailand has emerged as one of the world's fastest growing economies, said Wood Mackenzie Consultants Ltd., Edinburgh, with growth topping 10%/year.

Thailand's energy needs will be met 67% by oil, 20% by gas, and 13% by coal and hydroelectric sources in 1993. By 2000, the analyst forecast, oil will fall to 62%, gas will rise to 28%, and coal and hydroelectric will slip to 10%.

Thai gas production will rise 25% from 1992 to average 1 bcfd by 1995. Including production from new discoveries, production could rise to 1.5 bcfd by 2000, up almost 90% from the 1992 level.

Increased gas flow output in the mid-1990s will be due

largely to development of Gulf of Thailand fields.

Shortfall

By 1998, production from Gulf of Thailand fields will not be enough to offset the decline in today's fields. Wood Mackenzie said Thailand will need to import more than 1 bcfd by 2005 in the absence of future discoveries in the country.

Counting "necessarily tentative" estimates of future discoveries, about 750 MMcfd will need to be imported by 2005.

"There are no fields in Thailand we would classify as probable developments at present," Wood Mackenzie said.

The analyst estimates total recoverable oil reserves in Thailand at 320 million bbl, of which 47% remains. Total Thai gas reserves are placed at 6.3 tcf, of which 68% remains.

Gas production from today's fields and those under development will peak at 1.2 bcfd in 1996 and remain above the 1.1 bcfd mark until the turn of the century. Future discoveries could add 320 MMcfd by 2000.

Pipelines

Petroleum Authority of Thailand (PTT) expects to complete in 1996 a 36 in. pipeline from Unocal Corp.'s Erawan field to Mab Ta Phud in Rayong province. This will parallel an existing line, doubling Gulf of Thailand export capacity to 1.6 bcfd.

Gas will enter the pipeline from three main sources: Unocal fields, Bongkot, and Pilong field in the Joint Development Area (JDA) under negotiation between Thailand and Malaysia.

PTT is laying a 110 mile 32 in. line with 700 MMcfd capacity from Bongkot to Erawan and a 24 in. diameter line to take the gas to Khanon, 100 miles away on the mainland, where a gas treatment plant will be built to fire a power station.

A further $700 million, 28 in. pipeline is planned for completion in 1997 to take gas 70 miles from Mab Ta Phud to the Ban Pakong power station.

Imports

The Thai-Malay JDA in waters claimed by both countries was near agreement, leading Wood Mackenzie to predict drilling may resume in 1994. The firm estimates reserves of 3 tcf of gas and 200 million bbl of oil, with commercial gas production likely before 2000.

Total won exclusive exploration rights to about 2.5-3.5 tcf of undeveloped gas reserves in the Gulf of Martaban offshore Myanmar.

The most likely gas market is the industrial west coast of Thailand. Total could have 250 MMcfd available for export by 2000.

An extension of Malaysia's Peninsular Gas Utilization pipeline is to be complete in 1996, passing near the Thai border at Perlis. The pipeline will tap about 27 tcf of gas off Malaya.

PTT plans to start talks to move 300 MMcfd from Malaysia via Perlis by the turn of the century.

PTT expressed interest in taking gas by pipeline from Viet Nam's offshore Dai Hung field development, which holds about 400 million bbl of liquids and 2.5-3 tcf of associated and nonassociated gas.

The 300 mile distance from Dai Hung to PTT pipeline infrastructure makes imports from Viet Nam unlikely before 2000.

Unocal spending

Unocal Thailand and partners expect to spend as much as $1.75 billion developing natural gas off Thailand the next 5 years.

That's in addition to $3.3 billion Unocal and partners have spent off Thailand.

Unocal expects to drill as many as 50-60 wells/year to maintain its Offshore Thailand gas production of at least 800 MMcfd.

Unocal Thailand Pres. Brian Marcotte said his company's focus is to maintain a secure gas supply to Thailand the next 10-15 years, with a target of 900 MMcfd by 2000.

Reserves growth targeted

The company is trying to boost its gas reserves, estimated at 7 tcf of combined proved, probable, and possible reserves.

That accounts for an increasing emphasis on onshore Thailand prospects.

Unocal started a four well drilling program on its concessions in Northeast Thailand that covers a total area of 9,550 sq km.

The first well spudded is in Mukdahan province.

Northeastern prospects have big potential but drilling conditions are more difficult, Marcotte said, adding that the onshore emphasis does not signal a shift away from gulf potential.

Unocal also is interested in Gulf of Thailand acreage in the joint development area offered by Thailand and Malaysia and in disputed waters claimed by Thailand, Cambodia, and Viet Nam. The latter three countries are in the process of resolving their disputed offshore territorial claims, with U.S. companies' involvement in Viet Nam still problematic.

Bongkot field

Total and partners have brought on stream giant Bongkot natural gas field in the Gulf of Thailand off Thailand at a rate of 150 MMcfd.

Plans called for production to jump to 200 MMcfd in January 1994 and 250 MMcfd in January 1995. Reserves are estimated at 2 tcf. Development, to be complete in early 1994, began in August 1992 and entails drilling 27 wells.

Bongkot gas is delivered via pipeline operated by state owned PTT Exploration and Production (Pttep) to the Bangkok and Khanom areas, intended mainly to feed Thai power plants.

VIET NAM

CAPITAL: Hanoi
MONETARY UNIT: Dong
REFINING CAPACITY: N.A.
PRODUCTION: 122,500 b/d
RESERVES: 500,000 million bbl

Exploratory drilling off Viet Nam was set to snowball a midyear 1993, with 28 wells to be spudded compared with only two in 1992, Wood Mackenzie Consultants Ltd., Edinburgh, predicted.

About 70 exploration wells were predicted to be drilled during 1993-97, of which more than three-fourths will be drilled in the South Con Son basin. Seismic survey activity was expected to remain buoyant, with acquisition of 30,000 line km of 2D and 3D data covering 2,000 sq km planned in 1993.

Liquids production slipped to an average 92,000 b/d in 1993 from 1992's peak of 96,000 b/d, Wood Mackenzie said. All Viet Nam's oil comes from one field, Bach Ho, placed on stream by the Russian/Vietnamese combine Vietsovpetro in June 1986.

Vietnam's petroleum reserves are estimated at 900 million bbl oil equivalent, of which a little more than 10% has been produced by the end of 1992.

Most remaining reserves are said to lie in Dai Hung field awarded for development in January 1993 to a group led b

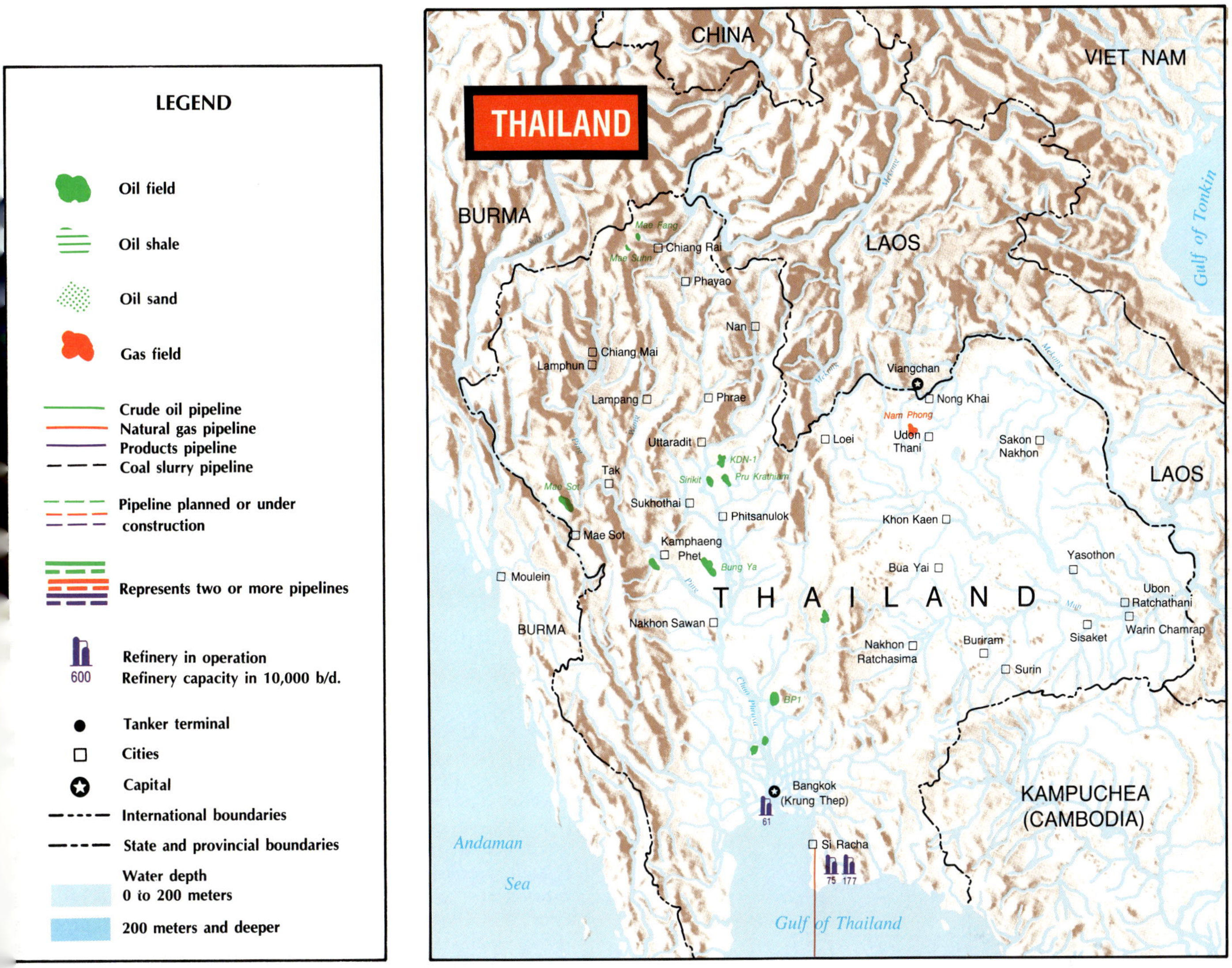

BHP Petroleum Pty. Ltd., Melbourne. BHP estimated Dai Hung reserves at 800 million bbl of light, sweet crude and 2-3 tcf of gas.

Wood Mackenzie estimated reserves in Dai Hung and Rong fields as 400 million bbl and 50-100 million bbl of oil, respectively, with Dai Hung also holding 1 tcf of gas

Vietsovpetro's plans to develop Rong field are thought to favor a jack up rig converted for production, with recovered oil to be stored on a tanker prior to export. The $250 million project is expected on stream in 1995.

Assuming Dai Hung and Rong field developments proceed as planned, the analyst predicted Viet Nam's liquids production could reach 210,000 b/d before 2000.

Contract extensions

Wood Mackenzie counted 23 production sharing contracts active with Viet Nam, a record high for that country. Ten of them were awarded in second half 1992, covering areas relinquished by Vietsovpetro in the South Con Son basin.

The Dai Hung development contract was the only one awarded in 1993, although Wood Mackenzie said state company Petrovietnam granted extensions to some existing contracts.

Petro-Canada won a 6 month extension on Blocks 03, 12A and 20, while Sceptre Resources Ltd., Calgary, gained a 1 year extension on Block 111. Wood Mackenzie said Cairn Energy plc, Edinburgh, Enterprise Oil plc, London, and International Petroleum Ltd., Dubai, were negotiating with Petrovietnam to extend their contracts.

Now that most of Viet Nam's prospective acreage is under license, Wood Mackenzie did not expect 1992's level of acreage awards to be repeated soon.

However, Petrovietnam was expected to invite bids for remaining offshore blocks, including deepwater acreage in the eastern South Con Son basin, after the end of 1993.

Block 05-1 in the South Con Son basin is likely to generate the most interest because it borders the Dai Hung field block. Block 05-1 is said to contain the undrilled Blue Dragon prospect, reputed to be of a size similar to Dai Hung.

1993 wildcats

Eight Vietnamese wildcats were drilled in 1993, of which three found hydrocarbons. BP Exploration Operating Co. Ltd. is said to have had the most notable successes, with its 1 Lan Tay and 1 Lan Do gas strikes on Block 06.

The 1 Lan Tay was temporarily abandoned in March 1993 after a blowout, while results of drillstem tests in 1 Lan Do were not been revealed.

Petronas Carigali Sdn. Bhd., the operating arm of Malaysia's state oil company, found oil with its 1X Jade on Block 02. The well is said to have tested 100-300 b/d of oil with a gravity of less than 10° and was abandoned as noncommercial.

Wood Mackenzie said BP was the most active foreign explorationist in Viet Nam in 1993 with eight exploration wells scheduled.

BHP planned four appraisal wells in Dai Hung field in the second half of 1993. IPE

Middle East

IRAN

CAPITAL: Tehran
MONETARY UNIT: Rial
REFINING CAPACITY: 1.162 million b/cd
PRODUCTION: 3.64 million b/d
RESERVES: 92.86 billion bbl

Iran since 1991 has been willing to work with international companies in upstream oil and gas projects within the formerly isolationist Islamic republic. Deals, however, have been scarce. The much heralded welcome of international capital has been intermittent at best.

President Hashemi Rafsanjani recognizes the need for foreign investment in Iran and has made normalization of international relations a goal. The country has a rapidly growing population exceeding 60 million, widespread damage from an 8 year war with Iraq, and fiscal difficulties that hamper reconstruction.

So Rafsanjani, who became president after the 1989 death of revolutionary leader Ayatollah Khomeini, has attempted to open a range of opportunities in his country to foreign capital. Many of those opportunities are based on oil reserves estimated at some 93 billion bbl and gas reserves of 700 tcf.

But a politically and culturally powerful clergy, led by Ayatollah Ali Khamenei, successor to Khomeini as revolutionary leader, remains hostile to non-Iranian, non-Islamic influences and resists efforts to open the economy. Its criticisms of the government's accommodations toward outsiders gain clout as domestic economic pressures grow.

Apparently in response to such opposition, the government has kept foreign oil companies, especially American, at some distance. Foreign service and supply companies, however, are working in Iran.

Whatever its difficulties in attracting international capital, the Iranian petroleum industry suffers no chronic shortage of skilled technicians. State owned National Iranian Oil Co. (NIOC) has been able to rebuild war damaged refineries and restore some of the production capacity lost since the Iraniar revolution of 1978-79.

Refinery modernization is under way, although the pace i measured. Additions to refinery cracking capacities wi change Iran's potential net trade position. The country i recent years has had to rely on imports to cover a ne products deficit estimated in 1992 at 34,000 b/d. In 1992, imported gasoline, kerosine jet fuel, and diesel fuel an produced surplus fuel oil, naphtha, and other products.

Recent capacity additions were expected to give it a ne product surplus estimated at 230,000 b/d in 1993, mostly fue oil. That surplus was expected to grow to perhaps 355,000 b/ in 1994 and 422,000 b/d by 1995, giving the Iranian refinin industry significant potential for investments in upgradin capacity.

Product demand in 1992 amounted to 874,000 b/d. Th total includes 135,000 b/d of motor gasoline, 133,000 b/d c kerosine jet fuel, 297,000 b/d of diesel, 219,000 b/d of fuel oi and 90,000 b/d of other products, Iranian refineries provide an estimated 120,000 b/d of motor gasoline, 75,000 b/d c kerosine jet fuel, 218,000 b/d of diesel, 329,000 b/d of fuel oi 4,000 b/d of naphtha, and 93,000 b/d of other product Demand growth is expected to average 4%/year through 200 about double the worldwide rate.

Current operations

Iran has disclosed a hefty oil discovery. Oil Ministe Gholamreza Aghazadeh said the discovery adds an estimate 7 billion bbl of oil to the country's proved reserves base.

The find is in southern Khuzestan province in the Darkhc vein region, south of Ahvaz. The minister said it is light crud and drilling is under way to define the field.

NIOC is stepping up drilling and gas injection to rais production capacity. The petroleum ministry recently est mated production capacity at 4.3 million b/d and said would increase to 4.5 million b/d during 1993.

Iran produced nearly 6 million b/d before the Islam revolution. Crucial to production capacity additions has bee

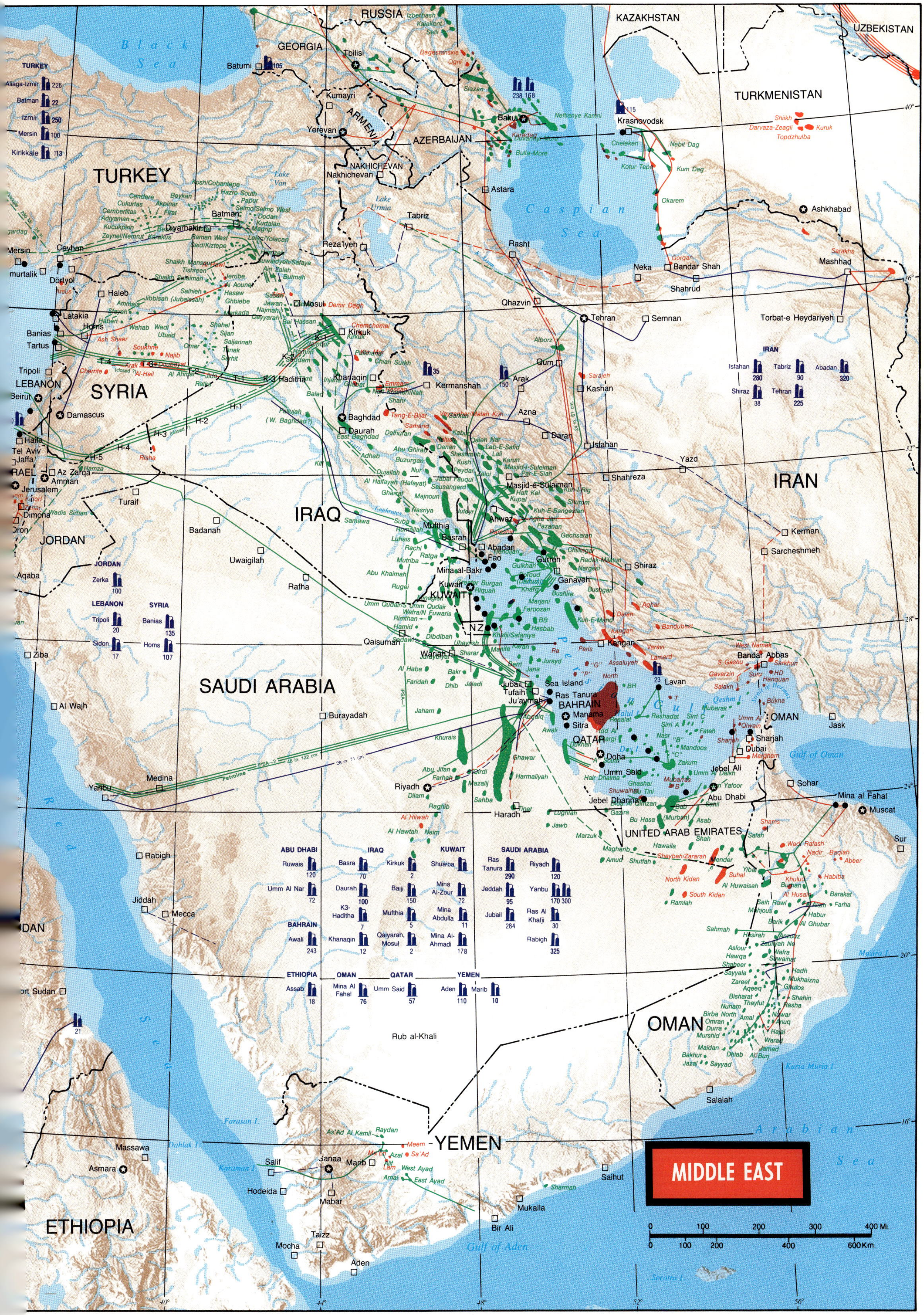
MIDDLE EAST
RUSSIA
GEORGIA
KAZAKHSTAN
UZBEKISTAN
TURKMENISTAN
AZERBAIJAN
ARMENIA
NAKHICHEVAN
TURKEY
SYRIA
LEBANON
JORDAN
IRAQ
IRAN
KUWAIT
SAUDI ARABIA
BAHRAIN
QATAR
UNITED ARAB EMIRATES
OMAN
YEMEN
ETHIOPIA
Black Sea
Caspian Sea
Red Sea
Gulf of Oman
Arabian Sea
Gulf of Aden
Rub al-Khali
Batumi
Tbilisi
Kumayri
Yerevan
Baku
Krasnovodsk
Ashkhabad
Mashhad
Astara
Tabriz
Rasht
Tehran
Qazvin
Semnan
Neka
Bandar Shah
Shahrud
Torbat-e Heydariyeh
Qum
Arak
Kashan
Kermanshah
Azna
Daran
Isfahan
Yazd
Shahreza
Kerman
Sarcheshmeh
Shiraz
Masjid-e-Sulaiman
Ahwaz
Abadan
Basrah
Fao
Mina al-Bakr
Ganaveh
Kangan
Bandar Abbas
Lavan
Jask
Mosul
Kirkuk
Khanaqin
Baghdad
Daurah
Haditha
Diyarbakir
Batman
Ceyhan
Mersin
Dörtyol
Haleb
Latakia
Banias
Tartus
Homs
Tripoli
Beirut
Damascus
Haifa
Tel Aviv
Jaffa
Jerusalem
Amman
Az Zarqa
Dimona
Aqaba
Turaif
Badanah
Uwaigilah
Rafha
Qaisumah
Kuwait
Ras Tanura
Jubail
Manama
Sitra
Doha
Umm Said
Abu Dhabi
Jebel Ali
Dubai
Sharjah
Mina al Fahal
Muscat
Sohar
Sur
Salalah
Riyadh
Haradh
Burayadah
Medina
Yanbu
Jiddah
Mecca
Al Wajh
Ziba
Rabigh
Sanaa
Marib
Salif
Hodeida
Mabar
Taizz
Mocha
Aden
Mukalla
Bir Ali
Saihut
Massawa
Asmara
TURKEY
Aliaga-Izmir 226
Batman 22
Izmir 250
Mersin 100
Kirikkale 113
IRAN
Isfahan 280
Tabriz 90
Abadan 320
Shiraz 38
Tehran 225
JORDAN
Zerka 100
LEBANON
Tripoli 20
Sidon 17
SYRIA
Banias 135
Homs 107
ABU DHABI
Ruwais 120
Umm Al Nar 72
BAHRAIN
Awali 243
IRAQ
Basra 70
Kirkuk 2
Daurah 100
Baiji 150
K3-Haditha 7
Mufthia 5
Khanaqin 12
Qaiyarah, Mosul 2
KUWAIT
Shuaiba
Mina Al-Zour 72
Mina Abdulla 11
Mina Al-Ahmadi 178
SAUDI ARABIA
Ras Tanura 290
Riyadh 120
Jeddah 95
Yanbu 170 300
Jubail 284
Ras Al Khafji 30
Rabigh 325
ETHIOPIA
Assab 18
OMAN
Mina Al Fahal 76
QATAR
Umm Said 57
YEMEN
Aden 110
Marib 10
0 100 200 300 400 Mi.
0 100 200 400 600 Km.

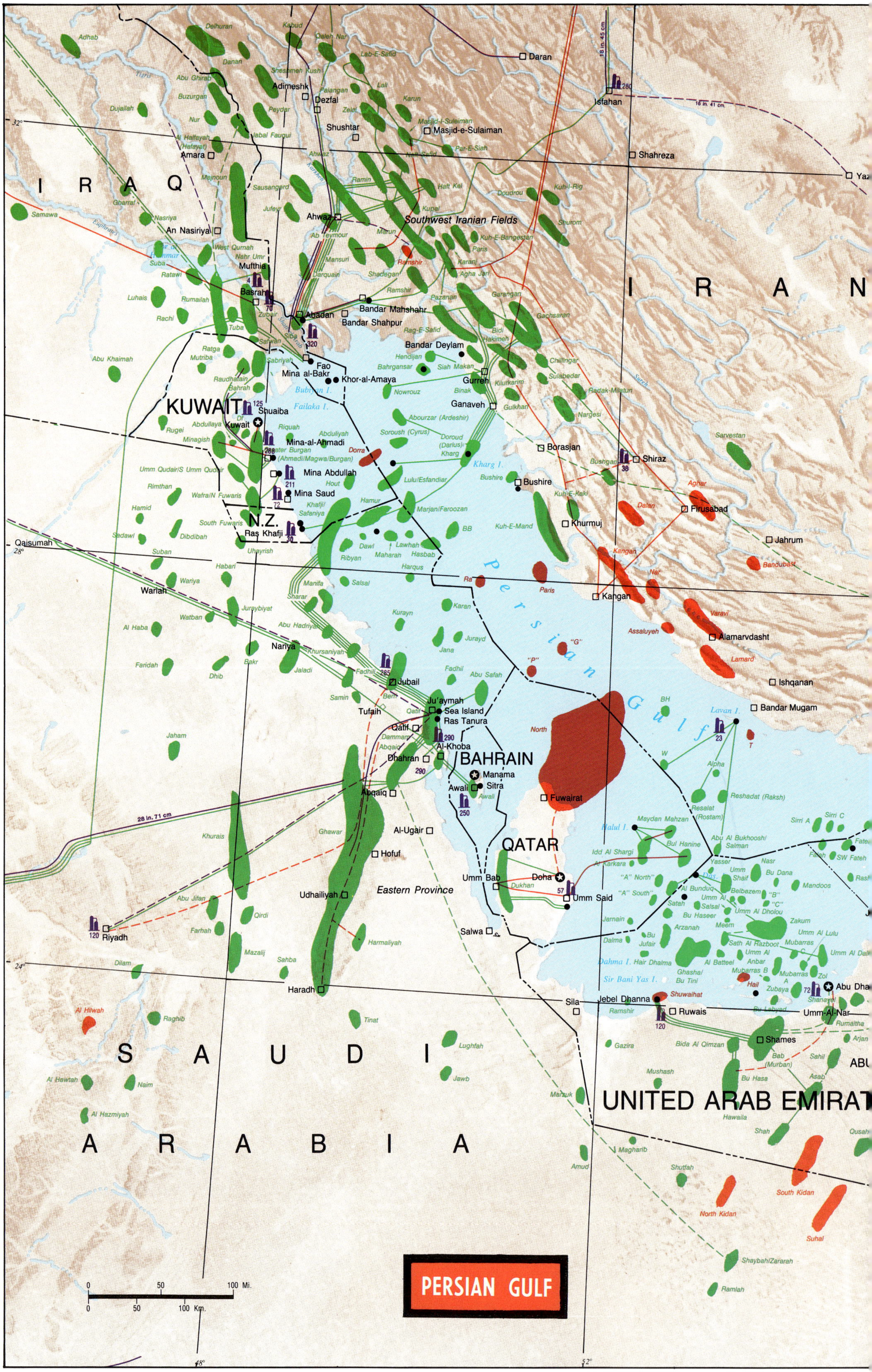
PERSIAN GULF
IRAQ
IRAN
KUWAIT
N.Z.
SAUDI ARABIA
BAHRAIN
QATAR
UNITED ARAB EMIRATES
Persian Gulf
Southwest Iranian Fields
Eastern Province
Daran
Isfahan
Adimeshk
Dezful
Shushtar
Masjid-e-Sulaiman
Shahreza
Amara
An Nasiriya
Ahwaz
Mufthia
Basrah
Abadan
Bandar Mahshahr
Bandar Shahpur
Bandar Deylam
Fao
Mina al-Bakr
Khor-al-Amaya
Gurreh
Ganaveh
Shuaiba
Kuwait
Mina-al-Ahmadi
Mina Abdullah
Mina Saud
Ras Khafji
Borasjan
Shiraz
Bushire
Firusabad
Khurmuj
Jahrum
Kangan
Alamarvdasht
Ishqanan
Bandar Mugam
Qaisumah
Warlah
Nariya
Jubail
Ju'aymah
Sea Island
Ras Tanura
Tufaih
Qatif
Al-Khobar
Dhahran
Manama
Sitra
Awali
Abqaiq
Al-Ugair
Hofuf
Fuwairat
Umm Bab
Doha
Umm Said
Udhailiyah
Salwa
Riyadh
Haradh
Sila
Jebel Dhanna
Ruwais
Abu Dhabi
Umm-Al-Nar
Shames
28 in. 71 cm
0 50 100 Mi.
0 50 100 Km.

LEGEND

- Oil field
- Oil shale
- Oil sand
- Gas field
- Crude oil pipeline
- Natural gas pipeline
- Products pipeline
- Coal slurry pipeline
- Pipeline planned or under construction
- Represents two or more pipelines
- Refinery in operation
- 600 Refinery capacity in 10,000 b/d.
- Tanker terminal
- Cities
- Capital
- International boundaries
- State and provincial boundaries
- Water depth 0 to 200 meters
- 200 meters and deeper

a steady program of natural gas pipeline construction, which facilitates gas lift operations.

NIOC operates eight refineries with total capacity as of Jan. 1, 1994, of slightly less than 1.2 million b/cd. That total includes an estimated 254,600 b/d at the Abadan refinery, 60% of which was damaged during the war with Iraq. Before the war, Abadan had crude distillation capacity of 630,000 b/d.

The company started up its new Arak refinery, 230 km southwest of Tehran, with throughput of 90,000 b/d vs. capacity of 150,000 b/d. Crude oil is delivered via truck, pending completion of a pipeline. In October 1993, throughput was expected to reach 320,000 b/d, and a 40,000 b/d fluid catalytic cracker is due on stream there.

The company is building or planning refineries at Bandar Abbas (240,000 b/d) and Chah Bandar (300,000 b/d). Start-up of the Bandar Abbas plant was reported to be delayed 1 year to 1995. Financing problems apparently have slowed construction.

A number of oil companies have conducted negotiations with NIOC about participating in Iranian upstream ventures. A common sticking point in those discussions is said to be allowed rates of return. NIOC and other branches of the government in 1991 sponsored a petroleum conference in Isfahan that signaled Iran's intentions to become more active than it had been in international petroleum projects. Soon after that meeting, Iran announced it had signed a letter of intent with Japan Petroleum Exploration Co. for drilling in the Strait of Hormuz. A final agreement has not materialized.

Shortly before 1993 opened, ENI announced its Saipem subsidiary and Technologie Progetti Lavori SpA of Italy were beginning a development program in South Pars gas field in the Persian Gulf under an agreement with NIOC. Saipem completed basic engineering design and is arranging financing through export credit agenices.

Meanwhile, Iranian officials show strong interest in the Central Asian states of the former Soviet Union and have entered preliminary agreements for pipeline and other projects.

Elsewhere, Iran in January 1993 started up a 100,000 b/d crude oil pipeline that connects the port city of Bandar Abbas with Kerman province to the north. From Kerman the line extends to Central Isfahan province, where it connects with the nationwide pipeline network. Another pipeline, which will augment supplies to Zanjan province in the northwest, also was commissioned during January 1993. During 1992, oil shortages led to sharp criticism from the public because Iran is one of the world's major oil exporters.

IRAQ

CAPITAL: Baghdad
MONETARY UNIT: Dinar
REFINING CAPACITY: 347,500 b/cd
PRODUCTION: 435,800 b/d
RESERVES: 100 billion bbl

Iraq's Oil Projects Co. started up a $20 million, 25 km products pipeline linking Baghdad with storage tanks in the eastern province of Dialalh. Iraq plans eventually to hook all its reserviors into a pipeline network. Most products move via truck to regional storage depots.

Iraq said its rebuilt southern ports of Khor Al-Zubair, with 21 piers, and Umm Qasr, with two terminals, are back on stream. The Al-Bakr terminal on the northern tip of the Persian Gulf was returned to service during August 1992 at a cost of about $16 milliion and is restored to full loading capacity of 1.6 million b/d. The Khour Al-Omayyeh port, also at the northern end of the gulf, is to be restored as well.

An adviser to Iran's minister of petroleum told a London conference in 1993 that Iraq can increase productive capacity to 3.5 million b/d by 2000 from the current 2 million b/d.

Iran destroyed most of Iraq's seaborne export facilities during the Iran-Iraq war in the early 1980s. About 600,000 b/d of capacity remains. There are three onshore export pipelines: a 1.65 million b/d capacity line to the eastern Merditerranean via Turkey, one with similar capacity to the Red Sea via Saudi Arabia, and one to the Mediterranean across Syria.

The Syrian route was closed by the Syrian government during the Iran-Iraq war and has been virtually abandoned. The Saudi government is unlikely to encourage Iraqi exports across its territory, having gained market share from embargoes on Iraqi production.

That leaves the Turkish route. This crosses Kurdish controlled territory in Iraq, now enjoying protected independence from Baghdad after Saddam Hussein's oppression of the Kurds. Although Turkey wants Iraqi oil to flow its way, the pipeline would be vulnerable without a settlement between Iraqi Kurds and the central governement in Baghdad.

ISRAEL

CAPITAL : Jerusalem
MONETARY UNIT: Shekel
REFINING CAPACITY: 208,000 b/cd
PRODUCTION: 200 b/d
RESERVES: 3,351,000 bbl

Israel and Egypt have agreed to study feasibility of a $1 billion pipeline to take gas from the Nile Delta to Israel via the Gaza Strip.

The two governments have exchanged letters of understanding for gas supply and electric power grid links. Israel's engery minister is said to be hoping for a 25 year contract to supply Egyptian gas for power generation. The Egyptian government reportedly said it has insufficient natural gas to export, but potential discoveries might provide a source.

The project would allow Israel to postpone a $200 million investment in a pipeline running from a Tel Aviv suburb to Eilat on the Red Sea.

In other action, Isramco Inc., Tel Aviv, chartered the Benreoch semisubmersible rig from Ross Offshore AS, Oslo, and was scheduled to drill a wildcat off Israel in January 1994. According to Norwell Ltd., Aberdeen, Scotland, appointed by Isramco to manage the project, 1 Yam Yafo will be the first offshore wildcat drilled in the region since the late 1980s.

The Benreoch charter term is for 5 months. Isramco has an option to extend the contract for two more wells. One would be a 60-80 day reentry and workover of a well drilled in the late 1980s on the same license. The other would be a second wildcat on the block.

Givot Olam Oil Ltd., Jerusalem, planned to drill on Israeli License 242 and looked for a drilling contractor to start work as 1994 neared. The company intends to drill one well to 4,000 m and may drill a second to 5,500 m.

LEBANON

CAPTIAL: Beirut
MONETARY UNIT: Lebanese Pound
REFINING CAPACITY: 37,500 b/cd
PRODUCTION: 0
RESERVES: 0

The ministry of Industry & Petroleum completed its first offshore survey in more than 20 years.

Geco-Prakla, Crawley, U.K., acquired 500 line km of data off nothern Lebanon using the Bin-Hal 511 survey vessel.

Geco-Prakla planned to present results at a series of meetings scheduled early in 1994. The ministry was to announce exploration licensing arrangements at those seminars.

OMAN

CAPITAL: Muscat
MONETARY UNIT: Omani Riyal
REFINING CAPACITY: 76,000 b/cd
PRODUCTION: 774,800 b/d
RESERVES: 4.7 billion bbl

Exploration activity is stirring in Oman. Occidental of Oman Inc. tested an oil discovery on its 1.6 million acre Suneinah block in Northwest Oman. And three independents plan a wildcat on an undrilled block offsetting Oxy's acreage.

Oxy's 1 Wadi Latham flowed at a rate of 2,439 b/d of 40° gravity oil and 2.2 MMcfd of gas through an 80/64 in. choke from a 201 ft section of lower Cretaceous Shuaiba limestone at 6,719-6,920 ft in a medium radius horizontal hole. Oxy operates the well with a 65% interest, with Finland's Neste Oy holding the rest.

The site is near the southern boundary of the block about 20 miles east of Al Barakah field, discovered by Oxy in 1991. Oxy produces about 32,000 b/d of oil from Al Barakah and Safah fields. Safah is a 1983 discovery.

Oxy drilled horizontal wells in developing both fields. The most recent producing well drilled in Safah attained a medium radius horizontal section of 4,944 ft, a record in the Middle East, the company said.

Meantime, Arakis Energy Corp., Vancouver, B.C., and Aegis Energy Inc., Lafayette, La., Signed letters of intent with American Exploration Co., Houston, to participate in a joint venture wildcat on American's Jebel Aswad block, also known as Block 15.

The combine's 1 Jebel Aswad, which was scheduled to spud in November 1993, will test middle Cretaceous Marib and Shuaiba. American said drilling to a proposed 11,500 f will take 50-60 days at a gross cost of about $4 million. American acquired Conquest Exploration Co. in February 1991, which had a production sharing agreement covering the 346,000 acre block. A unit of American will operate the well, while Arakis and Aegis will have 50% and 20% working interests, respectively.

Study scheduled for Omani—Indian gas pipeline

Oman has formed a group to study construction of a 1.8-. bcfd subsea pipeline to transport Omani gas to India. If built the $4 billion Oman-India line could play a key role in India' plan to press ahead with economic liberalization.

The project is part of an Omani campaign to create new outlets for its gas reserves, to which more than 20 tcf hav been added with recent deep discoveries at Barik, Sai Nihayda, and Saih Rawl.

The proposed gas line is to be laid in three segments alon one of these routes:

- A direct, deepwater route of about 900 miles, consisting either a single 38 in. line or two 28 in. lines form Sur, o Oman's eastern coast across the Gulf of Oman and Arabia Sea to a point on India's western coast.
- A shallow water route consisting of about 1,250 miles 42 in. pipe from a point near Sohar on the Batinah coas north across the mouth of the Persian Gulf, to India along th continental shelf off Iran and Pakistan.

Obstacles to pipeline construction along either route a formidable. Water depths along the direct route in son places exceed 3,000 m, and the pipeline would have to cross subsea mountain range as well as the mouth of the Indu River. The longer but more shallow route would requi

SAUDI ARABIA

CAPITAL: Riyadh
MONETARY UNIT: Riyal
REFINING CAPACITY: 1,614,000 b/cd
PRODUCTION: 7.981 million b/d
RESERVES: 258.703 billion bbl

The first project in a decade long program to expand and upgrade Saudi Arabia's refining network was scheduled to go in the field during October 1993 at Saudi Arabian Marketing & Refining Co.'s (Samarec) Yanbu refinery. It will be the first step in the conversion of this basic hydroskimmer to a classic reformulated fuels refinery.

Projects at other refineries in the kingdom will follow at a pace that could put $3.5-4.5 billion worth of work on site by the end of first quarter 1994. Completion of the Samarec projects is planned for 1997, the company says.

Saudi Arabian Oil Co.'s (Saudi Aramco) Ras Tanura refinery is also a major target of the long term upgrading program, which has been designated the Kingdom refinery upgrading program (KRUP). The new units at Ras Tanura have been sized and the technology selected. In mid-1993, managing contractor Brown & Root Braun was finalizing a lump sum turnkey contract at its Alhambra, Calif., offices covering design, procurement, and construction for approval by Saudi Aramco's executive committee. No official details had been released by summer 1993 on the specific configuration or timing of this revamp. The picture emerging is that KRUP will be stretched out more than originally anticipated. But the initial investment still makes this one of the world's largest single, active programs and will put Saudi Arabia on the road to big league refining. This is only the down payment on a master plan that calls for a total outlay of $16 billion during the next 10-12 years.

A large part is to go to futher expansion and upgrading of the Ras Tanura refinery plus investments in three joint venture refineries in the kingdom. Two of them have relatively good conversion capability, but a third is a large hydroskimmer that is ripe for upgrading.

The master plan is part of a two pronged strategy that is making the world's premier holder of oil reserves a power in international refining. In additon to domestic upgrading and expansion, the other element is joint refining ventures outside Saudi Arabia, so far in the U.S. and Korea. A Japanese partnership, with new refineries, is also a strong possibility.

The American partnership with Texaco Inc., Star Enterprise, got Saudi Aramco an interest in U.S. service stations and 615,000 b/d of capacity in three refineries. In South Korea, Saudi Aramco has acquired a 35% interest in Ssangyong Oil Refining Co., which operates a 145,000 b/d topping refinery at Onsan. Current action focuses on Samarec refineries at Yanbu, Riyadh, and Jeddah and Saudi Aramco's Ras Tanura refinery. Samarec doesn't operate the Ras Tanura plant but coordinates its output and handles its domestic and export product distribution. Samarec has selected Foster Wheeler Corp. as its managing contractor.

Licensors for the new process units at the Yanbu, Riyadh, and Jeddah refineries have been selected, definitive scoping has been completed, and cost estimates prepared on a ±25% basis. Cost estimate at Yanbu is $1.8 billion ($1.35-2.25 billion), Riyadh $541 million ($406-676 million), and Jeddah $478 million ($359-598 million).

Samarec will kick off the Yanbu upgrading project with a fast track project to debottleneck the crude unit. It will raise capacity to 300,000 b/d from 170,000 b/d.

Saudi Aramco is proceeding cautiously with its upgrading program. The company says its Ras Tanura refinery will be rebuilt to the 530,000 b/d level it had before a fire in 1990. The complex, long term replacement project will include a fluid catalytic cracker (FCC) and a hydrocracker.

Saudi domestic consumption has been running at 850,000 b/d and increasing at rates of 3-5%/year.

In 1993, Saudi Aramco's executive committee evaluated several options including increasing capacity of the Ras Tanura hydroskimmer to 530,000 b/d plus adding, among other things, a 90,000 b/d visbreaker, a 110,000 b/d FCC, a hydrocracker, 75,000 b/d of alkylation capacity, and state of the art control and utility systems.

The committee decided that the initial upgrade would not include a new crude unit, thus maintaining capacity at 300,000 b/d. Downstream units, though, would include a UOP/Unocal hydrocracker and UOP continuous catalytic reformer (CCR), but not an FCC. The decision will not preclude Ras Tanura from later boosting capacity to 530,000 b/d with the matching conversion capacity, which would include the FCC.

Work at Ras Tanura consists of three packages. The first is for the north refinery and includes water treatment, instrument air, power generation, the CCR, and a hydrotreater. The second package will cover the south refinery and includes the 40,000 b/d hydrocracker, a hydrogen plant, and a visbreaker. Each package is estimated to cost $500-600 million. The projects will be spread over 5 years, although normally such projects require 4 years, a contractor estimates.

The third package will cover a new distributed control system for the entire refinery, which is estimated to cost $200 million.

Across the kingdom from Ras Tanura on the Red Sea is the 300,000 b/d Petromin Petrola Rabigh Refinery Co. refinery at Rabigh. This plant was orginally supposed to be upgraded with Ras Tanura in Phase 2 of the upgrading program. The second phase was to begin 3 years after Phase 1, which covers the three Samarec refineries. The Ras Tanura timetable has apparently been moved up, essentially merging Phases 1 and 2.

But the Rabigh refinery appears to have dropped out of the upgrade picture for now. It is a joint venture of Samarec and Greece's Petrola International. It has no catalytic reformer or conversion capacity and therfore offers plenty of opportunity for upgrading. This refinery started up in 1990 after many delays and then had a fire in 1991 that cut capacity to its current level from 325,000 b/d.

Horizontal drilling taps more oil

Horizontal wells have eliminated gas and water coning problems and increased oil production to improve ultimate recovery in several large Saudi fields. Horizontal wells have produced oil at rates two to 10 times that of offset vertical wells. As horizontal drilling becomes more common, the cost of horizontal wells will decrease to near that of vertical wells in the Middle East.

Saudi Aramco is moving forcefully into horizontal drilling. The technical advances contributing to horizontal well success include new polycrystalline diamond compact bits, conventional bits modified for high speed motor drilling, steerable positive displacement motors, measurement while drilling, logging while drilling, innovative drilling fluids, and downhole adjustable stabilizers.

In Saudi Arabia, 13 horizontal wells have been drilled in carbonates and 8 wells in sandstones, with another 8 wells under way during first quarter 1993.

The first two horizontal wells were completed in carbonate reservoirs offshore in 1991. Saudi Aramco says it has several major horizontal well applications:

- Control of coning and edge water influx in relatively thin remaining oil column areas.
- Improvement of waterflood sweep efficiency.
- Production improvement in thin and tight reservoirs.

Each of these is also part of Saudi Aramco's plan for total reduction in development costs.

During the next 10 years, Saudi Aramco's horizontal drilling program will make up half of the company's total drilling budget.

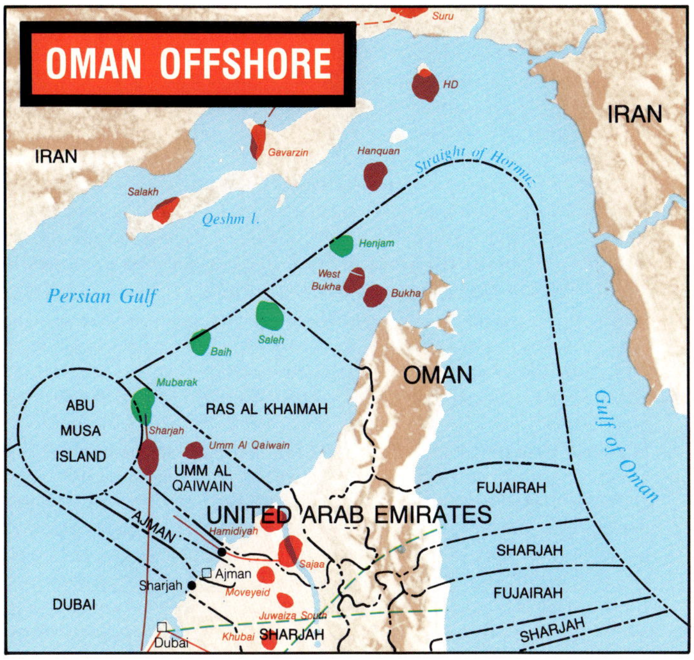

It says a 5-10% improvement in oil recovery by application of horizontal drilling technology would translate into a 12.5-25 billion bbl increase in reserves in its fields.

Horizontal drilling technology, Saudi Aramco says, has yielded numerous benefits to date for the company. It has improved well capacity by 150-400%, reduced the total number of wells by 30%; reduced total drilling, flow line, and facilities costs by 20-25%; reduced water and gas handling costs by up to 50%; improved recovery of original oil in place by 5-10%; and deferred installation of artificial lift and gas compression facilities.

Most of Saudi Aramco's horizontal drilling experience is derived from directional work on offshore platforms in giant Safaniya, Zuluf, Marjan, and Berri fields. As of 1992, Saudi Aramco drilled successful horizontal wells in Berri, Zuluf, Abqaiq, and giant Ghawar fields.

Because of the relative immaturity of many of Saudi Aramco's fields, the company can apply horizontal technology very early in its fields.

Well results disclosed in new onshore area

Saudi Aramco has detailed test rates for three wells related to its Medin oil, gas, and condensate discovery in Northwest Saudi Arabia.

The field is the first discovery on the northern Red Sea coastal plain.

Saudi Aramco plans further drilling and seismic surveys to further delineate extent of reserves.

The 1 Medin wildcat discovered the reservoir in November 1992 and flowed 45 MMcfd of sweet gas and 1,300 b/d of 39° gravity condensate through a 1 ¾ in. choke with 1,100 psi wellhead pressure from tertiary Miocene limestone at 7,300-7,732 ft.

Delineation well 2 Medin, spudded Nov. 30, 1992, began production tests Jan. 14, 1993. It cut the same reservoir 200 ft higher in the structure than 1 Medin. It tested 55 MMcfd of gas and 1,900 b/d of 58° gravity condensate through a 1 ¾ in. choke with 1,200 psi wellhead pressure. The 3 Medin well, spudded Jan. 23, 1993, on Mar. 4, 1993, flowed 2,300 b/d of 30° gravity sweet crude oil and 19 MMcfd of sweet gas through a 1 in. choke with 1,315 psi wellhead pressure from an interval at 7,760-7,845 ft.

Hydroskimming refinery operated by Saudi Arabian Oil Co. (Saudi Aramco) at Ras Tanura will be upgraded as part of Saudi Arabia's refinery modernization program. Photo courtesy of Saudi Aramco.

UNITED ARAB EMIRATES

CAPTIAL: Abu Dhabi Town
MONETARY UNIT: Dirham
REFINING CAPACITY: 192,500 b/cd
PRODUCTION: 2.189 million b/d
RESERVES: 98.1 billion bbl

Natural gas projects are moving toward center stage in the emirates. Abu Dhabi, U.A.E.'s biggest oil producer, is preparing to boost liquefied natural gas exports along with oil productive capacity.

Dubai is increasing its capacity to import gas from neighboring Sharjah, where an Amoco Corp. affiliate is increasing exploration around a recent gas/condensate discovery. Dubai's growing gas needs result in part from a pilot miscible gas injection project that was scheduled to begin in offshore Fateh oil field in spring 1993.

A third LNG processing train on Abu Dhabi's Das Island is scheduled to start up in 1994. The train, which Abu Dhabi Gas Liquefaction Co. Ltd. (Adgas) calls the world's biggest, will double Das Island's production capacity to nearly 5 million metric tons/year. Adgas is building the train to meet obligations under an expanded contract with Tokyo Electric Power Co. Inc. The orginial contract calls for delivery of 2 million tons/year of LNG and 800,000 tons/year of LPG through 1995.

Air Products & Chemicals Inc. is the cryogenic technology licensor for the third train, Chiyoda of Japan the main contractor. The facility will have three steam turbine driven refrigeration compressors, each developing 36,000 kw. The third train will liquefy 430 MMcfd of gas. Some of the gas will come from nonassociated Permian Khuff reserves in offshore Abu Al Bukhoosh field on the border with Iran. Total Abu Al Bukhoosh (Total-ABK) operates the field.

Khuff gas had been considered for gas lift in Abu Al Bukhoosh. Under current plans, all will go to Das for liquefaction. Khuff nonassociated gas also is under development in Umm Shaif oil field operated by Abu Dhabi Marine Operating Co. (ADMA-OPCO). Some Khuff production is on stream and is being injected for pressure maintenance. Umm Shaif Khuff supplies will supplement Abu Al Bukhoosh volumes to Das. ADMA-OPCO produces 450,000-500,000 b/d of oil from Umm Shaif and Lower Zakum fields. Late in 1992 it commissioned a 37 km, 36 in. oil trunk line from the Umm Shaif complex to Das Island and linked a new platform to an existing 30 in. trunk line to the island.

Umm Shaif associated gas and nonassociated gas from the Uweinat gas cap, along with associated gas from Lower Zakum field, feed the original LNG trains at Das Island. State owned Abu Dhabi National Oil Co. (Adnoc) owns 60% of ADMA-OPCO. Other owners are British Petroleum Co. plc 14.67%, Total 13.33%, and Japan Oil Development Co. 12%.

Most of Abu Dhabi's offshore oil production capacity additions are taking place in Upper Zakum field, where Zakum Development Co. (Zadco) has boosted the platform total to 84 and continues to drill. Upper Zakum crude travels 50 km to Zirku Island for stabilization, storage, and shipping. Adnoc owns 88% of Zadco, Japan Oil Development 12%.

In addition to production from Upper Zakum, Zadco handles production from Umm Al Dalkh and Satah fields, which were developed by a separate joint venture of Adnoc and Japan Oil Development called Umm Al Dalkh Development Co. (Udeco). Zadco absorbed Udeco's activities in 1988.

Umm Al Dalkh production, in which Adnoc owns an 88% interest and Japan oil Development 12%, moves to Zirku via upper Zakum facilities. Production from Satah field—Adnoc 60%, Japan Oil Development 40%—goes directly to Zirku. Remaining oil and gas production capacity additions are taking place onshore, where Abu Dhabi Co. for Onshore Operations is the operator.

Dubai enhanced oil recovery pilot

Dubai Petroleum Co. (DPC) scheduled start-up of injection operations in April 1993 for a 3 year miscible gas pilot enhanced recovery project in offshore Fateh field. The project will use about half of Fateh's production of rich natural gas, which will be injected intermittently under pressure to mix with oil in the upper Cretaceous Mishrif formation. Injected water will alternate with gas.

DPC estimates Fateh's recovery factor at more than 50%. Production attributable to a waterflood that began in 1972 has been about twice that of primary production. Enhanced recovery might add a recovery factor increment of about one-half that of the waterflood. If successful, the pilot will be commercial by itself. EOR pilot work so far has included a 3D seismic survey and installation of a small platform, designated ZH. In addition, DPC has drilled five wells, two for production and three for observation. Three existing wells, one a former producer, will inject gas for the pilot.

Under current plans, 3 years of evaluation and whatever construction may be required will follow 3 years of pilot production. DPC, however, can advance the schedule.

Fateh gas production now goes by pipeline to Jebel Ali, an industrial area south of the city of Dubai, by way of a 90 km pipeline. After processing by Dubai Gas Co., dry gas returns to the field as necessary for fuel. Jebel Ali industrial facilities use most of the production. Commercial Fateh gas injection would use all the field's current gas flow and require supplementary supplies.

Dubai can import as much as 300 MMcfd of gas from Amoco Sharjah Oil Co.'s onshore fields in Sharjah. The gas travels from Amoco's. Shalco gas processing plant through a 24 in. pipeline to Jebel Ali. In January 1993, Dubai began imports of gas from Mubarak field off Sharjah. The gas goes to Jebel Ali through a 90 km, 16 in. pipeline from a new platform in the field, operated by Crescent Petroleum Co. International Ltd. The pipeline, completed in August 1992, has capacity of 100 MMcfd.

In July 1992, DPC and McConnell Dowell Middle East commissioned a 60 km, 20 in. pipeline to Jebel Ali from ARCO Dubai Inc,'s onshore Margham retrograde condensate field. With a 16 in. parallel pipeline already in place, the new pipeline boosts deliverability of Margham production to Jebel Ali to 500 MMcfd. The new line's capacity is 300 MMcfd. Margham liquids are extracted at the Margham gas processing plant and exported via Jebel Ali. Dry gas is reinjected into the formation but is available for industrial use when needed. The pipeline, owned by Dubai Supply Authority, was commissioned within 100 days of project approval.

Sharjah active

Amoco's expanding exploration program and the start of exports of offshore gas to Dubai have lifted activity in Sharjah. The 2 Kahaif gas/condensate discovery that Amoco reported late in 1992 is on trend with the company's Moveyeid and Sajaa gas and retrograde condensate fields to the north and ARCO's Margham field to the south in Dubai.

The discovery flowed at rates as high as 73 MMcfd of gas and 1,615 b/d of condensate through a $^{56}/_{64}$ in. choke with 5,200 psi flowing tubing pressure from 700 ft of Thamama pay.

The productive onshore Sharjah-Dubai trend is a thrust belt that runs parallel to the Persian Gulf coast, analogous to the Overthrust Belt of the U.S. West. Amoco, which holds its Sharjah acreage under concession agreements, drilled the Khaif strike based on 3D seismic data interpretation. It's part of a drilling program that began late in 1991 and has so far included seven wells in Sajaa and Moveyeid fields, the two exploration wells, and a Kahaif delineation well. The two budgeted wildcats are part of the program.

Before the current drilling program began, there had not been a well drilled in Sharjah in about 6 years.

Sajaa and Moveyeid fields produce 600 MMcfd of gas and 38,000 b/d of condensate. In additon to the export line to Dubai, the fields have pipeline connections to the Emirates General Petroleum Co. system and Ras Al-Khaimah.

The Shalco LPG plant, with nameplate inlet capacity of 500 MMcfd, extracts 13,000 b/d of propane and butane, which with condensate are exported to the Far East from a terminal at Hamriyah. Amoco is considering an expansion of the LPG plant to a nameplate capacity of 700 MMcfd and has let contract to Fluor Corp. to study costs.

Offshore, Crescent Petrileum shut down Mubarak field in November 1992 to accommodate installation of the platform now producing gas for export to Dubai. The platform has facilities to separate and stabilize condensate and to dry and compress gas. It caps a development program based on Mubarak's Thamana gas/condensate reservoir and including the drilling of two horizontal wells.

The first Thamama horizontal well, completed in mid-1992 reached total depth of 15,591 ft with an 870 ft horizontal section. Crescent drilled the second well directionally to 17,400 ft, of which 800 ft was horizontal in Thamama pay.

The well flowed more than 45 MMcfd of gas and 6,300 b/d of condensate with wellhead flowing pressure of 2,181 psi on a restricted choke. Production potential is 80 MMcfd.

VAM®, the ideal Premium Connection, has what it takes to follow your directions. Whether you are planning horizontal drilling programs, or wells with severe deviations, look to VAM® for the muscle to get you to TD safely—and with cost-saving, trouble-free running.
"The VAM® System" is more than a wide array of advanced thread designs, it is your assurance of broad technical and product support from a worldwide network of plants—the largest and most comprehensive in the tubular industry.
"The VAM® System" also provides R&D offices and laboratories working together in several countries, field engineers ready to assist your teams, and convenient repair shops for the fast delivery of accessories.
Vallourec's cutting-edge technology is constantly striving to meet your application and economic needs be developing new and improved tubular products.
vallourec industries
Oil and Gas Market Division
130 rue de Silly
92100 Boulogne-Billancourt, France
Tel: (33-1) 49.09.35.00
Telex: Valre 633 601 PTV+
Fax: (33-1) 49.09.37.13
CERTIFICAT 1993/1349
ISO 9001
AFAQ
VAM®
WE FOLLOW YOUR DIRECTIONS
Our Technology Makes the Difference
OGD-941

Latin America

ARGENTINA

CAPITAL: Buenos Aires
MONETARY UNIT: Austral
REFINING CAPACITY: 709,485 b/cd
PRODUCTION: 574,800 b/d
RESERVES: 1.6 billion bbl

Argentina continued its rapid pace of petroleum industry privatization.

The Argentine government received bids totaling $4.2 billion for the restructured, privatized assets of Argentina's gas industry, reported N.M. Rothschild & Sons Ltd., London.

A group led by British Gas plc bought the distribution company that supplies gas to Buenos Aires, and Nova Corp., Calgary, leads a group that was buying 70% of Argentina's northern natural gas pipeline system.

Gas privatization

The privatization of Argentina's state gas utility, Gas del Estado, was completed late in 1992. The utility had been divided into two national gas transportation companies and eight regional distribution companies.

BG's group secured for $300 million a 70% share in the regional Distribuidora de Gas Metropolitana, which supplies 1.7 million consumers in Buenos Aires. Gas de Buenos Aires Norte, which supplies 850,000 consumers, was sold to Repsol Petroleo SA unit Gas Natural for $155 million.

British Gas has a 41% interest in the group and acts as operator. Partners are Argentine conglomerate Perez Companc 25%, Astra Cia. Argentina de Petroleo SA 20%, and Invertrad SA, an Argentine steel company, 14%.

As many as one third of the remaining 30% of the shares in Metropolitana will be allocated to Gas del Estado employees, British Gas said, while the balance of shares will be held by the Argentine government for later sale.

Nova purchase

The Nova group was to pay $248 million for a 70% interest in Argentina's northern gas pipeline system. Nova was to pay $37.5 million for its 15% share in the group.

Other members are Soc. Comercial del Plata and Techint Cia. Tecnica International, both of Argentina, and Malaysia's Petroliam Nasional Bhd.

Nova participates through its Novacorp International Consulting Inc. and Novacorp International SA units. The latter has been named technical operator of the 2,485 mile system that includes 12 compressor stations.

Enron Corp. submitted the winning bid of $561 million for a 70% interest in Transportadora de Gas del Sur SA, operator of the southern Argentina pipeline system.

Northwest basin awards

Meantime, international companies were acquiring Northwest basin assets, capitalizing on the privatization of YPF and tender offerings by the country.

A group led by Pluspetrol SA was awarded through international tender the 4.2 million acre Palmar Largo exploration tract in the Northwest basin, previously operated by YPF.

The Palmar Largo Unit holds proved reserves of 17 million bbl of crude and production of 7,400 b/d.

In a separate deal, Norcen International acquired a 100% interest in a 3.09 million acre block off Tierra del Fuego in southern Argentina. The block will be the first in Argentina in which Norcen acts as operator.

Ampolex bid

Ampolex Ltd., Sydney, Australia, successfully bid for a joint interest in the Agurague Unit in the Northwest basin near the Bolivian border.

The assets were purchased from YPF as part of its privatization.

The Aguarague Unit covers 17,000 sq km, contains three gas/condensate fields including Aguarague field, and pro-

SOUTH AMERICA
ARGENTINA
No. REFINERIES
Zamora 1
Compana 1
Galvan 1
Bahia Blanca 1
Buenos Aires 1
Duran 1
Dock Sud 1
LaPlata 1
Lujan deCuyo 1
Plaza Huincul 1
Quilmas 1
San Lorenzo 1
0 100 200 300 400 Mi.
0 200 400 600 Km.
MEXICO
BELIZE
GUATEMALA
HONDURAS
EL SALVADOR
NICARAGUA
COSTA RICA
PANAMA
CUBA
HAITI
DOMINICA REPUBLIC
PUERTO RICO
JAMAICA
TRINIDAD AND TOBAGO
Caribbean Sea
VENEZUELA
GUYANA
SURINAME
GUIANA
COLUMBIA
ECUADOR
PERU
BRAZIL
BOLIVIA
PARAGUAY
URUGUAY
CHILE
ARGENTINA
TIERRA DEL FUEGO
Pacific Ocean
Atlantic Ocean
Cabaiguán
Santiago de Cuba
Bonao
Port-au-Prince
Santo Domingo
Kingston
St. Croix
St. John's
Fort de France
Bridgetown
Point Fortin
Port of Spain
Caracas
Puerto La Cruz
Maracaibo
Barranquilla
Cartagena
Santa Marta
Colón
Panama
Managua
San José
Georgetown
Paramaribo
Cayenne
Bogota
Medellin
Cali
Quito
Esmeraldas
Tumaco
La Libertad
Talara
Iquitos
Lima
La Pampilla
Conchan
La Paz
Cochabamba
Santa Cruz
Sucre
Arica
Manaus
Belém
Fortaleza
Recife
Maceió
Aracaju
Salvador
Brasilia
Belo Horizonte
Vitoria
Campos
Rio de Janeiro
São Paulo
Santos
São Sebastião
Curitiba
Sao Francisco do Sul
Laguna
Canoas
Porto Alegre
Rio Grande do Sul
Asunción
Montevideo
Buenos Aires
La Plata
San Lorenzo
Cordoba
Mendoza
Santiago
Valparaiso
Concepción
Bahia Blanca
Puerto Rosales
Comodoro Rivadavia
Rio Gallegos
Rio Grande
Cape Horn
Falkland I.
Golfo de San Matías
BRAZIL
ALAGOAS
SERGIPE
BAHIA
Maceió
Aracaju
Camacari
Aratu
Salvador
0 50 100 Mi.
0 50 100 Km.
BRAZIL
Campos
Albacora
Marlim
Namorado
Enchova
Pampo
Linguado
Badejo
0 50 100 Mi.
0 50 100 Km.

SOUTH AMERICA, SOUTHERN PORTION

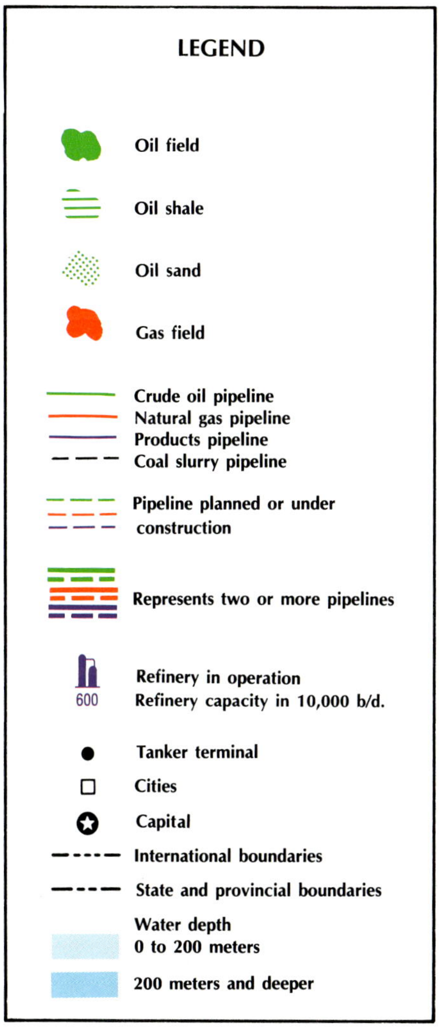

duces 160 MMcfd of gas and 9,000 b/d of liquids.

Ampolex's bid was $261 million for a 100% interest in the assets. It will own a 20-25% interest in the project.

Plans for the unit include a 7 year, $50 million exploration program.

Refinery modernization

Shell Cia. Argentina de Petroleo SA let a $15.7 million turnkey contract to Foxboro Co., Foxboro, Mass., for the second phase of the control room and field instrumentation modernization project at its 120,000 b/d Buenos Aires refinery. The contract includes three I/A Series system control nodes to manage crude unit, hydrotreater, Platforming, and FCC plants, plus intelligent transmitters and interfaces to programmed logic controllers and the plant information system.

BRAZIL

CAPITAL: Brasilia
MONETARY UNIT: New Cruzado
REFINING CAPACITY: 1.3 million b/cd
PRODUCTION: 631,300 b/d
RESERVES: 3.6 billion bbl

Petroleos Brasileiro SA continues to increase production and advance deepwater technology in the Campos basin off Brazil.

Petrobras placed on stream the first of four wells in the pilot early production project for Bijupira and Salema oil fields in the Campos basin off Rio de Janeiro. Petrobras used new technology to lay a hybrid rigid/flexible pipeline to transport crude from the two fields that set a world record for deepwater pipelines and resulted in significant cost savings.

Meantime, rising production from the Campos basin's Albacora and Marimba fields has pushed Brazilian oil production to a record high.

In addition, Petrobras drilled another sizable oil discovery in the basin.

Bijupira, Salema on stream

Petrobras started production from the Salema/Bijupira pilot system at 7,500 b/d in a project that cost about $83 million.

Plans called for production to build to 20,000 b/d of 28-30° gravity oil and 8.75 MMcfd of gas by January 1994. The two fields lie in water depths of 1,640-2,460 ft.

Petrobras is using the upgraded Petrobras XIII semisubmersible as the SS-20 floating production system (FPS) in Bijupira field to handle production from both fields. The FPS is installed in 2,066 ft of water.

Some Salema/Bijupira oil is shipped from the FPS via pipeline manifold installed in Corvina field to a fixed platform in shallower water in Cherne field using a 17 km, 8 in. pipeline. Petrobras laid the line with an innovative hybrid rigid/flexible recovery/lay method from the FPS to the Corvina pipeline manifold.

The rest of Salema/Bijupira crude moves via pipeline manifold in Corvina field to a nearby single point mooring buoy.

Salema/Bijupira gas moves through a 10 in., 21 km pipeline to the Marimba field platform.

Reserves are estimated in Bijupira at 120 million bbl of oil and 56 bcf of gas and in Salema at 35 million bbl and 28 bcf.

Production record

Petrobras' crude oil production reached 710,232 b/d in October 1993 with start-up of more wells in Marimba and Albacora fields.

That record was expected to fall in succeeding months, when Caravelas field in the Santos basin starts up. Initial production is expected to be 10,000 b/d.

Petrobras achieved the record production level by increasing productivity of Campos basin fields, which produced at 97.1% of capacity, highest in the state owned petroleum company's history.

Brazilian production rose by about 10% during the 12 months ended in September 1993, or an increase of more than 65,000 b/d.

At the same time, Petrobras jumped its reserves estimate by about 4% during first half 1993. Proved reserves in water depths to 1,000 m totaled 4.6 billion bbl of oil equivalent (BOE). If probable reserves in water depths of 1,000-2,000 m are considered, that number jumps to 11 billion BOE.

Campos discovery

Petrobras drilled its 4 RJS 396D wildcat about 10 km from giant Marlim field and 100 km offshore in 2,296 ft of water.

The discovery well's productive capacity was estimated at 3,000 b/d of oil.

Pay sands are at 8,200-256 ft. Initial estimates pegged potential reserves at 20 million BOE.

It was the 1,000th well Petrobras had drilled in the offshore Campos basin since exploratory drilling started there in 1971.

Parecis Plateau basin wilcat

Petrobras spudded the first wildcat in the unexplored Parecis Plateau basin of Mato Grosso and Rondonia states of Southwest Brazil.

Site is on the large Fazenda Itamarati cotton plantation, about 160 km northwest of Cuaiba. The activity stirred complaints about the lack of royalties for private landowners in Brazil by plantation owner and frequent Petrobras monopoly critic Olacir de Moraes. Petrobras has budgeted $8 million for Parecis Plateau basin exploration.

Amazon gas processing plant

Petrobras started up the first gas processing plant in the Amazon region of Brazil.

Propak Systems Ltd., Airdrie, Alta., under an $8.8 million Petrobras contract, installed the plant to handle liquids associated with natural gas produced from Rio Urucu oil and gas field.

Start-up operations began early in 1993.

Petrobras found the field in 1986 in the Solimoes basin of Amazonas state, about 600 km southwest of Manaus, and has spent more than $150 million on exploration and development and related facilities.

The environmentally sensitive Urucu region is one of the most dense, remote jungles in the world.

Production, market concerns

Urucu oil production averages about 11,000 b/d.

Urucu gas was 100% flared at first. Reinjection began in 1991, but excess deliverability of about 10 MMcfd has meant some flaring persists.

Petrobras was considering options for marketing Urucu gas, with a productive capacity of about 35 MMcfd, including a possible pipeline to Porto Velho for power generation.

More likely is a low temperature liquefied natural gas terminal along the Solimoes River that could tie in the much bigger gas reserves in the nearby Jurua River area, pegged some years ago at 282 bcf. Solimoes LNG could be transported via river vessel, avoiding the costly and environmentally contentious jungle pipeline.

Petrobras in 1989 estimated Rio Urucu's reserves at 85 million bbl of oil equivalent.

BOLIVIA

CAPITAL: La Paz
MONETARY UNIT: Boliviano
REFINING CAPACITY: 45,250 b/cd
PRODUCTION: 20,300 b/d
RESERVES: 108 million bbl

BHP Petroleum (Bolivia) Inc. took a farmout on Pan Andean Resources plc's Chapare exploration block in Central Bolivia.

BHP initially takes a 30% stake in the 3.5 million acre tract, with Pan Andean, London, retaining 20% and state owned Yacimentos Petroliferos Fiscales Bolivianos (YPFB) holding 50%.

YPFB awarded a 30 year operations contract on the block to Pan Andean in May 1993. The Chapare block offsets acreage operated by Maxus Bolivia Inc., declared commercial with 30,000 b/d of oil production planned by yearend 1994.

Pan Andean said there were four discoveries in four wells drilled on the Maxus block and YPFB acreage within 40 km of the Chapare block.

The farmout commits BHP to a $6 million seismic survey and data processing program in the first 2 years. BHP planned to begin acquiring seismic data in August 1993, with the aim of completing a 400 line km program by yearend 1993.

Mamore Block commercial

Maxus Energy Corp., Dallas, declared its Mamore Block in Bolivia commercial following long term tests of the 1 and 2 Surubi wells, which were drilled in 1992 into a combined 209 ft of net pay.

Maxus drilled a third well on the block and was preparing another drillsite. Maxus holds a 100% interest in the block, 96 miles northwest of Santa Cruz. Production was to begin in 1993 after pipeline and marketing arrangements were complete.

CHILE

CAPITAL: Santiago
MONETARY UNIT: Peso
REFINING CAPACITY: 164,615 b/cd
PRODUCTION: 14,300 b/d
RESERVES: 300 million bbl

Refineria de Petroleo Concon, Concon, Chile, let contract to Advanced PetroGas Systems Inc., Houston, to engineer, design, and fabricate a modularized 30 metric ton/day sulfur recovery unit at its 62,700 b/cd refinery at Concon.

Unit start-up was scheduled for second quarter 1994.

Strait of Magellan operations

Coastal Petroleum NV began bunkering operations for clients using the Strait of Magellan off southern Chile by offering 245,000 bbl of storage capacity for marine gas oil at Punta Arenas and Puerto Gregorio.

Deliveries were available by truck, pipeline, or vessel.

Intermediate and bunker C fuel oil are available.

Counting the new sites in Chile, Coastal offers bunkering services at 25 ports throughout the Americas.

COLOMBIA

CAPITAL: Bogota
MONETARY UNIT: Peso
REFINING CAPACITY: 248,850 b/cd
PRODUCTION: 449,800 b/d
RESERVES: 1.9 billion bbl

Development of Colombia's Cusiana and Cupiagua oil fields by BP Exploration Co. (Colombia) Ltd. depends strongly on export pipeline capacity.

There is limited pipeline capacity short term. So the fields will be developed in two phases.

About 10,000 b/d was being produced on long term test in 1993. Oil moves through an Empresa Colombiana de Petroleos (Ecopetrol) 12 in. pipeline running from Araguanay to the Covenas terminal on the Caribbean Sea coast.

Production was to stay at this volume until mid-1994.

The Cusiana acreage is one of several blocks BP holds in the Northeast Colombia.

Cusiana field lies near a 40,000 b/d Ecopetrol pipeline from Araguanay, which meets the 75,000 b/d Central Llanos pipeline at El Porvenir. This in turn joins the 150,000 b/d Oleoducto de Colombia (ODC) pipeline at Vasconia and continues northwest to Covenas.

To step up Cusiana oil flow production to 150,000 b/d by the end of 1995, as Phase 1 of development, BP plans to lay a 20 in. line from Cusiana to El Porvenir and a 30 in. loop to upgrade the Central Llanos section of the ODC pipeline to 220,000 b/d. First phase pipeline costs are estimated at $300 million.

Beyond 1995, BP is considering three options: laying a new pipeline, further upgrading of the Araguanay-El Porvenir-Covenas route, or tying into Occidental Petroleum Corp.'s 230,000 b/d Cano Limon-Covenas pipeline, which passes north of Cusiana. Pipelines for Phase 2 development will cost more than $1 billion gross, BP said.

Production

Before Cusiana's discovery, Colombia had reserves of less than 2 billion bbl of oil and 3.5 tcf of gas.

BP estimated Cusiana oil reserves at 1.5 billion bbl and Cupiagua at 500 million bbl and is not putting a figure on gas reserves because gas is recycled to aid oil production.

Juan Maria Rendon, president of Ecopetrol, said Cusiana and Cupiagua field reserves will at least double oil and gas reserves for Colombia. Although Cusiana gas has not been booked as reserves, a BP official confirmed Rendon's estimate of at least 3.5 tcf gas reserves in Cusiana.

UNITED STATES
TEXAS
LOUISIANA
New Orleans
Houston
Corpus Christi
Laredo
Brownsville
MacAllen
El Paso
Ciudad Juarez
Celulosa de Chihuahua
CHIHUAHUA
Chihuahua
Delicias
Cuidad Camargo
Jiménez
MEXICO
SINALOA
Culiacan
DURANGO
Durango
Gomez Palacio
Torreón
Parras
Saltillo
COAHUILA
Nueva Rosita
Sabinas
Nuevo Laredo
Monclova
Monterrey
Apodaca
Cadereyta
NUEVO LEON
Reynosa
Matamoros
San Fernando
TAMAULIPAS
Soto la Marina
Ciudad Victoria
ZACATECAS
Zacatecas
Mateahula
SAN LUIS POTOSI
San Luis Potosi
Mazatlán
NAYARIT
San Blas
Chicura
Canuch
Jayabo
Aguascalientes
Lagos
Leon
Guanajuato
GUANAJUATO
QUERETARO
Querétaro
Irapuato
Salamanca
Celaya
Guadalajara
Puerto Vallarta
JALISCO
Autlán
La Barca
COLIMA
Manzanillo
Tecoman
Lago de Cuitzeo
Morelia
MICHOACAN
Toluca
MEXICO
Mexico City
Cuernavaca
Tula
Pachuca
HIDALGO
Tianchinal
Poza Rica
Tuxpan
Naranjos
Ebano
Ciudad Madero
Tampico
Gulf
Golfo de Campeche
Apizaco
Tlaxcala
Puebla
PUEBLA
Azcapotzalco
Orizaba
Veracruz
Alvarado
VERACRUZ
Tierra Blanca
Coatzacoalcos
Minatitlan
Cardenas
Villahermosa
Cd. Pemex
TABASCO
Xicalango
INSET SHOWN
CHIAPAS
OAXACA
Oaxaca
Tehuantepec
Juchitán
Salina Cruz
Golfo de Tehuantepec
GUERRERO
Chilpancingo
Acapulco
Pacific Ocean
Maloob
Bacab
Kutz
Ixtoc
Akal
Nohoch
Kanaab
Chac
Luch
Abkatun
Takin
36 in. 91 cm
Chiapas-Tabasco area
Gulf of Campeche
Xicalango
Dos Bocas
Cunduacan
Ojiacaque
Santa Ana
Tupilco
Mecoacán
El Golpe
Tepotzingo
Puerto Ceiba
San Roman
Mango
Usumacinta
TABASCO
Samaria
Luna
La Venta
5 Presidentes
Tasajeras
Zaragoza
Coatzacoalcos
Nanchital
Cardenas
Villahermosa
Iride
Bitzal
El Cobo
Ciudad Pemex
Jose Colomo
Chilapilla
Vernet
Cactus
Sitio Grande
Rio Nuevo
Paredon
Minatitlan
Nuevo Teapa
El Plan
Tecuanapa
Mundo Nuevo
Giraldas
Cacho Lopez
Tepate
Artesa
Sabancuy
Copano
F. Nacional
Sarlat
Zapotero
Agave
Sunuapa
VERACRUZ
CHIAPAS
C. Nanchital
0 25 Mi
0 40 Km.

Cusiana's gas reserves were enough to trigger Ecopetrol interest in developing a national gas pipeline network. Rendon said Colombia's gas reserves may reach 8 bcf when Cusiana, Cupiagua, and Piedemonte field delineation is complete.

In mid-1994 two 40,000 b/d oil production units were to be installed on a site near the 2 Buenos Aires well pad, with a third to be installed, if required, in third quarter 1995.

Production will jump to 50,000 b/d in mid-1994, rising to 130,000 b/d in 1995 and 150,000 b/d by yearend. BP has not decided Phase 2 production rates, although Ecopetrol said this could be 600,000-800,000 b/d of oil. Development of Cusiana and Cupiagua has cost $2/bbl gross to date, BP said.

Piedemonte exploration

BP suspended the 1 Tamara wildcat on the Piedemonte block as a tight hole early in 1992. The company has a 10% interest in the 1 Volcanera which was drilled on adjacent acreage by operator Maxus Energy Corp.

Two BP wells were to be spudded on the 2,000 sq km Piedemonte block in 1993. The first, 1 Tacare, was to be started early at midyear. A second,

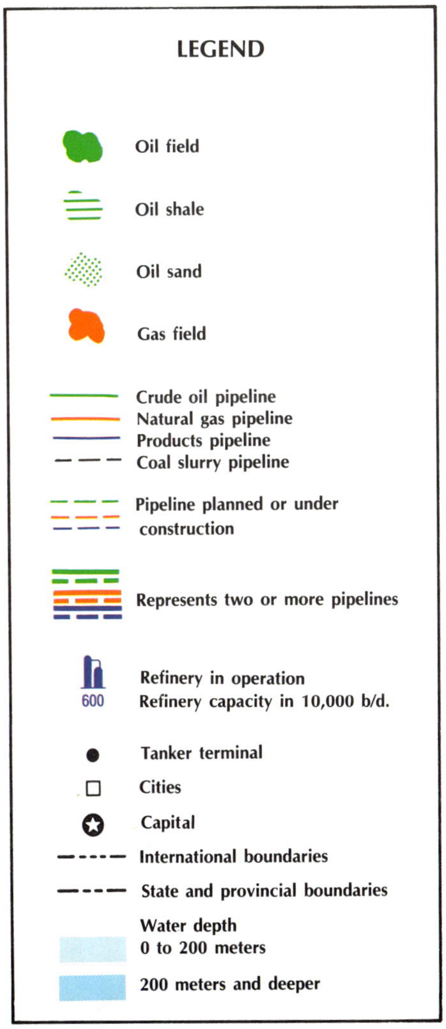

so far unnamed, was to be drilled in the southwest part of the block later in the year.

A $50 million/year budget has been set for Piedemonte exploration, allowing for two wells/year to the end of 1995. Early Cusiana wells cost about $25 million to drill, but Hayward said improved knowledge has helped halve well costs.

Garnet Resources activity

Garnet Resources Corp., Houston, was poised to step up its exploration and development campaign on a 344,000 acre tract in Colombia's Putumayo basin.

Garnet's plans in 1993 on Santana Block B included asking state oil company Empresa Colombiana de Petroleos (Ecopetrol) to declare Mary and Miraflor fields commercial and drilling four development wells in Toroyaco and Linda, Santana B's two commercial fields.

In addition, Argosy Energy International, Garnet's operating subsidiary in Colombia, since December 1992 has been preparing the first of two wildcat drillsites north-northeast of Miraflor field.

Possible commerciality of Mary and Miraflor fields received a boost in 1992 when tests of the 2 Mary delineation well indicated the reservoir likely will produce 5,300-5,500 b/d of oil from four zones. Three of the zones tested were new pays for Mary field and the fourth zone a confirmation of the pay in the 1 Mary discovery, about ½ mile the east.

Santana B drilling

Drilling was slated early in 1993 on the 3 Toroyaco development well. The rig was to drill continuously in Toroyaco and Linda fields, possibly completing the 3 Linda, 4 Toroyaco, and 4 Linda development wells before yearend 1993.

Garnet estimated 4 Toroyaco would be on line by June 1993, adding 1,500-1,700 b/d of oil to Santana B production.

At yearend 1992, combined Toroyaco and Linda production amounted to about 3,000 b/d. After starting commercial production from the fields in April 1992, Garnet's cumulative output by yearend reached 714,000 bbl, including 150,000 bbl net to Garnet.

In addition to development drilling in Toroyaco and Linda, Garnet expected to spud during first quarter 1993 the 1 North Miraflor wildcat. Drillsite was about 4 km north of the 1 Miraflor discovery well.

Mary, Miraflor

Garnet completed a feasibility study and preliminary engineering of development plans to get under way once Ecopetrol declares Mary and Miraflor fields commercial.

Two pipeline projects are key to expanding Santana B production.

Garnet intends to build an 18 km, 8 in. line to gather Mary and Miraflor crude and ship it across the Caqueta River for delivery into a 9 km transfer pipeline in Toroyaco and Linda fields. Laying the pipeline, including the Caqueta River crossing, was expected to take 12-15 months. Capacity can be expanded with additional pump horsepower to about 40,000 b/d.

CUBA

CAPITAL: Havana
MONETARY UNIT: Peso
REFINING CAPACITY: 280,000 b/cd
PRODUCTION: 25,000 b/d
RESERVES: 100 million bbl

Cuba's first licensing round in 30 years has cast new light on geology and recent exploration results in the country.

It is part of a privatization program aimed at reviving an economy that has been struggling since 1989 to make up for lost assistance from the former Soviet Union. Much of that help came in the form of subsidized crude oil and petroleum products.

In 1989, Cuba used an average of 228,400 b/d of products, according to the East-West Center, Honolulu. The country imported 172,000 b/d of Soviet crude and 98,000 b/d of products. By the last quarter of 1991, almost no Soviet oil was available to Cuba.

The National Assembly of Popular Power in 1992 changed the constitution to make private and foreign investment legal and to open the country to foreign trade.

The government now can shift land, sugar mills, factories, and other former state property to the private sector "when this leads to economic and social development and does not affect the economic, social, and political purposes of the state."

Cuba's transportation system is crippled by shortages of fuel and spare parts. Fuel problems also hurt agriculture and manufacturing.

What's offered

In February 1992, state owned Cubapetroleo SA (Cupet) described oil and gas licensing terms during promotion presentations held in London and Calgary.

U.S. companies can't invest in Cuba due to a trade embargo.

The country is offering production sharing contracts for 11 blocks—three offshore, seven onshore, and one onshore and off. Block sizes range from 1,400 sq km to 6,000 sq km.

Companies and groups face no restrictions in the number of applications they can submit to Cupet.

Contracts will have maximum terms of 25 years with exploration periods of 4 years. The state will take 11% of production and tax net profits at the rate of 30%.

Simon Petroleum Technology Ltd. (SPT) of the U.K. is selling data packages under a contract with Cupet.

Cuba has signed eight contracts with international oil companies during the past 3 years.

What's been found

Cuba produces 15,000-18,000 b/d of mostly 11-35° gravity crude oil from five fields, although oil gravity reaches as high as 60.5° in Motembo field.

Gas production averages 1.5 MMcfd.

About 25 fields have been discovered since exploration began in 1881, most along the northern coast. The largest is Varadero field, where oil in place is estimated at more than 1.1 billion bbl.

The island nation has two main geological provinces. Most discoveries have been in the northern province in pre-Upper Cretaceous Campanian pays.

The southern province is part of the Caribbean plate.

About 1,600 wells have been drilled in Cuba, most of them old and shallow. They are in areas covering less than 5% of the country's prospective area.

Since 1960, the year after the revolution that brought President Fidel Castro to power, drilling has totaled more than 1.7 million m of hole, according to Cupet.

Seismic surveys shot during the period exceed 70,000 line km, 50% of the work offshore.

ECUADOR

CAPITAL: Quito
MONETARY UNIT: Sucre
REFINING CAPACITY: 148,000 b/cd
PRODUCTION: 340,800 b/d
RESERVES: 2 billion bbl

Ecuador continued to grapple with efforts to privatize its petroleum sector after disclosing its plans to withdraw from the Organization of Petroleum Exporting Countries.

One of OPEC's smallest members, Ecuador in 1992 said it would leave the group in March 1993, citing high membership costs and minimal benefits. Industry observers also noted at the time Ecuador's plans to sharply boost production this century might run afoul of its OPEC quota.

Political controversy was stalling efforts to implement a new reform oriented hydrocarbon law in Ecuador that would open the country's petroleum sector to greater participation by foreign companies and privatize state petroleum companies, including Petroleos del Ecuador (Petroecuador).

New law status

A draft of the new hydrocarbon law circulated in Ecuador for a number of months and was to be submitted to that country's congress in the fall of 1993.

The new law was to set the stage for a sweeping restructuring of Ecuador's oil and gas industry under a master plan developed by ICF Resources Inc., Fairfax, Va., for a special Ecuadorian commission. That plan outlined steps needed upstream and downstream for a combined outlay of $4 billion to boost Ecuadorian oil and gas production, improve the country's crude transportation and refining infrastructures, and maximize petroleum revenues.

Oil authorities prepared several drafts of reforms to Ecuador's hydrocarbon law in hopes of submitting a final text to the congress for its approval before calling the next bidding round. In the draft proposals are alternatives to Ecuador's existing service contract system, including association and concession agreements.

But at yearend 1993 no draft had been submitted to congress, nor does it seem likely to be submitted in the near future, given political circumstances.

Further, the political cauldron has been stirred anew with fresh charges of environmental damage from oil industry operations in Ecuador's Oriente jungle region. The latest controversy entails a $1 billion class action lawsuit on behalf of Ecuadorian Indians alleging long term environmental pollution by Texaco Inc. during its 20 year operatorship of the region's key producing region.

While decisions on Petroecuador's future role in the industry is still pending reforms to the hydrocarbons law, the state oil company continues to maintain a dominant presence in the industry, from exploration to domestic marketing, including steps called for under the master reorganization plan.

Meantime, Petroecuador had selected 13 blocks for the next bidding round, tentatively set for January 1994. Oil Minister Francisco Acosta said that, failing congressional approval of the reforms, the planned bidding would be carried out under present service contract regulations. These have enabled so far a dozen foreign oil companies to carry out exploration since 1985. The successful ones are starting to produce oil in their blocks. Among the key start-ups in 1993 were projects operated by units of Occidental Petroleum Corp. and Ste. Nationale Elf Aquitaine.

Master plan

Underlying the tone of the reforms is a master plan developed by ICF that found about $4 billion must be spent on Ecuador's petroleum sector during 1994-2005 to achieve goals the government identified.

That breaks out to more than $2.5 billion for exploration and development, $971 million for expanding Ecuador's pipeline system, and $473 million for upgrading the refining sector.

Generally, it called for:

- Almost doubling oil production to more than 650,000 b/d in 1997, subsequently declining to 250,000 b/d by 2008. This would come mainly from infill drilling and waterflooding in Shushufindi, Sacha, and other large fields. CO_2 flooding and development of Ecuador's Pungarayacu tar sands would also be considered. Projections include undiscovered oil reserves pegged at 600 million bbl. In addition, offshore Amistad gas field would be developed, with resulting gas production intended to back out fuel oil use in the Guayaquil region.
- Expanding the SOTE oil trunk pipeline capacity to 600,000 b/d by 1996 to transport increased volumes of Oriente crude to Esmeraldas for export. Heavy crudes would be blended in the Oriente with lighter crudes to meet the minimum viscosity requirements for pipeline transport. Additional crude lines from Santo Domingo to La Libertad and directly to the trunk line from the southern Oriente, plus use of existing lines in Colombia and Peru, also are possible options.
- Upgrading the Esmeraldas refinery as soon as possible by adding a coker, another fluid cat cracker, and other units in order to reduce surplus volumes of lower value No. 6 fuel oil. Crude processing would increase to meet expected increases in domestic product demand, expected to rise to more than 175,000 b/d by 2008 from slightly more than 100,000 b/d in 1993. The La Libertad refinery would be maintained at low throughput levels through the 1990s. An alternative new refinery could be considered at Santo Domingo or Quito.

Reserves, output climb

Stepped up exploration and development by Petroecuador and foreign service contractors boosted Ecuador's oil reserves and production.

As of Dec. 31, 1992, total remaining reserves were pegged by the National Directorate of Hydrocarbons at 3.2 billion bbl. Of this, 1.95 billion bbl underlie fields under production and 1.3 billion bbl in fields not yet brought on production or fully developed. For the fields under production, that breaks out as Petroecuador 1.9 billion bbl, City Investing 12 million bbl, and Tripetrol SA 39 million bbl. Of those fields not producing or fully developed, that breaks out as Petroecuador 577 million bbl, Maxus group 228 million bbl, Oxy 198 million bbl, ARCO 162 million bbl, Oryx Energy Co. 121 million bbl, Elf group 24 million bbl, and Braspetro SA 8 million bbl.

Production has climbed steadily in 1993. By the end of September 1993, output reached 351,559 b/d, an increase of 6,000 b/d from August's level and 10,994 b/d from the average output during January-August. Much of the increase is due to new oil production by four operators that signed contracts since the mid-1980s.

Output is expected to increase further in 1994 to about 405,000 b/d. The increase will come mainly from Oxy's Block 15, where production is expected to reach 30,000 b/d, and from Maxus Energy Co.'s Block 16, where production will start at about 14,000 b/d by February 1994. Average crude gravity in 1994 is expected to drop to 26.3° from 27.9° in 1993, due in largely to the Maxus heavy oil development.

Oil output will continue to grow until 1999, when it is expected to reach a peak of 447,000 b/d. From then, output will start to dwindle to about 147,000 b/d in 2011.

Petroecuador activity

Petroecuador is quickly reviewing its exploration and development plans up to 1996, when its current authority expires.

As a first measure towards a more efficient management of exploration and development activities, Petroecuador's affiliates Petroproduccion and Petroamazonas, which so far have acted independently, have been merged into one unit. The new unit Petroproduccion will be responsible in the future for all exploration, development, and production activities in all of Petroecuador's areas.

Of the 13 blocks Petroecuador was to offer in the January 1994 tender, nine are located in the Oriente west of the blocks under exploration and/or development, along a north-south axis from the Colombian to the Peruvian borders.

Four blocks will be around the Gulf of Guayaquil.

Petroecuador's 1993 exploration campaign resulted in the discovery of oil reserves estimated at 460 million bbl in the Isphingo-Tambococha area of the Oriente.

By the end of June 1993, Petroecuador's oil output reached 330,449 b/d.

Pipeline work planned

To cope with the additional production expected, Petroecuador plans to expand the Lago Agrio-Balao trunk pipeline to 450,000 b/d.

A new, parallel line to the existing one will be built. It will be buried at points along the eastern side of the Andes, to avoid damage from earthquakes. In addition, a third line is under consideration from Villano to Baeza, to transport oil from several blocks in the Oriente.

The proposed 171 km, 18 in. pipeline will begin in ARCO's Villano field, where ARCO has discovered about 176 million bbl of heavy oil, and will extend to a pump station at Baeza on the Lago Agrio-Balao trunk pipeline. From this point the trunk line will be expanded to about 450,000 b/d from its present 325,000 b/d capacity.

MEXICO

CAPITAL: Mexico City
MONETARY UNIT: Peso
REFINING CAPACITY: 1.5 million b/cd
PRODUCTION: 2.7 million b/d
RESERVES: 51 billion bbl

Pemex's contribution to the Mexican economy is imposing. As the nation's largest taxpayer, with 1992 sales of $20.6 billion, Pemex provided $16.8 billion to the Mexican government's income in 1992 through hydrocarbon extraction duties, a special production and services tax, and a valued added tax. Including domestic sales and exports, Pemex accounts for a third of public sector revenue in Mexico.

Pemex is the world's sixth largest producer of crude oil and in first quarter 1993 was the fourth biggest exporter of crude to the U.S. after Saudi Arabia, Venezuela, and Nigeria. As Mexico's largest exporter, Pemex pulls in more than 30% of Mexico's export earnings, averaging slightly more than $8 billion/year in recent years. In 1982, Pemex provided 79% of Mexican export earnings with hydrocarbon export revenues of $16.6 billion.

Despite extensive cuts in its work force, Pemex still is considered the largest enterprise in Latin America. Peak employment in 1986 swelled to 210,000, but by yearend 1992, Pemex's payroll had been trimmed to 125,000. Insiders say more cuts are on the way.

About half of Pemex's crude oil output serves domestic consumption, and half is exported. About 58% of Mexico's oil exports go to the U.S., with the remainder broken out as Spain 17%, Japan almost 7%, Canada and France 2% each, and other nations combined 14%.

Pemex's role

Under Article 27 of the 1917 Mexican Constitution, and other laws, the government through Pemex maintains monopoly control of exploration and extraction of all hydrocarbons in Mexico, refining of crude oil, and the production of basic petrochemicals.

Despite constitutional and political constraints, Pemex under President Carlos Salinas de Gortari has opened new doors to foreign investment, either within the traditional bounds of Mexican law or under new interpretations of old rules.

Mexican law, for instance, dictates that only Pemex may produce basic petrochemicals, loosely defined as compounds derived after one conversion from the hydrocarbon feedstock. In 1986, 60 petrochemicals were considered basic and thus off limits to private sector involvement. The Mexican government—notably under Salinas—has chopped the number of basic petrochemicals to 8.

A list of secondary petrochemicals similarly has shrunk from 67 to 13 compounds. Mexican law allows private companies to produce this product category provided they are no more than 40% foreign owned. Chemicals not on either list carry no restrictions on foreign ownership.

The changes already have brought tangible results. In April 1993, Pemex signed a pact with Valero Energy Corp. and a Spanish construction firm to build a $350 million plant to produce methyl tertiary butyl ether, an oxygenate that Pemex in 1989 began adding to its gasoline to help mitigate severe urban air pollution in the country. The Ixhuatlan, Veracruz, complex is to have capacity of 500,000 metric tons/year when complete in 1995.

Growing cooperation

Among other examples of the quickening pace of growing cooperation between Pemex and the private sector:

- Pemex in 1992 imported an average 250 MMcfd of natural gas from a variety of U.S. suppliers to supply northern Mexico—an arrangement that for Pemex is less costly than transporting natural gas from Campeche Sound fields. Mexican petroleum industry analyst George Grayson, a government professor at William & Mary College, estimated Pemex's natural gas imports could grow to 300 MMcfd by yearend 1993. A shortage of capital has sparked speculation that Pemex soon will sign turnkey service contracts with U.S. firms to develop natural gas in Mexico's north, Grayson contends.
- In February 1993, Pemex formed a 50-50 joint venture with Shell Oil Co. to own and operate Shell's 225,000 b/d Deer Park, Tex., refinery, where a major upgrade was under way. Under the joint venture, Shell can sell Pemex as much as 45,000 b/d of unleaded gasoline. For its part, Pemex secured access to a refinery that can process more than 100,000 b/d of Mayan heavy crude, which accounts for 67.5% of Mexico's crude oil exports.
- Of 55 wells being drilled in Campeche Sound, about a third of the work is to go to private companies. Triton Engineering, Houston, in 1992 drilled a well for Pemex under a turnkey, lump sum contract in 127 days, compared with 249 days usually needed by Pemex. Sonat Corp., Houston, and Mexican partner EPN Arval won a bid in late 1991 to drill six wells for Pemex.
- In February 1992, Pemex announced a major change in strategy for its 60 intermediate petrochemical plants. Obsolete plants will be closed, some will be privatized through international open bidding, and others may be transformed to

Austria
and Abu Dhabi state oil companies teamed up to cooperate on crude oil supplies, technical development, and international exploration. The move stems from the purchase of a 20% interest in OMV AG by International Petroleum Investment Co. of Abu Dhabi.

Pakistan
is offering a 20% interest in state owned Mari Gas Co. Ltd. The offer is called for under an agreement with World Bank's Investment Finance Corp. that requires IFC to provide $19.5 million to drill 14 wells to boost Mari's natural gas productive capacity to more than 400 MMcfd. Of the 14, three more wells remain to be drilled by December.

Tankers

Venezuela's
Petroleos de Venezuela SA units PDV Marina and Bitor are considering whether to build special tankers to ship Orimulsion boiler fuel. The 40,000 dwt tankers will cost $40 million to build, Venezuelan state news agency Venpres reported.

Pipelines

TransCanada PipeLines
started this year's Ontario pipeline and compression construction program. It will lay 51 km of pipeline, replace 14 km more, build a compressor station, and add compression at six sites. The work will cost $200 million (Canadian).

Centana Energy Corp.
closed the first part of its purchase of certain assets from MND Energy Corp. The trade includes MND pipeline assets in Southeast Texas, including the Spindletop gas storage facility. Centana is a unit of Panhandle Eastern Corp., Houston, while MND is a unit of Mitchell Energy & Development Corp., The Woodlands, Tex.

U.S. Federal Energy Regulatory Commission
action cleared the way for NorAm Energy Corp. (formerly Arkla Inc.), Shreveport, La., to set up an unregulated natural gas gathering business. FERC affirmed that upon spindown of NorAm Gas Transmission Co.'s gathering facilities into a NorAm unit, these facilities will be exempt from FERC jurisdiction after certain conditions are met.

FERC
approved settlements that resolve Order 636 transition cost issues for Natural Gas Pipeline Co. of America and Texas Eastern Transmission Corp. Texas Eastern can collect all costs up to a designated spot price and $330 million of costs over that. Natural Gas Pipeline's settlement provides for a $408 million cap on recovery of gas supply realignment costs, amortized over 4 years.

FERC
granted preliminary approval to SunShine Interstate Transmission Co. to lay a 145 mile gas pipeline to Florida from Mississippi. The line will tie into SunShine Pipeline Co.'s 545 mile intrastate system in Florida (OGJ, Mar. 14, p. 38).

Premier Transco Ltd.,
a unit of British Gas, and planned operator of the Scotland to Northern Ireland gas pipeline currently under design, awarded engineering of the offshore section to Snamprogetti Ltd. through European Marine Contractors. Snamprogetti will provide detailed engineering of the 24 in. subsea portion, which extends to landfall in Northern Ireland near Skemaghan Point on Islandmagee from near Portnaughan Bay, Scotland.

Pipelaying
is to begin this month on a $27.5 million (Australian), 300 km Northern Territory natural gas pipeline linking the main Amadeus-Darwin trunk line with MIM Holdings' McArthur River lead and zinc mine near Borroloola on the western coast of the Gulf of Carpinteria. The work is to take 30 weeks. Gas will be used as fuel to generate power for the mine under a 20 year contract from Northern Territory Power & Water Authority.

France's Coflexip
plans a $55 million (Australian) flexible pipe manufacturing plant at Perth to support future work in Southeast Asia. Western Australia's government and Coflexip signed an $8.5 million government aid pact for the plant, under which Coflexip will receive $5 million for construction, $2 million for erecting a heavy crane at Fremantle wharf, and free 10 year use of 3 hectares of adjacent land. Start-up is slated for July 1995.

Indonesia
started up a $97.67 million, 221 km products pipeline to a Jakarta depot from its Cirebon, West Java, refinery. The line moves fuel oil, kerosine, and diesel oil.

Exploration

Viet Nam
will allow the venture of APN Petroleum Ltd. 55% and operator Occidental Petroleum Corp. 45% to drill a wildcat to test the Ung Trang structure on the 815,000 acre Block 04-3 in the South China Sea. Drilling is to start in July.

Lasmo
is developing an early production system to start output from its 1 Revancha and 1 Venganza oil discoveries in Colombia's Espinal block. Production is planned via Lasmo's nearby Purificacion field, where a 35,000 b/d capacity pipeline is under construction.

British Gas plc's
Savi Ragha Well Block 34 wildcat near Dera Ismail Khan, Pakistan, flowed 16 MMcfd of gas and 700 b/d of condensate from an 82 m thick pay zone. The well, BG's first in Pakistan, will be deepened to projected depth of 3,000 m. The wildcat is about 35 km from Dhodak gas/condensate field, slated for start-up in September.

Anadarko Petroleum Corp.
unit Anadarko Algeria Corp. and partners' 1 East Berkine oil discovery on Block 404 in the Sahara Desert of Algeria cut 165 ft of net Triassic pay (OGJ, May 23, p. 89). It's the sixth wildcat and the fourth discovery in Anadarko's Algerian program. The well is to be drilled to 15,900 ft during the next 2 weeks.

Research

Energy Research Clearing House,
The Woodlands, Tex., seeks proposals for cooperative funding in 1995 for research projects on exploration and production technologies deemed most needed by its members. Topics to be considered are reservoir and production engineering, enhanced oil recovery, geology, and geophysics.

U.S. Department of Energy
and Gas Research Institute, Chicago, agreed to synchronize their natural gas R&D and demonstration (RD&D) efforts. DOE and GRI, the two largest gas RD&D funding organizations in the U.S., long have collaborated on projects.

Elf Petroleum Norge AS
opened a marine research center beside a fjord at Randaberg near Stavanger, No way. The 40 million kroner ($5.5 milli Elf Akvamiljo center will study biodeg tion of hydrocarbons, effects of oil po on marine life, and ways to cle cuttings. Elf believes it is the first f company to open a research lab

Table 2

Classification of surface geochemical prospecting methods

Survey type	Media sampled	Target
1. Geochemical methods		
a. Direct methods:		
Hydrocarbon		Paraffin compounds
(1) Instantaneous	SA W A	
(2) Integrating	S SA W A LS R	
Bitumens (fluorescence spectrometry)	S W DS LS	Aromatic compounds
b. Indirect methods:		
Vitrinite refflectance	S R	Carbonaceous material
Alteration haloes	S R	Fe, Mn, Calcite
Chelates, trace elements	S R DS LS	Br, Cu, Pb, Zn, U
Delta carbonate	S R	Carbonates
Microbiological indicator	S DS	HC-consuming, bacteria
Soil salts	S W	$CaCl_2$, gypsum, calcite siallite, SO_4^-, HCO_3^-, Cl^-
Helium, radon, hydrogen	SA W	He, Rn, H_2
Associated gases	SA W	N_2, CO_2, NH_3, SO_2, O_2
Carbon, oxygen isotopes	S W	C, O_2, calcite, CH_x
Trace metals, halogens	S W DS	I, Ni, V
Biogeochemistry	Plants	Trace elements
2. Geophysical methods		
a. Indirect methods:		
Radiometric, gamma-ray spectrometry (by air, ground)	S R	Radionuclides of K, U, Th
Magnetics, micromagnetics (by air, ground)	S R	Magnetic minerals
Magnetic susceptibility (by ground, laboratory)	S R C	Magnetic minerals
Induced polarization (by ground)	S R	Resistivity & chargeability
Electromagnetics, VLF, & multifrequency (by air, ground)	S R	Conductive "chimney"
Magnetotelluric (ground)	S R	Resistive "chimney"
Radar (soft targets) (by ground)	A	Hydrocarbon gases

Types of sample media: S-soil, SA-soil air, W-water, DS-drainage sediment, A-air, R-rock, LS-lake and ocean sediment, C-cores & hole cuttings.

tive-setting.

Table 2 lists many direct and indirect techniques used in the past. For each geochemical (or geophysical) survey method used, various media are available for sampling as shown (second column). The third column shows the geochemical *target,* a petroleum-induced signature suspected to exist near the surface.

Techniques are termed *direct* if they identify one or more component hydrocarbon gases or liquids present in the media sampled. Direct methods employ techniques that suggest the nature of aromatic, naphthenic, or paraffinic compounds in the soils, and employ variations of gas chromatography, fluorescence spectroscopy, and mass spectrometry.

Sample media available include gases or liquids in the following: air, soil or subsoils, ground and surface water, drainage sediments, and lake and ocean sediments. Hydrocarbons detected by direct survey methods may include methane, ethane, and heavier alkanes, paraffins, alkenes, and multiple-ring aromatic compounds that include bitumens, and sulfur compounds.

Bitumens are high molecular weight components of oils. Their aromatic components fluoresce when exposed to ultraviolet light. Fluorescent compounds are detected in small quantities from soils and drainage sediments. Aromatic compounds can be used to "fingerprint" oil. Specific fluorescence spectra are matched with those data known to characterize oils produced nearby to establish oil "fingerprints." The technique is applicable for identifying multiple pays or multiple oil sources. In the opposite sense, fluorescence can suggest environmental contamination such as agricultural pesticides.

Techniques are termed *indirect* hydrocarbon sensing techniques if they are designed to sense a secondary product caused by the upward movement of hydrocarbons. Indirect techniques cannot sense hydrocarbons directly. As an example, radiometric "low" anomalies may occur where carbon dioxide has contributed to calcite recementation in the soil zone.[5] Induced polarization methods[5] can detect resistivity and chargeability contrasts caused by abundant calcite and/or pyrite in the subsoil area. Magnetic surveys can detect magnetite-like mineralization thought to be related to leaking oil field brines. The microbiological indicators of oil and gas might be termed quasi-direct sensing methods.

Alkali-earth elements, potassium, calcium, manganese, and magnesium, detected by assays can be deposited in some soils and leached from others by unusual geochemical environments. These environments are suspected to exist above and laterally-adjacent to the surface projection of oil pools. Arguably, situations exist where some indirect indicators could act as quasi-direct indicators. Trace elements and associated gases are examples.

Besides being direct indicators in some settings, hydrogen and helium also act as *structural* indicators.[7] Because of their small molecular size, workers speculate these elements more easily escape through faults and fractures than paraffinic and aromatic hydrocarbons, suggesting that a compound's relative geochemical mobility should be considered. Helium's ultimate source is from radioactivity decay of earth radionuclides, particularly uranium. Noble gases, helium, neon, and radon act as structural indicators because of their nonreactivity.

Table 3 compares several groups of geochemical methods used for recent onshore surveys against pertinent characteristics of the method. At least nine methods represent direct hydrocarbon indicators. Three others provide a semblance of the "fingerprinting" ability described. Thirteen techniques that display integrative advantages include seven direct and six indirect techniques. Because of their low cost per unit of area, several indirect methods are used for wide-area or reconnaissance surveys. Many methods perform well in both reconnaissance surveys (wide-area) or detail surveys (small area).

An important distinction for direct methods—*instantaneous* or *integrating*—is used here to classify methods. By design, integrating methods collect data over longer periods, or remove any short-term fluctuations. In the opposite sense, free soil gases as helium and hydrocarbons record significant fluctuations over the short periods. Techniques that purposely detect short term fluctuations are termed instantaneous techniques.

Do not overlook the advantage of large sampling volume. A large volume is better than a small volume when sampling earth phenomena. Magnetic, resistivity, and induced polarization geophysical surveys are good examples of geophysical techniques that measure potential-field responses for large volumes of rock to varying depths.

Table 3

Characteristics of geochemical methods

	Hydrocarbon methods							Other associated							Gamma-ray spectrometer (radiometric)			Pathfinder elements			Electrical methods			Magnetics			Other
METHODS	Free gases in soil	Adsorbed soil gas	Headspace gas	Dissolved gas in water	Stream sediment	Bitumen methods	Sorbent technique	Iodine	Microbiological indicators	Carbon Isotopes	N_2, O_2, CO_2, NH_4	Helium/hydrogen	Radon	Neon/argon	Total count (on foot)	Airborne	Truck-borne	Associated & chelatable	Alkali earths	Partial extraction of metals	Spectal & phase induced polarizatoin/resistivity	Spontaneous potential	Multifrequency EM (airborne)	NURE reprocess	Airborne, gradiometer	Ground	Spectral imagery
CHARACTERISTICS																											
1. Direct hydrocarbon indicator	x	x	x	x	x	x	x		?																		
2. Indicrect hydrocarbon indicator								x	x		x	x	x		x	x	x	x	x	x	x	x	x	x	x	x	x
3, Petroleum finger-printing						x	x			x																	
4. Integrates over time		x	x	x	x	x	x	x													x	x	x	x	x	x	
5. Tectonic-structural analysis indicator												x	x	x				?	?	?							
6. Reconnaissance	x	x	x	x	x	x	x	x	x	x	x	x	x	x	x	x	x	x	x	x	x	x	x	x	x	x	x
7. Detail method	x	x	x			x	x	x	x	x	x	x				x	x	x	x	x	x	x					
8. Large sampling volume/area of dispersion				x	x	?										x	x				x	x	x	x	x	x	
9. Cost per unit area																											
Low	x	x	x	x	x	x	x	x	x		x	x	x		x	x	x	x	x	x		x	x	x	x	x	x
Moderate	x	x	x			x	x			x		x		x							x				x		
High										x											x						

? Uncertain or possible in some cases

Project design

The design of an effective program means making decisions about:

- sample material,
- sampling pattern (spacing, sample media, depth),
- sample preparation,
- analytical procedure, and
- criteria for interpretation.[8]

To make these decisions requires some assumptions about the following factors:

1. Dispersion, or mobility characteristics of hydrocarbon compounds,
2. Local environmental constraints that influence dispersion processes,
3. Target size and its effect on sample density and sample spacing,
4. Availability of sample material,
5. Analytical capability and logistical considerations.

For example, target size can influence the program cost. The larger the target, the fewer samples are needed to detect it. As a rule of thumb, an anomaly cannot be confirmed with fewer than six samples occurring within the anomaly limits. Conversely, to establish higher confidence in the results, at least 75% of the samples should be located outside the anomaly to define what constitutes the local and regional anomaly threshold limits.

Evaluate logistical constraints. In high cost areas, it may be an advantage to "double or triple" sample but analyze only one third of the samples. Other samples can be retained for backup purposes only for anomaly follow-up, if necessary. In this example, a fill-in survey has been possible without remobilizing to the field.

Orientation surveying

Orientation studies consist of a series of preliminary experiments to decide optimum survey procedures and are completed in one of four ways:[9]

(1) by traveling to the field area to systematically collect samples and pertinent information;

(2) by performing a literature search;

(3) by performing a theoretical orientation;

(4) by performing no orientation (should be avoided).

The field orientation survey.

A classic orientation survey comprises field sampling and analysis around a known field. A scenario consists of preliminary field sampling and observation aimed at determining the existence and characteristics of hydrocarbon anomalies. This information allows selecting adequate prospecting techniques and for deciding the factors/criteria for interpreting geochemical data. Evaluate those factors to determine several factors:

(a) contrast between anomaly and background,

(b) the most suitable survey method(s),

(c) optimum depth of sampling,

(d) shape and extent of an anomaly,

(e) reproducibility of results,

(f) possibility of contamination, and

(g) sampling schemes, sampling media, and between-sample spacing.

Decisions about orientation surveys should be made by the company and not a contractor or laboratory. To complete or omit the orientation procedures is a first critical milestone for application of geochemical surveys.

The orientation survey is generally simple. It consists of one or two traverses and one or more vertical soil profiles. The work is conducted over a known oil field that is geologically similar to the proposed survey area (or possibly connected to it) and then continued outside the known productive area into background. Traverse lengths may be 2-10 miles. The spacing must be ⅛ to ¹⁄₁₀ mile between samples, or closer in the absence of data to indicate width of production. Fully-developed recent discoveries are good examples for orientation study.

New, shut-in discoveries are best.

The taking of shallow, deep (12 ft), and shot-hole samples is ideal; sampling both on-field and off-field is critical. The use of multiple techniques during the orientation is imperative. Expenditures ranging from $2,000-40,000 and the taking of up to 800 samples in large orientation surveys are reasonable. As a rule, over-surveying during orientation study ensures sufficient data to organize reconnaissance and detail surveys properly. Often the orientation survey area can be made a part of the larger survey anticipated.

The literature (orientation) study.

If it is impractical to visit the field area to conduct an orientation survey ahead of the main survey, significant information may be obtained by review of previous work, communications with contractors, and professionals familiar with this or similar areas.

Theoretical orientation.

This highly speculative approach is based on application of theoretical models, some basic principles of exploration geochemistry, and assumptions as to the surficial geology/geomorphology and climate in the survey area. This approach is not recommended.

Reconnaissance and detail surveying

Reconnaissance and *detail* refer to the study area size and the nature of the objectives set. Some practitioners refer to four categories: frontier, reconnaissance, detail, and development.

The design stage involves setting of objectives, tentative choices of field and laboratory techniques anticipated, and requires customer input on the following:

1. type and areal dimensions of largest and smallest trap (field) anticipated,
2. number of oil or gas objectives expected,
3. their depths,
4. known fields in area if any,
5. general geology,
6. extent of area under study—large (basin size), medium (several townships, a county, a fairway or trend), small (one to a few square miles),
7. possible sources of contamination,
8. input from orientation study,
9. objective(s) sought, and
10. level of detail and confidence limits demanded from results.

Reconnaissance surveys.

Size (or area) and level of detail are considerations for production surveying. Reduce very large basin-sized or fairway-sized areas to manageable units by reconnaissance surveying. Therefore, the use of indirect methods can exceed the use of direct methods.

The purpose of reconnaissance work is to search a relatively large area to call attention to local interest areas where eventual follow-up (detail) work will occur. The spacing and types of samples are chosen to detect, but not necessarily outline the favorable provinces or fairways, etc., at a low cost compared to detail surveys.

Systematic sampling across unknown ground should accomplish two objectives:

1. eliminate unprospective ground, and
2. identify prospective areas or targets sought.

In reconnaissance surveys, the first objective is most important and the other is of secondary importance; in detail surveys, the opposite applies.

Specifically, objectives for most reconnaissance surveys correspond to one of the objectives set out in Fig. 1 (A, 1-3). Reasons for surveying are numerous and reconnaissance surveys may also be used to accomplish one or more of the several additional goals listed below:

- To determine if unexplored basins (regions) are petroliferous in character and subsequently rank areas by this data;
- To define fairways and rank prospective areas before seismic surveys;
- To eliminate areas of low or no potential before land acquisition;
- To identify areas for further follow-up by detail surveys; and
- To evaluate prospective geochemical characteristics along seismic lines.

Detail surveys.

The purpose of detail surveying is to identify or confirm drillable objectives.

The confidence in drawing conclusions lies at a higher level than those of reconnaissance surveys. Between-sample spacing is usually less for detail work than reconnaissance work and, if unknown, can be decided using the semi-variogram calculation[10] from previous work. Background, local, and regional anomaly thresholds (geochemical anomaly contrast) must be maximized and readily identifiable. Anomaly limits must be seen.

Results must distinguish between false and valid anomalies at low levels of contrast. Sinclair's probability graphs[11] may be an aid over simple statistics to establish these limits. Objectives of most detailed surveys are outlined in Fig. 1 (B. 4-7).

The Author

Boleneus

David Boleneus has used geochemical and geophysical methods in exploration for petroleum and minerals for 16 years. He organized surface prospecting surveys of six fields for Brown and Ruth Laboratories to compare the success of 10 (direct and indirect) methods. He assembled a "User Manual for Surface Geochemical Methods in Petroleum Exploration" for Conoco Inc. He has 10 years' experience in oil provinces of the western U.S. with Sohio Petroleum Co. in Denver and other companies. He holds geology degrees from Eastern Washington University and Louisiana State University and has earned a master's degree in business administration.

References

1. Groth, P.K., and Groth, L.W., Bibliography for surface and near-surface hydrocarbon prospecting methods, APGE, 2nd edition, Denver, Colo., 1994.
2. Northeast Research Institute Inc., The Petrex fingerprint technique, Sales and informational brochure, NERI, Denver, Colo., 1992, 18 p.
3. Potter, R.W., Harrington, P.A., Silliman, A.H., and Viellenave, J.H., The significance of geochemical anomalies to hydrocarbon exploration, abs. AAPG Hedberg Research Conference, Apr. 24-28, 1994, Vancouver, B.C.
4. Klusman, R.W., Soil gas and related methods for natural resource exploration, John Wiley & Sons, 1993, 483 p.
5. Hansen, D.A., Radiometrics, in Practical Geophysics for the Exploration Geologist, R. Van Blaricom, ed., Northwest Mining Assn., Spokane, Wash., 1980, pp. 2-38.
6. Oehler, D.Z., and Sternberg, B.K., Seepage-induced anomalies, "false" anomalies, and implications for electrical prospecting, AAPG Bull., Vol. 68, No. 9, 1984, pp. 1,121-45.
7. Jones, V.T., and Drozd, R.J., Predictions of oil and gas potential by near-surface geochemistry, AAPG Bull., Vol. 7, No. 6, 1983, pp. 932-952.
8. Rose, A.W., Hawkes, H.E., and Webb, J.S., Geochemistry in mineral exploration, Academic Press, 2nd edition, London, 1979, 657 p.
9. Thomson, I., Getting it right, in Exploration geochemistry, design and interpretation of soil surveys, J.M. Robertson (ed.), Vol. 3, Reviews in Economic Geology, Economic Geology Publishing Co., El Paso, Tex., 1987.
10. Clark, I., Practical geostatistics: Applied Science Publishers, London, 1980, 129 p.
11. Sinclair, A.J., Probability graphs, Assn. of Exploration Geochemists, Rexdale, Ont., 1976, 76 p.

Fig. 1

APPLICATIONS FOR SURFACE GEOCHEMICAL SURVEYS

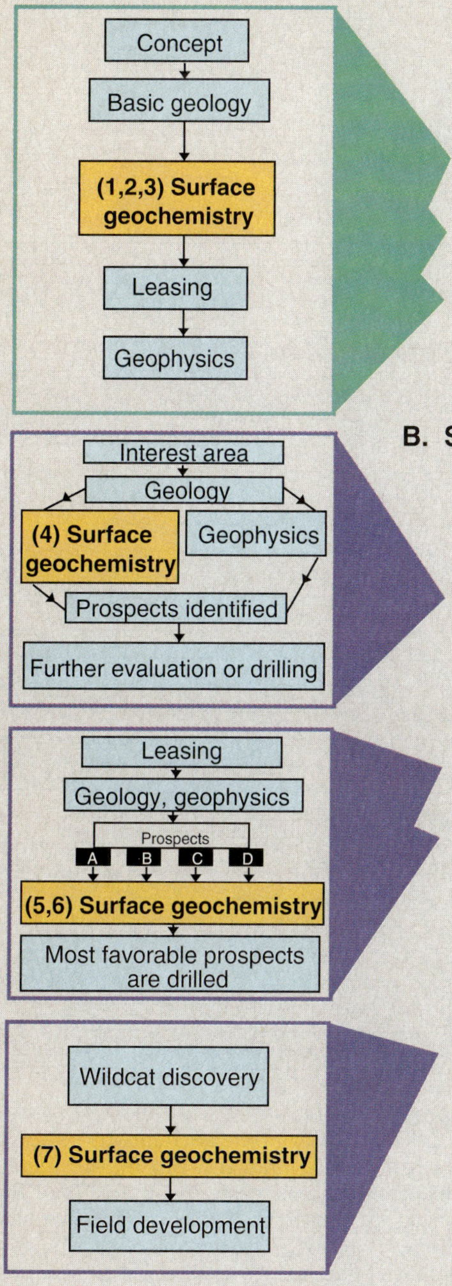

A. Surface geochemistry as a reconnaissance exploration tool:

Geochemistry can be applied to LARGE areas--such as basins, concessions, and plays. It is applied in "series" fashion with the more costly tools used later.

Objectives:

1. To identify interest areas, based on gross hydrocarbon character, at a low cost per square mile.
2. To select most valuable concessions; focus exploration in frontier areas.
3. To differentiate oil-prone from gas-prone or thermally-mature versus immature areas.

B. Surface geochemistry as a detail exploration tool:

Geochemistry can be applied to known interest areas previously identified by other methods (geophysics, leasing, and geochemistry). It is usually applied in "series/or parallel" fashion with other tools.

Objectives:

4. To identify drilling prospects; usually used in conjunction with geophysics.

5. To rank drilling prospects on the basis of their most favorable geochemical characteristics.
6. To assist a drilling partner decide to participate/not participate in one or more of a series of wells.

7. To define the areal extent of an oil accumulation to assist orderly field development.

OGJ

ment. It further represents that only surface geochemistry reveals important information about hydrocarbon charge, hydrocarbon type, and seepage of hydrocarbons to the surface.

Surveys involve the following stages:

(1) objective-setting and planning (design),

(2) orientation (information gathering about the sampling environment) and choice of methods,

(3) survey sampling (field, laboratory) and follow-up,

(4) data presentation and analysis, and

(5) interpretation.

Total time involves about one to four months. Companies and individuals who begin by selecting contractors solely on cost criteria may unknowingly forgo participating in the first two stages and sacrifice flexibility during the process.

Choice of methods

Selecting surface geochemistry methods normally occurs after objec-

Guidelines for surface geochemical surveying

David Boleneus *Registered professional geologist* *Spokane, Wash.*

This article classifies the various methods employed and provides guidelines useful in design of orientation, reconnaissance, and detail surface geochemical surveys for oil and gas exploration.

It is based on a user manual for surface geochemistry in petroleum exploration the author assembled in 1989.

Surface geochemistry is perhaps better termed surface prospecting because it employs both geochemical and geophysical techniques to detect petroleum, its constituents, or products subsequently caused by reactions near the earth's surface. It should not be confused with geochemistry of petroleum source rocks.

Introduction

Surface geochemistry provides a tool to map the surface expression of oil or gas accumulations, regardless of the type of trap or type of hydrocarbon present.

Surface geochemistry generally cannot discriminate between shows or marginal or economic quantities of petroleum. However, surface geochemistry represents a powerful petroleum indicator technique.

More than 20 different surface geochemical methods are known by which surface expressions can be measured. Further complicating their understanding, the protocols of each of these methods are perhaps equally variable.

Surface geochemistry is becoming more widely used because of its low cost, ease of application, and improvements in technology in the last 10 years. Individuals and companies often prefer to locate positive indications of petroleum on their land. An American Association of Petroleum Geologists-affiliated society has published an extensive bibliography on the subject,[1] and relevant studies on drilling successes of geochemical prospects are now available.[2,3]

Table 1

RELATIVE EFFECTIVENESS IN REDUCING RISK

	Geology	Seismic	Source geochemistry	Surface geochemistry
Source				
1. Present	High		High	
2. Thermally mature	Low		High	High
Seal				
3. Present	High			
4. Effective				Low
Reservoir				
5. Present	High	Low		
6. Porosity; type	Low			Low
7. Multiple pays	High	Low		High
Trap				
8. Timing to migration	High		High	
9. Preserved		High		
10. HC charged	Low			High
11. HC types	Low		High	High
Surface features				
12. Mineral alteration	Low			Low
13. Seepage of petroleum				High
Migration				
14. Extent & significance			High	Low
Sampling density				
15. By technique	Low to high	High	Low	High

☐ Areas where surface geochemistry can provide the most significant contributions.

Use of surface geochemistry is premised upon no reservoir being ideally sealed, leading one to deduce that imperfect seals are always present. The conclusion could be drawn that petroleum reservoirs leak (seepage). Then, from the time of field origin, the soils overlying a field represent natural, long-term, concentrating media of petroleum. This concentration often exists in parts per million although still within the detection limit of the new generation laboratory instruments. Surface geochemistry is successful because empirical study about microseepage shows that hydrocarbons are displayed in many forms directly (vertically) above oil and gas fields.[4]

Surface geochemistry should be used only if it can reduce the risk or finding costs of oil and gas. Surface geochemistry is a tool, like seismic, gravity, basement magnetics, geologic mapping, and drilling and logging technologies.

Beyond data assembled about the six attributes that explorers commonly use to describe petroleum occurrences—source rock, reservoir rock, structure, seal, thermal history, migration, and entrapment—surface geochemistry contributes two additional attributes.

Table 1 represents the relative contributions by geology, seismic, source rock studies, and surface geochemistry to describe the petroleum environ-

Detailed microbial surveys help improve reservoir characterization

James Tucker, Daniel Hitzman *Geo-Microbial Technologies Inc. Ochelata, Okla.*

Application of the Microbial Oil Survey Technique (MOST) as a reservoir characterization tool expands a surface geochemistry technique from exploration into the production industry.

The three examples presented show how tightly spaced microbial sampling grids over producing fields are effective tools to delineate reservoir drainage patterns associated with and related to reservoir heterogeneities. Using this surface microbial information with geology and geophysics is a powerful combination for identifying infill target and stepout potential within existing fields.

Distinct microbial population patterns identified in these surveys are:

1. Decreased microbial populations over or adjacent to producing wells indicating hydrocarbon withdrawal drainage patterns,
2. Linear high and low variations of the gas withdrawal patterns which illustrate the compartmentalization of the reservoir; and
3. Exploration type, broad microbial highs which indicate either missed horizons or stepout potential in the field.

New approach

The MOST technique used in this survey was first developed by Phillips Petroleum Co. in the 1950s.[1]

The technique is based on analyzing shallow surface soil samples for a specific suite of microorganisms which indicate the presence of light hydrocarbon gases. Where elevated clusters of these specific microbial populations are identified, anomalous hydrocarbon microseepage from oil and gas reservoirs is indicated.[2 3 4]

Fig. 1

MICROBIAL VALUES INDICATE RESERVOIR DRAINAGE PATTERNS

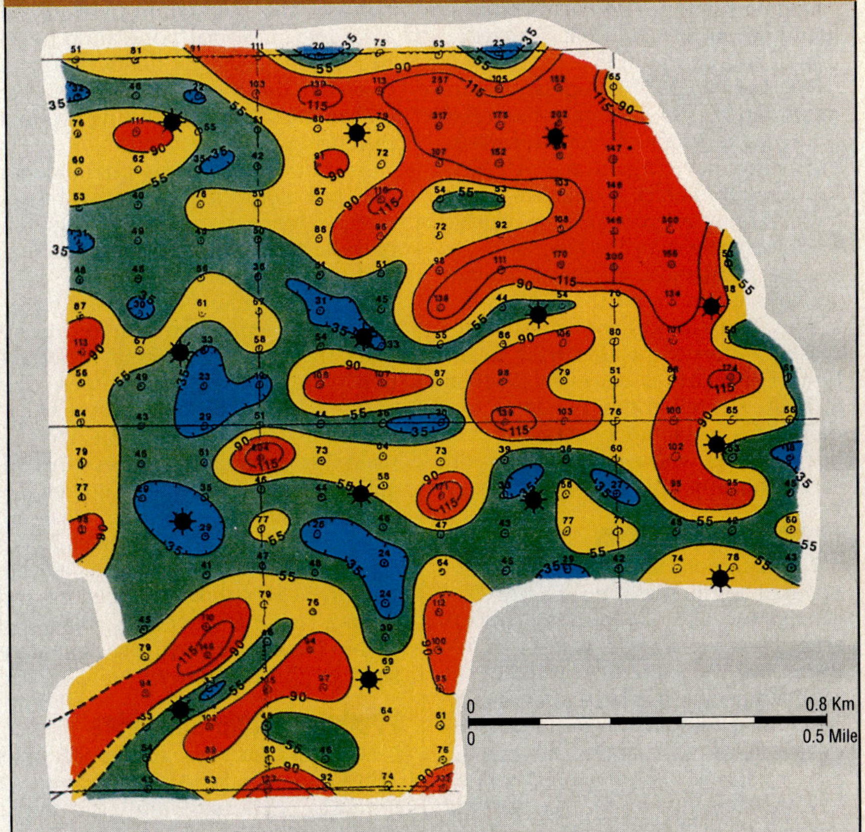

Part of Sacramento basin, Calif.

In the past MOST has been used primarily for exploration reconnaissance surveys, for ranking and comparison of wildcat prospects, and for hydrocarbon microseepage measurements of new and offset prospects.[5 6 7] This new application of MOST within producing fields broadens the ability of surface geochemical surveys to identify reservoir heterogeneities and bypassed pockets of reserves.

Anomalous hydrocarbon microseeps are identified by observing MOST population concentrations and distribution patterns within a survey area. There is a direct, positive relationship between the light hydrocarbon concentrations in the soils and

Fig. 2

RESERVOIR DRAINAGE PATTERN, S.E. OKLAHOMA

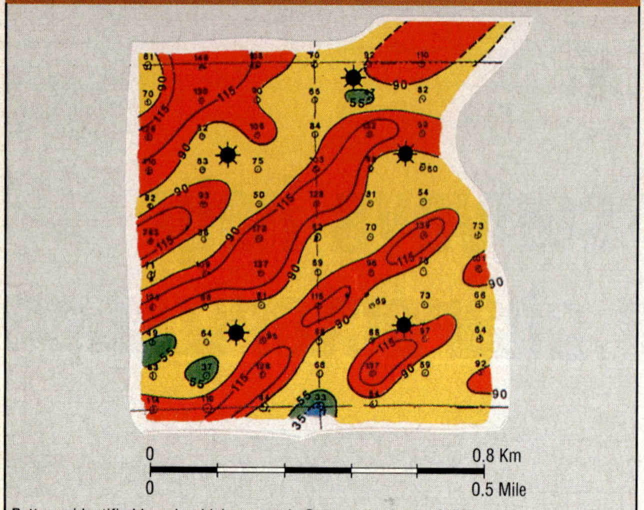

Patterns identified by microbial surveys in Sacramento basin, Calif.

OGJ

Fig. 3

RESERVOIR HETEROGENEITY PATTERNS

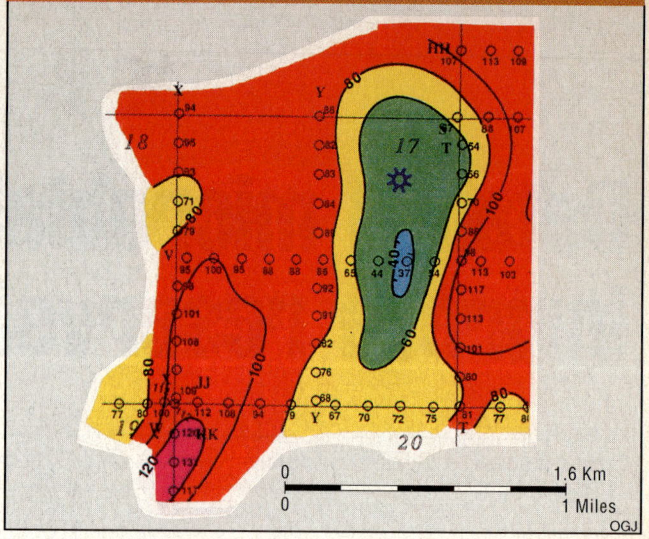

OGJ

these microbial populations. Surface contamination of produced oil and changing soil types do not affect these light hydrocarbon indicating microbial population distributions.

The specific MOST populations are measured by standard microbiological screening techniques for hydrocarbon-indicating microorganisms. A selective growth media is utilized which cultures only microorganisms capable of utilizing light hydrocarbons.

After incubation, microbial colonies are counted and reported as Microbial Values. Soil samples normally are processed within 48 hr after field collection. However, a new stabilization technique extends this period to one month for exploration areas or producing fields where express transport services are not available.

Production patterns

Hydrocarbon microseepage is dependent upon pressurized reservoirs driving the light hydrocarbon microseepage upward. The pattern of reduced microbial counts adjacent to producing wells has been a commonly observed phenomenon for older producing fields. Over some well-drained gas reservoirs, the microbial values have been found to even be anomalously low.

The phenomenon of apparent microseepage shutdown over producing fields is thought to be due to a change in the drive mechanism controlling microseepage. When a well is brought into production, the drive mechanism changes from a vertical, buoyancy driven force to horizontal gas streaming to the pressure sinks created around producing wells.

When this occurs, microseepage greatly decreases or ends and microbial populations at the surface decline rapidly. This change in drive mechanism and microbial population densities can be used to define reservoir drainage direction, radius, and heterogeneities around existing wells in producing fields.[8]

In *undrilled* areas this phenomenon will not occur and there will be a direct relationship between high microbial populations, microseepage, and potential reservoirs.

Rapid alterations in light hydrocarbon microseepage are required to explain the observed microbial population patterns identified in these surveys. Although more stingent research is required to completely explain this phenomenon, initial investigations suggest the migrations of gases and liquids within the crust are controlled by solid earth tides.[9] This seems to be the most logical mechanism for controlling the opening and closing of microfracture pathways within the crust allowing rapid microseepage to the surface.

Producing field examples

Microbial surveys conducted over active and abandoned fields identified drainage patterns, infill locations, stepout potential, and perhaps redrill potential of repressured abandoned fields.

The hydrocarbon microseepage trends will be affected by the actual hydrocarbon withdrawal, reservoir heterogeneities, and bypassed pockets of hydrocarbons.

Sacramento basin

In this example from one of the larger gas fields, all of the predicted microbial patterns have been identified (Fig. 1).

Clusters of *low* microbial populations over or adjacent to many of the producing wells identify the reservoir drainage radius around the wells. Some of these patterns are more radial while others were elongate. The elongated patterns are probably related to reservoir heterogeneities within the reservoir.

Areas of *high* microbial activity which were found between the producing wells indicate pockets of unproduced hydrocarbons within the field. These microbial highs were found to be either surrounded by radial withdrawal patterns or as elongated pods (Figs. 1, 2).

The first is probably related to yet undrained portions of the producing reservoir, while the second type appears to be related to reservoir fragments isolated from the producing well by reservoir heterogeneities. The second type would be potential targets for infill drilling.

Additionally, large, continuous clusters of elevated microbial values, in the northeast (Fig. 1), are similar to those associated with exploration programs. These indicate untapped reservoir potential and are either missed horizons for recompletions or flank potential for stepout drilling.

Southeast Oklahoma

In this example a large, elongated microbial low was identified within a larger area of microbial highs (Fig. 3).

Since most of our domestic microbial surveys are collected by clients, the results were presented to the client prior to GMT's knowledge of the production history of the survey area.

DOE OIL RECOVERY PROGRAM CLASS III PROPOSALS

Proposer	Project title	Site
	NEAR TERM PROPOSALS	
Texas Tech University, Lubbock, Tex.	Characterization, development of Spraberry Trend using horizontal drilling—a multidisciplinary approach	Spraberry Trend near Midland, Tex.
City of Long Beach, Calif., Dept. of Oil Properties	Boosting waterflood reserves with improved reservoir characterization, management	Wilmington field, Long Beach, Calif.
University of Texas at Arlington	Increased waterflood recovery by improved injection, completion technology	N. Jameson field, Colorado City, Tex.
Pacific Operators Offshore Inc., Santa Barbara, Calif.	Trilateral horizontal well to optimize recovery, reserves from multiple turbidite reservoir offshore	Carpinteria field off Santa Barbara County
Lomax Expl. Co., Inland Resources Inc., Salt Lake City	Reservoir characterization for waterflood demonstration in lacustrine slope/basin sandstones, Green River basin	Duchesne field Gilsonite Unit near Roosevelt, Utah
S-Cal Research Corp. and Joyce A. Kostura, San Rafael, Calif.	Improve Midway-Sunset steam injection economics	Midway-Sunset field, Kern County, Calif.
University of Utah, Salt Lake City	Reactivating idle lease by applying steam drive in low dip slope/basin reservoir	Midway-Sunset field, Kern County, Calif.
University of Houston, Tex.	Seismic-guided development of a Leonard time turbidite reservoir	Terry County, Tex.
Allied Oil & Tool Co., Scottsdale, Ariz.	Ultradeep reservoir opening in wells to extend life	Poso Creek, Jasmin, and Antelope Hills fields near Bakersfield, Calif.
Bureau of Economic Geology, University of Texas-Austin	Advanced reservoir characterization, simulation, and production optimization strategies in Delaware basin	Geraldine Ford and W. Ford fields, Orla, Tex.
	MID-TERM PROPOSALS	
Bayou Choctaw Inc., Metairie, La., and SMK Energy Corp., Tyler, Tex.	Bayou Choctaw field slope/basin clastic recovery program	Bayou Choctaw field near Plaquemine, La.
S-Cal Research Corp., San Rafael, Calif.	Improve steam injection economics	Midway-Sunset field, Kern County, Calif.
Shell Western E&P Inc., Bakersfield, Calif.	S. Belridge diatomite steam drive	S. Belridge field near McKittrick, Calif.
Improved Petroleum Recovery Consultants, Yorba Linda, Calif.	Improving layered reservoir steam drive economics by drilling seismic guided horizontal producers, slim hole injectors	Midway-Sunset field, Kern County, Calif.
City of Long Beach, Calif., Dept. of Oil Properties	Advanced reservoir characterization, thermal production technologies to increase heavy oil reserves	Wilmington field, Long Beach, Calif.
Orla Petco Inc., Midland, Tex.	Secondary CO_2 recovery optimized with advanced reservoir modeling and 3D seismic flood front monitoring	E. Ford field near Orla, Tex.
City of Long Beach, Calif., Dept. of Oil Properties	Integrated approach to improved reservoir characterization, waterflood management	Long Beach Unit, Wilmington field, Long Beach, Calif.
Strata Production Co., Roswell, N.M.	Advanced technologies for improved recovery from slope/basin clastic reservoir	Nash Draw field Brushy Canyon pool Eddy County, N.M.
Unocal Corp.-Coastal Calif. Business Unit, Ventura, Calif.	Multizone frac from extended reach horizontal wells in tight Repetto turbidites	Santa Clara field, Santa Barbara Channel off California
Chevron Production Co., Bakersfield, Calif.	Advanced Antelope shale reservoir characterization to learn CO_2 EOR viability in Monterey siliceous shale	Buena Vista Hills near Taft, Calif.
Oryx Energy Co., Dallas, Tex.	Dynamic early development, characterization, management of Discoaster B	Neptune field, deepwater Gulf of Mexico off Louisiana
Shell Western E&P Inc., Bakersfield, Calif.	Use of advanced steam, horizontal well technology in heavy oil reservoirs	Midway-Sunset field, near Maricopa, Calif.
ARCO Western Energy, Bakersfield, Calif.	Economic recovery of oil trapped at fan margins using high angle wells, multiple hydraulic fracturing	Yowlumne field near McKittrick, Calif.
Texaco Inc., Denver, Colo.	Diatomite oil recovery project	McKittrick field near McKittrick, Calif.
Bureau of Economic Geology, University of Texas-Austin	Integrated multidisciplinary application of advanced reservoir characterization and management	Inglewood field, Los Angeles County, Calif.
Future Enhanced Oil Recovery Inc., San Diego, Calif.	Technical, operation issues of a downhole steam generation system	Wilmington field, Long Beach, Calif.
McFarland Energy Inc., Santa Fe Springs, Calif.	Study of Ten Section field	Ten Section field near Bakersfield, Calif.
Castaic Lake Water Agency, Castaic, Calif.	Public-private association extending an oil field's economic limit by turning produced water to a resource	Placerita field near Santa Clarita, Calif.
Michigan Technological University, Houghton, Mich.	Recovering bypassed oil in three California basins using horizontal drains and microbial enhanced oil recovery	Richfield, Russell Ranch, and Bellevue West oil fields
Osyka Producing Co., Houston	Integrated study to boost oil recovery in Leonardian Spraberry slope/clastic reservoirs	Pegasus, Parks fields near Midland, Tex.
Parker & Parsley Development Co., Midland, Tex.	Advanced reservoir characterization, evaluation of CO_2 gravity drainage in naturally fractured Spraberry reservoir	Spraberry Trend near Midland, Tex.

well. Favorable conditions for improved recovery include 5,000 psia reservoir pressure, 300 million bbl OOIP with less than 4 million bbl recovery, and limited development by nine producing wells.

Team members are with Texas A&M University, University of California-Berkeley, University of Southern California, and Halliburton Energy Services.

California thermal

Shell Western and Maurer Engineering Inc., Houston, will try an advanced steamflood of 13.5-15.8° gravity oil in a shut-in Miocene Monarch sand reservoir in Midway-Sunset field over 5½ years. The $18 million project would test modifications to the steam assisted gravity drainage (SAGD) process described in Canadian engineering literature.

Ten slim hole vertical wells would be used to inject steam more cost effectively into the base of the Monarch sand aimed at a more rapid oil response in horizontal production wells.

Improved Petroleum Recovery Consultants proposed a 7 year project in Midway-Sunset field's Upper Miocene Marvic reservoir. Team members include Unocal Corp., its Brea, Calif., laboratory, and University of California-Berkeley.

Elements include use of slim holes rather than the usual high cost cased holes for steam injection, use of in-situ seismic guidance based on analysis of drill bit signals for locating the thin shale interbeddings, the drilling of oscillatory horizontal production wells to optimally pierce the shale barriers and maximize exposure to the producing formation.

A 7 year thermal project proposed in the Tar zone on the southern half of Fault Block IIA in Wilmington field involves City of Long Beach, Tidelands Oil Production Co., Long Beach, University of Southern California, and David K. Davies & Associates Inc., Kingwood, Tex.

An existing steamflood has been relatively inefficient due to inadequate characterization of the heterogeneous turbidite geology, high permeability thief zones, low gravity oil, and nonuniform distributions of remaining oil.

The team plans to develop 3D deterministic and stochastic geologic models and reservoir simulations, steam injection/production via new horizontal wells, and hot water alternating steam (WAS) drive.

The project would also involve installing an insulated, subsurface harbor channel crossing to supply steam to an island location and use of a novel alkaline water/steam completion technique to control well sanding and cut completion costs.

The public abstract supplied to DOE for the mid-term project proposed by S-Cal Research Corp. and Joyce A. Kostura is similar to the abstract for the same proposers' near term project. The 63 month mid-term proposal involves 13° gravity oil that requires steam injected at high rates through horizontal drainholes at the base of dipping reservoirs and gas/surfactant for improved sweep.

The proposal points out that 12,000 vertical wells have been involved in steam projects, many of which are now shut-in.

DSG proposal

Future Enhanced Oil Recovery Inc.'s Wilmington field downhole steam generator proposal also involves Universities of Houston and Arlington, Tex., and Viscosity Control Systems, San Fernando, Calif.

The proposal envisions studying or integrating engineering and geologic data, water chemistry, mineralogical data, crosswell seismic, and analogous sands in outcrop. Sedimentological and reservoir models will be developed, a downhole steam generator and auxiliary equipment will be fabricated and tested, and a huff-n-puff pilot carried out.

A second budget period could involve a steamflood with a new injection well and four producers.

Immiscible CO_2

Chevron Production Co. proposed a 5½ year project to demonstrate immiscible CO_2 flooding in stripper category Miocene Antelope/Monterey shale in Buena Vista Hills field, where primary recovery is estimated at 6.5% of OOIP.

Reservoir characterization will involve high resolution core analysis including profile permeameter, oil fingerprinting, borehole imagery, modern wireline logging, and multiscale fracture characterization. Cross-well seismic imagery, including new attenuation imaging techniques for fracture characterization, will be applied to monitor sweep efficiency.

In comparable rock, laboratory displacement tests and associated computer simulations showed CO_2 recovery in core to be 50-60% and indicated reservoir recoveries in the range of 5-15% of OOIP.

> *One group proposed a 6 year project to implement microbial enhanced oil recovery combined with horizontal producing wells in Upper Miocene turbidite sands in three California basins.*

Produced/potable water

Castaic Lake Water Agency, ARCO Western Energy, and Kennedy/Jenks Consulting, Ventura, Calif., proposed a 25 year project to convert water produced from Placerita oil field of Los Angeles County into potable water.

The Ventura basin field is under cyclic and continuous steam injection, and water handling is second the costliest operation after steam costs. The field's water cut is 94%, so operators are treating and handling more than 50,000 b/d of water. Current water disposal is underground injection below the producing zone's oil-water contact.

The project aims to add reserves by reducing water cut by permanently removing water from the reservoir, cutting costs for injection and chemicals for water treatment, and increasing thermal efficiency.

Proved reserves of oil and water sufficient to last 20 years must exist to support a full scale produced water treatment facility. A pilot scale plant would have to demonstrate feasibility for a later 48,000 b/d treatment plant for the Santa Clarita Valley.

Microbial EOR plan

Michigan Technological University, University of South Florida, Tampa, and Magness Petroleum, Placentia, Calif., proposed a 6 year project to implement microbial enhanced oil recovery combined with horizontal producing wells in Upper Miocene turbidite sands in the San Joaquin, Los Angeles, and Cuyama basins.

Technologies to be developed are designed to benefit small to medium sized independents by demonstrating how to design and implement horizontal drains, initiate and manage efficient recovery programs, and evaluate the feasibility of using microbial EOR technologies.

The group plans to characterize 15-30 San Joaquin basin fields to develop a detailed, 3D picture of Stevens and Monterey formations.

More California plans

ARCO Western Energy and others would drill a 56° well to obliquely penetrate the fan margin area of a reservoir in Yowlumne field. Exploitation of oil in fan margins becomes uneconomic due to lateral thinning of individual sand lobes and lateral deterioration of reservoir properties.

Multiple hydraulic frac jobs are to be run in the high angle well to effect vertical communication between layers and provide 1,100 ft of formation penetration. Micro-seismic logging and other methods will be used to determine hydraulic fracture azimuths.

The well's artificial lift system will involve advanced electric submersible pumping equipment to improve reliability and maximize production rate. As modeled, the fractured slant well should replace three conventional vertical wells.

The team includes Scotia Group Inc., Dallas, Dowell-Schlumberger, University of California-Santa Barbara, and California Division of Oil & Gas.

City of Long Beach proposed to team with others to improve reservoir characterization and waterflood management in the Upper Miocene Puente/Lower Pliocene Repetto formations in Wilmington field's Long Beach Unit.

Waterflooding implemented at unit start-up involved well completions over vertical intervals as thick as 1,000 ft with open holes gravel packed behind slotted liners. This made it difficult to waterflood all the individual sand layers uniformly.

The 5½ year project would involve improving understand of fault locations and their sealing nature, stratigraphic continuity of sand bodies, lateral variation in permeability, and continuity and extent of individual flow units. Drilling horizontally with continuous coiled tubing, chemical methods of profile modification, realigning waterflood patterns, redrilling producers as injectors, and adding infill producers may be tried.

Other participants are

ARCO Long Beach Inc., California State Lands Commission, Colorado School of Mines, and Thums Long Beach Co.

Bureau of Economic Geology and Stocker Resources Inc., Los Angeles, a subsidiary of Plains Resources Inc., Houston, over 6 years will try to locate and produce a large share of the more than 413 million bbl of oil remaining in Pliocene Vickers-Rindge reservoirs in Inglewood field.

Plans include an alkaline-surfactant-polymer flood, an as yet untested multilateral injection well supported by slim hole injection wells targeted to flood discontinuous flow unit compartments, and innovative profile modification treatments including polymer gel treatments, foam-diverted acid stimulation, and a detrital-blocking technique to further re-engineer the existing waterflood to improve lateral and vertical sweep efficiency.

A main objective of McFarland Energy Inc.'s proposed 3 year Ten Section field study is to devise and implement strategies to produce bypassed, uncontacted reserves.

Discovered in 1936 with 214 million bbl OOIP and 249 bcf OGIP in Miocene Stevens sands, the field has produced 76 million bbl and 208 bcf.

The field has 105 idle Stevens wells, and 17 wells are producing 460 b/d of oil, 174 Mcfd of gas, and 3,292 b/d of water.

Stevens zone II, the primary oil reservoir, is the main target. The study would involve detailed reservoir characterization and reservoir simulation and other work that might improve waterflooding, such as polymers, in situ gelation for profile control, horizontal well(s), injection strategies, and simulation assisted reservoir management.

Team members are Intera Information Technologies Inc., Denver, and Ten Section Gas Services Inc., Los Angeles.

California diatomite

Two projects are aimed at oil in California's diatomaceous earth in Kern County.

Texaco Inc. units and Davy International, Pittsburgh, hope to demonstrate Texaco's Diatomite Solvent Extraction Process (TDSEP) on Monterey diatomite in McKittrick field.

They plan to design, build, and operate a 300 b/d pilot plant to demonstrate TDSEP over 42 months. The pilot plant operation is intended to confirm laboratory investigations from the 1980s and provide data to design and buld a commercial size plant.

Shell Western E&P Inc. and others proposed a 51 month, continuously optimized diatomite steam drive on 15 acres Shell operates in South Belridge field. Steaming would start in early 1995 and continue 4 years. Existing steam facilities would be used, but new infill wells would be drilled and existing wells would be converted to complete the steamflood patterns.

Recovery efficiency in diatomite in the San Joaquin basin have been low due to extreme reservoir heterogeneity in the form of variable oil saturations and gravities, very low permeabilities, natural fractures, and induced fracture reorientation.

The project has been designed to reduce uncertainties, mitigate steam breakthrough, maximize heat injection without extending hydraulic fractures, and assess the impact of steam injection on subsidence related well failures. Other participants are Crutcher-Tufts Production Co. and Santa Fe Energy Resources Inc. of Bakersfield and University of California-Berkeley.

West Texas Spraberry

Two proposals involve coaxing more oil from West Texas Spraberry reservoirs.

Osyka Producing Inc. and others propose to interpret 2D and 3D seismic data over Pegasus field to enhance resolution and imaging at the Spraberry level to delineate fracture zones and/or enhanced porosity zones.

Descriptive reservoir characterization will entail drilling and coring two wells through Spraberry, descriptive core analysis, petrophysical and petrographic analysis, calibration of logs to core data, and developing a calibration technique for old electric logs.

Another proposal involves adding oil reserves by purging Placerita field of formation water and processing it into potable water at the ultimate rate of 48,000 b/d for use in the Santa Clarita Valley.

These results and integrated geological and engineering data will be used in 3D reservoir simulation to characterize fracture patterns. Techniques designed at Pegasus will be demonstrated through recompletions and more infill drilling at Pegasus and Parks fields.

Team members on the 6 year project are Principle Energy Development, Cypress, Tex., and Mobil Exploration & Producing U.S. Inc.

A group led by Parker & Parsley Development Co., Midland, may try CO_2 flooding in a Spraberry reservoir. Spraberry, with its extensive natural fractures, has not been CO_2 flooded.

Based on theoretical and experimental work combined with an analysis of the successful CO_2 pilot in the naturally fractured Midale reservoir, plans for reservoir characterization followed by a Spraberry CO_2 pilot have been formulated.

Team members are New Mexico Petroleum Recovery Research Center, Socorro, N.M., Epic Consulting, Calgary, Alta., Lincoln Elkins, Lakewood, Colo., and University of Texas-Permian Basin, El Paso.

Spraberry, covering 2,500 sq miles, is the world's largest areal reservoir. Spraberry produced 18.575 million bbl from 7,587 wells in 1993. Parker & Parsley operates 2,400 Spraberry wells and drilled 249 in 1993.

Other Permian basin

Orla Petco Inc. and Great Western Drilling Co. would start carbon dioxide flooding in the first year of an 8 year project in Permian Ramsey (Delaware) sand at 2,700 ft in East Ford field, Reeves County, Tex. Extensive reservoir characterization has been completed.

Preliminary work with synthetic seismograms and time-average equations to estimate seismic velocity and acoustic impedance in Ramsey as a function of oil, water, and CO_2 saturation indicates that reflection methods can map CO_2 flood fronts, the proposers said. CO_2 is to be supplied by means of an innovative alliance with a major company.

Strata Production Co. leads a 5 year proposal in Nash Draw Brushy Canyon field of Eddy County, N.M., to develop a control area using standard reservoir management and compare its performance with that of an area developed using advanced management methods.

Other participants are Petroleum Recovery Research Center, Institute for Improved Oil Recovery, geophysicist Dr. Bob Hardage, Scott Exploration Inc., and Pecos Petroleum Engineering Inc.

A reservoir model integrating production, petrophysical, geostatistical, geophysical, and geologic data will be built. Various combinations of vertical and horizontal wells would be drilled and selective completions performed to optimize production performance. ▪

PUBLICATIONS

Guidebook 25: Waulsortian Mounds and Reservoir Potential of the Ullin Limestone ("Warsaw") in Southern Illinois and Adjacent

Areas in Kentucky, *by Zakaria Lasemi, Janis D. Treworgy, Rodney D. Norby, John P. Grube, and Bryan G. Huff, published by Illinois State Geological Survey, 615 E. Peabody Dr., Champaign, Ill. 61820-6964. $6 plus shipping.*

The presence of Waulsortian-type carbonate mud mounds within the Ullin is of interest because such mounds have potential for reservoir development. This is supported by recent discoveries of prolific hydrocarbons in the Ullin limestone, in part associated with the Waulsortian mound facies.

The carbonate mounds appear morphologically similar to Silurian reef mounds in the basin, except for the lack of frame-building organisms.

Regional Geologic Framework of the Cretaceous, Offshore Mississippi, Open File Report 21, *by A. John Warner, published by Mississippi Office of Geology, P.O. Box 20307, Jackson, Miss. 39289-1307. 40 p., 1 plate, $8 plus handling.*

This is a study of the deeply buried Cretaceous rocks rocks beneath coastal Mississippi and state coastal waters. It briefly discusses 15 Cretaceous formations, presents paleogeographic maps for Selma and Hosston, and includes regional correlations.

ARGENTINA

Yacimientos Petroliferos Fiscales AS let a contract to a joint venture of Veritas Energy Services Inc., Calgary, and Union Geofisica Argentina SA for seismic surveys in Argentina.

The $10 million contract, with acquisition starting in late July, calls for 1,250 line km of 2D Vibroseis work in the Neuquen basin and 600 line km in the Chaco-Parana basin.

ALBERTA

Talisman Energy Inc., Calgary, may develop an expanded horizontal program aimed at oil in Upper Cretaceous Belly River in the Warburg area southwest of Edmonton.

The company completed a horizontal well flowing more than 500 b/d in mid-June from about 3,150 ft true vertical depth during initial tests of Belly River, which has not been seen as a prime candidate for horizontal drilling because of concerns about formation damage during drilling.

Careful design of the drilling and completion systems enabled Talisman to drill 2,300 ft of horizontal section, the company said. It plans to drill one more horizontal well before deciding on the expanded program.

Talisman also had a strong vertical well program in the area, where it has consolidated properties acquired from Encor Energy Inc. with Warburg area properties of its own.

Each of 16 vertical wells drilled the past two years is averaging 100 b/d of light, sweet crude. Most extended Belly River pools. These vertical wells and the horizontal well have added about 2.4 million bbl of oil reserves for Talisman. Three more stepouts are to be drilled this summer.

BR. COLUMBIA

Talisman Energy Inc., Calgary, expects to drill a record eight to 10 wells this year in the mountainous Monkman area of Northeast British Columbia.

It had three rigs drilling in the area in mid-July, one of which had reached a Triassic reservoir.

Talisman 50-50 with Amoco Canada Petroleum Co. Ltd. gauged 47 MMcfd of gas at the Burnt c-61-A/93-0-8 well 5 miles from a tie-in to the Sukunka dehydration facility. Acid gas content of 8.4% is one of the lowest in the area.

The well, Talisman's 17th in the area the past 5 years, is a major step towards projected production of 150 MMcfd of gas with Westcoast Transmission's Pine River expansion starts up in November 1994, Talisman said. Talisman's Monkman area production averaged 81 MMcfd in 1993.

Burnt c-61-A is the longest reach hole Talisman has drilled. By building location 2,550 ft offset from ultimate bottomhole location and taking advantage of new directional drilling technology, Talisman said it saved $1 million in construction costs. The $6.75 million drilling cost includes $2 million for lease construction.

U.S. GULF

Marathon Oil Co., operator, BP Exploration, and Shell Offshore Inc. are reviewing data from an apparent discovery on Green Canyon Block 244.

The 1 Green Canyon 244, in 2,761 ft of water, encountered 300-400 ft of net oil and gas pay, with the majority of net oil pay in one sand. Fluid samples indicated a high quality, 36.5° gravity crude, Marathon said. Total depth is 16,212 ft measured depth.

The block is part of an exploratory unit that includes Green Canyon blocks 200, 201, 244, and 245. Unit interests are held by Marathon, BP Exploration, and Shell Offshore Inc. one third each. The discovery well is 12 miles southeast of Shell's Bullwinkle platform on Green Canyon Block 65.

Matrix Oil & Gas Inc., Covington, La., plans to drill more development wells this year in West Cameron Block 192 field, discovered in 1954.

Matrix's B8 OCS-0190 well in the field flowed 4.47 MMcfd of gas from perforations at 9,507-52 ft true vertical depth. It was placed on production June 19. Matrix and Santa Fe Minerals Inc. own the field 50-50.

LOUISIANA

Las Colinas Energy Corp., Irving, Tex., and two other independents plan to start acquisition on Phase I of the Lacassine Refuge 3D seismic survey in Cameron Parish in late August.

The second phase of the 52 sq mile survey is to be acquired in 1995. Universal Seismic Acquisition Inc. has the Phase I contract.

MICHIGAN

Exploration is on tap to extend northern Michigan's shallow Devonian Antrim shale gas play 130-160 miles south of the main existing production fairway (OGJ, May 30, p. 97).

Amoco Production Co. has staked 11 Antrim exploratory tests in seven southern Michigan counties.

The sites to be drilled are in 35-2n-3w, 35-4n-5w, and 14-4n-6w, Eaton County; 10-5n-8w, Ionia County; 29-2n-1e, Ingham County; 20-4n-3e and 11-4n-5e, Livingston County; 9-5n-4e and 8-6n-3e, Shiawassee County; 34-6n-6e, Genesee County; and 29-7n-11e, Lapeer County, Petroleum Information reported.

Amoco is pressing horizontal drilling in Ordovician Prairie du Chien reservoirs in Michigan.

The company intends to horizontally redrill its Clayton field discovery well in Arenac County at 10,800 ft true vertical depth, PI reported. The original completion in 1986 was for gas and condensate from Glenwood and Prairie du Chien.

Seven Clayton field wells produced 16.5 bcf of gas and 897,800 bbl of condensate through 1993, PI noted.

Amoco also staked an 11,500 ft horizontal Prairie du Chien well in South Buckeye gas/condensate field of Gladwin County. Amoco recently finished redrilling a horizontal well in this area and is drilling another from surface.

Exploratory drilling results, directions in Kirgizstan

M.U. Usuvaliyev
K.T. Toktonaliyev

Structural exploration drilling began in Kirgizia in 1962. During 1976-82 it was concentrated in the southwest part of the Fergana depressions, where two oil fields were discovered, Beshkent and Togap.

Both the Beshkent and Togap folds flatten out with depth, and on the top of the Paleogene sediments they are expressed as structural noses that plunge to the west at 5-10° at Togap and 15-20° at Beshkent. The axis of the fold on the top of the Paleogene is shifted 600-900 m to the north. Even though these folds have identical structure, the composition and quality of the oil is markedly different.

Beshkent, Togap fields

Beshkent field was discovered by Well No. 5. A flow of oil at 5.6 cu m/day was recovered from Oligocene Stratum III in the 1,809-20 m interval. Formation pressure is 236 atm, and density of the oil is 0.864. Content in percent is sulfur 0.2, asphalt tar 28.0, silica-gel tar 12.9, and paraffin 11.6. Structural exploration wells 8 and 9 yielded oil at 1.5 and 7 cu m/day, respectively.

Stratum III in Beshkent field consists of dense, fine-grained clayey sandstone 15-18 m thick. Net pay is 5 m. Porosity is in the 15-25% range.

Togap field was discovered by structural-exploration Well No. 1. A flow of oil at 5.7 cu m/day was recovered from Stratum III in the 1,310-20 m interval.

Formation pressure is 115 atm, and density of the oil is 0.873. Content in percent is sulfur 0.27, asphalt tar 32, silica-gel tar 19.5, and paraffin 17.4.

In Togap field Stratum III consists of sandy siltstone with beds of fine-grained sandstone. Thickness is 10-15 m, and net pay is 3.5 m. Porosity is in the 13-21% range.

Other drilling

A profile of five wells was drilled on the Tamchin and Vtoraya Tamchin structures. With depth these two features flatten out and pass into a single fold on the top of the Paleogene sediments—Tamchin proper.

Five additional wells were drilled to study the flanks of the Tamchin structure; these showed that on the Paleogene the fold has shifted its position somewhat to the west. Well 10 yielded 0.2 cu m/day of oil and water from Paleogene Stratum IV of the Rishtan beds. Density of the oil is 0.867. Content in percent is sulfur 25, asphalt tar 28, silica-gel tar 10.3, and paraffin 10.5. The productive stratum is a sand-silt rock, and porosity is 18.25%. Clay content is 24-60%.

A total of 12 wells were drilled in four areas: Severnyy Togap, Murgun, Shark, and Ayzar.

Structural and parametric wells were drilled in the north border area of the Kara-Tau (southwest part of the Fergana depression) to determine reservoir quality of the rocks and to trace facies changes. Three parametric wells were drilled at distances of 7-10 km from one another. It was found that the Permo-Triassic and Jurassic drop out of the section, and that thin (50 m) Cretaceous sediments rest on eroded metamorphosed Paleozoics—and Paleogene on the east.

There are also discordant breaks in deposition between the Ancient Quaternary and the Neogene Baktriy beds, the Neogene Brick-Red formation, and the Oligocene Sumsar beds, the Sumsar beds and the Eocene Green Clay formation, and Paleogene and Cretaceous, and the Cretaceous and Paleozoics.

Practically no marker horizons are present in the Neogene because of pinchouts due to breaks in deposition. This situation necessitates expenditure of excessive amounts on deep drilling because of discordance between the Neogene and Paleogene.

1990s exploration

In the coming years structural-exploration drilling will be carried out largely in Southwest and East Fergana and then in the north of Kirgizia.

Several structures on the south flank of the Kara-Tau high (Karagan, Staro-Sulyuktin, Ketmenkul, Sarydzhar, and Dashtakin) are primary drilling targets. A profile of drilling is planned for the Tashravat-Murgun downwarp, which is located between the Kara-Tau and Mayday-Tau highs.

Another target is the Zambarych group of structures of Southwest Fergana,

Extracted from Petroleum Geology, Vol. 28, No. 5/6, published by Dr. James Clarke, P.O. Box 71, McLean, Va. Translated from Povysheniye effektivnosti geologo-razvedochnykh rabot na neft' i gaz v Kirgizskoy S.S.R., Frunze "ILIM," 1990.

which were found in 1948. Seventeen structures were mapped, nine of which are in Kirgizia. The oil-gas potential of this region is unclear. However, taking into account the proximity of the Karagach and Togap-Beshkent fields, this region may be highly favorable.

V.A. Babakhyan (1948-50) mapped 19 local structures in the Uzgen-Kurshab depression of East Fergana. He described most of these as structural noses. The Paleocene sediments were assessed as without prospects, but the Lower Cretaceous and Jurassic were considered as favorable. Babakhyan thought that the Kurshab group of structures—Digmay, Zerger, Tokubay anticlines, and Dzhilandin, Kondubin, Shurabashat structural noses—were prime targets.

Two exploration wells were drilled in the Dosmet area of the Uzgen downwarp in 1981-82. A flow of water with dissolved gas was recovered from Jurassic Stratum XXIII in the 984-1,009 m interval. The gas is largely methane. This indicated favorability for the Jurassic as well as the Permo-Triassic.

Future targets

Targets for drilling over the next 20 years are the Alay, Talas, Vostochno-Chuy, Issuk-Kul, and Naryn-Chatyrkul depressions.

The Paleogene section of the Alay depression appears to be the same as in the Fergana depression. This depression joins the Cis-Kun-Lun oil-gas downwarp of China. Three anticlinal structures are known, which geological studies indicate are drilling targets.

The Talas depression is joined on the northwest by the Chu-Sarysuy depression of Kazakstan. Its oil-gas potential has not yet been studied. Several local structures are present, but the sedimentary section has not yet been studied. It is possible that this depression is similar to the Chu-Sarysuy.

In the Vostochno-Chuy depression there is interest in pinchouts to the north of Upper Paleozoic and Lower Cenozoic deposits.

The west part of the Issuk-Kul Valley has not been drilled, although salt is known in the Neogene section, and thick Upper Paleozoic clastics are present in the mountain frame. Methane is present in formation water of the Carboniferous and Paleogene-Neogene.

The Naryn-Chatyrkul depression was connected with the Fergana depression before uplift of the Fergana Range beginning in the late Jurassic. Drilling has penetrated Jurassic and unmetamorphosed Paleozoics in the section. Their presence beneath thick Cenozoic deposits suggests that they have oil-gas potential. ▪

DOE tallies Class III oil recovery field projects

Here are details from mid-term proposals submitted as part of the U.S. Department of Energy's Class III oil recovery field demonstration candidate projects.

All of the proposals emphasize dissemination of project details so that the results, if successful, can be applied widely in similar reservoirs. Project results will also be fed into a national petroleum technology transfer network.

Class III near term proposals will be profiled in a later article.

Mid-term proposals

Most Class III mid-term proposals involve projects of 5-6 years' duration, but one is to last 25 years.

Flex-Gulf of Mexico

Oryx Energy Co., TGS/Calibre, and Western Geophysical proposed a project involving Viosca Knoll Block 826 in 1,900 ft of water 135 miles east-southeast of New Orleans.

The Neptune development has 11 Upper Miocene reservoirs in structural-stratigraphic traps surrounding a salt dome.

In such reservoirs multimillion barrel discoveries often occur in stacked, highly stratified Miocene turbidite sandstones that require optimum wellbore placement to exceed marginal economic thresholds. Oryx said its site is constrained by these economic realities but offers significant recoveries if the lateral continuity of the turbidite sand bodies can be properly characterized and managed.

The Discoaster B reservoir, with an estimated 23 million bbl recoverable, is the target of accelerated reservoir characterization to coincide with initial development drilling.

The program is to integrate high resolution 3D seismic and 3D amplitude versus offset data, sequence stratigraphy, turbidite facies data derived from cores and wellbore imaging, production data from two initial development wells, and geostatistical analysis and reservoir simulation.

Gulf Coast

Bayou Choctaw Inc. and SMK Energy Corp. proposed to apply advanced reservoir characterization techniques in Bayou Choctaw field, Iberville Parish, La.

The 5 year, $16.7 million proposal notes that technology constraints in the field are lack of seismic data, few modern open hole or cased hole logs, no conventional cores, and no available sidewall cores. Drilling and production practices by previous operators were dictated by offset drainage protection, not reservoir management.

The companies would rework/recomplete existing inactive wellbores using modern technologies, acquiring 3D seismic data to define reservoirs, and drill new wells to tap them.

They postulate that this program will add a minimum of 30 million bbl of oil reserves and 3,000 b/d of production in the field. Success could be applied to more than 134 salt domes in Louisiana alone.

California offshore

Unocal spearheads a 5 year proposal involving Lower Pliocene Repetto turbidite sequences LP-K, L, M, and N accessed from Platform Gilda in Santa Clara field of the Ventura basin in Santa Barbara Channel federal waters.

The project would use high angle wells with multiple hydraulic fracs, including activities that have not been attempted off California. The friable Repetto has in-situ porosity of 5-30%, and permeability averages 3 md, with extreme vertical variation in both properties.

A Wilmington field project would involve installing an insulated, subsurface channel across Los Angeles harbor to supply steam to an island location.

With oil gravity averaging 20° and viscosity 2 cp at 175° F. and 5,000 psia in situ, pressure communication is limited to within a few hundred feet of a well. Modern well logs and 3D seismic data are available, but only 60.5 ft of core was taken from a single well.

Development of Lower Repetto from Platform Gilda costs as much as $4 million/

particles (two neutrons and two protons bound together) are dangerous if they find their way into the human body.

The new long-range alpha detectors can detect contamination from as far as several yards away.

The two basic detectors that Los Alamos scientists have designed are airflow and electrostatic. In the airflow long-range alpha detector, a fan transports the ion pairs into a chamber where they are collected on a metal screen. A computer controlled electrometer detects the electric charge as a small current, which is converted into a measure of the radioactivity of a sample.

The electrostatic detector uses a charged field to transport the ion pairs created by alphas to the signal plane.

Los Alamos is developing the liquid waste monitor for its radioactive liquid waste treatment facility. The instrument will be used to monitor radioactive liquid waste as it enters the treatment facility. Further improvements would allow this detector to monitor the processed liquid before it leaves the plant. If the liquid is below regulatory limits for radionuclides, it could be discharged without further treatment.

Rocks for teachers

The North Carolina Geological Survey (NCGS), several state agencies, science education coalitions, and mineral and crushed rock companies, are cooperating to distribute basic earth science resources to public elementary and middle schools. Donations from the mining industry plus funding from the North Carolina Mining Commission, the Education and Industry Committee for Earth Science, and the Math Science Alliance, are enabling NCGS to assemble 1,400 rock kits for teachers.

Each kit is representative of North Carolina's geology and contains 19 bagged and labeled rock and mineral specimens, a streak plate, a geological map of North Carolina and an instruction guide. The guide, which was developed with the Math/Science Nucleus, a 10-year-old program involved in earth-science education at the elementary level, contains data sheets about each specimen, 19 hands-on activities, and a section on the geology of North Carolina.

Training workshops for teachers will be organized on request.

Museum exhibit traces evolution's family tree

The American Museum of Natural History, New York, launches its 125th anniversary celebration this year with the opening of two newly renovated and restructured fossil halls. The renovation of the museum will restore the building's architectural detail and exterior vistas, and the floor plan of the exhibit will show visitors the evolution of life.

The two halls are the first of six that will be redone by 1996. The halls are structured to show evolution as it unfolded and interrelationships of living things, as opposed to traditional exhibits that trace evolution through time. The path through the new exhibit is patterned as an "evolutionary family tree," with paths branching off to represent the development of new evolutionary characteristics, such as new features in a skeleton or new ways of reproducing. Each branch leads to an alcove containing both fossil and modern representatives of closely related animals.

The mammal wing will feature actual fossils, with casts used only to fill gaps in the museum's collection. Also, a group of special fossils will be available for visitors to handle.

For more information, contact the American Museum of Natural History, Central Park West at 79th St., New York, 10024-5192. Phone: (212) 769-5800.

Exhibits trace evolutionary family tree.

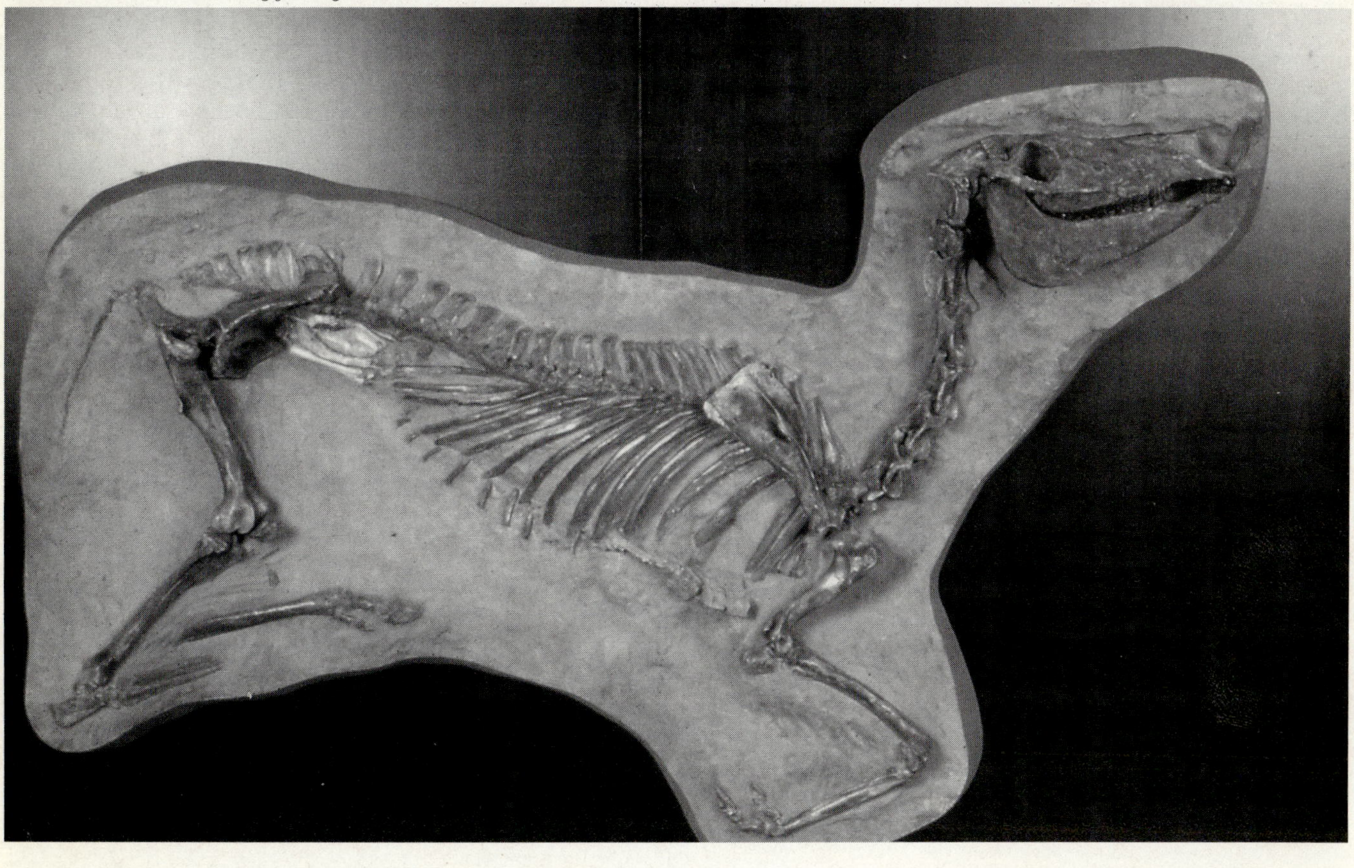

AMERICAN MUSEUM OF NATURAL HISTORY, NEW YORK

AGI testifies on geoscience funding

The American Geological Institute presented testimony to four congressional appropriations subcommittees concerning funding geoscience programs for FY 1995. In oral testimony and written statements, AGI Director of Government Affairs Craig M. Schiffries stressed the payoffs from federal investments in R&D on geological hazards.

"The disastrous earthquake that struck Los Angeles Jan. 17, 1994, and the devastating floods that inundated the Midwest in 1993 provide powerful reminders of the vital role the geosciences play in an ever-growing range of national goals. Recent earthquakes and floods have resulted in more than $10 billion in emergency supplemental appropriations and even larger private losses. The societal benefits of geoscience R&D on earthquakes and other geologic hazards extend to such areas as housing, transportation, commerce, agriculture, and human health and safety."

AGI's testimony to the Subcommittee on Interior and Related Agencies chaired by Rep. Sydney Yates (D-Ill.) and to the corresponding Senate subcommittee chaired by Sen. Robert Byrd (D-W.Va.) concerned FY 1995 budget requests for the U.S. Geological Survey and Fossil Energy R&D for the Department of Energy.

"Geoscience R&D," Schiffries noted, "is a key component of the Domestic Natural Gas and Oil Initiative that was announced by the Clinton administration Dec. 9, 1993. While research on renewable resources and energy efficiency is critical to America's future, continued research on fossil energy is no less important. The societal benefits of fossil energy R&D extend to such areas as economic and national security, job creation, capital investment, and reduction in the trade deficit."

USGS budget

A Transition Team within the USGS presented a vision for the 21st century and formulated policy options for achieving it. According to the team's report, "The mission is evolving rapidly to include a greater emphasis on integrated analyses of Earth's environment, hazards, and resources to assure sustained global health, welfare, and prosperity." The FY 1995 budget request for USGS is $583.7 million, an increase of 1.0 percent relative to the FY 1994 current plan.

"AGI urges Congress to support a resumption of real growth for the entire USGS and to begin redressing the long-term negative growth of the Geologic Division. Funding for many USGS programs that are directly relevant to national priorities would fail to keep pace with inflation. For example, funding for geologic hazards surveys would decline in constant dollars despite the recent flurry of earthquakes in California. Likewise, the budget for marine and coastal geologic surveys would be level-funded despite the coastal damage inflicted by recent hurricanes and winter storms. Moreover, the budget for energy resource surveys would also fail to keep pace with inflation despite the program's relevance to the Domestic Natural Gas and Oil Initiative."

DOE fossil energy R&D

"The budget request for the Department of Energy's fossil energy R&D programs reflects priorities articulated in the administration's Domestic Natural Gas and Oil Initiative. The FY 1995 budget request for fossil energy R&D is $469.4 million, an increase of $38.7 million, or 9.0 percent, relative to FY 1994 appropriations. Substantial increases in funding for natural gas R&D (59.7 percent) and petroleum R&D (42.1 percent) would be partially offset by decreases in coal R&D (23.5 percent) and other programs (12.1 percent).

"Premature abandonment of productive domestic oil fields is one of America's most serious energy problems," Schiffries said. "The development and application of advanced technologies could add an estimated 10 to 30 billion barrels of crude oil to current domestic reserves. This incremental domestic supply would be from existing fields, rather than from new fields opened in environmentally sensitive regions. Two-thirds of the oil found in the United States will remain unrecovered if only conventional production methods are used."

NSF budget

In testifying before the House Subcommittee on VA, HUD, and Independent Agencies May 5, and before the corresponding Senate Subcommittee May 20, Schiffries emphasized the importance of geoscience research and the success of the National Science Foundation.

"Geoscience research and information enhance society's ability to make wise policy decisions on resource development, environmental protection, natural hazards reduction, waste disposal, and land-use planning. It is also important to emphasize the need to maintain the health of the basic science on which applications and policy decisions must be ultimately based. AGI urges Congress to reaffirm its commitment to American preeminence in science and technology by fully funding the fiscal 1995 NSF budget request of $3.20 billion."

Funding geoscience research

"The rationale for supporting geoscience research and education has never been stronger," Schiffries told Chairman Rep. Louis Stokes (D-Ohio). "Global climate change, natural disasters, energy resources, and water quality issues are reported daily by the news media. Geoscience research plays an increasingly important role in an ever-growing range of scientific and societal problems, and federal investments in geoscience research should be increased accordingly. The proposed budget represents a modest investment in the future of our nation and our planet. AGI urges Congress to fund fully the fiscal 1995 budget request for the NSF Directorate for Geosciences."

Geoscience education

Stressing the need for NSF-sponsored education programs to Sen. Barbara Mikulski (D-Md.), Schiffries said, "A basic understanding of Earth is required for citizens to make informed decisions about the delicate balance between resource use and environmental protection. At the K-12 level, geoscience education has been largely underemphasized and underfunded for 40 years. We cannot afford to let that situation persist for another 40 years.

"We recommend that NSF and the geoscience community jointly develop and implement strategic plans for improving geoscience education and achieving parity with other scientific disciplines.

"Ethnic minorities are severely underrepresented among geoscience majors and graduate students. Women are also underrepresented among geoscience majors, and the problem is substantially worse at the graduate level. Continued NSF leadership is needed to increase recruitment and retention of underrepresented groups through scholarships and mentoring and to develop strategies to address underlying factors that prevent greater participation of those groups."

Copies of AGI testimony are available from Craig Schiffries, AGI, 4220 King St., Alexandria, Va. 22302-1507.

operate as joint ventures in restructuring to take place this year.

E&P

Pemex Exploration & Production is responsible for exploration and extraction of oil and gas, field gathering pipelines, and crude export facilities.

Of Pemex's $40 billion in assets, Pemex E&P received 55% under the reorganization. The unit took 58% of the Pemex budget in 1993. After falling for a decade, official Pemex crude oil reserves figures have stabilized since 1991 and stood at 44.4 billion bbl at the start of 1993.

Pemex's official figure for total hydrocarbon reserves, including crude oil, condensate, and natural gas, has hovered around 65 billion bbl of oil equivalent (BOE) since 1990, down from 72 billion BOE a decade ago.

In 1992, Pemex's crude oil production averaged 2.67 million b/d, or about 8,000 b/d less than in 1991, when production hit levels not seen since 1984.

Pemex crude exports averaged 1.37 million b/d in 1992 and generated about $7.4 billion in revenues for the year, about the same as in 1991, or about 89% of total Pemex exports. Pemex 1992 crude exports broke out as 67.5% Maya, 21% Isthmus, and 11.5% Olmeca. Mexico uses Isthmus grade crude for most domestic consumption.

Refining

Pemex Refining controls production and marketing of all refined products in Mexico.

It owns and operates Pemex's seven refinery complexes, more than 85 refined products distribution terminals, and crude and products pipelines. This division got 20% of Pemex's assets and 27% of the 1993 budget.

In 1992, Pemex refineries processed 1.57 million b/d of crude, natural gas liquids and condensate, and secondary process liquids, an 0.8% decline from 1991.

The more than 3,000 filling stations that display the Pemex logo are not part of Pemex Refining. They are private concessionaires that earn a commission based on the quantity of fuel sold. Domestic fuel prices and taxes are set by the Secretariat of Finance Ministry (Hacienda), the Commerce & Industrial Development Secretary (Secofi), and Pemex.

Gas, basic petrochemicals

Pemex Gas & Basic Petrochemicals manages supply and marketing of LPG, natural gas, and basic petrochemicals in Mexico.

It owns and operates Pemex's gas processing plants, LPG pipelines and distribution terminals, natural gas transmission pipelines, and all basic petrochemical plants. The division was allocated 12% of Pemex's assets.

Pemex produced 3.58 bcfd of natural gas in 1992, a 1.4% decline from 1991 and the second consecutive year of slight drops in production. The southern zone accounted for 54% of natural gas production, Campeche 33%, and the northern zone 13%.

Pemex processed 3.16 bcfd of natural gas at nine plants in 1992, a 3.7% decline from the previous year.

In all, Pemex subsidiaries produced more than 19 million tons of petrochemicals in 1992 and recovered 646,000 tons of sulfur from crude oil.

Petrochemicals

Pemex Petrochemicals owns and operates Pemex's 60 secondary petrochemical plants within 10 petrochemical complexes.

The unit received 10% of Pemex's assets.

It produced 13.7 million tons of intermediate petrochemicals in 1992, selling 5.6 million tons domestically and exporting 1.2 million tons. Pemex claims its biggest petrochemical complex, La Cangrejera in Veracruz, ranks fourth in size in the world with a 4.3 million ton/year capacity.

PERU

CAPITAL: Lima
MONETARY UNIT: Inti
REFINING CAPACITY: 184,250 b/cd
PRODUCTION: 126,300 b/d
RESERVES: 381 million bbl

Petroleos del Peru predicted a 10% increase in Peru's oil production in 1993 and a further increase in 1994.

Petroperu also forecast a sharp increase in drilling in Peru for 1993.

After several years of declining oil production, a gradual turnaround in the cash strapped country's petroleum industry is largely tied to its progress in privatization. The government in 1992 began a campaign to privatize all state owned companies by the end of July 1995.

The government has decided to postpone until 2000 plans to develop the giant Camisea gas field. Peru does not have a market big enough to justify the project's $1.8 billion investment and will continue to expand its electric power grid with oil fired plants.

The government also dropped a proposal to sell Camisea gas to Brazil as infeasible.

Among other developments in Peru's petroleum industry:

- Occidental Petroleum Corp. signed a risk service contract with Petroperu to explore the 2.2 million acre Block 4 in the Maranon basin east of the Andes.
- Peru's privatization commission let a 20 year contract to Petrotech International Corp. of Delaware to operate a former Belco Petroleum Corp. offshore block currently operated by Petroperu unit Petromar SA, slated for liquidation under the privatization program.
- Great Western Resources, Houston, signed a 30 year exploration and development contract with Petroperu for Block 65 in the Maranon basin.
- Petroperu prequalified companies from the U.S. and Mexico bidding under an international tender to develop Aguaytia gas field and operate two small producing oil fields in the north central jungle.
- Petroperu expected to call tenders for additional toll contracts to operate its La Pampilla and Talara refineries, similar to the 1 year contract it has implemented with Marc Rich Co., Zug, Switzerland, for processing crude at La Pampilla.

Petroperu forecasts

Petroperu estimated Peru's oil production would rise to 127,200 b/d in 1993 and 133,200 b/d in 1994 from 115,900 b/d in 1992.

Most of the 1993 increase will come from jungle fields operated by Oxy and offshore fields operated by Petromar.

The 4% increase in 1994 is to come mainly as a result of development drilling in Petroperu's Pavayacu-Corrientes and Chambira fields in the jungle.

Peru's oil production profile began to improve in fourth quarter 1992 after Oxy and Petromar boosted output.

Petroperu expected 15 wildcats to be drilled in Peru in 1993, compared with only two—both by Petroperu—in 1992. Of those 15, Petroperu was to drill five wildcats—two on the north coast and three in the northern jungle.

Private companies would drill the remaining slate of wildcats. Grana y Montero were to drill two on the northern coast, Petromineros one in the northern jungle, Vegsa two on the northern coast, Mobil Oil Corp. one in the northern jungle, Eurocan Ltd. two in the central jungle, and Advantage Resources Inc. two in the northern jungle. In addition, Petroperu planned to drill 52 development wells, mainly on the northern coast. It expected contractor companies, including Petromar, to drill 93 development wells in 1993.

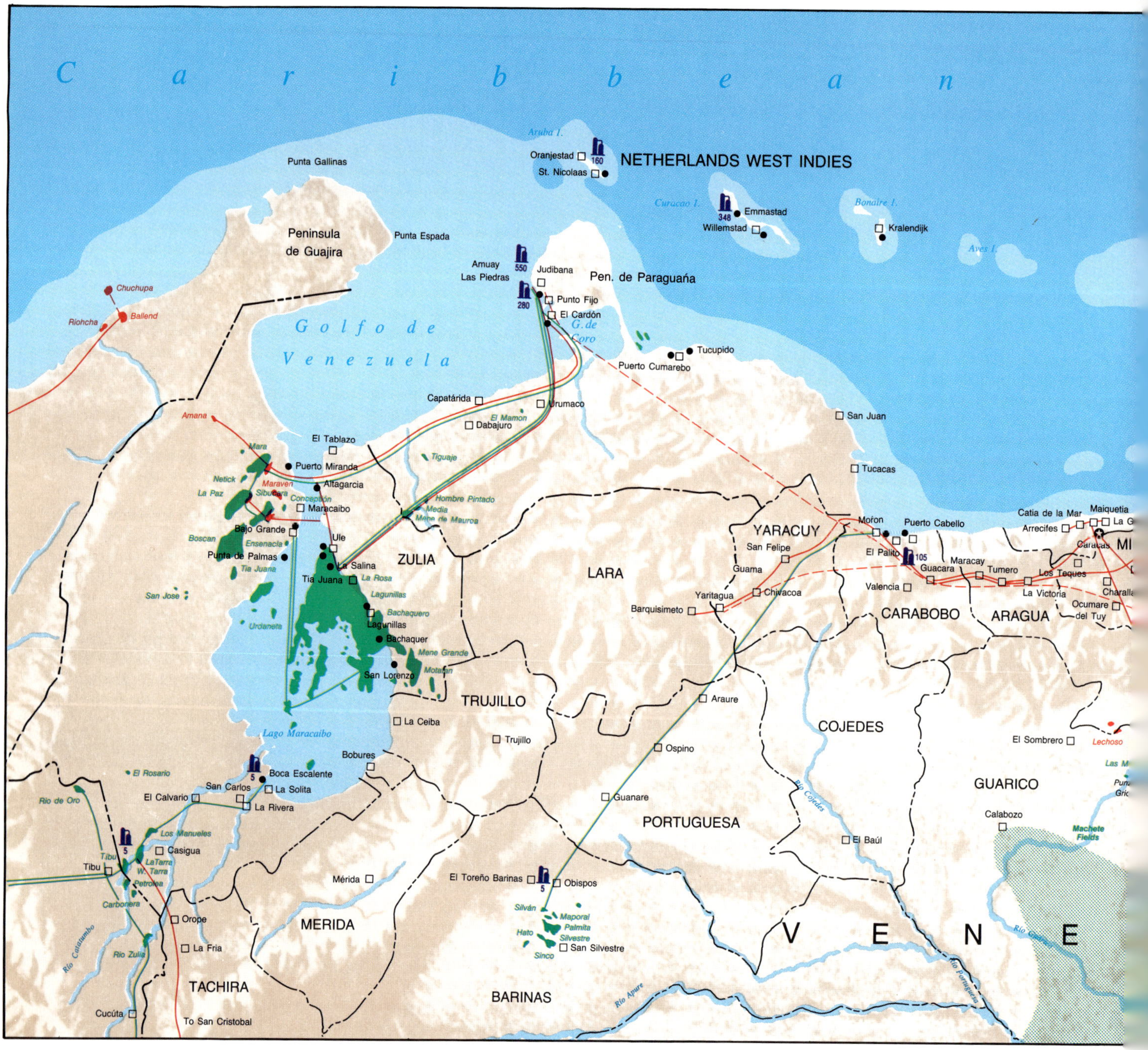

Privatization strides

Peru continues to make strides in privatizing its petroleum sector.

Effective Nov. 18, 1993, the country had a new state petroleum company, Perupetro SA, set up to oversee new and existing contract responsibilities of Petroleos del Peru SA (Petroperu), which is to be privatized in 1994. Until privatization is complete, Petroperu will continue to operate its oil field, refining, transportation, and marketing assets.

Petroperu Nov. 16, 1992, signed its final contract, a 30 year exploration and production agreement with a Peruvian subsidiary of Petrotech International Corp., a unit of McAlistair Towing & Transportation, New York., covering Block Z-2B off Peru's northern coast.

In addition, Petroperu's board approved terms of contracts with Maple Gas Corp., Dallas, for the $150 million development of Aguaytia gas field and operation of Maquia and Agua Caliente oil fields in Peru's central jungle.

TRINIDAD & TOBAGO

CAPITAL: Port of Spain
MONETARY UNIT: Dollar
REFINING CAPACITY: 305,000 b/cd
PRODUCTION: 119,500 b/d
RESERVES: 466 million bbl

Trinidad and Tobago's petroleum sector remained at a crossroads in 1993.

While heavily reliant on oil and gas for domestic energy consumption and hard currency export earnings, the small Caribbean island nation faces some tough choices in reviving its hydrocarbon sector in the 1990s.

Exploration and production of crude oil have stagnated in recent years, and domestic refinery utilization remains low at 36%. However, substantial natural gas reserves in Trinidad and Tobago offer the promise of a burgeoning natural gas

based economy with an eye to liquefied natural gas and gas based petrochemical exports.

Any solutions will involve considerable outlays by the government as well as a sizable infusion of capital by foreign companies. Therein lie some of the hard choices.

Those are among the findings of a study by East-West Center (EWC), Honolulu.

Oil, gas roles

The petroleum industry in Trinidad and Tobago is the bedrock of the nation's economy, notes EWC. Petroleum alone accounted for 31% of the country's 1990 gross domestic product.

Oil exports that year accounted for 66% of trade revenues. Including petrochemicals, hydrocarbon exports accounted for almost 85% of export earnings. Domestic energy consumption is heavily weighted toward natural gas.

Oil sector

With proved reserves estimated at 521 milllion bbl, Trinidad and Tobago's crude reserves could be depleted in less than 10 years at current production levels, EWC said.

Oil & Gas Journal estimated Trinidad and Tobago crude oil and condensate reserves at 572.6 million bbl as of Jan. 1, 1993, and 1992 average production at 139,100 b/d. EWC put the country's potential reserves at 2.9 billion bbl, notably offshore.

After peaking at 230,000 b/d in 1978, EWC noted, the country's crude production fell to 160,000 b/d in 1983, rose to an average 176,000 b/d in 1983-85, and then slipped again, to 151,000 b/d in 1988 and 146,000 b/d in 1991. Production averaged 142,000 b/d the first 6 months of 1992, EWC said.

About three fourths of Trinidad and Tobago's oil production comes from offshore fields. Amoco Trinidad Oil Co., the only major foreign petroleum company currently producing significant volumes of oil in the country, accounted for about half the crude output in 1990. Premier Consolidated Oilfields Ltd., London, produced less than 1,000 b/d that year.

All of Trinidad and Tobago's crude exports of 71,000 b/d went to the U.S. in 1991. The country in 1991 imported about 35,000 b/d of crude for its refineries, a sharp jump from 18,000 b/d in 1990. That doesn't include about 70,000 b/d of Venezuelan crude processed in Trinidad and Tobago for Maraven SA, an affiliate of Venezuela's state petroleum company Petroleos de Venezuela SA (Pdvsa). Meantime, Cuba has approached Trinidad and Tobago about the possibility of buying crude to replace lost supplies from Russia.

Foreign investment prospects

Prospects for foreign investment in Trinidad and Tobago's petroleum sector remain mixed, although the new government has promised to reexamine its energy policy and create a better environment for foreign investors, EWC said.

The analyst notes that although nationalist sentiments in Trinidad and Tobago were not as strong as in other major Latin American producers in the 1970s, the government acquired all of Shell Oil Co.'s assets and part of Texaco Inc.'s and Tesoro Petroleum Corp.'s assets, using them to create its first state oil company, Trinidad and Tobago Oil Co. (Trintoc) at a time of rising oil prices.

Texaco and Tesoro pulled out of the country in 1985, when the government acquired Texaco's 225,000 b/d Point a Pierre refinery and onshore producing properties and Tesoro's 49.9% share of Trinidad-Tesoro Co. With the latter purchase, the government created its second state petroleum company, Trinidad and Tobago Petroleum (Trinopec) in 1985.

"The withdrawal of these companies and the company's complicated hydrocarbons tax regime have had a negative impact on the government efforts to attract foreign investment in the oil sector," EWC said. In addition to a petroleum profits tax (PPT) that foreign companies must pay, the government in 1981 introduced a supplemental profits tax (SPT), which has especially drawn criticism from foreign investors who consider the rates too high.

The administration of Patrick Manning, elected in 1991, has taken steps to merge Trintoc and Trintopec and form a single state company. Under a first stage, the two companies' exploration and production departments were merged late in 1992 and further streamlining and consolidation was to occur in early 1993. In addition, the government took steps in 1992 to reform the tax regime by simplifying and cutting SPT rates.

However, EWC noted, oil exploration results have been disappointing. And unlike the high quality of current crude production, new finds are likely to be heavier and higher in sulfur content. That makes it difficult for domestic refineries, with their limited secondary processing capability, to accommodate the newer crudes.

Refining status

EWC blames a combination of poor management, low profits, reduced U.S. fuel oil demand, and high oil prices during the early 1980s for slashing Trinidad and Tobago's refining capacity to 260,000 b/d by 1985 from 456,000 b/d in 1980.

Since them, refining capacity has rebounded to 305,000 b/d, although utilization may be as low as 40-56%, depending on how much capacity is leased to foreign companies.

Trintoc owns both refineries and produced about 100,000 b/d of products in 1991 with a crude throughput of about 110,000 b/d. Trintoc leases part of its refinery capacity to Pdvsa, which processed 30,000 b/d of crude in Trinidad and Tobago in 1990, shipping all products to Venezuela. Pdvsa processed 70,000 b/d of crude in Trinidad and Tobago in 1991-92. Nigeria also is considering refining some of its crude in Trinidad and Tobago under a technical agreement between the two countries.

Trinidad and Tobago exported 84,000 b/d, or 83%, of the refined products it produced in 1991. In 1990, Trinidad and Tobago shipped 20% of its products exports to the U.S. It also exported liquefied petroleum gas, kerosine, diesel, and fuel oil to Barbados, diesel to Brazil, Costa Rica, and Dominican Republic, gasoline, kerosine, diesel, and fuel oil to Guyana, fuel oil to Jamaica, LPG to Panama, and gasoline, kerosine, and fuel oil to Suriname, among others. Gasoline accounted for 48% of Trinidad and Tobago's domestic consumption of 18,000 b/d of refined products in 1991. Other products consumption broke out as diesel 20%, LPG 9%, kerosine 8%, and fuel oil 3%. Because natural gas fuels most of the nation's power plants, 99% of fuel oil is exported.

Refining problems

EWC notes Trinidad and Tobago's refining industry is based on aging facilities and infrastructure in need of rehabiliation.

"Its refining capacity remains simple, with a cracking to distillation ratio of only 9%, a rather low level that is unable to meet increasing demand for ligher products in international markets. In 1991, more than 50% of Trintoc's refined petroleum products was fuel oil. The country therefore must upgrade its facilities in order to be able to meet the more stringent fuel specifications in the U.S. and to increase flexibility to cater to other product export markets."

A major dilemma for Trinidad and Tobago's petroleum sector, EWC contends, is that while there is substantial underutilized refining capacity in the country, the government is relying on crude exports for foreign exchange earnings. Thus it would be difficult for the government to pursue higher refinery utilization and expand crude exports at the same time. Among options EWC thinks the government could pursue are to:

- Rationalize refining capacity by closing down one of the refineries so as to continue focusing on crude exports.
- Reduce crude exports or increase crude imports, thus diverting more oil into its refineries and thus boost product exports.
- Seek more processing contracts with such countries as Venezuela and Nigeria to utilize excess capacity.

Natural gas: bright spot

Natural gas remains the bright spot in Trinidad and Tobago's petroleum outlook.

Gas resources remained mostly underdeveloped until the late 1970s, when the government began efforts to diversify the country's energy mix. Gas development accelerated during the 1980s when oil production began to decline and the search for oil instead turned up substantial gas discoveries.

With proved reserves EWC put at 9.2 tcf, the nation ranks fourth in Latin America after Venezuela, Mexico, and Argentina. It also ranks fourth in gas production with 1991 output of 558 MMcfd. OGJ estimated Trinidad and Tobago gas reserves as of Jan. 1, 1993, at 8.7 tcf and average production through October 1992 at 443 MMcfd. Even at a conservative estimate of about 9 tcf, Trinidad and Tobago could produce at current levels for 45 years. In addition, a Trintopec report puts the country's probable and gas reserves at 22.5 tcf. The biggest foreign holders of gas reserves in Trinidad and Tobago are Amoco, British Gas plc, Deminex, and Occidental Petroleum Corp. Amoco plans to spend as much as $300 million to develop West-East Queen's Beach and North-Southeast Galeota gas fields off eastern Trinidad to meet a new gas supply contract with National Gas Co. EWC noted the two fields have a combined productive capacity of 480 MMcfd.

VENEZUELA

CAPITAL: Caracas
MONETARY UNIT: Bolivar
REFINING CAPACITY: 1.2 million b/cd
PRODUCTION: 2.3 million b/d
RESERVES: 63.3 billion bbl

Venezuela's Corpoven SA, a unit of Petroloes de Venezuela SA (Pdvsa), late in 1992 signed the first two operating agreements with private companies chosen during the second round of bidding on reactivation of marginal oil fields.

Nine winning bids were announced. Alirio Parra, Venezuela's minister of energy and mines, said expected investment of $860 million during 1993-96 from the nine successful bidders should increase production in the areas to more than 140,000 b/d in the first 3 years of operation.

Lagoven SA said it expects to recover 800 million bbl of crude oil from reactivated marginal fields in three areas it administers under the program. Lagoven disclosed production and investment details for contracts to be signed with winning bidders Total and Royal Dutch/Shell group. Maxus Energy Co., a third winning bidder for a Lagoven area, disclosed its plans under the program. Lagoven said it expects Total, Shell, and Maxus to spend almost a combined $4 billion during the 20 year life of their operating contracts.

Operating agreements

An operating agreement was signed with a group including Norcen International Ltd. 45%, Argentina's Cia. Naviera Perez Companc SA 45%, and Servicios Corod de Venezuela SA 10%.

The group will operate the Oritupano-Leona block in Anzopategui and Monagas states. The block holds 15 fields with cumulative production of 255 million bbl and remaining proved reserves estimated at 169 million bbl.

The Norcen group's initial 3 year work program carries a capital commitment of $150 million. Work during the initial period will focus on production enhancement, including work-overs and well repair, installation of added production facilities, and drilling. Norcen expects the initial phase to yield increased production of 15,000 b/d.

A group made up of Argentina's Tecpetrol SA, Astra, Australia's Ampolex (U.S.A.) Inc., and Chile's Sipetrol SA now has an operating agreement covering the Quiamare-La Ceiba block in Anzoategui. The block holds six fields with cumulative production of 71 million and estimated remaining reserves of 119 million bbl.

Teikoku operations

Japan's Teikoku Oil Co. Ltd. was successful bidder on the Sanvi-Guere block about 300 km southeast of Caracas. It expected to sign a formal contract, Kyodo News Service reported. Cumulative production from the block, which was produced during 1955-91, is about 34 million bbl.

The new area is close to the East Guarico block won by Teikoku in 1992. Teikoku took over operations on the East Guarico block, also known as Unit 9, in early 1992. Pdvsa said Teikoku is the first Japanese company to conduct oil operations in Venezuela. The East Guarico block was the only block operating when bidding began. The block includes about 16 fields, and when Teikoku took over production was about 1,370 b/d from 27 wells.

Total plans

Total predicted incremental recovery would reach about 168 million bbl of oil from the 53,957 acre Jusepin area in northern Monagas state of eastern Venezuela.

During 1938-89, Jusepin produced about 194 million bbl of 31° gravity crude, or about 14% of original oil in place (OOIP) estimated at 1.3 billion bbl. Total forecast production at 300 b/d by the third year of activity, to climb to 43,000 b/d by the end of the 13th year of the contract.

Shell plans

Shell de Venezuela plans total outlays of $1.33 billion in the West Urdaneta area of Lake Maracaibo in western Venezuela.

It estimates incremental production at 240 million bbl of oil. The 411,582 acre producing area yielded 34 million bbl of crude during 1974-93. That represents about 4% of OOIP pegged at 780 million bbl. Shell expects initial production will reach 8,000 b/d in the first year and 47,000 b/d by the end of the 7th year of the contract. It plans to spend more than $18 million the first 3 years.

Maxus plans

Maxus Energy Corp., Dallas, was successful bidder with Otepi Consultores SA on the Quiriquire block and will post a letter of credit for $34 million for its 95% stake. The company's 3 year work commitment was to start in 1994 with a study of fields that are producing 800-1,000 b/d. The program includes a 3D seismic program over potential deep prospects. Maxus plans to start a drilling program in 1995.

Hunt for new markets

Pdvsa was targeting new markets in Asia and Latin America for its crude oil and refined products under a new 10 year business development plan.

Pdvsa Pres. Gustavo Roosen sees relatively slim prospects for growth in the U.S. and Europe, where it is already well established. Accordingly, the company must seek new markets with strong growth potential in order to place the 3.1 million b/d of crude and products it expects to be exporting by 2003.

As part of this push, Pdvsa is reviewing the possibility of laying a pipeline from Venezuela to Panama, which would give it better access to Pacific ports, Roosen said. Roosen noted the concept has been around for some time but is not a priority for now.

Foreign investment key

Pdvsa's new 10 year plan continues to rely on substantial foreign investment in its oil, gas, petrochemical, and coal sectors.

Pdvsa hopes to develop crude oil productive capacity of 4 million b/d by about 2000, up from 2.8 million b/d currently. Oil production is projected to reach 3.6 million b/d and exports 3.1 million b/d in 2003. Pdvsa exports of crude and refined products in first half 1993 averaged 2.03 million b/d at $14.57/bbl. That puts the petroleum export total at about $5.4 billion, about 11% below the government target. For 1993, the government was targeting 2.075 million b/d at $15.70/bbl for a total of $11.9 billion vs. $11.2 billion in 1992.

Heavy oil projects

Venezuela's congress approved two heavy oil projects involving foreign investors and a combined cost of more than $4.8 billion.

The surprise, last minute vote came only 3 days after the congress approved the $5.6 billion Cristobal Colon liquefied natural gas export project, representing the biggest energy project in Venezuela's history and the first direct equity investment by foreign companies in the country's upstream petroleum sector since nationalization in 1976.

Total project

A joint venture of Total 40%, Maraven 35%, and a combine of Itochu Corp. and Marubeni Corp. 25% plans a project estimated to cost more than $3.1 billion that would produce 114,000 b/d of 9° gravity, high sulfur crude from the Zuata region of the Orinoco belt in Anzoategui state.

The venture would use delayed coking and hydrodesulfurization processes in Venezuela to yield 100,000 b/d of 31° gravity, 0.06 wt % sulfur content crude.

The extra heavy crude would be transported via a 210 km pipeline to a delayed coking plant to be built near the Caribbean port of Jose, site of existing petrochemical and gas processing plants.

Conoco project

A joint venture of Conoco 49.9% and Maraven SA 49.9% plans a $1.7 billion project to produce 120,000 b/d of 9° gravity, 3.7 wt % sulfur content crude, also from the Zuata region.

The venture would use delayed coking to yield about 102,000 b/d of 20° gravity, 2.7 wt % sulfur crude.

The partners would place total output with Conoco refineries at Lake Charles, La., and Ponca City, Okla. IPE

Africa

ALGERIA

CAPITAL: Algiers
MONETARY UNIT: Dinar
REFINING CAPACITY: 530,000 b/cd
PRODUCTION: 750,000 b/d
RESERVES: 9.2 billion bbl

Algeria is increasing its capacity to export natural gas in order to reinforce its strong position in the growing international market.

The country's reserves are estimated at more than 3.6 trillion cu m. Algerian energy and development policy is based on a rational exploitation of this resource.

A liquefied natural gas (LNG) pioneer, Algeria has one of the world's most important LNG production capacities. With a location encouraging export to nearby countries, Algeria has an important place in the world natural gas market and an exclusive role within its trading region.

This requires development of transport capacity. Accordingly, Algeria has built the Transmediterranean Pipeline (Transmed) linking Algeria to Italy and in 1993 launched construction of the 530 km Maghreb/Europe (GME) pipeline to Spain.

In 1992, natural gas represented 63.9% of total primary hydrocarbon production of 162.4 million metric tons of oil equivalent (TOE). Production net of reinjection operations reached 107.4 million TOE, of which natural gas represented 45.9%.

Algeria marketed 49.4 million TOE of natural gas, of which 34.4 million TOE was exported in gaseous and liquefied forms.

The export total accounted for 10.5% of the 1992 world market. Algeria held 24.2% of the international LNG market and 6.2% of the pipeline export market.

Natural gas represented 40.1% of Algeria's hydrocarbon exports in 1992, LNG alone 22.6%. If all byproducts of LNG are assumed to be exported, natural gas in its two states represents 41.1% of total hydrocarbon exports (35.3 million TOE), LNG and its byproducts alone 23.7%.

A growing role

Algeria plans to increase its international trade in natural gas in coming years. The effort will especially focus on southern Europe.

Sonatrach aimed at increasing natural gas exports from 35.4 billion cu m in 1992 to 60 billion cu m.

To meet this objective, the national company started refurbishing its LNG facilities. From 1997 on, LNG plants in Arzew and Skikda will be able to produce as much as 30 billion cu m/year.

Sonatrach's four LNG plants operate at less than capacity rates because they need steam. At Arzew, liquefaction capacities are 10.5 billion cu m/year each in the GL 1Z and GL 2Z plants and about 2 billion cu m/year in the GL 4Z plant. The GL 1K plant at Skikda has liquefaction capacity of 8 billion cu m/year.

All the LNG plants produce byproducts such as LPG ethane, helium, nitrogen, and gasoline. They produce their own electricity and steam, consuming about 13% of inlet gas streams in the process.

Sonatrach accounts for these gas use volumes, as well as for gas consumed in transportation of gas from producing areas to LNG plants and other facilities, as "autoconsumption," totaling 208.3 trillion BTU in 1992.

At Arzew, Sonatrach operates two LPG separation plants, GP 1Z with a capacity of 4 million tons/year of LPG and GP 2Z with capacity of 600,000 tons/year of LPG.

The GME pipeline under construction will be able to carry 7.2 billion cu m/year at completion of its first phase in 1996. This capacity will be increased to 18.5 billion cu m/year in its final phase.

In addition, Sonatrach has decided to increase capacity of the Transmed pipeline to 24 billion cu m/year.

AFRICA
A F R I C A
FRANCE
ROMANIA
C.I.S.
FORMER YUGOSLAVIA
ITALY
BULGARIA
PORTUGAL
SPAIN
GREECE
ALB
TURKEY
Black Sea
Caspian Sea
Lavéra
Genoa
Trieste
Madrid
Rome
Lisbon
Istanbul
Ankara
Athens
Strait of Gibraltar
Ceuta
Oran
Algiers
Maison Carree
Skikda
Bejaia
Arzew
Bizerte
Tunis
La Skhirra
TUNISIA
MALTA
Rabat
Melilla
Mohammedia
Casablanca
Sidi-Kacem
Safi
MOROCCO
Essaouira
Tenerife
LaPalmas
Canary I.
El Aaiún
Hassi R'Mel
Hassi Messaoud
ALGERIA
Tripoli
Mediterranean Sea
CYPRUS
SYRIA
LEBANON
ISRAEL
JORDAN
IRAQ
IRAN
Tehran
Baghdad
Damascus
Kuwait
Alexandria
Cairo
Suez
Benghazi
Tobruk
Marsa el Brega
Ras Lanuf
Sider
LIBYA
EGYPT
Riyadh
SAUDI ARABIA
Villa Cisneros
MAURITANIA
MALI
NIGER
Nouakchott
SENEGAL
Dakar
GAMBIA
GUINEA BISSAU
GUINEA
Conakry
SIERRA LEONE
Freetown
LIBERIA
Monrovia
COTE D'IVOIRE (IVORY COAST)
Abidjan
GHANA
Accra
TOGO
Lome
PEOPLE'S REPUBLIC OF BENIN
Porto Novo
Lagos
Bamako
Ougadougou
BURKINA FASO
Niamey
NIGERIA
Kano
Kaduna
Abuja
Makurdi
Warri
Port Harcourt
Bonny
CHAD
Lake Chad
N'Djamena
SUDAN
Khartoum
Port Sudan
Red Sea
YEMEN
DEMOCRATIC YEMEN
Assab
DJIBOUTI
Djibouti
ETHIOPIA
Addis Ababa
SOMALIA
Mogadishu
CENTRAL AFRICAN EMPIRE
Bangui
CAMEROON
Victoria
Yaoundé
Malabo
EQUA. GUINEA
Bata
Sao Tome
Cape Lopez
Port Gentil
Libreville
GABON
CONGO
Gulf of Guinea
Brazzaville
Pointe Noire
Kinshasa
Muanda
CABINDA
Ango-Ango
ZAIRE
UGANDA
Kampala
KENYA
Nairobi
Mombasa
RWANDA
Kigali
Bujumbura
BURUNDI
TANZANIA
Zanzibar
Dar es Salaam
Lake Victoria
Lake Rudolf
Lake Albert
Lake Tanganyika
Luanda
ANGOLA
Lobito
ZAMBIA
Ndola
Lusaka
MALAWI
Lilongwe
Zomba
Moroni
Comoro I.
ZIMBABWE
Salisbury
Umtali
Beira
PEOPLE'S REPUBLIC OF MOZAMBIQUE
Mozambique Channel
MADAGASCAR
Tamatave
Tananarive
NAMIBIA
Windhoek
Walvis Bay
BOTSWANA
Mmabatho
Thohoyandou
Pretoria
Johannesburg
Maputo
SWAZILAND
Sasolburg
LESOTHO
Durban
REPUBLIC OF SOUTH AFRICA
Umtata
Cape Town
Cape of Good Hope
South Atlantic Ocean
Indian Ocean
0 300 600 Mi.
0 200 400 600 Km.

Big investment seen required for southern Africa

Africa needs massive financing to develop its oil and gas potential, says Tei Mante, director, Sub-Saharan Africa, International Finance Corp., Washington, D.C.

And there will be stiff competition from the former Soviet Union and other regions around the world vying for foreign capital and technology.

The World Bank estimated that $50-100 billion will be needed between now and 2000 for oil and mineral exploration and development in Africa, Mante said.

He spoke at Sub-Saharan Oil and Minerals Conference in Cape Town in late 1993, where ministers and representatives of governments in southern Africa gathered to describe their petroleum investment rules and objectives to private oil and mining companies.

The conference attracted the largest gathering of African energy ministers in many years—maybe ever—said a spokesman for the conference organizer. Better than expected attendance reflected a high interest in African oil and mineral opportunities.

Petrochemicals, natural gas

As in a growing list of regions elsewhere, African nations are welcoming investors in upstream and downstream operations. Special emphasis is being placed on developing downstream manufacturing capabilities in many countries of southern Africa.

"Petrochemical manufacturing holds the key to development," Chief Don Etiebet, Nigeria's minister for petroleum and mineral resources, told the conference.

Investors are welcome in petrochemical manufacturing, he said.

Nigeria plans to build three methyl tertiary butyl ether plants, for example, to produce product for export, said Etiebet. Private companies will be involved in this construction program. A key reason for diversifying downstream is to reduce crude oil market risk.

Another thrust of countries in the region is to recover and utilize much more of current natural gas production, the bulk of which is currently flared. One estimate is that 74% of the gas produced in southern Africa is flared.

In Nigeria, development of natural gas is just beginning, Etiebet told the conference. He pegged Nigeria's gas reserves at 100 tcf.

Chevron Corp. has two gas utilization proposals in Nigeria. In partnership with Nigerian National Petroleum Corp., it plans a $570 million project to recover 170 MMcfd of associated gas now being flared. The gas will be gathered from offshore platforms, liquids extracted, and dry gas sold for domestic use.

Another proposal to lay a gas pipeline to fuel power plants in Ghana is under discussion.

Worldwide demand for natural gas could grow twice as fast as oil demand, said William Edman, managing director of Chevron Nigeria Ltd., Lagos.

As in other regions, one reason for the emphasis on natural gas in Africa is its environmental benefit. But Albina Africano, Angola's minister of petroleum, told the Cape Town conference, "Oil is also seen as beneficial in Africa because it replaces wood."

Kerosine and gasoline are the best intermediate alternatives for wood, she said.

Financing development

Financing oil and gas development in southern Africa won't be easy, but governments in the region are reassessing their investment rules with an aim to ensure an environment that will attract foreign capital.

The potential from an oil standpoint is considerable, Robert Angel, chief executive of South Africa's Engen told the conference. Subequatorial Africa has the potential for significant growth in oil demand, and regionally there is a good upstream/downstream balance.

Angel estimated total petroleum investment in the region at $100 billion. Currently, 63% is held by multinational companies, 18% by governments, and 19% by private African companies. He expects a shift away from government ownership as private investment is encouraged.

Angel outlined a plan of action for developing southern Africa's petroleum potential:

- Encourage private African companies to participate in development.
- Continue privatization efforts to make organizations cost competitive.
- Make company organization and performance transparent to public and potential investors.
- Develop technical and business skills.
- Develop a blueprint for regional cooperations that will optimize the use of limited capital.

In closing the conference, South African President F.W. de Klerk said the years of isolation are over, and international investors and explorers can now return to South Africa.

He cited plans by South Africa's oil company, Soekor, to offer exploration licenses on offshore blocks in second quarter 1994 as part of an effort to encourage foreign participation in South Africa's oil industry. IPE

Local markets

The local market consumes 15 million TOE/year of natural gas—65.8% of primary energy consumption. This is expected to increase in coming years to 19 million TOE/year as the industrial sector, which accounts for 15% of Algerian demand but now is in recession, recovers.

The national distribution network has sufficient capacity to handle the increase.

Sonatrach's major Algerian oil and gas customers, all state concerns, and their activities include Sonelgaz, electricity and gas; Naftec, refining; Asmidal, fertilizer; and Enip, petrochemicals.

Naftec returns some of its products to Sonatrach for exports and markets the rest through a distributing company, Naftal.

Maghreb-Europe gas line

Algeria's state petroleum company Sonatrach let a lump sum contract worth more than $450 million to Bechtel Corp. to engineer and construct the 330 mile Algerian segment of the Maghreb-Europe gas pipeline.

When complete in 1996, the Maghreb-Europe pipeline system will be the first large diameter gas transmission link from Morocco to the European continent.

Spanish gas company Enagas in 1993 was to award construction contracts on the Moroccan and Spanish portions of the pipeline.

The total Maghreb-Europe pipeline system will run from Algeria's Hassi R'Mel gas field in the Sahara desert, across northern Morocco, and under the Strait of Gibraltar to Cordoba, Spain, where it will tie into the Spanish gas transmission pipeline grid. The system could be expanded to transport gas to Portugal, France, and Germany.

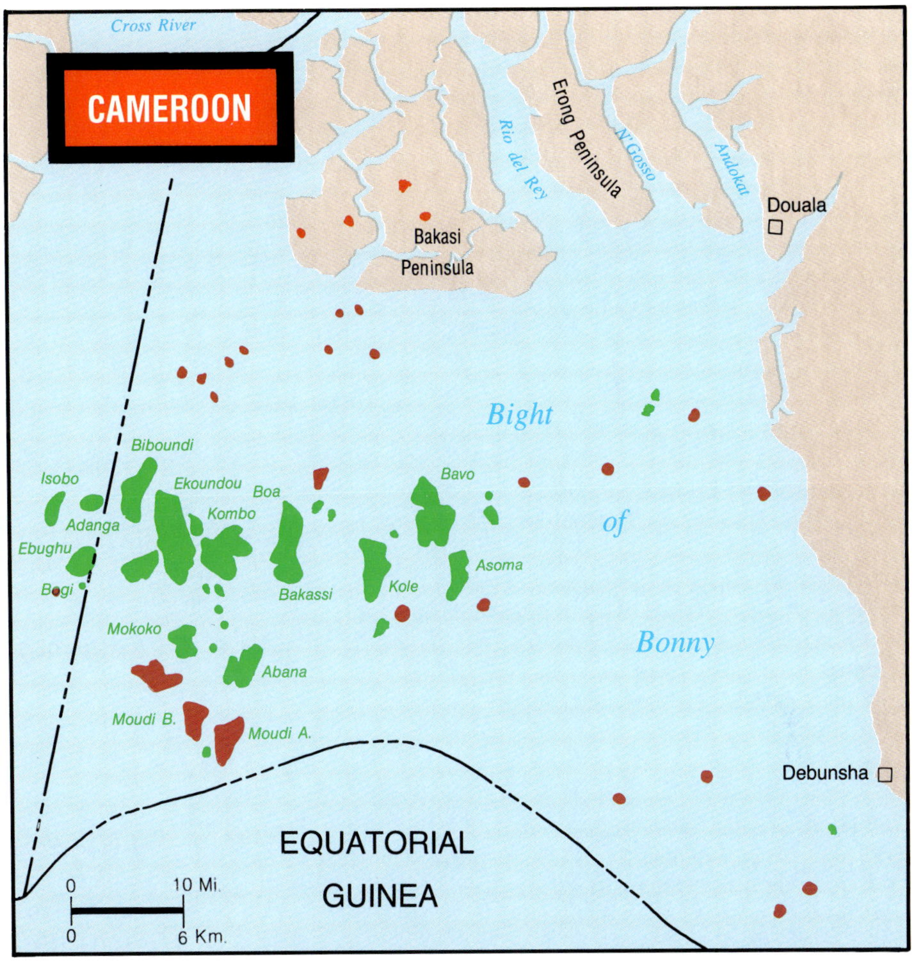

Armel Morvan, Bechtel's manager of the construction project, said construction camps were to be set up in summer 1994, and construction is to begin in the fall. He said terrain along the pipeline route is rugged desert, mostly rock, with temperatures reaching more than 110° F. in the summer and dropping below freezing in winter. The workforce at peak construction is expected to total about 2,000.

Mike Thiele, Bechtel senior vice-president and manager of the company's pipeline and industrial business, said the Maghreb-Europe line at first will have capacity of about 700 MMcfd of gas. System capacity will ramp up to about 1.8 bcfd as compression is added. In addition to engineering and construction of the Algerian portion of the pipeline, Bechtel's contract includes procurement of materials and components and installing a fiber optic telecommunication and control system along the pipeline route.

Bechtel Financing Services Inc. arranged financing for the project through the U.S. Export-Import Bank and France's Coface and Germany's Hermes export credit agencies.

Second licensing round

Algeria's Sonatrach disclosed details of a second exploration licensing round.

Thirty-one blocks were offered covering more than 200,000 sq km in a round designed to license acreage near producing fields as well as frontier territory.

Mohamed Babaghayou, Sonatrach deputy general manager of exploration and research, spelled out new bidding procedures designed to improve on 1992's first round.

"Experience has shown the industry is thirsty for information on Algeria," Babaghayou said. "We also learned that bidding procedures must take account of planning and budgeting requirements of oil companies."

Bethioua GL2/Z LNG complex in Algeria is the site of a 3 year, $500 million upgrade by M.W. Kellogg that got under way in 1992. File photo.

WEST AFRICA
GHANA
TOGO
BENIN
NIGERIA
CAMEROON
EQUATORIAL GUINEA
RIO MUNI
GABON
CONGO
CABINDA
ZAIRE
ANGOLA
Atlantic Ocean
Kaduna
Abuja
Bida
Yola
Ilorin
Baro
Ibadan
Makurdi
Donga
Lome
Porto Nova
Lagos
Tema
Seme
Salt Pond
Tano
Meren
Okan
Forcados
Warri
Ewan
Oguta
M'Bede
Ea
Kolo Creek
Pennington
Port Harcourt
Alesa-Eleme
Ekulama
Okubie
Idoho
Kole
Victoria
Douala
Yaounde
Malabo
Sanaga Sud
Kribi
Ebolowa
Bata
Oyem
Libreville
Sao Tome and Principe
Booue
Cape Lopez
Port Gentil
Lambarene
Odombo
Ikenge
Rabi-Kounga
Gamba
Totou
Ivinga
Mossaka
Lucina
Tchibala
Tchibala S.
M'Bya N.
M'Bya S.
Yanga-Sendji
Pointe Noire
Loango
Pointe Indienne
Sendji
Emeraude
Takula
Malongo
W. Malongo
Kungulo
Cabinda
Mibale
Moanda
Essungo
Lele
N'Zombo
Quinguila
Cabeca Da Cobra
Maleva North
Mavanga
Sulele
Brazzaville
Kinshasa
Zaire River
Ambriz
E. Lombo
Luanda
Cacuaco
Mulenvoss
Benfica
Uacongo
Barra do Cuanza
Galinda
Tobias
Nova Gaia
Lobito
12 in. 31 cm
16 in. 41 cm
6 in. 15 cm
10 in. 25 cm
Atlantic Ocean
Cape Lopez
Lopez N.
Lopez S.
Clairette
Port Gentil
Anguille N.E.
Anguille
Anguille S.W.
Anguille S.E.
Pt. Gentil SM
N'Tchengue
N'Tchengue Ocean
M'Bega
Alewana
Animba
Doree
Ozouri
Octopus
Pageau
Girelle
Torpille
Pelican
Obando
Baliste
Ayol
Grondin
Olende
Oguendjo
Mandoros
Gonelle
Barbier
Breme
Batanga
Gongoue
N. Tombenyoni
Pointe Weze
Mporaloko
GABON
Loanda
Bilape
Chikawar
Odombo
Ongindjo
Assewe
Ikenge
Omboue
0 25 Mi.
0 40 Km.
0 100 200 300 Mi
0 100 200 300 400 Km.

Second round procedures

Here are second round procedures:

- Companies buying data packages were to have 4 months—June 1993 to September 1993—to evaluate the data.
- In October 1993, companies wishing to bid were required to submit a letter of intent to Sonatrach indicating chosen blocks.
- Blocks of interest to more than one company were to enter a bidding round opening Dec. 1 and closing 6 months later.
- Direct negotiations would begin if only one company shows interest in a block.
- Sonatrach was to make detailed presentations before the end of May 1994 about its legal and contractual requirements to each company applying for acreage.

Two companies were offering data packages: Integrated Exploration & Development Services Ltd., Tetbury, U.K., and TGS-Calibre Geophysical Co., Houston.

IEDS was marketing packages on Tindouf basin Blocks 301 through 308, 358, and 363 through 365, Dahar Depression Blocks 413 through 415b, Melrhir and Chotts basin Blocks 108 and 128b, Southern Dahar Depression Blocks 443b and 444, and Ghadames basin Blocks 210 and 235b.

TGS was marketing data on Blocks 319, 348, and 349 in District 8 and Blocks 103, 116, and 120 in District 9. These districts in western Algeria include the Bechar and Timimoum basins and the Atlas fold belt. TGS said several wells in the area had good shows of oil and gas.

Sonatrach hopes to raise drilling activity to 55 wildcats/year by 1995 and 100 wildcats/year by the end of the century. In 1993, Sontrach intended to drill 27 new field wildcats on its own and 14 in association with foreign companies.

ANGOLA

CAPITAL: Luanda
MONETARY UNIT: Kwanza
REFINING CAPACITY: 32,100 b/cd
PRODUCTION: 498,400 b/d
RESERVES: 1.5 billion bbl

The civil war in Angola was expected to trim the country's oil production for 1993 by about 6% from 1992.

Angolan production will rebound in 1994 and reach a new high of 715,900 b/d in 1996, says Wood Mackenzie Consultants Ltd., Edinburgh.

The resurgence will be driven by a recovery in exploration and appraisal drilling in 1993, which is likely to be maintained for 5 years.

Wood Mackenzie expected the Cabinda enclave to produce an average 303,500 b/d in 1993, or 58% of Angola's total. Production off mainland Angola will total 216,400 b/d, with onshore production contributing an average 1,600 b/d.

Production from Block 2 off mainland Angola was shut down twice by fighting. From Jan. 18 to Feb. 8, 1993, Unita forces occupied the Soyo supply base and Quinfuquena terminal. This forced producers to change crude export routes. From May 31 to July 15, 1993, some Block 2 production facilities were shut down as they came under fire from land-based Unita rebels. Before the closures, crude oil was piped ashore from Block 2 fields to the terminal and back out again to an offshore loading point.

Production forecast

The Quinfuquena terminal was bypassed and oil was exported from the original loading buoy adjacent to Essungo field and a new buoy near East Lombo field. Nevertheless, Block 2 production for 1993 was forecast to be down 21% from 1992.

All Block 2 fields were producing normally, said Wood Mackenzie, apart from Essungo and Raia which remained vulnerable to land-based attack. Block 1 Safueiro field was shut in as production is exported via Essungo field.

Onshore oil production was severely curtailed, with most fields shut down since late 1992. Only the small Kwanza basin fields were producing, the analyst said. 1993's onshore production was expected to be only 1⁄20th of the 1992 total. Assuming onshore fields are reopened in mid-1994 and that Block 2 fields produce normally, Wood Mackenzie predicted total Angolan production would average 564,500 b/d in 1994.

The Angolan success rate is predicted to decline in the next 5 years because of the large number of frontier wells planned. Satellite accumulations off Cabinda and in Blocks 2 and 3 are thought likely to buoy the figures.

Platform contracts

Cabinda Gulf Oil Co. let contracts worth a combined $100 million to AMEC Process & Energy Ltd., London, to build two platforms for North N'Dola and South Sanha fields off Angola. They will be built at Durban for completion in August 1995.

They will be the first platforms built in South Africa for export. North N'Dola platform will have a 2,200 short ton topsides and 1,700 short ton jacket, while South Sanha will have a 2,400 short ton topsides and 1,500 short ton jacket.

FPSO use

Angola's Sonangol national oil company will use Oceaneering International Inc. division Oceaneering Production Systems' Ocean Producer floating production, storage, and offloading system in Block 4 about 17 miles off northern Angola. The FPSO will receive, process, store, and offload

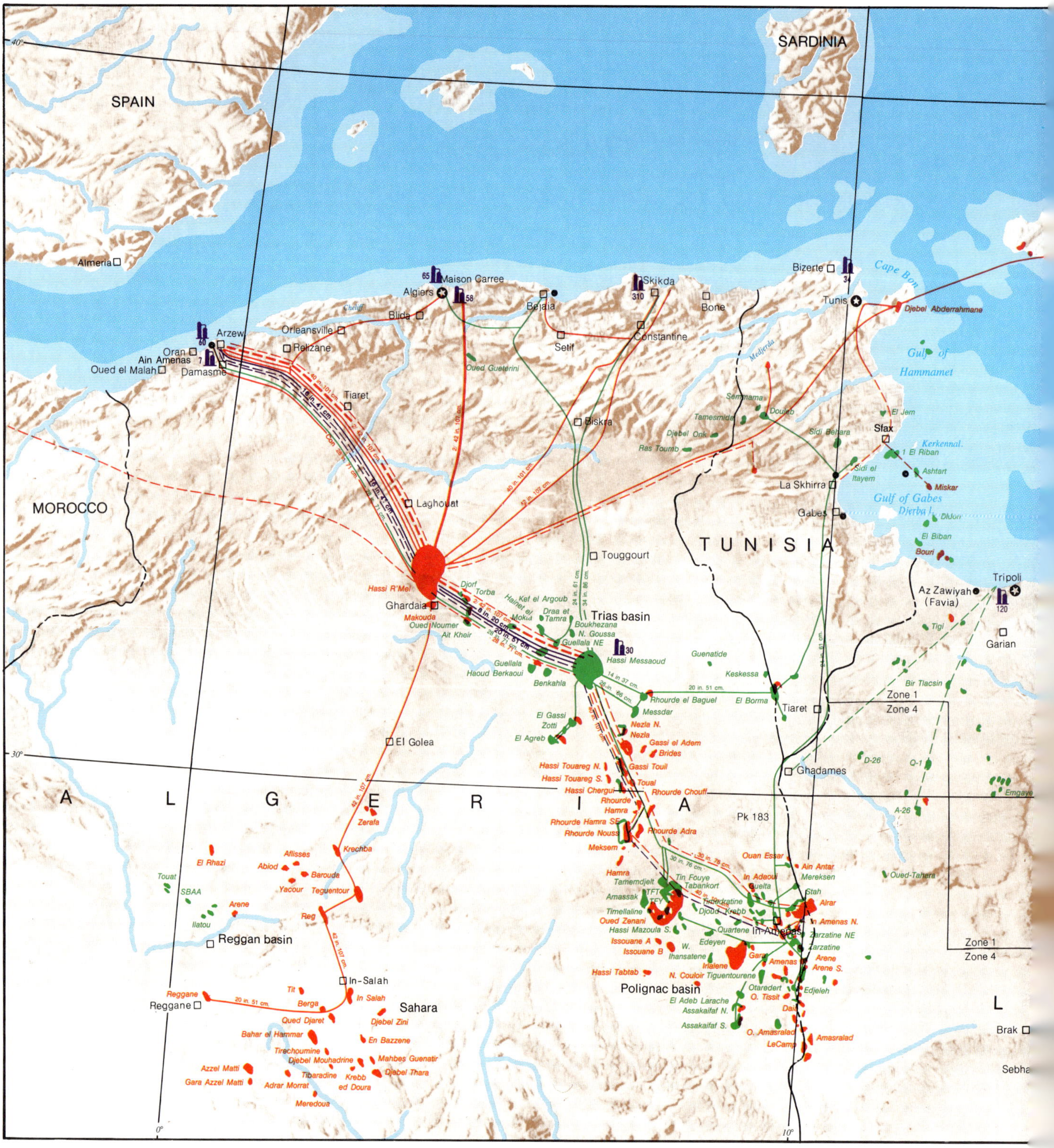

crude produced from two seabed wells. Contract length is 1 year, which was expected to start early in January 1994, with options for two 1 year extensions.

Platform, pipeline contract

Cabinda Gulf Oil Co. Ltd. let contract to Offshore Pipelines Inc., Houston, to build platform and pipeline facilities off Cabinda, Angola.

OPI will build, ship, and install the 1,566 ton, four pile N'Sano A platform. It will lay a 3.5 mile, 16 in. crude line to Takula GS-P platform from N'Sano A, a 1.7 mile, 12 in. crude line to Takula O platform from N'Sano A, and a 1.7 mile, 6 in. gas lift line to Takula O from N'Sano A.

Cabinda Gulf also let contract to OPI to build two risers and install 4.2 miles of 12 in. pipeline to GS-Mike platform from Tak-O platform.

CAMEROON

CAPITAL: Yaounde
MONETARY UNIT: CFA franc
REFINING CAPACITY: 42,000 b/cd
PRODUCTION: 118,800 b/d
RESERVES: 400 million bbl

Exxon Co. International hired Gulf Interstate Engineering Co., Houston, for preliminary engineering services for the

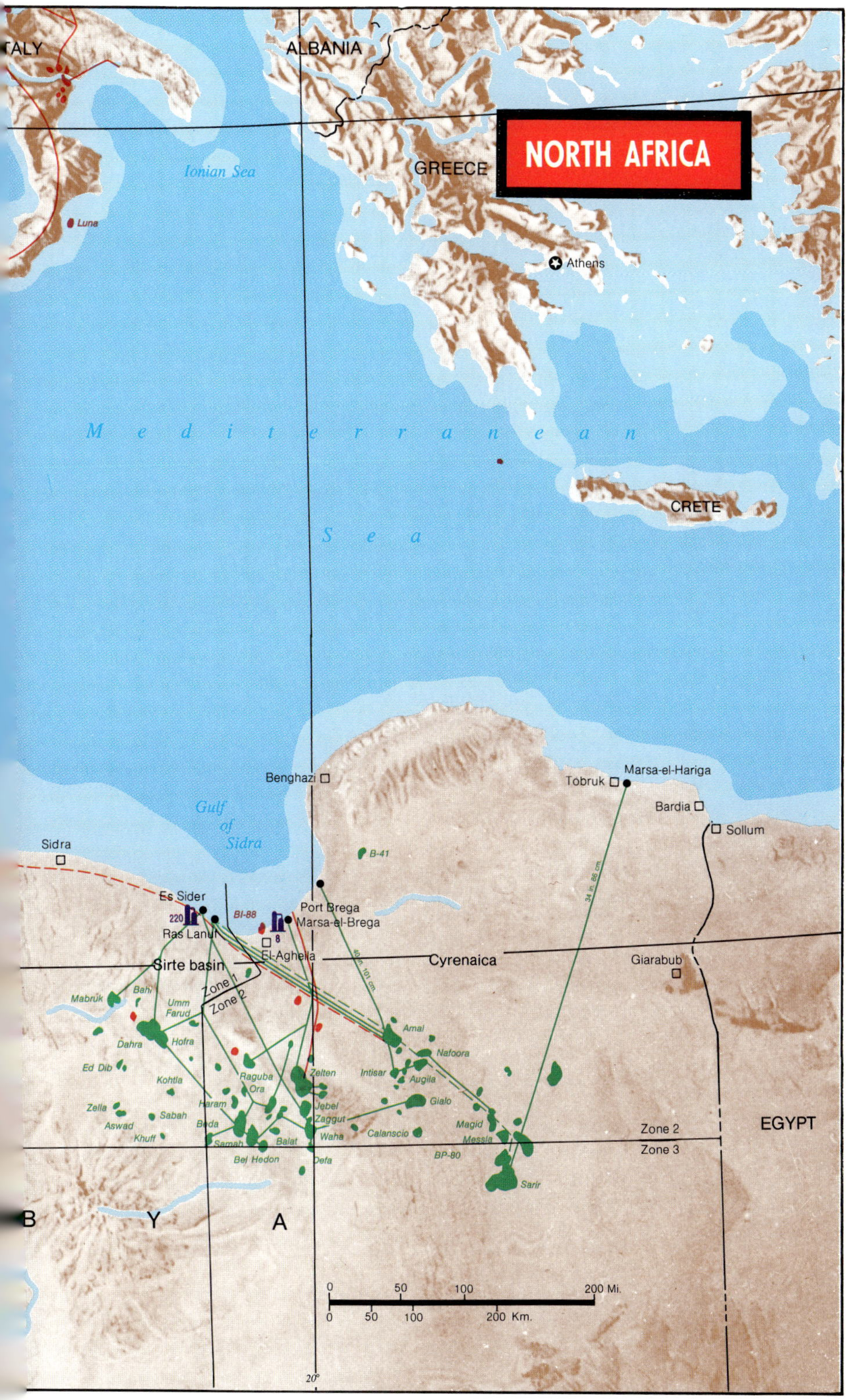

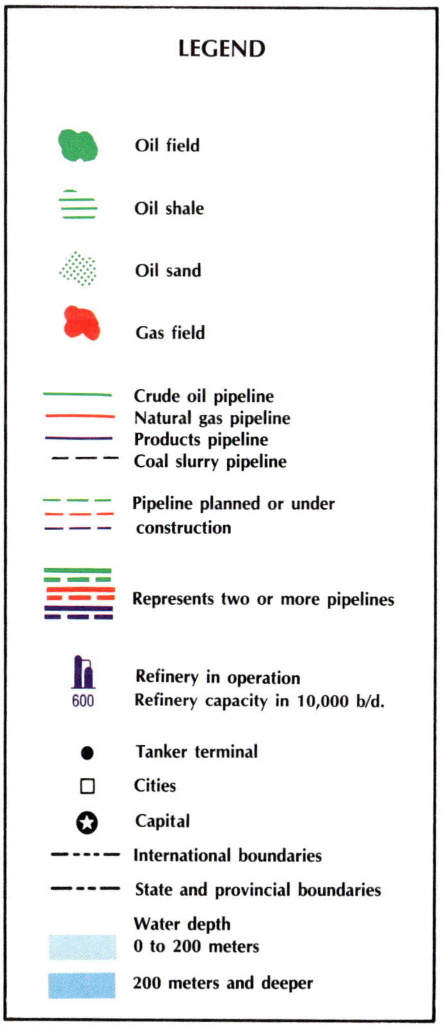

Chad-Cameroon pipeline system in central western Africa.

Gulf's contract is part of a feasibility study for a 1,000 km crude oil pipeline from near Doba, Chad, to an export terminal on the coast of Cameroon.

CONGO

CAPITAL: Brazzaville
MONETARY UNIT: CFA franc
REFINING CAPACITY: 21,000 b/cd
PRODUCTION: 200,000 b/d
RESERVES: 830 million bbl

Elf Congo let a 1.53 billion franc ($260 million) contract to a combine of Bouygues Offshore and ETPM for two platforms to develop N'Kossa oil field off Congo.

The contract covers design, construction, and installation of two drilling platforms to be installed in 170 m of water. Each will have a 10,000 metric ton jacket, including 4,000 tons of piles and conductors, and a 2,000 ton deck. It also covers equipment procurement and laying associated pipelines. Delivery is set for July 1995.

In addition, Bouygues and affiliate Bouygues Offshore will design and build a 300 million franc ($51 million) barge to serve as N'Kossa's floating production vessel (FPS). The barge will be 220 m long, 45 m wide, and 15.5 m high and house 30,000 tons of equipment and 160 workers. The 100,000 ton displacement barge will be anchored off Congo in 180 m of water and will be the first single hull, post-tensioned concrete FPS. Delivery is set for Mar. 30, 1995.

Offshore wells

Ste. National Elf Aquitaine let contract to Ross Offshore AS, Oslo, to drill two wells off Congo.

The Pelerin rig was to spud the first well in September 1993. The fixed part of the contract covers 8 months, after which Elf has an option to drill two more wells.

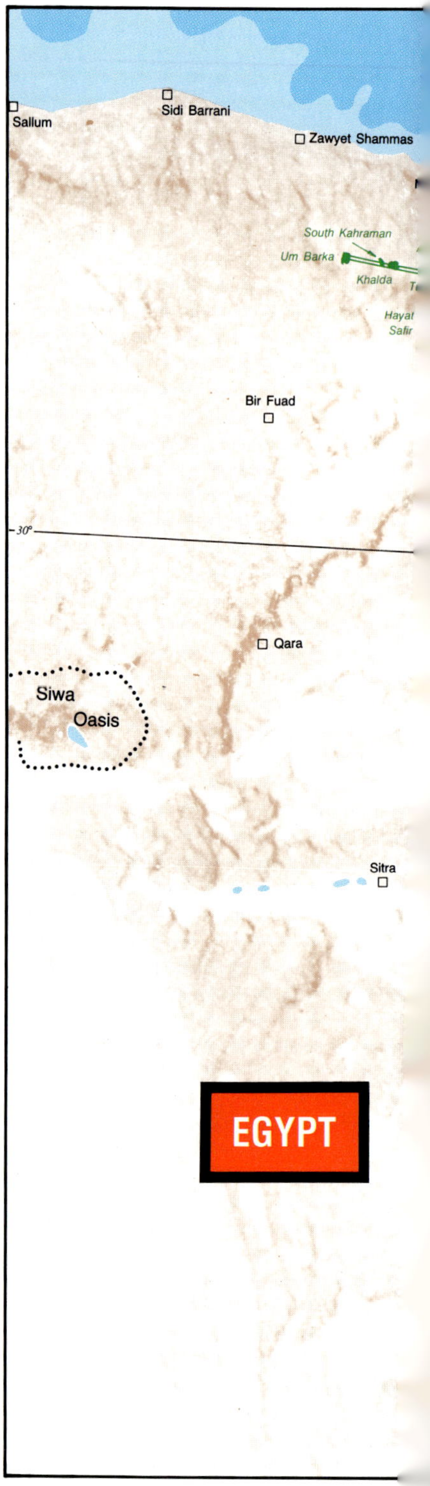

EGYPT

CAPITAL: Cairo
MONETARY UNIT: Pound
REFINING CAPACITY: 532,153 b/cd
PRODUCTION: 894,300 b/d
RESERVES: 6.3 billion bbl

Shell International Petroleum Co. Ltd. found oil and gas for the second time on its Obayed permit in Egypt's Western Desert, Middle East Economic Survey reported.

Shell's 3 Obayed flowed 3,000 b/d of 45° gravity crude oil and 5.7 MMcfd of gas. Shell estimated the well could raise field reserves to 3.1 tcf from 1.4 tcf, MEES said.

Baltim concession activity

International Egyptian Oil Co. (IEOC), a unit of Italy's Agip SpA, and Amoco Corp.'s Amoco Egypt Oil Co. found gas and condensate in 144 ft of water on the Baltim concession, 120 miles north of Cairo in the Mediterranean Sea.

Their discovery well flowed 31.6 MMcfd of gas and 586 b/d of condensate through a ¾ in. choke with 2,706 psi flowing wellhead pressure from Messinian Abu Madi pay sands above 12,835 ft measured depth.

Operator IEOC and Amoco Egypt each own a 50% interest in the well.

Relube plant

Egypt's Enppi of Cairo let contract to KTI Group for basic engineering and critical design support for Alexandria Petroleum Co.'s Relube plant.

The 30,000 metric ton/year waste lube oil refining plant will be built next to an existing complex. Completion is scheduled for spring 1995.

EQUATORIAL GUINEA

CAPITAL: Malabo
MONETARY UNIT: CFA franc
REFINING CAPACITY: N.A.
PRODUCTION: 4,000 b/d
RESERVES: 3.6 million bbl

Among the oil and gas producing countries on the coast of West Africa is Equatorial Guinea, a collection of territories that has decided it wants a piece of the action.

Equatorial Guinea comprises the island of Bioko off the coast of Cameroon, the Rio Muni coastal enclave sandwiched between Cameroon and Gabon, and the tiny island of San Antonio, far southwest of Bioko in the Atlantic Ocean.

The capital is Malabo on Bioko Island. Here President Obiang Nguema Mbasogo, an army general, oversees what a British Foreign Office official euphemistically calls "a single party regime."

EMS Ltd., Oxted, U.K., landed the job of organizing Equatorial Guinea's first exploration licensing round, with bidding expected to close in December 1993 at the earliest. All unlicensed prospective acreage will be offered.

Virgin territory

Don Holmes, managing director of EMS, explained that the Rio Muni area is virtually virgin territory. Six wells were drilled during 1968-91, the last three by Ste. Nationale Elf Aquitaine. Elf held a major portion of Offshore Rio Muni acreage until 1992, when it pulled out after fulfilling commitments.

London independents Hamilton Oil Co. Ltd. and Enterprise Oil plc took over the block nearest the border with Gabon, where some acreage is in dispute. Hamilton acquired 590 line km of seismic data, but will not drill two commitment wells until a border is agreed on. Adding spice to the dispute, in Gabon's recent licensing round Elf won some acreage now licensed to Hamilton/Enterprise.

Available Rio Muni concessions are split into seven offshore blocks extending about 100 km offshore on the continental shelf and slope and five onshore with total area of 13,000 sq km. Water depths of offshore blocks reach 6,500 ft. Onshore blocks are available only in the coastal sedimentary basin. Basement rocks cover most of the interior.

Bioko production

Three blocks off northern Bioko, near producing fields off Nigeria and Cameroon, are already under license. Walter International Inc., Houston, holds one block containing Equatorial Guinea's only producing field, Alba. It produces 4,000-5,000 b/d of condensate, according to EMS.

Two blocks are held by United Meridian International Corp., Houston, a combine of Conoco Inc., British Petroleum Co. plc, and Den norske stats oljeselskap AS.

United Meridian won acreage in mid-1992 and has not begun drilling.

Free acreage around Bioko is not yet split into blocks. Existing licenses are defined by 15 min grids covering about 750 sq km.

Norman Hempstead, EMS consulting geophysicist, said Rio Muni is an area of complex geology. Six wells along a 160 km coastline provide only a preliminary assessment. Recent surveys found many potential trapping mechanisms, he said, while three wells found source rocks and identified potential reservoirs.

GABON

CAPITAL: Libreville
MONETARY UNIT: CFA franc
REFINING CAPACITY: 17,300 b/cd
PRODUCTION: 296,700 b/d
RESERVES: 730 million bbl

Twenty exploration blocks with a combined area of more than 60,000 sq km (23,160 sq miles) were on offer in the seventh licensing round opened by Gabon's Ministry of Mines, Energy and Hydraulic Resources.

Data were available from Geco-Prakla Exploration Services, Crawley, U.K., in three packages: North Gabon, South Gabon, and Deep Offshore Gabon.

The North Gabon package includes information on Blocks A-93, C-93, F-93, H-93, I-93, J-93, K-93, and M-93, which are onshore except for J-93, K-93, and part of M-93.

South Gabon data cover Blocks N-93, O-93, P-93, Q-93, R-93 and S-93, which are near-shore concessions with water depths to 500 m except on R-93, where depths range from 50 to 1,000 m.

Deepwater offshore concessions are Blocks B-93, D-93, E-93, G-93, L-93, and T-93, which have water depths of 200-2,000 m.

Closing date for receipt of seventh round applications at the ministry offices in Libreville was July 31, 1994.

Wood Mackenzie Consultants Ltd., Edinburgh, said all blocks in the round have been offered before. The ministry was said to be excited by the potential of the southernmost blocks after recent 3D seismic surveys by Amoco Corp.

In January 1993, Amoco acquired the Phenix concession, which lies between Block S-93 and the Congolese border, out of round. Results on this block and the adjoining block over the border in Congo, also operated by Amoco, are said to be encouraging.

The ministry said while Gabon's shallow, postsalt, northern and central offshore areas are comparatively well explored, the same cannot be said for presalt and postsalt sections onshore and southern presalt offshore areas. Deepwater blocks are said to be undrilled.

Thirteen exploration blocks were offered in 1992's sixth licensing round. Bids were required by June 30, 1993, after which first awards were scheduled for late 1993 or early 1994.

COTE D'IVOIRE (Ivory Coast)

CAPITAL: Abidjan
MONETARY UNIT: CFA franc
REFINING CAPACITY: 62,800 b/cd
PRODUCTION: 500 b/d
RESERVES: 50 million bbl

Cote d'Ivoire opened an international bidding round at yearend 1993 for production sharing contracts (PSCs) on three blocks that lie mainly in the Gulf of Guinea.

Qualified offers resulting in PSCs signed before Apr. 30, 1994, would be allowed special incentives designed to speed oil and gas development.

The tender is part of a plan through which Cote d'Ivoire hopes to encourage domestic economic growth by stimulating enough oil and gas production to become energy self-sufficient.

Cote d'Ivoire imports enough crude to feed Ste. Ivoirienne de Raffinage's 59,000 b/cd refinery in the capital city of Abidjan. About one-third of the plant's refined product output is needed to meet domestic demand.

Cote d'Ivoire in December 1992 amended laws covering petroleum PSCs and service contracts to liberalize terms and take a more active role in developing its oil and gas resources. Combined with recent tax law revisions, the new legal framework is intended to spur investment in petroleum exploration and development.

To help Cote d'Ivoire achieve its economic goals, each company achieving commercial production under a PSC awarded in the tender will be required to sell 10% of its share of profit oil or gas to Petroci at 75% of a calculated market price. Losses occurring because of the 25% price differential may be recovered from future production as cost oil or gas.

Blocks offered

Tracts included in Cote d'Ivoire's offshore tender lie within the Abidjan margin on the republic's continental shelf.

All of Cote d'Ivoire's oil and gas exploration and development has occurred on the margin, where the shelf is underlain by a thick sedimentary section extending far offshore. All told, 78 exploratory and 32 development wells have been drilled on the Abidjan margin and 39,456 line km of 3D seismic and 31,720 line km of 2D seismic collected.

Water depths on the blocks range to 2,000 m, but most of the prospective oil and gas areas are in less than 200 m of water:

- Block 93-C1-01 is a 5,796 sq km irregularly shaped tract on which about 26,562 line km of 3D seismic data and 2,256 line km of 2D data have been collected.
- Block 93-C1-02 is a 4,012 sq km tract. About 2,943 line km of 2D seismic data have been collected on the block.
- Block 93-C1-03 is a 5,833 sq km tract on which about 3,912 line km of 2D seismic data are available.

Within Block 2, but not included in the current tender, is Belier field, a 1974 discovery productive from upper Cretaceous lower Senonian sandstones at 1,885-2,005 m below sea level.

In Block 3, but not being offered, are Foxtrot and Espoir fields, discovered in 1981 and 1979, respectively. Foxtrot hydrocarbons occur in Cretaceous upper Albian sandstones at 2,100-2,503 m below sea level, while Espoir is productive from Albian sandstones at 1,800-2,136 m below sea level.

All the blocks also hold several undeveloped oil and gas discoveries.

Offshore prospects

Much of offshore Cote d'Ivoire's oil and gas prospectivity is defined by seismic and well data. Hydrocarbons within the coastal basin are trapped mainly in rift habitat and postrift or drift habitat.

Primary rift objectives include intra-Albian and upper Albian traps, while postrift/drift opportunities are mainly Cenomanian to Maastrichtian. Main Abidjan margin reservoirs are found in Maastrichtian, lower Senonian, and Albian intervals.

Presence of sand is the main factor affecting Albian trapping. Extensional fault blocks are the primary Albian tectonic style, and possible traps are linked to sealing by normal faults. Areal closures are about 15 sq km, with vertical closures of 80 m. Down to basement faults are predominant, with throws of as much as 1,000 m.

Some of the most obvious Cenomanian traps have been drilled, including block edge folds superimposed on Albian paleo highs that gain relief as a result of differential compaction. Although more limited than at Albian depths, Cenomanian areal closures are significant.

Sandy intervals are found mostly in lower Senonian strata, and depth maps show strong attenuation of structural closures, compared with Albian levels. Possible lower Senonian prospects are present on the Belier-Assinie ridge, provided stratigraphy can be mapped accurately and stratigraphic controls understood.

Maastrichtian sand bodies detected so far are typically channel depositions, with cumulative thicknesses of individual beds amounting to as much as 85 m and average porosity of 20%. Structural control of Maastrichtian objectives is limited to undulations superimposed on top of Albian ridges.

Most of the prospective Maastrichtian area is associated with a delta system active in the Cenomanian and lower Senonian.

PSC requirements, incentives

As set out in the 1992 law and further defined in an April 1993 decree, many PSC terms covering the three offshore tracts are negotiable.

Perhaps foremost, PSCs which were signed before Apr. 30, 1994, need not have included a signature bonus. Production sharing contractors will not have to pay royalties on production and will be exempt from all taxes except the national income tax.

Cote d'Ivoire reduced its income tax rate to 35% from 50%.

Bids received before the deadline were to be opened in the order received. Negotiations for acreage could begin after review of any bid, in which case the government would announce whether serious talks were in progress and whether bids still were being accepted.

Cote d'Ivoire's new PSC includes an exploration period led by one 3 year phase and followed by two additional 2 year phases, a 2 year appraisal period, and a 25 year exploitation period. Extensions may be granted during all three periods, including:

- As many as 90 days during an exploration phase to complete a production test or abandon an exploratory well.
- During appraisal, an additional 2 years to appraise an oil discovery and 4 years for a gas discovery.
- Two extensions during development and production, the first for as long as 10 years.

Twenty-five percent of each block must be relinquished at the end of the initial 3 year exploration phase and another 25% after the first 2 year additional exploration phase.

Production sharing contractors will be required to carry Petroci through the exploration period with a 10% interest. In case of a commercial discovery, Petroci may increase its interest to as much as 30%.

If at declaration of a gas field's commerciality a market is not available, the contractor will be allowed to extend his rights for as long as 5 years.

A jet wash-down crew sets up a 7 m deep shooting pattern for a seismic survey near Port Gentil, Gabon. Photo by Michel Davalan, Elf Gabon, a subsidiary of Elf Aquitaine Production.

PSC biddable terms

Bidders are expected to propose minimum work commitments and estimated costs of the work program. Exploratory wells drilled in excess of phase requirements may be carried forward into the first phase following.

Other biddable items include percentages of cost oil, profit oil splits, production bonuses, and annual training budgets. Proposed cost recovery percentages may vary according to water depth and daily production volumes. PSC applicants will be required to bid production bonuses based on threshold oil volumes of 30,000 b/d, 50,000 b/d, and 100,000 b/d.

Shares of profit oil or gas are negotiable based either on specified production increments and water depths or on the ratio of cumulative net income divided by cumulative investments. Splits of profit oil produced in less than 200 m of water should vary within the following production windows: 0-10,000 b/d, 10,001-20,000 b/d, 20,001-30,000 b/d, and more than 30,000 b/d. In water more than 200 m deep, splits should differ if production ranges from 0-25,000 b/d, 25,001-50,000 b/d, and more than 50,000 b/d.

Shares of profit gas from fields in less than 100 m of water will vary if gas production is 0-75 MMcfd, 75-150 MMcfd, and more than 150 MMcfd. In water more than 100 m deep, shares of gas should be proposed for production of 0-150 MMcfd and more than 150 MMcfd.

Contractors' shares of profit oil or gas may be calculated either before or after payment of income taxes.

Joint venture

Global Natural Resources Inc., Houston, unit GNR (Cote d'Ivoire) Ltd. agreed to a joint venture with United Meridian Corp., Houston, and World Bank's International Finance Corp. covering exploration and development of hydrocarbons on Block CI-11 off Ivory Coast.

Two wells on the block found oil and gas and flowed oil at commercial rates on drill stem test in the 1980s. Under a 1 year initial exploration term, the venture will drill two wells on the structures tapped by the earlier two wells, with the first to have spudded in the fourth quarter 1993.

LIBYA

CAPITAL: Tripoli
MONETARY UNIT: Dinar
REFINING CAPACITY: 348,400 b/cd
PRODUCTION: 1.4 million b/d
RESERVES: 22.8 billion bbl

PanCanadian Petroleum Ltd., Calgary, will earn a 15%

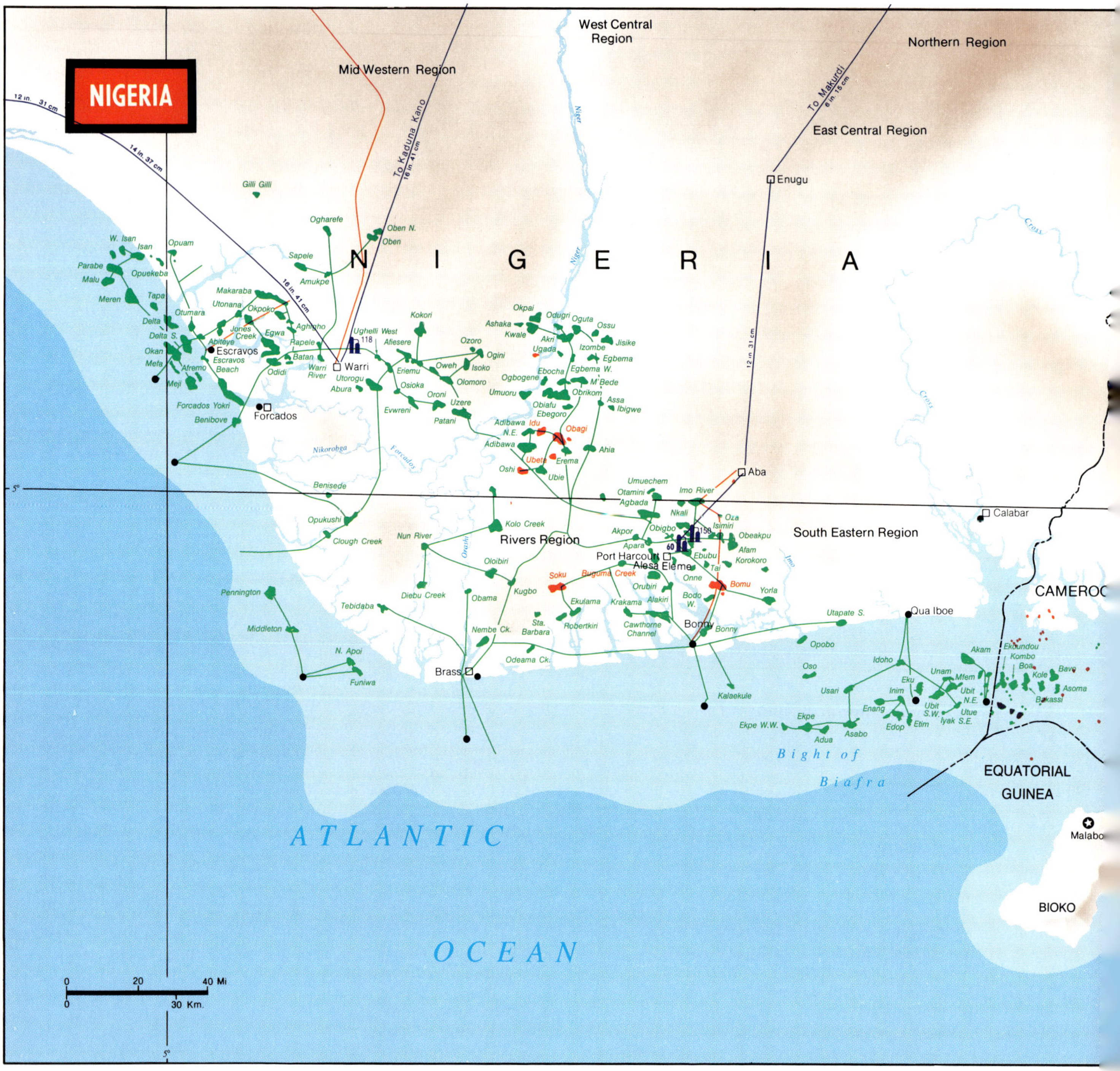

working interest in 3.9 million acres in Libya under an agreement with Fina Exploration Libya BV.

The agreement covers blocks NC 170, NC 171, and NC 172 in the Sirte basin. The work commitment calls for acquiring 8,500 line km of seismic data and drilling 12 wells. Drilling was to start at midyear 1993. Westcoast Petroleum Ltd., Calgary, and Chieftain International Inc., Edmonton, signed a similar agreement with Fina in 1992.

NAMIBIA

CAPITAL: Windhoek
MONETARY UNIT: Rand
REFINING CAPACITY: N.A.
PRODUCTION: N.A.
RESERVES: N.A.

Norsk Hydro AS spudded its first well off Namibia late in 1993 in 500 m of water on Block Q1911.

This was the first of three commitment wells.

Operator Hydro holds 40% interest in the block with partners Den norske stats oljeselskap AS and Saga Petroleum AS, each holders of 30% interest.

Statoil said information on drilling results will be restricted.

Occidental work

An Occidental Petroleum Corp. unit was to conduct on site work in Namibia under an agreement with Brilund Ltd. unit Etosha Petroleum, then acquire an assignment to explore Etosha's 35 million acre concession.

Oxy was to acquire 100% interest in the area for $286,000 and drill one wildcat.

Etosha was to receive added payments of as much as $300,000 if Oxy continues after drilling the first well.

Oxy will hold all rights to discoveries, with Etosha

receiving a 4% overriding royalty interest and as much as $2 million.

Nopec contract

National Petroleum Corp. of Namibia (Namcor) let an exclusive contract to Nopec AS, Naersnes, Norway, to conduct a multiclient 2-D seismic survey off Namibia and market and license the data.

The survey was to cover 10,000 line km of regional seismic from the Angola border to 22° 30' in the south.

Acquisition was to have started early in 1993 and be completed in 5 months. Full data were to be available by yearend 1993.

The agreement included marketing and licensing 1,160 line km of 2-D seismic data acquired by Namcor in 1992.

NIGERIA

CAPITAL: Abuja
MONETARY UNIT: Naira
REFINING CAPACITY: 433,250 b/cd
PRODUCTION: 1.9 million b/d
RESERVES: 18 billion bbl

Nigerian National Petroleum Corp. (NNPC) and Chevron Oil Corp. will install an oil production platform off Ondo state, western Nigeria, to boost production by about 15,000 b/d.

Estimated cost of the project is $150 million.

Meantime, Chevron inaugurated its 19th Nigerian oil field, Opuekeba, in the Escravos concession. Initial production is 16,000 b/d, and production is expected to peak at 30,000 b/d. Estimated reserves are 46 million bbl. Idama, Inda, Abighorodo, and Belema/Belema North oil fields were to start production late in 1993, adding 65,000 b/d to Nigeria's production.

Offshore work

Texaco Overseas (Nigeria) Petroleum Co. Unlimited hired Offshore Pipelines Inc., Houston, to engineer, fabricate, transport, install, and commission two offshore structures and three pipelines in North Apoi and Okubie fields off Nigeria.

In December 1993, OPI was to begin fabrication of the North Apoi J three pile well protector platform and the Okubie A four pile platform to be installed in 42-50 ft of water. The three associated pipelines include 0.9 and 8 miles of 6 in. size and an 8 mile, 10 in. line.

Production sharing contracts

DuPont E&P Nigeria Ltd. and NNPC signed a production sharing contract for Oil Prospecting License 220 in the Southeast Niger Delta deepwater area. The block covers 2,421 sq km in 150-1,100 m of water. Contract terms call for DuPont to commit $54 million to a work program during the first 6 years of the license.

Agip SpA unit Nigerian Agip Oil Co. signed a production sharing contract (PSC) with NNPC covering exploration on two deepwater blocks off Nigeria. The acreage involves 4,700 sq km in 200-1,200 m of water 60-100 km from shore in the Gulf of Guinea. Agip will be operator of a joint venture with Exxon Corp. and Amoco Corp.

Mobil Corp. unit Mobil Producing Nigeria Unlimited signed a production sharing contract with NNPC calling for Mobil to explore 565,000 acre Block OPL 221 off Nigeria. Fifty miles offshore and 10 miles southwest of current Mobil production, it lies in 600-3,000 ft of water. Geologic and geophysical work were scheduled. Operator Mobil holds a 40% interest in the joint venture, NNPC 60%.

Exxon Corp. affiliate Esso Exploration & Production Nigeria Ltd. signed a production sharing agreement with NNPC to explore 500,000 acre Block 9, 47 miles off Nigeria. The agreement marks Exxon's return to upstream operations in the country.

NNPC awarded production sharing contracts to Elf Nigeria Ltd., a unit of Ste. Nationale Elf Aquitaine, for three onshore blocks in the Benue basin and two blocks in 200-1,000 m of water in the Atlantic Ocean south of the OML 100 permit area where Elf Nigeria began producing oil in Afia and Odudu fields in March 1993. Elf Nigeria is to pay exploration and development costs on the new tracts in return for undisclosed shares of oil production.

Production start-up

Elf Petroleum Nigeria Ltd. began oil production from Afia field in 60-75 m of water 60 km off Nigeria, the first of four fields on production license OML 100 to start up. Others are Odudu field, IME, and Edikan.

The four hold reserves estimated at 150 million bbl. By mid-1994, they were to be producing about 60,000 b/d of low sulfur crude into a 250,000 dwt floating storage tanker.

Development will entail four producing and one processing platform and 40 wells, including three horizontal wells in Odudu. Interests are Elf 40% and Nigerian National Petroleum Corp. 60%.

Exploration activity

Nigeria awarded five exploration licenses to Shell Petroleum Development Co. (SPDC) covering deepwater Blocks 212 and 219 and three onshore blocks in the Benue basin.

The offshore acreage lies in water depths ranging from 200 m to more than 1,500 m. SPDC created a new company, Shell Nigeria Exploration & Production Co. Ltd., to conduct exploration and production operations on the offshore blocks, where it plans extensive 3D seismic surveys.

Aztec Energy Corp., Dallas, Abacan Resources Corp., Calgary, and Abacan International Resources Management Inc. were to participate in exploration of three contiguous blocks of more than 2 million acres total in the Benin basin of Nigeria and Benin. Oil prospects lie in several Tertiary and Cretaceous formations. Two concessions of 500,000 acres each are in Nigeria, one offshore and one onshore, and the third is 1 million acres onshore in Benin. The companies plan an initial $10 million exploration program.

Refinery agreements

Gulf-Petro Trading Co. Inc., Dallas, and subsidiary Petrec International signed joint venture agreements with NNPC, Lagos.

One called for Gulf-Petro Refining Ltd. to build a 10,000 b/d portable refinery at Bonny, Nigeria, with work to start in late June or early July 1993.

The other agreement is for cleaning all crude tanks and waste water facilities at NNPC refineries at Warri, Port Harcourt, and Kaduna, Nigeria

Ammonia plant

The key ammonia plant in National Fertilizer Co. of Nigeria's fertilizer complex at Onne, Rivers state, Nigeria, is turning out product at a rapid clip.

It has averaged 1,000 tons/day of production since it went on stream in 1987, passing the 2 million ton cumulative mark late in 1992. That is 112% of planned yearly production capacity at an on stream factor of 99%. Both figures are believed to be world records, said M.W. Kellogg Co., Houston. Kellogg led an international group that designed, engineered, built, and started up the plant.

SOUTH AFRICA

CAPITAL: Pretoria/Cape Town
MONETARY UNIT: Rand
REFINING CAPACITY: 364,038 b/cd
PRODUCTION: 4,100 b/d
RESERVES: 41 million bbl

In recent years a number of small oil fields have been discovered by South Africa's Soekor in the Bredasdorp basin—mainly in deepwater submarine fan-channel complexes.

However, the greater potential is for gas. Several gas fields have already been discovered.

The offshore basins of the Republic of South Africa lie along the southern margins of the African plate. They originated in the middle to late Jurassic during the breakup of southern Gondwana.

South Africa's first natural hydrocarbon production came on stream in 1992 to feed the Mossgas synfuels plant near Mossel Bay on the south coast. A recent independent review confirmed that significant potential remains—primarily in the Orange, Bredasdorp, and Pletmost basins but especially in the unexplored Southern Outeniqua basin, which holds the potential for giant oil fields.

South Africa's infrastructure is well developed with, by African standards, an exceptional road network, modern international airports, and several harbors capable of handling all but the largest ships.

Development of the region's energy potential would do much to contribute to stability of the subcontinent through generation of wealth in general and sorely needed jobs in particular.

The energy demand, facilities, and skills are all in place—the geology is the last remaining area of uncertainty for a successful project.

Mossel Bay platform

South Africa's Mossgas let contract to Genesis Engineering Consultants, London, for preliminary engineering of two expansions of the F-A natural gas production platform in Mossel Bay.

The expansions cover subsea tieback of as many as six satellite wells and addition of export gas compression to extend the life of F-A field. Mossgas produces synthetic oil products from Mossel Bay gas.

New company

South Africa's Sasol Ltd. and AECI Ltd. of Johannesburg were to merge their petrochemical and plastics interests to form a company with $784 million/year in sales, London's Financial Times reported.

First project for the venture will be conversion of AECI's PVC production from carbide feedstock to using Sasol's ethylene.

SUDAN

CAPITAL: Khartoum
MONETARY UNIT: Pound
REFINING CAPACITY: 21,700 b/cd
PRODUCTION: 2,000 b/d
RESERVES: 300 million bbl

A Vancouver, B.C., independent and its Sudanese partner's development plan has been initially approved by the government of Sudan. The plan calls for an initial production of 40,000 b/d from Heglig and Unity oil fields in Sudan.

Arakis Energy Corp. and the private Sudanese company State Petroleum Corp. (SPC) want to begin the first commercial hydrocarbon production in the destitute, war torn country.

They are picking up where Chevron Corp. left off after years of grappling with an ambitious, costly—and ultimately futile—effort to export crude oil from Sudan.

After finding almost 300 million bbl of oil in Sudan during the early 1980s, Chevron scuttled a $2 billion project to export 50,000 b/d of Sudanese crude in 1986. It drilled 90 wells and sank more than $1 billion into the project.

But it dropped the plan, citing the 1986 collapse of oil prices and concerns over security after repeated guerrilla attacks delayed work.

Chevron and Sudan signed a memorandum of understanding in 1992 allowing third parties to develop the fields it found in southern Sudan. Private Sudanese interests later acquired Chevron's interests in the two concessions that cover 42 million acres and include the discoveries.

Arakis and SPC in 1993 began acquiring development rights to the two blocks.

What's involved

The Arakis-SPC plan calls for a first phase to produce 40,000 b/d from Unity and Heglig oil fields on the Muglad block by mid-1995.

The oil is to be shipped by pipeline to a proposed 60,000 b/d refinery at Kosti for local use. An undisclosed Japanese group has proposed building the refinery at Kosti and extending the proposed pipeline to Port Sudan on the Red Sea.

Under a second phase, Sudanese oil production could ultimately ramp up to 300,000 b/d, thus supporting a crude export project.

Sudan imports about 60,000 b/d of crude oil and refined products.

Arakis-SPC is involved only in the first phase, upstream portion of the project.

The company also was negotiating funding options for the project and is likely to pursue loans from the World Bank, International Monetary Fund, and other international banks.

That may pose a problem for Sudan, which is in arrears with IMF, Arakis said.

The company also said the government is committed to pay world prices for crude from the two blocks.

Plans call for setting up 20 person operating camps at each production site, with field operations expected to begin in January 1994.

Oil could be produced as early as January 1994, with production trucked to an existing 1,200 b/d topping plant at Muglad.

Project details

Arakis estimated a cost of $140 million for wellhead equipment, an oil gathering system, and surface equipment to produce the initial 40,000 b/d of oil.

Proved producible oil reserves are pegged at 280 million bbl, or about 20% of the 1.4 billion bbl original oil in place (OOIP) identified to date on the two blocks.

Of that reserves total, 240 million bbl are attributed to Unity and Heglig fields on the Muglad block and 40 million to Adar/Yale fields on the Melut block.

Lifting costs were estimated at $3/bbl, with finding and field development costs placed at 60¢/bbl.

Under a tentative agreement, Arakis as operator will earn a 70% net revenue interest until payout, with the government getting a 30% royalty. After payout, the split will be 50-50.

The revival of development plans for the Sudanese fields is justified, Arakis contends, by improved oil prices compared with the $8/bbl medium gravity African crudes were getting in 1986 and a significant lessening of rebel activity in Sudan.

Unity field

An initial 25,000 b/d will be produced from 19 completed wells in Unity field.

These wells produce from multiple fault-separated oil zones.

The main producing area in Unity field is productive from 10 of 11 identified sands.

Individual flow rates of more than 2,000 b/d/zone have been gauged from six of the 11 pay zones during drill stem testing of appraisal wells, which were drilled to 6,500 ft.

A test produced 3,000 b/d of oil for 30 days from just one of seven pay zones in the 5 Unity well.

Other pay zones will be tested and placed on production later.

The rest of the wells in the field will produce 1,200-2,000 b/d initially from one zone, with several pay zones in each well yet to be drill stem tested and placed on production.

Heglig field

Heglig field has 17 wells that will contribute 15,000 b/d to initial production. Drill stem tests have indicated an average production of 2,400 b/d of 29-36° gravity crude from either the Bentiu Unit 1 or 3 zones.

Heglig area OOIP has been estimated at 555 million bbl, of which 90 million bbl is classed as proved. Porosity of the pay zones runs about 25-30%.

The initial production facility design to serve both fields will be capable of processing 50,000 b/d of oil.

The crude, which has a high bubble point, will be handled by conventional oil field technology. A hot oil pump sized to the pipeline gathering system will facilitate pigging the entire gathering system, which will hook up to a field gate serviced by a 20 in. pipeline linked to the refinery.

Arakis has in hand $5 million of the $30 million it needs for its commitment to the project for detailed engineering and drill stem testing. It will try to raise the remainder on equity markets.

TUNISIA

CAPITAL: Tunis
MONETARY UNIT: Dinar
REFINING CAPACITY: 34,000 b/cd
PRODUCTION: 100,100 b/d
RESERVES: 1.7 billion bbl

Exploration and development was gathering momentum off Tunisia in 1993.

British Gas plc started work on a £492 million ($738 million) gas field development project in Tunisia that will help the country become self-sufficient in natural gas.

Meantime, a 50-50 combine of operator Union Texas Petroleum Holdings Inc. and Louisiana Land & Exploration Co. signed an oil and gas exploration permit off Tunisia.

Tunisia's gas fields have been thought too small to develop as a source of gas exports. Domestic requirements have been met mainly by Algerian gas arriving via the Trans-Mediterranean pipeline.

BG Miskar plans

BG intended to develop Miskar field to supply gas for electricity generation and distribution. A long term supply contract has been signed with Ste. Tunisienne du Electricite et du Gaz (STEG).

The contract commits British Gas to supply STEG with as much as 160 MMcfd of gas for 5 years after production begins in mid-1995, and 200 MMcfd thereafter.

Miskar field was discovered in 1975 on the Amilcar concession.

It lies 75 miles off Tunisia in 225 ft of water in the Gulf of Gabes.

BG estimates field reserves at 1 tcf.

Three platforms will be used to develop Miskar field. Twelve wells will produce via wellhead platform to an adjacent processing platform. A flare boom platform will be installed nearby.

Development drilling was under way with the J.T. Angel jack up rig operated by Reading & Bates Corp., Houston.

Produced gas will be exported via subsea pipeline to the Hannibal onshore processing plant at Sfax. Here gas will be prepared for industrial and domestic consumption, while liquids will be stripped out for sale.

Other BG Tunisia

BG is 100% owner and operator of Miskar. Its other Tunisian interests are as operator in Amilcar 50%, West Kerkennah 49%, and Roumedia 50% and as a minority interest owner in South Kerkennah 15%.

Amilcar offshore permit is said to contain several smaller gas fields along with Miskar.

These will be developed as Miskar production declines. BG expects Miskar to provide secure supplies for 20 years.

West Kerkennah onshore/offshore permit includes two producing concessions. The onshore Gremda/El Ain area holds five oil wells producing a total 2,500 b/d. West Gremda field lies within the city of Sfax. Two wells produce on El Hajeb/Guebiba permit at a combined rate of 200 b/d of oil.

In 1992 BG discovered Cercina oil field near the Kerkennah Islands on the West Kerkennah block. This was under development, with first oil expected by yearend 1993 at 4,000-5,000 b/d.

Cercina crude oil will be exported via offshore loading into tankers. Peak production of 10,000 b/d is expected in 1995.

BG also said it expected to produce 7,000 b/d of oil from Rhemoura field on the West Kerkennah permit beginning in fall 1993.

BG bought into South Kerkennah license in 1991, acquiring the 15% stake of Springfield Oil Services Inc., Tunis.

The permit is operated by Amoco Corp.

Roumedia permit is the largest single concession in Tunisia, covering 2,300 sq km. BG acquired the permit in 1992.

Union Texas/LL&E permit

The Union Texas/LL&E permit calls for a 4 year exploration program on the Ramla block.

Ramla is a 1 million acre block about 80 miles offshore in the Gulf of Gabes and about 180 miles southeast of Tunis.

Water depths average about 200 ft.

Union Texas Maghreb Inc. and LL&E Tunisia Ltd. each will bear 50% of exploration costs. In the event of development of a discovery, state owned petroleum company L'Enterprise Tunisienne D'Activities Petrolieres (ETAP) will have the right to participate with as much as a 50% interest.

The companies have agreed to acquire seismic data and drill one well during the permit's 4 year term. Permit extensions are available under certain conditions. IPE

Commonwealth of Independent States

C.I.S.

CAPITAL: Moscow
MONETARY UNIT: Ruble
REFINING CAPACITY: 9,845,939 b/cd
PRODUCTION: 7.847 million b/d
RESERVES: 57 billion bbl

Russia's crude oil production slide is expected to continue during 1994, while its natural gas flow will turn around from a slump.

That was the key finding of a draft report by Russia's Association for the Fuel & Energy Complex (Asofec) to a plenary meeting of the former Soviet republics' Intergovernmental Council on Oil & Gas. Other delegates at the meeting called for cuts in taxes and duties they said were hindering bilateral relations, although such prospects are dim in light of most republics' economies remaining in shambles.

Several delegates noted problems between Russia and other former Soviet republics over nonpayment of debts for petroleum supplies. Russia recently has cut off supplies to a number of other C.I.S. members for lack of payment of petroleum debts. The council was established in 1993 by Armenia, Belarus, Georgia, Kazakhstan, Kyrgyzstan, Moldova, Russia, Tajikistan, Ukraine, and Uzbekistan. Also attending as observers were Hungary, Slovakia, and Czech Republic.

Asofec's president said the association predicts Russian crude production in 1994 will average 6.54 million b/d, down from a forecast 6.8-7 million b/d during 1993. That compares with peak of 11.4 million b/d in 1988.

Natural gas flow is expected to increase by 350-385 bcf in 1994 to 22.75-23.1 tcf. Natural gas production, previously a bright spot in Russia's energy production picture, declined during first half 1993 from the same 1992 period.

At the same time, Russian production of refined products is expected to slip to 4.5 million b/d, down from 5.09 million b/d expected during 1993. Asofec said refined products supplies will cover Russian demand but will not be sufficient to cover other former Soviet republics' demand.

Some delegates at the meeting called Asofec's predictions too optimistic. Asofec pointed out that while Russia's oil production decline can't be stopped during the next 2 years, it could be slowed significantly through cooperation and investment by all former Soviet states in Russia's crumbling petroleum infrastructure.

Delegates to the meeting sought to establish improved relations among council members. Asofec called on oil and gas enterprises of the council nations to form more joint ventures among its members. It said a preferred approach would entail service contracts, thereby allowing direct agreements without a preliminary lender process and possible conflicts with Russian law.

Asofec also called for coordinated legislation governing development of natural resources, taxes, and accounting procedures.

Most delegates from other former Soviet republics cited an urgent need to catch up with payment of debt to Russia for hydrocarbon supplies. C.I.S. countries owe Russia more than $10 billion for oil and gas supplies, equal to 53% of the payment for 1 year's deliveries, Asofec said.

Belarus noted at the meeting it paid 60% of its liabilities for Russian petroleum purchases. In late August 1993, Belarus's take of Russian gas supplies was slashed to 423.6 MMcfd from the typical 1.094 bcfd because of nonpayment for the gas. After the partial payment, Russia boosted Belarus's gas supplies to 600.1 MMcfd.

Similarly, Estonia's imports of Russian natural gas fell to about 52 MMcfd in first half 1993 from about 119 MMcfd during the same period in 1992. Estonia stepped up payments and currently pays about $2.27/Mcf.

Russian gas enterprise Gazprom noted Belarus's payment of $30 million helped stave off a threatened strike by western Siberian gas workers. However, Gazprom noted,

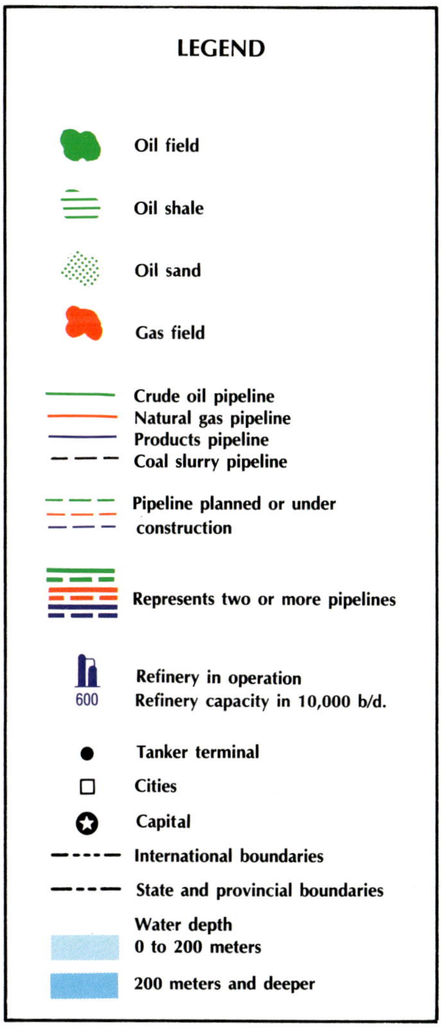

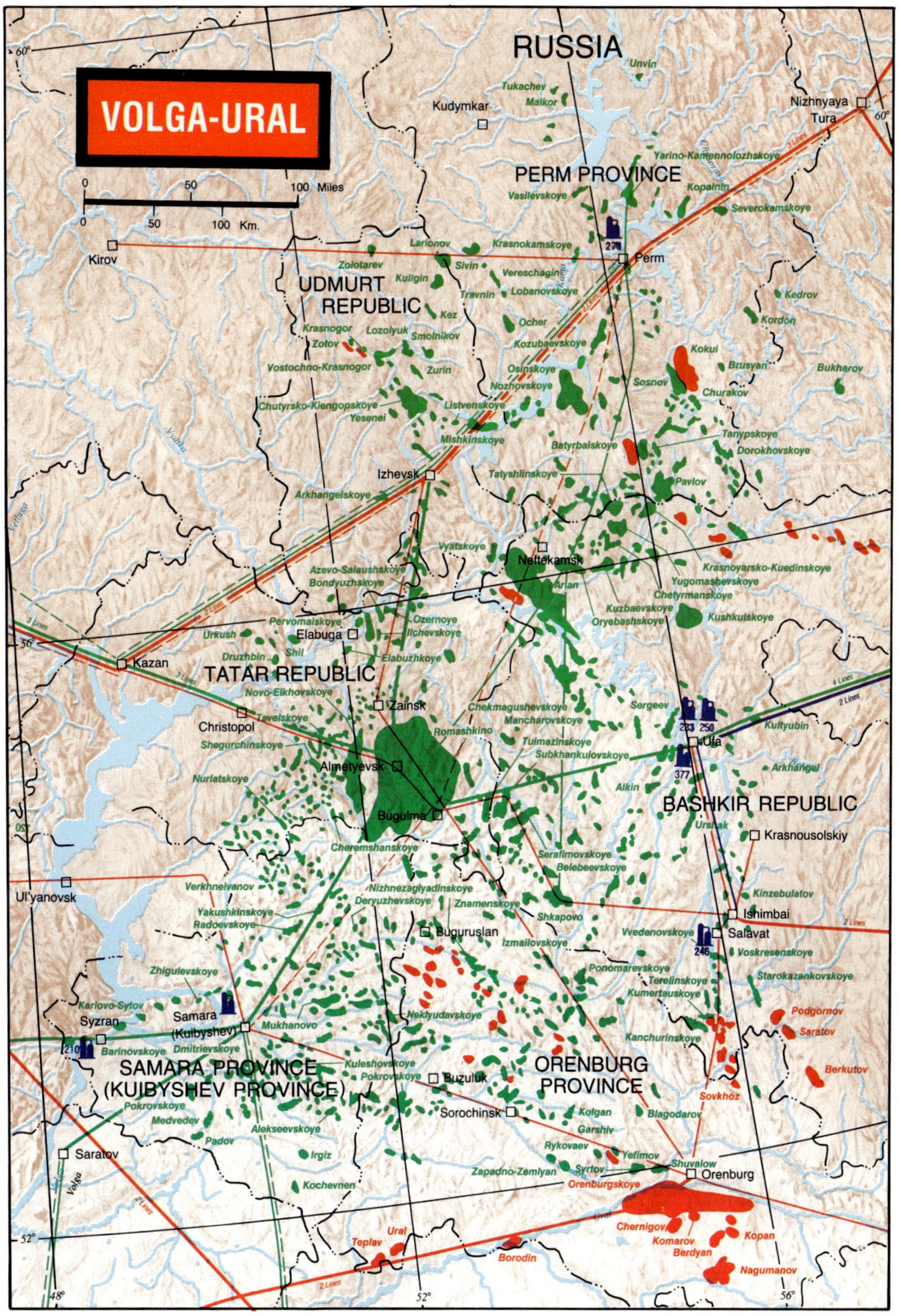

one concern still owes, at last report, $100 million to the Yamalo-Nenets autonomous region alone.

Ukraine, with a fuel debt to Russia of $1 billion and mounting, said it will cover all its liabilities. However, the C.I.S. republic in August 1993 paid only $26 million instead of $65 million as promised. In late August 1993, Russia cut gas shipments to Ukraine by 25%. Ukraine had been receiving about 4.2 bcfd. By remaining in arrears, Ukraine faces the threat of a further 25% cut if the bills aren't paid.

Russian gas production hits hydrocarbon slide

The last remaining stable sector of Russia's energy industry is getting shaky. Gas flow, which did not follow the sharp decline in Russian oil, coal, and refinery production in 1988-92, slumped in 1993.

The nation's gas production in the first 6 months of 1993 was down less than 2% from the same 1992 period. But June-August 1993 flow skidded about 5% from the same 1992 period.

Flow from western Siberia's supergiant Urengoi field, considered the world's largest gas field, has started to fall below plan. Urengoi has until recently accounted for as much as 45% of Russia's gas production. Preliminary estimates show that Urengoi, astride the Arctic Circle, likely will provide 273 billion cu m (9.64 tcf) of gas in 1993, once official tallies for the year are made. Moscow planners had called for 230 bcm (9.88 tcf).

Urengoi's 1994 production also is expected to fall short of target by another 5 bcm (176.5 bcf). Officials say the field's reserves are still big enough to maintain peak flow, but reservoir pressures are dropping. Only nine of 15 required compressor stations have been built.

Moscow had hoped that giant Pestsovoye field, just north of Urengoi, would be able to make up for Urengoi's production shortfall. But Pestsovoye development will require 100 billion rubles at today's current value "and even more when our new round of inflation is taken into consideration," the Moscow newspaper Delovoi Mir declared. "There is nowhere to obtain this amount of money either now or in the foreseeable future," Delovoi Mir said.

"This chronic lack of funds is one of the main reasons development and exploitation of Pestsovoye have been postponed from 1995, as originally planned, to 1997 at the earliest." The newspaper said the delay in placing Pestsovoye on stream means a loss of about 27 bcm (953 bcf)/year in gas flow, along with about 30% as much crude as

COMMONWEALTH OF INDEPENDENT STATES (C.I.S.)
POLAND
RUSSIA
LITHUANIA
LATVIA
ESTONIA
FINLAND
BELARUS
UKRAINE
MOLDOVA
CRIMEA
GEORGIA
ARMENIA
AZERBAIJAN
IRAN
AFGHANISTAN
TURKMENISTAN
UZBEKISTAN
KAZAKHSTAN
KIRGIZSTAN
TADZHIKISTAN
Barents Sea
White Sea
Black Sea
Caspian Sea
Aral Sea
Lake Balkhash
Lake Ladoga
Gulf of Finland
Sea of Azov
Arctic Circle
Kola Pen.
Kanin Pen.
Mangyshlak Pen.
Timan-Pechora fields
Volga-Ural fields
Saratov fields
Volgograd fields
Krasnodar fields
Orenburg fields
Emba fields
Kalmyk fields
Dagestan fields
Fergana fields
Warsaw
Kaliningrad
Klaipeda
Liepaya
Riga
Tallinn
Tartu
Vyborg
St. Petersburg (Leningrad)
Petrozavosk
Arkhangel
Naryan-Mar
Pechora
Ukhta
Kaunas
Vilnius
Mazheikai
Pskov
Novgorod
Kirishi
Valdai
Cherepovets
Torzhok
Gradno
Minsk
Polotsk
Brest
Lvov
Brody
Dolina
Nadvornaya
Chernovtsy
Mozyr
Smolensk
Bryansk
Kaluga
Moscow
Yaroslavl
Ryazan
Vladimir
Kiev
Vinnitsa
Kishinev
Unecha
Orel
Yelets
Poltava
Kremenchug
Kharkov
Odessa
Kherson
Zaporozhye
Zmiuev
Kursk
Voronezh
Michurinsk
Nizhniy Novgorod (Gorkiy)
Kirov
Ostrogozhsk
Lisichansk
Lugansk
Donetsk
Sevastopol
Simferopol
Yalta
Rostov
Novorossiisk
Tikhoretsk
Krasnodar
Tuapse
Sochi
Stavropol
Volgograd
Saratov
Alexandrou
Gai
Syzran
Samara (Kuibyshev)
Ul'yanovsk
Kazan
Izhevsk
Zainsk
Neftekamsk
Perm
Ufa
Buzuluk
Salavat
Ishimbay
Orenburg
Uralsk
Magnitogorsk
Chelyabinsk
Kurgan
Sverdlovsk
Nizhnyaya Tura
Serov
Tyumen
Tobol
Khanty Mansi
Aktyubinsk
Orsk
Atbasar
Astrakhan
Atyrau (Guryev)
Kulsary
Beineu
Grozny
Makhachkala
Batumi
Tbilisi
Yerevan
Nakhichevan
Reza'iyeh
Tabriz
Baku
Astara
Rasht
Aktau (Shevchenko)
Krasnovodsk
Kungrad
Khiva
Tehran
Bandar Shah
Semnan
Shahrud
Ashgabat (Ashkhabad)
Bukhara
Chardzhou
Samarkand
Mashhad
Isfahan
Daran
Mazar-i-Sharif
Shibarghan
Dushanbe
Tashkent
Chimkent
Namangan
Fergana
Bishkek (Frunze)
Alma-Ata
Karaganda
Balkhash
Lake Issyk Kul'
Kolguyev
Samotlor
0 100 200 300 Mi.
0 100 200 300 400 Km.

Urengoi produces and 10% as much condensate.

Delovoi Mir suggested development of western Siberia's far northern gas fields might be hastened by creating a joint stock company controlled by Russian government agencies but including foreign investors.

Russian oil exports to show little effect

Russian oil exports to countries outside the C.I.S. probably won't be large enough to have a major effect on the world market during the next two decades, an official study indicates. The Moscow newspaper Business World Weekly cited Russian Ministry of Fuel & Energy data projecting the nation's total exports of crude and products, including deliveries to states, at 3.1-3.8 million b/d in 2010. That compares with 5.9 million b/d in 1990, an expected 2.34-2.6 million b/d in 1995, and 2.3-3 million b/d in 2000.

Of the total crude and products exports projected for 2010, 1.8-2.24 million b/d likely will go to non-C.I.S. countries, down from 2.7 million b/d in 1990. Estimates are 940,000 b/d to 1.1 million b/d for 1995 and 1.1-1.7 million b/d in 2000.

Russian crude and products exports to other states are expected to be 1.4-1.5 million b/d in 1995, 1.2-1.3 million b/d in 2000, and 1.3-1.4 million b/d in 2010. By comparison, Russian crude and products deliveries to other republics in the C.I.S. were 3.2 million b/d in 1990.

Russia's State Statistical Committee admits that it erred in reporting earlier that the nation's oil exports declined during the first 9 months of 1992. According to a more recent Russian report, the nation's crude export volume to non-C.I.S. customers rose 14% to more than 1.3 million b/d in 1992 if deliveries by nongovernment organizations are added to those announced for the state oil firm Nafta Moskva.

The Business World Weekly survey showed that Russian crude and condensate production in 2010 likely will be 7.4-8 million b/d, far below the 1990 figure of 10.36 million b/d, 9.22 million b/d in 1991, and 7.85 million b/d in 1992.

A further decline in Russian oil flow to about 7 million b/d in 1993 and 6.34-6.8 million b/d in 1995 is expected. Production is seen rising to 6.6-7.2 million b/d in 2000 and 7.4-8 million b/d in 2010.

Eastern Russia's resources tallied

Russia's Academy of Sciences has reported new data on estimated hydrocarbon resources and reserves in eastern Siberia and the Russian Far East. Potential gas resources in the area are believed to total about 32 trillion cu m (1.13 quadrillion cu ft). Of this, eastern Siberian "commercial" gas reserves (roughly proved/probable) total 944 bcm (33.3 tcf), including 654 bcm (23.1 tcf) in Krasnoyarsk territory and 290 bcm (10.2 tcf) in Irkutsk province.

Potential Russian eastern Siberian and far eastern oil resources were placed at more than 8 billion metric tons (58.4 billion bbl) and commercial oil reserves at 3.1 million tons (2.3 billion bbl), including 133 million tons (971 billion bbl) in Krasnoyarsk territory and 180 million tons (1.31 billion bbl) in Irkutsk province.

Potential far eastern gas resources in the Sakha republic (Yakutia) and Sakhalin province were projected at 16.7 trillion cu m (589.5 tcf). Commercial gas reserves in the Sakha republic were estimated at 959 bcm (33.9 tcf) and in Sakhalin province at 615 bcm (21.7 tcf).

Potential oil resources in the Russian far east were placed at 3.76 billion tons (27.4 billion bbl) and commercial reserves at 324 million tons (2.37 billion bbl). Commercial far eastern oil reserves included 131 million tons (956 million bbl) in the Sakha republic and 193 million tons (1.41 billion bbl) in Sakhalin province.

The academy emphasized that oil and gas deposits are very unevenly distributed in eastern Siberia and Russia's Far East. It said that geological exploration organizations expect to increase commercial hydrocarbon reserves in the region "manyfold."

Kazakh liquids, gas reserves

Kazakhstan's 160 fields hold reserves totaling a combined 15.56 billion bbl of oil, 62.9 tcf of gas, and 6.01 billion bbl of condensate, data compiled by Kazakh Scientific Research Geological Exploration Petroleum Institute (Kaznigri) show.

The head of Kaznigri's department of reserves and resources told a Geneva sysmposium 100 fields lie in the republic's onshore pre-Caspian basin with reserves totaling 9.4 billion bbl of oil, 58 tcf of gas, and 5.98 billion bbl of condensate. He said potential reserves for the pre-Caspian basin amount to 50 billion bbl of oil, 15 billion bbl of condensate, and 280 tcf of gas, including about 15 billion bbl of oil in offshore reservoirs.

Another Kaznigri spokesman said 80% of potential resources lie below 16,000 ft. Deep reservoirs and sour hydrocarbons are the main hindrances to increased production.

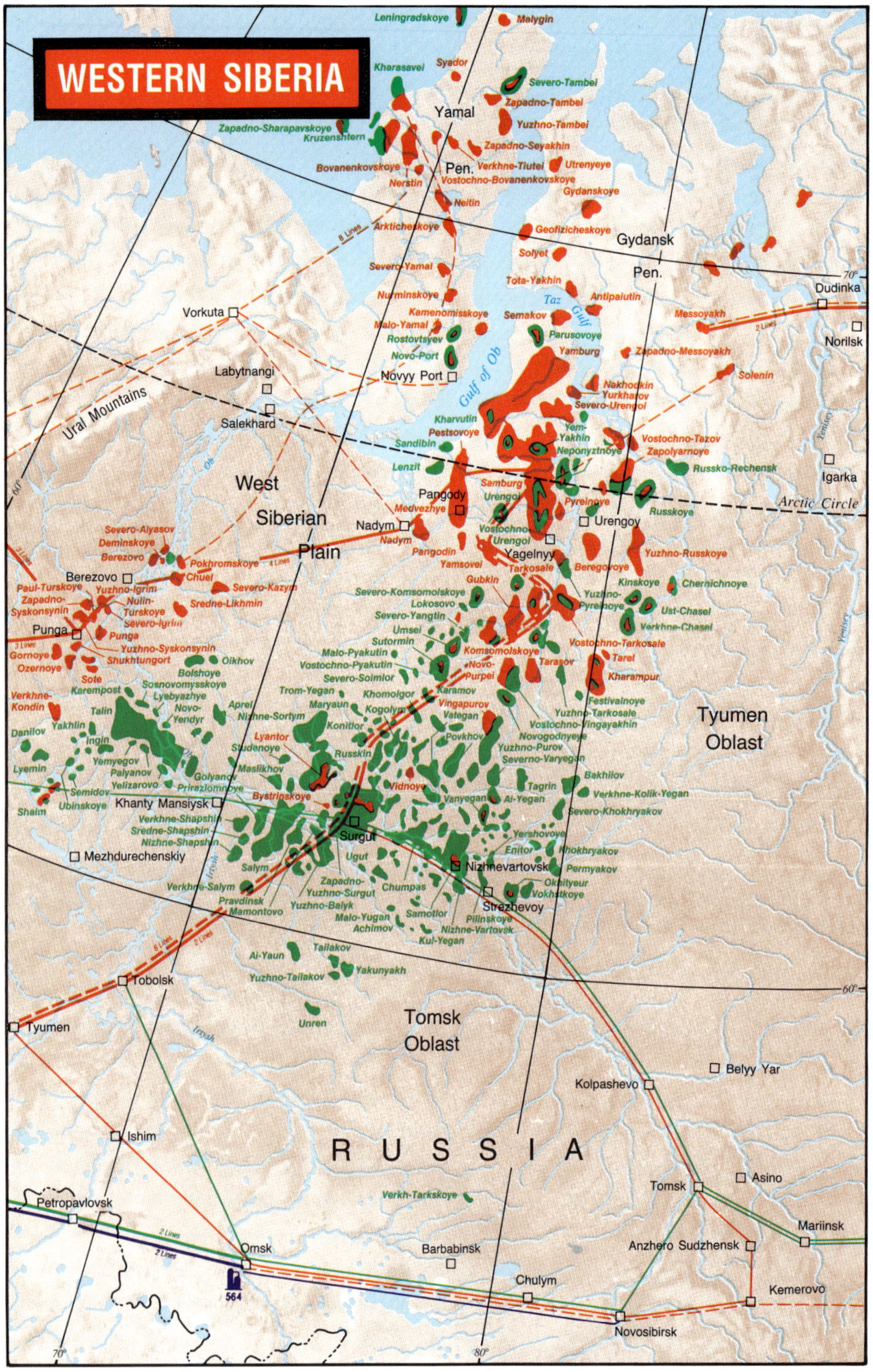

Petroconsultants SA, Geneva, estimated Kazakhstan's 1992 oil and condensate production averaged 511,500 b/d, while gas flow averaged 505 MMcfd. Most Kazakh oil comes from the Mangyshlak and Buzachi peninsulas with total 1992 production averaging 243,250 b/d, down 13% from 1991, most recent figures show.

Oil production from the Kazakh pre-Caspian basin averaged 229,200 b/d in 1992, up 2.4% from 1991. That included 17,800 b/d from Karachaganak field, currently being studied for gas development by British Gas plc and Agip SpA.

Production figures for Tengiz field were not broken out, but Petroconsultants estimated an average 65,000 b/d in 1992. Turgay basin production total averaged 25,470 b/d in 1992, up 9% from 1991. The company said the overall Kazakh production decline was due to poor performance in old Mangyshlak-Buzachi fields and fluctuation in Karachaganak production. Those could not be fully offset by increased production from Tengiz field and the Turgay basin.

Latvia eyes possible Baltic Sea resources

Latvia, a former Soviet republic, has expressed optimism over possible development of oil resources on its Baltic Sea shelf. But whatever prospects may exist for Latvia to become self-sufficient in petroleum are faint at least during the rest of this century and highly dubious for the more distant future. It is believed there is no commercial hydrocarbon production anywhere in the former Soviet sector of the Baltic, and Latvia's "oil fields" may be only potentially productive structures.

Latvian press reports claim "oil fields" have been found in the Baltic near the border with Sweden. They say Latvia's prime minister has met with Swedish and Finnish officials to discuss development of the deposits. "Special Latvian government representatives consulted with American firms regarding exploration and development of the fields," said a dispatch from Latvia's capital, Riga, published in Moscow's Izvestia newspaper.

"The main difficulty," it said, "is that Latvia has no clear laws to protect its interests in offshore oil development involving foreign partners." Riga sources concede Latvian oil production prospects are "vague." Latvia apparently has no onshore commercial oil or gas production and must depend largely on imports from Russia.

Moscow reports have referred to discovery of small oil deposits on Sweden's Gotland Island west of Latvia. But the Gotland finds are believed to be marginally commercial at best. The C.I.S. did not report successful Baltic drilling near the boundary with Sweden. An upper Ordovician onshore strike at about 2,700 ft near the town of Kuldiga, northeast of the port of Liepaya, was described by the Soviets as an "oil field," but no commercial production has been reported there.

Friction between Moscow and Riga over a pipeline shipping refined products to Latvia's Baltic coast has been a major threat to Latvian petroleum supplies. This pipeline from Samara, formerly Kuibyshev, on the Volga River to the Latvian port of Ventspils was shut down by Russia in 1992. Talks were held to resume flow through the line, which is the subject of an ownership dispute between Russia and now independent Latvia.

The line's construction was financed by Russia when Latvia was part of the Soviet Union. Reexports of Russian crude and products were the largest category of Latvian foreign trade during the last half of 1992. Suggestions have been made that a joint stock company be formed by Russian and Latvian interests to operate the pipeline. But Latvian authorities complain the proposed enterprise favors Russia.

While still insisting that it owns the pipeline, Russia has

placed itself in a difficult position by shutting down the line to Ventspils. It has had to deliver petroleum products by railroad to Ventspils at much higher cost.

Latvia is concerned that Russia will expand or build new Baltic oil loading facilities in Kaliningrad province or St. Petersburg, which could make its use of Ventspils unnecessary. But this would require considerable time and put a further strain on Moscow's tight capital construction budget.

Russia discloses possibilities for offshore oil and gas E&D

Russia has disclosed more data on hydrocarbon resources, field development plans, and project priorities it expects to employ in preparing for a "new epoch" of petroleum exploitation on its huge far northern continental shelf. A Moscow press analysis of studies conducted by the Morneftegaz (Offshore Oil & Gas) Research Institute concedes the Russian Federation may be ill prepared to tap the estimated 50 billion tons (365 billion bbl) of oil equivalent believed to lie beneath arctic and subarctic seas.

With the nation plagued by severe political, economic, financial, ethnic, and separatist problems, raising the $100 billion (at 1992 prices) deemed necessary to achieve Morneftegaz's ambitious long term goals will be extremely hard. But the program to accelerate offshore hydrocarbon development in environmentally hositle sectors of Russia's far nothern seas is dictated by urgent need, not choice.

Moscow was forced to place more emphasis on the arctic shelf when Caspian Sea oil and gas production representing almost 99% of total offshore hydrocarbon production of the C.I.S. fell into the hands of now fiercely independent Azerbaijan and, to a much smaller degree, Turkmenistan when the U.S.S.R. collapsed in 1991. Moreover, Moscow lost large portions of southern subsea territory when the former Soviet Union disintegrated.

Despite the shrinkage in areas under Moscow's control, Russia reigns over more continental shelf than any other country. The Russian Federation reckons the size of its remaining shelf at 6 million sq km (2.32 million sq miles). Moscow figures that amounts to 20% of the world total. Most of Russia's continental shelf lies north of the Arctic Circle.

Delovoi Mir quoted official sources as saying that even with Russia's loss of dominion over extensive proved and potentially petroliferous subsea territory to other C.I.S. republics, it still holds 87% of all offshore hydrocarbon resources previously held by the U.S.S.R. It said an estimated 13 billion tons (94.9 billion bbl) of oil and 52 trillion cu m (1.84 quadrillion cu ft) of gas resources lie on Russia's continental shelf.

Those figures admittedly are very speculative because, Delovoi Mir said, only 8% of Russia's shelf has been explored for hydrocarbons. Other studies say less than 1% of Russia's continental shelf has seen exploratory drilling, and less than 2% of the nation's potentially petroliferous seabed has been extensively explored. Unfortunately for Russia, whose urgent need is more crude reserves, most of the hydrocarbons found in its best explored areas are gas/condensate. Three supergiant gas/condensate fields have been found in the Barents and Kara seas.

Ukraine now owns most petroleum prone sectors of the Black Sea and the neighboring Sea of Azov. Azerbaijan, Kazakhstan, and Turkmenistan control the biggest and most valuable undeveloped petroliferous portions of the Caspian. The three non-Russian republics are trying to attract foreign firms into joint ventures for exploration and development of their sectors of the Caspian without asking Moscow's consent.

The only shelf areas with significant petroleum producing potential that Russia retains in the southern part of the C.I.S. are along the northwest coast of the Caspian and the eastern end of the Sea of Azov. Russia's remaining portion of the Black Sea shelf is regarded as having little possibility for important exploration success.

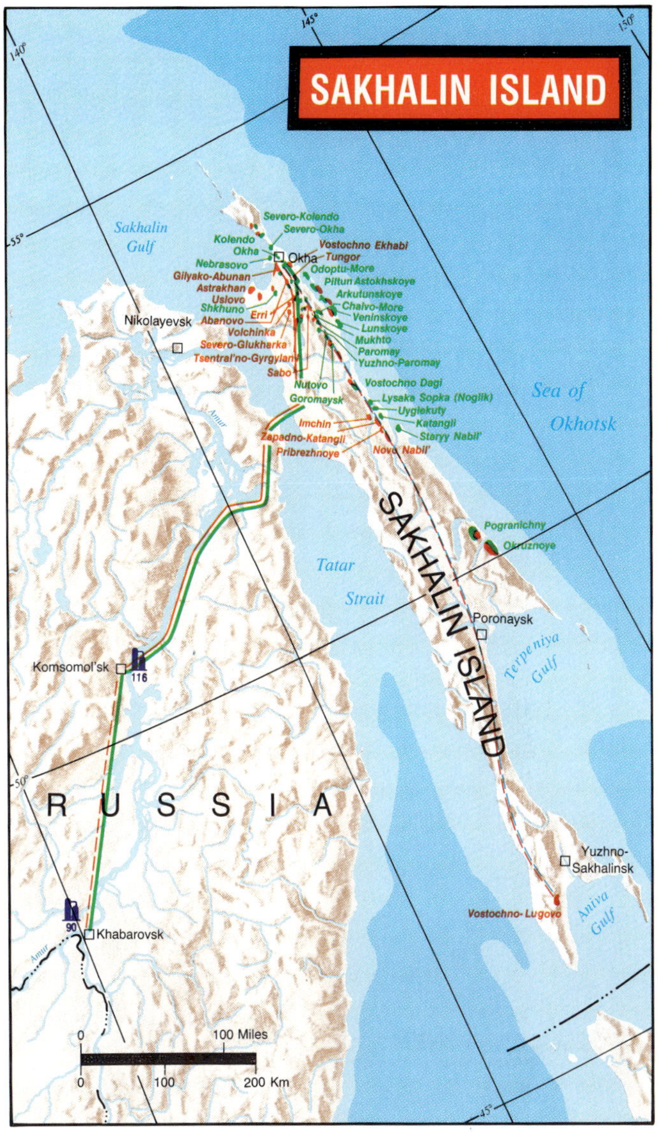

Prospective areas

Russia's Caspian Sea shelf off Astrakhan province, the Kalmyk republic to the southwest, and Dagestan, which lies between Kalmyk territory and Azerbaijan's northern border, is modestly prospective for petroleum discoveries. Some large, but mostly small, oil and gas fields are near the coast of Astrakhan province, the Kalmyk republic, and Dagestan.

One Dagestan field, Inchkhe-More, believed at first to be small and only marginally commercial, extends into the Caspian. Lying in no more than 20 m (66 ft) of water, Inchkhe-More is more recently reported to hold 50 million tons (365 million bbl) of crude and condensate and 90 bcm (3.18 tcf) of gas reserves.

The non-Russian portion of the Caspian shelf, representing at least 75% of the total area, probably has an even higher percentage of the sea's undiscovered oil and gas. On the Sea of Azov's Russian held eastern coast, a gas field, Beisugskoye, lies offshore. Farther north, the Baltic Sea is not highly rated for its hydrocarbon potential.

Only a small sector of the C.I.S.'s Baltic shelf off Kaliningrad province belongs to Russia, with the rest owned by Lithuania, Latvia, and Estonia. One recent, and perhaps too optimistic, report estimated Baltic oil resources on the

ГОСУДАРСТВЕННАЯ НЕФТЯНАЯ КОМПАНИЯ

НАФТА МОСКВА

STATE OIL COMPANY

NAFTA MOSCOW

VVO "Sojuznefteexport" dating back to 1931 as an import-export organization for crude oil and petroleum products was reformed into the state oil company "NAFTA MOSCOW" on the 29th of May 1992.

According to its legal regulations "NAFTA MOSCOW" carries out versatile activities the main of which are the following:

- **Export, import and reexport of crude oil and products of refining and petrochemical processing;**
- **Barter operations;**
- **Investments and any kind of business related with production, processing, store, transportation and marketing of crude oil and petroleum products;**
- **Participation in originating and running enterprises, companies, banks;**
- **Stock–exchange operations with goods and securities;**
- **Consulting on legal and financial matters as well as mediatory services.**

The company operates both on the territory of Russian Federation and outside its borders. Early in 1994 the state oil company "NAFTA MOSCOW" was a shareholder in 29 Russian and foreign enterprises.

For the fruitful activities and economic cooperation the organization was awarded the international prizes:

"GOLD MERCURY INTERNATIONAL AWARD" - Gold Mercury International Association, Rome, Italy, 1980.

"EUROMARKET AWARD – 1993" – EEC Euromarket Award, Brussels, Belgium, E.M.R.C.

Address: 32/34 Smolenskaja–Sennaja sq. Moscow 121200, Russia
Phones: 253-94-88, 253-94-89
Cable: NAFT MOSCOW
Telex: 411148 A, B, C, D, E
Fax: 244-22-91

shelves of these three nations at 300 million tons (2.19 billion bbl). A small Baltic oil field was found off Russia's Kaliningrad province in 1984, but development was abandoned, purportedly for ecological reasons.

Delovoi Mir now says two small fields, not one, with total crude reserves of 9 million tons (65.7 million bbl) have been found off Kaliningrad province. It disclosed that Russian and German firms conducted a feasibility study that may lead to joint development of the fields.

Deepening Russia's troubles in trying to become an important offshore hydrocarbon producer is its loss of a large portion of industrial facilities that produce platforms and equipment for Caspian fields. The Morneftegaz Research Institute noted that Russia can no longer count on using the big facility near Azerbaijan's capital of Baku that built deepwater (by C.I.S. standards) drilling and production platforms.

Ukraine owns the Okean shipyard at Nikolayev, near the Black Sea coast northeast of Odessa, said to be capable of building jack-up and semisubmersible rigs as well as drillships. However, Russia has experience building mobile rigs at the Vyborg shipyard near St. Petersburg on the Gulf of Finland. Work on jack ups, semisubmersibles, and drillships, some of them intended for use in arctic waters, was carried out at Vyborg in cooperation with the Finnish firm Rauma-Repola Oy.

At Astrakhan, near the northern end of the Caspian, Russia still has facilities that built jack ups and semisubmersibles, often using foreign parts. This shipyard also installed topsides sections and repaired mobile rigs operating in the Caspian and assembled large foreign-made floating cranes and a big foreign-built pipelay barge towed down the Volga River in sections.

A shipyard at Nikolayevsk-on-Amur in the far east can build topsides components for platforms. It also is involved in design work on ice resistant fixed platforms slated for use in the Sea of Okhotsk off Sakhalin Island.

Russia eyes loss to others in C.I.S.

Russia may lose billions in foreign oil exploration and development dollars to other countries in the C.I.S. That could happen if it does not move quickly to curb administrative nitpicking, bureaucratic infighting, and deep seated official and civilian xenophobia.

Such is the opinion of Moscow observers who predict that major western oil companies will shift their interest and funds elsewhere in the C.I.S. if Russia continues to reject or delay potentially lucrative exploration and profit sharing production agreements. Moscow's influential Izvestia, which strongly supports President Boris Yeltsin's economic reforms, warned that terms offered to Russia by foreign oil firms contain the most favorable conditions the nation can expect.

Izvestia urged Moscow leaders to emulate Kazakhstan's more realistic bargaining attitude and not risk greater alienation of western oil companies who are disillusioned by sudden changes in Russian laws, policies, and contract provisions that reflect increasingly greedy demands.

Russian bickering over agreements with France's Elf Neftegaz regarding exploration in the lower Volga region may be giving Kazakhstan a huge advantage in obtaining foreign investment, Izvestia said. The newspaper reported Elf is moving swiftly to determine whether its concession in Kazakhstan's Aktyubinsk province contains big subsalt oil reservoirs that may rival supergiant Tengiz field in the extreme southeast part of the pre-Caspian depression. Izvestia said some geologists believe the pre-Caspian depression and adjacent areas of the Caspian Sea shelf could contain seven times as much oil as Tengiz, which has recoverable reserves estimated at 1 billion tons (7.3 billion bbl) or more.

Russia's Supreme Soviet recently approved Elf's oil exploration project in Volgograd and Saratov provinces northwest of the pre-Caspian depression. But foot dragging by Moscow, and especially local authorities in the two provinces, has put implementation of the lower Volga venture substantially behind Elf's work pace in Kazkhstan.

Izvestia quoted geologists as estimating that Aktyubinsk province, where the Elf unit is exploring, holds 817 million tons (5.96 billion bbl) of crude reserves. It asserted that Elf's concession in the Temir district, south of Atkyubinsk city, could account for about 120 million tons (876 million bbl). Temir pay is believed to lie at 4,000-6,000 m (13,123-19,685 ft) beneath thick salt strata Kazakh officials say the republic lacks the resources to explore, let alone develop, the area's oil.

"Elf plans to invest $380 million in its Temir concession during the first 5 years of exploration," Izvestia reported. "The production sharing contract puts all of the risk on the French firm, and it won't cost Kazakhstan a single dollar if oil isn't found.

"For Kazakhstan and other former Soviet republics, this type of agreement with foreign investors is optimal. No C.I.S. member state, including Russia, can afford such a huge risk for wildcatting. Kazakh officials in the republic's ministry of geology and the state stock company Temirmunai, which are Elf's partners in the Temir venture, regard the deal as a 'huge plus,' offering greater advantages than Russia has obtained from foreign investors."

C.I.S. production plan and gas/oil pipeline network

Rosneftegasstroi, Moscow, Chairman Gennadi Shmal outlined a plan to overcome the C.I.S. oil and gas crisis, originally proposed by the Institute of Energy Problems of the Russian Academy of Sciences. "It should be stressed that these figures, although optimistic, are quite possible," he said. "The production volumes depend on a number of factors, the main of which is the volumes of investments in oil production and gas industry and the execution of large construction programs."

Russia's policy on oil and gas exports to C.I.S. members will be considerably different from that for other countries, he said. "During the period under discussion, Russia will keep the favorable balance in the exchange of power resources with C.I.S. members.

Only Kazakhstan will export fuel to Russia, in the form of coal and oil.

"In order to perfect the structure of exports, the share of power raw materials will be decreased along with the growth of the share of more expensive products of its processing." Foreign countries want to use more Russian oil and gas, Shmal said, with increased reliability of supply. The effective utilization of credits for developing and reactivating projects, including main pipelines, is the most important factor in developing economic cooperation between Russia and distant foreign states.

The C.I.S. gas supply network involves 200 gas and condensate fields, 140,000 miles of main gas pipelines, 44 underground storage sites with total capacity of 2.83 tcf, 907 compressor stations with 3,850 compressors and 50 million kw capacity, 4,500 gas distribution stations, and six gas processing plants.

The main gas pipelines have 1,339 water crossings totaling 540 miles, 2,472 railway crossings, and 5,700 road crossings.

Until 1989, about 14,000-16,000 miles/year of pipeline were built, including 6,000-10,000 miles of main lines. During 1985-90, about 28,200 miles of gas pipeline were laid. Even in 1991, when spending for pipeline construction was cut, 2,700 miles of main pipeline were laid, including 2,200 miles of gas pipeline. Another 1,200 miles of distribution and 2,200 miles of gathering pipelines were laid.

In 1992, 3,200 miles were laid, Shmal said. In 1993, a considerable growth in pipeline construction was planned, with about 7,800 miles to be laid.

Russia, the largest C.I.S. member, planned to place 3,381 miles of gas pipeline in operation during 1993, including 1,612 miles of large diameter trunk line. Its gas distribution enterprise estimated it would need $114-123 million/year to provide more gas to users. The republic is emphasizing construction of small diameter lateral gas lines using polymers instead of steel to serve smaller towns that were bypassed by major trunk lines only a small distance away.

Shmal said, "Even according to world standards this program of pipeline construction is rather impressive." There are 78,000 miles of 40-56 in. pipelines in Russia. By 1990, 56 in. lines accounted for 83.4% of the Russian gas network. The longest 56 in. pipeline runs 2,800 miles from Urengoi via Pomary to Uzhgorod.

A system of 48 in. and 56 in. pipelines was laid from the northern fields of western Siberia: Medvezhje, Urengoi, and Yamburg. It has total capacity of 21.2 tcf/year. In the last 20 years, about 11,000 miles of the gas supply system have been rehabilitated, with 90% of this work involving substitution of pipe by laying new sections nearby. This testifies to the lack of diagnostic techniques to locate damaged sections and carry out repair work, Shmal said.

Up to 1,000 miles of gas pipeline and 600 miles of oil pipeline are rehabilitated each year, with work considerably behind demand. However, 30-40% of all main pipelines may be in need of intrapipe diagnostics.

More than 98% of C.I.S. crude oil moves by pipeline, along with a big volume of products. Total length of crude oil and products pipelines in the C.I.S. is 56,000 miles, Shmal said.

However, different figures are supplied by JP Kenny & Partners, Staines, U.K. More than 90% of current oil production moves through the Transneft network. Total pipeline length is 43,500 miles, with about 600 pump stations and storage capacity of more than 100 million bbl. Trunk lines make up about 40,000 miles of network, Kenny said, with an average diameter of 32 in. By far the largest system is in Russia.

There are also 370,000 miles of field gathering and distribution lines, mainly the responsibility of regional oil producers, and 19,000 miles of product lines, largely operated by Transnefteproduct. Transneft's role has changed dramatically with the move to a market economy. Its original role as buyer, seller, and transporter of oil has been replaced by that of a shipping company, charging oil producers a tariff for transportation. Ruble tariffs are charged for republic and interrepublic delivery, with hard currency required for export.

As an example of Russia's new-found cooperation with the West, the Polar Lights joint venture during winter 1992-93 started welding double joints of arctic grade, coated steel pipe for a 12 in. pipeline to move oil from Ardalin oil field, north of the Arctic Circle in Russia's Timan-Pechora basin, to a Komineft trunk line at Kharyaga, Russia.

Joint venture partners Conoco Timan-Pechora Ltd. and Arkhangelskgeologia during 1993 completed slightly more than half the 40 mile, 40,000 b/d line. Pipe-laying was scheduled to resume in winter 1994, with completion of the Ardalin line expected by July 1994 in time to start moving oil during second half 1994.

Planned projects flowing from breakup of the U.S.S.R. include laying 680 miles of 28-32 in. pipeline to ship oil from western Kazakhstan to Kumkol. In this way oil will reach Kazakh refineries at Chimkent and Pavlodar.

Siberian gas is to replace supplies from Turkmenistan to the southern regions of Russia. This will require a 56 in. pipeline to be laid from Cheboksary to Izobilnoye. Large underground gas storage sites in Ukraine and other regions are now separated from the Russian pipeline system, requiring construction of underground storage units in Russia. A major problem lies in providing enough crude oil to refineries in areas of high use of petroleum products.

Reconstruction and development are also taking place in the opposite direction, Shmal said. Turkmenistan, Kazakhstan, and Azerbaijan are working out plans to export gas. One plan involves shipping crude oil from Tengiz to Novorossiisk for further export by tanker. The intent is to boost exports from Tengiz to 150-220 million bbl/year in the first decade of the next century.

Supply of oil from Azerbaijan's Azeri field will reach 125 million bbl/year by 2000. Talks were conducted to route Azeri oil to Novorossiisk via Grozny and Tikhoretsk or collect it in Komsomolsk or Voznesensk for joint shipment with oil from Kazakhstan and Russia. The latter plan would use today's oil pipelines with slight changes.

Turkmenistan's government plans to transport natural gas from Dauletabad field to Turkey. One option is to lay a 56 in. pipeline with 990 bcf/year capacity across the Caspian Sea from north to south. The possibility of laying a gas pipeline from Turkmenistan to China also is under discussion.

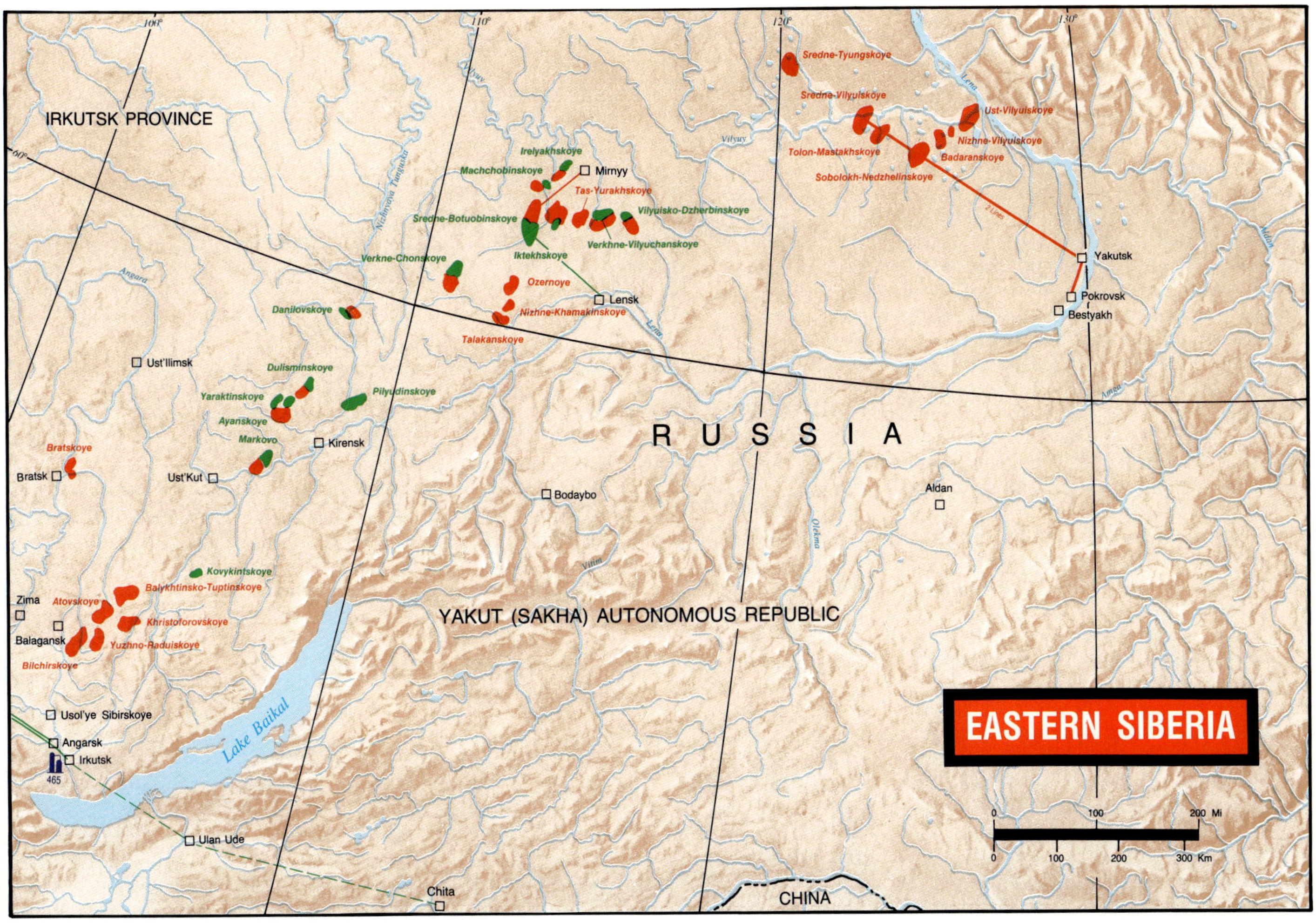

"The preservation and stoppage of uncontrolled disintegration of oil and gas construction organizations are the most important tasks during the transitional period in pipeline transport," Shmal said. Long term credits for oil and gas construction companies are required to build capital, even though inflation rates are high.

First arctic LNG project eyed for W. Siberia's Yamal Peninsula

Russia is trying to figure whether it is practical to liquefy part of the huge gas production foreseen for western Siberia's Yamal Peninsula. It would be the world's first arctic LNG project.

Likely LNG markets are the republic's northwest sector and tanker deliveries to other customers. Use of arctic sea lanes to transport Yamal LNG would not change Moscow's plans to transmit most of the peninsula's production by pipeline across the Ural Mountains to the central part of European Russia and to countries beyond the nation's western border.

But advocates of the Yamal LNG plan emphasize that employment of LNG tankers would allow Russia to deliver gas to the world market "regardless of the political situation in the C.I.S."

Izvestia reported Yamal gas would move by pipeline from Kharasavei field to an offshore liquefaction plant placed on large barges anchored in shallow water. LNG would be transferred to tankers from the "artificial island," which also would serve as a terminal for incoming equipment and materials to develop onshore fields.

Russian economists and maritime officials based in Murmansk said the main components of this complex could easily be moved from one offshore Yamal site to another as required when the peninsula's gas fields go on production. They believe the entire system could become operational in 3-4 years, in time for use by 1997, the target date for starting development of nearby gas deposits.

In downstream C.I.S. projects:

- Kazakhstan republic formed Kazakhgaz holding company to aid downstream development of Karachaganak oil and gas field. Plans call for a refinery to be built at Uralsk to process 3.5 million tons/year of crude and condensate and 6 bcm/year of gas. Long term plans call for a complex to be built there to produce 400,000 tons/year of petrochemicals.
- Ukraine's government issued a decree calling for the Drogobych refinery, southwest of Lvov, to build units during 1993-98 that can produce 825,000 tons/year of unleaded gasoline, 75,000 tons of solid paraffin, and 300,000 tons of industrial lubricants. Financing is to be obtained from the plant's export profits and foreign credits.
- Germany's BASF AG and Russia's Gazprom formed a joint venture to build a 1.2-1.5 billion Deutschemark ($700-800 million) petrochemical complex at Novyy Urengoy, Siberia. The plant will use gas produced locally as feedstock to manufacture as much as 340,000 tons/year of ethylene and 300,000 tons/year of polyethylene. Construction is scheduled to begin in 1995. The plant's output is intended for sale in Russia.
- British Petroleum Co. plc opened a lubricants plant in Baku. Downstream division BP Oil Co. said the project is part of a growing cooperation with upstream operations. It will help establish BP's brand name in a region where the company is known only as an explorer.
- Turkmenistan plans to revamp its two aging refineries. The Krasnovodsk refinery by the Caspian Sea and Chardzhou refinery in eastern Turkmenistan have a com-

bined 150,000 b/d capacity.

- Pennzoil Caspian Corp. let a $25 million contract to Creole Production Services, Houston, to supply a 150 MMcfd gas processing plant for its Guneshli field development off Azerbaijan in the Caspian Sea.

Creole said the plant is part of a comprehensive gas gathering and compression system for Guneshli and adjacent fields. The plant will be fabricated and assembled at Shreveport, La.

The project was scheduled for completion during third quarter 1993.

- Zangas, a Russian state company that operates the Stavropol gas terminal, agreed with Turkish private concern Tekfen Holding to build a $115 million polypropylene plant at Stavropol. Construction of the 100,000 ton/year unit will take 2 years.

The scheme will be financed by sale of products on international markets.

- Kazakhstan chose a Japanese group including Mitsui & Co., Mitsubishi Corp., and Tokyo Engineering Corp. to build a 120,000 b/d refinery on the Mangyshlak Peninsula on the northeast coast of the Caspian Sea.

The plant will refine crude from the Tengiz area and is expected to cost about $1 billion.

- Russia's refinery output in 1992 declined sharply from 1991. Automotive gasoline production was down 74,000 b/d (10.5%), diesel fuel 140,000 b/d (10.7%), heavy fuel oil 182,000 b/d (10.3%), and kerosine 46,000 b/d (16.9%).

Products exports dropped 23% to 540,000 b/d, and their value slipped 7% to $4.69 billion from $5.034 billion.

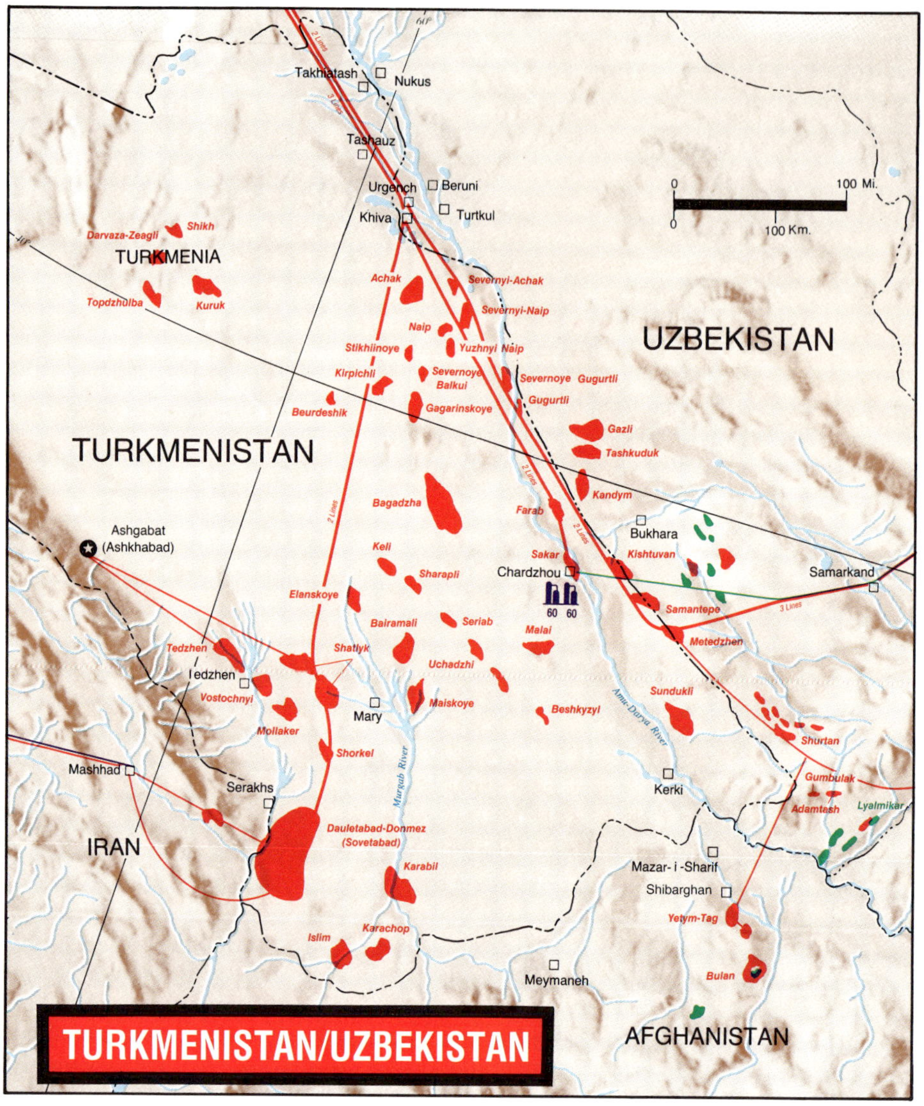

TURKMENISTAN/UZBEKISTAN

Environmental rules growing tougher

Environmental rules in the C.I.S. are getting more stringent for foreign groups trying to strike oil and gas deals.

Growing green awareness among grassroots groups in the C.I.S. and high expectations of western wealth and environmental expertise are among several factors fueling the trend.

C.I.S. environmental requirements are mounting for foreign companies, even though member countries are relatively lenient when applying rules to domestic firms. Russia, Kakakhstan, and other former Soviet republics want foreign investment to help develop hydrocarbons and other resources and spur economic growth.

At the same time, they want their environments to be protected from further damage, said Bernhard Metzger, manager of an environmental survey conducted in Kazakhstan by Arthur D. Little Inc. (ADL) consulting company, Cambridge, Mass. ADL bills its survey as the largest by western entities.

"There is a growing green awareness in the C.I.S. among watchful citizens, advocacy groups, and nongovernment agencies that want to improve current conditions and have high expectations of western industries," Metzger said. "They know the West is sophisticated and wealthy. They expect the best, even though environmental enforcement of their own business has been less stringent." As a result, C.I.S. members are starting to demand comprehensive environmental studies and top notch environmental protection programs by foreign companies before granting rights to develop natural resources.

In addition, new environmental protection laws and a pending environmental impact assessment law are expected to have big implications for companies wanting to do business in the C.I.S.

To be successful in the C.I.S., Metzger said, western industries must be sensitive to local customs and cultural idiosyncracies, ready for sudden political changes, and responsive to evolving rules. "Any company wanting to invest there should understand how seriously they are taking the environmental performance of outside developers," he advised.

Metzger based his conclusions on results of a preacquisition environmental survey of Kazakhstan's 60 sq mile Karachaganak gas field for a joint venture of units of British Gas and Agip.

The survey established a baseline reading of pollution to identify possible problems before the two companies signed an agreement to help Kazakh officials continue exploring and developing the field.

As part of the study, ADL gathered data on environmental characteristics and socioeconomic conditions in a several hundred square mile area around the field. ADL

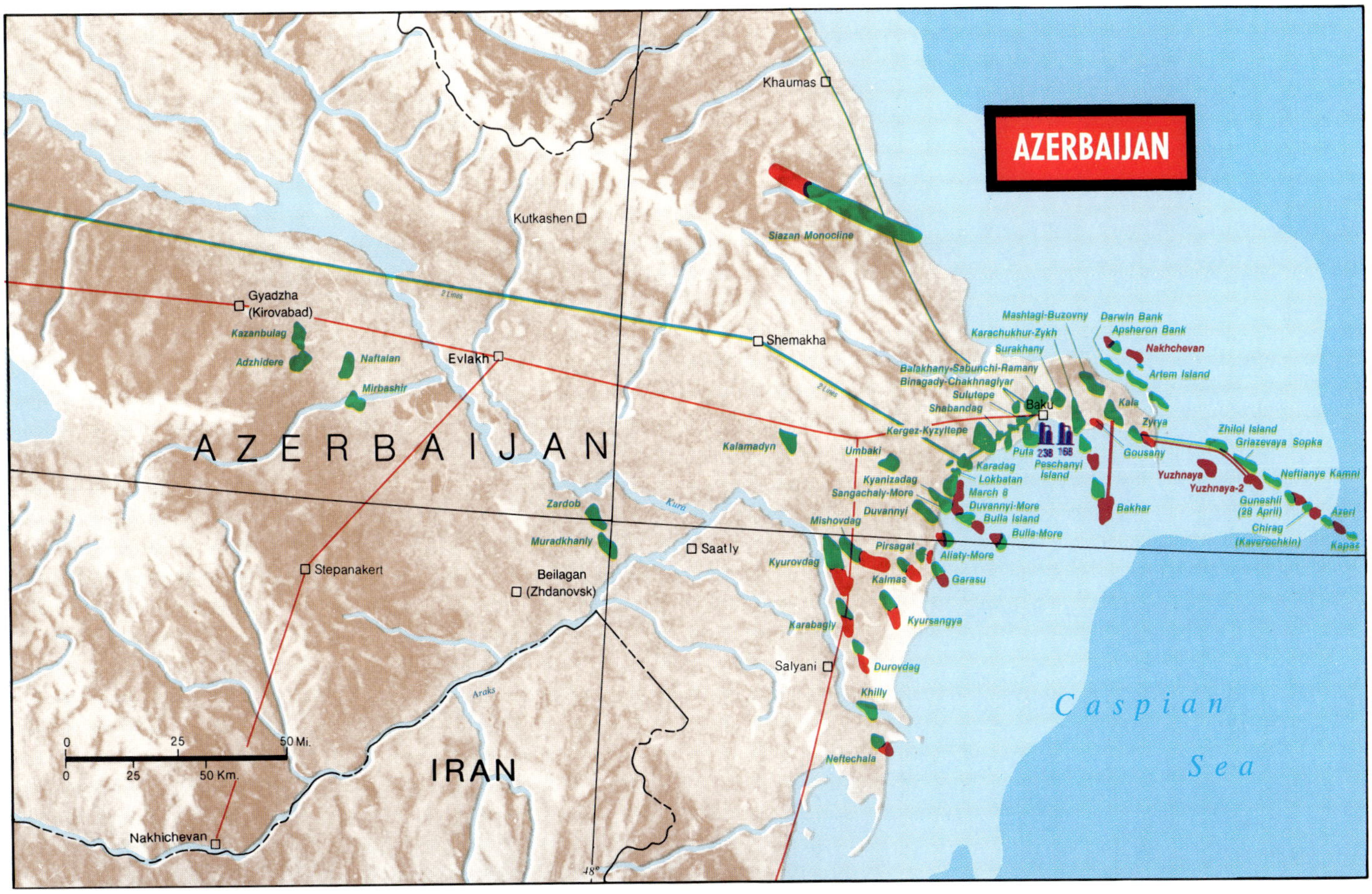

The Polar Lights joint venture in mid-1993 loaded more than 150 items in Houston for shipment to Russia in preparation for the start of development drilling in the Ardalin complex of the Timan-Pechora basin. Photo courtesy of Conoco Inc.

and about a half dozen Russian, British, German, and Italian subcontractors analyzed hundreds of soil and water samples at a Moscow laboratory and a mobile lab flown in from the West to assess environmental damage resulting from past and current field operations in the area and surrounding villages.

In addition, ADL used a western mobile air monitoring lab to track air quality and a satellite communications system linking three continents to manage equipment deployment and staff logistics.

Despite some serious but episodic air pollution events, the ADL team found the environmental impact of previous drilling operations was "much less serious than any of us expected," Metzger said. However, the survey found that many years of intense farming surrounding Karachaganak had caused serious environmental damage, including heavy soil erosion and contamination with pesticide residue, destruction of natural habitats, and reduced biodiversity.

Metzger said ADL's study showed companies willing to obey environmental mandates will find acceptance by the locals and backing of international financing organizations.

"But if you fail to take into ac-

Drilling action during Sakhalin Island feasibility study conducted by a consortium of companies. Photo courtesy of Marathon Oil Co.

count local regulations, the movement toward more local control and local players in the game, the danger of appearing too know-it-all, and the hard environmental line they are making westerners toe, you won't get your permit and you may never understand why."

Eximbank funding Russia's U.S. oil equipment purchases

The U.S. Export-Import Bank has signed a pact with the Russian Federation to provide financing for Russia's purchase of $2 billion of U.S. oil, gas, and petrochemcial equipment and services.

Eximbank said the accord, capping more than a year of negotiations, will help Russia revitalize oil and gas production and processing facilities and ultimately increase Russian exports and foreign exchange revenues.

The World Bank was expected to waive certain regulations and allow Russia to establish escrow accounts required by the agreement.

Standard documentation for individual transactions under the accord were to be completed in mid-1993, allowing Eximbank to begin approving individual transactions by September 1993.

Under the framework, hard currency revenues from export sales of Russian oil and gas will be deposited in a foreign escrow account, which will provide security for Eximbank supported financing and cover payments of principal, interest, fees, and the like.

The Russian Ministry of Fuel & Energy will decide which potential transactions by its oil and gas producing associations will be included under the agreement. U.S. suppliers should contact the production associations directly to discuss potential transactions.

Eximbank's board will consider financing for each transaction case by case.

Talks on a U.S.-Russia oil and gas agreement began in May 1992. At an April 1993 summit in Vancouver, B.C., Presidents Clinton and Yeltsin discussed the plan and agreed to press for speedy conclusion as part of an international assistance package for Russia. The two sides agreed Apr. 13, 1993, on the major principles of the agreement.

They have not disclosed details of the pact. IPE

PARMELEE WRENCH CO. INC.

126 Jersey Street   Harrison, N.J. 07029   U.S.A.   Telephone: (201) 483-3232   Fax: (201) 484-3621

TURN ANY PIPE IN A TIGHT PLACE, IDEAL FOR USE ON HIGHLY POLISHED PIPE

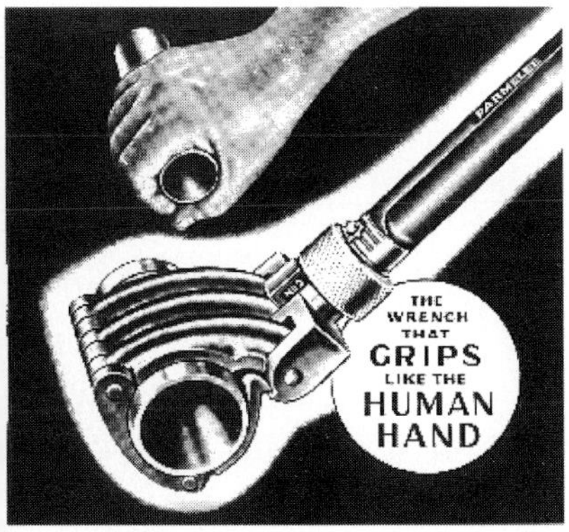

The Parmelee Pipe Wrench is designed for use on highly polished or any kind of pipe and for working in close quarters. It carries an instant appeal to mechanics because of its unique and simple principle. The handy ratchet action of the Parmelee gives a new grip with each pull of the handle. The Wrench has no teeth - consequently it cannot bite into the pipe. Its grip is sure and certain for its girth completely surrounds the pipe it grips, just as the thumb and fingers of a person's hand would surround it if they were gripping the pipe.

Wherever cylindrical tubes or rods have to be made up by threading, the Parmelee Pipe Wrench is the easiest and best tool to use.

All sizes of Parmelee Pipe Wrenches are furnished with lock nuts which operate from girth end. The handle, girth and lock bolt on all sizes, except no. 4, are of drop forged steel, while the No. 4 is cast of high carbon alloy steel. All sizes are toughened and hardened by scientific heat treatment.

In addition to the standard girth sizes listed here, we can also furnish additional sizes of girths, including metric sizes.

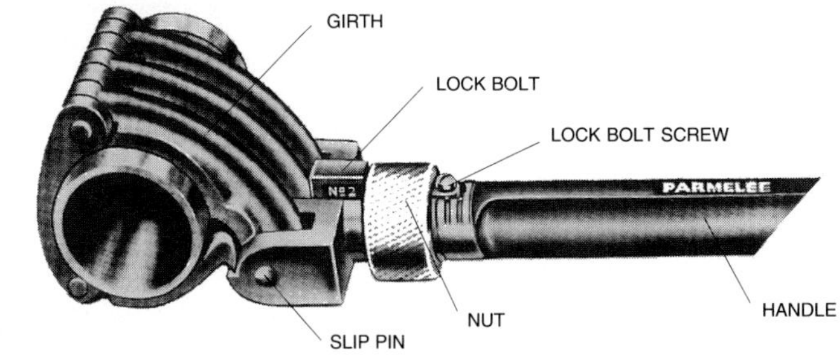

Available parts for #1 Wrench

Parmelee No.	O.D. (inches)	I.P.S. (inches)	Girth width (inches)	Description	Parmelee No.	O.D. (inches)	I.P.S. (inches)	Girth width (inches)	Description
H1	-	-	-	10" Handle	G1 1-3/8	1 3/8	-		Girth
G1 7/8	7/8	-		Girth	G1 1-13/32	1 13/32	-		Girth
G1 15/16	15/16	-		Girth	G1 1-7/16	1 7/16	-		Girth
G1 1	1	-		Girth	G1 1/2 IPS	-	1/2	2	Girth
G1 1-1/16	1 1/16	-		Girth	G1 3/4 IPS	-	3/4		Girth
G1 1-1/8	1 1/8	-	2	Girth	G1 1 IPS	-	1		Girth
G1 1-3/16	1 3/16	-		Girth	L1	-	-	-	Lock bolt
G1 1-7/32	1 7/32	-		Girth	N1	-	-	-	Adjusting nut
G1 1-1/4	1 1/4	-		Girth	P1	-	-	-	Slip pin
G1 1-5/16	1 5/16	-		Girth					

Available parts for #2 Wrench

Parmelee No.	O.D. (inches)	I.P.S. (inches)	Girth width (inches)	Description	Parmelee No.	O.D. (inches)	I.P.S. (inches)	Girth width (inches)	Description
H2	-	-	-	20" Handle	G2 2-1/8	2 1/8	-		Girth
G2 1	1	-		Girth + liner	G2 2-5/32	2 5/32	-		Girth
G2 1-1/16	1 1/16	-		Girth + liner	G2 2-3/16	2 3/16	-		Girth
G2 1-1/8	1 1/8	-		Girth + liner	G2 2-7/32	2 7/32	-		Girth
G2 1-3/16	1 3/16	-		Girth	G2 2-1/4	2 1/4	-		Girth
G2 1-7/32	1 7/32	-		Girth	G2 2-9/32	2 9/32	-		Girth
G2 1-1/4	1 1/4	-		Girth	G2 2-5/16	2 5/16	-		Girth
G2 1-9/32	1 9/32	-		Girth	G2 2-3/8	2 3/8	-		Girth
G2 1-11/32	1 11/32	-		Girth	G2 2-7/16	2 7/16	-		Girth
G2 1-3/8	1 3/8	-		Girth	G2 2-1/2	2 1/2	-	3 1/8	Girth
G2 1-7/16	1 7/16	-		Girth	G2 2-9/16	2 9/16	-		Girth
G2 1-1/2	1 1/2	-		Girth	G2 2-5/8	2 5/8	-		Girth
G2 1-9/16	1 9/16	-	3 1/8	Girth	G2 2-11/16	2 11/16	-		Girth
G2 1-5/8	1 5/8	-		Girth	G2 2-3/4	2 3/4	-		Girth
G2 1-21/32	1 21/32	-		Girth	G2 2-13/16	2 13/16	-		Girth
G2 1-11/16	1 11/16	-		Girth	G2 2-7/8	2 7/8	-		Girth
G2 1-3/4	1 3/4	-		Girth	G2 1 IPS	-	1		Girth
G2 1-25/32	1 25/32	-		Girth	G2 1-1/4 IPS	-	1 1/4		Girth
G2 1-13/16	1 13/16	-		Girth	G2 1-1/2 IPS	-	1 1/2		Girth
G2 1-7/8	1 7/8	-		Girth	G2 2 IPS	-	2		Girth
G2 1-29/32	1 29/32	-		Girth	L2	-	-	-	Lock bolt
G2 1-15/16	1 15/16	-		Girth	N2	-	-	-	Adjusting nut
G2 2	2	-		Girth	P2	-	-	-	Slip pin
G2 2-1/16	2 1/16	-		Girth					

Available parts for #3 Wrench

Parmelee No.	O.D. (inches)	I.P.S. (inches)	Girth width (inches)	Description	Parmelee No.	O.D. (inches)	I.P.S. (inches)	Girth width (inches)	Description
H3	-	-	-	20" Handle	G3 3-1/8	3 1/8	-		Girth
G3 2-3/16	2 3/16	-		Girth + liner	G3 3-3/16	3 3/16	-		Girth
G3 2-1/4	2 1/4	-		Girth	G3 3-1/4	3 1/4	-		Girth
G3 2-9/32	2 9/32	-		Girth	G3 3-5/16	3 5/16	-		Girth
G3 2-5/16	2 5/16	-		Girth	G3 3-3/8	3 3/8	-		Girth
G3 2-3/8	2 3/8	-		Girth	G3 3-1/2	3 1/2	-	3 1/8	Girth
G3 2-1/2	2 1/2	-		Girth	G3 3-5/8	3 5/8	-		Girth
G3 2-9/16	2 9/16	-	3 1/8	Girth	G3 3-11/16	3 11/16	-		Girth
G3 2-5/8	2 5/8	-		Girth	G3 2 IPS	-	2		Girth
G3 2-21/32	2 21/32	-		Girth	G3 2-1/2 IPS	-	2 1/2		Girth
G3 2-11/16	2 11/16	-		Girth	G3 3 IPS	-	3		Girth
G3 2-3/4	2 3/4	-		Girth	L3	-	-	-	Lock bolt
G3 2-7/8	2 7/8	-		Girth	N3	-	-	-	Adjusting nut
G3 3	3	-		Girth	P3	-	-	-	Slip pin
G3 3-1/16	3 1/16	-		Girth					

Available parts for #4 Wrench

Parmelee No.	O.D. (inches)	I.P.S. (inches)	Girth width (inches)	Description	Parmelee No.	O.D. (inches)	I.P.S. (inches)	Girth width (inches)	Description
H4	-	-	-	37" Handle	G4 5-9/16	5 9/16	-		Girth
G4 3-1/4	3 1/4	-		Girth	G4 5-5/8	5 5/8	-		Girth
G4 3-7/16	3 7/16	-		Girth	G4 6	6	-		Girth
G4 3-1/2	3 1/2	-		Girth	G4 6-7/16	6 7/16	-		Girth
G4 3-5/8	3 5/8	-		Girth	G4 6-5/8	6 5/8	-		Girth
G4 3-3/4	3 3/4	-		Girth	G4 8	8	-	3 5/8	Girth
G4 3-7/8	3 7/8	-		Girth	G4 9	9	-		Girth
G4 4	4	-		Girth	G4 3 IPS	-	3		Girth
G4 4-1/8	4 1/8	-	3 5/8	Girth	G4 3-1/2 IPS	-	3 1/2		Girth
G4 4-1/4	4 1/4	-		Girth	G4 4 IPS	-	4		Girth
G4 4-1/2	4 1/2	-		Girth	C4 5 IPS	-	5		Girth
G4 4-3/8	4 3/8	-		Girth	G4 6 IPS	-	6		Girth
G4 4-7/16	4 7/16	-		Girth	L4	-	-	-	Lock bolt
G4 4-5/8	4 5/8	-		Girth	N4	-	-	-	Adjusting nut
G4 4-3/4	4 3/4	-		Girth	P4	-	-	-	Slip pin
G4 5-3/8	5 3/8	-		Girth					

Iron pipe size wrench sets

Parmelee No.	Handle size No.	I.P.S. girths sizes (inches)	Girth width (inches)	Weight (lbs)	(oz)
C1	#1	1/2 3/4 1	2	4	0
C2	#2	1 1 1/4 1 1/2 2	3 1/8	16	8
C3	#3	2 2 1/2 3	3 1/8	3 1/8	10
C4	#4	3 4	3 5/8	3 5/8	8

OTHER SIZES UPON REQUEST

Europe

ALBANIA

CAPITAL: Tirana
MONETARY UNIT: Lek
REFINING CAPACITY: 40,000 b/cd
PRODUCTION: 8,000 b/d
RESERVES: 165 million bbl

OMV (Albanien) Exploration GmbH became 100% owner and operator of the Rodoni 1 Block off Albania by buying a 50% interest from Deminex.

The block was awarded 50-50 to the two companies in 1991.

OMV plans to drill a 10,000 ft wildcat on the block within 3 years.

AUSTRIA

CAPITAL: Vienna
MONETARY UNIT: Schilling
REFINING CAPACITY: 210,000 b/cd
PRODUCTION: 24,500 b/d
RESERVES: 94.73 million bbl

Austria diversified its energy supply by taking its first natural gas from Norway late in 1993.

OMV, Austria's state oil and gas company, will import 300 cu m/year from the Norwegian North Sea, inceasing the volume to 1 billion cu m/year in 2002 with the option to buy another 500 million cu m/year at that time.

OMV received the gas through a 348 mile, 32 in. line from Emden to Ludwigshafen, Germany, on the Austrian border. The line is rated at 290 MMcfd, expandable to 775 million.

In 1988 OMV contracted with Statoil, Norway's state oil company, to buy 35 bcf/year of gas for 20 years starting in 1993.

Previously, OMV met 20% of Austrian demand with domestic gas, and has been inporting 80%—about 6.2 bcf/year—from Russia.

OMV also launched a program to restructure the company toward oil and gas and away from chemicals and to sell its pharmaceutical chemcials unit.

It approached Middle East national oil companies and large western international oil firms to offer them a long term investment in OMV.

The company's fertilizer operations have come under severe competition from eastern European imports at a time when demand has dropped. OMV was closing 1 million tons/year of fertilizer capacity to reduce production to 600,000 or 700,000 tons/year.

BELGIUM

CAPITAL: Brussels
MONETARY UNIT: Franc
REFINING CAPACITY: 607,000 b/cd
PRODUCTION: 0
RESERVES: 0

A British Petroleum Co. plc products tanker was hit by a bulk carrier outside the port of Antwerp June 3, 1993, killing several crewmen and spilling about 2,000 bbl of petroleum.

Belgian Refining Corp. completed rebuilding a 25,000 b/d straight run distillate desulfurizer at its 90,000 b/d Antwerp refinery after a June 1992 fire.

State owned Distrigaz continued to negotiate for added gas supplies from Russia, Algeria, Holland, and Norway.

The government increased taxes on inland use of high sulfur fuel oil to $16/ton, encouraging barge users to switch to low sulfur fuel oil, which is taxed at $7.50/ton.

BP Chemicals sold its 120,000 metric ton/year polyethylene plant at Antwerp to Neste Chemicals.

BULGARIA

CAPITAL: Sofia
MONETARY UNIT: Lev
REFINING CAPACITY: 300,000 b/cd
PRODUCTION: 1,000 b/d
RESERVES: 15 million bbl

Texaco Exploration Offshore Bulgaria Ltd. became the first western company to claim a hydrocarbons discovery in the Black Sea off Bulgaria.

Texaco's 1 Galata flowed 34 MMcfd of gas through a 1¼ in. choke from pay at 3,400 ft. The well, which was spudded in June 1993, was the first of two commitment wells. A second well was scheduled to be spudded early in 1994.

The discovery well was drilled to 8,700 ft by Diamond Offshore's Ocean Liberator semisubmersible operating in 115 ft of water on Block 3, 19 miles southeast of Varna. Drilling was serviced from a supply base at Balchick, 25 miles north of Texaco's local office at Varna. Equipment and services were shared in a cost saving agreement with fellow operator Union Pacific Resources Inc., which was the first western company to complete a well, on its Block 1 concession.

The general manager of Texaco's frontier exploration department said more drilling is required to establish the commerciality of the discovery.

Block 3 license partners are Texaco 20%, Enterprise Oil Exploration Ltd. 20%, OMV (Bulgaria) Exploration GmbH 10%, and Bulgarian Committee of Geology & Mineral Resources 50%.

Ocean Liberator is the only western semisubmersible operating in the Black Sea. The rig was scheduled to be released by Texaco to drill two wells for British Gas plc on Block 4.

Block 4, known as Kamchia, is said by British Gas to hold a gas/condensate discovery. Drilling was scheduled to begin during fourth quarter 1993.

Elsewhere, Bulgaria shut down its Burgas refinery in October 1993 after Russia suspended oil supply. Iran offered to supply 70,000 metric tons of crude to help ease the country's fuel shortage. Bulgaria imports 98% of its 12 million ton/year crude requirement, almost all of it from Russia. The nation's three refineries have a combined capacity of 300,000 b/cd.

Meantime, Mobil Corp. licensed Bulgarian state company Neftochim to use its low pressure xylene isomerization (MLPI) technology to produce the main raw material for polyester fibers and films. Neftochim will modify a unit to use the technology at its Burgas refinery.

Using Mobil's ZSM-5 catalyst, MLPI converts orthoxylenes and metaxylenes into paraxylene isomer, a basic building block used to produce polyester fibers and films. It is Mobil's first such licensing agreement in eastern Europe.

Bulgarian discovery

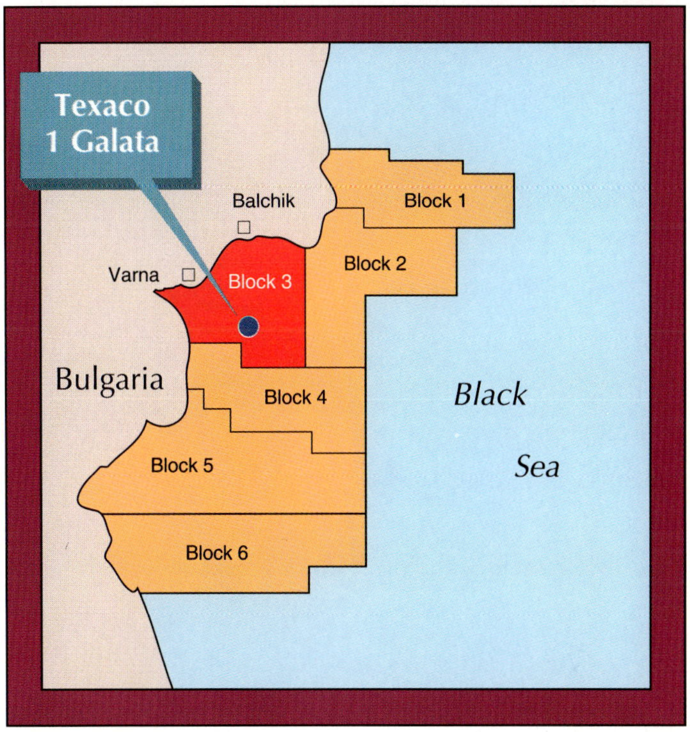

CZECH REPUBLIC

CAPITAL: Prague
MONETARY UNIT: Koruna
REFINING CAPACITY: 307,250 b/cd
PRODUCTION: 2,000 b/d
RESERVES: 15 million bbl

Czechoslovakia ceased to exist Jan. 1, 1993, and was succeeded by two independent states: Czech republic and Slovakia. Czech and Slovak leaders decided during summer 1992 to divide their country. The international boundary between the successor states is the internal boundary that separated them as constituent republics of Czechoslovakia.

Slovnaft, Bratislava, and Itochu Corp., Tokyo, plan a petrochemical joint venture in the Czech republic. The agreement calls for a 3 month feasibility study of expanding and renovating Slovnaft's polypropylene complex at a cost of $40-50 million.

Slovnaft is undergoing privatization, and plans call for scrapping one of two plants and expanding capacity of the other to 80,000 metric tons/year.

Czech republic's Chemopetrol Litvinov and Kaucuk Kralupy formed Mero-IKL Ltd., Prague, to lay and operate a crude oil pipeline to Kralupy from Ingolstadt, Germany. The pipeline will be about 340 km long with nameplate capacity of 220,000 b/d. Plans include expanding the line to 330,000 b/d after 2000. The line will supply the Czech refining industry and provide an alternative to Russian supplies. Construction is to be complete in 1995.

DENMARK

CAPITAL: Copenhagen
MONETARY UNIT: Krone
REFINING CAPACITY: 183,500 b/cd
PRODUCTION: 162,500 b/d
RESERVES: 779.936 million bbl

The Danish energy ministry reported Denmark was nearing self-sufficiency in energy production.

Energy imports, including natural gas, were dropping as the country's share of North Sea production increased.

Danish fishermen were demanding compensation from the offshore oil industry, claiming that their businesses had been disrupted by offshore oil and gas exploration and production. They claimed a 1982 law requires the fishing industry to be consulted before offshore oil developments are approved. They asked that the government lift the ban against fishing within a 500 m radius of offshore platforms.

The fishermen also complained many of the fish netted near Tyra field are contaminated by oil.

Danish Premier Poul Nyrup Rasumssen disclosed Denmark will begin talks with Greenland in 1994 on whether to give the province full control of its minerals and natural

Location No. Refineries
United Kingdom
South Killingholme 1
Fawley 1
South Humberside 1
Ellesmere Port 1
Coryton 1
Port Clarence 1
Shell Haven 1
Stanlow 1
Scotland
Grangemouth 1
Dundee 1
Wales
Milford Haven 3
Llandarcy 1
Belgium
Antwerp 4
France
L'Orcher 1
La Mede 1
Mardyck 1
Reichstett-Vandenheim 1
Donges 1
Feyzin 1
Grandpuits 1
Fos sur Mer 1
Port Jerome 1
Notre Dame de Gravenchon 1
Berre l'Etang 1
Petit Couronne 1
Dunkirk 1
Lavera 1
Netherlands
Rotterdam 2
Pernis 2
Amsterdam 1
Vlissingen 1
Italy
Livorno 1
Porto Marghera 1
Sannazaro, Pavia 1
Taranto 1
Falconara, Marittima 1
LaSpezia 1
Augusta, Siracusa 1
Frassino, Mantova 1
Busalla 1
Priolo Gargallo 1
Rome 1
Milazzo 1
Gela 1
Sarroch 1
Trecate, Novara 1
Priolo 1
Cremona 1
Germany
Merseburg 1
Hamburg 1
Heide 1
Wesseling 1
Burghausen 1
Godorf 1
Harburg Grasbrook 1
Neustadt 1
Neustadt-Donau 1
Karlsruhe 2
Harburg 1
Zeitz 1
Leuna 1
Woerth 1
Schwedt 1
Vohburg/Ingolstadt 1
Geisenkirchen 1
Lingen 1
Salzbergen 1
Wilhelmshaven 1
NORWAY
SCOTLAND
UNITED KINGDOM
N. IRELAND
IRELAND
WALES
ENGLAND
DENMARK
NETHERLANDS
BELGIUM
GERMANY
LUXEMBOURG
FRANCE
SWITZERLAND
SPAIN
PORTUGAL
ANDORRA
CORSICA
SARDINIA
ITALY
MOROCCO
ALGERIA
TUNISIA
North Sea
North Atlantic Ocean
English Channel
Bay of Biscay
Mediterranean Sea
Ligurian Sea
Gulf of Cadiz
Strait of Gibraltar
Orkney I.
I. of Man
Bantry Bay
Skagerrak
Minorca I.
Majorca I.
Iviza I.
Pantelleria I.
Mongstad
Bergen
Oslo
Karmoy
Stavanger
Lysekil
Göteborg
Fredericia
Kalundborg
Copenhagen
Malmo
Skaelskor
Flotta
Fraserburgh
St. Fergus
Forties
Ekofisk
Grangemouth
Glasgow
Belfast
Dublin
Whitegate
North Tees
Liverpool
Manchester
Birmingham
Milford Haven
London
Fawley
Wytch Farm
Bacton
Groningen
Schoonebeek
Amsterdam
Rotterdam
Antwerp
Brussels
Dunkerque
Wilhelmshaven
Brunsbuttel
Heide
Schwerin
Rostock
Hamburg
Bremen
Hanover
Berlin
Magdeburg
Bonn
Eisenach
Erfurt
Jena
Leipzig
Gera
Frankfurt-am-Main
Suhl
Karlsruhe
Vohburg
Ingolstadt
Munich
Baden-Wurttemberg
Notre Dame de Gravenchon
Port Jerome
Le Havre
Rouen
L'Orcher
Petit Couronne
Vernon
Gargenville
Paris
Coulommes
Grandpuits
Nancy
Brest
LeMans
Donges
Nantes
Tours
Blois
Orleans
Vierzon
Reichstett-Vandenheim
Mardyck
Basel
Zurich
Bern
Geneva
Roussines
Chazelles
Ambes
Bordeaux
Vichy
Lyon
Feyzin
Grenoble
Parentis
Auros
Toulouse
Lacq
Lourdes
Bayonne
San Sebastian
Vizcaya
Bilbao
Vitoria
Logrono
Burgos
Valladolid
La Coruna
Oporto
Madrid
Zargoza
Barcelona
Tarragona
Castellon de la Plana
Valencia
Puertollano
Cabo Ruivo
Lisbon
Sines
Huelva
Cordoba
Rota
Cadiz
Malaga
Algeciras
Almeria
Cartegena
Fos sur Mer
Marseille
La Mede
Lavera
Turin
Milan
Lacchiarella
Porto Maghera
Trieste
Koper
Savona
Genoa
Ravenna
Rome
Porto Torres
Cagliari
Palermo
Rabat
Oran
Arzew
Algiers
Skikda
Tunis
Ashtart

resources.

Greenland, which has home rule status, wants to use oil and gas and mineral leasing revenues for its domestic budget.

AS Dansk Shell was upgrading its 62,000 b/d Frederica refinery to add a gas oil hydrodesulfurization unit and sulfur handling units.

The German firms Ruhrgas and BEB Erdgas signed an agreement with Dansk Naturgas AS to boost gas supplies to 90 bcf/year from the current 15 bcf/year. Ruhrgas will take two thirds and BEB one third of the gas, receiving it at Ellund on the German/Danish border.

Dansk Olieror plans to increase the capacity of its oil pipeline from the North Sea fields to Frederica in Jutland to 200,000 b/d from 180,000 b/d in 1994.

FINLAND

CAPITAL: **Helsinki**
MONETARY UNIT: **Markka**
REFINING CAPACITY: **200,000 b/cd**
PRODUCTION: **0**
RESERVES: **0**

Russia in 1993 was negotiating a trade pact for delivery of more natural gas to Finland.

The deal hinged on construction of underground storage in Finland and extension of Finland's gas grid west to the city of Turku and other areas along the Gulf of Bothnia.

Moscow, in turn, wanted to buy more Finnish mobile drilling rigs for use in the Barents Sea as well as other Finnish equipment and supplies for onshore oil and gas work.

Finland imported 87.5 bcf of Russian gas in 1992.

Meanwhile, Neste doubled capacity at its Raikkola gas pipeline compressor station on the Finnish-Russian border to 3 million cu m/year. Cost was $20 million.

Neste Gas opened a new LPG underground storage cavern and terminal at Tornio, a city north of the Arctic Circle and near Finland's northeastern border with Sweden.

The cavern, which can hold 42,000 tons of propane, was being filled with LPG from the North Sea and from Neste's refineries.

Neste, Esso, Shell, Mobil, and smaller Finnish companies agreed to a protocol that will minimize the risk of environmental damage from refining, storage, distribution, and marketing operations in Finland.

The accord follows European Community environmental standards and allows companies to develop their own methods to limit pollution.

Meanwhile, Conoco said it will enter Finnish marketing with its Jet service station subsidiary. Its goal is a 6% market share in 3-4 years.

The Finnish Barents group, which is participating in planned development of Russian Barents Sea fields, said the project will start 2-3 years later than expected.

The group, which includes Neste and Repola Offshore, had planned to start drilling in 1995 and producing gas about 2000. It blamed political and economic problems in Russia for the delay.

Neste closed its Porvoo refinery for 5 weeks in 1993 for a $40 million upgrade. It added a desulfurization unit to the fluid catalytic cracking unit to lower the sulfur content of gasoline and middle distillates.

Neste was building a $6 million, 100,000 metric ton/year purified benzene plant in Espoo.

FRANCE

CAPITAL: **Paris**
MONETARY UNIT: **Franc**
REFINING CAPACITY: **1,861,430 b/cd**
PRODUCTION: **54,400 b/d**
RESERVES: **177.434 million bbl**

The French government was moving toward deregulation of its gas and electric power sectors during 1993.

It was studying ways to introduce more flexibility into the monopolies of Gaz de France and Elecricite de France and make them conform to the European Community's deregulation policy.

That would require scrapping the companies' import/export monopolies, opening distribution to third party access, and introducing unbundling of production, transportation, and distribution functions.

The government launched a program to privatize Elf Aquitaine, the state owned oil company. It owned 51% of Elf and planned to sell all but 10-15% in 1994, raising about $8.5 billion.

Elf and Rhone-Poulenc, the chemicals group, were among four companies to be privatized first in the government program of selling off 21 companies. The other two companies were banks.

The government owned 43% of Rhone-Poulenc and planned to sell all of that.

Gaz de France reportedly was negotiating to buy more gas from Norway. It already was due to buy 8-9 billion cu m/year by 2000.

The Industry Ministry reported consumption of LPG as a substitute for gasoline or gasoil fell 40% since

EASTERN EUROPE
LITHUANIA
RUSSIA
KALININGRAD
BELARUS
POLAND
UKRAINE
GERMANY
CZECH REP
SLOVAKIA
MOLDOVA
AUSTRIA
HUNGARY
ROMANIA
SLOVENIA
CROATIA
BOSNIA AND HERZEGOVINA
YUGOSLAVIA
SERBIA
MONTENEGRO
ALBANIA
MACEDONIA
BULGARIA
TURKEY
GREECE
ITALY
SICILY
Baltic Sea
Adriatic Sea
Black Sea
Aegean Sea
Sea of Marmara
Copenhagen
Smolensk
Kaunas
Vilnius
Kaliningrad
Krasnoborskoye
Ushakovskoye
Molodechno
Borisov
Minsk
Gdansk
Rostock
Reinkenhagen
Kolberg
Hamburg
Stettin
Grodno
Schwedt
Bromberg
Baranovichi
Pripyat basin
Rechitsa
Vishanskoye
Davydovskoye
Ostashkovichskoye
Tishkovskoye
Rechitskoye
Ivatsevichi
Mozyr
Roxforde
Brandenburg
Berlin
Frankfurt-an-der-Oder
Magdeburg
Golzau
Poznan
Plock
Glinik
Warsaw
Kobrin
Pinsk
Brest
Ovruch
Fallstein
Rybaki
Staakow
Cottbus
Grünberg
Krotoszyn
Lodz
Volkenroda
Mühlhausen
Salzwedel-Peckensen
Langensalza
Schwarzheide
Leipzig
Jena
Zeitz
Erfurt
Eisenach
Suhl
Gera
Dresden
Breslau
Radom
Pulawy
Lublin
Kovel
Zhitomir
Kiev
Lutsk
Rovno
Dubno
Berdichev
Skvira
Kielce
Sandomierce
Rozwadow
Neisse
Cosel
Litvinov
Zaluzi
Kralupy
Prague
As
Kolin
Pardubice
Trzebinia
Oswiecim
Bochnia
Grobla
Partynia
Lubaczow
Jaroslaw
Rudki
Brody
Staro-Konstantinov
Tarnow
Jaslo
Jedlicze
Czechowice
Gorlice
Drogobych
Stryy
Borislav
Dolina
Lugskoye
Ternopol
Vinnitsa
Dashava
Kadobnensk-Kalusz-Grabovsk
West Ukrainian fields
Waidhaus
Pilsen
Ostrava
Stahov
Pribor
Zukow
Martin
Smilov
Mikova
Olka
Strazske
Nadvornaya
Starynj
Bitkov
Kosmachskoye
Kosov
Chernovtsy
Krasnoilsk
Trebic
Brno
Vacenovice
Hodonin
Gbely
Kosice
Uzhgorod
Solushsk
Vohburg
Munich
Linz
Stefanov
Malacky
Zvolen
Nitra
Sahy
Nyirbogdany
Miskolo
Leninvaros
Satu-Mare
Baia-Mare
Transylvania
Iasi
Ugeny
Kishinev
Vienna
Schwechat
Bratislava
Sala
Pedemes
Demjen
Mezokeresztes
Hajduszoboszlo
Nadudvar
Biharnagybajom
Debrecen
Salzburg
Sopron
Komarom
Szöny
Budapest
Százhalombatta
Szandaszolos
Oradea
Cluj
Zaul-de-Cimpie
Bogota
Tauni
Bozna
Sarmasel
Teleac
Sulpacu Barcau
Lucacesti
Darmanesti
Mierourea
Singeorgiu
Saros
Noul-Sasesc
Bicaz
Zemes
Stanesti
Bacău
Gheorghiu-Dej
Graz
Nagylengyel
Lovaszi
Budafop
Zalaegerszeg
Nagylonizsa
Pannonian basin
Ulles
Algyo
Pusztafoldvar
Battonya
Szeged
Arad
Sibiu
Brasov
Boldesti
Moreni
Galati
Reni
Izmail
Maribor
Lendava
Petisovci
Legrad
Molve
Babocsa
Pécs
Velebit
Novi Knezevac
Timisoara
Ljubljana
Verona
Padua
Venice
Trieste
Koper
Zagreb
Lepavina
Pitomaca
Sandrovac
Popelana
Benicanci
Dugo Selo
Sombor
Ada
Mokrin
Srpska Crnja
Medja
Srbobran
Begejci
Resita
Petroseni
Gura-Ocnitei
Frasinu-Mislea
Cimpina
Ploiesti
Teleajen
Buzău
Fäurei
Midia
Lebada
Navodari
Constanta
East Swan
Rijeka
Sisak
Janja Lipa
Gakovo
Obod
Bizovac
Novi Sad
Boka
Gospodjinci
Nikolinci
Mramorak Selo
Ticleni
Pitesti
Brazi
Bilteni
Leonardi
Budisteni
Borzetsi
Bucharest
Po
Correggio
Minerbio
Bosanski Brod
Nova Gradiska
Pancevo
Belgrade
Turnu-Severin
Craiova
Oltenia
Arges
Pula
Ivana
Porto Garibaldi
Amelia
Ravenna
Cesenatico
Riccione
Rimini
Bologna
Dugi
Zadar
Giurgiu
Ruse
Tolbukhin
Kavarna
Varna
Tjulenovo
Florence
Sarajevo
Dolni Dabnik
Chiren
Gigen
Pleven
Gorni Dabnik
Devetak
Beli Izvor
Vraca
Staro Orjachovo
Falconara
Ancona
Split
Aleksinac
Nis
Otmanli
Burgas
Sofia
Stara Zagora
Vis Island
Pristina
Bellante
Teramo
Cellino
Pescara
San Stefano
Dubrovnik
Plovdiv
Maritsa
Edirne
Orestias
Civitavecchia
Rome
Fiumicino
Valle Cupa
Alliano
Vasto
Portocannone
Cupello
S. Salvo
Chieuti
Skopje
Ripi
Petroliana-Fornesina
Monte Stillo
Candela
Gaeta
Tirana
Cerrik
Stalin
Ballaj
Divjaka
Seman
Marinze
Fier
Patos
Kucove
Bubullima
Qytezi
Gorisht
Drashovice
Philippi
Kavala
Alexandroupolis
Prinos
S. Kavalla
Thessaloniki
Naples
Pietralunga
Bari
Brindisi
Taranto
Grottole
Ferrandina
Rotondella
Nova Siri
Pisticci
Ballsh
Vlorë
Kastoria
Grevena
Finiq
Sibari
Larissa
San Leonardo
C. Cimiti
Arta
Delfi
Aspropyrgos
Athens
Eleusis
Corinth
Lavinia
Palermo
Milazzo
Messina
Mazara del Vallo
Marinella
Marsela
Lippone
Gagliano
San Nicola
Bronte
20 in. 51 cm
24 in. 61 cm
40 in. 101 cm
56 in. 142 cm
18 in. 45 cm
26 in. 66 cm
28 in. 71 cm
0 100 200 Mi.
0 100 200 300 Km.

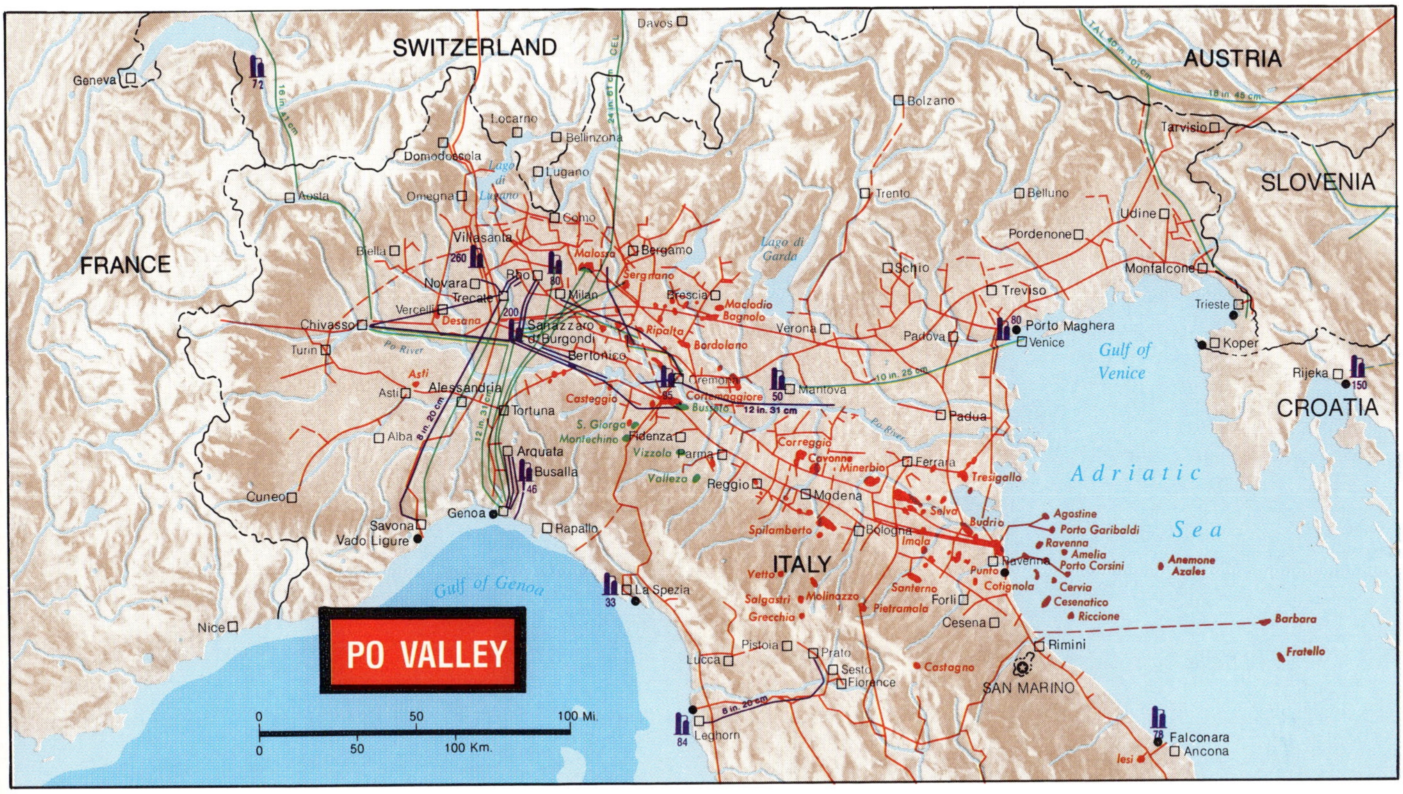

1985 due to its higher costs. The ministry recommended lowering taxes for LPG used in vehicles.

Gaz de France acquired Shell France's heating oil business, doubling its market share to 10%.

Refining shakeout

France's refining industry was undergoing a shakeout caused by massive environmental expenses and a changing products market.

Some changes would have been caused anyway by surplus refining capacity and weak products demand growth in Europe.

But accommodating new environmental mandates and their corresponding petroleum product specification changes will cost French refiners $7.7 billion during the 1990s.

In 1992 French refiners posted their lowest refining margins in a decade, booking combined after tax operating losses of $152 million.

Refiners' average gross margins on Brent blend crude were about $2.28/bbl, below the breakeven point, and French marketing margins were the lowest in Europe in mid-1992.

French refineries are designed to largely produce gasoline, but now must cope with a market in which diesel fuel demand growth is 6-7%/year, far outpacing gasoline demand growth.

At the same time, demand for heavy fuel oil is depressed because of the government bias toward nuclear power plants, which now supply 75% of the nation's power.

France is the largest consumer of diesel fuel in Europe, using about 400,000 b/d, compared with about 350,000 b/d of leaded and unleaded gasoline.

French refiners can meet most of the domestic demand for gasoline but only 80% of the demand for diesel.

And because demand for heavy fuel oil is flat, refiners cannot compensate for the increasing diesel demand by using more of the middle of the barrel.

Total restarted its La Mede refinery in Provence in April 1993 after repairing severe damage from a November 1992 fire and explosion. It said the accident was caused by a leak in the catalytic cracker gas plant, which released about 10 tons of liquefied gases.

Total resumed negotiations with Saudi Aramco over a possible Saudi stake in Total's French downstream operations. Talks had begun in 1991 but were shelved. Similar talks with Elf broke down because the Saudis considered the price was too high.

GERMANY

CAPITAL: Bonn
MONETARY UNIT: Deutschemark
REFINING CAPACITY: 2,265,060 b/cd
PRODUCTION: 61,200 b/d
RESERVES: 449.314 million bbl

Natural gas pipeline construction again dominated Germany's energy developments in 1993.

The U.K. North Sea's Markham field began delivering gas to Wintershall Gas, a subsidiary of Germany's Wintershall and Russia's Gazprom. It was the first North Sea gas moved into Germany.

Wintershall is buying 3 billion cu m of Markham gas during 10 years, shipping it to the continent through the Wintershall operated North Sea West Gas transit line.

The gas is landed at Den Helder in Netherlands, and Gasuine moves it to the Dutch-German border into the recently completed Midal line.

Markham gas initially will fill a new 1.5 billion cu m storage cavern at Rehden in Northwest Germany.

Ruhrgas AG laid a 90 mile gas pipeline link between Werne and Wetter in Central Germany.

It was part of a new pipeline network to supply gas from western Germany, along with gas from Russia, to the former East Germany.

The 48 in. section tied into a hub of Ruhrgas' pipeline grid at Werne, allowing Norwegian, Dutch, and German natural gas to be supplied under a 70 billion cu ft/year contract with Erdgasversorgungs GmbH.

Ruhrgas previously completed a 40 in. line about 110 miles from Schluctern to Vitzeroda, shipping West German

gas to the Thuringia region of East Germany. The Werne-Schluctern-Vitzeroda link cost $630 million. The Werne-Schluctern branch added a major north-south route to Ruhrgas' network, which totals 5,600 miles of line.

Wintershall was laying an 80 mile, $148 million gas pipeline to link Hamburg with the north-south German Midal gas line Wintershall and its Russian partner Gazprom were building.

Ruhrgas will store as much as 35 bcf of gas in depleted Breitbrunn/Eggstatt gas field near Rosenheim, southern Germany, starting in 1996. The project, which will cost as much as $300 million, will have a maximum production rate of 17.6 MMcf/hr and an injection rate of 8.8 MMcf/hr.

Verbundnetz Gas AG began laying a $410 million, 115 mile, 42 in. pipeline connecting Berlin with the West German grid.

The line, between Salzwendel and Bernau, north of Berlin, will enable Norwegian imports to meet demand in the Berlin, Brandenburg, and Saxony-Anhalt regions.

Other developments

Germany's environmental minister recommended that the country's annual auto tax, based on engine displacement, be changed to a tailpipe gas emissions based system.

The ministry noted that even through German carbon emissions dropped 14.5% in 1987-92, carbon emissions from autos were increasing.

The transportation minister said that after a Jan. 1, 1994, fuel tax increase, no more fuel tax hikes are in prospect.

Petrolchemie und Kraftstoffe AG, Schwedt, was building a methyl tertiary butyl ether (MTBE) complex and butane isomerization unit at its Schwedt refinery.

The 84,000 metric ton/year (2,000 b/d) MTBE plant will process the stream from a fluid catalytic cracking unit. It is due completion in late 1994 and the isomerization unit in mid-1995.

Deutsche BP completed work on a 1 million ton/year catalytic cracker and deisohexanizer at its Raffinerie Vohburg-Ingolstadt. The units will enable the plant to make more gasoline.

GREECE

CAPITAL: Athens
MONETARY UNIT: Drachma
REFINING CAPACITY: 395,500 b/cd
PRODUCTION: 11,300 b/d
RESERVES: 41 million bbl

Greece joined the international trend toward privatization.

It disclosed plans to privatize two downstream companies in the first sale of state firms under a petroleum industry deregulation program.

On the auction block were the Hellenic Aspropyrgos refinery at Athens and the EKO-Chemicals refining/petrochemical complex plant at Thessaloniki.

The 125,000 b/d Aspropyrgos refinery is the second most advanced in the Mediterranean area and can meet future sulfur level regulations for fuels.

The EKO complex contains a less sophisticated 70,000 b/d refinery with PVC, caustic, chloride, and ethylene units. A network of 700 gasoline stations is included.

Aspropyrgos and EKO combined have more than half of the Greek refined products market. Their sale was planned by mid-1994.

Also, the government sent parliament a bill that would allow international oil companies to explore for oil and gas again in Greece.

A state firm has a near monopoly on exploration, but the government planned to offer large oil firms 100,000 sq km for exploration, 70% of which was offshore.

The acreage is not expected to include regions in the northern Aegean Sea that are disputed by Greece and Turkey.

The government was considering legislation that would limit increases in products prices. It was responding to public complaints that product prices in Greece did not decline after world oil prices fell in 1993.

Athens Public Gas Enterprise plans to spend $300 million during 10 years to construct a gas distribution network in the Attica suburbs of Athens.

Athens is due to begin receiving supplies of Russian gas from a pipeline under construction from the Bulgarian border.

HUNGARY

CAPITAL: Budapest
MONETARY UNIT: Forint
REFINING CAPACITY: 241,500 b/cd
PRODUCTION: 31,700 b/d
RESERVES: 139.031 million bbl

Hungary's Mecseki Szenbanyak FA and Canadian Fracmaster Ltd., Calgary, formed Methane Master Kft, a joint venture that has acquired exclusive rights to explore for and produce coalbed methane in the Mescek basin of Hungary. Estimated reserves to a depth of 1,500 m are 3 tcf.

Fracmaster said there are substantial reserves below that depth. Investment in the project is to exceed $75 million. Drilling under a 35 year concession was scheduled to start Mar. 1, 1993. Full scale development, scheduled to start in January 1994, will include drilling 300 wells.

Hungary's state owned MOL Oil & Gas Co. has announced plans to modernize and expand its network of gasoline stations this decade. The MOL-2000 project would cost about $188 million. Hungary has about 1,100 filling stations, of which 195 belong to foreign firms or joint ventures. MOL expects the number of Hungarian gasoline stations to reach 1,500-2,000 by 2000 and has discussed plans to install filling stations in Romania and Ukraine with local partners.

Hungarian state holding company AV Rt named M.M. Lazard Freres VC of France privatization adivser to MOL. The French company is to set the date, conditions, and sales method for a minority stake in MOL. Hungary will retain majority ownership.

Conoco Inc. opened its first Jet retail outlet at Kenderes, 90 miles southeast of Budapest. The company planned to open two more stations during summer 1993. Products on sale are leaded and unleaded gasoline, diesel fuel, two stroke engine fuel, and home heating oil.

IRELAND

CAPITAL: Dublin
MONETARY UNIT: Irish pound
REFINING CAPACITY: 52,500 b/cd
PRODUCTION: 0
RESERVES: 0

Ireland signed a treaty with the U.K. Department of Trade and Industry governing regulation of a gas pipeline to transport supplies from the British grid to Ireland.

The pipeline will extend from Moffat, West Scotland, across the Irish Sea to Ballough, north of Dublin on Ireland's east coast. Irish state utility Bord Gas Eireann is to be owner and operator of the line.

The $248 million line will consist of a 170 mile offshore section and a 70 mile onshore section in Scotland between the Irish sea and the Scottish grid at Moffat.

The treaty guarantees access for links to Northern Ireland and the Isle of Man if required. Northern Ireland can take up to 280 MMcfd and Isle of Man 25 MMcfd.

The line will enable Ireland, dependent on offshore Kinsale Head and Ballycotton field gas, to either import or export gas.

The gas board also signed an $8.3 million contract with British Gas for a 1 year supply.

Ireland was negotiating with Russian officials to help it operate the state owned 52,500 b/d refinery and terminal at

English Channel
Cherbourg
Bay of the Seine
Antar 12 in. 31 cm
Le Havre
57
144
Port Jerome
N.D de Gravenchon
Rouen
L'Orcher
309
Caen
Petit-Couronne
158
65
125
Lisieux
Bernay
Beauvias
Gargenville
Seine
3 lines
Vernon
Paris
Valenciennes
Mons
Charleroi
Maubeuge
36 in. 91 cm
36 in. 91 cm
BELGIUM
GERMANY
Charlesville
LUXEMBOURG
Luxembourg
Aisne
Reims
Marne
Meuse
Metz
Bellange
12 in. 31 cm
Nancy
Epinal
Saône
Gulf of St Malo
Brest
Coulommes
Chaunoy
Grandpuits
98
Brie-Chailly-Chartrettes
Valence
St. Martin-de-Bossenay
Villemar
Gisy-les-Nobles
Troyes
Rennes
Vern-sur-Seiche
Laval
Le Mans
Lorient
Vannes
Antar 12 in. 31 cm
Chateaubriant
16 in. 41 cm
Paris Basin
St. Firmin-des-Bois
Chuelles
Chateaurenard
215
Donges
Angers
F R A N C E
Blois
Auxerre
Montbard
Briare
Nantes
Tours
Loire
20 in. 51 cm
Vierzon
Dijon
Cholet
Besancon
Bourges
Nevers
Les Sables d'Olonne
Parthenay
Chatellerault
Issoudun
Beaune
Fontenay-le-Comte
10 in. 25 cm
Poiters
Chauvigny
Chateauroux
Autun
Dole
Cressier
60
Chalon
B a y
Niort
La Rochelle
Roussiness
18 in. 45 cm
Moulins
Lons-le-Saunier
SWITZERLAND
Revigny
Sugeres
Vienne
24 in. 61 cm
Vindecy
10 in. 25 cm
o f
Macon
16 in. 41 cm
Lake Geneva
Saintes
Royan
Vichy
Roanne
Bourg
Gonova
Les Ancizes
Annemasse
Le Verdon
Chazelles
Lyon
Annecy
B i s c a y
34 in. 86 cm
Clermont-Ferrand
130
Feyzin
Vienne
Pauillac
Issoire
Chambery
Perigueux
Tulle
St. Etienne
Ambes
Brive
24 in. 61 cm
Bergerac
Rhone
Lavergne
Bordeaux
12 in. 31 cm
18 in. 45 cm
Aquitaine
Dordogne
Allier
Grenoble
Cazaux
Lugos
Mothes
Marmande
Valence
ITALY
Parentis
Cabeil-Ychoux
Lucats
Auros
Decazeville
Lot
N. Mimizan
Basin
Agen
Villefranche
Rodez
Mont-de-Marsan
16 in. 41 cm
24 in. 61 cm
Garonne
8 in. 20 cm
Tartas
Montauban
Dax
Lussagnet
240
Vizcaya
Orthez
Avignon
L'Isle
Bilbao
San Sebastián
Bayonne
Lacq
Lacq sup
Toulouse
Castres
Pont d'As
St. Faust
St. Marcet
Proupiary
Gallician
Vitoria
Oloron
Tarbes
Rousse
Meillon
Mazeres
Lourdes
Montpellier
Frontignan
Manosque
Berre l'Etang
Sete
Fos-sur-Mer
100
La Mede
26
Lavera
136
180
Marseille
Pierrefitte
Pamiers
Carcassonne
Narbonne
Besiers
Toulon
St. Girons
Logroño
Gulf of Lions
Ebro
Andorra
ANDORRA
Perpignan
Huesca
La Junquera
Barbastro
M e d i t e r r a n e a n
Monzon
S e a
Balaguer
Gerona
Manresa
Zargoza
Palamos
Lerida
Tarrega
Igualada
Fraga
Tarrasa
Mataró
SPAIN
21
80
Barcelona
Fayon
Tarragona
Alcaniz
Angula
Dorado
Montanazo
Salmonete
Casablanca
Tarragona
Castellon
Vinaroz
Amposta
FRANCE
Teruel
Castellon de la Plana
102
0
50
100 Mi.
0
50
100 Km.
MINORCA
Valencia
MAJORCA
IBIZA

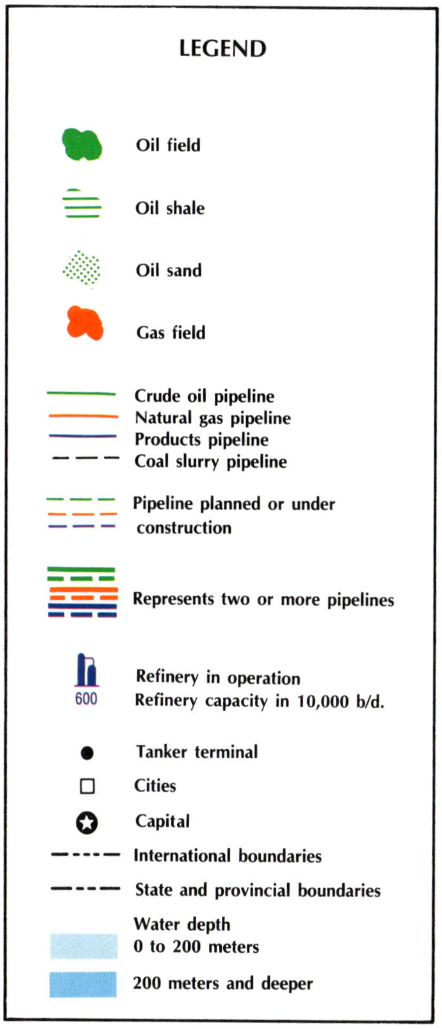

Whitegate in County Cork.

For several years, the Irish government has been looking for a foreign partner to modernize and assume operations at the refinery.

The government postponed a frontier licensing round to give offshore operators more time to acquire and process seismic data.

The frontier round will offer 128 blocks in the Erris and Slyne troughs about 45 miles off Ireland's western coast.

The government had set new licensing terms and tax measures in 1992 in an effort to boost offshore exploration.

ITALY

CAPITAL: Rome
MONETARY UNIT: Lira
REFINING CAPACITY: 2,262,000 b/cd
PRODUCTION: 86,000 b/d
RESERVES: 620.5 million bbl

Italy's state oil and gas companies may be sold as one company, Soc. Italiana Idrocarburi, in a new plan to privatize the state energy/chemicals group Ente Nazionale Idrocarburi (ENI).

ENI, the parent company of Agip, installed a new slate of managers and drafted a business plan that called for it to withdraw from all nonoil and gas activities during 1993-97. It planned to sell 50-60 subsidiaries.

Besides Agip and Agip Petroli, functions to be sold were to include gas company and distribution division Snam, design and engineering contractor Snamprogetti, and drilling/construction contractor Saipem.

The government planned to fully privatize Agip and put sale of the company back on track.

The sale was delayed in 1993 following corruption scandals that resulted in the arrest of hundreds of top Italian business and government officials, including some key ENI executives. Most of the charges involved bribes and kickbacks.

The government planned to sell 15% of Agip stock at first, after some affiliate companies were streamlined to make the package more attractive.

It was pressing ENI for faster privatization, both to cut losses and to fill government coffers with the sale of assets.

Adriatic production

Agip SpA was starting commercial oil and gas production from Aquila field in the Adriatic.

The company said the field, in 2,788 ft of water, holds the deepwater record for commercial production and the deepwater record for a horizontally drilled well section.

It also said the field has the biggest flow rate for an Italian oil well, with preproduction testing of a delineation well gauging 10,000 b/d of oil and 3 MMcfd of gas.

For full development of the field, Agip was considering installing a tension leg platform or a subsea development tied back to a floating production storage system. Development costs were estimated at $106 million.

Agip drilled the discovery well in 1981 and estimated potential reserves are 400 million bbl of oil and 2.8 tcf of gas. Agip planned to drill two more wells in the field to bring production to 30,000 b/d, equal to a third of the company's production in Italy.

Fina Italiana SpA tested a rare oil discovery in southern Italy.

The 1 Tempa Rossa discovery cut a gross oil column of more than 1,000 m and produced 16-22° gravity oil at a combined rate of 7,630 b/d from three zones. A confirmation on an adjoining block cut 800 ft of pay and produced about 3,500 b/d.

An Agip group found oil near Potenza in southern Italy. The 1 Cerro Falcone wildcat flowed 600 b/d of 36° gravity oil in Potenza province. It was the latest in a string of discoveries in the area.

Refining outlays

Italian refiners planned to spend $7.2 billion in the 1990s, most of it to develop cleaner fuels and take plant-specific environmental mitigation measures to conform to new European Community regulations.

Italian refiners generally lagged some of their counterparts in Europe and North America on environmental spending because they faced a constant margin squeeze as a result of product prices remaining under tight government controls.

In 1992 the government began to implement price deregulation in line with EC directives. At the same time, the government was enforcing stiffer emissions rules to improve air quality in urban areas.

Unione Petrolifera, the Italian refiners association, said the most pressing concern for Italian refiners in the 1990s is rising demand for more environmentally benign products.

Their chief concerns will be reducing benzene levels in gasoline and sulfur levels in gas oil and heating oil. At the same time, Italian refiners will have to increase the market share of high value-added products to satisfy expected growing demand for lighter products.

Refiners also must streamline distribution and marketing systems to reach EC standards. Unione Petrolifera said refinery investments by 2000 will focus on making cleaner fuels for electrical power generating plants, to improve the quality of gasoline, for distribution and marketing operations, and to replace and revamp plants and equipment.

Kutwait Petroleum International closed its 85,000 b/d refinery in Naples after a compressor broke down. The products terminal remained open.

The company had decided to close the 50 year old refinery soon anyway because of its deteriorating condition and inefficiency.

ISAB's 240,000 b/d refinery at Santa Panagia Bay in Sicily was shut down for 3 weeks following a fire.

Anonima Petroli Italiana planned to build a 220,000 kw cogeneration unit at its 63,000 b/d refinery at Falconara. Completion is due in 1996.

The government was considering banning tankers from the Venice lagoon and requiring construction of a subsea oil pipeline from Trieste to the Porto Marghera refinery at Venice. It also banned its oil, gas, and chemical tankers from the Bonifacio Straits between Corsica and Sardinia, following a similar ban by France. Tankers traversed the straits 1,400 times in 1992.

Natural gas trade

Russia's Gazprom will ship $2 billion worth of gas to Italy's ENI in exchange for equipment to modernize its transportation network.

ENI unit Snam will take delivery of the gas for sale in Italy, while Snamprogretti will supply Russia pipe and machinery to upgrade pipelines and Nuovo Pignone will supply gas turbines for compressor stations. Gazprom supplies ENI with almost 500 bcf/year through two pipelines that cross Ukraine, Slovakia, and Austria. The lines operate near capacity, but throughput can be increased.

In exchange for the equipment, Gazprom will ship Snam all incremental volumes resulting from the upgrade. Industry officials estimated the added Russian imports at 200 bcf/year.

A loop of the trans-Mediterranean gas pipeline was expected to be partially completed and carrying 7 billion cu m of gas in 1994 and 13.5 billion by 1996.

The existing 12.5 billion cu m/year line has been operating since 1983. The new line will run parallel to the first one, except for the stretch crossing the Straits of Messina and a segment north of Rome.

MALTA

CAPITAL: Valletta
MONETARY UNIT: Pound
REFINING CAPACITY: 0
PRODUCTION: 0
RESERVES: 0

Malta awarded an exploration and production sharing contract covering the 1,340 sq km block VII off southern Malta to Shell Malta exploration BV and Nimir Petroleum Co. Malta Ltd. in a 50-50 venture.

The frontier block is in water depths to 1,300 ft. Shell began 3D seismic surveys on the acreage.

For several years, Malta has sought its first oil production.

The island lies on a trend between the two largest producing fields in the Mediterranean: Sicily's Vega field and Libya's Bouri field.

The government issued a production sharing contract to Amoco Mediterranean Petroleum Co. and Agip International BV for a 837,000 acre spread about 60 miles off Malta's southern coast.

The contract covers Blocks 3 and 43 and calls for seismic acquisition and drilling at least one well during 3 years. Amoco is operator with a 60% working interest, while Agip holds the rest.

Meanwhile, Amoco drilled a wildcat in License Area 3 north of Malta where it operates two blocks with BHP Petroleum Co.

East of the island, Texaco Exploration Malta Inc. holds concessions in the strait between Malta and Sicily near Italy's Vega oil field.

NETHERLANDS

CAPITAL: Amsterdam
MONETARY UNIT: Guilder
REFINING CAPACITY: 1,183,500 b/cd
PRODUCTION: 45,700 b/d
RESERVES: 132.086 million bbl

Despite opposition from environmental groups, the government allowed gas drilling to resume in the Waddenzee mudflats area, believed to contain 75 billion cu m of gas.

Norway's Statoil was considering two possible landing sites in Netherlands, near the towns of Lauwersoog and Den Helder, for a new gas pipeline from North Sea fields to Europe.

But the Lauwersoog site would require the line to cross the Waddenzee mudflats.

Utilities belonging to the Dutch Electricity Generating Board started a 250,000 kw integrated coal gasification combined cycle power plant at Buggenum. It is believed to be the largest of its kind in the world.

Shell Nederland Raffinaderij BV began a $2 billion rennovation of its 400,000 b/d Pernis refinery near Rotterdam.

It planned to place a $1.5 billion hydrocracker and oil gasification plant on stream in 1997 to upgrade the plant's product mix and spend another $500 million on emissions control measures during 10 years to meet Dutch limits on sulfur dioxide and nitrous oxides.

The refinery processes 50 kinds of high sulfur crude, but the market is demanding more lighter products.

The new hydrocracker and gasification plant will not substantially change refinery capacity. The hydrocracker will convert heavy, high sulfur, middle distillates to light, desulfurized products such as naphtha or low sulfur diesel and kerosine.

The gasifier will convert heavy residues to synthesis gas, some of which will be broken down to provide hydrogen for the cracker.

The rest will fuel a new steam and power generation plant.

Mobil Chemical Co. will build a $35 million, 4.4 million gal/year plant in Amsterdam to make synthetic lubricants base stocks.

Construction was due to be complete late in 1995. The plant was sited next to Mobil's product distribution terminal in the West Haven section of Amsterdam.

Mobil started up an 11,000 metric ton/year polypropylene plant in Kerkrade, 150 miles south of Amsterdam, increasing its European capacity by 25%.

NORWAY

CAPITAL: Oslo
MONETARY UNIT: Krone
REFINING CAPACITY: 287,500 b/cd
PRODUCTION: 2,247,300 b/d
RESERVES: 9,283,760,000 bbl

Public sentiment in Norway was running against joining the European Community, even through the neighboring countries of Sweden and Finland were likely to join.

One of the major problems was that an EC directive would prevent Statoil from giving preferential treatment to Norwegian firms for contracts.

Norway was expected to decide in 1995 whether to lay a fourth pipeline to move North Sea gas to continental Europe.

In service is a Zeepipe line to Belgium and the Norpipe line to Germany. The Europipe line was being laid to deliver Troll field gas to the continent in 1996.

Meanwhile, Storting (parliament) was considering legislation to allow offshore oil companies to participate in sales of gas from the Norwegian North Sea. Under existing rules, the oil firms produce gas but the Norwegian gas negotiating committee (GFU) markets it.

Prime Minister Gro Harlem Brundtland praised GFU but called for "a closer link between resource management and sales activities." Storting planned to study ways to link gas fields to sales contracts earlier in development.

Before a contract is concluded, the Ministry of Industry and Energy will decide which gas fields will supply the contracted gas, Brundtland said. GFU and the licensees will provide plans to the ministry, and licensees will have the opportunity to negotiate sales along with GFU.

State owned Den norske state oljeselskap AS (Statoil) was reconsidering its plan to build a methanol plant at Tjeldbergodden, using associated gas from Heidrun oil field as feedstock.

Statoil said the initial $360 million cost of the methanol plant now was likely to be much higher. It had planned for the plant to go on stream in 1996, with Statoil owning 82% and Conoco 18%. Conoco is developing Heidrun field.

Statoil merged its petrochemical division with the Finnish firm Neste Oy's petrochemical business. A petrochemical firm will be started, with each company owning 50%.

Statoil signed an agreement with a group of Norwegian companies to develop a system to recover volatile organic compounds (VOCs) emitted during loading of shuttle tankers. The test program will cost $5.5 million.

An estimated 700,000 bbl/year of oil vaporizes during offshore buoy loading of more than 400 million bbl of oil from Norwegian fields. The project will seek to cut VOC emissions by about 500,000 bbl/year.

Norway's latest attempt to attract exploration interest in the Barents Sea was a disappointment.

Only two Barents production licenses were awarded in the 14th round of licensing in 1993. The government said it will ask oil companies what incentives they need to increase exploration of the area.

Only 15 other exploration licenses were awarded, with 11 in the North Sea and four in the Norwegian Sea.

Norway was considering tightening its petroleum tax laws to reduce capital outflow from the Norwegian oil industry.

Under tax reforms introduced in 1992, a 78% tax on production income encourages foreign companies to fund developments with debt equity, against which tax allowances can be claimed. The government said companies are using the opportunity to repatriate funds to partner companies by transferring reserve funds into share capital.

The government was considering regulating the capital structure of petroleum companies, new rules limiting debt ratios, limits to or abolition of tax allowances on debt interest payments, and increased tax rates.

POLAND

CAPITAL: Warsaw
MONETARY UNIT: Zloty
REFINING CAPACITY: 333,000 b/cd
PRODUCTION: 3,400 b/d
RESERVES: 36.793 million bbl

Natural gas demand is set to flourish in Poland.

Cambridge Energy Research Associates (CERA), Paris, identified three pointers to a steady growth market.

Six million out of more than 13 million Polish homes are connected to a gas supply, but this is a recent development. Gas contributes only 8.5% to primary energy use.

Gas price subsidies have been virtually eliminated in Poland, providing internal cash resources and an investment framework for new gas supplies to Europe.

Poland's government is aware of the potential, holding a second oil and gas exploration licensing round that closed Apr. 12, 1994, for 27 blocks in the Lublin area of the Polish Lowlands and in the Carpathian Mountains southeast of Krakow. This follows 1992's first round for conventional and coalbed methane prospects.

Polish Oil & Gas Co. (POGC), Warsaw, participated in nine blocks near Lublin, while operatorship of Ciecierzyn and Melgiew gas fields near Lublin were offered for joint development with POGC.

The Polish government plans to increase the natural gas portion of primary energy supply to 12% by 2000. This will increase natural gas consumption to 450 bcf in 2000 from 300 bcf in 1992.

CERA said Poland's gas production could increase substantially. Production was 90 bcf in 1991. This is expected to double by 2000 via increased flow from existing fields and from new discoveries.

Even with this increase, domestic production will provide no more than 40-50% of Polish demand, making imports essential. Russia sent nearly 200 bcf to Poland in 1992, but diversification of supply is now a priority.

Four options are under discussion. A western route is planned to take 35 bcf of North Sea and Russian gas from the German grid in Saxony into Siberia starting during summer 1993.

A northern pipeline from Denmark via the Baltic Sea could carry British, Norwegian, or Danish gas starting in late 1997. An eastern link to the existing Northern Lights pipeline could bring gas from Tyumen province and the Barents Sea.

More fancifully, a southern option involves an LNG plant on the Croatian island of Krk to receive gas from Algeria. Ethnic conflicts in the region and high costs are among hurdles that make this the longest shot of the four.

Coalbed methane acreage awarded

Poland has awarded coalbed methane development rights on 60,000 acres in the Upper Silesian basin to a unit of McCormick Energy Inc., Houston.

McCormick Poland is one of two companies to receive coalbed methane concessions during summer 1993 from Poland's Ministry of Environmental Protection, Natural Resources, and Forestry. The ministry in June 1993 signed a deal with Amoco Poland Ltd. covering exploration and

LOWLANDS
50 Mi.
50 Km.
North Sea
Heligoland Bight
Philips 34 in. 86 cm
Philips 36 in. 91 cm
20 in. 51 cm
30 in. 76 cm
Penzoil 36 in. 91 cm
Dan
Forbes
Gordon
West Sole
Viking
Indefatigable
Broken Bank
Deborah
N. Hewitt
Hewitt
Sean
Lemon Bank
Placid
Bacton
Kotter
Logger
ENGLAND
Ameland-N.
West Frisian I.
East Frisian I.
Borkum
Juist
Norderney
Langeoog
Spiekeroog
Schiermonnikoog
Terschelling
Vlieland
Texel
Wadden Zee
Ijssel Meer
Markerwaard
Northeast Polder
E. Flevaland
S. Flevaland
Delfzijl
Groningen
Hoogezand
Leeuwarden
Harlingen
Emden
Norden
Aurich
Jever
Leer
Amsterdam
Haarlem
Ijmuiden
Leiden
The Hague
Rotterdam
Pernis
Utrecht
Amersfoort
Hilversum
Zwolle
Ommen
Hengelo
Enschede
Arnheim
Nijmegen
Zutphen
Winterswijk
Lingen
Salzbergen
Munster
NETHERLANDS
Over Flakkee
Schouwen
N. Beveland
Walcheren
Vlissingen
Goes
Breda
Tilburg
Oosterhout
Eindhoven
Helmond
Venlo
Kleve
Dinslaken
Gelsenkirchen
Duisburg
Essen
Dortmund
Hamm
Dusseldorf
Cologne
Bonn
Wesseling
Godorf
Westphalia
Zeebrugge
Oostende
Brugge
Nieuwpoort
Dunkerque
Terneuzen
Axel
Antwerp
Turnhout
Gent
Mechelen
Brussels
Leuven
Diest
Hasselt
Liege
Verviers
Namur
Charleroi
Mons
Kortrijk
Lille
Cassel
Denain
FRANCE
BELGIUM

SWEDEN
Copenhagen
Malmo
Arhus
Fredericia
Kalundborg
D E N M A R K
Korsor
Nyborg
Skaelskor
Baltic Sea
Flensburg
Kiel Bay
Fehmarn
Schwedeneck
Kiel
Preetz
Great Plon Lake
Schleswig
Heide
Warnau
Plon
Boostedt
Lubeck Bay
Holstein
Bramstedt
Lubeck
Stralsund
Rostock
Reinkenhagen
Kolberg
Koslin
Wismar
Wolin
Schwerin
Mölln
Hamburg
Harburg
Dibbersen
Reitbrook
Sinstorf
Potrau
Mechelfeld-S.
Harburg Grasbrook
Volkensen
G E R M A N Y
Stettin
Walez
Achim
Lower Saxony Basin
Verden
Bomlitz
Wittenberge
Schwedt
P O L A N D
20 in. 51 cm
Landsberg
Ahrensheide
Hademstorf
Eystrup
Luben
Wietze
Eldingen
Orrel
Hankensbuttel
Wittingen
Knesebeck
Nienhagen
Hohne
Vorhop
Ehra
Landesbergen
Hanigsen
Lachendorf
Misburg
Lehrte
Gifhorn
Lieferde
Bokeloh
Hannover
Eddesse
Oelheim
Wolfsburg
Braunschweig
Meerdt
Sarsted
Kemme
Molme
Oberg
Helmstedt
Bad Salzdetfurth
Hohenassel
Salzgitter
Roxforde
Brandenburg
Berlin
Potsdam
Rudersdorf
Frankfurt-an-der Oder
Magdeburg
Rybaki
Fallstein
Staakow
Guben
Zary
Zagan
Schwarzheide
Kassel
Merseburg
Leuna
Leipzig
Volkenroda
Muhlh
Langensalza
Gorlitz
Eisenach
Erfurt
Jena
Zeitz
Dresden
Zittau
Bad Hersfeld
Gera
Karl Marx Stadt
Zaluzi
CZECH REP.
Suhl
Kralupy
Prague
Plauen
Karlovy Vary

development of coalbed methane on a 120,000 acre Upper Silesian tract in the same vicinity.

The ministry's most recent coalbed methane agreement is one of three new contracts between McCormick Energy affiliates and Polish parties. The deals allow McCormick to drill coalbed wells on public and privately owned land and use production from the wells to heat and generate electricity for mines in southern Poland.

McCormick Poland plans to drill nine wells on the ministry concession and adjoining acreage during an evaluation phase. Full development could lead to drilling of as many as another 160 methane wells and installation of compressors, pipelines, and other equipment.

A timetable for development was not disclosed. But based on data from 70 active coal mines and more than 1,000 well cores, wells in the concession area could produce more than 400 bcf of methane in the next 20 years, McCormick says. McCormick Poland also signed agreements with two mines near the concession area:

- Nadwislanski Spolka Weglowa SA granted the company methane development rights on a 5,000 acre tract mined by KWK Brzeszcze.
- Jastrzebska Spolka Weglowa SA awarded similar rights on a 2,800 acre tract at Krupinski mine.

The McCormick unit also is negotiating to expand methane development to acreage owned by other Polish coal mines.

In addition, another affiliated company, McCormick Power Co., has agreed to install cogeneration units at Krupinski mine. The units are to be fueled by methane otherwise vented as part of coal companies' mine degasification programs. McCormick expects to sign similar contracts at several other mines.

Poland has a growing market for gas, which now supplies only 8% of the country's total energy needs. Imports account for about two thirds of Poland's gas supply.

Low BTU gas due study for development

Gas/condensate fields in western Poland, undeveloped because of high nitrogen and hydrogen sulfide contents, are to be studied for development as part of a novel power generation project.

Drake Resources Ltd., London, in collaboration with POGC, will study the feasibility of producing more than 400 bcf of low BTU gas and about 50 million bbl of condensate. Drake estimated capital outlays of at least $50 million in a project designed to supply at least 50,000-75,000 kw of electrical power from a central generating plant.

Target reserves lie in 12 fields in two groups southwest and west of Poznan, and 10-90 km east of the German border. Enough wells have been drilled, tested, and suspended in all the fields to begin development. The gas has nitrogen contents of 58-90% and hydrogen sulfide levels ranging from a trace to about 1%.

Drake bases its development plan on three key features:

- Use of combined cycle gas turbines to generate electricity from produced gas without the need for nitrogen removal.
- Multiple field development projects to yield economies of scale and enable separation and selective blending of gas components to improve heating value with further enrichment, if necessary, by blending in methane from a local trunk line carrying high BTU gas.
- Environmental benefits, in comparison with Poland's largely coal based power generation, which include improved efficiency and reduced emissions.

Power generation in Poland currently is fueled 70% by hard coal, 10% brown coal, 13% imported oil, and 7% by natural gas, half of which is imported.

Drake reckons its project will show up to 45% gas conversion efficiency, compared with 30% efficiency at best for existing coal based plants. Ash hanling and airborne particulates will not be a problem, and sulfur dioxide emissions will be minimized. Carbon dioxide emissions will be only 25-35% of those from existing coal fired plants generating the same amount of power.

The feasibility study was scheduled for completion during 1993. The fields have been split into two groups to stage the development. Total production from the seven fields in Phase 1 is estimated at 40-80 MMcfd of gas and 700-1,700 b/d of liquids.

If the study reveals that further hydrocarbons must be added to the produced gas to meet fuel gas standards, the study will be expanded to include development of other fields. If the gas still falls short of requirements, methane could be imported from the Polish gas trunk line network.

Once the study has been completed, and if the project proves commercially viable, power generation could follow within 2 years, Drake said.

Elsewhere, Poland's Industry & Trade Ministry said four European companies and one U.S. company will invest $300 million during 5 years to privatize and modernize Poland's gas industry. The ministry identified them as BOC of Britain, Linde of Germany, AGH of Sweden, Hydrogas of Norway, and Liquid Carbonic of the U.S.

Neste Oil Poland Ltd. has opened the first service station in Poland to be built by a western company. The station, at Zakroczym, 35 km from Warsaw is open 24 hr/day and accepts payment in local currency or by major international credit cards. Customers can buy leaded gasoline and diesel produced in Poland, and unleaded and leaded fuels produced by Neste Oy in Finland.

Poland's Mazowieckie Zaklady Rafineryjne i Petrochemiczne late in 1993 let contract to Costain Oil, Gas & Process Ltd,. Manchester, U.K., for supply and construction of a cryogenic offgas recovery unit to be installed at its Plock plant. The unit will process gas for use in other processes on the site. Start-up is scheduled for yearend 1994.

Poland's Mazovian Refinery & Petrochemical Works let contract to KTI Group BV, Zoetermeer, Netherlands, for a study and engineering of MZRiP's Ethylene Plant II debottlenecking and modernization program at Plock. KTI will assess performance of the plant and recommend changes needed to increase capacity to 360,000 tons/year from 300,000 tons/year.

PORTUGAL

CAPITAL: Lisbon
MONETARY UNIT: Escudo
REFINING CAPACITY: 294,000 b/cd
PRODUCTION: 0
RESERVES: 0

Portugal's efforts to establish a domestic pipeline grid ran aground in 1993, forcing a delay in the proposed project.

The government was working on a legal framework that would permit formation of a group to build and operate a domestic gas network.

The government will require operators to maintain a 20 day strategic gas reserve rather than share a reserve with Spain.

Portugal decided to link up with the Maghreb-Europe gas pipeline for most of the supplies for a domestic gas grid, with only small volumes from Algeria through a planned LNG terminal at Setubal rather than the other way around.

In addition to the Setubal terminal, the project calls for a 248 mile, high pressure pipeline to Braga and installation of distribution lines and storage facilities.

Projected start-up was 1995, with Portuguese gas demand estimated at 350 MMcfd.

Gaz de France was to have been the lead company in the project but failed to reach agreement with the Portuguese government. Negotiations reportedly broke down over take or pay terms.

After that failure, a number of foreign companies sought to participate in the project with domestic firms. Foreign participation was to be limited to a 49% interest.

The government also was working to privatize the nation's fuel network, as required by EC guidelines. EC had granted Portugal a 1 year reprieve from that requirement on the grounds that the privatization process was putting state owned Petrogal at a competitive disadvantage against foreign firms.

Parliament approved the privatization of CNP through direct sale. It had been losing money for several years.

Neste Oy, which leases CNP's Sines petrochemical complex for $30 million/year, was expected to buy CNP if the government assumed a large portion of CNP's debt.

Ste. Nationale Elf Aquitaine refinery at Grandpuits, France, is one of three the French state oil company owns in France. Photo by G. Houlbreque, courtesy of Elf.

SLOVAKIA

CAPITAL: Bratislava
MONETARY UNIT: Koruna
REFINING CAPACITY: 193,400 b/cd
PRODUCTION: Not available
RESERVES: Not available

Slovakia was formed when Czechoslovakia ceased to exist Jan. 1, 1993, and was succeeded by two independent states: Czech republic and Slovakia. Czech and Slovak leaders decided during summer 1992 to divide their country. The international boundary between the successor states is the internal boundary that separated them as constituent republics of Czechoslovakia.

Novacke Chemicke Zavody (NCZ), a Slovak chemical company based 140 km northeast of Bratislava in Slovakia's Upper Nitra Valley, retained Wright Killen & Co., Houston, to help secure partners to form a joint venture during the final steps of NCZ privatization. Formed in 1940, NCZ specializes in PVC and allied products, calcium carbide, and organic chemicals.

The company is vertically integrated within the PVC business and has diversified to build a significant presence in European calcium carbide markets.

SPAIN

CAPITAL: Madrid
MONETARY UNIT: Peseta
REFINING CAPACITY: 1,283,000 b/cd
PRODUCTION: 18,400 b/d
RESERVES: 22.518 million bbl

The government passed legislation ending the distribution monopoly of state owned Cia. Arrendataria del Monopolio de Petroleos SA (Campsa) on gasoline and diesel made in Spain but retained the right to set maximum prices for products.

Campsa was limited to operating the existing products pipeline network. Its tanker trucks and service stations were divided between Campsa shareholders Repsol, Petronor, Cepsa, and BP Oil.

Design work began in 1993 on the Strait of Gibraltar section of the planned Maghreb-Europe gas pipeline from Algeria to Spain.

Twin 20 in. lines will carry gas from Sonatrach's Hassi R'Mel field in Algeria to the gas grid in southern Spain. Start-up is due in 1995.

The Greek tanker Aegean Sea spilled 523,264 bbl of oil late in 1992 off Northwest Spain after running aground at a harbor entrance off La Coruna. About half of the spilled crude burned in a fire that followed the accident, and heavy seas helped disperse the rest.

A similar accident occurred at the same spot in 1976. The Spanish government urged Repsol to build a deepwater terminal on the coast.

Refining changes

Spain's downstream petroleum sector weathered major changes in 1992-93, largely due to its participation in the European Community.

During 1990-91, much progress was made in liberalizing the oil sector, beginning with the breakup of Campsa.

That led to development of new retail outlets, a new price system, and involvement of multinational companies in Spanish refineries.

The emphasis during the rest of the 1990s will be on consolidation as the semipublic company Repsol SA presses privatization and restructuring, with ambitious expansion plans upstream and downstream.

Repsol, which dominates the Spanish oil market, is 65% owned by the state. It plans to fully privatize by the end of the century.

Repsol's restructuring is intended to free more capital for an aggressive expansion in international operations, as well as to help consolidate its downstream presence in Spain.

Royal Dutch/Shell asked Spain to return assets that were seized from its Spanish subsidiary Shell Espana SA when the government created an oil monopoly in 1927. The assets include various terminals, service stations, and other land.

Shell argued that with the dismantling of the Campsa monopoly, the reason for the takeover no longer is valid, and it is due some compensation.

SWEDEN

CAPITAL: Stockholm
MONETARY UNIT: Krona
REFINING CAPACITY: 427,500 b/cd
PRODUCTION: 0
RESERVES: 0

The future of Sweden's OK Petroleum, which operates two domestic refineries, was very much in doubt in 1993.

After debating whether to buy the whole firm or sell its interest, Finland's Neste Oye decided to sell its 24% share. The other two owners are the Swedish government with 24% and KF Group with 52%.

Later, OK Petroleum said Middle East oil companies expressed an interest in buying the company's refineries and storage facilities. OK's fuel distribution and gasoline marketing operations were not for sale.

The Swedish subsidiary of Statoil, the Norwegian state oil company, was buying British Petroleum's 240 service stations in Sweden, making Statoil the biggest marketer in Scandinavia with 700 outlets.

OK Petroleum planned to build an isomerization unit at its Scanraff refinery south of Gothenburg to make unleaded gasoline for the domestic market. It will go on stream in 1994.

Lysekil Refining was upgrading the 28,000 fluid catalytic cracking unit at its 200,000 b/d Lysekil refinery. Completion was due in 1994.

Swedish regulators planned to introduce a system of dyes for diesel fuel to prevent deliberate misfueling, due to varying energy taxes for middle distillates.

SWITZERLAND

CAPITAL: Bern
MONETARY UNIT: Franc
REFINING CAPACITY: 132,000 b/cd
PRODUCTION: 0
RESERVES: 0

The Tiefengas Konsortium Swisspetrol/Sulzer (TGK) was seeking a partner to drill a 23,450 ft well to test potential gas reservoirs in the upper Jurassic to the Triassic.

It said seismic data and geochemistry indicate deep gas in the north alpine overthrust belt of Switzerland.

Swissgas began taking another 100 million cu m/year of gas from Gasunie of Netherlands, bringing its total to 600 million/year through the trans-European gas pipeline.

The pipeline was working to loop its line between Wallbach on the Swiss-German border to Daeniken in canton Solothurn.

Shell continued to upgrade its 60,000 b/d refinery at Cressier. The $125 million project was necessary to meet the government's requirement that diesel fuel contain no more than 0.05% sulfur by 1994.

TURKEY

CAPITAL: Ankara
MONETARY UNIT: Lira
REFINING CAPACITY: 713,000 b/cd
PRODUCTION: 75,800 b/d
RESERVES: 487.67 million bbl

Turkey agreed to construction of a $1.4 billion export pipeline to move Azerbaijani crude from Baku to a Mediterranean terminal at Ceyhan.

The 660 mile pipeline will carry as much as 800,000 b/d of crude from Caspian Sea fields through Iran and the Republic of Nachichevan.

The terminal at Ceyhan, operated by the Turkish state oil company Botas, can handle two 350,000 dwt and two 150,000 dwt tankers at the same time.

Another project on the books would move gas from Turkmenistan and Kazakhstan to Europe via Turkey.

The Turks favor a 3,000 mile, dual 56 in. pipelines stretching 300 miles from the gas fields of Turkmenistan, through Armenia and Turkey. Capacity would be 1.4 tcf/year and construction cost would be more than $15 billion.

Turkish Petroleum Refineries Corp. started a 16,500 b/d hydrocracker at its 224,000 b/d Izmit refinery on the Aegean Sea and a 14,500 b/d hydrocracker at its 113,000 b/d Kirikkale refinery in Central Anatolia. They cost a combined $500 million.

The company awarded a $300 million contract for construction of a complex at the Izmit refinery. The project includes a 23,000 b/d unicracker as well as upstream and downstream process units. The complex is to start up by mid-1996.

Mobil sold a 56% interest in two liquefied petroleum gas marketing companies in Turkey for $256 million to Aygaz and Koc Holding.

Tatneft entered a joint venture with Garnet Resources Corp. to explore Garnet's Akseki, Egridir, and Isparta licenses in Southwest Turkey. They total 2.66 million acres in the Taurus Mountains.

UNITED KINGDOM

CAPITAL: London
MONETARY UNIT: Pound
REFINING CAPACITY: 1,866,640 b/cd
PRODUCTION: 1,866,200 b/d
RESERVES: 4,554,260,000 bbl

Parliament accepted a petroleum revenue tax reform package that did not include a transitional period.

Oil companies complained that the changes would improve prospects for current development projects and maintaining production from aging fields but hurt new exploration.

The action cut the petroleum revenue tax on current producing fields to 50% from 75% effective July 1, 1993, and abolished the tax on fields approved for development after Mar. 16, 1993. However, after Mar. 16 the government would no longer allow exploration costs to be reclaimed against profits of producing fields.

Britain's Monopolies and Mergers Commission, after a year long study, recommended that British Gas' transportation and merchant functions be split.

It said BG's domestic supply monopoly should be reduced by 1997 and total free market competition be in place by 2002, with BG's merchant functions being split off as a separate company by Mar. 31, 1997.

It said the reduced monopoly should be restricted to users of 150 MMBTU/year, down from 250 MMBTU. Price ceilings would be relaxed to compensate for the loss of business.

The U.K. government awarded its first out-of-round onshore licenses in 1993. The awards were part of a procedure designed to expedite exploration of acreage where companies have identified oil or gas prospects.

Eukan Energy Ltd. and affiliate Altwood Petroleum Ltd. got four blocks that contained gas discovery wells abandoned in the 1970s and 1980s.

Exploration/ production

The Energy Ministry approved development of three oil and gas fields on Block 110/13 in the Liverpool Bay area of the Irish Sea.

Hamilton Oil Co. Ltd., operator, will develop Douglas, Hamilton, and North Hamilton fields at a cost of about $1.5 billion.

A gas terminal will be built at Point of Ayr, North Wales. Gas will be supplied beginning in December 1995 to a power generating plant to be built by Powergen plc at Connah's Quay.

Oil production is due to begin in November 1995, with offshore loading on the Douglas field platform. It will rise to 70,000 b/d.

Field development will involve three platforms in Douglas field, providing drilling, production, processing, and accommodation.

Hamilton and North Hamilton fields will be developed using unmanned satellite platforms tied back to Douglas field. The operator may also develop Lennox field on Block 110/15 through an unmanned platform.

Hamilton estimates oil reserves in Douglas and Lennox fields at 150-200 million bbl. Gas reserves for Hamilton, North Hamilton, and Lennox are a total of 1.2 tcf.

Powergen will begin taking 300 MMcfd for its power plant beginning in mid-1996.

Powergen bought a 3.9% interest in the offshore fields, leaving Monument Oil & Gas plc with a 20% interest, Hamilton 46.1%, Lasmo plc 30%.

BP Exploration began producing 15,000 b/d from the offshore extension of Wytch Farm oil field in Dorset.

The 80 million bbl offshore Sherwood reservoir is produced from extended reach wells drilled from an onshore drilling pad 2.4 miles away. At least two more wells were planned.

Production from the onshore portion of Wytch Farm field began in 1979 Production during 1993 amounted to 80,000 b/d of oil.

BP considered various options to produce the offshore field, including construction of an artificial island in Poole Bay. It chose directional drilling from an onshore site because it was the most economic option, although environmentalists had objected to creation of an artificial island.

Forty gas production, supply, transporation, and marketing companies use Theddlethorpe terminal on Lincolnshire's North Sea coast. The terminal has undergone an expansion program by operator Conoco (U.K.) Ltd.

BP drilled a major discovery in the Faero-Shetland basin in the Atlantic Ocean west of the Shetland Islands.

The company's Block 204/24a-2 wildcat flowed 3,800 b/d of 25° gravity oil through a 1 in. choke. BP's initial estimate of reserves is 250-500 million bbl.

The well cut 160 ft of oil bearing sand in drilling to 9,295 ft in 1,600 ft of water. It followed a 1990 well drilled 2.8 miles to the east that cut only a thin oil and gas column and was not tested.

Because of deep water and a harsh environment, development will be costly. BP found Clair field 75 miles east of the new discovery 20 years ago, but it still has not been developed.

Gas terminal expansion

Capacity of Theddlethorpe gas terminal on the Lincolnshire coast was expanded to 2.6 bfgd. The terminal processes gas arriving through four pipelines from 12 offshore fields. Average processing rate is expected to be 1.4 bcfgd, or about one fourth of Britain's demand.

The terminal, built in 1972 to handle gas from Viking field, is operated by Conoco (U.K.) Ltd., which owns a 50% interest. BP Exploration Operating Co. holds the other 50%.

The expanded terminal processes gas from Viking, Lincolnshire Offshore Gas Gathering System, Pickerill, and Caister Murdoch System complexes.

Mobil began a $7.5 million construction program to expand its Scottish Area Gas Evacuation (SAGE) system terminal at St. Fergus, Scotland. The expansion will enable the terminal to receive initial gas shipments from Brae and outlying fields in the North Sea.

The existing treatment train processes 400 MMcfd of gas from Mobil's Beryl field. The second train will bring capacity to 1.15 bcfgd. Gas sales from North, South, Central, and East Brae will start in October 1994 at a rate of 270 MMcfd, rising to 350 MMcfd in 2003. After processing at St. Fergus, gas will enter the British Gas grid.

Shell U.K. Ltd. let a $23 million contract to revamp a gas oil hydrodesulfurizer at its 262,000 b/d Stanlow refinery. Construction will begin in 1994. IPE

The Neddrill 6 semisubmersible drills a Britannia field appraisal well. Photo courtesy of Conoco.

North Sea

North Sea operations in 1993 were plagued by storms.

Even so, operators continued to log progress in placing new fields on stream.

Early in the year, gales and high seas showed how vulnerable North Sea operations are to weather.

In January a tanker went aground in the Shetland Islands during a storm, production in the Norwegian and U.K. sectors was disrupted, and construction work was halted. Further storms disrupted installation in February.

A tidal wave of U.K. development program completions followed the 1992 peak in development spending. Seventeen U.K. offshore fields went on stream during 1993, compared with Norway's four, Netherlands' three, and Denmark's two.

Scottish Enterprise, Aberdeen, estimated U.K. development spending at £4.7 billion ($7.05 billion) in 1992 and £4.4 billion in 1993 ($6.6 billion). After this peak, U.K. capital outlay is predicted to tail off, falling to less than £3 billion ($4.5 billion) in 1997.

The agency sees 1993 as the development spending peak for western Europe as a whole, at £12.4 billion ($18.6 billion). This will fall to £10.9 billion ($16.35 billion) in 1994 and £10.1 billion ($15.15 billion) in 1995.

U.K.'s new fields rallied British oil production, pushing it back above the million b/d mark. Norway continued to dominate the North Sea in terms of oil flow, however, with production hitting a record average 2.47 million b/d in July 1993.

While analysts predicted a slow decline for the North Sea as a whole, one large oil discovery west of the Shetland Islands and another off mid-Norway raised hopes of more strikes in quiet corners of the region.

Britain's government caused a storm of its own in March 1993, when oil industry tax reforms forced U.K. operators to reevaluate their assets. This resulted in development projects being rescheduled, programs to extend the lives of older fields, and a slide in U.K. exploration and appraisal drilling.

Although the boom days are over for the North Sea, some operators are convinced that reduction of development and operational costs could yield opportunities in the region for years to come.

Here are selected highlights of 1993's operations in the North Sea:

Storm damage

Storms early in January 1993 caused the U.K.'s biggest oil spill.

On Jan. 5 the Liberian registered Braer tanker was passing between the Orkney and Shetland Islands in a Force 10 gale.

The engine failed and the tanker began to drift. Attempts to attach lines to the ship to enable tugs to tow it out of danger were foiled by high seas. Braer struck rocks just after 11 a.m.

The 89,000 dwt vessel was carrying more than 600,000 bbl of crude oil. Rough weather kept salvage crews from working and prevented aircraft from spraying dispersant.

For 3 weeks salvage teams and cleanup crews were thwarted, until the vessel could be surveyed by divers for the first time Jan. 24. There was no cargo or bunker oil left on board.

Meanwhile, winds gusted to 100 mph in the northern and central North Sea. Oil producers had to shut down fields that relied on offshore loading.

Worst hit was Norway's Den norske stats oljeselskap AS, which lost production from its main fields, Statfjord and Gullfaks, several times during January and February.

Wood Mackenzie Consultants Ltd., Edinburgh, said Statfjord, Gullfaks, and Snorre fields had to defer a combined 7 million bbl of oil and natural gas liquids production in January.

Mobil North Sea Ltd. had to halve U.K. Beryl field production to 55,000 b/d in mid-January but said storage capacity would enable output at this level until weather improved.

Nevertheless, Beryl suffered most on the U.K. side, with average production for January at 65,000 b/d, compared with 110,000 b/d in December 1992.

The Shell-Esso combine, with Shell U.K. Exploration & Production as operator, saw production from Kittiwake field average 8,000 b/d in January, compared with 30,000 b/d in December 1992. Production vessels in Buchan and Donan fields, which both average about 12,000 b/d, were pulled out of fields for several days by operator BP Exploration Operating Co. Ltd.

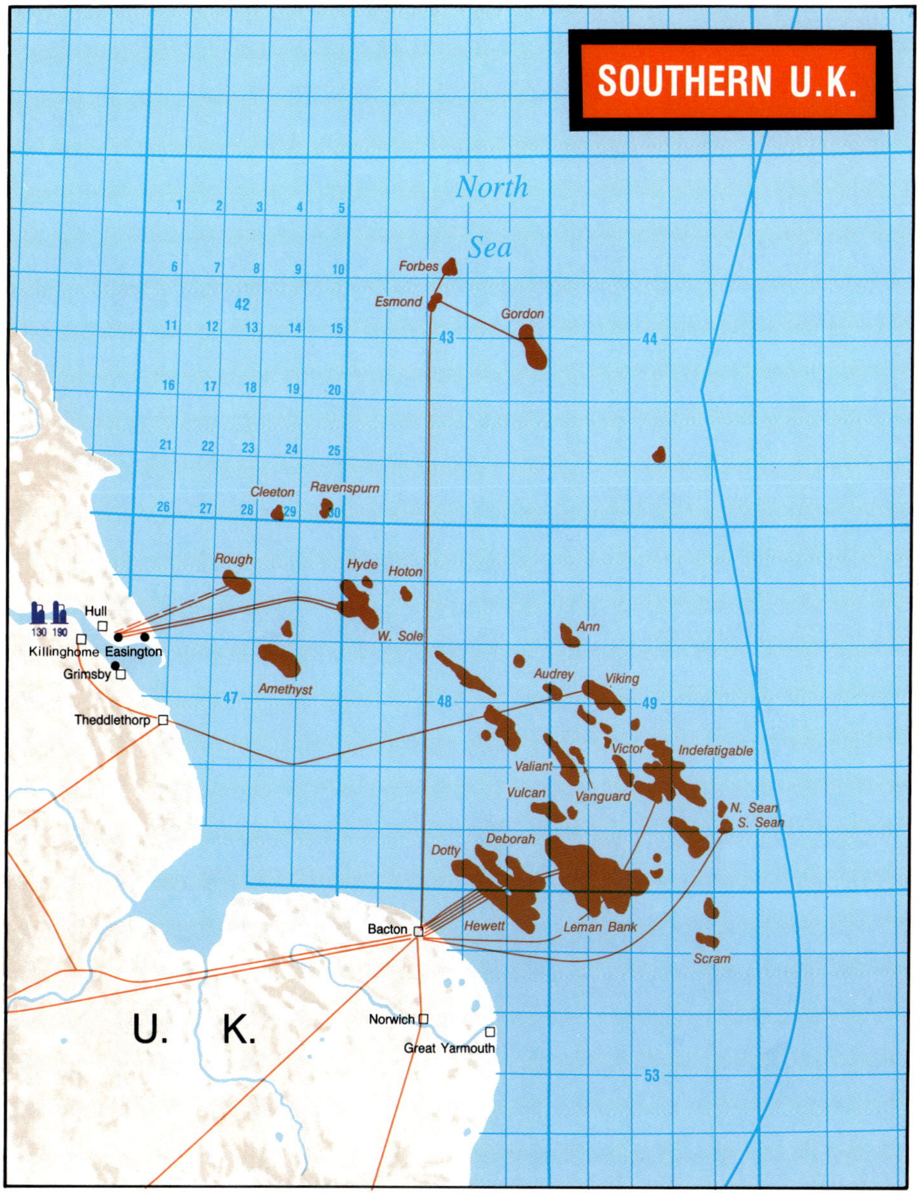

East Sleipner

Among new fields placed on stream in 1993, development of Norway's East Sleipner gas field was the most pressured.

Operator Statoil laid its reputation on the line with a promise to deliver East Sleipner gas to European buyers Oct. 1 as scheduled, despite the sinking of the original gravity base structure for the platform in 1991.

East Sleipner was chosen as the first field to provide gas under the Troll sales contract of 1986. This deal committed Norwegian gas producers to supply 35 tcf, worth 700 billion kroner ($100 billion) at current prices, to Europe during almost 30 years.

On Sept. 29 the last of four production wells was tested to determine production rate. On deadline day, the target field flow of 370 MMcfd was achieved without the need to call on gas from other Norwegian fields.

First production from East Sleipner also gave the Zeepipe trunk line its first commercial use. Most of East Sleipner's gas was sent 830 km to Zeebrugge, Belgium, where it joined Europe's gas pipeline grid. The rest went through the existing Statpipe line to Emden in Germany.

Before East Sleipner began commercial gas production, delivery of condensate to Statoil's gas terminal at Karsto, north of Stavanger, began Sept. 19.

Statoil reckons East Sleipner development cost 15 billion kroner ($2.1 billion). Despite the loss of the platform base, that was less than the original 1986 estimate of 18 billion kroner ($2.5 billion).

Scott

In the U.K. North Sea, Amerada Hess Ltd. placed Scott field on stream Sept. 2, 1993. It is the biggest British oil field to be developed this decade.

Amerada Hess figures Scott will meet 10% of U.K. oil needs from reserves of 450 million bbl of oil and 287 tcf of gas.

Among main considerations in Scott development was the need to reach production start-up and production plateau quickly to improve field economics. First oil flow amounted to about 30,000 b/d. Amerada Hess then worked to hit a 180,000 b/d plateau within a month.

To achieve first oil quickly, the company reduced the period of offshore construction, hookup, and commissioning. A two platform development program was chosen. Processing, drilling, and production facilities are located on one platform, bridge-linked to a utilities and quarters platform.

The production platform jacket was installed on Block 15/22 in September 1992 using the DB102 crane barge. The utilities platform jacket was installed alongside in March 1993.

Twelve topsides modules were installed throughout April 1993, with a combined weight of 32,000 metric tons. The largest of these was the 10,360 metric ton process module. The last lift, on May 1, 1993, was of the two bridges linking the platforms.

Scott's use of the North Sea's most extensive subsea development allowed early production through predrilled wells. First production was from one of three well clusters tied back to subsea manifolds.

At plateau production, each well is designed to produce at least 25,000 b/d of oil. With seven wells on stream, the platform is designed to process about 200,000 b/d of oil.

Draugen

Norske Shell AS became the first company to produce oil commercially in the Norwegian Sea when Draugen field

NORTH SEA
Atlantic Ocean
SCOTLAND
N. IRELAND
IRELAND
WALES
Kirkwall
Flotta
Wick
Nigg Bay
Cromarty Firth
Inverness
Fraserburgh
Peterhead
Cruden Bay
Aberdeen
Dundee
Glenrothes
Grangemouth
Edinburgh
Finnart
Ardyne Point
Hunterston
Ardrossan
Glasgow
Outer Hebrides
Inner Hebrides
Firth of Clyde
North Channel
Londonderry
Belfast
Lough Neagh
Carlisle
Solway Firth
I. of Man
Lancaster
Heysham
Morecambe Bay
Liverpool Bay
Liverpool
Eastham
Stanlow
Ellesmere Port
Bradford
Manchester
Birmingham
Irish Sea
Galway Bay
Shannon
Dublin
Limerick
Wexford
Cork
Whitegate
Bantry
Bantry Bay
Kinsale Head
St. George's Channel
Cardigan Bay
Anglesey I.
Milford Haven
Angle Bay
Pembroke
Llandarcy
Neath
Cardiff
Bristol
Bristol Bay
Bath
Gloucester
Celtic Sea
Exeter
Plymouth
Bournemouth
Fawley
Weymouth
Kimmeridge
English Channel
0 50 100 Mi
0 50 100 Km.
Complete offshore supply and service base
Offshore supply and service base
Specialized service base

NORWAY
SWEDEN
DENMARK
NETHERLANDS
BELGIUM
GERMANY
FRANCE
POLAND
CZECH REP.
LUXEMBOURG
North Sea
Baltic Sea
Skagerrak
Heligoland Bight
Kiel Bay
Lubeck Bay
Strait of Dover
Oslo
Stockholm
Dalaro
Nynasham
Mongstad
Sotra
Bergen
Karsto
Stavanger
Sola
Valloy
Tonsberg
Slagentangen
Lysekil
Goteborg
Arhus
Esbjerg
Fredericia
Kalundborg
Copenhagen
Malmo
Skaelskor
Kiel
Heide
Brunsbuttel
Cuxhaven
Bremerhaven
Hamburg
Harburg
Grasbrook
Wilhelmshaven
Emden
Bremen
Lubeck
Schwerin
Rostock
Stettin
Schwedt
Berlin
Frankfurt-an-der-Oder
Magdeburg
Hannover
Misburg
Cottbus
Leipzig
Zeitz
Dresden
Jena
Gera
Erfurt
Eisenach
Suhl
Prague
Groningen
Amsterdam
The Hague
Rotterdam
Pernis
Vlissingen
Antwerp
Brussels
Feluy
Gent
Liege
Zeebrugge
Dunkerque
Calais
Boulogne
Dieppe
Denain
Valenciennes
Dortmund
Essen
Duisburg
Dusseldorf
Dinslaken
Gelsenkirchen
Cologne
Bonn
Godorf
Wesseling
Raunheim
Frankfurt am Main
Mannheim
Speyer
Woerth
Karlsruhe
Volkingen
Metz
Luxembourg
Lingen
Emsland
Salzbergen
Schoonebeek
Great Yarmouth
Ipswich
Felixstowe
Kent
Bacton
Troll
Frigg
Ekofisk
Forties
Brent
Heimdal
Sleipner
Statfjord
Gullfaks
Oseberg
Valhall
Tyra
Dan
Gorm
Indefatigable
Leman Bank
Viking
Hewett

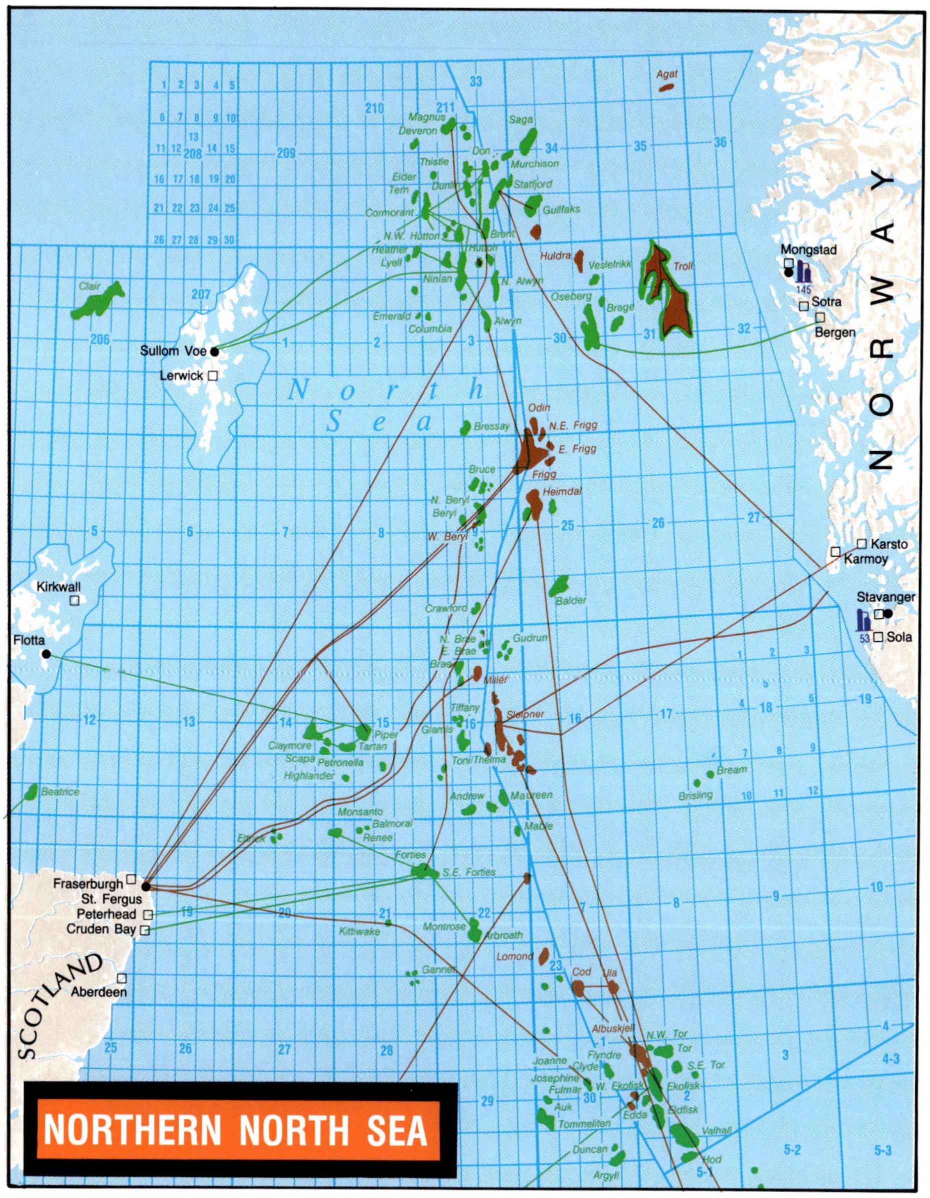

went on stream Oct. 19, 1993. Draugen reserves are 580 million bbl of oil and 100 bcf of gas.

The field lies in Block 6407/9, where water depth is 800-900 ft. The platform was towed out of Stavanger May 3, 1993. The 830 km journey northward took 10 days and was the longest tow of a fixed platform.

Because of lack of infrastructure in the Haltenbanken area, Shell developed Draugen using a monotower platform with oil storage in the base to allow offshore loading into shuttle tankers.

First oil came from a subsea well 9 km from the platform. A second subsea well 3 km away was to be placed on stream when tests were complete on the first well. Combined production from the two wells is expected to be 25,000-40,000 b/d.

The first of four production wells was about to be drilled from the platform. These will help boost production to a plateau of 95,000 b/d in mid-1994. The platform has slots for 10 production wells.

Three water injectors have been drilled from the platform, as well as a gas injector. Three more water injectors were planned, the first of which was drilling ahead.

The platform's concrete gravity base houses seven oil storage cells which can hold 11 days' output—more than 1 million bbl. Oil will be exported in shuttle tankers leased from development partner Statoil.

Although gas will be used for power generation on the platform, most of it will be reinjected. Future development of gas reserves depends on plans to lay a gas export pipeline from the Haltenbanken area.

Only one other field, Heidrun on Block 6507/7, is so far under development off mid-Norway. Operator Norske Conoco AS expects first oil from Heidrun in 1995.

Heidrun reserves are 750 million bbl of oil and 1.6 tcf of gas. A concrete tension leg platform has been chosen for the development, with offshore loading of oil for export.

Gryphon

Kerr-McGee Oil (U.K.) plc placed Gryphon oil field on production Oct. 14, 1993, less than 10 months after receiving development approval from the U.K. government.

Use of the U.K.'s first purpose built, permanently moored, production vessel enabled the £265 million ($400 million) project to be developed 2 years earlier than originally planned.

Gryphon field lies on U.K. North Sea Block 9/18b. It is expected to yield 96 million bbl of 22° gravity oil during 15-20 years. Peak production was expected to be 50,000 b/d in mid-1994.

Eight subsea production wells, five of which are extended reach, will be tied back to the Gryphon A vessel. Four water injection wells, an injection water source well, and a dual service gas well also will be used.

The first well produced more than 10,000 b/d of oil. A second well, placed on stream Oct. 19, 1993, is expected to hike production to 25,000 b/d.

Kerr-McGee said a floating production, storage, and offloading vessel was preferred to a fixed platform because of flexibility, cost effectiveness, and ability to accelerate oil production from the field.

The floater, designed to withstand a 100 year wave while moored, can hold 525,000 bbl of oil and process 60,000 b/d. Offloading to a shuttle tanker can take place at a 600,000 b/d rate.

As many as four shuttle tankers will be available for Gryphon under a contract with Statoil. Most of the time only one shuttle tanker will be required.

Everest-Lomond

Amoco (U.K.) Exploration Co. installed Lomond gas field platform topsides on Block 21/21 in January 1993, completing the major infrastructure for a £1.2 billion ($1.8 billion) development of Everest and Lomond fields via the new Central Area Transmission System (CATS) gas pipeline.

Development work on Everest and Lomond was slowed by storms more than any other North Sea project. February's 30 knot winds and 35 ft waves caused the Safe Supporter accommodation vessel to be pulled off anchor alongside the Lomond platform during hookup.

The vessel's anchors were damaged, requiring it to be towed to Bergen for repairs. A team of 100 workers continued hookup work while stationed on West Omikron jack up

North Sea fields starting production in 1993

Field	Operator	Block	Start-up	Reserves: Oil (Million bbl)	Gas (bcf)	Condensate (Million bbl)
NORWAY						
Embla	Phillips	2/7a	May 12	50	100	—
Brage	Norsk Hydro	31/4a	Sep 24	290	60	—
E. Sleipner	Statoil	15/9	Oct 1	130	1,900	—
Draugen	Shell	6407/9	Oct 19	580	100	—
U.K.						
Piper B*	Elf Enterprise	15/17	Feb 3	172	14	—
Lyell	Conoco	3/2	Mar 22	31	—	—
Chanter	Elf Enterprise	15/17	Apr 26	5	30	5
Everest	Amoco	22/10	May 5	—	900	35
Bruce	BP	9/9b	May 18	220	2,600	—
Saltire	Elf Enterprise	15/17	May 23	130	100	—
Lomond	Amoco	21/21	Jul 14	—	600	20
Guinevere	Mobil	48/17a	Jul 15	—	70	—
Lancelot	Mobil	48/17b	Jul 15	—	220	—
Hudson	Amerada Hess	210/24a	Jul 19	87	—	—
Orwell	ARCO	50/26a	Aug 3	—	250	—
Hyde	BP	48/6	Sep 1		140	—
Scott	Amerada Hess	15/21a	Sep 2	450	290	—
Caister	Total	44/23a	Oct 2	—	300	3.5
Murdoch	Conoco	44/22a	Oct 2	—	320	2.2
Ann	Phillips	49/6a	Oct 2	—	130	—
Gryphon	Kerr-McGee	9/18b	Oct 14	96	—	—
NETHERLANDS						
Horizon	Unocal	P/9	Aug 17	38	—	—
F/3	NAM	F/3	Oct 1	75	210	—
F/15	Elf	F/15	Oct 19	—	400	—
DENMARK						
Regnar	DUC	5505/17	Sep 27	3	—	—

*Remaining reserves figure before redevelopment. Original reserves were 1 billion bbl of oil and 120 bcf of gas.

Sources include Wood Mackenzie Consultants Ltd. and Arthur Andersen & Co.

U.K. North Sea fields likely to be developed by 2003

Field	Operator	Block
1994-98		
Andrew	BP	16/28
Arkwright	Amoco	22/23
Armada	British Gas	22/5 & 16/29
Beinn	Marathon	16/7
Birch	Lasmo	16/12
Blenheim	ARCO	16/21
Britannia	Chevron/Conoco	16/26 &15/30
Captain	Texaco	13/22
Enoch	BP	16/13
ETAP	BP	22/24
Fife	Amerada Hess	31/26
Forth	BP	9/23
Guillemot	Shell/Esso	21/25
Janice	Phillips	30/17
Keith	Hamilton	9/8
Mallard	Shell/Esso	21/19
Pelican	Shell/Esso	211/26
Pine	Lasmo	16/12
South Piper	Elf Enterprise	15/17
Stirling	Sun	16/21
Thelma	Agip	16/17
West Brae	Marathon	16/7
1999-2003		
Braemar	Marathon	16/7
Clair	BP	206/8
Elgin	Elf Enterprise	22/30
Erskine	Texaco	23/26
Franklin	Elf Enterprise	29/5
Grant	Total	3/14
Ross	BP	13/29
Skua	Shell	22/24

Source: Grampian Regional Council

rig, which arrived to complete tie-backs to the platform.

Then a Danish tanker ran aground near the mouth of the River Tees, close by the shore approach of the CATS pipeline, through which Everest and Lomond gas would be exported.

The 97,000 dwt Freja Svea tanker snagged an anchor in 60 mph winds and 30 ft waves. Although the engine was engaged, the crew could not keep the vessel from running aground.

Amoco thought the lost anchor may have damaged the CATS pipe. Once the storm died down, a sonar survey of the inshore stretch of CATS showed no damage.

Amoco sent first gas through the CATS pipeline May 9, 1993, to Teesside, England, to power an Enron Corp. power generating station.

Lomond gas went on stream in July 1993, delayed by 2 weeks.

Piper B

From a safety viewpoint, the event of the year in the North Sea was the return to production of Piper field. Safety systems on the new Piper platform are seen as the benchmark of the new U.K. safety regime.

Elf Enterprise Caledonia Ltd. resumed production from Block 15/17 Piper field, almost 5 years after a fire and explosion damaged the original platform beyond repair and killed 167 men.

Redevelopment of Piper was tied in with development of nearby Saltire and Chanter fields, which Elf Enterprise took over, along with Piper, upon purchase of Occidental Corp.'s North Sea assets in 1991.

Elf Enterprise estimated Piper held 172 million bbl of oil and 14 bcf of gas from original reserves of 1 billion bbl of oil and 120 bcf of gas.

Wood Mackenzie estimated Piper redevelopment would cost £850 million ($1.3 billion), compared with original capital costs of £710 million ($1.1 billion) for the first platform, which began production in December 1976.

Redevelopment using Piper Bravo platform has brought advantages over the original structure, said Elf Enterprise Managing Director Michel Romieu: "Now we have new facilities, which are more competitive and can be further automated to make development costs among the best in the market."

Piper operating costs have been reduced from £5 ($7.50)/bbl with the old platform to £3 ($4.50)/bbl at present. This is £0.7 ($1.05)/bbl less than average operating costs on the U.K. shelf.

Romieu expects Piper operating costs to drop to £2.50 ($3.75)/bbl in 1994.

Elf Enterprise did not escape the effects of storms. The company had to halt production briefly from Claymore and Scapa fields off the U.K. when the Cam Sentinel standby vessel lost power and collided with the Tharos accommodation vessel on Mar. 28, 1993.

U.K. gas basin

The Caister-Murdoch system in southern North Sea

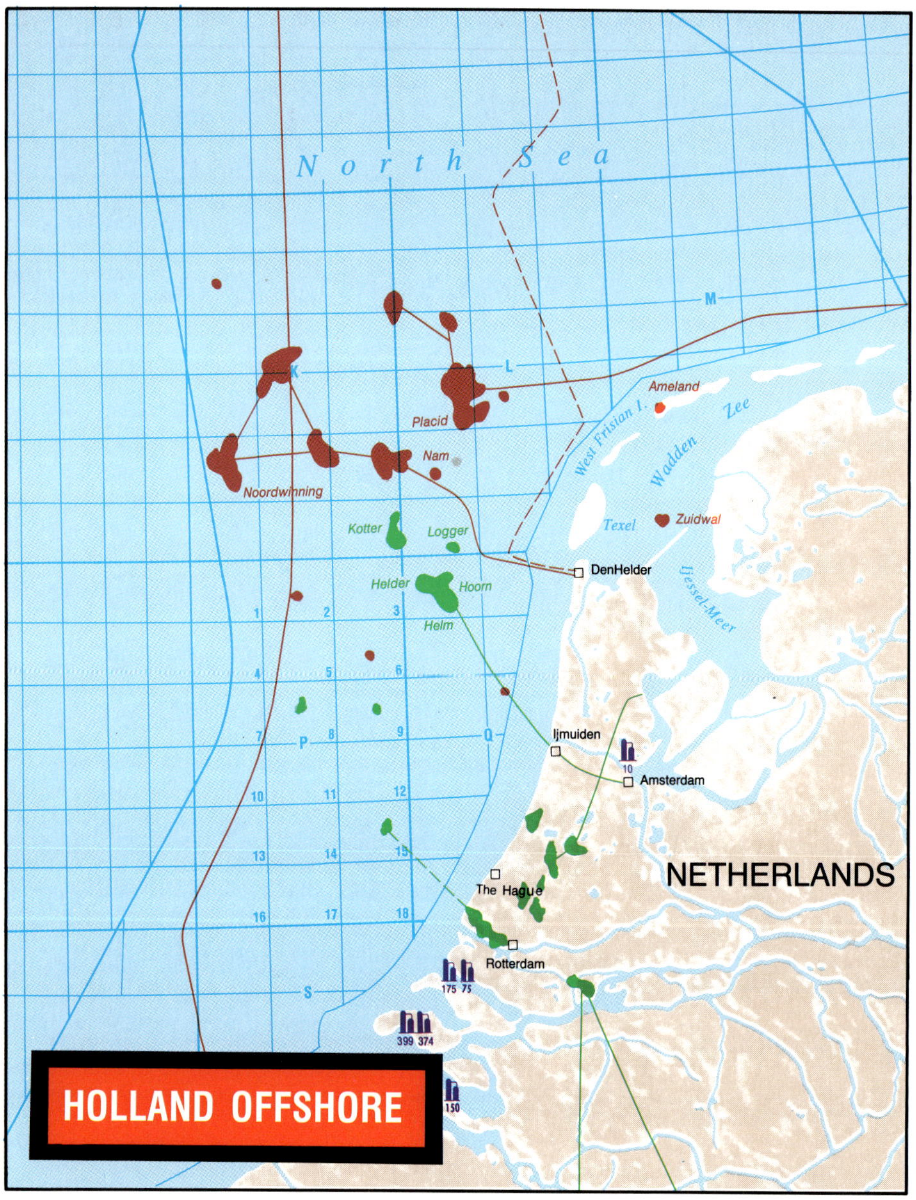

Quadrant 44 went on stream during 1993, opening a new area in the U.K.'s southern gas basin.

Fabrication of the two unmanned platforms was completed in March. Tow-out of Total's Caister platform to Block 44/23a and Conoco's Murdoch platform to Block 44/22a took place early in April 1993.

Caister and Murdoch gas production began Oct. 2, 1993, with flow expected to reach a 300 MMcfd peak in 1994. Gas moves by pipeline to Conoco's Theddlethorpe, U.K., terminal for delivery to independent gas suppliers. The line has a capacity of 750 MMcfd, which will enable it to handle production from future developments in the area.

Conoco has one discovery west of Murdoch field and was in discussions with Shell over development of Ketch and Schooner fields as satellites.

In other activity, ARCO British Ltd. placed Orwell gas field on stream Aug. 3, 1993, in U.K. Block 50/26a. Orwell is the longest subsea tie-back development in the world, with the Orwell manifold lying 34 km away from the Thames host platform.

Dutch action

Nederlandse Aardolie Maatschappij BV (NAM), Shell-Esso's Dutch joint venture, brought F/3 field on stream on Oct. 1, 1993, but saw no more than pilot production because of pipeline problems.

F/3 gas is sent ashore through the Nogat pipeline, but construction at the Nogat terminal curtailed gas production.

F/3 was developed using Netherlands' first concrete gravity base structure. This supports a drilling-/production deck and bridge-linked accommodation deck. The base's honeycomb cells can store 190,000 bbl of oil. A separate offshore loading platform is nearby.

Unocal Nederland BV began oil flow from Horizon field on Block P/9 using a 16 slot wellhead/process platform with capacity of 20,000 b/d. The platform is bridge-linked to a jack up rig providing utilities, accommodation, and helipad.

Oil is sent 30 miles by pipeline to Helder platform on Block Q/1 for further processing. From Helder the oil goes through the main trunkline from the block to a terminal near Amsterdam. Horizon was aptly named, because horizontal drilling developments since discovery of the field in 1982 made it viable. Five wells were being placed on stream to afford plateau production of 18,000 b/d early in 1994.

Amoco in August 1993 completed installation of a central processing platform in Block P/15 gas field. The platform will take gas from four satellite wellhead protector platforms and three subsea completions.

P/15 production was expected to start late in 1993 at about 470 MMcfd, with gas being sent 25 miles to shore via a new 26 in. diameter pipeline to a new gas terminal on the Hook of Holland.

Danish fields

Dansk Undergrunds Consortium, the joint venture of operator Maersk Olie & Gas AS, Texaco Denmark Inc., and Shell, placed two fields on stream off Denmark.

Regnar field is a small oil reservoir on Block 5505/17, which was brought on stream in late September 1993. Developed with a single subsea well tied back to Dan field, it is the sector's first subsea field.

Another small oil field, Valdemar on Block 5504/7, went on production in October 1993. Valdemar was developed using an unmanned wellhead platform to handle production from three wells for processing in Tyra field.

Development drilling was under way from Gorm, Skjold, and Dan field platforms. DUC has been drilling in the oil zone in Tyra field. The Danish Energy Agency forecasts that Danish oil production would total 60 million bbl in 1993 and peak at 62 million bbl in 1994, largely on account of development drilling. By 1997, however, production will fall to 55 million bbl for the year.

Other developments

Several of the batch of new U.K. oil fields will export production through BP's Forties pipeline. BP said Forties pipeline will be able to handle 1 million b/d when upgrading and expansion are complete.

BP shut down the Forties oil pipeline system in late March to tie in the Unity riser platform and 470,000 b/d of

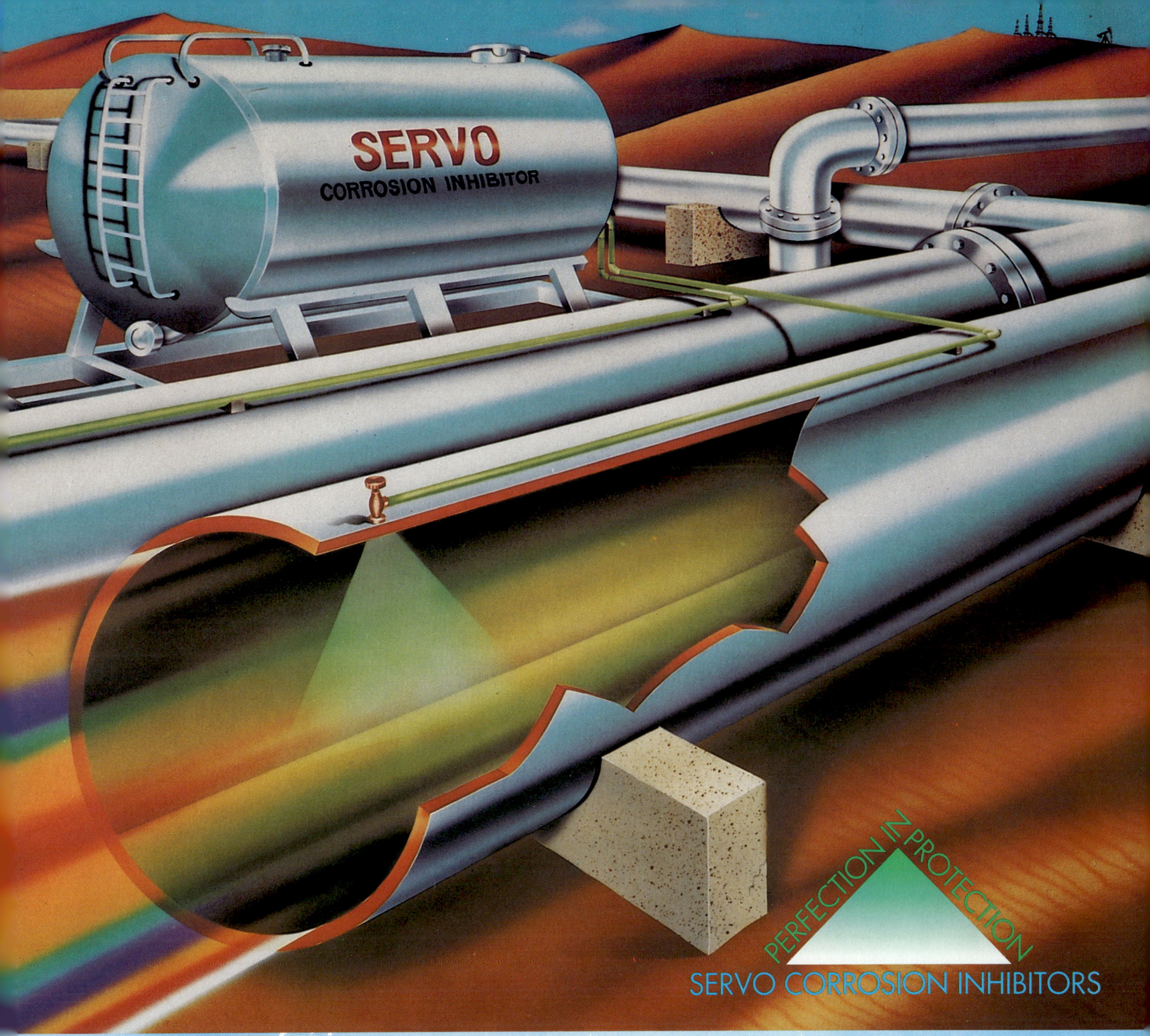
SERVO
CORROSION INHIBITOR
PERFECTION IN PROTECTION
SERVO CORROSION INHIBITORS

production at Kinneil terminal, near Edinburgh.

The 23 day shutdown interrupted the flow from eight fields—Arbroath, Balmoral, Brae, Buchan, Forties, Glamis, Miller, and Montrose—causing 9 million bbl of production to be deferred. First new throughput for Forties came from Amerada Hess's Scott field. BP expected Agip U.K. Ltd. to send first oil from Tiffany field, expected at about 70,000 b/d, through Forties in mid-November 1993, followed by Toni oil in December 1993. In June 1993, Marathon Oil U.K. Ltd. completed installation of the 28,500 metric ton East Brae platform on U.K. Block 16/3a. First oil was expected mid-December 1993, with exports via Forties rising to 115,000 b/d by 1995.

Chevron U.K. Ltd. launched a floating storage unit (FSU) for Block 16/26 Alba field at Astano SA's Ferrol shipyard in Spain. The 125,000 dwt tanker left the yard in late March 1993. Sea trials took place from then until July, after which the vessel was towed to Alba field, arriving in August. In June 1993, Chevron saw the Alba platform installed. A 2.7 km pipline bundle was towed out to Alba field in September. It was installed by Rockwater Ltd., Aberdeen, using remotely operated vehicles.

In the Norwegian sector, Elf Aquitaine Norge AS began predrilling four Froy field development wells in Block 25/5. The Treasure Saga semisubmersible rig spudded the first well early in June 1993 and was to drill until April 1994. The field is due on stream in January 1995.

Elf also completed installation of subsea equipment on three wellheads comprising Lille Frigg field in Norwegian Block 25/2a. The three high temperature/high pressure wells were tied back during the summer. Lille Frigg was due on stream January 1994.

Nelson

Shell installed topsides for Block 22/6a Nelson field in a 6 day operation during early August 1993. The installation required eight lifts, which were carried out in two phases due to interruption by bad weather.

On Aug. 8, 1993, the 9,875 metric ton deck was lifted onto the jacket by contractor Heeremac using the DB102 crane vessel in a 4 hr operation. That evening the 2,120 metric ton drilling module was lowered into place. Next day the 1,950 metric ton living quarters and drilling derrick were installed. Then the weather worsened and the installation team had to choose between more lifts or starting to weld the deck to the jacket.

"Having completed the four heaviest lifts, poorer weather was forecast, which would have hampered further lifting operations. So it was decided to begin lifting," said Simon Manktelow, installation engineer at Shell. After welding was finished Aug. 11, 1993, the utilities skid, exhaust tower, derrick top section, and boom were installed in 5 hr.

Nelson was to begin production in March 1994, adding 135,000 b/d to Forties throughput. Block 22/11 Nelson field is one of the largest U.K. developments of the 1990s. Proved and probable reserves are put at 480 million bbl of oil and 85 bcf of sales gas. Enterprise Oil plc will assume operatorship of Nelson after first oil. This marks the arrival of Enterprise as the newest production operator in the North Sea.

Discoveries

With companies in Norway and U.K. alike coming to terms with maturity of the region, BP and Statoil sprang surprises by disclosing their biggest North Sea discoveries in years.

BP announced that appraisal of a Block 204/24a discovery west of the Shetland Islands showed reserves of 250-500 million bbl of oil. This was said to be BP's biggest U.K. find in 5 years.

"The discovery has the potential to open a significant new play in the West of Shetlands province, although there is substantial further work to be done to evaluate its size," said John Browne, managing director of BP Exploration.

BP was leading a group looking to develop Clair field, 75 miles northeast of the latest discovery. And while BP unveiled its new discovery, the U.K. Department of Industry revealed that BP and Shell had been awarded five more blocks near the discovery on a 50-50 basis.

BP completed its largest 3D seismic survey to guide future West of Shetlands exploration drilling. About 2,040 sq km, equal to 10 North Sea blocks, was covered.

Onboard seismic and navigation data processing were said to have enabled early interpretation of data. Also, selection of well location was made within weeks of data being acquired. The 204/20-1 well was spudded by the Ocean Valiant rig Sept. 14, 1993, on a drilling program expected to take 35 days. The new test, 35 km northeast of the 1990 202/24-1 discovery well, is on a block owned 50-50 by BP and Shell and operated by BP. Days later, Statoil announced a 440 million bbl oil discovery in 1,300 ft of water on Block 6608/10 off mid-Norway. Later named Norne, this strike was said to be Statoil's biggest in 8 years.

Then in June 1993, Statoil said a Norne appraisal well completed in the previous February suggested 100,000-200,000 b/d of oil could be produced from relatively few wells. Another appraisal well was likely to be drilled during the winter. Statoil said it intends to work toward submission of a Norne development plan to the Norwegian government in 1994. Monohull and semisubmersible production systems were being studied, with production capacity of 120,000-150,000 b/d.

At least two wells were planned by Statoil on a 14th round block north of Norne. Kyrre Nese, Statoil's senior vice-president, exploration and production, said heavy exploration is planned in the Norne region. IPE

MANNESMANN
DEMAG

Your partner for all offshore compressors – Mannesmann Demag Process Compressors

Our strengths . . .
- more than 60 years of experience in manufacturing centrifugal compressors
- a finely graded system of standardised and proven equipment
- packaging of skids, modules and single lift units up to 400 tons
- modelling by up-to-date computer techniques and data bases
- modern powerful test beds even allowing class I testing with original GT drives
- a sophisticated quality assurance system according to ISO 9001
- a well qualified offshore team managing your inquiries and orders

. . . your benefits
- a fully developed machine design offering a safe and economical operation
- tailored compressor concepts with high reliability for all applications
- short installation, commissioning and start-up periods
- minimised weights, dimensions and therefore low investment costs
- a sound judgement of the aerodynamical and mechanical performances
- a high degree of quality and availability of all compressors and ancillaries
- a project specific planning and customised and quick order processing

Close to you – world-wide

Together with our own sales and service organisation specialists we are close to you all over the world.

For more information on our products and local representatives please contact

Mannesmann Demag Verdichter
Wolfgang-Reuter-Platz
D-47053 Duisburg
Phone (+49) 203/605-1
Fax (+49) 203/61061
Telex 855855

mannesmann technology

8752-001

Platform of NINIAN, Central North Sea

Export gas compressor with dry gas seals for Chevron NINIAN, Central North Sea, p = 141 bar

Four reinjection compressors, p = 176 bar, for the DRAUGEN platform of Norske Shell AS

Export gas compressor, p = 105 bar, with dry gas seals for the Sole Pit platform of Shell

Compressor for gas gathering on the F3-FB-1 platform of NAM, p = 121 bar

Chronology of events for 1993

JANUARY

THE MARKET: The near-month futures price for light sweet crude on the New York Mercantile Exchange (Nymex) begins the year at $19.64/bbl, up 52¢/bbl from the first week of 1992. The Nymex near-month natural gas price begins the year at $1.71/MMBTU, up 39¢/MMBTU from a year earlier.
OPEC: OPEC Pres. Alirio Parra of Venezuela tours member countries seeking production restraint to strengthen crude prices. OPEC's 1992 output averaged 24.38 million b/d, up 1.1 million b/d from 1991 and the highest level since 1980.
DRILLING: Baker Hughes reports that the number of active rigs in the U.S. averaged 721 in 1992, the lowest yearly average in more than 50 years.
PRODUCTION: BHP and Petronas Carigali plan to develop Dai Hung (Big Bear) oil field off Viet Nam. Production could reach 250,000 b/d by 2000...New financing rescues the $5.2 billion (Canadian) Hibernia oil field development project off Newfoundland following pullout of 25% interest holder Gulf Canada. Other partners, except Petro-Canada, increase interests, and Murphy Oil joins the project.
DOWNSTREAM: Shell plans a $2 billion upgrade, including a $1.5 billion hydrocracker and oil gasification plant, at its Pernis refinery near Rotterdam.
OIL SPILLS: The 89,000 dwt Braer tanker runs aground in the southern Shetland Islands, spilling much of its cargo of 607,000 bbl of crude...The 255,000 dwt Maersk Navigator and 96,550 dwt Sanko Honour, both carrying crude, collide and catch fire off northwestern Sumatra.
NATURAL GAS: AGA projects U.S. gas consumption of 20.7 tcf in 1993 vs. 20 tcf in 1990, which would be the sixth straight yearly increase.
LNG: Australia's North West Shelf LNG operation starts up its third Burrup Peninsula LNG train, raising capacity to 6 million metric tons/year from 4 million tons/year.
TRANSPORTATION: Iran starts up a 100,000 b/d crude oil pipeline from Bandar Abbas to Kerman province.
GOVERNMENT: California Coastal Commission grants a conditional permit for interim tankering of 50,000 b/d of crude from giant Point Arguello field off California...Philippines approves a plan to privatize the state oil company...Bill Clinton becomes President of the U.S. He nominates Hazel O'Leary secretary of Energy...Italy's Parliament approves a plan to privatize ENI...Trade barriers fall in the European Community.

FEBRUARY

THE MARKET: The Nymex crude price flutters around $20/bbl before and after an early-month OPEC meeting. The Nymex gas price rises to $1.81/MMBTU.
OPEC: OPEC agrees to reduce its group production quota by 1 million b/d from the first quarter to 23.582 million b/d for the second quarter and to make the new quota effective a month earlier than usual.
PRODUCTION: North Sea's Piper oil field resumes production nearly 5 years after an explosion and fire damaged the original platform and killed 167 men.
DOWNSTREAM: API says U.S. refinery distillation capacity dropped by 200,000 b/d in 1992 to 15.5 million b/d...Malaysia and Taiwan sign an agreement that clears the way for construction of a $1.26 billion, 150,000 b/d refinery in Sarawak by a joint venture of Chinese Petroleum, Petronas, and Pacific Resources.
GOVERNMENT: India proposes to convert state owned Oil & Natural Gas Commission (ONGC) into a private limited company and to reorganize it and Oil India Ltd....President Clinton proposes a 25.7¢/MMBTU energy tax with an extra 34.2¢/MMBTU on crude oil.
EXPLORATION: China opens for bidding onshore areas covering 420,000 sq km in nine provinces, including the promising Tarim basin.
TRANSPORTATION: Sonatrach starts work on the 431 km first phase of a 988 km LPG pipeline from Alrar-Rhourde-Houss in southern Algeria to Hassi R-Mel.

MARCH

THE MARKET: OPEC production cuts don't live up to the February agreement but seem to balance the market. Nymex prices hover around $20/bbl for oil and $2/MMBTU for gas.
FORMER SOVIET UNION: Elf Aquitaine receives approval for what it claims is the first oil and gas production sharing contract by a western company in Russia, covering 18,000 sq km in the Volgograd and Saratov regions.
GAS PROCESSING: Shell Canada begins processing gas at its $1 billion, 300 MMcfd Caroline project in Alberta.
GOVERNMENT: The U.K. announces tax changes for the North Sea that will lighten the burden on producing fields but reduce writeoffs from exploration...Greece announces plans to privatize the Hellenic Aspropyrgos refinery in Athens and the EKO-Chemicals complex at Thessaloniki.
LNG: Lagoven, Shell, Exxon, and Mitsubishi sign a joint venture agreement for the $4 billion Cristobal Colon project to develop gas fields off Venezuela and build a 6 million ton/year LNG plant.

APRIL

THE MARKET: Nymex crude prices stay above $20/bbl, while gas prices spurt to $2.60/MMBTU near the end of the month.
FORMER SOVIET UNION: Chevron and Kazakhstan sign agreements that enable development to begin of Tengiz and Korolev oil fields...Overseas Private Investment Corp. approves $150 million in financial aid to Polar Lights, a 50-50 venture of Conoco and Russia's Arkhangelskgeologia production association, which plans development of Ardalin complex fields in the Timan-Pichora basin...Benton Oil & Gas completes a Russian oil pipeline and expects its joint venture, Geoilbent, to start production from Western Siberia's Gubinskoye oil and gas field in the third quarter.
PRODUCTION: Shell/Esso plans to spend $1.95 billion to redevelop Brent oil field in a project that will extend life of the U.K.'s largest oil field by at least 10 years...Venezuela's cabinet approves a $1.7 billion project by Conoco and Maraven to produce and upgrade heavy oil from the Orinoco belt...Arakis Energy Corp. of Vancouver, B.C., and State Petroleum Corp. of Sudan plan to develop Heglig, Unity, and Adar/Yale oil fields in Sudan, which Chevron discovered but relinquished...Norway's production of crude and condensate averages a record 2.44 million b/d.
COGENERATION: The Teesside power station, called the world's largest gas-fired, combined-cycle cogeneration plant, reaches full operation in England. It can generate 1,875,000 kw of electricity and includes a 300 MMcfd gas processing plant.
GOVERNMENT: Pres. Clinton pledges to reduce carbon emissions from combustion of fossil fuels to 1990 levels by the year 2000.
TRANSPORTATION: Colombia's Ecopetrol says it will seek tenders for construction of a north-south trunk pipeline that will anchor a nationwide natural gas grid.
LNG: Pertamina plans a $580 million seventh train at its Bontang LNG complex in East Kalimantan.

MAY

THE MARKET: The Nymex crude price slips to $19.92/bbl, the gas price to $2.09/MMBTU.
TRANSPORTATION: Saudi Aramco finishes an expansion of its 1,200 km Petroline crude oil pipeline to Yanbu to 5 million b/d from 3.2 million b/d.
DOWNSTREAM: Petronas, British Petroleum Chemicals, and Idemitsu Petrochemicals plan to build Malaysia's first petrochemical complex, including a 320,000 ton/year ethylene plant and 200,000 polyethylene plant, at Kertih...Marubeni of Japan agrees with Uzbekistan's state oil company to build a $1 billion,

100,000 b/d refinery.
FORMER SOVIET UNION: Tengizchevroil, the joint venture of Chevron and Kazakhstan, ships its first crude oil to market...Purneftegas Oil Producing Association of western Siberia wins approval for a $174 million loan from European Bank for Reconstruction and Development.
GOVERNMENT: Canada's House of Commons approves the North American Free Trade Agreement, which aims at lowering barriers to trade among Canada, Mexico, and the U.S....Argentina begins selling stock in Yacimientos Petroliferos Fiscales.

JUNE
THE MARKET: The Nymex crude price drifts to $18.73/bbl at month's end. The Nymex gas price stays at about $2.10/MMBTU.
OPEC: Except for Kuwait, OPEC members agree to extend the second quarter group quota of 23.58 million b/d into the third quarter. Kuwait's assertion of a 2.16 million b/d quota for itself makes the group quota 24.14 million b/d.
LNG: Oman LNG LLC Co. forms to build and operate a 5 million ton/year liquefaction plant in Muscat.
GOVERNMENT: U.S. President Clinton abandons his proposal for a BTU tax with an extra levy on oil.
DOWNSTREAM: Saudi Arabia decides to merge Samarec, state-owned operator of three refineries, into Saudi Aramco, state-owned oil producer and operator of one refinery, making Aramco the world's largest integrated oil company.

JULY
THE MARKET: The crude price slide becomes a plunge on news that the UN may soon lift its embargo on export of Iraqi oil. OPEC schedules, then postpones, an emergency meeting. The Nymex crude price sinks below $18/bbl. The gas futures price stays above $2/MMBTU.
GOVERNMENT: U.S. Congress passes a budget bill that includes a 4.3¢/gal tax on transportation fuels.
FORMER SOVIET UNION: Export-Import Bank agrees to provide financing for a $2 billion purchase by Russia of oil, gas, and petrochemical equipment and services from the U.S....International Finance Corp. agrees to lend $10 million to Russia's Vasyugan Services Enterprises for a plan to increase oil and gas production in western Siberia's Tomsk region...Barclay's Bank of London provides $1.1 billion in financing to renovate the Novokiybyeshev refinery.
DOWNSTREAM: A joint venture of Shell Oil and Pemex starts work on a $1 billion, 2 year upgrade of its Deer Park, Tex., refinery and petrochemical complex, which will be partly supplied by heavy Maya crude from Mexico.

AUGUST
THE MARKET: The Nymex crude price languishes below $18/bbl most of the month. The gas price holds at $2.20/MMBTU.
OPEC: OPEC's ministerial monitoring subcommittee recommends that the current production ceiling of 23.582 million b/d be continued into the fourth quarter. OPEC's July output is estimated to have been 24.64 million b/d.
GOVERNMENT: Pdvsa completes its second round of bidding for contracts to produce oil and gas from marginal fields in Venezuela...Britain's Monopolies & Mergers Commission recommends that the transportation and merchant functions of British Gas be split.
PRODUCTION: Chevron makes its first lifting of Point Arguello crude at the Gaviota marine terminal in California following 4 years of controversy over tankering.
OIL SPILLS: Marine Spill Response Corp., a private spill response contractor for 77 oil shippers organized after the 1989 Exxon Valdez spill, begins operations in the U.S.

SEPTEMBER
THE MARKET: In a jittery market, the Nymex crude price drops below $17/bbl then jumps back above $18/bbl as traders worry about OPEC resolve, Russian politics, and prospects for resumption of Iraqi exports. The gas price slips below $2.20/MMBTU.
OPEC: OPEC sets a group quota of 24.5 million b/d for the fourth quarter of 1993 and first quarter of 1994 and pursuades Kuwait to accept a quota of 2 million b/d.
GOVERNMENT: India, despite a 1 day antiprivatization strike, decides to allow foreign companies to conduct speculative seismic surveys...Pemex, having dismissed 120,000 employees—half its workforce—since 1989, discloses plans to further cut staff to 50,000....U.K. gas regulator Ofgas says British Gas can't raise rates in Britain to offset exploration losses abroad.
DOWNSTREAM: National Petroleum Council says U.S. refiners will have to make capital investments totaling $36.5 billion—$5.5 billion more than the industry's book value—to meet environmental requirements during 1991-2000...Huntsman Corp., a 50-50 joint venture of Huntsman Financial Corp. of the U.S. and Consolidated Press Holdings of Australia, agrees to buy Texaco Chemical Co. for $1.06 billion...Kuwait National Petroleum Co. places its Mina Abdullah refinery, damaged during the Iraqi invasion of 1990, back on stream at a diminished capacity of 230,000 b/d.
EXPLORATION: Phillips and partners announce a discovery on Ship Shoal South Addition Block 349 off Louisiana on a structure imaged beneath a thick layer of salt with 3D seismic technology.
PRODUCTION: Canadian Oxy and partners open Yemen's Masila project, which includes 10 oil fields, production capacity of 120,000 b/d, 138 km of pipelines, and port facilities.

OCTOBER
THE MARKET: The Nymex crude price ends the month at $17.40/bbl, while the gas price rises to $2.35/MMBTU.
GOVERNMENT: FERC meets its goal of restructuring gas pipeline rates under Order 636 before the heating season, having approved restructuring plans of 76 transmission pipelines...Jean Chretien is elected prime minister of Canada as the Canadian Liberal party sweeps elections in the House of Commons and ousts the Progressive Conservative government.
LNG: Abu Dhabi Gas Liquefaction Co. completes an agreement to double to 5 million metric tons/year exports of LNG to Japan's Tokyo Electric Power Co. when its Das Island liquefaction plant expansion starts operations in 1994.
PRODUCTION: North Sea production hits a record-high 4.94 million b/d average for the month.

NOVEMBER
THE MARKET: Oil prices fall to their lowest levels since before the 1990 Persian Gulf crisis as demand lags expectations and new production comes on stream. OPEC holds an emergency meeting but decides not to call for production cuts below those set in September. The Nymex crude price drops more than $1/bbl on Nov. 29 to close at $15.31/bbl, its lowest level since June 10, 1990. The gas price stays above $2.30/MMBTU.
FORMER SOVIET UNION: Eight companies agree in principle with Azerbaijan to develop Chirag and Azeri fields in the Caspian Sea under a production sharing contract worth $7 billion...Ukraine awards its first offshore license, granting Crimean Petroleum Co. exploration and production rights to a prospect in the Black Sea.
GOVERNMENT: The U.S. House and Senate approve the North American Free Trade Agreement among the U.S., Canada, and Mexico.
DOWNSTREAM: Kuwait National Petroleum restarts its Shuaiba refinery, the last of its three war-damaged refineries to resume operations.

DECEMBER
THE MARKET: The plunge in oil prices continues. The Nymex price drops to $14.57/bbl Dec. 3, its lowest level since Nov. 23, 1988. Brent crude for January delivery also hits a 5 year low at $13.27/bbl Dec. 7. The Nymex gas price drops below $2/MMBTU.
GOVERNMENT: IPAA asks the Clinton administration for relief from low oil prices...Saskatchewan prepares to end at yearend a 10 year royalty holiday for new oil and gas wells...EPA proposes to require that 30% of the oxygenates used in U.S. reformulated gasoline come from renewable sources, namely ethanol.
PRODUCTION: Exxon starts production from its controversial, $2 billion, expanded Santa Ynez Unit development off California.

Schematics show N. Sea production facilities

A large number of small satellite fields tied into existing production facilities have helped create a complex production network in the central North Sea.

Although this network lowers development costs, development of discoveries often is delayed until production through existing facilities falls far enough to make room for additional production streams.

As 1994 approached, more than 50 discoveries were thought to have potential of being phased into the production network by 2000.

The accompanying schematic diagrams, covering the region between Hod field near Ekofisk in the south and Magnus field in the north, are not to scale and do not provide facility orientation on a platform. They do, however, show main components of production facilities and an approximate orientation between fields.

A companion map for each of the seven areas locates the fields on the U.K. and Norway grid systems. Discoveries without published development plans as of September 1993 are not included on the map.

To keep the diagrams simple, all flow and control lines between facilities are not included. If several flow lines connect the same facilities, only one line is shown.

Northern North Sea pipelines and areas diagramed

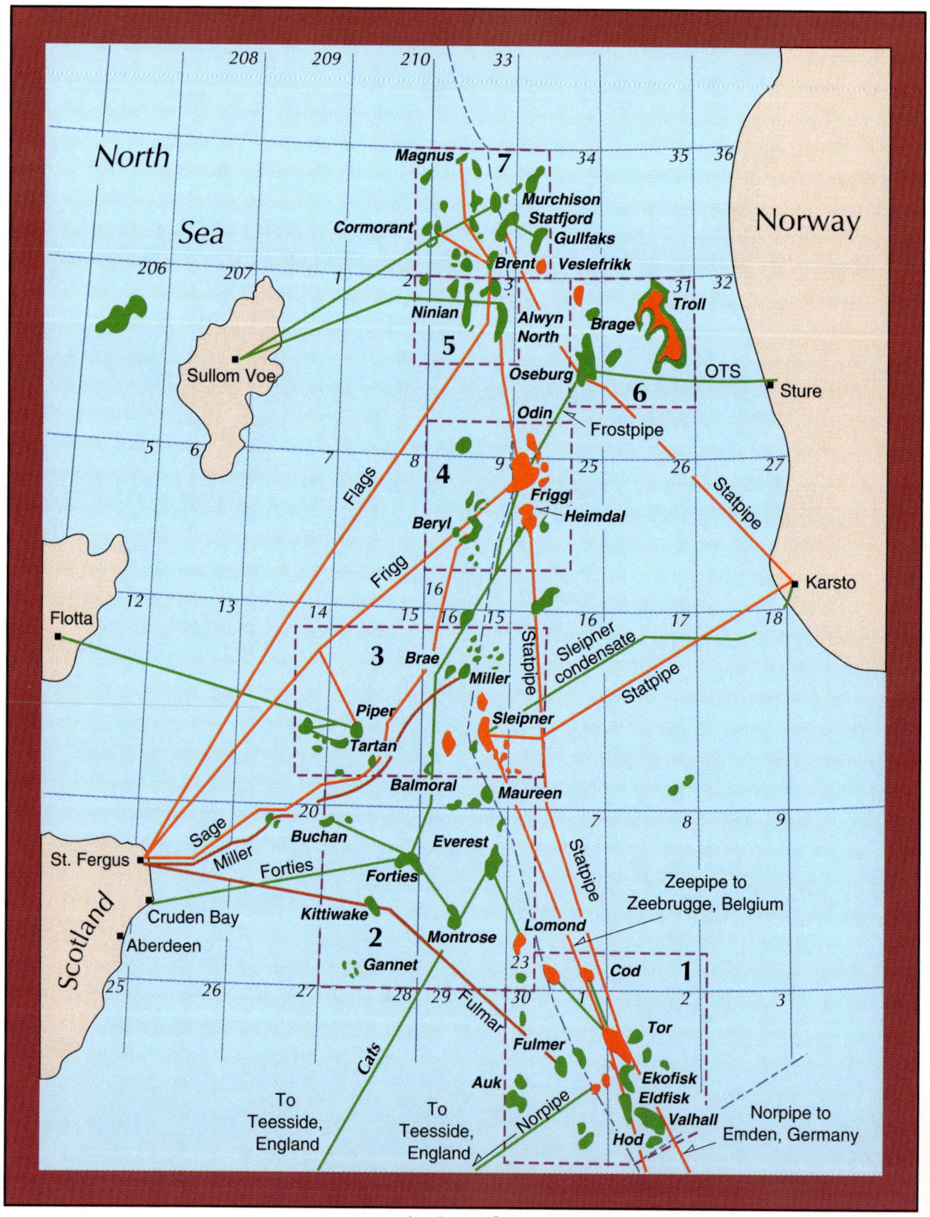

New pipelines take shape

A number of new gas and oil pipelines have been installed or were being installed to accommodate new fields off the U.K. and Norway. They join an increasingly complex pipeline system moving oil and gas to the U.K. and continental Europe.

The 36 in. Brent/South Cormorant and 36 in. Ninian pipelines feed Sullom Voe terminal in Scotland. Each line has a capacity of 1 million b/d.

BP Exploration Operating Co. Ltd. operates the Ninian lin, while Shell U.K. Exploration & Production Ltd. is operator for the Brent/South Cormorant line. Both lines were commissioned in 1978.

Oil from the Piper area moves through a 30 in. pipeline to Flotta in the Orkney Isles. Elf Enterprises Caledonia Ltd. operates to 500,000 b/d line, commissioned in 1976.

Area 1 (vicinity of Ekofisk)

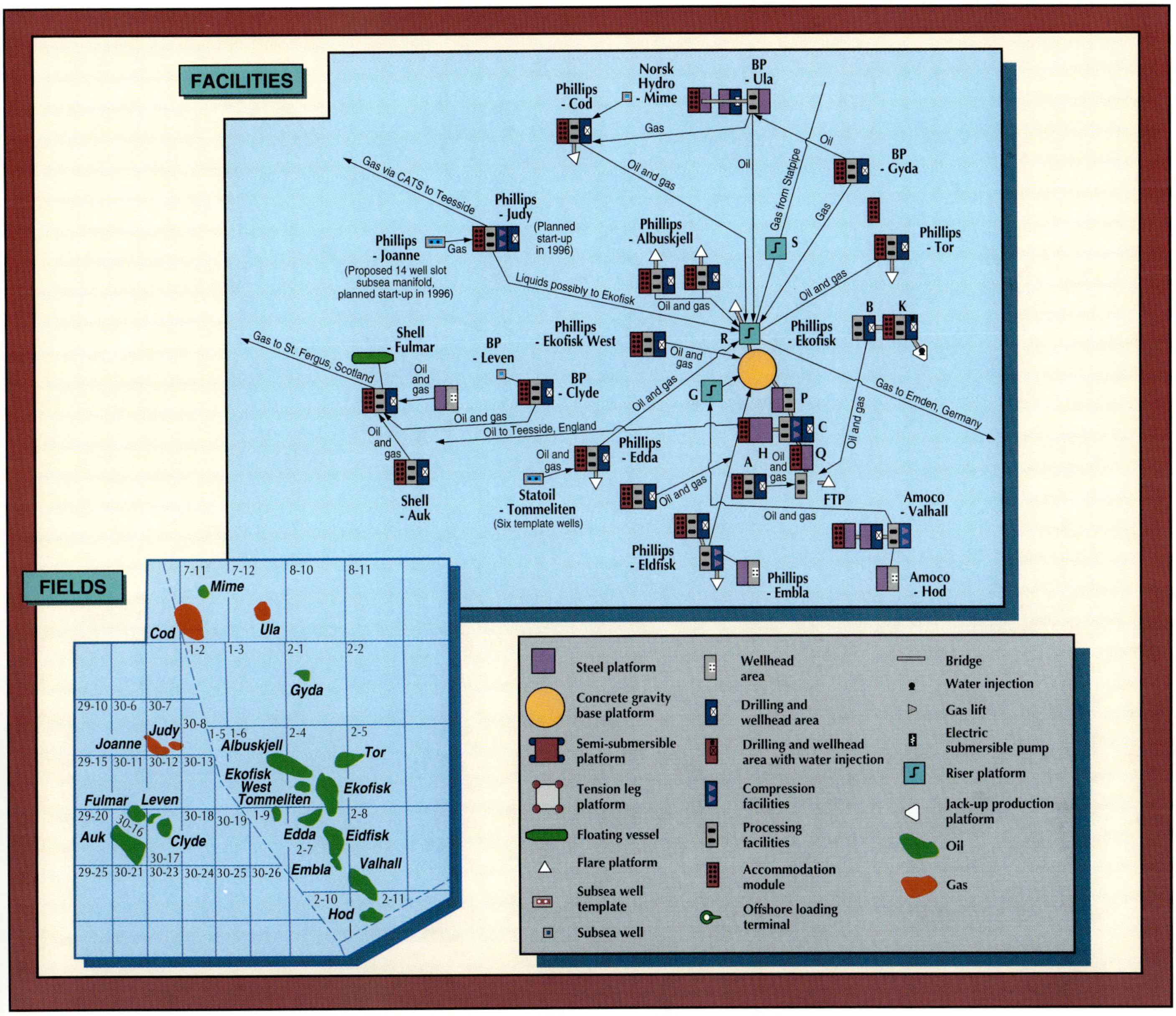

In 1992, a 36 in. line replaced the 32 in. Forties pipeline, doubling capacity to about 1 million b/d. BP is the line operator. Tied into the Forties sytem are lines from Brae, Buchan, and Montrose field areas.

The remaining oil pipeline to the U.K. is Norpipe, operated by Phillips Petroleum Co. Norway. The 34 in. line, connecting the Ekofisk field complex with Teesside, U.K., has a capacity of about 700,000 b/d. The line was commissioned in 1975.

The 28 in. Oseberg Transportation System (OTS) transports oil from Oseberg field to Sture, Norway. Norsk Hydro Produksjon A.S. operates the 700,000 b/d capacity line. First oil arrived in Sture on Dec. 20, 1988.

Frostpipe, operated by Elf Petroleum Norge AS, is an 18 in. line designed to move liquids from Frigg to Oseberg field. Capacity is about 100,000 b/d.

Another pipeline moving liquids to Norway is the Sleipner condensate line. Capacity of the Den norske stats oljeselskap A.S. operated 20 in. line is about 200,000 b/d.

The St. Fergus area is site of the main gas pipeline terminal for the U.K. North Sea. The five U.K. gas pipeline systems are Flags (Far North Liquids and Associated Gas System), Frigg, SAGE (Scottish Area Gas Evacuation), Miller, and Fulmar.

The 20 in., Shell operated, Flags gas and NGL line's first deliveries took place in 1982 from Brent field. Inlet capacity is about 950 MMcfd.

Total Oil Marine U.K. operates two 32 in. lines from Frigg to St. Fergus. The lines were placed in service in August 1978. Capacity is about 1.2 bcfd.

Deliveries through the SAGE line started in October 1992. This Mobil North Sea Ltd. operated line moves rich gas from Beryl and other fields. Initial deliveries were about 265 MMcfd. Beryl gas was previously reinjected, but the field is in a blowdown stage.

The BP operated, 30 in. Miller pipeline line was installed in 1992.

The Fulmar line, operated by Shell, started up in 1986. The 20 in. system transports about 250 MMcfd.

Amoco U.K. Exploration Co.'s CATS pipeline went on stream in 1993. Capacity is about 1.4 bcfd through the 36 in. gas line from the Everest/Lomond area to Teesside.

The Statpipe system ships gas from fields in the northern North Sea to Ekofisk, where the gas enters the Norpipe line to Emden, Germany. Statoil is operator of

Area 2 (vicinity of forties)

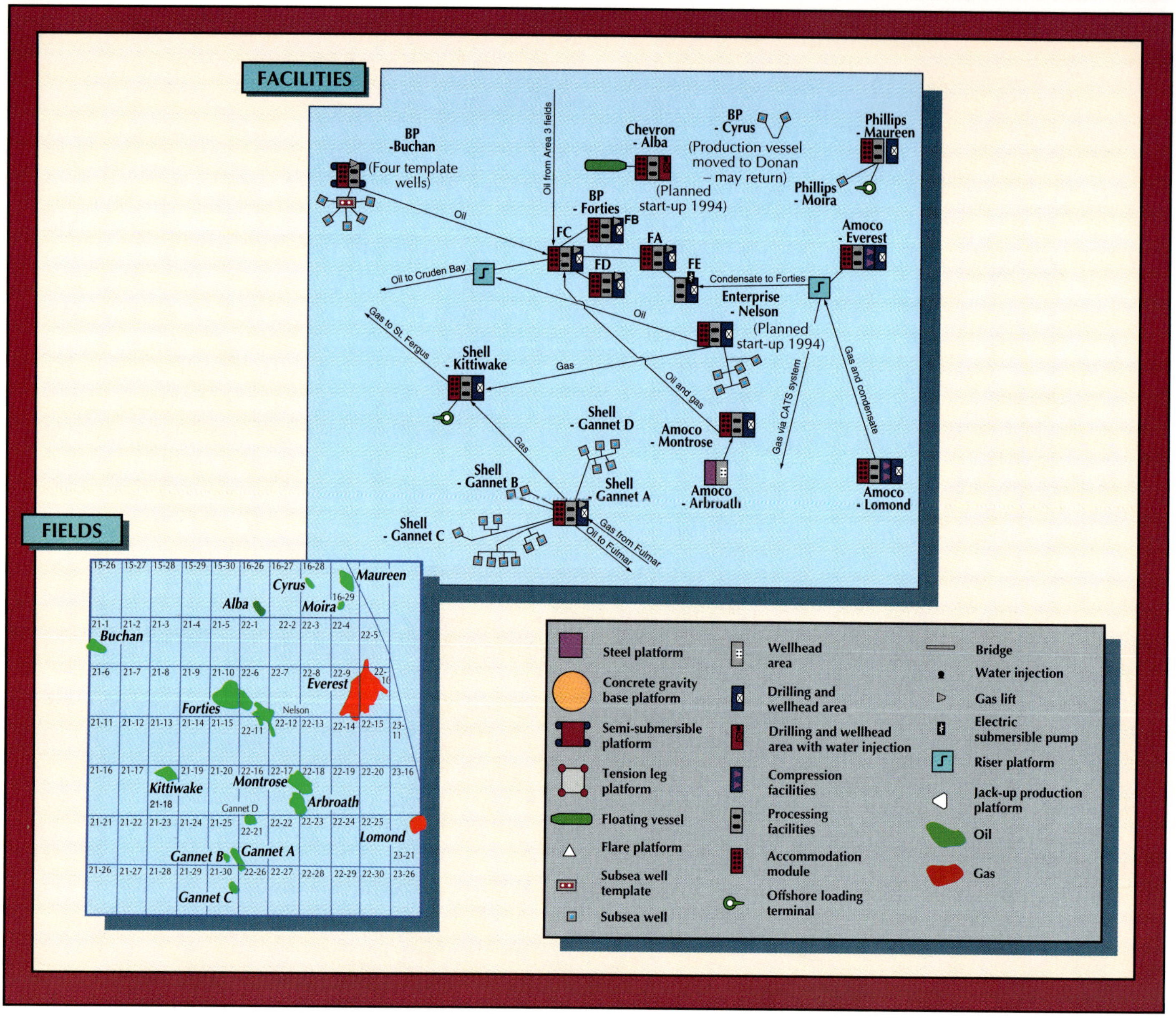

Statpipe.

The 28 in. and 36 in. system includes two riser platforms and a terminal at Kårstø, Norway, where gas liquids are separated. Operations started in 1985. The dry gas section between the riser platform in Block 16/11S and Ekofisk field has a capacity of about 1.7 bcfd

Statoil's Zeepipe, also completed in 1993, connects Sleipner field with Zeebrugge, Belgium. One leg of the line will connect Sleipner with the Statpipe riser platform in Block 16/11S. Another leg of Zeepipe (not shown on map) will connect Sleipner with Troll field via Kollnes, Norway. Completion of this leg is expected in 1996. Zeepipe has a capacity of about 1.2 bcfd.

Europipe (not shown on map) is a 40 in. line under construction as 1994 approached. The line is being laid from the riser platform on Block 16/11S to Germany, largely paralleling the Statpipe/Norpipe line. Completion date in 1995 for the 1.2 bcfd capacity line will coincide with sales commitments of Troll area gas. Statoil is the operator of Europipe.

The 36 in. Norpipe line from Ekofisk to Emden started service in September 1977. Phillips operates the 2 bcfd capacity line.

Area 1

Release in late 1992 of the Deep Sea Pioneer floating production vessel, a converted semisubmersible drilling rig, eliminated from the schematic Hamilton Oil Co. Ltd's Argyll, Duncan, and Innes fields in U.K. Block 30/4. The fields were abandoned in 1993.

Argyll first produced June 11, 1975. It was the North Sea's first field to use a converted semisubmersible for production.

Ekofisk area development includes seven fields. Not shown are four booster platforms on the pipeline to Teesside and Emden. Phillips Petroleum Co. Norway operates Ekofisk, a 1969 discovery and the first oil field found in the Norwegian North Sea.

In 1971, Ekofisk became the first North Sea field to produce commercial oil. The Gulftide jack up rig provided initial production facilities for 18 months. Use of concrete construction for a platform, a huge oil storage tank, is another first at Ekofisk.

At yearend 1993 field facilities were being modified to increase water and gas injection capacity. Installation of a Converted jack up rig is planned for added water

Area 3 (vicinity of Beryl and Frigg)

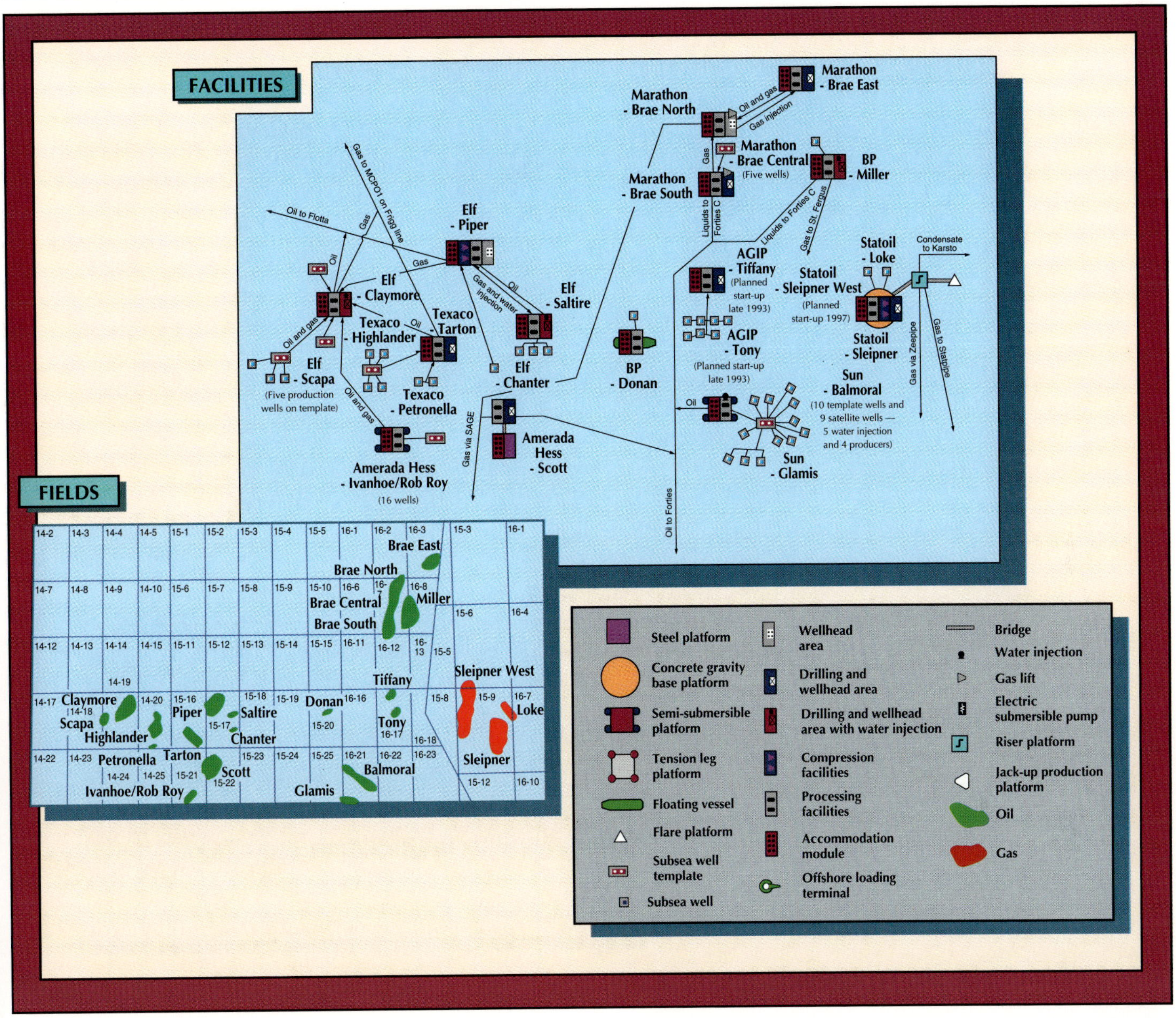

injection capacity. Also, a new production platform was planned to increase safety measures.

Estimates are that Ekofisk will still be producing after the production license expires in 2011. Phillips plans to submit a proposal that is estimated to cost $3-4 billion to change out existing facilities and allow production beyond the current license term.

Auk field, operated by Shell, was the second oil field to start production from a fixed platform off the the U.K. Placed on stream in 1975, the field was near depletion in 1993, although facility upgrades and changing from tanker loading to a pipeline to Fulmar were designed to extend the field's life.

In Shell's Fulmar field, a submerged turret loading (STL) system includes a purpose-built 95,000 dwt tanker and will allow all-weather loading. Installed in 1993, the STL system replaces a conventional floating storage vessel.

Area 2

In Area 2, Forties field facilities are expected to last at least until 2010 or about 35 years from the initial platform installation in June 1974.

Production started in September 1975 from the BP operated field. About 25% of the estimated 2.5 billion bbl of recoverable oil remained to be produced at yearend 1993, mostly by artificial lift with either gas lift or electric submersible pumps.

Elsewhere in the area, the oil price collapse in 1986 forced Shell to postpone development of Kittiwake and Gannet fields until a lower cost design was complete.

Kittiwake was developed as a stand alone facility using a slim-line jacket. Production started in 1990. Facility costs were reduced by installing the refurbished offshore loading terminal from Auk field and reusing accommodation modules originally intended for the Auk platform.

A revised development plan for Gannet field included minimum manning and facilities, single lift installation, and more subsea technology. Production started in October 1992.

The first phase of Chevron U.K. Ltd.'s Alba field development included a turret moored floating storage vessel. Second phase development in the southern part of the reservoir will either have a second platform or use subsea technology.

Nelson field is one of the largest oil fields among recent

Area 4 (vicinity of Piper and Brea)

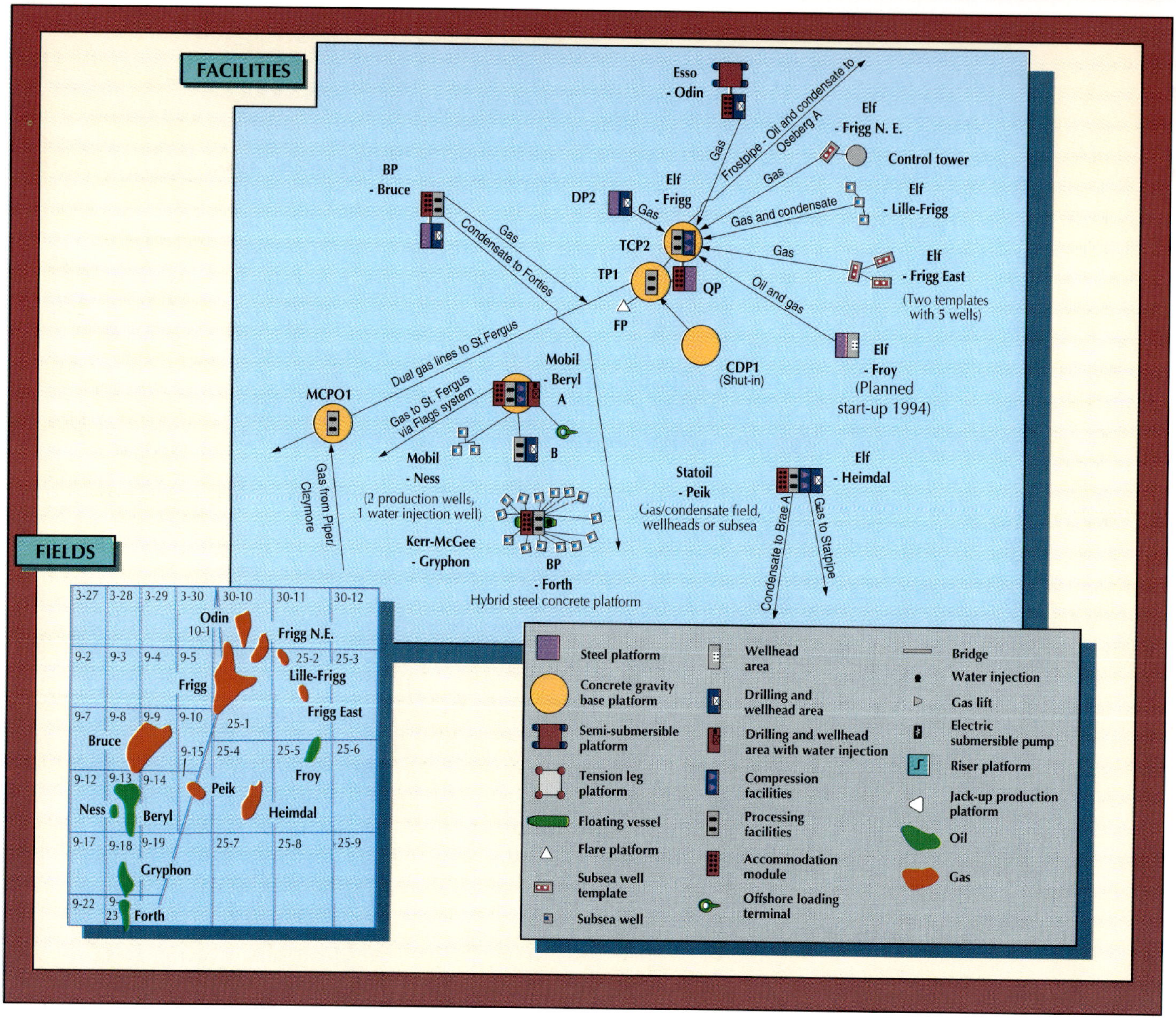

North Sea discoveries. Proved and probable reserves are about 480 million bbl of oil and 85 bcf of gas. Production is expected to start early in 1994. Shell was operator through the field's development stage, while Enterprise Oil plc will be operator during the production stage.

Two horizontal wells were on production in Cyrus field, operated by BP. Production started in April 1990 via the Seillean production vessel, a 42,000 dwt tanker. In 1992, the vessel moved to test the Donan subsea well in Block 15/20.

Amoco U.K. Exploration Inc.'s operated Arbroath field produces through an unmanned facility tied into Montrose field. Arbroath, discovered in 1969, went on production in 1990.

Everest and Lomond gas fields are covered by a contract that commits about 300 MMcfd to a Teesside power generating plant via the CATS line.

Area 3

Piper B platform replaces the original Piper field platform that was destroyed by an explosion and fire in 1988. Oil production from Piper B started in February 1993. The satellite fields Saltire and Chanter also went on stream early in 1993.

Statoil's East Sleipner platform was installed in June 1993. It replaces a concrete gravity base that sank in August 1991 during trials. Two subsea wells went on stream in August 1993.

East Brae field production is forecast to start in December 1993. North Brae gas will be injected into East Brae, while plans are to start sales of East Brae gas to the SAGE system in October 1994. Marathon Oil U.K. operates all four Brae fields.

BP's Miller field started producing in June 1992. Water injection was also commenced.

When Balmoral field started producing in 1986, some of the firsts for the Sun Oil Britain Ltd.'s project included:

- A purpose built semisubmersible production vessel
- A combination of working drilling derrick and flare tower.
- Deployment of a glycol dehydration system on a floating vessel.
- Large scale application of remotely controlled subsea chokes for production and gas lift.
- First field in the North Sea to gravel pack highly deviated (more than 60°) wells.

Area 5 (vicinity of Ninian)

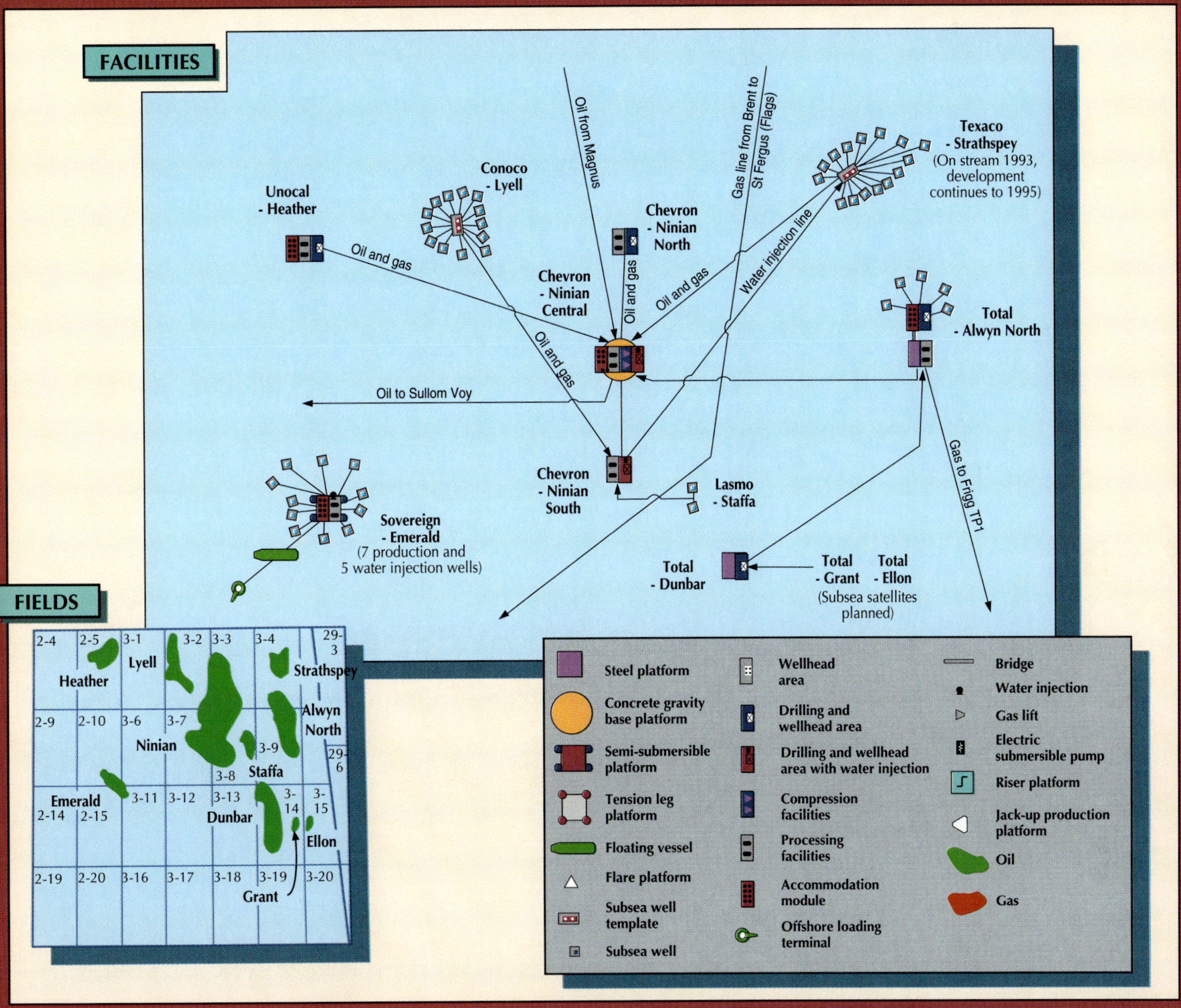

Amerada Hess Ltd.'s operated Scott field, among the larger oil to go on stream in the U.K. North Sea in the last decade, is to reach a plateau of about 180,000 b/d. Estimated reserves are 450 million bbl of oil and 287 bcf of gas.

Area 4

As gas production declined in Elf Petroleum Norge AS's operated Frigg field, a number of satellite fields were placed on stream. Frigg started production in 1977 and since 1987 has had a production decline. Production is likely to end in 1997.

Esso Norge AS's Odin is another field that is likely to cease production by 1997. At the end of 1992, about 90% of Odin's estimated 1 tcf of recoverable gas had been produced. A floating utility vessel supplies the simple production platform with utilities.

Mobil Oil North Sea Ltd. started oil production from Beryl field in 1976. To increase oil recovery, gas was reinjected. With completion of the SAGE pipeline to St. Fergus, gas sales began in October 1992.

The Tentech 850C floating, production, storage, and offloading vessel Gryphon Alpha serves Kerr-McGee's Gryphon field. Production started in October 1993.

BP's Bruce field began producing in August 1993. Bruce is the largest gas field in the U.K. Reserves are estimated at 2.6 tcf and 220 million bbl of liquids.

Area 5

Ninian platform, operated by Chevron, ties in a number of smaller fields. Ninian started producing in 1978. Its facilities have been upgraded to conform to North Sea safety requirements and allow other fields to use the production facilities.

The first three fields to take advantage of this availability are three subsea developments: Lasmo's Staffa, Conoco's Lyell and Texaco North Sea U.K.'s Strathspey.

Staffa went on production in 1992, Lyell in April 1993. Production from Strathspey was scheduled to commence in the last quarter of 1993.

The subsea wells in Sovereign's Emerald field are tied back to the Alibaba converted semisubmersible rig, renamed Emerald Producer. Oil is exported by shuttle tankers from a floating storage vessel. Production started in late 1990.

Area 6 (vicinity of Oseberg and Troll)

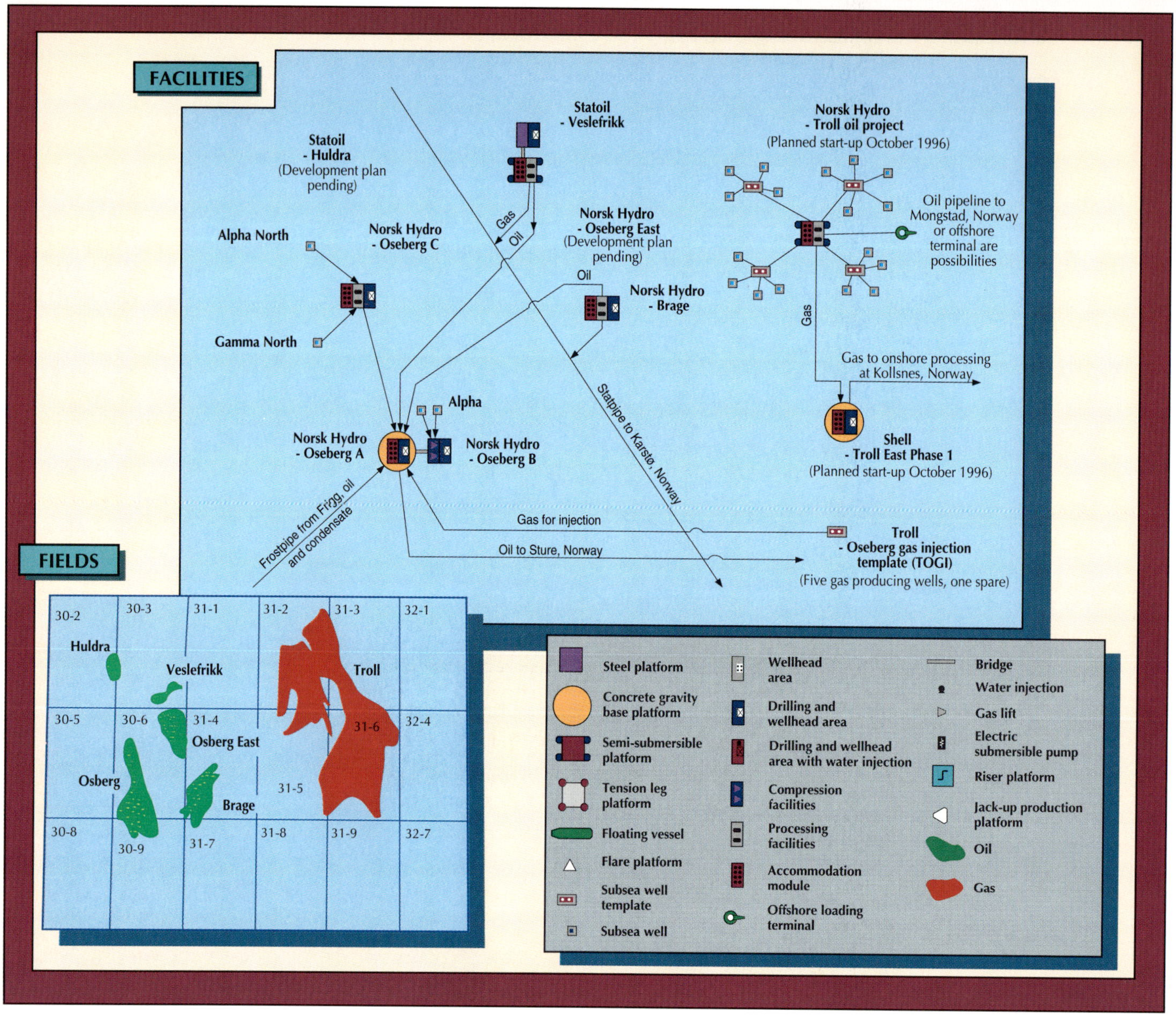

Area 6

To improve oil recovery, gas from wells on a subsea template in Troll field is injected in Oseberg field, operated by Norsk Hydro. Production from the first phase of the project, Platform A, began in 1988. The second phase, Platform C, was on stream in 1991. Estimated 1993 production was 450,000 b/d of oil.

Statoil's operated Veslefrikk field has a wellhead platform bridge-linked to processing and accommodation units on the converted West Vision semisubmersible. Production, which started in 1990, was about 65,000 b/d of oil in 1993.

Brage field went on production in 1993 and is expected to reach a rate of about 85,000 b/d. Norsk Hydro operates the field. Brage associated gas will act as an extra guarantee for Troll gas contract volumes. The gas will be shipped to Ekofisk for storage.

Troll field is being developed in several phases. Norsk Hydro in 1993 was developing Troll's oil leg in the west part of the field with subsea completions. Shell is operator for the development of Troll East gas that will have a Condeep SP4 platform. Both projects are expected to start production in 1996. A third phase, West Troll gas, may commence production in 2005.

Estimated recoverable reserves are about 46 tcf and 400 million bbl of oil. Production during Phase 1 is expected to be 20 bcfd.

Area 7

Area 7 contains most of the large gravity base platforms in the North Sea. These are in Gullfaks, Statfjord, Brent, Dunlin, and South Cormorant fields. Two of the North Sea's tension leg platforms also are in this area: Snorre and Hutton.

Although Statoil's operated Statfjord field has been producing since November 1979, production has increased because of extended reach and horizontal drilling. The upgrade program has added about 500 million bbl to recoverable reserves and increased production to more than 700,000 b/d of oil.

Production from two subsea satellites fields, North and East Statfjord, is planned for 1994.

Production from Statoil's Gullfaks field started in 1987. Horizontal wells from Gullfaks and subsea developments such as Tordis will be used to develop a number of small prospects in the area.

Area 7 (vicinity of Brent and Statfjord)

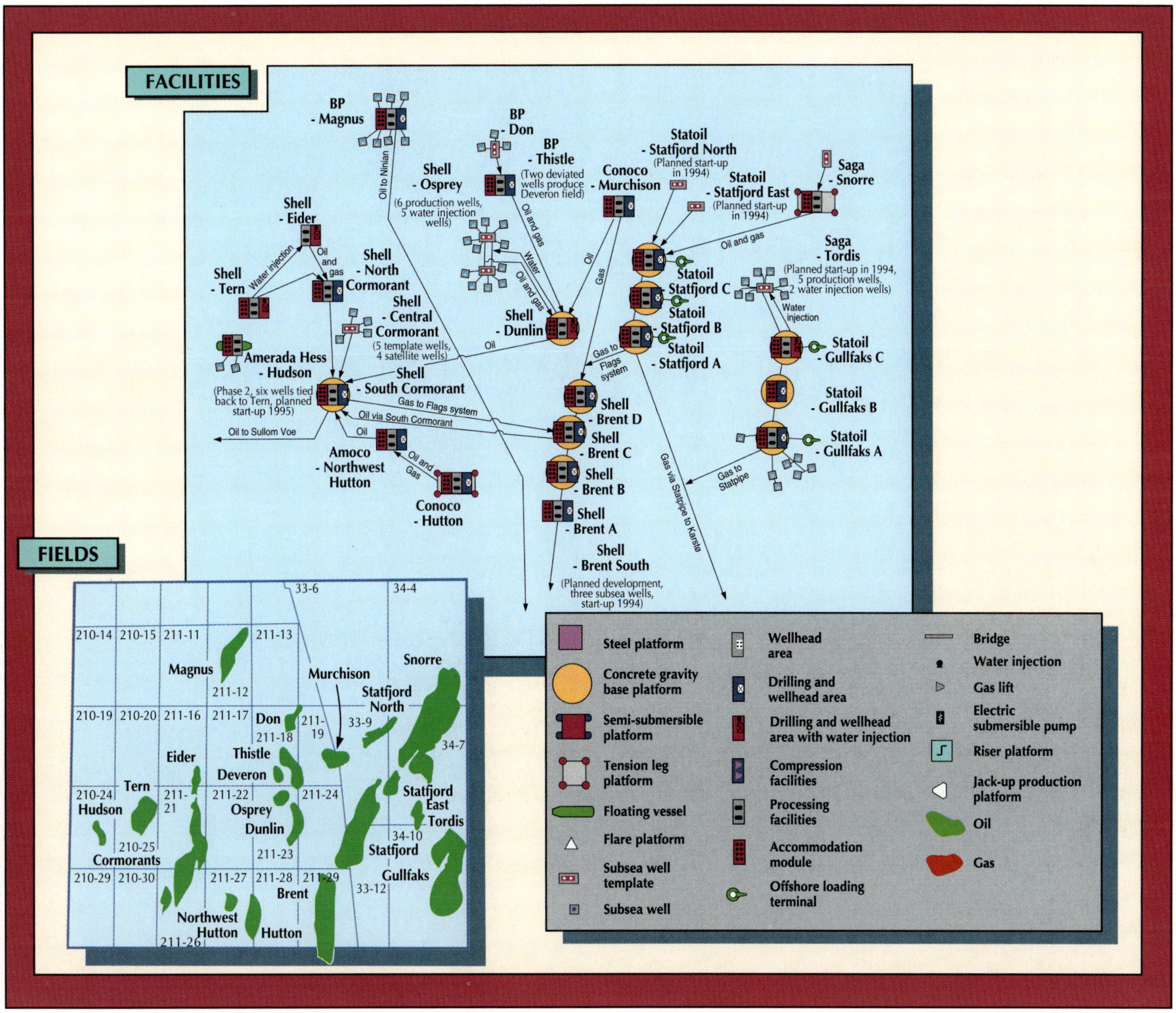

The price of oil from Shell's Brent field is used as the base North Sea crude oil price, although Brent is no longer the main oil producing field in the North Sea. Brent production in 1993 was less than half of the field's peak flow of 416,000 b/d in 1985-86.

To recover more oil and gas, gas processing modules on three of Brent's platforms will be replaced with ones that operate at lower pressures. The $1.95 billion refurbishment during 5 years will make gas the more important production stream from Brent and extend the field's life another 10 years to about 2008.

Shell's Dunlin is another old platform—production started in 1978—that is used to produce satellite fields. Shell's Osprey subsea development has a processing module on the Dunlin platform, and water for injection is supplied from Dunlin field. Osprey production started in 1991.

BP's Don field was tied into production facilities in Thistle field in 1989. Production from Thistle field, also operated by BP, started in February 1978 and has declined to about 30,000 b/d of oil from a peak of 139,000 in 1982.

Magnus field, also operated by BP, has been on production since August 1983.

Shell's Eider platform is an unmanned unit that began producing in November 1988. Water injection began in 1989.

Production from Tern field, also operated by Shell, started in June 1989 with gas lift starting in September 1989 and water injection in February 1990.

Shell's South Cormorant platform has been producing since 1979. North Cormorant was brought on in 1982 and Central Cormorant in 1983. Water is injected in all three fields.

Conoco's Hutton was the North Sea's first tension leg platform. Production from it started in August 1984. Pipelines from Hutton are tied into Amoco's Northwest Hutton platform that started producing in April 1983.

Hudson field, operated by Amerada Hess, went on stream in July 1993 using the Petrojarl I production vessel. In 1994, subsea wells will be tied back to the Tern platform.

Production from Saga Petroleum AS's Snorre tension leg platform began in August 1992. Estimated reserves are about 750 million bbl of oil and 250 bcf of gas. Peak production in 1994 is expected to be 190,000 b/d. IPE

E&D technology integration increasing

Integration of exploration and development technologies is penetrating oil company culture more deeply as cost cutting and staff reduction requirements grow more urgent in the 1990s.

Few major new oil and gas provinces have been found since major discoveries in the 1960s and 1970s, and most productive basins are reaching or approaching maturity.

Nevertheless, world oil reserves equal about 45 years of supply at current demand levels, compared with 30 years in 1973. This increase has resulted in part from:

- Technological breakthroughs in the understanding of reservoir complexity and behavior.
- A more integrated approach to management of hydrocarbon assets in existing fields.
- Greater emphasis on energy conservation.

As Shell Oil Co. points out, more than three fourths of annual additions to world oil reserves are obtained by optimizing recovery from existing fields, while the rest comes from discoveries.

Increasingly, says Shell International Petroleum Mij.'s Michael J. Pink, "The explorer, petroleum engineer, and production engineer must evaluate together the options for exploration drilling to ensure that in case of success, the discovery has the potential of being developed economically."

This article treats the integration of exploration and development technologies in settings that mainly involve research and development, geophysical, land, exploration, drilling, engineering, development, and production programs in various parts of the world (Fig. 1).

It also spells out the role workstations and other computers have played in helping professionals collect and manipulate the vast data volumes they need to conduct integrated activities.

Use of integration in offshore projects

Several examples surfaced during 1993 of integration—and the need for it—in offshore projects.

Shell Offshore Co. made big internal and external changes to meet Gulf of Mexico operating challenges while cutting costs, Robert L. Howard, vice-president of U.S. operations, told the American Association of Petroleum Geologists (AAPG).

Formerly separate exploration and development departments that rarely communicated are being distilled into engineering and geoscience "skill pools."

Exploration personnel are capitalizing on drilling technology and applying economic analyses earlier in the life of projects. Production personnel are making heavier use of 3D seismic surveys and other geophysical methods.

Howard said Shell is forming strategic alliances with service and supply companies and signing more incentive and turnkey contracts to take advantage of service company expertise.

Integrated team approach

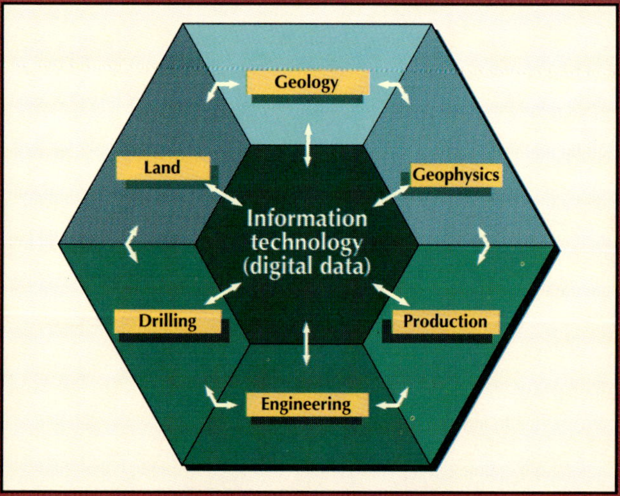

Source: Phillips Petroleum Co.

He noted that unprecedented levels of cooperation will be needed to exploit discoveries in the Gulf of Mexico deepwater areas, where Shell and others have drilled more than 200 wells since 1985 in 1,500 ft of water or more (Fig. 2).

Because of staggering costs, of the 31 discoveries industry has made since 1984 in gulf deep water only six have been or are being developed Fewer than 5% of the leases have been evaluated.

Howard suggested several integrated approaches to deepwater developments:

- Operators can join forces in building early infrastructure with the notion of longer lifespan and development of more than one oil or gas reservoir. This reduces front end risk capital and makes some nearby accumulations commercial.
- Seismic contractors could focus on blocks of heavily leased acreage around existing discoveries, where clients are obvious and easier to group. That would reduce or eliminate risk capital associated with speculative shoots and allow contractors to concentrate on quality rather than quantity.
- Drilling contractors can plan rig types and availability around corridor geology and bathymetry and work to achieve longer term contracts based on corridor development scenarios in a "share the risk" mode.

Where integration could aid deepwater developments

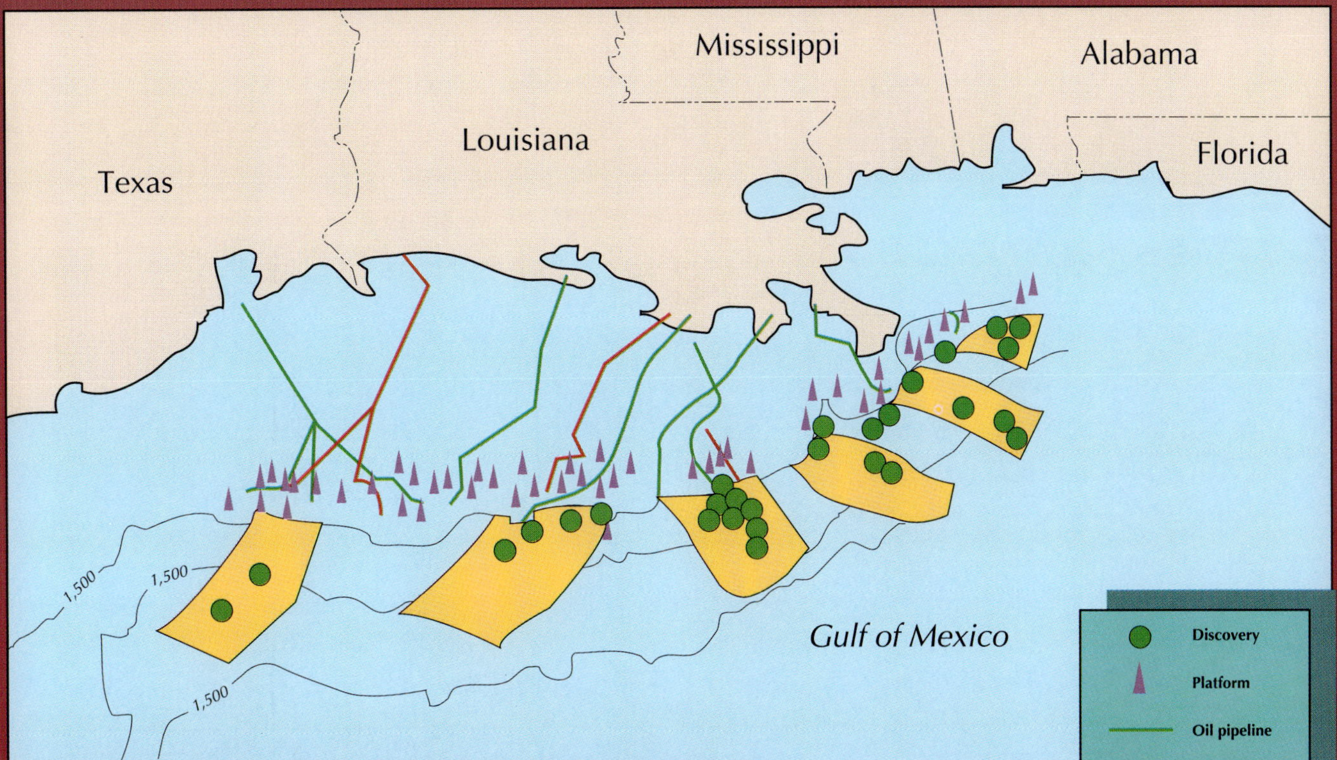

Deepwater discoveries can be grouped on the basis of geology, nearness to each other, and nearness to shelf infrastructure within 15 miles of the 1,500 ft bathymetric contour, along six corridors in which synergy between projects seems possible and desirable.

Source: Shell Oil Co.

Nine field North Sea plan requires cooperation

One program entailing high levels of cooperation is the Eastern Trough Area Project (ETAP) to develop nine Central U.K. North Sea oil and gas fields.

ETAP involves a cluster of nine fields on nine blocks with interests held by eight companies. Five of the fields are quite small, but the payoff from all nine is a significant 650 million bbl of oil equivalent that includes 1.2 tcf of gas.

ETAP will involve only a single processing and export platform, with the other fields tied back as unmanned process platforms or subsea developments.

Scrapping of the U.K. Petroleum Revenue Tax on new developments sharply reduced project costs, but integration of disciplines and cooperative aspects is important as well.

BP Exploration Co. Ltd. will operate ETAP on behalf of Shell-Esso U.K. plc, Agip (U.K.) Ltd., Fina Exploration Ltd., Hamilton Oil Co. Ltd., Murphy Petroleum Ltd., Lasmo plc, and Monument Oil & Gas plc.

Such cooperation between operators and contractors is needed to reverse a downward spending trend on Northwest Europe's continental shelf, said Norman Chambers, managing director of Brown & Root Marine Europe and Africa.

He said companies need to pool their data on reservoirs and work out capital and operating costs that will make developments viable. A few companies are working together, but the reach and scope of the alliances should be extended into a collaborative approach across the region, Chambers said.

Integration maturing in R&D activities

Oil company research and development operations have formed integrated teams to mirror reorganizations of their in-house clients.

For instance, several large companies have tapped technology developed by computer hardware and software companies and geophysical contractors to improve development and production efficiencies.

Many major companies are organizing in-house research teams to work with multidisciplined operating and business unit teams, Phillips Petroleum Co. said. In the case of computer based reservoir visualization, the line between production and exploration is nearly rubbed out.

There is still a strategic need for companies to have unique skills and develop technological areas that can provide a competitive advantage. But the trends are for companies to harvest more technology from outside sources or to rely more on jointly funded projects involving groups of oil companies, service companies, and research organizations and colleges.

Phillips manages integrated teams

Phillips said interdiscipline teams within its operations find and develop hydrocarbon reserves profitably using state of the art computing and communications.

New developments in computer technologies have allowed management practices to change and become more effective, the company said.

The interdiscipline, integrated team of the 1990s represents effective management of human resources. Application of new technologies by the integrated team represents effective use of physical resources.

Phillips continues to aggressively support the team and its technologies as part of its business strategy.

The integrated team can consist of geology, geophysics,

In many ways, the computer terminal has replaced the conference table for interdiscipline teams (Fig. 3). Photo courtesy of Phillips Petroleum Co.

and engineering staff dedicated full time to a specific project. An example of this at Phillips is a dedicated exploitation team in Woking, U.K., assigned to 4.7 tcf Hewett field in the North Sea.

As an alternative to the dedicated approach, an ad hoc team can be assembled with a specific exploration or exploitation problem to solve that requires staff from any or all of the individual disciplines in drilling, production, reservoir engineering, geology, geophysics, land, and information technology. An ad hoc team of this type can support a full time, dedicated team.

Phillips team members can be located in various offices worldwide. Corporate research and development staff based in Bartlesville, Okla., are frequently assigned to division office teams but, as a result of modern communication links, are not required to work in that office.

The integrated team is not new to Phillips. Examples of such teams can be traced to the drilling and production of the Bartlesville sands in Northeast Oklahoma in the early 1900s.

What makes the integrated team of the 1990s unique is the technology that supports the team and in many cases even drives the team to greater levels of integration. The modern, integrated team is just as likely to meet at a computer terminal as at a conference table (Fig. 3).

Management can bring the right people together to form an integrated team, but technology can enable the team to work to full potential.

Supercomputers' role in integrating disciplines

Massive number crunching supercomputers have left their mark on integrated teams mainly through the disciplines of geophysics and engineering, Phillips said.

Advancements in 3D seismic acquisition and processing are closely linked to the supercomputers. The use of 3D seismic in exploitation has drawn geology, geophysics, and engineering closer together.

Concurrently, the supercomputer enabled reservoir engineering simulators to more accurately—and with greater detail—model primary field performance and secondary and enhanced oil recovery and to design development strategies for new fields.

Information for the simulator's enhanced detail in the reservoir geological model is frequently provided by 3D seismic data. The previous paper and pencil technologies allowed each discipline to work its piece of the problem, put the results on a map, graph, or cross section, and pass the results along to the next discipline.

When Phillips processed the first in-house 3D survey on a Cray 1M supercomputer and ran the first reservoir simulation in 1984, the paper and pencil era began a rapid decline and the era of team interaction began a rapid ascent. To date more than 100 3D surveys have passed through Phillips' Cray computers.

Use of workstations, software described

The interactive workstation also has been evolving and multiplying to help analyze mountains of data.

Initially, output from supercomputers and mainframe computers was in hardcopy format. This limitation restricted interpretation to only a part of the total 3D data volumes.

Interactive workstations eliminated the need for paper and pencil and allowed interpretation to move from the batch mode to the truly interactive domain and to fully utilize all of the data.

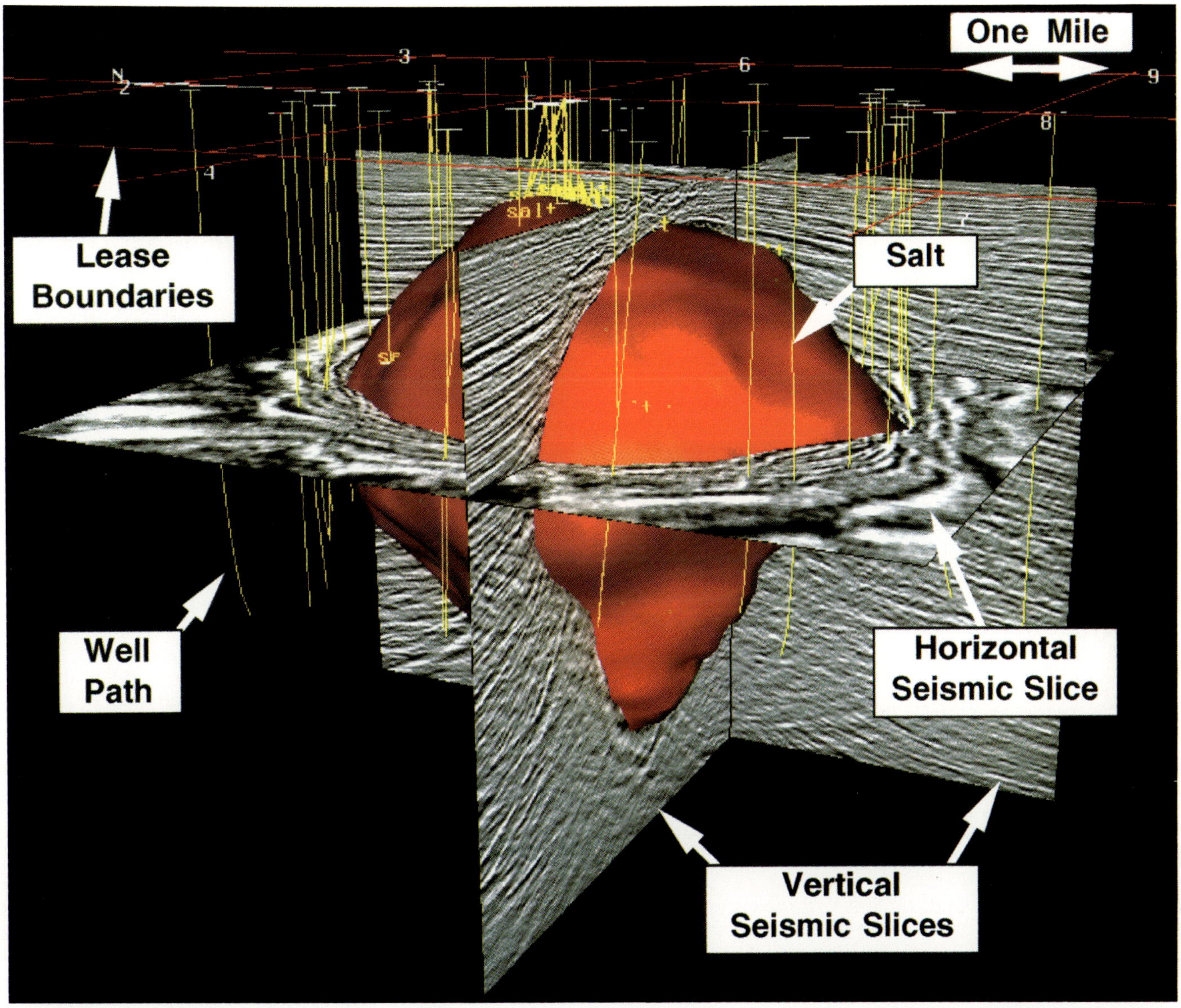

Visualization techniques blend many types of information in this depiction of a Gulf of Mexico salt dome (Fig. 4). Graphic courtesy of Phillips Petroleum Co.

In 1980 Phillips had fewer than 10 interactive workstations. In 1993 the Phillips scientific community had about 1,000 workstations.

Every major project undertaken by Phillips interdiscipline teams will involve many sessions on several interactive workstations.

Examples are a predrilling analysis on a 486 PC, a simulation run on an IBM RS-6000, a Gulf of Mexico subsalt visualization on a Silicon Graphics, a velocity analysis on a DEC, geologic cross section and map generation on a Sun, map drafting on a MicroVAX, or petrophysical well log analysis on an HP computer.

For instance, images of complex subsurface geometries like a Gulf of Mexico salt dome can be enhanced with lease boundaries, wellbore paths, vertical and horizontal seismic sections, and a highlighted salt interface (Fig. 4).

Recent workstation developments include interfacing several machines locally in a physical cluster or using networks and software to link central processing units (CPUs) from various sites into a virtual cluster.

These clusters provide near-supercomputer capabilities but allow data to reside and be processed at a local office. The workstation has become the indispensable workhorse of the interdiscipline team.

Some of the more significant in-house software used by the integrated teams are engineering production analysis, reservoir simulation visualization, drilling history data acquisition and reporting, petrophysical analysis and borehole visualization, subsidence modeling, basemap generation through graphic retrieval, 3D seismic poststack depth conversion, 3D seismic visualization, 3D seismic acquisition modeling, and seismic lithology inversion.

Discovery to abandonment

Shell noted that integrated exploration and production activities need to be based on a clear vision that a multidisciplinary approach must prevail from field discovery to abandonment.

And exploration and subsequent discovery of potential commercial hydrocarbons in the Cook Inlet and North Slope areas of Alaska prompted an integrated team effort among Phillips' geoscience and engineering staff.

Technology has played an integral role in each issue that affects the exploration decision tree. Smaller teams are solving tasks that were once performed by larger organizations and at the same time processing more data in greater detail.

The Sun and Hewlett Packard workstations are the main computer resource for the geoscientist. For example, all electronic log data, both conventional density, sonic, dipmeter, and the more advanced tools like the formation

How slimhole drilling can cut cost, waste

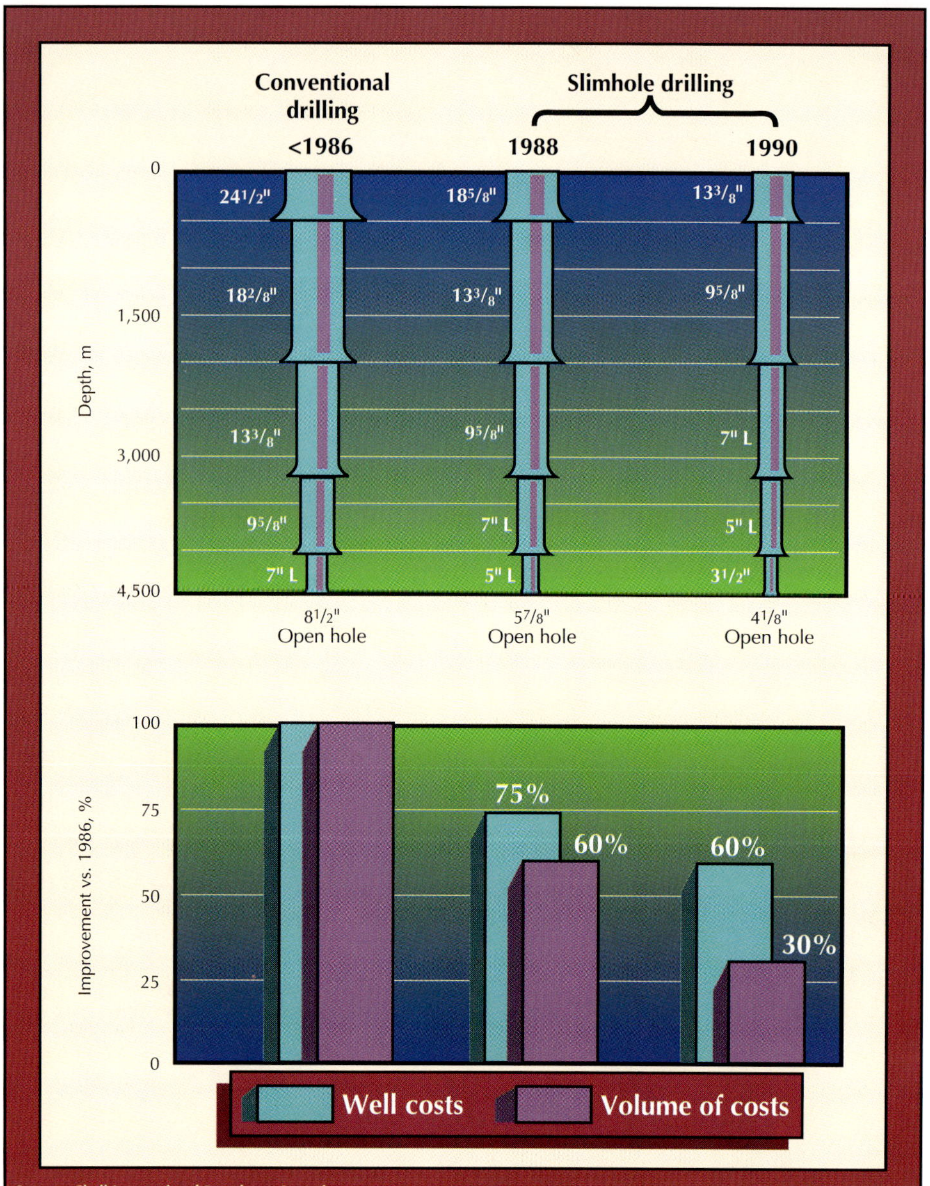

Source: Shell International Petroleum Co. Ltd.

microscanner (FMS) are analyzed on local workstations.

Satellite links on location provide transmission of log data directly from the wellsite to the analyst's desktop computer. A unique feature of the HP workstation log analysis system allows professionals at different sites to work together on the same project interactively and at the same time. Satellite and land line communications on the rig provide direct links to the mainframe-based daily drilling record system and electronic mail. The daily drilling information is entered from the rig and can be uploaded directly to the mainframe drilling history system. This information is then instantly accessible worldwide.

Tight holes have security levels added, and the information is available only to persons who need to know. Therefore, from any type of computer connected to the Phillips computer network, a team member can access data or mail from a rig drilling in the middle of the Beaufort Sea or Cook Inlet.

Those capabilities are available for Phillips locations around the world wherever there are reliable communication lines.

The Alaska projects have proven to be a working example of true team effort since their beginnings. Through the leadership of the Phillips Alaskan exploration team based in Houston, supported by corporate staffs, the first well in many years was drilled on the North Slope in 1990.

Since that time, North American partnership, development geology, and research and services staffs have formed an integrated development team to support all activities. Close cooperation with production, land, and facilities groups has proven very successful. Technology continues to play a vital role in facilitating these efforts.

Technology integration boosts drilling capabilities

Two particular advances—horizontal drilling and slimhole drilling—have had an immense effect on oil production and soon may prove to be a valuable addition to explorationists' efficient pursuit of new reserves.

AAPG anticipates an estimated potential target for horizontal wells worldwide to be about 8 billion bbl of recoverable oil.

In the venture operated by Shell's Nigerian affiliate, about 2 billion st-tk bbl of original oil in place lies in thin oil rims that can be exploited only with horizontal wells.

Shell companies drilled more than 100 horizontal wells during 1986-91 in a wide variety of reservoir types and geological settings. The process has matured such that the technology and experience are now available to drill, log, core, test, and complete ever longer horizontal sections at decreasing cost.

The productivity improvements of horizontal wells vary from two to five times that of vertical wells, Shell noted, and estimates of future use of the technique suggest an expansion reminiscent of the growth of 3D seismic.

Slimhole drilling technology allows a hole as small as 4 in. in diameter to be drilled to 5,000 m, leading to cost reductions in rig size, location area, casing, drilling mud, cement, and transportation.

Cost savings are 25-40% compared with conventional drilling, and the technique has obvious environmental advantages (Fig. 5).

Application of these drilling techniques in conjunction with focused exploration technology help Shell companies make fruitful economic development decisions and sustain their businesses. Shell also notes that companies face significant technical challenges when they drill very deep prospects in proven areas.

Drilling operations at 5,000-8,000 m under pressures in excess of 750 bar and temperatures as high as 210° C. have the potential to add several hundred billion barrels of oil equivalent to world hydrocarbon reserves.

With current technology and operating practices it is uneconomical to produce much of this oil and gas, but Shell believes the size of the reward appears large enough to warrant further attention.

Seismic stratigraphy solves drilling problem

Geophysical data may provide a solution to a potentially dangerous drilling problem, notes Peter Trabant, a Houston consulting oceanographer.

Deepwater drillers in the Gulf of Mexico have encountered serious drilling problems that stem from the presence of geologically young, unconsolidated sands. These sands provide an excellent conduit for high pressured interstitial fluids when pierced by the drillbit.

Casing is not always installed when the sands are penetrated, and mud weights required to overcome formation pressures cannot be used because they may fracture the sands and cause rapid loss of circulation.

A partial solution to this problem may be found by mapping shallow seismic sequence stratigraphy, with emphasis on depth, thickness, and types of sand bearing facies or systems tracts.

In addressing this particular drilling problem and related foundation engineering properties of near seafloor sediments on the upper continental slope, it is vital to understand the origin and nature of the shallow deposits. These characteristics may be determined prior to drilling by applying the concepts of sequence seismic stratigraphy in interpretation of high resolution seismic data routinely acquired for hazards studies.

Trabant spelled out the application of the seismic stratigraphic concept as a basis for interpretation of Pleistocene sedimentary sequences to depths of 3,000 ft on the continental slope in the northern Gulf of Mexico.

The concept of seismic stratigraphy is based on observations that sedimentary deposits are cyclic in nature and the observed cyclicity frequently appears to be a worldwide phenomenon attributable to relative changes in sea level.

The problem-causing shallow sands on the continental slope occur within the same types of strata that produce high quality petroleum reservoirs at depth and are detected to a large extent using the seismic sequence stratigraphic method.

The deepwater sands are the result of rapid sedimentation during low-stands in sea level when the shoreline lies along the outer continental shelf edge. Depositional sedimentary units or facies resulting from this phenomenon include the basin floor and slope fan complexes of the seismic sequence stratigraphic model developed since the 1960s at Exxon Production Research by Peter Vail and others (Fig. 6).

Detailed interpretation and mapping of the shallow seismic sequence stratigraphy allow an operator to determine presence, age, and location of sand-prone facies well in advance of drilling.

Mapping of high porosity, sand-prone systems tracts on the upper continental shelf can be accomplished from high resolution, multifold hazards survey data. Such survey data should be acquired via a seismic source with frequency components above 250 hz. Processing of the multifold field data must be done for optimum resolution and include migration.

The method is applicable worldwide with minor modifications to account for differences in morphology and sedimentary setting.

Lowstand sand distribution*

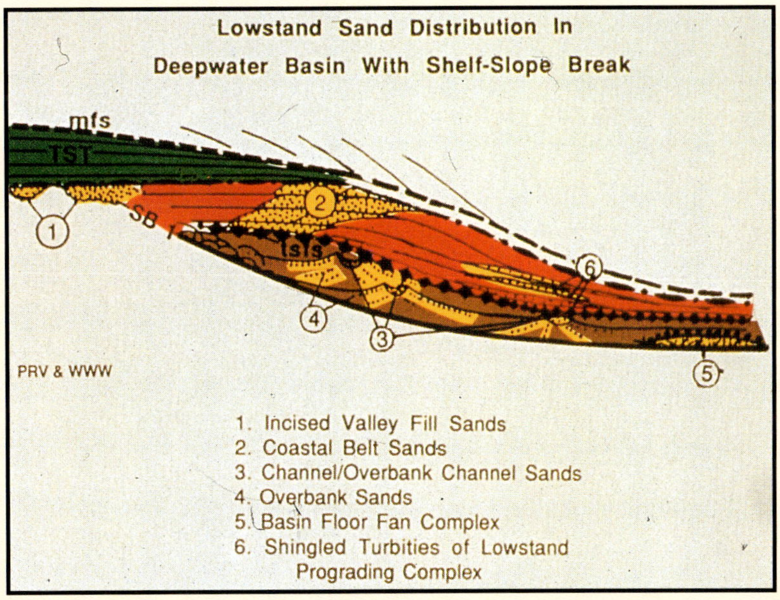

*In deepwater basin with shelf-slope break
Source: Trabant, adapted from Vail and Wornardt,1990.

Integration seen vital in West Texas reef play

Petrie Parkman & Co., Denver, cited integration of data as a primary determinant of success or failure in a Pennsylvanian Canyon reef oil exploration play in the horseshoe atoll area of West Texas.

The analyst described the play as one of the hottest in the U.S. Lower 48 states. Many independent operators are using the combination of affordable 3D seismic and computer aided exploration technology to target relatively small reefs that have eluded explorationists armed with less sophisticated methods.

The company said 3D seismic had been acquired over more than 750 sq miles in the greater horseshoe atoll area from early 1990 through early 1993.

But Petrie Parkman said integration of all available seismic and geologic data was key to arriving at a comprehensive subsurface model.

Explorationists who merely rely on bumpology—assuming every seismic bump is a pinnacle reef with little regard for underlying geological conditions—will produce disappointing results.

Without a proper framework, a seismic bump could be mistaken for a reef when in fact it is caused by an underlying structure that gives a false impression of a build-up, the analyst said.

Properly acquired and processed 3D seismic data can go far beyond simply delineating the geometry of a reservoir container, in this case a pinnacle reef. When merged with subsurface geological control, seismic data can help predict the quality of reservoir rock within the container, the analyst said.

A liquefied natural gas tanker carrying a cargo of Indonesian LNG heads into a Japanese port. The two countries are the world's biggest exporter and importer, respectively, in the surging LNG trade. Photo courtesy of Pertamina.

World LNG trade keeps on climbing

World trade in liquefied natural gas, already climbing rapidly, could double or triple in the next 20 years.

Estimates are that industry may have to spend as much as $40-60 billion in new facilities and expanded fleets to meet projected LNG demand by 2010, Oil & Gas Journal reported in mid-1993.

Fueled by soaring demand in the Asia-Pacific region, the surge in LNG trade may overwhelm projections of the industry's ability to accommodate that growth. Some industry analysts told OGJ they saw shortfalls on the horizon in terms of LNG plant capacity and LNG tanker fleets.

Underlying those possible shortfalls were current and projected prices for LNG, tied to a consensus view that oil prices will remain relatively low for the foreseeable future. Developers are unlikely to press costly grassroots LNG projects without the prospect of substantial price increases. And prospective LNG customers balk at signing contracts for speculative supplies at prices needed to justify those investments while energy prices are expected to remain relatively flat.

The result, OGJ said, likely will be a shakeout of the many project proposals that emerged in recent years. That could leave LNG far short of fulfilling its market potential absent some preferential treatment by governments on both sides of the supply/demand equation.

Dominant players

The Asia-Pacific region's LNG producers and consumers will continue to dominate world trade in the fuel.

But new LNG export players were emerging as 1994 approached, notably in the Middle East, Latin America, and Africa. On the consuming side, questions remained whether the U.S. will live up to its trumpeted potential and whether Europe will be a major new growth market.

There will be no lack of other prospective LNG supply sources beyond expansion of traditional suppliers, notably Algeria, Abu Dhabi, Alaska's Cook Inlet, Brunei, Libya, Indonesia, Malaysia, and Australia. Among prospective sources are Qatar, Oman, Nigeria, Iran, Russia's Sakhalin Island, Indonesia's Natuna Island, Australia's Withnell Bay, Alaska's North Slope, Norway, Papua New Guinea, Argentina, Trinidad and Tobago, Venezuela, and Yemen.

As 1994 approached, the clamor was growing worldwide for new carbon and other energy taxes as well as a host of new regulations to curb oil and coal use, the so-called dirty fuels. Such steps ostensibly were intended to combat global warming and other claimed environmental and energy security ills the dirty fuels were said to inflict.

Passage of such new taxes could amount to an indirect subsidy for all natural gas supplies by discouraging investment in competing fuels, especially in electrical power generation. Thus, with consuming nations increasing the cost of oil and coal through new regulations and taxes, LNG was headed for a bigger advantage in world energy trade.

Accordingly, the environmental push that was driving much of the projected demand for LNG may achieve what expectations of ever increasing oil prices in the late 1970s and early 1980s did not achieve: buoyant long term economics of new world scale LNG projects.

Some analysts contended that increasing politicization of energy markets will be the engine that drives LNG growth in the coming years. Governments seeking to reduce reliance on oil imported from the Middle East and/or to accommodate environmentalists' agendas will be the key to viability of future projects by placing priority on LNG as a fuel of preference.

Demand outlook

Analysts agree on a growth trend for LNG. They just differ on degree, OGJ reported.

Shell International Gas Ltd.'s (SIG) most conservative outlook calls for world natural gas demand of 2.5 trillion cu m in 2010. SIG Managing Director J.R. Williams at Gastech '93 in Paris predicted as much as $800 billion to $1.1 trillion of combined capital spending by the natural gas industry will be needed to meet this increased demand.

SIG expects world demand for LNG to reach 140-200 million tons/year by 2010, with the biggest increase occurring in nations that currently import LNG. However, Williams said only about one third of the roughly 30 LNG projects proposed or under speculation to meet that expected demand growth will come to fruition because of fierce competition among project promoters.

Industry has invested almost $50 billion in current dollars in the world LNG infrastructure. Even at that, daily world trade in LNG represents only about ⅒th of the peak daily consumption of natural gas in the U.S.

France's Cedigaz sees total international gas trade jumping to 550 billion cu m in 2010 from more than 400 billion cu m in 2000 and 320 billion cu m in 1991. That's out of a total world consumption projected to climb to as much as 3.1-3.5 trillion cu m in 2010 from 2.6-2.7 trillion cu m in 2000 and 2.1 trillion cu m in 1991.

LNG trade, while accounting for only 24-30% of all world gas trade and only about 3-4% of world marketed gas production, nevertheless will continue to be the fastest growing segment of the gas market. LNG demand has grown an average 20%/year since 1970.

Cedigaz estimates world LNG trade will almost triple in 20 years, jumping to 184-212 billion cu m in 2010 from 116-130 billion cu m in 2000 and 77 billion cu m in 1991.

Drewry Shipping Consultants Ltd., London, predicts world LNG trade will climb to 170 billion cu m by 2010, a 136% jump from 1990's level, compared with a 66% increase in pipeline gas trade to 410 billion cu m in the same period.

Asia-Pacific demand

The Pacific Rim accounts for almost two thirds of the LNG transported and three fourths of the LNG delivered in the world.

World LNG trade rose 3.5% in 1992 to 59.5 million metric tons (79.2 billion cu m gasified), of which the Far East accounted for 73.4%, compared with 73.6% in 1991, according to data from Groupe International des Importateurs de Gas Natural Liquefie (Giignl). Indonesia accounted for 39% of all LNG exports in 1992, followed by Algeria 24%, Malaysia 12%, Brunei 9%, and Australia almost 8%, Giignl estimated.

Japan accounted for almost 65% of all LNG imports in 1992, followed by France with 11.5%. Along with Japan, South Korea and Taiwan are expected to be the major new sources of LNG import growth, and with Japan imported almost 42 million tons of LNG in 1991.

The three Asian countries could absorb LNG imports of as much as 63-66 million tons by 2000 and as much as 91-100 million tons by 2010, said Ichiro Yokose, managing director, fuels, Mitsubishi Corp. He made the prediction in a paper at an oil and gas conference in Dubai early in 1992.

How world LNG demand has mushroomed

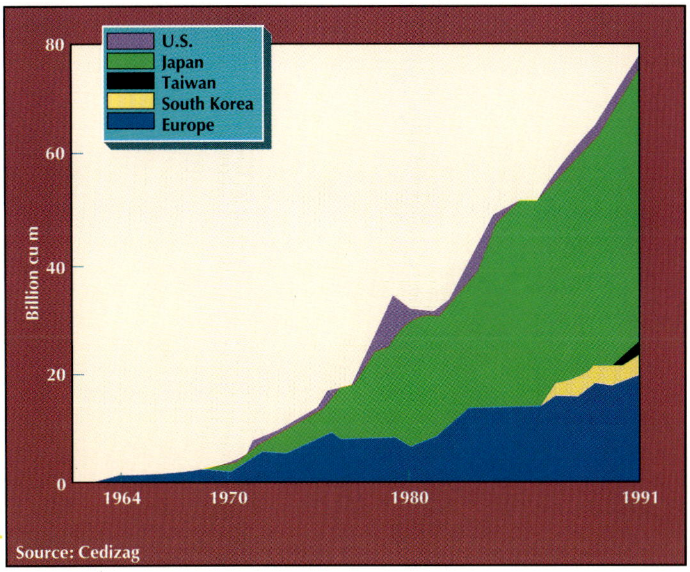

Source: Cedizag

Yokose predicted incremental demand to 2000 in the three nations will be covered by expansion of existing supply sources in Australia, Malaysia, and Abu Dhabi and new projects in Qatar, Oman, and Sakhalin Island. The big question is what new supply sources will come to the fore to meet the roughly 50% further increase to 2010, Yokose said.

He contended the big driver in the LNG market, Japanese demand, will be affected by:

- The outlook for electrical power demand.
- Delays in construction of nuclear power plants.
- Increases in town gas use.
- The Japanese government's policy to diversify energy sources among coal, oil, gas, and alternate fuels such as methanol.
- The current gloomy overall economic condition, which is likely to continue in the years to come.

Other demand

In North America, few analysts see prospects of the U.S. becoming a significant LNG importer in the short to medium term.

Gas Research Institute (GRI), Chicago, in 1992 lowered its forecast of U.S. LNG imports in 2010 by 25% from 1.2 quadrillion BTU. The previous forecast had suggested U.S. LNG terminal capacity will be strained after 2000 and a new terminal will be needed by 2010.

GRI also raised its estimates of U.S. gas production and U.S. imports of Canadian gas in 2010 but cut its estimates of gas acquisition prices, including a drop in projected LNG import prices to $3.74/MMBTU in 2010 from $4.87/MMBTU it projected in 1991.

Because the pipeline load factor in the U.S. is relatively low at about 70%, gas storage and peak shaving LNG facilities will become the solution to improving load factors for pipelines, said A. Pastuhov, president, AVP Corp., South Dartmouth, Mass.

"Until load factors are significantly improved, I do not believe (U.S.) LNG imports will increase...during the next decade," Pastuhov said in a paper written for presentation at Gastech '93.

"It is possible, however, that in the event either the Venezuelan project gets a new life or that LNG out of Trinidad becomes a reality, some of the output from

World suppliers and buyers of LNG*

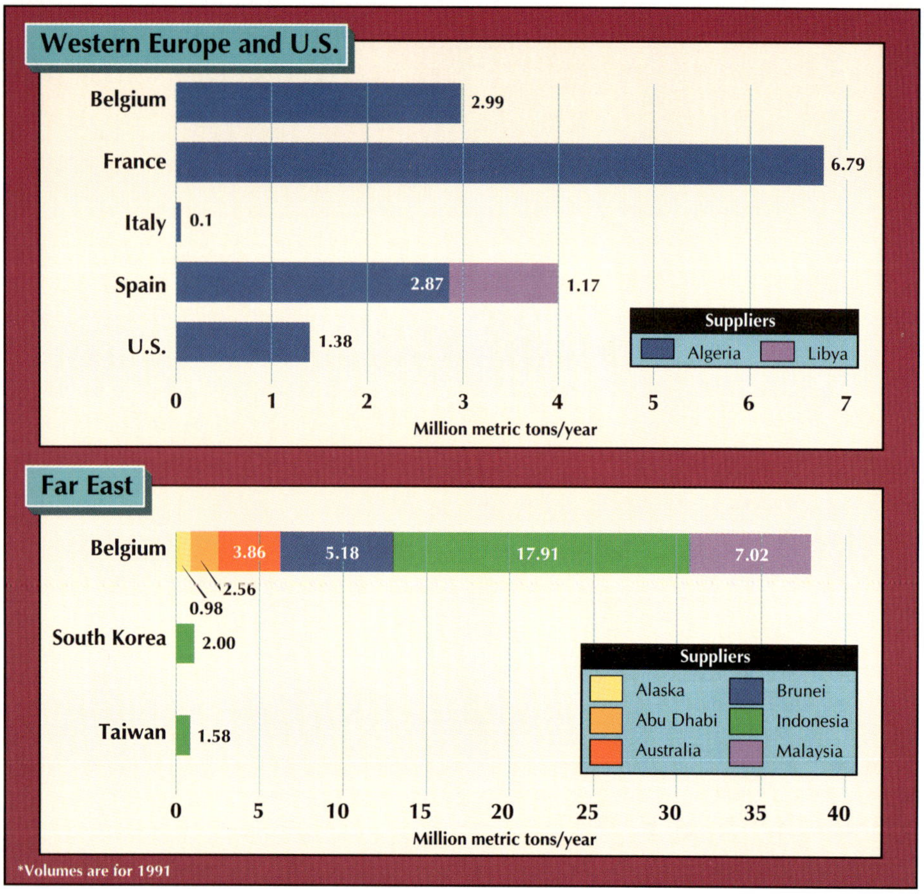

either project or both may come into U.S. terminals such as Cove Point and/or Elba Island.

"I do not believe those imports will be for baseload supplies, but rather such imports will be more for winter supplies, especially if the pipeline load factor improves with time."

Cedigaz said conditions are ripe for LNG to boost its share of Europe's gas market. However, uncertainties persist over the cost of new LNG supplies relative to world market gas prices and how effectively gas will back out electrical power as a result of environmental initiatives.

Cedigaz predicted European LNG imports will more than double by 2010, claiming 10-12% of Europe's gas market by then. While traditional suppliers Algeria and Libya will account for significant increases in LNG supplies to Europe, other suppliers will be needed after 2000. Cedigaz looks to Nigeria to fill some of the gap by 2000, with longer term European LNG supplies coming from Qatar, Oman, Iran, and even Norway.

Other countries could join the list of LNG importers. Institute of Gas Technology's LNG Observer pointed to India's consideration of several pipeline and LNG import projects to accommodate an expected spurt in natural gas demand for environmental reasons. High sulfur coal provides more than half of India's primary commercial energy consumption.

One such project under consideration was being pushed by an Enron Corp. group to build a 2 million kw gas fired power plant on India's west coast to be fueled by LNG. Enron signed a letter of intent to buy 2.5 million tons/year of LNG from Qatar's Ras Laffan LNG project beginning in 1997-98.

LNG vs. pipelines

Perhaps the most critical factor in demand for LNG will be the relative status of gas pipelines, OGJ reported.

Cedigaz cited these factors in its claim that balancing of natural gas markets by pipeline is reaching its limits:

- The growing distance between reserves and markets implies that pipeline transportation may be physically, technically, or economically infeasible.
- Most major traditional pipeline exporting countries are reaching the limits of their export capacity. No major onshore exporter has effectively emerged on international markets since the late 1970s.
- Importing countries want to diversify supply sources. But because the range of pipeline exporters is narrow, reliance on more distant imports via LNG is ultimately necessary.
- The proliferation of remote gas markets, which have good potential but are far from major gas pipeline networks in Asia and Europe, also tends to favor establishment of LNG receiving terminals instead of pipeline connections to these networks. The modular, phased nature of an LNG project also lends itself to these developing markets.
- Increasing capability of power utilities to purchase natural gas at a higher price and in significant volumes could create a new class of LNG buyers owning their own terminals, especially in Europe, similar to importing Japanese utilities.
- Increasing awareness of political instability affecting pipeline deliveries, such as gas pipeline supply cutoffs in the C.I.S.

LNG tanker outlook

Critical to LNG project viability is shipping, said, Drewry, which estimated tankers will account for at least one third of the $3.5-4.5 billion cost of a new grassroots project.

Drewry noted there is almost no secondhand market for LNG carriers and estimated newbuilding prices soared to $270-290 million for a 125,000 cu m vessel in 1993 from $120 million at yearend 1986.

The London shipping consultant said newbuilding prices will rise slightly in the near term because:

- World shipbuilding capacity will remain constrained by the limited number of yards with experience in building LNG carriers and by orders for other vessels.
- Subsidies will continue to be phased out.
- Shipbuilders, notably in Japan, will face labor shortages and higher costs.

As a result, shipowners will seek to maintain their vessels as long as possible, rendering obsolete the current standard of 20-25 years, assuming government regulations aren't a barrier.

Shippers also will continue to find ways to trim LNG transportation costs, which Drewry estimated at anywhere from 27-37¢/MMBTU for existing Algeria-Europe trade to $2.27/MMBTU for a proposed Qatar-Japan project. IPE

Oil spill response, such as this deployment of containment boom in Prince William Sound off Alaska following the Exxon Valdez oil spill, will be a critical test of how the petroleum industry is judged on environmental issues in the years to come.

Industry's oil spill response capability crucial test in '90s

The petroleum industry's ability to cope with oil spill response and cleanup is critical to whether it prospers in the years to come.

After all, it was the 1989 Exxon Valdez supertanker oil spill in Alaska's Prince William Sound that ushered in a new era of heightened environmental awareness. With that new awareness came a surge in onerous new environmental regulation that ultimately will cost industry tens of billions of dollars in added capital and operating costs. Much of that added regulatory cost can be seen in the provisions of the 1990 U.S. Oil Pollution Act (OPA).

The Valdez spill and its aftermath promises to continue to rouse controversy for years to come.

The first judicial settlements under OPA came in 1993 and involved whopping penalties assessed for two spills off Washington state.

There were significant measures industry took to improve spill response and cleanup efforts in 1993. One, a direct product of OPA, was the official start-up of the Marine Spill Response Corp.(MSRC), U.S. industry's cooperative offshore oil spill response/cleanup program. Another was a crucial oil spill test burn off Newfoundland that was designed to show the efficiency of burning spilled oil at sea while monitoring that procedure's environmental effects.

At the same time, industry continued to grapple with concerns over the public's perception of how it is prepared to cope with oil spills and some of the more stringent provisions of OPA, including insurance requirements that some say could virtually shut down many small operators in the U.S. Gulf Coast region.

Exxon Valdez spill still controversial

Almost 5 years after its occurrence rocked the petroleum industry and revitalized the U.S. environmental movement, the Exxon Valdez tanker oil spill off Alaska continued to stir controversy.

Conflicting reports abound over whether there is long term damage to the Prince William Sound ecosystem resulting from the Mar. 24, 1989, spill. The worst U.S. oil spill occurred when the tanker laden with Alaskan North Slope crude struck a submerged reef and spilled 258,000 bbl of oil.

Since then, Exxon Corp. has spent more than $3.5 billion for cleanup, fines, and further environmental mitigation, monitoring, and studies related to the spill. However, government scientists at recent conferences disclosed studies they contend show long term, significant damage to the sound.

Exxon in April 1993 launched a counteroffensive, disclosing results of studies it funded that it claims show no credible scientific evidence of long term damage. At the same time, the company blasted as flawed the government's data on assessing environmental damage to the sound and charged that test samples from the sound were mishandled. Both contributed to exaggerated damage claims, Exxon said.

Exxon presented details of its studies at an American Society for Testing and Materials (ASTM) symposium in Atlanta.

Meantime, Prince William Sound still shows lingering effects from the Exxon Valdez oil spill, witnesses told the U.S. House merchant marine committee in early 1993.

But recovery has been so rapid that there is more controversy over how to use $900 million in natural resource recovery funds that Exxon paid than over how badly species are suffering.

Exxon: no long term damage

Exxon said results of studies by oil spill researchers from academia and private environmental consultants funded by Exxon verify that while the spill had acute short term effects on the environment, there is no credible scientific evidence of long term damage.

Researchers studied the presence and persistence of oil in the environment, recovery of seabirds, and the health of key commercial fisheries populations.

"These sophisticated studies produce conclusions that refute the broadly communicated claims of long term and dire impacts on the environment," Exxon said.

Exxon sought to refute some government claims ahead of the ASTM meeting, specifically claims that:

- Certain Gulf of Alaska murre colonies may take as long as 70 years to recover from the spill. Studies by John Wiens, a U.S. ornithologist, show murre populations are at historic levels and the birds are reproducing normally, Exxon said.
- Significant portions of the sound are contaminated today with high levels of oil comparable to those the day of the spill. "In fact, comprehensive shoreline surveys conducted by state, federal, and Exxon scientists show that by the spring of 1992, oil was present only on 0.2% of the 3,000 miles of shoreline in the sound," Exxon said. "Biological recovery of affected species has been rapid and in most cases is nearly complete."
- The herring fishery in the sound has been devastated. A Battelle Marine Science Laboratories study found the oil did not have significant effects on the sound's herring population and the fishery continues to support record harvests, Exxon said.

Exxon claims faulty data used

Exxon said it used chemical fingerprinting to determine that some of the hydrocarbons found in test samples from the sound came from naturally occurring oil seeps upstream and from marine diesel fuel.

It also claimed frequent instances of lab contamination of the samples, contending lab contamination was responsible for about half the suspect substances found in tissue samples from wildlife tests in 1989-90. Texas A&M University, which conducted most of the lab testing for the National Ocean and Atmospheric Administration (NOAA) under a $2 million contract, denied Exxon's claims.

NOAA officials said Exxon's claims were plausible and the agency has begun to review its database on the spill. However, the Seattle Times newspaper quoted a NOAA official in Seattle as charging Exxon with putting up a "smokescreen" that does not make "any difference in the big picture in regards to damage."

Exxon scientist Hans Jahns, acknowledging the extent of short term damage from the spill, said, "The issue is the claim of continuing long term damage," the Seattle Times reported. "That doesn't agree with our data, and it doesn't agree with oil spill literature. One year after the spill, as expected, there was not much exposure to Exxon Valdez oil."

Committee testimony reflects dispute

The House committee was told nature is doing the main job of restoration so well that scientists and environmental groups are urging that most of the settlement money be used to acquire upland habitat, including land and timber rights in the sound and adjacent areas. About $27 million in acquisitions are planned.

Commerce Sec. Ronald Brown disclosed $25 million more would be spent in 1993 for land acquisition.

Brown conceded, "There has been criticism regarding the progress to date in using the settlement money to restore the sound region. Natural resource restoration is a complicated and time consuming process, but I am committed to expediting the process to the greatest extent possible."

Rep. Gerry Studds (D-Mass.), committee chairman, said, "Some species such as bald eagles appear to have recovered reasonably well, while others may take many more years to return to anything approaching normalcy and in fact are still suffering from contact with lingering oil on beaches, in eelgrass beds, and in the food chain."

Elenore McMullen, chief of the native village of Port Graham, said, "Since the spill there is simply less life in Prince William Sound. Ducks are rarely seen. Seals are difficult, if not impossible to find. Sea otters are scarce. Even the pink salmon run has gotten smaller."

Studds said the settlement money "must not become a slush fund for unending research studies, nor should it be used to pay for locally popular projects that have little, if anything, to do with the spill or sound.

"While the trustee system established under the terms of the settlement is cumbersome and exhausting, that system must ensure that the funds are used for their proper purpose, the restoration of the environment damaged by the oil."

Coast Guard OPA update

Adm. William Kime, U.S. Coast Guard commandant, told the committee his agency is making continued progress in implementing OPA, the tanker safety legislation passed in response to the accident.

He said the Coast Guard, Alaska, the petroleum industry, and others are negotiating with Russia to conduct an offshore in situ test of the burning of spilled oil as a response technique.

He said Alyeska Pipeline Service Co. "has accumulated the most extensive spill response equipment inventory in the world and they now routinely exercise it."

That gear includes five 210 ft emergency response vessels, 50 other response vessels and skimmers with a cumulative recovery rate of 44,000 bbl/hr, and more than 200,000 ft of containment boom.

James Hermiller, Alyeska president, said Trans-Alaska Pipeline System owners have spent more than $237 million since 1989 to enhance oil spill prevention and response capability in the sound and will spend another $73 million this year.

He said, "All laden outbound tankers and all partially laden inbound tankers are escorted in the sound by two specially fitted vessels to assist in safe navigation and to provide immediate assistance in the event of a tanker problem or an oil spill."

Big fines assessed for Washington spills

The U.S. Justice Department in February 1993 hit two oil companies blamed for separate spills in Washington state with a total of $14.7 million in penalties and cleanup costs.

Texaco Refining & Marketing Inc. and U.S. Oil & Refining Co. agreed to consent decrees in U.S. District Court at Seattle to resolve suits brought by Justice, the Environmental Protection Agency, and the Coast Guard.

They were the first judicial settlements under OPA.

Justice said Texaco spilled about 5,000 bbl of oil near and into Fidalgo Bay Feb. 22, 1991, when a pump failed at its refinery near Anacortes. Justice also sued Texaco for environmental violations stemming from later smaller spills at the refinery.

Under the decree, Texaco is required to complete the

Offshore Newfoundland oil spill test burn

Source: Environment Canada

spill cleanup at a cost of about $8 million, install state of the art spill prevention equipment costing $800,000, pay the government's cleanup cost of $125,000, and pay a $480,000 civil penalty to the U.S. in addition to a $20,000 fine paid earlier to Washington state.

U.S. Oil spilled more than 14,000 bbl of oil in Tacoma Jan. 6, 1991, when a pipe burst during the unloading of a tanker. Some of the oil entered Commencement Bay, part of Puget Sound.

Under its decree, U.S. Oil is required to complete cleaning up the spill at a cost of $4 million, install and maintain state of the art spill prevention equipment costing $800,000, pay the government's $60,000 in removal costs, and pay a federal fine of $425,000 in addition to a $45,000 civil penalty paid the state.

U.S. Oil also must file records regarding its financial condition because the settlement takes the company's financial condition into consideration.

Myles Flint, acting assistant attorney general for environment and natural resources, said, "We could not have achieved these settlements under previous law." OPA allows a fine of as much as $1,000/bbl of oil spilled.

Rear Adm. A.E. Henn, Coast Guard chief of marine safety, security, and environmental protection, said, "These judicial settlements, the first under OPA, signal a new era in marine environmental protection when polluters who violate federal law can expect substantial financial penalties."

Virginia spill settlement

Star Enterprise, owned by Texaco and Saudi Refining Inc., agreed to pay a $2.75 million fine to Virginia for an oil spill from a Northern Virginia pipeline terminal.

The Virginia Water Control Board must now approve the Star Enterprise settlement or sue in court.

The Fairfax City tank farm is estimated to have leaked more than 172,000 gal into the ground under a residential neighborhood.

The settlement also requires Star to pay $800,000 to hook up water lines to 100 homes that now use well water and pay for studies on how the leak is affecting public health.

Star settled a lawsuit with a number of homeowners in 1992, agreeing to buy their houses and pay them for medical costs.

It is making improvements at the tank farm costing $3.3 million and expects to spend $100 million on the cleanup effort.

Oil spill test burn successful

A major crude oil spill test burn off Newfoundland was successfully concluded in early August 1993.

Preliminary results suggest burning oil is a practical and sound way to reduce the environmental effects of marine oil spills.

The experimental burn was the first controlled burn of crude oil in open sea to record the effects of burning oil emissions on air and water quality. A group of 25 government agencies and industry associations led by Environment Canada (EC) conducted the test burn. The 1 day burn took place 40 km east of St. John's, Newf., under favorable weather conditions Aug. 12.

Test burn rationale

The test was intended to generate data needed to help serve as a yardstick for determining when to resort to burning in accidental oil spills. Involved were oil spill specialists from the U.S., U.K., Norway, France, Sweden, Finland, and the Baltic republics of the former U.S.S.R.

The agencies were expected to make their first analyses of the oil burn experiment available in February 1994.

EC Project Manager Merv Fingas noted, "Extensive tests have shown that emissions from controlled burns do not pose risks to environmental health, and now we will see if this is also the case at sea."

EC Emergencies Science Coordinator Greg Halley said, "Burning oil enables you to take advantage of the window of opportunity you may have in the first days of the spill. You could have the majority of the oil burned before weather conditions deteriorate.

"Mobilizing skimmers and storage containers for mechanical recovery is much more complicated and time consuming than deploying a fire resistant boom. That's why burning oil is such an attractive option. It eliminates delays and removes the oil very efficiently."

Halley does not expect burning to replace the traditional boom and skimmer approach because burning is feasible only in open sea or near uninhabited land. It simply offers another option for oil spill response, he noted.

Test burn details

Once wind, waves, and ocean currents fell within EC safety thresholds, a heavy duty boom was laid to serve as a backup containment and recovery system for the spilled oil.

A boom to contain the burning slick was laid 150 m ahead of the backup boom. Then a Canadian Coast Guard (CCG) offshore supply vessel began discharging Alberta sweet crude into the fire boom.

The vessel maintained a speed of about 0.5 knots to maintain oil thickness at the 2-3 mm level required to sustain combustion on the water. Once 10% of the oil was released into the boom, a Helitorch ignited the fire, and the oil discharge continued until the volume reached a total of 304 bbl. The burn lasted 90 min., 30 min. longer than expected. A second identical burn also took longer than expected, consuming 182 bbl. Cleanup of remaining oil traces with absorbent pads and skimmers followed the second burn. The two fires burned 99% of the discharged oil.

Emissions sampling

During the test burn, two remote control sampling boats followed the boom, each equipped with 40 measuring instruments.

Behind them, a blimp tethered to a floating platform and operated by the U.S. National Institute of Standards and Technology was positioned in the resulting smoke plume to sample soot concentrations and carbon compounds.

Another vessel positioned 500 m downwind from the fire sampled particulate size, polynuclear aromatic hydrocarbons (PAHs), volatile organic compounds (VOCs), metals, carbonyls, carbon dioxide, carbon monoxide, sulfur dioxide, and nitrogen oxides.

The U.S. Environmental Protection Agency sent a remotely operate underwater vehicle directly below the slick to videotape the burn and record water temperatures.

EC remote controlled helicopters flew directly above the burning oil to sample air emissions from the fire. Wide samples extracted from rotor blades were used to indicate PAH concentrations in the soot.

Unlike low temperature incineration, which produces PAHs, the test fire, burning at about 800° C., was expected to destroy most PAHs present in the original oil. Further, VOC concentrations in the smoke were expected to be lower than those emitted via evaporation of oil in open water.

Instruments also measured soot, particle analysis, plume trajectory, heat release rates, oil residue, and water column temperature, chemistry, and biology.

Enviromental mitigation steps

Because the oil in the boom was maintained at the thinnest possible level needed to sustain a fire, and because it was present less than 6 hr, the oil did not penetrate below the top ½ m of the water column.

The burn did not affect most commercial fish in the region because they are primarily bottom dwellers, and most fish eggs and larvae don't occur in the region in August.

Because of the limited scope of the test, the burn did not significantly affect seabirds or mammals.

Officials said the Newfoundland test was an eye-opener because it included real life effects of wind and waves.

Joe Mullin, a physical scientist with MMS, said it was surprising that the oil residue from the burn remained afloat.

"The fact that it didn't sink showed that all the 'nasties,' the volatile organic compounds and the lighter components, were gasified and burned off. We were left with something that won't put off toxic chemicals or vapors if left in the water."

Before concluding that burning oil at sea is an acceptable spill countermeasure, some scientists want to see results from a test burn conducted in warm water. U.S. Coast Guard's Bob Hiltabrand suggested taking samples 3-5 km behind the burn at sea level vs. the 1.6 km distance in the Newfoundland burn and using a different system for releasing oil into the boom that would more closely approximate an actual spill.

Hiltabrand said an experiment off the coast of Louisiana or Texas would take about 2 years to coordinate, and he figures it will be 4 years before the oil burning technique becomes operational in the U.S.

MMS at presstime was planning a workshop for spill response decisionmakers in January 1994. Further tests and lab experiments are needed to determine the effects of fresh and weathered crude oils and the burning of emulsions.

MMS's Mullin said, "It's up to each regional response team to decide if it wants to use this technology. The region in Alaska has embraced it already. Several regional response teams are considering it."

He noted that each region will have to identify sensitive areas where burning oil is off limits, such as a mangrove swamp or an area immediately adjacent to a refinery.

Comparisons with other burns

One of the first oil burning experiments was conducted in 1967, when the Torry Canyon tanker ran aground off England, spilling almost 700,000 bbl of crude into the ocean.

Using bombs, rockets, and other incendiaries, that effort failed because of a lack of technology to contain the oil, EC said.

That led to development of fire resistant booms and subsequent successful 24 hr burns in laboratory tanks in the mid-1980s.

In the first open water test burn, 12 bbl of crude burned in 30 min. off Norway in 1988, leaving only a 5% residue.

A test burn at the site of the Exxon Valdez oil spill off Alaska in 1989 involved about 400-800 bbl of North Slope crude that burned in only 75 min. It left an easily recovered, stiff residue of about 7 bbl, or the equivalent burn-off of 98% of the oil. Some officials have estimated an emergency oil burn response to the Exxon Valdez spill would have eliminated as much as 50% of the oil in just 2 hr, EC said.

EC had expected 98% of the oil discharged in the Newfoundland test to burn off, leaving about 8 bbl of residue after each burn.

Vice Project Manager Nick Vanderkooy, on loan from Canadian Marine Drilling Ltd., Calgary, conceded burn efficiencies probably will fall in real spills, notably with unfavorable weather and the possibility a slick might not be fully contained.

"It may drop to 50%, but you're still doing a heck of a lot better than you could with mechanical recovery," he said.

Vanderkooy noted an average of 10-15% of spilled oil typically is recovered using mechanical means.

Also participating in the test project management team with EC and EPA were CCG, U.S. Coast Guard, U.S. Minerals Management Service, U.S. Marine Spill Response Corp. (MSRC), American Petroleum Institute, Canadian Association of Petroleum Producers, and 3M Ceramics.

MSRC meets start-up target

MSRC, the largest U.S. offshore oil spill cleanup company, which was spawned in the wake of the 1989 Exxon Valdez tanker oil spill off Alaska, met its Aug. 18 start-up deadline.

That is the date the U.S. Coast Guard ordered companies that ship or produce oil in U.S. coastal waters to have the plans and capability for responding to major oil spills as required by OPA.

MSRC acts as a private offshore oil spill response contractor for 77 of those shippers, which are funding it through the Marine Preservation Association (MPA), Scottsdale, Ariz.

Burning of crude oil spilled at sea is gaining wider acceptance as a method of oil spill response. An August 1993 test burn off Newfoundland by a group of oil spill response organizations led by Environment Canada yielded important data in underscoring the technique's advantages. Photo at upper left shows an aerial view of the test burn. In photo at upper right, a remote controlled model helicopter samples emissions from the burn's smoke plume. Middle photos show views of the burn from the Canadian Coast Guard vessel Ann Harvey. Photo at lower left shows the burned oil residue in the fireboom following the completion of the first of two burns. Photo at lower right shows the second test burn from the M/V Casaco service vessel. Photos courtesy Environment Canada.

MPA members mostly are major integrated petroleum companies and tanker firms.

It has stationed equipment at 22 locations around the U.S. coast (See map, OGJ, Dec. 28, 1992, p. 22) and has trained its personnel. It will not operate off Alaska, where Alyeska Pipeline Service Co. has a parallel response capability.

MSRC has 435 employees, has spent $400 million on ships and gear, and will have an operating budget of about $90 million/year.

MSRC capabilities

MSRC is designed to make a best effort response to spills of 1,200-210,000 bbl in U.S. waters, not including the Great Lakes or inland waters.

The Coast Guard requires MSRC to be able to bring 10,000 bbl of cleanup capacity to the scene of a spill within 12 hr of the accident, 20,000 bbl within 36 hr, 40,000 bbl within 60 hr, and another 40,000 bbl thereafter.

Steve Duca, MSRC vice-president, stressed weather and other conditions often may prevent the firm from cleaning 100% of every spill, so that is not the goal.

"There is the expectation that the response company will clean up a spill without damage to the environment. That is not the case. We represent a major oil spill response resource for this country, but we are not going to be able to put every bit of toothpaste back into the tube."

Virtually all of the MSRC equipment is in place, and MSRC has waivers from the Coast Guard for gear that it temporarily lacks.

The last of MSRC's response personnel at command posts and in the field will continue training on their equipment. Also, the firm has been conducting a series of unannounced readiness drills. Duca said those have uncovered some defects "but no problem has been a showstopper."

MSRC is headquartered in Washington, D.C., but when spills occur, the general manager in one of five MSRC regions will direct the cleanup under supervision of the Coast Guard, using company equipment and employees and independent contractors if necessary.

The company's spill response capability is centered on sixteen 210 ft offshore response vessels. They carry boom, boom boats, skimming systems, oil/water separators, and can hold 4,000 bbl of recovered oil.

The 22 MSRC locations also are stocked with an array of equipment including barges, trucks, booms, skimmers, dispersants, and communications gear.

MSRC's first formal response

The company had its first formal response Aug. 11, 1993, even though it was not yet operational. A collision in Florida's Tampa Bay between a freighter and two products barges resulted in a spill of about 5,000 bbl of No. 6 oil from a barge. At the request of a member company, MSRC rushed 18,000 ft of boom to Tampa from its Miami center. About 2,500 ft was deployed. Within 2 hr, a response vessel also was dispatched toward the location.

Duca said, "MSRC has dealt with a number of very, very difficult issues in terms of acquisition and training to get to this point of readiness.

"We still have a lot of work to do in training and improving and integrating our operation to better serve our clients."

MSRC spill response status

OPA 90 became law Aug. 18, 1993, said David Usher, president of oil spill cleanup contractor Marine Pollution Control, Detroit. This was a short test period, but enough to show the long term value of OPA.

By October, more than 1,600 vessel response plans had been submitted to the U.S. Coast Guard for review, covering 5,800 vessels worldwide. Also, 88 contractors had applied to the Coast Guard for classification as oil spill response organizations, for which they need to demonstrate adequate standards.

"While Coast Guard approval is not required, it testifies to professionalism," said Usher. "The Coast Guard has given interim classifications to 70 contractors and final approval to 20. In addition, the Coast Guard has visited and inspected 200 sites."

Rear Adm. Henn said, "Despite our best efforts at prevention, it is inevitable that some oil will be spilled."

The response plan requirements of OPA 90 deal with the need for planning clean up operations before a spill occurs. This will be done through the national Preparedness for Response Exercise Program (PREP). The PREP guidance document is currently under development and was published in August 1993. Implementation date for PREP was Jan. 1, 1994.

Henn said large spills can overwhelm local response capability and require special planning. Such spills, dubbed "spills of national significance," are the subject of a plan under development, which will identify in detail the roles of federal, state, local, and private responders.

Tampa Bay spill response

In reponse to a recent 5,000 bbl oil spill in Tampa Bay, Fla., resulting from a collision between a freighter and two products barges, Usher said 18 contractors used 1,600 people at peak in the cleanup.

Equipment called out included 11 portable barges, 26 skid mounted vacuum transfer units, 18 skimmers, six rope mops, 86,000 ft of boom, 7,500 ft of inflatable offshore boom, a 16,000 bbl barge and mobile communications center, four vacuum trucks, more than 50 dump trucks, one helicopter, 19 pressure washers and more than 50 boats.

Usher said this response would not have been possible without OPA 90, with transporters, insurers, spill cooperatives and state and federal agencies working together to limit overall costs of the spill..

"On the negative side, the Coast Guard used its resources too long in offloading the barge," said Usher. While he admired the Coast Guard's "Shoot first, ask questions later" approach to spill response, Usher said this should not be at the expense of the private sector, which had been paid for by the tanker industry.

OPA insurance concerns

The U.S. Minerals Management Service in late 1993 began seeking a way to implement OPA's oil spill insurance requirements without harming the U.S. offshore petroleum industry.

Producers warned MMS its proposed rule, published in the Aug. 25, 1993, Federal Register, could force many small offshore operators to halt operations. The insurance industry said it could not underwrite all current offshore projects are the required $150 million of insurance each. The U.S. Coast Guard has a similar rule pending for tanker operations.

At presstime, the National Petroleum Council was to unveil a study on how the law could restrain offshore production, at Energy Sec. Hazel O'Leary's request. A Louisiana State University study found the proposed rule would literally shut down the oil and gas industry in Louisiana. If found that only 122 of the state's 2,000 oil and gas operators could afford insurance at $5/$1,000/year coverage and only 70 at $10/$1,000/year coverage.

Taylor Energy Co., New Orleans, said the insurance requirement would shut in 300,000 b/d of Gulf of Mexico production and force independents out of business. IPE

Eleo Maersk very large crude carrier is a double hull tanker design built by Odense Steel Shipyard Ltd. for Maersk Shipping AS. Photo courtesy of Maersk.

Tanker industry moves to clean up operations

As 1994 approached, the international tanker industry was logging significant strides in improving the safety and environmental awareness of its operations.

When Exxon Corp.'s Exxon Valdez tanker ran aground Mar. 24, 1989, in Alaska's Prince William Sound more than 250,000 bbl of oil spilled into the water.

The public outcry and political fallout that ensued spawned a shock wave that swept through the international oil industry.

With a string of mishaps following the Exxon Valdez spill, public scrutiny focused on the tanker industry even more sharply and found it lacking.

The U.S. government reacted strongly with passage of the Oil Pollution Act (OPA) of 1990, causing a second shock wave.

OPA defied worldwide thinking on tanker safety. It set narrow requirements and specified double hulls for new tankers, forced creation of a national spill response program in the U.S., and made tanker operators solely liable for spill damage.

While the industry was debating the rights and wrongs of those three rulings, two bigger tanker spills hit Europe in 1992.

The Aegean Sea and Braer tanker accidents forced the pace of European spill prevention and cleanup legislation.

Meantime, tanker owners were operating in generally weak markets for crude oil carriers throughout 1992, the latest year for which most data are available. Scrappings and newbuilding deliveries were on the rise.

Spot charter rates for crude carriers averaged considerably less than those of 1991. Rates for product carriers remained close to the depressed levels of second half 1991, when temporary stimulation from the Persian Gulf war had eased.

Spill liability

In 1993, Drewry Shipping Consultants Ltd., London, said the tanker industry had largely accepted the double

Double hull for Mobil's Eagle VLCC takes shape at Sumitomo Heavy Industries Ltd.'s Oppama shipyard in Japan. Later, the ship undergoes sea trials before Aug. 25, 1993, commissioning.

Mobil Corp. commissions double hull VLCC

Mobil Corp.'s Eagle, a very large crude carrier (VLCC), was commissioned in 1993 at Sumitomo Heavy Industries Ltd.'s Oppama shipyard in Japan as part of a growing number of double hull tankers in the world fleet.

Mobil said the 280,000 dwt vessel, built for Mobil Shipping & Transportation Co., is the first double hull vessel of her class owned by an oil company and the sixth double hull VLCC in the world.

She is 1,089 ft long and 190 ft wide. Cargo capacity is 2.2 million bbl.

Mobil's research, as well as studies by the National Academy of Sciences, found the double hull to be the best option in vessels of this size to safeguard the environment.

Construction began early in 1992.

The Eagle's double hull, along with other environmental and safety features of the vessel, exceed the new, more stringent tanker safety standards set by the U.S. Oil Pollution Act of 1990 and International Maritime Organization rules.

An evolution

Mobil said it developed and patented the design for the first double bottom tanker in 1968. Ten ships in Mobil's fleet have double bottoms.

The double hull is a further evolution of the double bottom design. It extends the double bottom up the sides of the vessel to provide a second steel barrier between the cargo and the water. Because crude oil is carried in the interior tanks, the outer steel skin could be penetrated without spilling cargo.

Between the inner and outer hulls of the Eagle is a space about 10 ft wide on the bottom and 13 ft wide on the sides. When the ship is fully loaded with crude oil, this space is empty. After the tanker has discharged its cargo, the protective space between the two hulls is filled with seawater as ballast for the return voyage.

Mobil said although the double hull adds significantly to the engineering complexity and cost of a VLCC, there is an equal increase in the vessel's safety and environmental security.

Eagle was destined to join Mobil's deepsea fleet of owned and chartered vessels that transport in excess of 650 million bbl/year of crude and products.

hull requirement set out in OPA but universally condemned the principle of unlimited liability for spills borne by shipowners and operators.

"One prominent shipowner has already described the trading of tankers to the U.S. as a game of Russian roulette," said Drewry.

"Some have simply stopped trading in black oils to U.S. ports in their own ships, while others are relying solely on chartering."

By introducing unlimited liability in U.S waters, Drewry argued that American authorities were in danger of creating a self-imposed oil embargo at a time when the country's dependence on oil imports was rising.

"The old principle of sharing responsibility for transportation between cargo owner and shipping company is not included in OPA to the detriment of the environment," said Andreas Ugland, chairman of International Association of Independent Tanker Owners (Intertanko), Oslo.

"This significantly reduces the incentive for charterers to choose the best operator and the best ship. This should be changed."

Intertanko and the U.S. Coast Guard reached agreement over independent operators' problems in securing oil spill response coverage on the U.S. West Coast.

Intertanko members holding contracts with National Response Corp. (NRC), New York, were told NRC could not provide a U.S. West Coast service.

They feared they may have to duplicate East Coast coverage currently in place in order to provide full U.S. response.

A meeting between Intertanko and the U.S. Coast Guard in late July 1993 resulted in a compromise. Intertanko members' vessels would stay more than 60 miles from the U.S. West Coast until they were entering port, and this they would do directly.

Oil field activity shares the desert with remnants of war in Kuwait. Here, Kuwait Santa Fe Rig 12 works over a well in Raudhatain oil field near the hulk of a damaged Iraqi tank.

Kuwait bounces back from Persian Gulf war

Kuwait revitalized its oil production so successfully, following the Persian Gulf war of 1990-91, that capacity at yearend 1993 was greater than before the Iraqi invasion.

A status report by a Kuwait Petroleum Corp. official in November 1993 detailed progress since Iraqi forces were ousted from Kuwait in February 1991:

- 111 oil wells drilled and more than 650 wells worked over. Only 110 wells were found intact after a spree of sabotage by retreating Iraqi soldiers.
- 18 of a total 26 oil gathering centers, all damaged in the war, repaired and returned to operation.
- Partial or full restoration of Kuwait's three refineries was expected to be complete in first half 1994.
- Repairs to marine export facilities were expected to be complete in first half 1994.

Kuwait bounced back with such vigor that government demands for increased production quotas caused headaches for the Organization of Petroleum Exporting Countries during 1993.

Trail of devastation

Iraqi forces invaded Kuwait Aug. 2, 1990, and occupied the country 7 months. They left a trail of more than 700 blazing oil wells, vast lakes of spilled oil from well blowouts as a result of sabotage, wrecked oil gathering centers, and damaged refineries and export terminals.

State owned Kuwait Oil Co. (KOC) said 1,080 wells were on production before the Iraqi invasion. Of 726 damaged wells in Kuwait proper, KOC estimated 706 were repairable, with 421 of those requiring workover.

At first KOC thought it would have to restore production largely by drilling new wells. Accordingly, the company began drilling before the last well fire was doused.

All 26 oil gathering centers in operation before the invasion were damaged or destroyed during the occupation. KOC originally planned to get 15 oil gathering centers back in operation immediately and later added three more.

P. Stuart Tholand, Bechtel Corp. senior vice-president and manager of Kuwaiti restoration projects, said to do this quickly KOC decided to return the gathering centers to their preinvasion condition.

"A lot of the gathering centers have been here a long time," Tholand said. Replacements for other destroyed gathering centers would follow, built to state of the art designs.

State owned Kuwait National Petroleum Co. (KNPC) has three refineries in Kuwait: Mina Al Abdullah, Mina Al Ahmadi and Shuaiba. The Iraqis only damaged control equipment at Mina Al Abdullah, while the control room and processing systems were damaged at Mina Al Ahmadi. Shuaiba refinery, on the other hand, was so badly damaged that KNPC debated abandonment of the wreckage in favor of construction of a new high grade plant.

After a first survey of devastation, firefighter Red Adair said it would take at least 2 years to cap the wild wells. However, 27 firefighting teams from around the world achieved the task in less than 8 months.

After Iraqi forces were ousted from Kuwait in February 1991, it was estimated that putting out well fires and returning production to prewar capacity would take as long as 5 years.

Emergency action

Bechtel Corp., Houston, won the project management contracts for emergency work and reconstruction of Kuwait's oil production facilities. Bechtel and KOC began planning emergency repairs in November 1990, while Iraqi forces still held Kuwait. The emergency phase of reconstruction was called Al Awda, Arabic for The Return.

In March 1991, only a few days after Iraqi forces retreated from Kuwait, Bechtel engineers began to lead damage assessment, planning, engineering, procurement and logistics, and construction operations.

The contractor arrived amid the ruins in March 1991 and completed its work in July 1993, at which point Kuwait's oil production capacity had reached 2.1 million b/d, almost the level before the war.

KOC returned to production sooner than anticipated, with 250,000 b/d used to meet local demand within 6 months of liberation.

KOC's target had been to return production to 2 million b/d. Production was up to 1.7 million b/d through 16 gathering stations by the end of 1992. The 2 million b/d capacity target was achieved in February 1993.

Bechtel said the first sabotaged well was capped in March 1991 and the last in November that year. Middle East Economic Survey (MEES) estimated in May 1993 Kuwait had spent more than $10 billion on reconstruction work.

Well repairs

KOC began to concentrate on workovers early in 1992, with jobs taking an average 8.5 days/well. KPC subsidiary Kuwait Santa Fe reportedly was averaging 30-40 wells/month.

Completion of the workover program was scheduled for mid-1993. The most challenging workovers were saved for last.

"The real difficult, nasty, junky wells we know we're going to have to spend some time on are not being addressed right now," said Gary L. Bauer, Kuwait Santa Fe general manager, in February 1993.

The well repair program completed in 1993 had two main stages: postcapping and workover.

Firefighters killed all but 30 or 40 of the burning wells without cementing them. Instead, they often dug down to undamaged casing and cut off the casing at that point. Then they installed new casingheads or whatever was handy to shut them in.

That was the condition of the wells when postcapping crews set to work.

"They take a power swivel and crane and they've got a BOP set up on a platform," explained Gary W. Casswell, Kuwait Santa Fe general drilling superintendent.

"They first kill the well, then nipple up the BOP and go in with the power swivel and do a manual backoff of the tubing. Hopefully, it backs off fairly close to the surface."

The postcapping crew then set a packer, reconditioned casing, and installed new spools and christmas tree. Nearly all the damaged wells required new wellhead equipment because of the intense heat and the amount of time some of the wells were left to burn.

After KOC scheduled a workover, "We come in, set a check valve, rekill the well if we have to, then nipple down the tree, nipple up the BOP, and start our workover," Casswell said.

A typical workover involved the crew pulling and checking tubing, performing whatever work was necessary downhole, reperforating, checking the integrity of the casing, then rerunning the production string.

"From there," Casswell said, "you're back on line. Probably within 2 days after we complete a workover they've already got a pipeline laid to it and are ready to go on production to a gathering center."

KOC mobilized 17 drilling rigs to drill new wells and work over damaged wells since the liberation of Kuwait in early 1991.

Reservoir damage

At yearend 1993, KOC had still to determine the extent of reservoir damage resulting from the 700-plus well fires, some of which burned for 9 months. Counting artificial lift wells in the Neutral Zone between Kuwait and Saudi Arabia, about 1,100 wells were damaged. Kuwait and Saudi Arabia share Neutral Zone production 50-50.

Wells returned to production were said to have greater water cuts than they did before the invasion. KOC said higher water cuts were deliberate, resulting from water production from edge wells under current reservoir management plans.

Abdul Al-Rabah, manager of KOC's production development group, said the company was deliberately enhancing water production, especially from edge wells, in an effort to keep the oil sweep uniform.

Most of Kuwait's oil fields, including the Burgan complex, have water drive reservoirs. Associated gas was being flared at gathering centers where condensate recovery units had not been returned to service.

New drilling

Among new wells planned by KOC was a pilot project involving drilling of Kuwait's first two extended reach wells.

Kuwait Santa Fe drilled the first two horizontal wells in Southeast Kuwait. Horizontal drilling was chosen to target tight, vertically fractured, lower Cretaceous Mauddud limestone.

The first horizontal well was drilled in Ahmadi field. Liner stuck in the 2,000 ft horizontal section, after which the well was reentered and drilled vertically.

A second well in Magwa field met with better results, returning initial production of 4,000 b/d through 1,800 ft of perforated liner.

For both the Ahmadi and Magwa wells, kickoff point was 4,100 ft and total measured depth about 6,100 ft. Angle was built up at 2°/100 ft of hole.

The Mauddud formation has 10-15% porosity, with permeability varying from 10 md to 100 md. The only other Mauddud formation well in Kuwait produces about 500 b/d.

As 1994 approached, Kuwait was considering granting U.S. companies concession rights in an oil field repossessed under a newly demarcated border with Iraq.

Kuwaiti Oil Minister Ali Al-Baghli said U.S companies could receive permission to produce oil from 11 wells in Rutqa field, which taps the same reservoir as Iraq's South Rumaila field. Before Iraq's invasion, Kuwait produced 10,000 b/d from Rutqa, while Iraq produced an undisclosed volume from wells farther north.

Exploration plan

From 1993 to 1998, KOC intends to explore for more oil

fields in Kuwait. The company's exploration strategy was said to involve 2D seismic surveying of most of the land and transition zones of Kuwait, 3D seismic surveys to provide more precise subsurface information on some oil fields, and drilling an adequate number of deep wildcats throughout Kuwait, targeting oil reservoirs in Jurassic and pre-Jurassic formations.

Kuwaiti Finance Minister Nasser Al-Rodhan told the Al-Seyaaseh newspaper the government was considering allowing foreign oil companies to participate in exploration and drilling with KOC.

London's Financial Times reported talks had begun between KOC and oil groups.

These offered Kuwait foreign expertise and, more interestingly in view of the country's $5 billion budget deficit in 1993, foreign capital.

British Petroleum Co. plc already held a contract with KOC to provide technical consultancy, while Union Carbide Corp. was involved in a joint venture to build a petrochemical plant.

MEES said Neutral Zone exploration will continue to focus on the Najma, Sargelu, and Marrat prospects, all Jurassic formations.

Neutral Zone operations

Deep drilling in the Neutral Zone was said to be resulting from encouraging discoveries in Kuwait during the 1980s.

Kuwait's oil ministry reported early in November 1993 an oil discovery in the Neutral Zone.

The South Umm Gudair DW-1 wildcat was reported by MEES to have discovered first oil in the Gotnia formation. The well flowed a maximum 6,144 b/d of 25° oil through a 1¼ in. choke. Drilling was carried out jointly by KOC and Texaco Saudia Inc.

Arabian Oil Co., a Saudi-Japanese-Kuwaiti joint venture, holds the rights to Neutral Zone offshore production.

KOC said Neutral Zone offshore production amounted to 350,000-400,000 b/d before the invasion. Installations were little damaged during the war and production quickly returned to normal.

Onshore production in the Neutral Zone was hit badly, though, with almost all 365 wells sabotaged by the Iraqis and left burning.

Ninety-five of the original wells were now producing at yearend 1993, although KOC reckoned it would take until 1996 to restore full production.

Cleaning up oil lakes

Operations were started by KOC in March 1991 to clean up oil lakes formed from damage to oil wells. Vacuum trucks and pipelines pumped recovered oil directly to Mina Al Ahmadi refinery and field treatment plants.

By late November 1993 about 17.3 million bbl of weathered crude had been recovered, and an estimated 4.7 million bbl remained.

Recovery was said to be taking place at about 40,000 b/d of weathered crude, with completion of lake depletion scheduled for March 1994.

MEES said several shipments of weathered crude had been made by November 1993, with total value estimated at more than $62 million.

Bechtel's Tholand said one problem with recovery of spilled oil was the variation in oil from lake to lake.

"No two lakes are the same in terms of content of salt and sand. Everything is different. It's a little difficult to know in each instance what you're going to get."

Tholand said recovery operations accelerated in summer when the crude oil was warm. Though careful to leave judgments on environmental damage to specialists, he noted that in general the oil had not penetrated very deeply into the sand.

Gathering centers

The second phase of repair work, known as Al Tameer from the Arabic for The Reconstruction, involved rehabilitating the oil fields' gathering and processing capacity and much of the infrastructure.

Kuwait originally had five gathering centers in the north: two in Raudhatain field, two in Sabriyah and one in Ratqa. There were three in the west, one in Minagish and two in Umm Gudair. There were 18 in the southeast: two each in Ahmadi and Magwa fields and 14 in Burgan.

The first gathering centers brought back into operation were in the Burgan complex. Two others were in Raudhatain and Sabriyah fields.

A heavily damaged gathering center in Umm Gudair field in western Kuwait was replaced by Daniel En-Fab Systems Inc., Houston, which shipped the equipment to Kuwait in modules.

Round-the-clock work brought total Kuwaiti production to 400,000 b/d by December 1991. By the end of June 1993, 18 of the original centers were back in operation, and production goals were being met.

A total 1,800 miles of new flow lines was laid to support rehabilitation of oil gathering centers. The first postwar oil was pumped through two of the original 26 gathering centers May 26, 1991.

KOC's crude handling capacity in surface installations is estimated at 2.83 million b/d, with plans under way to raise this to 3.65 million b/d by August 1996. Further projects under study would raise handling capacity still further to 4.45 million b/d by March 1997.

Bechtel planned to push oil storage capacity to 18 million bbl. By mid-1993 half the preinvasion storage capacity of 15 million bbl in two tank farms was available.

Al Awda/ Al Tameer tally

Upon completion of its restoration project, Bechtel calculated it had mobilized more than 16,000 workers in putting out well fires, stemming oil flow from damaged wells, and returning fields to production.

The contractor deployed 742 aircraft and ships to carry more than 520,000 tons of equipment. About 5,800 pieces of equipment shipped included bulldozers, cranes, trucks, front loaders, ambulances, and support vehicles.

Bechtel's tally showed that in only 8 months the emergency program capped 751 wells sabotaged during the Iraqi occupation and gulf war, including 647 wells that were on fire.

"Starting with no equipment or supplies, in a country devastated by war, KOC, the firefighters, and Bechtel accomplished what was thought to be an impossible task," said Don Gunther, Bechtel executive vice-president.

"This remarkable industrial achievement is an outstanding example of what men and women can do when they work together for the common good."

Bechtel estimated that 250 miles of oil and water pipelines were required for Al Awda. Total capacity of the water network was 25 million gal/day, delivered to fire sites. Each pump and hose could deliver 6,000 gal/min of water.

A total 360 lagoons were excavated, lined with plastic sheeting, and filled with as much as 1 million gal of water each to support firefighting.

In March and April 1991, four firefighting companies had teams mobilized to Kuwait. They came from Boots & Coots Inc., Red Adair Co. Inc., and Wild Well Control Inc., all of Houston, and Safety Boss Ltd., Calgary.

Between August and October 1991 another 23 teams, each made up of 10 firefighters and 20-40 support crew, joined firefighters already in Kuwait. They came from

Kuwait, Iran, China, Hungary, U.K., France, Canada, Romania, and Russia.

While fires were being controlled, unexploded ordnance had to be removed from 500 sq miles of land. Bomb disposal squads detonated more than 23,000 land mines, rockets, grenades, and unexploded bombs.

Access roads, work sites and drill pads had to be built. More than 250 miles of new roads were required. Construction involved more than 10 million cu m of sand/gravel mixture.

Huge support systems backed up Al Awda and Al Tameer. The communication system assembled for reconstruction included 23 satellite telephone networks, 4,500 telephones, and 2,000 portable radio phones.

Two medical emergency helicopters and a 40 bed hospital provided safety cover. About 100 medical staff operated from seven medical stations.

Catering too was on a grand scale. Six dining halls provided as many as 35,000 meals/day. A total 3.5 million meals were cooked during the firefighting campaign.

Refinery damage

A $3 million assessment of physical and economic damage inflicted on Kuwait's refineries, petrochemical plants, and service stations was conducted by Edward Rushton Son & Kenyon, London.

Commissioned by KNPC, the valuation was submitted to the Kuwaiti government as part of a claim for damage repairs. It was later submitted to the United Nations for war reparations. Edward Rushton said the total U.N. claim is likely to run into billions of dollars.

The firm said Kuwait's exploration and production facilities were mainly damaged willfully by Iraqi forces. Damage to refineries was said to have been caused largely by exchange of fire between Iraqi and allied coalition forces.

KNPC operated three refineries with combined capacity of 830,000 b/d before the invasion.

In September 1993 KNPC announced completion of repairs to Mina Abdullah refinery. Sami Fahd Al-Rashid, director of planning, told reporters all production units had been recommissioned after repairs. The refinery was said to be operating at full capacity of 230,000 b/d.

Al-Rashid said all units at Mina Al Ahmadi refinery also were to be operational by the end of 1993, except for one crude oil distillation unit which was to be on stream in October 1994. Mina Al Ahmadi was processing 185,000 b/d, compared with pre-invasion capacity of 400,000 b/d.

About 25% of storage capacity at the refineries was destroyed, either through direct damage to tanks or loss of connected pipelines.

Temporary repairs of damaged oil piers at Mina Al Ahmadi terminal and control systems at Mina Abdullah sea island were completed first to allow exports. Repairs of marine export facilities, including the oil export pier at Shuaiba terminal, were expected to be complete in first half 1994.

A marine consultant was comparing the cost of options to repair or refurbish existing marine facilities with building anew.

KNPC announced in November 1993 it intended to start production and export of unleaded gasoline to European specifications in 1997. Also, domestic production was to change over to unleaded products in line with a Gulf Cooperation Council plan for the region.

Shuaiba refinery

Worst hit of the refineries was Shuaiba.

KNPC listed damage as extensive damage to the second stage hydrocracker, substantial damage in one crude unit tower and associated equipment, partial damage to the crude oil vacuum distillation section and associated equipment, partial damage to the heavy oil unit, damage to the main interconnecting pipe rack, destruction of some fuel oil and kerosine tanks, light damage to a large number of products tanks, light damage to the Shuaiba pier and approach way, and destruction of one electrical substation.

KNPC completed recommissioning of Shuaiba refinery Nov. 17, 1993, about 2½ years after it was damaged in the war. Capacity was 130,000 b/d for the restored plant, compared with 200,000 b/d before the invasion.

The return to operation brought Kuwait's total refining capacity up to 530,000 b/d.

As Shuaiba went on stream, the Mina Abdullah refinery was said to be processing 230,000 b/d of crude oil and Mina Al-Ahmadi refinery 170,000 b/d.

KNPC said repairs to the 20 year old Shuaiba refinery took 20 months to complete at a cost of $85 million. Major repairs and reconstruction were required on four process units, preventive maintenance was carried out on 23 process units, and repairs were required on 36 petroleum products storage tanks.

Shuaiba Technical Services Manager Sa'd Abd Al-Wahhab Al-Sa'd told MEES a damaged heavy oil vacuum tower was repaired and its capacity restored to prewar level of 26,000 b/d.

The refinery's main isocracker was repaired, with capacity at 33,000 b/d compared with 44,000 b/d before the invasion.

Other units repaired included three 70 MMcfd hydrogen compression trains.

Two were in operation, while the third was idle as a backup unit.

KNPC expected its total refining capacity to reach 650,000 b/d by the end of January 1994 and 730,000 b/d by the end of 1994 or early 1995.

Export terminals

Kuwait's offshore Sea Island export terminal and single point mooring system were destroyed and the southern pier at Mina Shuaiba badly damaged in the war, although the northern pier was almost unscathed.

Exports resumed from the Sea Island terminal Sept. 6, 1992, following completion of repairs.

The terminal lies 12 km east of Mina Al Ahmadi's north and south piers. It has two berths of about 28 m water depth capable before the invasion of loading 19,500 tons/hr of oil.

A single point mooring buoy lies 2.4 km east of Sea Island. Its pipelines to onshore installations were cut during the war.

Mina Abdullah refinery's southern pier resumed operations in March 1992.

Third phase

In October 1993 KOC let a $100 million contract to Ralph M. Parsons Co., Pasadena, Calif., to provide program management services and further restore and increase oil production and processing in Kuwait.

MEES said the total installed cost of the third phase of restoration is likely to be $3-5 billion. This will involve more than 50 individual projects, including construction of desalting and water treatment units and two new gathering centers in Minagish and Umm Gudair fields in western Kuwait.

Work will include restoration of oil production and processing facilities; repair, reconstruction, and new construction of some buildings and utilities in Ahmadi township; maintenance support for KOC installations and computer systems; and engineering and construction of energy production facilities. IPE

World gas processing to continue growth

Gas processing, especially in the U.S., will recover from shrinking margins during the mid 1990s to remain profitable for the rest of the decade and into the next century.

And with new production, particularly in the Middle East and Africa, supplies for world trade will be ample for the rest of the decade, predicts Purvin & Gertz, Dallas consulting firm.

U.S. natural gas processing economics have improved dramatically from the low margins of the late 1980s. That improvement so far this decade has resulted from firmer NGL demand and prices and, until late 1992, from generally weak natural gas wellhead prices.

The outlook for NGL and gas prices for the remainder of

Commissioning in 1993 of the Sunray, Tex., gas processing plant near Amarillo by Maxus Energy Corp., Dallas, brought to 240-250 MMcfd Maxus' processing capacity in its U.S. Midcontinent region. The company also processes gas in Southeast Sumatra, Northwest Java, and the U.S. Gulf Coast. The $65.5 million Sunray plant, which features a state of the art cold box for producing helium, processes sweet and sour gas through separate trains. Photo courtesy of Maxus.

World LPG supply

Fig. 1

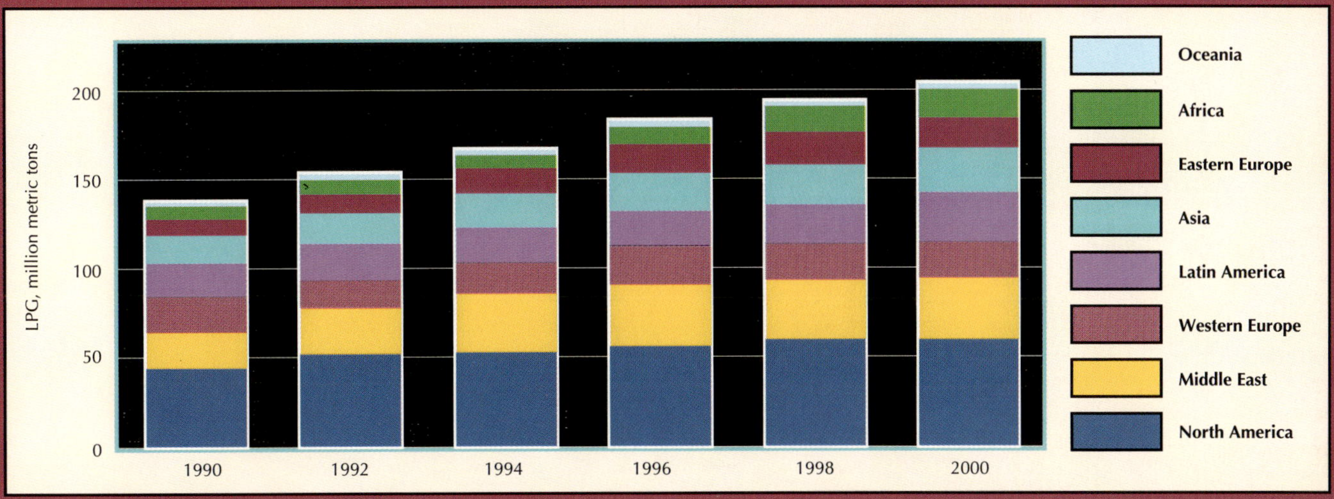

the decade indicates that gas processing margins will continue tight in the mid 1990s, as natural gas prices rise faster than NGL prices, then gradually improve.

Although gas processing will remain profitable throughout the decade, says the firm, several structural changes in the industry are under way or seem likely.

These include the disappearance of the so called gas bubble, a rapid increase in NGL feedstock requirements for olefin and methyl tertiary butyl ether (MTBE) manufacture, and emergence of alternative fuel mandates in developed areas of the world.

If gas prices rise too quickly, says Purvin & Gertz, profits from processing will be significantly squeezed.

On the other hand, petrochemical feedstock demand for NGL has increased in the early years of this decade. Whether NGL supplies will be adequate over the long term and how a possible shortfall could affect NGL prices are open to question.

Additionally, clean air and alternative fuel legislation appears to improve the prospects for expanded use of propane as engine fuel.

Postwar slowdown

The world LPG industry generally had ample supplies during most of the 1980s. Since the Persian Gulf war, however, growth in the world's LPG supplies has slowed relative to traditional premium captive demand growth.

This situation has led to tighter LPG supply/demand balances, excess shipping capacity, firmer prices, and a reduced need to develop price sensitive markets to use available supplies, Purvin & Gertz says.

With the U.S. likely to become more active in world LPG trade, North American gas processing will be influenced by emerging global trends.

Supply growth

World LPG supplies earlier this decade grew steadily, Purvin & Gertz data show. This improvement is highlighted by record LPG production in Saudi Arabia, renewal of supplies from Kuwait, and buildup of North Sea production.

By 2000, world LPG supplies will increase to nearly 203 million metric tons/day, or 6.6 million b/d (Fig. 1).

The largest increase in LPG supplies will be from the Middle East, where for the first half of this decade supplies have been almost totally related to crude oil production.

In the future, however, LPG production from the region will depend more on processing of nonassociated gas and gas/condensate reserves.

Saudi Arabia will continue to dominate LPG production from the region. By 2000, Saudi Arabia will expand its master gas system, which is currently operating at capacity.

Throughout the 1990s, Iran likely will significantly increase its LPG production with the full commissioning of its Bandar Imam facilities. Prospects for recovery of LPG from nonassociated gas production in Iran also will improve by 2000.

In 1995, Iraq should emerge as a major LPG producer and exporter, says Purvin & Gertz, assuming a resolution of the United Nations' sanctions and the completion of repairs or reconstruction of war damaged LPG production and export facilities.

Kuwait resumed LPG recovery in 1992 and reached full production levels consistent with crude oil rates by the first of 1994.

Qatar significantly increased LPG production in 1992 with commissioning of the first phase of its North field project. Prospects for additional LPG supplies have been enhanced by the probable implementation of an LNG project by early in the next century.

Development of a gas/condensate recycling project onshore and commissioning of a third train on Das Island will likely increase significantly LPG production from Abu Dhabi by 1995.

LPG supplies in western Europe will increase through the mid 1990s as they become available from rising gas/condensate flow in the North Sea.

Total LPG production in eastern Europe, including the former Soviet Union, will rebound as the region adjusts to economic restructuring and introduction of western capital and technology.

While LPG exports will gradually increase, most of the production increases will be absorbed by growing internal markets.

LPG production in Asia is expanding with growing refinery production, particularly in South Korea, and increasing supplies from gas processing, specifically in Malaysia.

During the second half of the 1990s, LPG supplies in Oceania will be further enhanced with development of the

LPG available for export: Middle East, Africa

Fig. 2

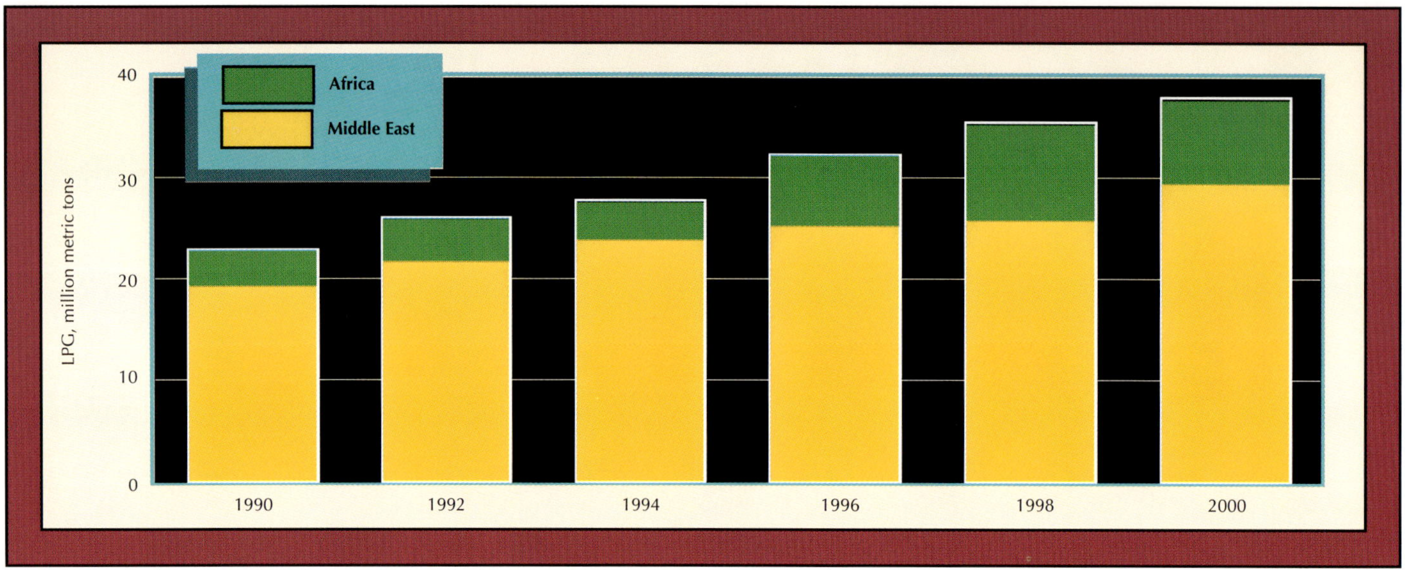

Australian Northwest Shelf LPG recovery project.

Latin American LPG supplies will be led by major gas processing expansions, particularly in Brazil and Venezuela. Increasing production in Argentina and Mexico also contribute to growing supplies in the region, Purvin & Gertz predicts.

Major production increases from new gas processing projects in Algeria and Nigeria will significantly increase LPG supplies in Africa during the latter half of the 1990s.

LPG supplies from the Middle East and Africa have dominated international trade in LPG. In 1992, exports from those two regions accounted for more than 95% of total large volume waterborne movements.

Future world trade also will be influenced by exports available from the Middle East and Africa. By the end of the century, waterborne trade from the two regions will increase by nearly 50% (Fig. 2).

While exports from the Middle East will continue to dominate world LPG trade, Saudi Arabia's role will diminish somewhat, as increasing domestic requirements reduce supplies available for export. Throughout the period, the United Arab Emirates will emerge as a more significant source of Middle East supplies.

By the end of the 1990s, LPG exports from Africa will increase substantially. Doubling of exports from Algeria and emergence of Nigeria as a major supplier will result in the region becoming more important to world LPG trade.

Market demand

Purvin & Gertz sees substantial growth for most world LPG markets (Fig. 3).

Average annual growth in the residential commercial sector of 3.5% should sustain that sector's current market share of about 46%. This outlook is supported by continuing strong demand in the developing countries of the Far East, Latin America, and Eastern Europe, including republics of the former Soviet Union.

Residential-commercial growth in North America, western Europe, and Japan will moderate to reflect the maturity of those markets.

Through 1993, the petrochemical sector accounted for about 23% of total world consumption. Growth in this sector will be substantial, dominated by increased butane use as MTBE feedstock.

Continued growth in propane use as olefin plant feedstock also will contribute to base petrochemical demand for LPG in most regions of the world. This is particularly true for the former Soviet Union where supply of LPG feedstocks to the petrochemical industry will be enhanced with completion of the trans-Urals pipeline project in the second half of the 1990s.

Purvin & Gertz says the expectation that future world LPG supplies will exceed premium market demand makes clear a continuing need to develop price sensitive markets, especially for propane. This condition will contribute to growth of LPG use in the petrochemical sector, particularly in the U.S. and Europe.

The Far East and Middle East also will participate in the clearing of world LPG supplies in new olefin plants with LPG feedstock flexibility. By 2000, use of LPG in the petrochemical sector will increase to a market share of about 28%.

With the exception of the engine fuel market, most all other world markets for LPG will grow at lower rates than residential-commercial and petrochemical sectors.

Increasing environmental pressures and clean air initiatives will enhance the growth prospects for LPG as engine fuel, says Purvin & Gertz. Engine fuel growth in any region will, however, be directly influenced by tax incentives.

Refinery use of LPG, mainly butane, will be affected by future specifications for motor gasoline, particularly in the U.S. Generally, lower vapor pressure specifications tend to reduce refiners' use of normal butane.

This reduction is offset somewhat by an expected increase in isobutane consumption as feedstock for alkylation and for internal production of MTBE. Overall, the market share for refinery use of LPG will decline slightly during the forecast period.

The greatest decline in LPG market share will occur in the industrial sector. Natural gas continues to penetrate this market sector, although growth in industrial LPG use remains likely for some developing countries.

Although projected supplies will keep pace with demand buildups, significant delays or cancellations of anticipated new supply projects could aggravate the current global tightness in LPG supply, particularly for butane.

This development could speed a growing trend of substituting propane for butane in LPG mix applications, as well as promote the use of alternative fuels in some Far East industrial LPG markets.

World LPG demand

Fig. 3

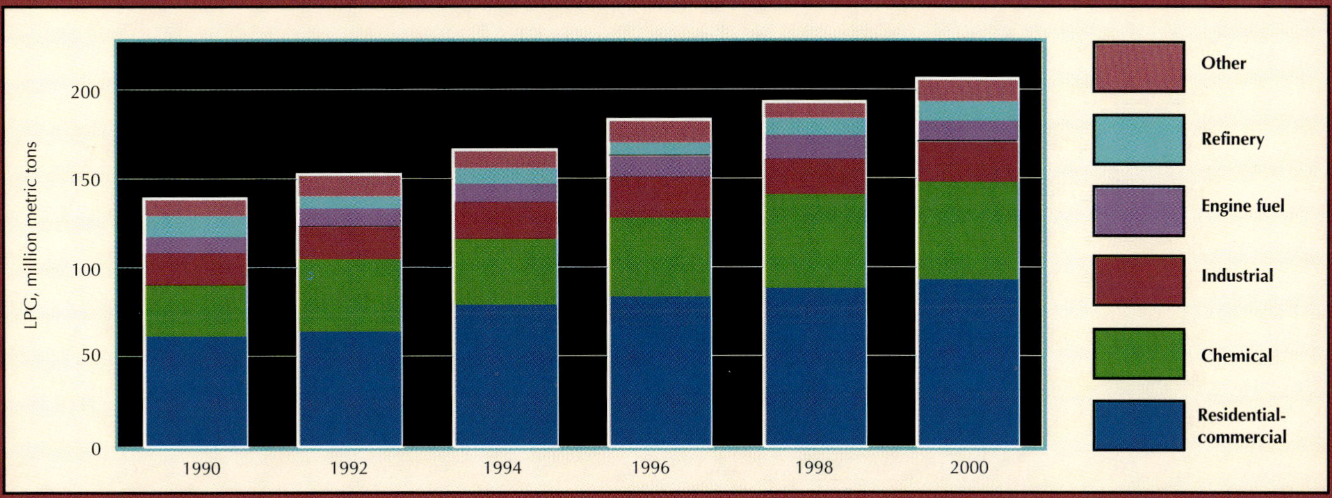

Outlook in the U.S.

Purvin & Gertz points out that in the first half of this decade, the U.S. natural gas industry has been in transition. Gas supply shifted from surplus to nearly balanced or even tight in some areas during certain times of year.

Gas demand has continued to grow, while U.S. drilling activity has been restrained at best—often depressed.

Although the U.S. has a relatively large potential gas reserve base available for development, says Purvin & Gertz, wellhead prices may be too low to spark enough drilling to keep pace with demand growth. If that occurs, gas demand could face tight supplies by the second half of the 1990s.

The outlook for natural gas prices and import levels through the middle of the decade suggests that natural gas supplies will continue to increase, but growth may slow toward the end of the decade.

Total Lower 48 natural gas supplies, including imports and nonconventional gas, will increase from about 19.7 tcf in 1992 to 21.8 tcf in 2000.

This view agrees generally with the National Petroleum Council, which in late 1992 released a study on the potential for natural gas supplies.[1]

Conventional Lower 48 dry gas production will increase from about 17.2 tcf in 1992 to about 17.9 tcf in 2000. Associated natural gas production will decline during this period as crude oil production in the Lower 48 continues to fall.

Nonassociated gas production rates will increase, more than offsetting the drop in associated gas production.

Reduction in associated gas volume is particularly significant to gas processing because associated gas is typically rich, which results in considerably higher NGL production rates per unit of gas volume compared with nonassociated gas.

This reduction, says Purvin & Gertz, will moderate somewhat the effect of increasing natural gas production rates on NGL supplies, particularly butane and natural gasoline.

U.S. NGL supplies

In the U.S., NGL is produced from natural gas in gas processing plants and from crude oil in refineries. Future NGL supply levels depend, therefore, on many complex variables that affect the natural gas, gas processing, and refining industries.

Total NGL supplies have increased significantly in the last several years. NGL production from gas processing has increased since 1990 mainly because of substantial increases in gas processing margins since the late 1980s.

Refinery production of NGL, mainly propane and butane, has increased because of increased conversion levels and reduced gasoline summertime volatility.

Total NGL supplies increased about 12% from 2.1 million b/d in 1990 to about 2.3 million b/d in 1992 (Fig. 4).

U.S. NGL production will continue to rise during the next several years, peak at about 2.4 million b/d around 1996, then gradually decline as production from gas processing moderates.

Ethane, propane prospects

Purvin & Gertz takes note of noticeable increases in U.S. ethane production earlier this decade due mainly to dramatic improvement in ethane extraction economics. Ethane extraction has been generally profitable in the 1990s, compared with marginal or unprofitable economics of the late 1980s.

As a result, ethane production increased from about 490,000 b/d in 1990 to about 570,000 b/d in 1992. In 1993, U.S. ethane production dipped as ethane rejection increased in response to reduced demand.

Overall, however, Purvin & Gertz believes ethane supplies will continue to expand during the rest of the 1990s, as ethane recovery levels slowly increase.

Anticipated ethane extraction margins should be high enough to encourage many gas processors to recover ethane.

Also contributing to a projected strength in ethane supplies through the mid 1990s is a gradual increase in natural gas volumes available for processing, says Purvin & Gertz.

Earlier this decade, U.S. propane production increased from about 920,000 b/d in 1990 to 1.2 million b/d in 1992. The rise resulted from increased extraction from natural gas as well as increased production from refineries.

Total U.S. propane production will continue to expand during the 1990s. Supplies from natural gas processing will continue to rise through the mid 1990s and then slightly decline late in the decade, Purvin & Gertz preicts.

The projected slow decline results from a relatively flat

U.S. NGL production

Fig. 4

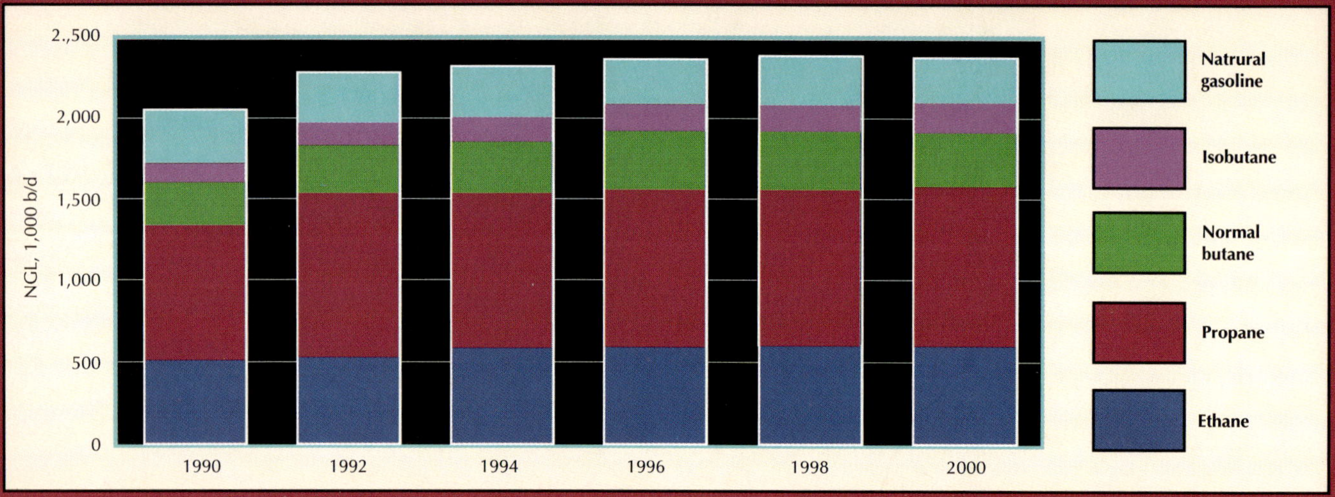

outlook for natural gas volumes available for processing and a general decline in average gas quality.

Refinery propane production will continue to increase during the 1990s.

This trend is despite U.S. clean air legislation that has forced most companies to modify refinery configurations to meet more stringent product specifications. The complex changes at most refineries may, as in reduced reforming severity, reduce LPG production.

But other changes will offset this yield shift, says Purvin & Gertz.

Butane's fortunes

In the early 1990s, normal butane supplies in the U.S. increased mainly because of reductions in summertime gasoline volatility. Refiners have been forced to back out large volumes of butane from gasoline blends during the summer.

As a result, most high conversion refineries have been oversupplied in butane during the summer. Normal butane supplies increased from about 270,000 b/d in 1990 to about 320,000 b/d in 1992.

Supplies from gas processing increased somewhat during the period, but most of the change resulted from the effect of lower gasoline volatility on refinery supplies.

Purvin & Gertz says normal butane supplies will increase through 1995, then remain relatively flat during the rest of the decade (Fig. 4).

Isobutane and natural gasoline supplies have increased since 1990 because of increased gas processing.

Supplies of isobutane and natural gasoline have continued to rise through the mid 1990s but will slowly decline due to the same shifts in gas volumes and quality that affect propane recovery.

Additional isobutane supplies via isomerization will be required to meet increasing demand.

U.S. demand for NGL increased earlier this decade, from about 2.2 million b/d in 1990 to about 2.4 million b/d in 1992 (Fig. 5). Many factors contributed to the increase.

Ethane, propane outlook

Through the first few years of this decade, says Purvin & Gertz, most U.S. demand for ethane is for ethylene manufacture.

In 1992, for example, U.S. ethane demand was about 570,000 b/d. Estimated ethane demand for ethylene plant feedstock was nearly 560,000 b/d.

Total U.S. demand for ethane has increased significantly since 1990 with the addition of NGL based ethylene plant capacity and expansion of domestic ethane supplies.

Unlike the late 1980s, when competitive ethylene feedstock economics favored heavier feedstock—LPG, naphtha, and gas oil—over ethane, ethane has been a competitive feedstock, and consumption will increase if added supplies continue to be available.

U.S. ethane demand will increase to 630-640,000 b/d by 2000, as domestic supplies gradually expand and Canadian imports increase. Demand would increase even further if the full potential of Canadian imports were realized.

Future demand for propane in the U.S. will be influenced mainly by growth in three markets: residential-commercial, petrochemical, and engine fuel. Demand will grow in all of those sectors during the 1990s.

Earlier this decade, propane demand grew from about 920,000 b/d in 1990 to slightly more than 1 million b/d in 1992. Most of the growth occurred in the residential-commercial and petrochemical sectors.

In addition, agricultural demand for propane increased sharply in 1992 because of a particularly strong crop drying season. In 1993, however, a similarly strong crop drying season did little to shore up propane prices.

On a weather adjusted basis, residential-commercial demand has generally increased since the mid 1980s. Propane demand growth in this sector continues to be regional with some markets experiencing relatively strong growth while others appear stagnant or in decline.

U.S. residential-commercial demand for propane will grow at moderate rates through the rest of the 1990s. Demand growth in this sector will average 1.5-2% during the period.

The demand for propane in petrochemicals is more variable than for most other end uses because of the highly competitive nature of the petrochemical industry and feedstock flexibility of many crackers.

Propane cracking rates are significantly altered in flexible olefin plants as feedstock cracking economics shift with changes in supply, demand, and price of feedstocks and ethylene coproducts. As a result, about 20-40% of propane demand in petrochemicals is price sensitive and can vary as feedstock economics change.

U.S. NGL demand

Fig. 5

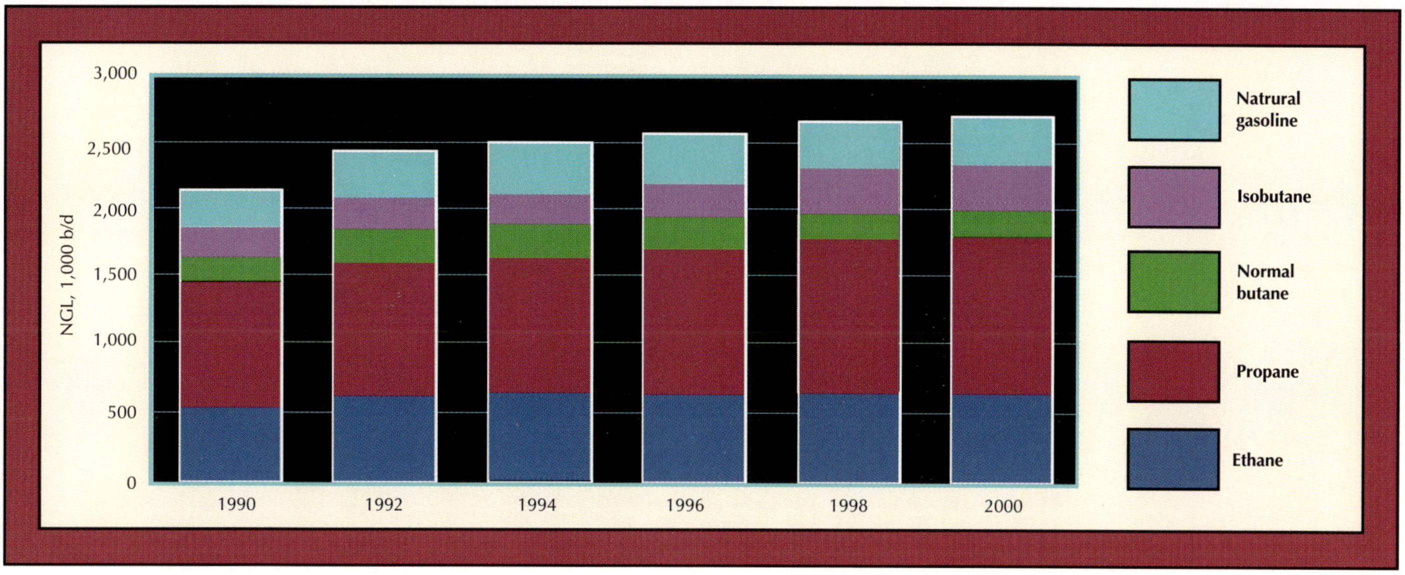

During the 1990s, says Purvin & Gertz, world supply of propane will increase faster than base demand.

Engine fuel demand for propane has declined from levels of the mid 1980s and is now relatively flat. With clean fuels and alternative fuels required by law, however, U.S. prospects for increased engine fuel demand for propane appear brighter in the 1990s.

Most of the demand for clean fuels wrought by federal and state legislation is being met with reformulated gasoline. Alternative clean fuels—LPG, compressed natural gas, and methanol—are playing a role in only a portion of the transportation fuels market, particularly in fleet applications.

Butane, fuel changes

Purvin & Gertz says that the most important factors influencing butane demand in the 1990s have been gasoline volatility changes and rapid development of MTBE production capacity. Summertime gasoline Rvp specifications were reduced in the late 1980s and early 1990s.

As a result, refineries earlier this decade had to reduce the volume of n butane included in summertime gasoline blends. Most high conversion refineries subsequently have had excess butane during summer months.

Refinery demand for n butane has declined significantly in the 1990s with many refiners now storing or selling excess supplies during summer months. As a result of this supply/demand shift, a price sensitive market for n butane developed for use as a seasonal feedstock in flexible olefin plants.

Total U.S. demand for n butane has declined significantly since the 1980s mainly because of gasoline Rvp reduction.

During the longer term, n butane has declined as gasoline Rvp levels are being further reduced in the mid 1990s and butane cracking in olefin plants is discontinued.

Demand for MTBE is rising rapidly mainly because of growth of reformulated gasoline consumption in the U.S. and minimum oxygen requirements included in the 1990 Clean Air Act amendments, says Purvin & Gertz.

Growth in gasoline demand requirements in other world markets such as the Far East and continuation of the phaseout of leaded gasoline also are contributing to MTBE demand growth.

Many companies decided to build new MTBE plants or expand existing ones to meet rising demand for the product. Several new MTBE projects based on refinery isobutylene supplies have been or are in the process of being added.

Supplies of refinery produced isobutylene are limited, however, forcing many companies to elect to build MTBE plants based on dehydrogenation of isobutane. Addition of new butane based MTBE plants has led to a dramatic increase in demand for isobutane and the need to expand isomerization capacity.

Total U.S. demand for isobutane will grow by an average of about 7%/year during the 1990s and reach about 380,000 b/d by the end of the decade. Most of the growth stems from expansion of MTBE feedstock requirements, but refinery demand for isobutane also will increase during the second half of the decade.

Natural gasoline

Natural gasoline is used in the U.S. petrochemical and refining industries. It can be cracked in any olefin plant with naphtha cracking flexibility.

Refineries can directly blend natural gasoline into motor gasoline or use it as feedstock for isomerization units and blend the isomerate into motor gasoline.

Historically, refinery applications have accounted for about 70-80% of total U.S. demand for natural gasoline. Natural gasoline demand has increased somewhat since 1990 in response to increased domestic supplies from gas processing and higher import levels.

In general, says Purvin & Gertz, natural gasoline demand will follow supply trends. As more supplies become available, additional gasoline blending can occur, particularly in winter months, and natural gasoline can replace alternate olefin plant feedstocks.

Conversely, if natural gasoline supplies decline, refiners can readjust gasoline blends and olefin manufacturers can use alternate feedstocks.

Natural gasoline demand will increase in the short to medium term, says Purvin & Gertz, as U.S. supplies from gas processing continue to rise. Demand will decline slowly after 1996 as natural gasoline production from gas processing peaks and begins to slide.

Reference

1."The Potential for Natural Gas in the U.S.," National Petroleum Council, December 1992. IPE

WORLDWIDE GAS PROCESSING

Capacities as of January 1, 1993, and average production

		MMcfd		Production 1,000 gpd (Average based on the past 12 months)								
Country	No. plants	Gas capacity	Gas through-put	Ethane	Prop.	Isobut.	Process or unsplit butane	LP-gas mix	Raw NGL mix	Debut. nat. gaso.	Other	Total products
United States												
Alabama	9	713.0	539.8	....	173.3	....	140.8	6.4	69.7	116.1	526.7	1,033.0
Alaska	3	6,375.0	6,322.0	....	2.3	....	16.0	....	3,175.0	....	....	3,193.3
Arkansas	4	908.0	549.5	3.6	7.9	3.6	4.9	....	26.7	12.9	....	59.6
California	28	999.0	636.3	....	318.9	50.8	175.5	65.6	296.6	70.0	19.2	996.6
Colorado	44	1,290.8	841.0	65.0	247.0	21.0	97.5	50.4	854.3	66.4	38.5	1,440.1
Florida	2	890.0	662.0	....	89.2	7.8	55.1	....	....	47.1	....	199.2
Kansas	21	4,872.0	3,662.6	409.0	981.0	108.1	247.9	484.5	1,082.5	228.1	369.4	3,910.5
Kentucky	3	141.0	120.3	....	....	....	....	....	200.3	....	....	200.3
Louisiana	68	18,338.4	12,169.3	1,680.7	1,328.4	447.5	553.9	395.4	6,179.5	868.8	374.5	11,827.7
Michigan	28	4,731.9	771.9	....	28.4	....	....	2.6	569.7	....	4.8	605.5
Mississippi	4	881.2	214.7	....	11.1	....	10.2	....	11.3	5.5	2.8	40.9
Montana	6	19.5	7.7	....	19.9	....	16.2	....	3.2	8.2	....	47.5
New Mexico	32	2,944.2	2,010.8	1,091.2	682.3	120.5	217.4	51.2	3,437.5	227.8	243.3	6,071.2
North Dakota	7	187.9	92.3	....	159.3	....	99.6	....	71.2	32.5	11.6	374.2
Ohio	1	20.0	8.8	....	....	....	....	....	8.8	....	....	8.8
Oklahoma	101	4,839.1	2,732.5	581.8	821.1	120.0	318.3	221.1	4,885.5	303.3	523.1	7,774.2
Pennsylvania	2	14.0	8.3	....	4.7	....	3.0	....	....	4.1	....	11.8
Texas	300	17,234.8	11,389.5	3,606.8	3,102.9	592.0	1,156.4	143.2	18,348.4	1,864.7	1,232.5	30,046.9
Utah	15	629.9	483.6	....	63.3	....	26.5	12.0	169.9	45.0	14.7	331.4
West Virginia	7	398.9	327.2	228.6	185.2	29.8	58.2	88.2	279.8	....	....	869.8
Wyoming	41	3,691.9	2,960.6	105.5	500.9	53.0	229.6	36.0	2,177.3	217.6	796.8	4,116.7
Total U.S.	**726**	**70,120.4**	**46,510.7**	**7,772.2**	**8,727.1**	**1,554.1**	**3,427.0**	**1,556.6**	**41,847.2**	**4,126.1**	**4,149.9**	**73,160.2**
Abu Dhabi	6	1,719.0	1,362.6	....	....	....	....	....	....	546.0	....	546.0
Algeria	4	5,335.2	4,268.2	....	....	....	....	632.8	....	....	5,449.2	6,082.0
Argentina	14	1,488.4	1,498.9	687.7	497.2	....	334.8	....	....	319.8	0.8	1,840.3
Australia	5	2,333.0	1,350.0	279.0	441.0	....	249.0	666.8	....	980.2	850.0	3,466.0
Austria	1	18.7	18.7	....	....	....	....	....	....	44.9	....	44.9
Bahrain	1	280.0	155.0	....	....	....	....	25.6	....	....	....	25.6
Bangladesh	2	240.0	178.0	....	....	....	....	....	....	....	8.2	8.2
Bolivia	2	474.0	177.3	....	1.4	2.5	....	161.7	....	61.5	136.9	364.0
Brazil	11	452.2	465.5	2.3	....	....	....	733.9	101.6	159.6	1.6	999.0
Brunei	2	1,082.0	777.4	....	7.9	....	....	....	....	239.4	....	247.3
Canada												
Alberta	492	23,918.1	15,311.4	6,186.3	3,323.7	1,050.9	851.3	1,189.8	4,161.2	3,866.0	254.5	20,869.7
British Columbia	20	2,848.1	2,151.4	....	395.4	....	179.4	3.5	50.4	251.5	189.5	1,069.7
Northwest Territories	2	219.9	28.3	....	....	....	....	....	....	....	....	....
Saskatchewan	30	353.2	257.6	....	13.5	....	6.9	12.9	104.2	11.1	18.2	166.8
Yukon	1	65.0	65.0	....	....	....	....	....	....	....	....	....
Chile	2	469.0	406.0	....	155.2	....	110.9	....	....	89.8	....	355.9
China, Taiwan	4	271.0	60.2	....	16.0	....	12.1	....	1.3	9.0	....	38.4
C.I.S.	22	3,071.9	2,457.5	1,270.8	387.2	693.7	2,417.1	8,490.6	....	....	410.8	13,670.3
Colombia	7	366.4	212.3	110.0	51.3	....	54.9	67.6	....	66.8	1.0	351.6
Denmark	1	450.0	240.0	....	....	....	....	....	....	....	100.0	100.0
Dubai	1	330.0	330.0	....	....	....	....	....	....	....	650.5	650.5
Ecuador	2	41.0	20.5	....	....	....	....	97.1	....	37.3	....	134.4
Egypt	12	1,915.0	1,782.0	....	....	....	....	1,236.0	1,020.2	575.4	160.0	2,991.6
France	1	570.0	563.0	....	139.0	....	132.0	....	....	270.0	....	541.0
Germany	3	366.0	274.0	....	....	....	....	....	....	....	....	....
Greece	2	16.6	10.7	....	....	....	....	....	40.7	....	....	40.7
Hungary	12	946.6	519.9	....	43.7	5.7	42.9	7.9	284.6	571.8	43.7	1,000.3
Indonesia	11	6,012.6	5,385.6	17.3	1,674.9	....	1,449.0	51.6	1,277.2	976.7	27.4	5,474.1
Italy	17	7,614.0	1,587.0	....	....	....	....	2.7	8.0	7.5	....	18.2
Kuwait	3	1,815.0	1,452.0	....	1,722.0	....	980.0	....	....	811.8	....	3,513.8
Libya	7	1,851.0	1,227.0	....	....	....	....	37.8	1,494.0	....	551.0	2,082.8
Malaysia	2	282.0	105.0	161.4	23.2	136.3	28.2	....	....	154.1	....	503.2
Mexico	11	4,649.0	3,561.1	7,176.6	....	....	....	7,897.1	....	2,976.6	8,022.5	26,072.8
Netherlands	1	440.0	140.0	....	....	....	....	....	....	....	6.0	6.0
New Zealand	6	1,441.0	807.0	....	39.0	....	15.7	252.7	....	31.6	860.0	1,199.0
Norway	2	950.0	537.0	....	682.0	108.0	256.0	....	290.0	193.0	....	1,529.0
Oman	8	826.0	309.4	....	....	....	....	....	202.0	43.0	....	245.0
Pakistan	7	1,395.0	308.7	....	....	....	....	53.8	64.9	....	9.1	127.8
Peru	2	70.0	37.0	....	3.4	....	1.0	5.6	28.9	....	4.0	42.9
Qatar	2	579.0	291.0	....	506.0	288.0	....	....	....	199.0	....	993.0
Saudi Arabia	5	4,300.0	....	....	....	....	....	....	....	....	....	....
Sharjah, U.A.E.	2	940.0	940.0	....	324.0	....	231.0	....	....	....	1,837.0	2,392.0
Spain	3	428.0	177.0	....	....	....	....	....	....	....	73.0	73.0
Thailand	2	850.0	550.0	....	....	....	....	400.4	8.4	84.0	....	492.8
Trinidad & Tobago	1	650.0	511.0	....	160.0	....	110.0	....	125.0	....	....	395.0
United Kingdom	11	8,485.0	979.2	....	332.0	....	291.0	....		1,748.5	49.1	2,420.6
Venezuela	14	2,913.0	2,557.0	213.3	197.4	....	113.4	231.3	3,274.6	171.9	....	4,201.9
Yemen	1	400.0	443.0	....	....	....	....	53.6	305.5	....	....	359.1
Total, outside U.S.	**780**	**96,530.9**	**56,836.4**	**16,104.7**	**11,136.4**	**2,285.1**	**7,866.6**	**22,312.8**	**12,842.7**	**15,497.8**	**19,714.0**	**107,746.1**
World Total	**1,506**	**166,651.3**	**103,347.1**	**23,876.9**	**19,863.5**	**3,839.2**	**11,293.6**	**23,869.4**	**54,689.9**	**19,623.9**	**23,863.9**	**180,906.3**

Methanol plant operated by Celanese Canada Inc. at Edmonton, Alta., can produced 750,000 metric tons/year. Methanol is one of the few bright spots in the current scene for petrochemical markets.

Near term global outlook weak for petrochemicals

The current deterioration of the petrochemical industry—particularly in the U.S. and western Europe—is a cause of great concern to operators and analysts alike.

Although the rapidly developing Asian market will continue to be a major factor into the next century, the immediate global outlook is for a weak market.

World polyolefins operating rates were expected to bottom out in 1993 despite persistence of robust capacity additions.

So predicted Chem Systems Inc. (CMI), Tarrytown, N.Y.

CMI sees harbingers of the next industry cycle in the U.S. economic recovery, a slowdown in new project planning, and a reduction in fixed costs.

An exception to the downturn in the global petrochemical industry is the continued strong demand for methanol, spurred largely in response to additives for cleaner burning automotive fuels.

Tecnon (UK) Ltd. in early 1993 predicted world methanol supplies will meet demand through 2001 except for slight shortfalls in 1993-94.

U.S. market outlook

CMI cited the U.S. petrochemical profitability index, a weighted average of cash margins for manufacturing petro-

Petrochemical profitability indexes

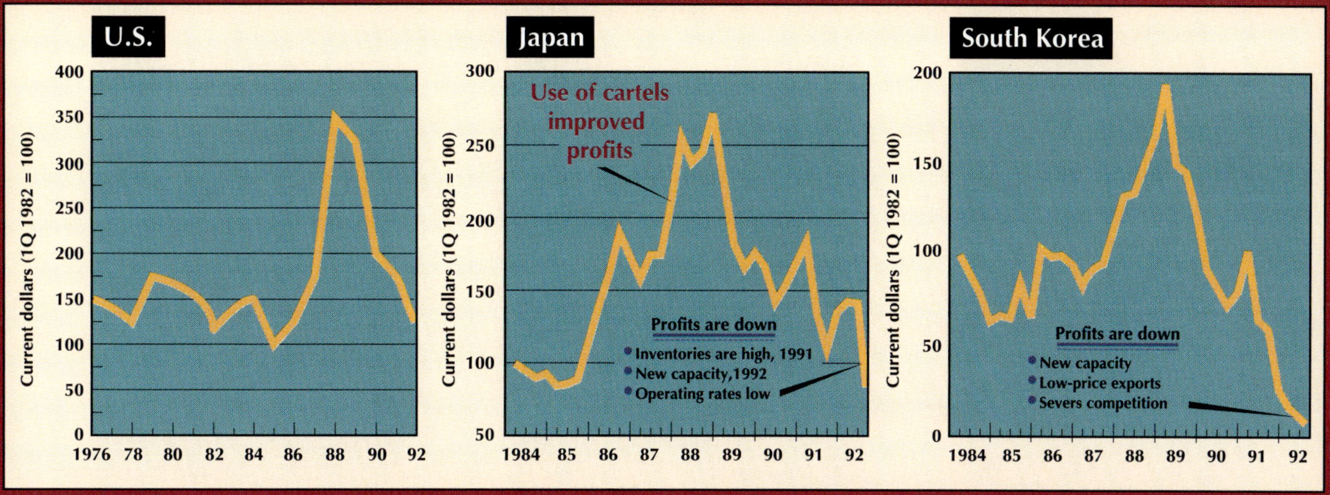

Source: Chem Systems Inc.

chemicals in the most efficient, "leader" plants, as underscoring the downturn the U.S. market has been in since 1988.

CMI Vice Pres. Bruce Pickover reported U.S. petrochemical profitability has reached a new peak about every 9 years.

The high profitability and consequent capacity expansions of the late 1980s led to falling margins, which in 1992-93 reached 1982-85 levels. The U.S. industry's current return is less than its 1981, 1983, and 1984 levels, although operating rates are about 10% higher. This pattern indicates a deterioration of industry structure—a cause of great concern, Pickover said.

The production of petrochemical derivatives was actually quite strong in 1992, despite a weak global economy. This strength led to large growth in basic building block production. The increase was not, however, caused by a growth in exports because of the weak dollar, as is generally believed. Additionally, U.S. demand has been strong, contrary to what one would expect in a weak economy.

The key to understanding demand growth is recent price trends. At the end of 1990 and throughout 1991, petrochemical prices dropped sharply as operating rates declined and refined products prices decreased after the Persian Gulf crisis was resolved.

But petrochemicals are now priced favorably, compared with competitive materials. Petrochemicals are even replacing paper, glass, and metals in many applications.

For most of 1992, natural gas prices, and consequently ethylene feedstock prices, increased sharply. And because margins were already at very low levels, these increases had to be passed on. Because of the increased product prices, companies replenished their depleted inventories.

Chem Systems anticipated limited production growth in 1993 despite the expected stronger economy. And exports probably will decline further as new overseas capacity comes on stream, says Pickover.

Furthermore, natural gas and natural gas liquids prices declined sharply in 1992, and some of these cost savings had to be passed on. Inventories are likely to be reduced again once companies see falling prices.

U.S. market structure

According to CMI, several factors differentiate the current U.S. industry downturn from previous cycles:

- Cost curves for most commodity petrochemicals have flattened considerably as licensing has led to a dissemination of many leading technologies.
- Leading producers have far less of a competitive cost advantage because petrochemicals prices are often set by marginal producers.
- Because of this reduced competition, fewer marginal companies are likely to leave the business because of a significant disadvantage in cost position.

Non-U.S. producers are gaining a substantial share of the U.S. market. The U.S. imports major quantities of commodity petrochemicals such as butadiene, propylene, cumene, ethylene glycol, methanol, styrene, and polyethylene, which puts downward pressure on prices and margins.

The U.S. is also experiencing a declining market for its exports. Asia is the largest importer of petrochemicals, accounting for one half to three fourths of the world's net exports. Production in East Asia was almost insignificant in the early 1980s, but South Korea has expanded its capacity dramatically, and by the mid-1990s, many other Asian countries will complete major complexes. Also, by 2000 China and India will have several facilities that will reduce their trade deficits considerably.

Capacity integration

There is also a global trend toward capacity integration. For example, high density polyethylene production was about half captive in 1984, but by 1994, it will be about 75% captive.

The problems facing the petrochemical industry have caused many companies to reposition themselves. Royal Dutch/Shell Group, and Exxon Corp., for example, have competitive strengths including refinery integration and scale of operation and plan to increase their shares in olefins and such primary derivatives as commodity polymers.

Through strategic alliances, companies can take advantage of such strengths on a global basis while minimizing the need to invest in foreign markets. And as competition increases in process and product technology, research and development costs can be shared.

Finally, Pickover said, alliances in the future may be structured to offer partners the option of acquiring or divesting the business as strategies change.

On the other hand, a number of companies are focusing on fewer core petrochemical businesses. Several companies have diversified away from commodity petrochemicals altogether and moved toward higher margin specialty

Outlook for world methanol

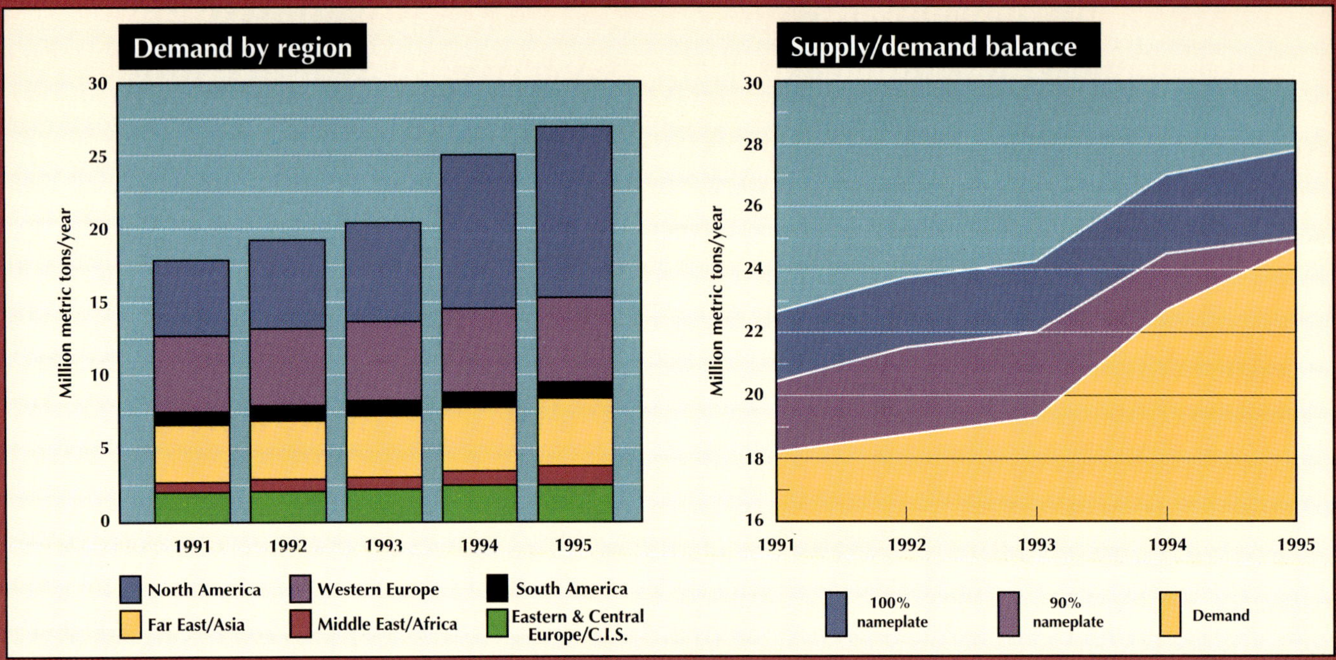

Source: Crocco & Associates Inc.

products.

U.S. product forecasts

CMI contends the most important factor affecting margins is capacity utilization. Other factors include global supply/demand, integrated trade, and international competition.

CMI expects some short term softening of ethylene prices as costs decline. On average, margins for cracking ethane and naphtha will be close, based on yearly average values. Consequently, ethane is expected to be the optimum feed on a return-on-investment basis, considering its lower fixed and working capital requirements.

Naphtha crackers will frequently produce higher margins, however, because these plants can switch feeds when conditions warrant.

Margins for cracking an ethane-propane mix, the most common feed, are expected to remain fairly constant over the next few years, Pickover said.

These margins will then begin to increase in 1995, reaching a peak of about 19¢/lb sometime during 1996-2000.

U.S. propylene

Because about half of the propylene used in the U.S. petrochemical industry is recovered from refineries, the price is set by its alternative use in alkylation, the largest refined product application.

At a given alkylate price, propylene values vary inversely with isobutane values.

Propylene prices in 1993 were currently extremely weak because alkylation values were somewhat low and isobutane values are high.

Other factors affecting propylene prices are:

- Heavier feed used in U.S. olefins plants.
- Improved catalytic cracking in U.S. refineries.
- Weak polypropylene production, partially because of reduced export markets.
- Weak propylene export market because of competition from Libya and South Korea.

In 1992-93, there was an oversupply of propylene from olefins plants, where heavier feedstocks were cracked, and from refineries, where new cracking catalysts produced more olefins. Demand was relatively weak. U.S. polypropylene production did not grow as sharply as expected because exports declined.

CMI projected that if ethylene prices cycle upward, propylene prices will follow, so that a reasonable ratio will be maintained.

U.S. aromatics/benzene

Factors affecting U.S. aromatics and benzene prices are:

- Toluene hydrodealkylation economics will continue to set benzene prices.
- Byproduct benzene production from olefins plants, refineries, and coal tar will be insufficient to meet demand.
- The 1990 U.S. Clean Air Act amendments will reduce reformate production.

As a result of decreased reformate production, toluene hydrodealkylation will still be needed to supply the deficit in benzene production from olefins plants, refineries, and coal tar operations.

CMI in 1993 predicted average margins for benzene via toluene hydrodealkylation this decade at only a few cents per gallon, including those times when plants cannot cover cash costs and discretionary units are not operated.

Aromatics and benzene prices are currently very low. They are expected to trend upward, however, with increasing gasoline values.

High density polyethylene and styrene prices, incidentally, are expected to remain weak the next few years, then increase during 1996-2000 because of cycling raw material costs and increasing margins.

Japan and Asia's boom market

Developments in Asia will continue to be an important influence on the global petrochemical industry for the remainder of the 1990s and into the 21st century.

Howard R. Blum, director of Chem Systems Japan Ltd., predicted higher growth in production and consumption of petrochemicals in Asia than for the rest of the world

North American methanol supply/demand

	Estimated	1,000 metric tons/year Forecast			
Demand	1991	1992	1993	1994	1995
Formaldehyde	1,561	1,580	1,599	1,631	1,668
DMT	177	177	178	179	178
Acetic acid	600	600	600	600	600
MTBE	1,617	2,090	2,480	4,000	5,000
Methyl methacrylate	195	195	198	198	198
Gasoline/fuels	11	17	40	78	130
Solvents	272	282	290	299	307
Others	1,082	1,157	1,176	1,191	1,210
Total demand	**5,515**	**6,098**	**6,561**	**8,176**	**9,291**
Nameplate production capacity	7,112	7,644	7,654	9,154	9,154
Capacity at 90%	6,401	6,880	6,889	8,239	8,239
Forecast surplus/ (shortage)* at 90%	886	782	328	63	(1,052)

Source: Crocco & Associates Inc.

Western Europe methanol supply/demand

	Estimated	1,000 metric tons/year Forecast			
Demand	1991	1992	1993	1994	1995
Formaldehyde	2,400	2,411	2,435	2,469	2,512
DMT	91	92	93	93	93
Acetic acid	367	365	365	365	575
MTBE	940	953	1,003	1,013	1,028
Methyl methacrylate	104	124	174	174	174
Gasoline/fuels	50	50	50	50	50
Solvents	164	167	172	174	174
Others	1,031	1,007	1,022	1,040	1,056
Total demand	**5,147**	**5,169**	**5,314**	**5,378**	**5,667**
Nameplate production capacity	2,970	2,970	2,970	2,970	2,970
Capacity at 90%	2,673	2,673	2,673	2,673	2,673
Forecast surplus/ (shortage)* at 90%	(2,474)	(2,496)	(2,641)	(2,705)	(2,994)

*Importability.

Source: Crocco & Associates Inc.

combined.

Total Asian ethylene capacity additions during 1991-96 will be about 6 million metric tons/year.

Japan has been the leading Asian petrochemical center, but Japanese industry restructuring and Japanese led investments have sparked production growth in other Asian countries.

The effect of Japan's current general economic recession on the petrochemical industry is magnified because inventories were built up in late 1991, and new capacity is coming on stream.

Chemical production in Japan has declined, and profitability has decreased sharply. Operating rates have been reduced, plants have been shut down, and start-ups have been postponed.

South Korea, Taiwan, China

Petrochemical production in South Korea has increased dramatically, making the country a major net exporter of polyolefins and styrene. CMI estimates that the South Korean capacity buildup has increased the industry downturn by 1 year.

One such example of the capacity increase is Han Yang Petrochemical Corp.'s Yoechun plant, which started up in December 1992.

South Korean consumption has increased sharply as well. Profitability, however, has fallen sharply because of the new capacity.

In Taiwan, after years of a relatively stagnant industry, petrochemical production and consumption are booming.

Factors causing this boom are a strong economy and increased demand for products in China. And with South Korean projects starting up, low cost raw materials are available.

Chinese Petroleum Corp.'s CPC 5 is proceeding toward a planned 1994 start-up, and Formosa Plastics Corp.'s (FPC) naphtha cracker complex is expected to come on stream in 1996. FPC, however, still has many problems, such as feedstocks and ports, to work out, which could cause delays in commissioning the plant. Plans are under way for a seventh complex led by a group of companies with USI Far East playing a prominent role, according to Blum.

Chinese imports have a tremendous effect on the regional as well as the global supply and demand balance.

Chinese imports were strong in 1992, partially absorbing South Korean output. In 1992, China's $36.2 billion chemical output exceeded 1991 levels by 12%, and imports and exports increased by 40-50% compared with 1991.

Although China has scaled back its plans, 12 petrochemical complexes were included in its 5 year plan, and several additional projects have since been proposed. China's program of political and economic reform may cause great changes to those plans in the coming years. CMI projects ethylene capacity in China to increase to 3.5 million metric tons/year by 2000.

Asean countries

In Indonesia, PT Tripolyta Indonesia's polypropylene plant is operating, and BP Chemicals Ltd.'s polyethylene plant is scheduled to start up shortly. The PT Chandra Asri complex is proceeding with foreign funding and is likely to be on stream in 1996 at the earliest.

In its Kepres 39 decree, however, the Indonesian government excludes projects requiring local financing and feedstock guarantees.

Malaysia is planning three complexes:

- A polypropylene plant is operating at Pasir Gudang, and ethylene and polyethylene plants are under construction.
- An LPG dehydrogenation unit is starting up at Kuantan. The propylene will be used for polypropylene and the isobutylene for MTBE.
- A natural gas based ethylene unit is planned at Terengganu, with the output to be used for polyethylene. This plant is unlikely to be on stream before 1996.

Petrochemical Corp. of Singapore has the largest petrochemical plant in the region with an ethylene capacity of about 450,000 metric tons/year. Mobil Petrochemicals International Ltd. is completing an aromatics complex in Singapore and plans to double the size of its olefins complex by 1996.

Thailand, as well, is planning a second olefins complex, based on naphtha feedstock and including an aromatics plant.

Asian trade patterns

For some products, such as polyolefins and polyester intermediates, Asia's trade deficit is expected to decline significantly.

But for others, such as vinyl chloride monomer, polyvinyl chloride, acrylonitrile, and styrene, the deficit will remain the same.

Because the region is not homogeneous, pricing is complex. Factors affecting prices include global and local costs and margins, operating rates, global and local trade patterns, and company strategies. CMI said the region is changing so fast that historical relationships may be only guidelines.

Although ethylene has been a global market, low prices and high shipping costs will make supplying ethylene to the region prohibitive during the next few years. Only South Korea, Japan, and possibly the gas based producers in Saudi Arabia and Qatar will supply Asia. The principal ethylene importers will be Taiwan and Indonesia.

The U.S. will be a large exporter of propylene to Asia in the near term, and the major price setting factor will be U.S. export price plus freight. Japan is also expected to export propylene. The main importing countries will be Taiwan, Indonesia, and South Korea.

The U.S., South Korea, and Japan will accommodate the Asian polyethylene deficit. Ethylene glycol will come from the U.S., western Canada, and Saudi Arabia. The U.S. also will supply ethylene dichloride and vinyl chloride monomer, primarily to South Korea, Taiwan, and Japan.

Polypropylene and styrene will be supplied largely by the U.S. and South Korca, and paraxylene will be supplied by the U.S. and Japan.

European markets

The European petrochemical industry is currently in crisis, according to John A. Philpot, managing director of Chem Systems Ltd. Every company has had its earnings slashed, and many have sustained heavy losses. Producers are experiencing a period when the average return to leader plants is approximately zero.

The recent steep drop in margins is linked with a mild downturn in operating rates since a peak in 1987-90. The tight market balance experienced during 1987-91 eased in 1992, when operating rates fell to 85%.

Demand, however, did not fall. The combination of new capacity and adverse trade swings caused the drop in operating rates.

The utilization problem in Europe has been mirrored in the Far East and Latin America. In contrast, rates in western Canada and the Middle East have remained at more than 90%, while U.S. utilization rates have been somewhere in between.

One product currently undergoing restructuring in Europe is polypropylene. Shell/Himont has led this process, and it is widely believed that BASF/ICI will follow. This leaves 15 producers, few of which have the capability and resources to match these ventures, said Philpot.

If all the speculation is to be believed, every company is talking to most others, and at least some further changes will occur in 1994. The key to remaining optimistic is believing that the actions being talked about in the polypropylene market will turn into reality.

It is highly likely, said Philpot, that during the next few years the modest recovery after 1994 will lead to another profitable cycle.

Worst over soon?

Although the polyolefins industry still is mired in a global downturn, CMI said several factors indicate the worst may soon be over:

- More companies are considering mergers, strategic alliances, acquisitions, technology licensing, or feedstock integration in restructuring strategies aimed at cutting costs and improving performance in the 1990s.
- Environmentalists, legislators, and polyolefins manufacturers and consumers are working together to solve solid waste problems, reducing the chances of environmental bans.
- Polyolefins sales were strong last year in 1992-93, in spite of a world economic recession.

CMI estimated linear low density polyethylene (Lldpe) and polypropylene makers in 1993 used about 79% of installed world manufacturing capacity, down from 83% in 1992. Lldpe operating rates are expected to rebound to 81% in 1994 and 88% by 2000, while polypropylene capacity utilization recovers to 82% in 1994 and 88% by the end of the century.

Similarly, high density polyethylene (HDPE) producers operated at about 81% of rated capacity in 1993—1 percentage point less than in 1992—84% in 1994, and 85% in 2000.

By contrast, operating rates of low density polyethylene (LDPE) plants will remain about stable through the end of the century, slipping 1 percentage point in 1993 to 87% and increasing to 89% in 1994 before falling to 86% in 2000.

Polyolefins outlook mixed

CMI said the outlook is mixed for polyolefins because some companies are depending on market growth to bring supply and demand into balance, while others are working on technological advances to extend the capabilities of certain materials.

For example, Lldpe capacity by 1995 will surpass 12.5 million metric tons/year, up from 6.9 million tons/year in 1990. If manufacturers could produce Lldpe with processing properties of LDPE, they could curb LDPE output by more than 2 million metric tons/year by 2000.

Because polyolefins capacity is being expanded in developed and developing countries, CMI expects interregional trading patterns to change significantly. Thirty new Lldpe manufacturers are expected to begin operating in 1990-95, including 14 in Asia.

As a result, producers in North America in 1995 will account for about 43.5% of world Lldpe capacity, down from 55.6% in 1990, western Europe 15.9%, down from 18.1%, and Japan 6.8%, down from 7.4%.

Meantime, countries in the Middle East and Far East, excluding Japan, by 1995 will have 8.8% and 15.9%, respectively, of the world's Lldpe production capability, down from 7.6% and 8.2% in 1990.

CMI also expects big changes in trading patterns on polypropylene markets, with combined interregional exports by 1995 decreasing to 915,000 tons from 1.7 million tons in 1990.

Methanol outlook

Results of a Tecnon study show that by 2001, methanol demand for methyl tertiary butyl ether and fuel uses will surpass demand for production of formaldehyde and other chemicals and solvents. Formaldehyde production is currently the largest consumer of methanol, followed by manufacture of chemicals and solvents.

Western Europe will continue to be the largest importer of methanol, followed by Japan and the U.S. The study predicts the region's methanol trade deficit will grow from 2.911 million metric tons/year in 1990 to 3.665 million tons/year in 2001.

During the same period, Japan's methanol trade deficit will grow from 1.654 million metric tons/year to 2.559 million tons/year, and the U.S.'s from 1.152 million metric tons/year to 2.328 million tons/year.

The Middle East/Africa region will continue to bolster its role as the largest methanol exporter. Exports from the region will increase from 2.053 million metric tons/year in 1990 to 4.43 million tons/year in 2001.

Canada's 1990 methanol exports of 1.489 million metric tons/year will grow to 1.836 million tons/year by 2001. And Central America/South America, now a small methanol exporter at 700,000 metric tons in 1990, will become a major exporter by 2001. This region's methanol surplus will grow to 2.34 million metric tons/year during the period. IPE

PennWell Maps the Petroleum World

From Southeast Asia to the Middle East, from the Former Soviet Union to South America, PennWell has been locating and plotting pipelines, oil and gas fields, plants, platforms, terminals, and more for over 25 years. Painstakingly researched and documented by PennWell's editorial and cartographic staff, we provide you "inside information," by surveying the operating companies themselves. PennWell Maps, leading the industry in providing the most comprehensive and detailed geologic and pipeline databases for the practicing petroleum professional.

Petroleum Market Intelligence

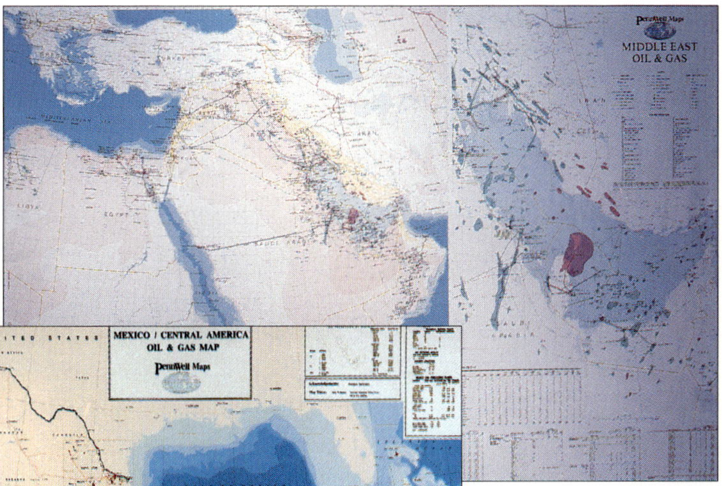

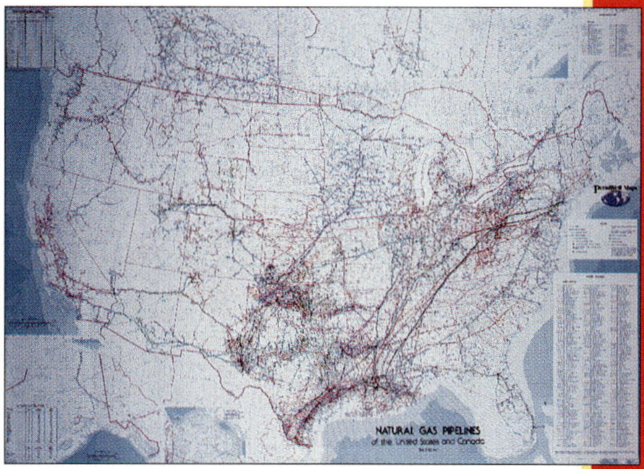

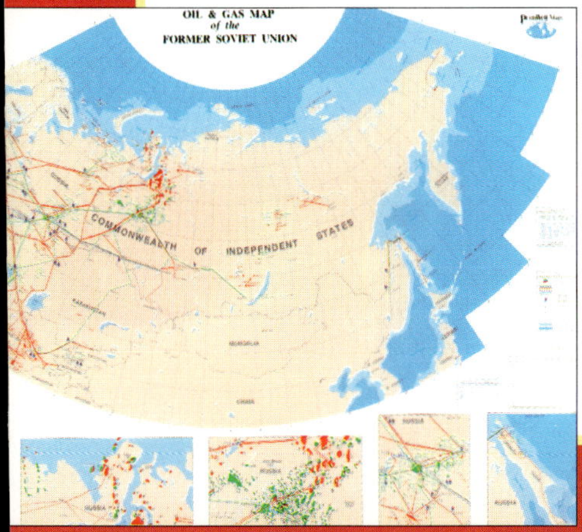

PennWell® Maps

1421 S. Sheridan Road
Tulsa, OK 74112 • USA
(918) 831-9421
Fax (918) 831-9555
Telex 211012

Yemen exploration interest still simmering

Yemen at yearend 1993 was preparing to extend exploration deadlines of oil concessions in the busy Shabwa area in the east central part of the country following poor exploration results.

Yemeni Oil Minister Salih Abu Bakr Bin Hussainoun said the picture of hydrocarbon prospects in Shabwa is incomplete despite the number of companies working there.

He said, "This is because the geology of Shabwa is complicated and the time provided for in the production sharing agreements, whether in the first or second exploration periods, is insufficient for the companies to complete their seismic, drilling, and evaluation commitments."

Middle East Economic Survey (MEES) reported that recent Shabwa drilling had yielded 23 dry holes and some minor oil and gas shows.

MEES said the ministry had started talks with Exxon Corp. to take over from Sun Co., which reportedly wanted to relinquish its blocks. It said Canadian Occidental Petroleum Ltd. had moved to its second exploration period.

Total group discovers oil in Block 10

Exploration led to a discovery during the year on another block in the East Shabwa area.

A group led by France's Total Aden Exploration & Production was proceeding with plans to determine commerciality of its Kharir oil strike in the northern portion of Yemen's East Shabwa Block 10 in Central Yemen (Fig. 1).

Yemen's expanding oil infrastructure Fig. 1

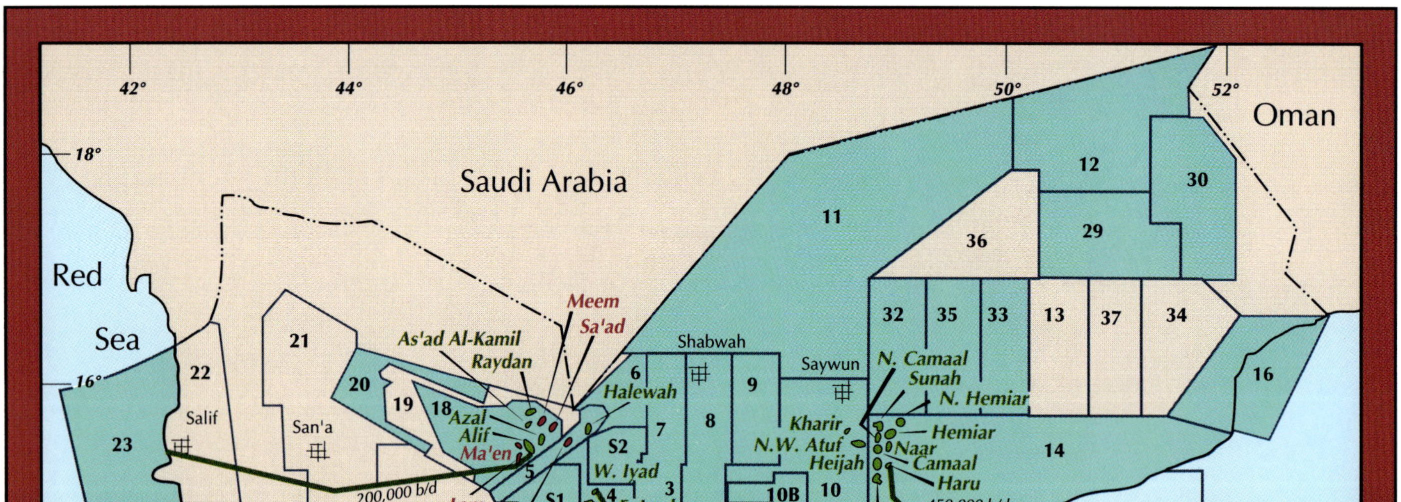

Where operators are exploring in Yemen

Fig. 2

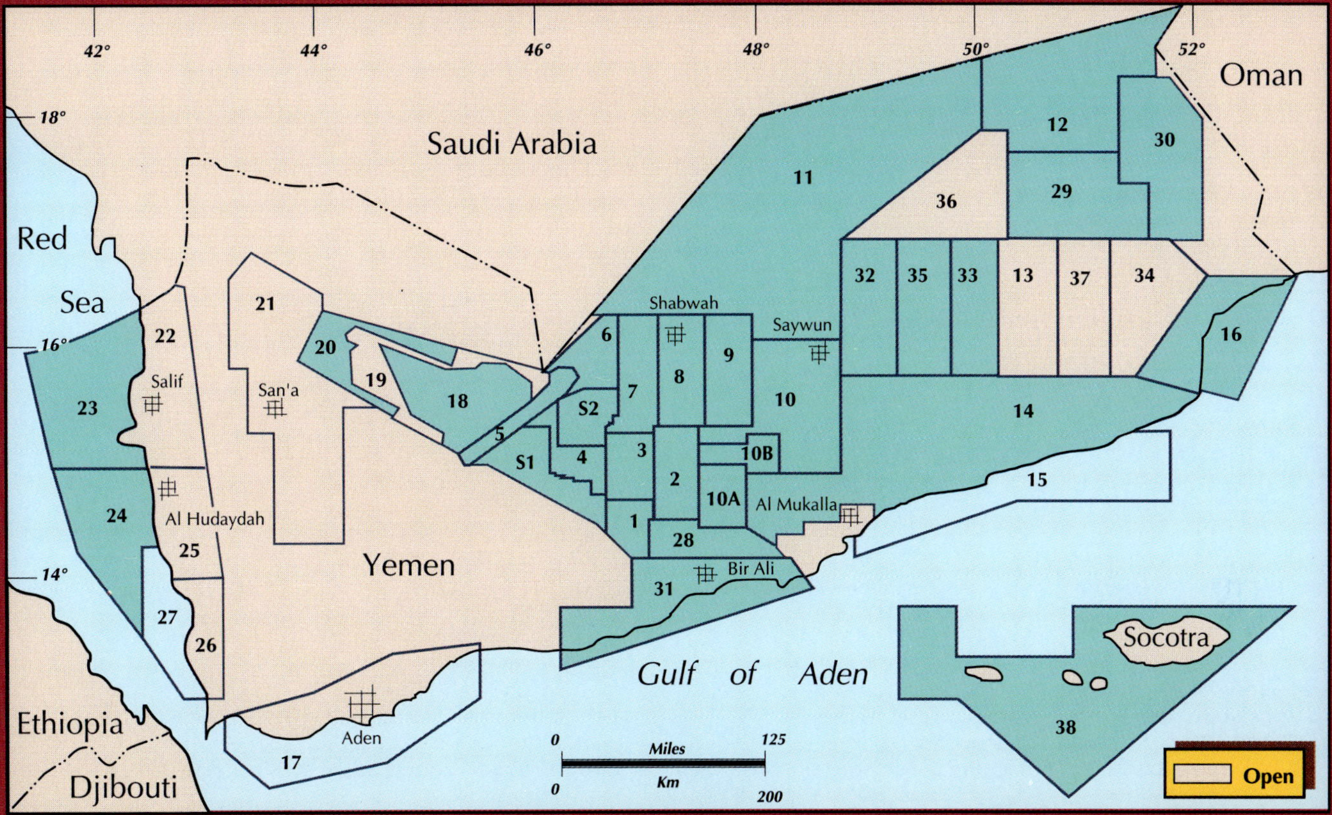

Block	Holders	Block	Holders	Block	Holders
1	Sun	9	Crescent, LL&E, British Gas	23	BP, Amoco
S1	Shell, Sodec, Pecten	10	Total, Unocal, Kufpec, Coplex, Command, Nomeco	24	Adair Oil International
2	Crescent, Clyde	10A	Total, Unocal, Kufpec, Coplex, Command, Nomeco	28	Adair Oil International
S2	Oxy, CCC	10B	OMV, Clyde, Ansan Wikfs	29	Nimir
3	Chevron, Agip	11	Elf	30	Petro-Canada, Sun, Lasmo
4	ARCO, Nimir	12	ARCO, Pecten	31	Lasmo
5	Total, Exxon, Hunt, Kufpec, Machinoimport, Zarubezhgeologia	14	CanOxy, Pecten, Consolidated	32	Clyde, Norsk Hydro, Ansan Wikfs, Oranje Nassau
6	BHP, Ministry of Oil & Mineral Resources	16	Nimir	33	Nimir
7	BP, Furnet, Amoco	18	Hunt, Exxon	35	Lasmo, Clyde, British Gas
8	BP, Furnet, Amoco	20	Phillips, IPL	38	British Gas

Interest owner Unocal Corp. said the group drillstem tested two appraisal wells that followed the 2 Kharir discovery, which in April flowed 12,000 b/d of 31° gravity oil with no water from four lower Cretaceous sands at 5,500-5,900 ft. Tests of the individual zones ranged from 350 to 5,700 b/d of oil.

The 1 Kharir flowed a combined 10,700 b/d of oil on four tests of Cretaceous pay at 5,622-6,824 ft. Rates were 300-4,800 b/d, and gravities were 30.5-34°.

East Shabwa Blocks 10 and 10A cover a combined 2.9 million acres. Participants planned to acquire at least 500 km of seismic data and drill at least two wildcats during 1994-95. The acreage is west of the Masila block estimated to hold reserves of more than 300 million bbl of oil.

Total operates the East Shabwa block. Partners are Unocal, Kuwait Foreign Petroleum Exploration, and a group of companies that acquired Pecten Yemen's interest near yearend 1993.

That group consisted of Nomeco Oil & Gas Co. and Australian companies Command Petroleum Holdings NL and Coplex Resources NL.

Seismic, drilling programs advance

Operators were carrying out work programs on many other blocks during the year (Fig. 2).

Baker Hughes Inc.'s count of active rigs ranged from five to eight per month during most of 1993.

A group led by Lasmo plc spudded several wells in late 1993 on the 7,395 sq km Hood Block 35.

The 1 Wadi Sana was programmed for 2,850 m or 50 m into basement, whichever is shallower. Main objective was Cretaceous Quishn clastics that produce in Masila Block 14 to the southwest. Secondary objectives were Lower Jurassic Kohlan and basement, which are productive in the Masila block and adjacent East Shabwa block.

The Wadi Sana drillsite is 27 km northeast of the CanOxy group's discoveries on the Masila block. Wadi Sana was to be the first of three wells on the Hood block.

North of Wadi Sana, the group drilled the 1 Qabr Hood well to 5,110 ft, ran logs, and observed fluorescence in some sidewall cores, Coplex said. The group ran a repeat formation test tool, cemented 9⅝ in. casing.

Qishn was drillstem tested at 3,348-70 ft and indicated

Yemen's geologic provinces

FIG. 3

Source: Gustavson Associates Inc.

tight. The formation was acidized, and oil and formation water flowed back. The well was pumped for 8 hr. Oil cut declined from 445 b/d to 23 b/d during the first 5 hr, and then water cut increased to 100%. The well was to be plugged.

The group next planned to drill 1 Washa, about 20 km northeast of 1 Qabr Hood, to test the hydrocarbon potential of Qishn, fractured carbonates, and basement. Planned total depth was 11,375 ft or 150 ft into basement. Drilling was to be completed in February 1994.

Hood Block interests are Lasmo 33%, British Gas 30%, Idemitsu 15%, Coplex 12%, and Hoodoil 10%.

A group led by Clyde Petroleum plc and Norsk Hydro was shooting seismic surveys on Block 32, north of and believed on trend with discoveries drilled by CanOxy and Total.

Louisiana Land & Exploration Co.'s LL&E Yemen Ltd. unit acquired a 17% interest in Yemen's 1.18 million acre Al-Hajar Block 9 operated by Crescent Petroleum (Shabwa) Inc. under a production sharing agreement (PSA) signed in January 1991.

Crescent had gathered, processed, and interpreted 1,200 line km of seismic data. Two wells were to be drilled on the acreage during the PSA's initial 4 year term ending in January 1995. British Gas Exploration & Production Ltd. also holds an interest in the block.

More operators line up to explore

New operators continued to acquire exploration interests in Yemen, and some blocks and interests were changing hands for various reasons.

Austria's OMV AG unit OMV (Yemen) Exploration GmbH took a farmout on Sarr Block 10B in the Masila area of eastern Yemen. OMV holds a 50% interest in the production sharing agreement, Clyde Petroleum plc 9%, and Ansan Wikfs (Hadramaut) Ltd. 41%.

OMV had the option to take over as operator. The drilling commitment is for one well within 2 years after the government ratified the agreement with Clyde.

Adair Oil International Inc., Tulsa, Okla., signed agreements to explore Blocks 24 and 28.

Block 24, the Alkathib area in the Red Sea off Hodeidah, covers 9,972 sq km. The work program, consisting of two 36 month phases, calls for 800 line km of seismic surveys and the drilling of two exploratory wells in phase one and 200 line km of seismic and one more well in phase two.

Block 28, the North Belhaf area, covers 3,771 sq km. The work program consists of two 30 month phases. Adair is required to shoot 1,500 line km of seismic surveys and drill two exploratory wells in each phase.

Block 28 borders to the southwest CanOxy's 6.8 million acre Masila Block 14, where nine oil fields have been discovered. Adair was negotiating with several oil companies, including U.S. companies not yet active in Yemen, that had expressed an interest in participating in the blocks.

Offshore, British Gas plc signed a memorandum of understanding with Yemen's petroleum exploration and production board to develop Socotra Block 38 in the Gulf of Aden. That gave British Gas exclusive rights to negotiate a PSA covering a 17,500 sq km area around the island.

BHP Petroleum Yemen Inc. acquired a 50% interest and operatorship of the 3,200 sq km Block 6 in North Central Yemen from Golden Spike Energy Yemen Ltd. Yemen's ministry retained the other 50%, having signed a PSA with Golden Spike in 1993. Required work program in the initial 3 year exploration period is 500 line km of seismic surveys and two wells.

Late in the year, India's state owned Oil & Natural Gas

Masila block stratigraphic column*

Fig. 4

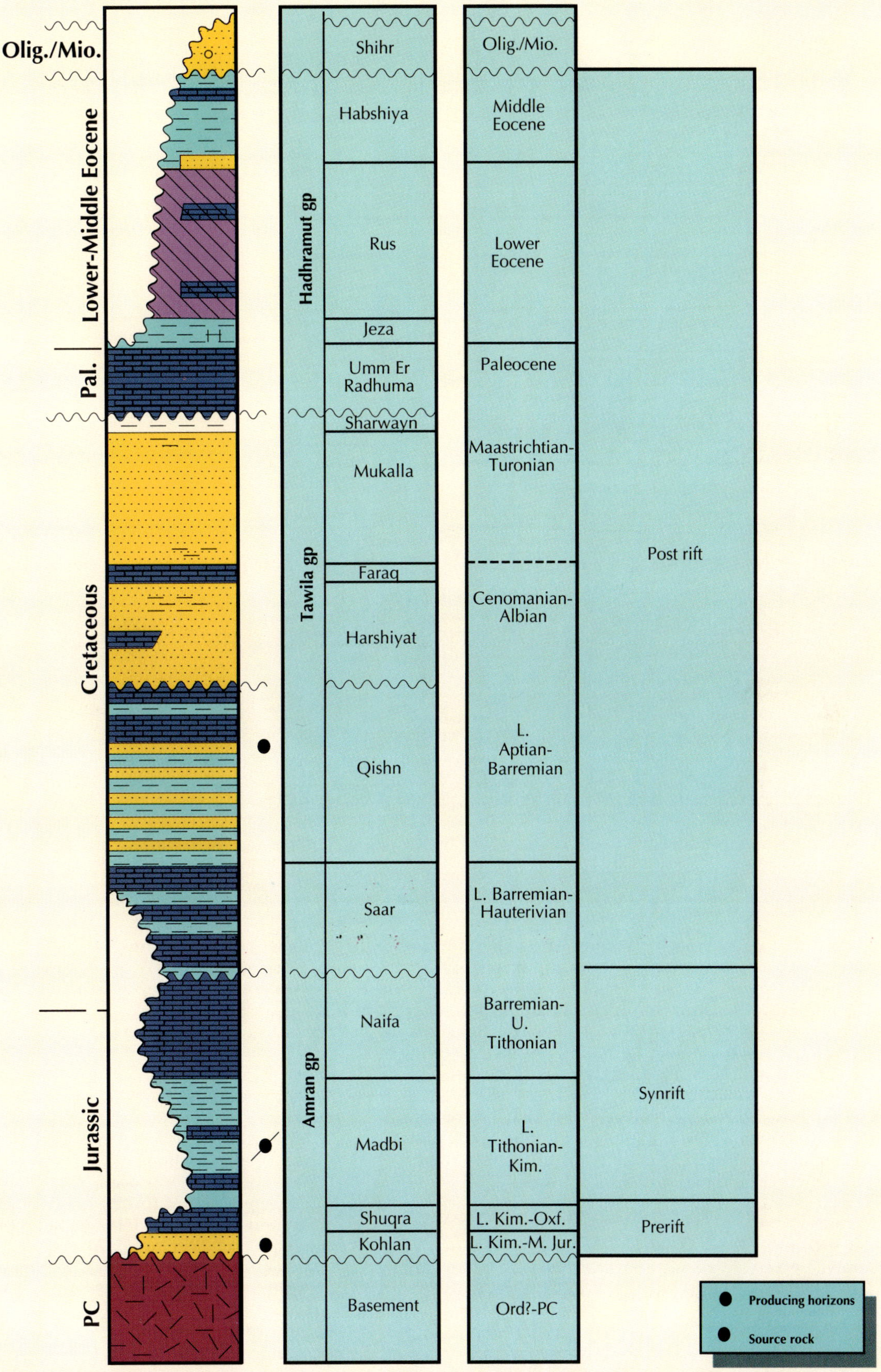

*Northwestern Masila contract area.

Source: Canadian Occidental Petroleum Ltd.

Sunah oil pool structural cross section Fig. 5

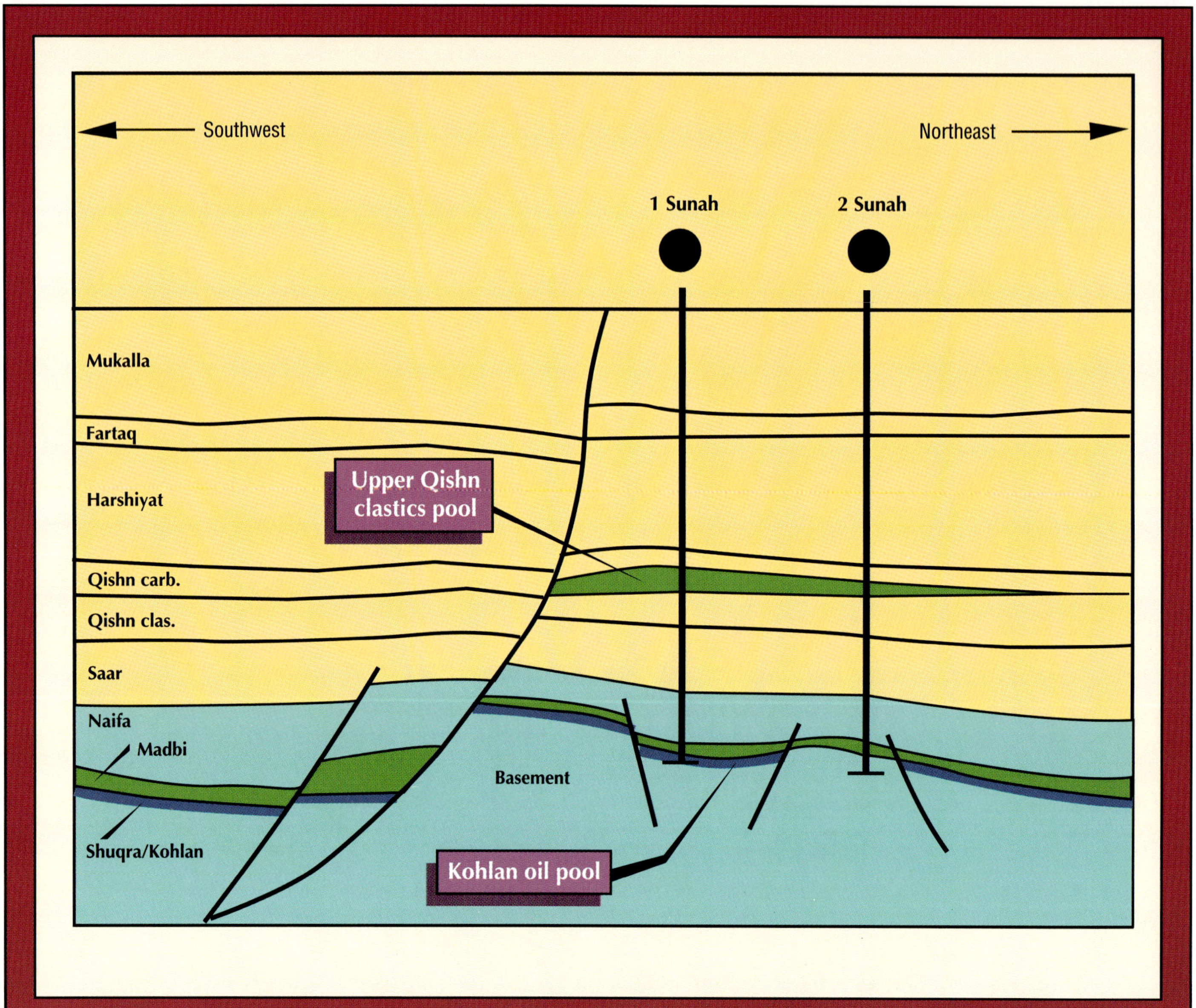

Source: Canadian Occidental Petroleum Ltd.

Commission began to focus on obtaining exploration rights outside India, including Yemen and elsewhere.

Meanwhile, Yemen in early 1993 declined to ratify the PSA of Red Eagle Resources Corp., Oklahoma City, covering Block 13 about 70 miles northeast of CanOxy's Masila area. Red Eagle refused to deposit $1 million in earnest money in the Central Bank of Yemen, saying there was no such requirement in the 90 page PSA.

Red Eagle in 1992 signed a PSA that would have committed it to spend about $50 million in two initial work phases, including acquisition of 1,200 line km of seismic data and drilling of three wildcats in about 30 months.

Oil production, reserves growing

Yemen's reserves soared to an estimated 4 billion bbl as of Jan. 1, 1994, compared with 550 million bbl in 1988 and none in 1983.

The CanOxy group officially inaugurated its Masila project Sept. 23, 1993. The project consisted of nine oil fields, production capacity of 120,000 b/d, 138 km of pipelines, and port facilities with 2.5 million bbl storage capacity and tanker loading berths.

Development and construction on the Masila block were completed in 17 months. CanOxy has a 52% interest in the group, which includes Shell Oil Co.'s Pecten Yemen Co. 20%, Occidental Petroleum Corp. 18%, Consolidated Contractors International Co. SAL 10%, and the Yemeni government.

The group made its first deliveries of light crude in July 1993.

The group delivered the oil to the Aden refinery from Masila fields in Block 14 in the Hadramaut region. Masila production was to rise to 126,000 b/d during part of third quarter and all of fourth quarter 1993.

Yemen had 19 oil fields capable of production or scheduled to be developed at yearend 1993. The country's production climbed to an estimated 228,000 b/d during September from about 180,000 b/d early in the year.

Geology of Masila area outlined

Maps and charts show the complex geology that resulted in the extension of exploration deadlines in East Central

Yemen, known as the Empty Quarter, is a hot center for exploration with about a dozen land rigs based in the country. Photo of courtesy Nabors Drilling International Ltd.

Yemen (Figs. 3 and 4).

CanOxy's first prospect to be drilled was at Sunah in the extreme northwest corner of its contract area (Fig. 5).

It was not the largest closure recognized but appeared to present the thickest Jurassic sedimentary section and was best located to receive hydrocarbons generated in and migrating from a thick Jurassic Madbi section within adjacent half grabens.

Objectives were postulated to be Kohlan sandstone sitting on basement and sealed by Shuqra limestones, sand lenses in the Madbi shales, porous and/or fractured limestones of the Jurassic Naifa formation, possible sandstones interbedded with shales in the Barremian Qishn formation, and, if present, an Alif sandstone equivalent.

The 1 Sunah had 4 m of oil bearing Kohlan sand and 6 m of pay in the uppermost sand of Qishn. The Kohlan sand flowed at a rate of 3,767 b/d of 36° gravity oil, and 30° gravity oil was recovered from the Qishn sand.

Tying the well with seismic required some changes in the original interpretation.

While the basal prerift sedimentary package was correctly identified, some departures were encountered in the overlying stratigraphy.

Of note was the identification of an early Cretaceous platform carbonate unit that was named the Saar formation. A fortuitous discovery was the degree and quality of the Qishn sand development, an interval thought most likely to be carbonate and shale dominated. The revised stratigraphy of the northwestern Masila Block is shown in Fig. 4.

The 1 Heijah well encountered no Jurassic sediments over a prominent basement high but found oil in two Qishn sands.

The 2 Camaal well, between the Sunah and Heijah discoveries, cut 29 m of net pay in a 94 m gross column within sands of the Qishn formation.

CanOxy said it recognized other potential reservoirs other than Qishn and Kohlan during its drilling program. An objective of its seismic campaign is to define subtle traps and identify new play concepts in the western part of the contract area.

Refinery expansions on the horizon

Yemen plans to maximize potential to refine its share of the crude produced in the country.

It plans a $350 million expansion to boost capacity of Aden Refining Co.'s 130,000 b/d Little Aden refinery in two stages to 200,000 b/d, then to 250,000 b/d.

It also hoped to construct a 100,000 b/d grassroots refinery at Mukalla at a cost of $300 million to handle its share of production from the Shabwa area.

Yemen Hunt Oil Co. also operates a 10,000 b/d topping plant at Marib.

Expatriate safety called into question

Several incidents occurred during 1993 that called into question the safety of expatriates working in Yemen.

Early in 1993, Yemen's oil minister told Hunt Oil Co. he could not guarantee the safety of its 300 expatriate staff following a spate of hijackings and kidnapings in Yemen. Hunt continued its operations, which accounted at the time for 170,000 b/d of Yemen's 200,000 b/d of oil production. Late in the year the U.S. ambassador to Yemen was kidnaped and later released unharmed.

Gas transport, export projects

Prominent in Yemen's future are efforts to develop and transport its natural gas reserves and line up domestic and export markets.

Enron Corp. signed a memorandum of understanding with General Gas Corp. of Yemen to help the country develop infrastructure needed to supply gas to domestic industrial markets and export LNG.

The project is to entail a 250 mile pipeline to an export terminal to be built on Yemen's coast.

Yemen and Enron hope to sign an implementation agreement by late 1994. IPE

Cook Inlet, Beaufort/North Slope finds revive Alaskan drilling action

A possible world class oil discovery in Alaska's Cook Inlet and a tantalizing find in the Beaufort Sea have revived hopes that Alaska may sustain its status as a premier petroleum province well into the next century.

A major exploration/development push is under way in Alaska's Cook Inlet basin, where an ARCO group is delineating Sunfish, an apparent giant field discovery and operators are trying to extend the life of aging offshore platforms (OGJ, May 24, p. 23).

In addition, the Kuvlum strike by an ARCO Alaska Inc. group off the Arctic National Wildlife Refuge has rekindled industry interest in the North Slope/Beaufort shelf region as a still highly prospective exploratory frontier. Although follow-up drilling failed to prove sufficient reserves needed for commerciality in the Beaufort—a level put at 1 billion bbl—Kuvlum pointed to the increasing likelihood of more huge structures containing oil on and off the North Slope.

Meantime, a surge of exploratory and delineation drilling elsewhere on Alaska's North Slope and in the Beaufort Sea has been led by ARCO and the Alaska unit of British Petroleum Co. plc. The two companies, the key North Slope operators, are seeking to forestall the day when North Slope oil production won't be enough to economically sustain the Trans-Alaska Pipeline System (TAPS).

Their efforts and others' have yielded a string of oil discoveries in recent years that may not approach Prudhoe Bay's supergiant status. However, because some sizable finds are within striking distance of the North Slope infrastructure, they could prove commercial if developed together.

In addition, using advances in exploration technology, operators continue to discover small—by North Slope standards—accumulations in the shadow of Prudhoe Bay area facilities.

At the same time, North Slope operators are pressing extensive efforts to extend the life of Prudhoe Bay, Kuparuk River, and other North Slope fields with enhanced recovery and other programs to tap more of the billions of barrels of oil that otherwise would be left unrecovered.

It adds up to an impressive E&D revival in a high cost state that until recently had seen an exodus of companies following offshore exploratory disappointments and plummeting oil prices in the 1980s.

Cook Inlet renaissance

Almost 36 years after the Cook Inlet basin gave Alaska its first significant oil discovery, the basin is bounding back with a burst of exploration and development.

Players in the renaissance are ARCO Alaska Inc., Phillips Petroleum Co., Unocal Corp., Shell Western E&P Inc., and Anchorage independent Stewart Petroleum Co.

ARCO, as operator for itself and Phillips, in 1993 kicked off a two or three well drilling program to delineate a deeper pay discovery that holds the potential of being the largest oil find in the prolific Cook Inlet basin.

The company estimates its Sunfish discovery it made in deeper pay in North Cook Inlet gas field in 1991 and confirmed in 1992 may hold gross reserves of as much as 750 million bbl of oil.

Sunfish discovery

ARCO drilled its 1 Sunfish discovery with Global Marine's Glomar Adriatic VIII jack up in state waters on projected section 12-11n-10w, 30 miles southwest of Anchorage. Sand at 12,160 ft yielded 1,100 b/d of 42° gravity oil and 1 MMcfd of gas through a $^{24}/_{64}$ in. choke.

The strike lies 1½ miles southwest of Platform Tyonek from which Phillips, as operator, developed North Cook Inlet gas field. The field, a 1962 discovery by Amoco Production Co. forerunner Pan American Petroleum Co., has produced 1.05 tcf from Sterling and Beluga sands at average depths of 4,200 and 5,100 ft.

In 1992, ARCO moved 1¼ miles south and slightly west of the discovery well to drill 1 North Foreland State on projected section 13-11n-10w. The 17,770 ft confirmation well flowed 3,610 b/d of high quality oil and 3 MMcfd of gas.

In addition to confirming the earlier find, the North Foreland well found two more, related hydrocarbon bearing zones. ARCO confirmed that the Sunfish and North Foreland pays are Tyonek equivalent of Tertiary age.

Phillips also joined the play as an operator in 1992, spudding 2 Sunfish in November as a delineation well from Platform Tyonek on projected section 6-11n-9w. The directional hole was planned to bottom 1¾ miles northwest of the surface location, or 2 miles northwest of 1 Sunfish, but mechanical problems intervened.

Sunfish plans

Work by ARCO-Phillips in 1993 targeted acreage acquired by the combine at Alaska's Sale 76 in January 1993. The two companies won all 15 tracts they bid on, paying bonuses totaling $64.4 million, or 98.6% of the sale's total high bonus bids of $65.3 million.

ARCO, as operator, planned to use Rowan Drilling Co.'s Rowan Gilbert Rowe, to drill 3 South Cook Inlet in about 100 ft of water on projected section 35-10n-11w. The drillsite lies about 13 miles southwest of the 1 North Foreland confirmation.

The projected 16,000 ft vertical well was to test Tract 14, which ARCO-Phillips acquired for a bonus of $12.8 million, or $2,502.67/acre. The bonus was the second highest of the sale.

ARCO planned to use the Glomar Adriatic VIII for a pair of delineation wells. The first was expected to be either a replacement for the troubled 2 Sunfish or another delineation, 2 South Cook Inlet on projected section 18-10n-10w, 10 miles southwest of the North Foreland confirmation. The well was to test Tract 10, the top dollar tract in Sale 76 with a bonus of $17.1 million, or $3,001.26/acre.

If the assignment is for 2 South Cook Inlet, the jack up on completion of that job—and depending on what the well finds—will move to 2 North Foreland or 1 South Cook Inlet, both of them between the North Foreland confirmation well and the 2 South Cook Inlet on a northeast-southwest trend.

Phillips also planned to drill 3 Sunfish from Platform Tyonek. Like 2 Sunfish, it was programmed to further delineate the Sunfish and North Foreland sands.

Along with delineation wells, ARCO conducted a 3D seismic survey summer over a 150 sq mile area in North Cook Inlet. In preparation for the survey during the approximately 22 ft tidal drop of Cook Inlet waters, ARCO in 1992 made a test run, successfully laying and retrieving cables from the boulder-strewn, glacial-scoured bed of the Inlet.

Results of wells drilled this summer will determine whether a 3D survey will be run in 1994 south of the two wells that have been successfully tested.

Sunfish prospect

The Sunfish discovery in Cook Inlet, which had largely been developed in the 1960s and early 1970s, had its inception with a new stratigraphic model developed by ARCO for the inlet.

"It was good old-fashioned geological work with biostratigraphy, samples, and logs," said Pamela Parks, ARCO Alaska's manager of South Alaska exploration. "And then we added to it technology that wasn't available 3 or 4 years ago."

Another factor, Parks said was ARCO's long association with the inlet. "We've had a continuous presence in the inlet since the 1950s, and we've continued working it."

The first trial of the stratigraphic model took place in 1990 with drilling of 1 Sturgeon on projected section 25-5n-17w, near Kalgin Island about 60 miles southwest of North Cook Inlet gas field.

"The stratigraphy worked, but the oil wasn't there," Parks said of the 7,206 ft hole. Although 1 Sturgeon was dry, the support given the stratigraphic model encouraged ARCO to move north to test the Sunfish prospect.

With delineation still under way, the earliest Sunfish could come on line would be late 1996, Parks said. Full development of the field could require as many as five more platforms to bolster the 15 presently in Cook Inlet. The discovery in North Cook Inlet does not seem to have implications for the lower Cook Inlet. "The lower Cook Inlet has a thin Tertiary section," Parks said. "Therefore it is not impacted by the Sunfish discovery."

Unocal perspective

In a pair of fields of more than 100 million bbl southwest of the Sunfish play, Unocal is another player in revitalization of Cook Inlet.

The company is an old hand in the inlet. Before Prudhoe Bay went on stream, Unocal was Alaska's largest producer, due mainly to flush production from McArthur River field.

"All of the development in the late 1960s was with 1960s technology," said Wylie R. Barrow Jr., Unocal's general manager for Alaska. "Generally it was with wide spacing and peripheral waterfloods. After 20 years of production, it was obvious well spacing was inefficient, and with modern seismic the geological structure is different from the original conception and requires more wells to develop it properly."

Another factor, Barrow said, is new technology's ability to drill extended reach and horizontal wells that enable the operator to do things that were not possible before.

"Underdevelopment, new geology, and new drilling technology," Barrow said. "The combination makes opportunity. That's really the basic premise toward redevelopment."

Granite Point

Unocal trained its sights on Granite Point, a 1965 discovery by Mobil that has produced 118.6 million bbl, and Middle Ground Shoal, a 1962 discovery by Pan American that has yielded 167.3 million bbl. Unocal had a foothold with a 25% interest in a state tract held jointly with Mobil 75%, on the south end of Granite Point field. The tract had been developed from Granite Point Platform operated by Mobil.

Two other platforms in Granite Point field—Bruce and Anna—and two in Middle Ground Shoal field—Baker on the north end and Dillon on the south end—were owned by Amoco 62.5%, Texaco 25%, and Chevron 12.5%.

Unocal bought Amoco's interest for an undisclosed price effective Sept. 1, 1990, and followed up with acquisition by trade of Texaco's interest and Chevron's, giving the company a 100% interest in Platforms Bruce, Anna, Baker, and Dillon. The company later tried to acquire Shell's interest in the midportion of Middle Ground Shoal developed by Platforms A and C, but the companies were not able to come to terms.

Granite Point oil field is a focal point for Unocal Corp.'s drilling program in Cook Inlet. This is Unocal's Platform Anna.

Unocal chose Granite Point as the proving ground, drilling four wells last year that boosted production from Granite Point platform to 5,200 b/d from 2,300 b/d. Two of the extended reach wells tapped the Tyonek formation for rates four to five times greater than conventional wells in the formation. Two other wells proved up production in the Hemlock, a deeper pay discovery.

In the wake of that successful program, Unocal disclosed a $60 million campaign to drill 12 more wells from Granite Point platform in a period of about 4 years. The wells are expected to add production of 7,000 b/d of oil and 4.9 MMcfd of gas. Plans call for adding a gas compressor on the platform to set the stage for development drilling to resume in 1994.

Chakachatna

A second, larger project on Unocal's revitalization list is Chakachatna, which envisions development of significant reserves from Platforms Bruce and Anna north of Granite Point platform in Granite Point field and Platforms Baker and Dillon in Middle Ground Shoal field.

Unocal plans to start Chakachatna rolling with an investment of $119 million.

"The $119 million is basically to drill a set of concept wells in the first phase and pay for a drilling rig and associated production facilities," Barrow said.

The first phase calls for drilling 20 development wells—four in 1993, 11 in 1994, and five in 1995. The wells are expected to increase production from the four platforms to about 11,000 b/d from the current 4,000 b/d.

The Chakachatna project, encompassing the four 100% Unocal owned platforms, could point the way to more such projects.

"There is more opportunity," Barrow said, citing Platforms Dolly Varden, Grayling, and Steelhead in McArthur River field and the Monopod platform in nearby Trading Bay field. "They all have possibilities. The future depends on drilling success, commodity price, and our ability to operate cost effectively. Those questions will be answered before the end of 1995.

"In a couple of years, it will be very interesting to see where the inlet stands. With the development Unocal is talking about and ARCO's projects, it's going to be exciting times."

In other activity, Shell plans a four well development drilling program on Platform C in its share of Middle Ground Shoal field. Drilling is scheduled to start late this summer and take about 9 months to complete. It is is part of a plan to maintain production.

Stewart's program

West of McArthur River field, Stewart is preparing to join Alaska's producing ranks as the first independent to develop production in the modern era of the state's oil industry.

The company in 1991 proved up a new offshore field, directionally drilling the discovery well from an onshore site in the West Foreland area about 65 miles southwest of Anchorage.

The drill pad was an abandoned air strip included in a 700 acre surface use contract negotiated with Salamatof Native Association Inc. The contract also allowed use of an existing road from the pad to Marathon Oil Co.'s Trading Bay production facility, 2½ miles to the north.

A subsurface agreement with Cook Inlet Region Inc. (CIRI) covered a 55 acre area within the Salamatof agreement area, allowing drilling through CIRI's subsurface estate en route to Stewart's offshore acreage and granting a subsurface easement through a pipeline corridor area along the west side of the road to Marathon's facility as established by the Salamatof agreement.

Stewart's 1 West McArthur River Unit, section 16-8n-14w, 13,742 ft measured depth, bottoming about 9,500 ft northwest of the surface location, found oil in the Hemlock, topped just below 9,200 ft.

The well was first tested in December 1991 and fully tested in June 1992, indicating a conservatively estimated initial production rate of about 3,400 b/d using a jet pump to assist the flow of 32-34° gravity, sulfur free oil.

Stewart placed the well on production in 1993, trucking the oil at first to Marathon's Trading Bay site, where Stewart is completing an offloading facility. Production will be limited by trucking capacity to about 3,000 b/d.

Stewart Pres. W.R. Bill Stewart, said, "We won't know the field's full potential until we run a pipeline."

Plans called for starting construction of an 8 in. 2½ mile pipeline in August 1993, with completion expected to take about 3 weeks.

Unocal will buy the crude and ship it through the Cook Inlet pipeline to the Drift River terminal for shipment by tanker to southern California refiners.

Along with the oil line, Stewart plans to lay a 4 in. gas pipeline in the same corridor.

Stewart's 6,300 acre West McArthur River Unit includes proved gas reserves in West Foreland gas field, discovered in 1962 by Pan American on the southern end of the structure tapped by Stewart's discovery well.

Pan American's 1 West Foreland, section 21-9n-14w, had a calculated absolute open flow of 16.45 MMcfd. It was shut in because of lack of market and has been retested several times since then. The onshore wellsite is presently held by Phillips.

Stewart spudded the first confirmation well in April 1993 with Grace Drilling Co.'s Rig No. 160. Its 2 West McArthur River Unit had a proposed 13,500 ft measured depth and 9,500 ft true vertical depth. Bottomhole location was about ⅔ of a mile south of the discovery well.

Development of the 6,330 acres included in Stewart's block will require about five production wells and two injection wells.

Estimated reserves are 50 million bbl. Gas reserves are estimated at a total 21.7 bcf underlying Stewart's 263 acre portion of the 640 acre participating area of West Foreland gas field. The field holds estimated reserves of 20 bcf and 13.5 bcf in undeveloped reserves within the Stewart block north of the gas field.

West McArthur River investment to date in operations by Stewart on behalf of itself, private Alaskan and non-Alaskan participants, and a South Korean refining company totals about $20 million, Stewart said. Estimated future investment totals $50 million for the Stewart acreage.

"It costs about $10 million for each completed well if you don't have major drilling problems," Stewart said.

The West McArthur River prospect was based on shows turned up by wildcats drilled in the 1960s and seismic that was reshot using today's more sophisticated stacking technique for processing and better data than that of earlier surveys in the first wave of exploration in the 1960s.

In the state's Sale 76 in January, Stewart, bidding unopposed, paid $20,767.77 or $6.57/acre for Tract 16, 6 miles north of the West McArthur River discovery well. The lease is on another prospect, Stewart said.

Kuvlum excitement

ARCO group's Kuvlum oil discovery in the Beaufort Sea at first sparked a flurry of excitement in U.S. exploration circles.

ARCO in summer 1993 estimated the Kuvlum reservoir could hold reserves of as much as 1 billion bbl or more and scheduled two delineation wells in the area. It also conducted a 3D seismic survey covering the 35,000 acre Kuvlum block.

The well that touched off the Kuvlum play was ARCO's 1 OCS Y-0866, which flowed 3,400 b/d of 34° gravity oil at a wellhead pressure of 300 psi with a 600:1 GOR. The well was drilled in 110 ft of water 60 miles east of Prudhoe Bay and 16 miles offshore from the western edge of the Arctic National Wildlife Refuge Coastal Plain.

In fall 1993, ARCO disclosed Kuvlum is not commercial as a stand alone development. ARCO drilled 2 Kuvlum and 3 Kuvlum to delineate the structure. The 2 Kuvlum, drilled to 11,125 ft about 3 miles southwest of 1 Kuvlum, found hydrocarbon bearing formations, including the reservoir tested in 1 Kuvlum. The 3 Kuvlum, 1 mile northeast of 1 Kuvlum, was drilled to 8,000 ft. It found hydrocarbons but not a commercial quality reservoir.

Undeterred, ARCO and Phillips slated a $22 million wildcat to test the nearby Wild Weasel prospect 9 miles offshore in the eastern Beaufort. ARCO's 1 OCS-Y-1597, drilled to 9,314 ft, failed to find commercial hydrocarbons.

It was after the Wild Weasel failure that ARCO said it would "stand back and reevaluate the eastern Beaufort."

If Kuvlum or Wild Weasel had measured up to the 1 billion bbl mark—with a marginal chance for significantly greater upside potential—regarded by ARCO as necessary to support development, the economics for other accumulations along the Alaskan coast could change. ARCO then said it would affect development prospects for the Point Thomson and Badami discoveries on the eastern North Slope. Davis said.

The former, a 1975 discovery by Exxon Corp. about 5

miles west of the ANWR boundary, is estimated to hold reserves of 7 tcf of gas and 300 million bbl of condensate. The latter, a 1990 discovery by Conoco Inc. 20 miles east of Endicott field, has yet to be delineated.

Kuvlum's nearness to ANWR raised the question of whether development of the field would increase pressure to open ANWR to exploration and development.

"Geologically, there isn't any connection between the two," Davis said. "I don't think developing Kuvlum would necessarily put any pressure on opening ANWR.

"If ANWR is opened, it doesn't mean all of the problems go away. It doesn't look like ANWR is going to be opened. The state has control of its resource. It's time to focus on state acreage. Most of the area north of the Brooks Range was explored in the 1960s and 1970s. Geophysical technology has changed considerably since then. We think it's time to explore there again."

Colville High

One area that is undergoing such exploration is the Colville High area along the Colville River delta west of Kuparuk River field, where six wildcats scattered over a large area have yielded oil.

The wells include three drilled by Texaco Inc. and one by Amerada Hess Corp. prior to ARCO's 1 Fiord in section 2-12n-5e and 1 Kalubik in section 11-13n-7e, both drilled in 1992. The Fiord well flowed oil from one interval at a rate of 1,065 b/d and from another, after hydraulic fracturing, at a rate of 180 b/d. The Kalubik flowed oil at a rate of 1,200 b/d from one interval and from another, after hydraulic fracturing, at a rate of 410 b/d.

During the 1992-3 winter season, ARCO drilled three wildcats related to the Colville High: 3 Kuukpik in section 22-13n-6e, ARCO-BP 1 Till in section 10-12n-6e, and ARCO-BP 1 Colville River in section 17-11n-6e. All were tight holes.

"We're aiming at separate accumulations," Davis said. "The job is to identify a critical number of accumulations which on their own would be subeconomic."

The goal is to find enough oil to warrant construction of a centralized processing plant at a site to which oil could be piped from as much as a 40 mile radius.

In addition to the wildcats related to the search for Colville oil, ARCO during the 1992-93 winter season drilled two wildcats in the central foothills area in a search for what Davis called "stand alones."

The wildcats were 1 Big Bend in section 24-3s-2w and 1 Tulaga No. 1 in section 14-4n-4e. Both tight holes were within a 40 mile radius of Umiat oil field, a 1946 discovery by the U.S. Navy, and East Umiat gas field, a 1963 discovery by BP Exploration (Alaska) Inc. Three wells were completed in the oil field, which has been shut in since 1951. Only one well was drilled in East Umiat. It was suspended.

Other exploratory wells drilled by ARCO during the 1992-93 winter season were 1 Jones Island in 28-14n-11e, 12 miles northeast of Milne Point oil field, and 3 North Prudhoe Bay State in section 25-12n-14e, a follow-up to the company's 4 West Beach State, a 1991 new pool discovery.

Results of the Jones Island well were not released. A 4 hr test of the Ivishak formation in 3 North Prudhoe Bay State flowed 35° gravity oil at a 3,000 b/d rate.

ARCO described production as comparable to the nearby West Beach field, which went on stream in April 1993, producing 3,000 b/d from the discovery well. West Beach is estimated to hold 10-65 million bbl of oil. Operator ARCO and Exxon hold equal interests in West Beach.

BP wildcat drilling

BP resumed wildcat drilling on the North Slope in 1993 with the company's largest exploration program in Alaska since 1985. Plans called for five wildcats—three operated by BP and two by ARCO.

BP kicked off the program with 1 Cascade in section 3-12n-11e, southeast of Milne Point field, spudding in late January with Nabors Drilling Co.'s Rig 28E and releasing the rig at the beginning of April. The initial program called for directionally drilling the hole to bottom about 800 ft southwest of the surface location. The program for a redrill called for bottoming about 700 ft west and slightly north of the surface location. Results were not released.

BeauDrill Kulluk arctic class semisubmersible drilling rig is on location drilling the Kuvlum discovery well in the Beaufort Sea off Alaska for ARCO Alaska Inc. and partners. Photo courtesy ARCO Inc.

The other two wells to be operated by BP are 1 Sag Southwest in section 2-11n-16e and 1 Yukon Gold in section 23-8n-23e. Sag Southwest was to be drilled in June from the main production island in Endicott field. Drilling at Yukon Gold was scheduled to begin in late October or early November.

BP owns a two-thirds interest in the prospect, Unocal one-third. The project is expected to cost about $15 million, including about $1.3 million for ice pad construction. Drilling and evaluation is expected to require 2-3 months.

Depending on results, a sidetrack to a second subsurface location may be drilled from the same pad the following winter or a second well could be drilled from another pad.

Drilling contractor for Yukon Gold is Pool Arctic Alaska.

BP participated in two wildcats drilled in 1993 by ARCO, as operator, in the Colville River Delta area west of Kuparuk River field: ARCO-BP 1 Till and ARCO-BP 1 Colville River.

And Exxon Corp. joined the North Slope winter campaign with a wildcat on Thetis Island, 6 miles northwest of Kuparuk River field, drilling the tight hole in section 19-14n-8e.

Badami appraisal

One of BP's first moves to evaluate a joint venture with Conoco Inc. and Petrofina Delaware Inc. in the Badami discovery area between Endicott field and the Point Thomson Unit ended in May 1993 with completion of a 3D seismic survey of the area by GECO-Prakla Inc.

Under terms of the joint venture, BP will operate the appraisal phase of the joint venture area, which includes 17 leases covering more than 70,000 acres about 20 miles east of Endicott field. BP's equity depends on the extent of appraisal and exploration work performed.

The 1 Badami discovery, drilled in section 9-9n-20e by Conoco and Petrofina in 1990, flowed 4,250 b/d. A second well was drilled in 1992, but results have not been revealed, and no reserve estimates have been released.

BP expects 1994 exploratory activity on the slope to be similar in scope to 1993. IPE

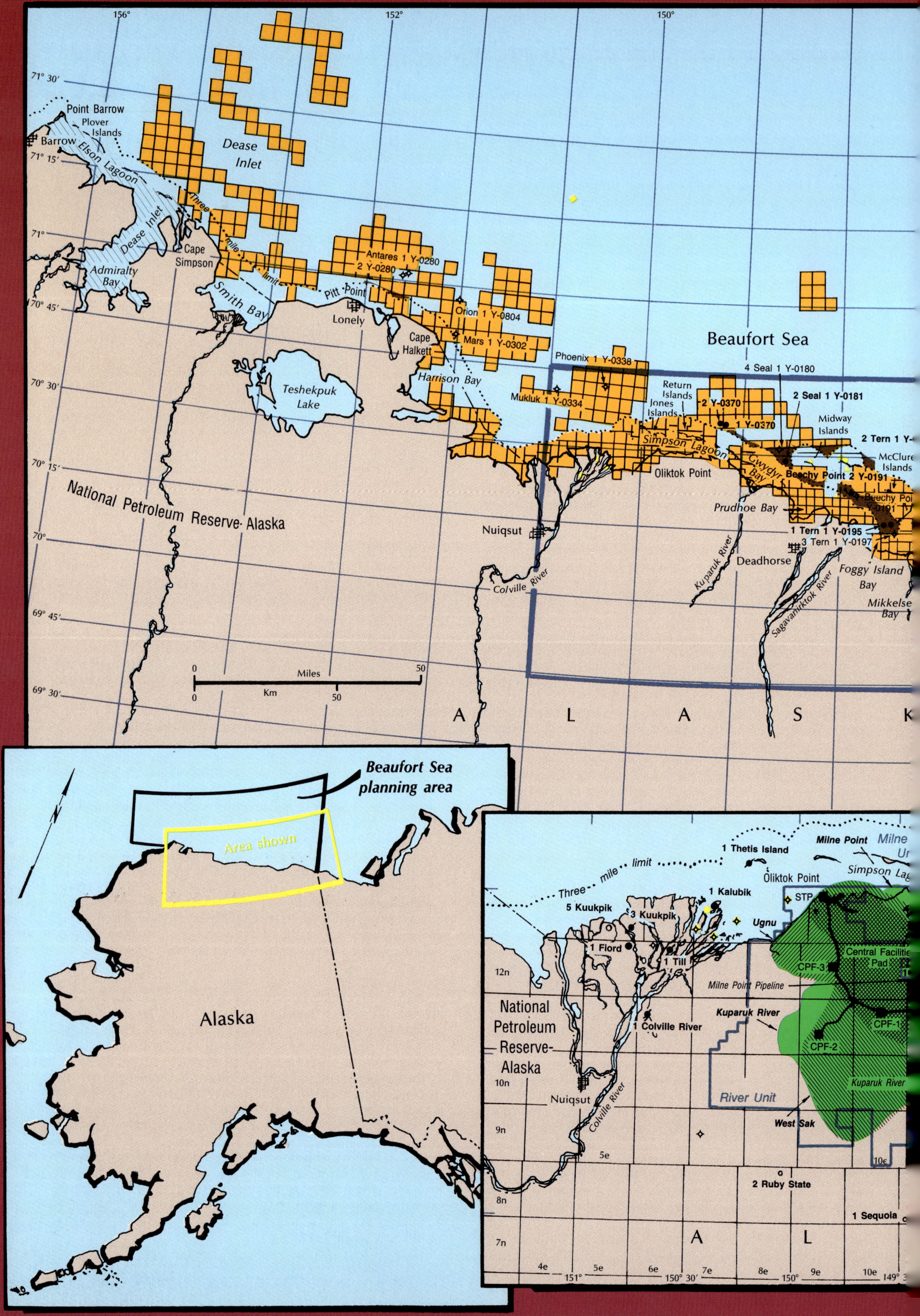
156°
152°
150°
71° 30′
71° 15′
71°
70° 45′
70° 30′
70° 15′
70°
69° 45′
69° 30′
Point Barrow
Plover Islands
Barrow
Elson Lagoon
Dease Inlet
Dease Inlet
Admiralty Bay
Cape Simpson
Three mile limit
Smith Bay
Pitt Point
Lonely
Cape Halkett
Harrison Bay
Teshekpuk Lake
Antares 1 Y-0280
2 Y-0280
Orion 1 Y-0804
Mars 1 Y-0302
Phoenix 1 Y-0338
Mukluk 1 Y-0334
Beaufort Sea
4 Seal 1 Y-0180
2 Seal 1 Y-0181
Return Islands
Jones Islands
2 Y-0370
1 Y-0370
Midway Islands
2 Tern 1 Y-
McClure Islands
Simpson Lagoon
Gwydyr Bay
Oliktok Point
Beechy Point 2 Y-0191
Prudhoe Bay
1 Tern 1 Y-0195
3 Tern 1 Y-0197
Deadhorse
Foggy Island Bay
Mikkelsen Bay
National Petroleum Reserve-Alaska
Nuiqsut
Colville River
Kuparuk River
Sagavanirktok River
0 Miles 50
0 Km 50
A L A S K
Beaufort Sea planning area
Area shown
Alaska
1 Thetis Island
Milne Point
Milne Ur
Three mile limit
Oliktok Point
Simpson Lag
1 Kalubik
STP
5 Kuukpik
3 Kuukpik
Ugnu
1 Fiord
1 Till
Central Facilities Pad
CPF-3
CPF-1
CPF-2
12n
Milne Point Pipeline
National Petroleum Reserve-Alaska
Kuparuk River
1 Colville River
10n
Nuiqsut
Colville River
River Unit
Kuparuk River
West Sak
9n
5e
8n
2 Ruby State
1 Sequoia
7n
A L
4e
151°
5e
6e
150° 30′
7e
8e
150°
9e
10e
149°

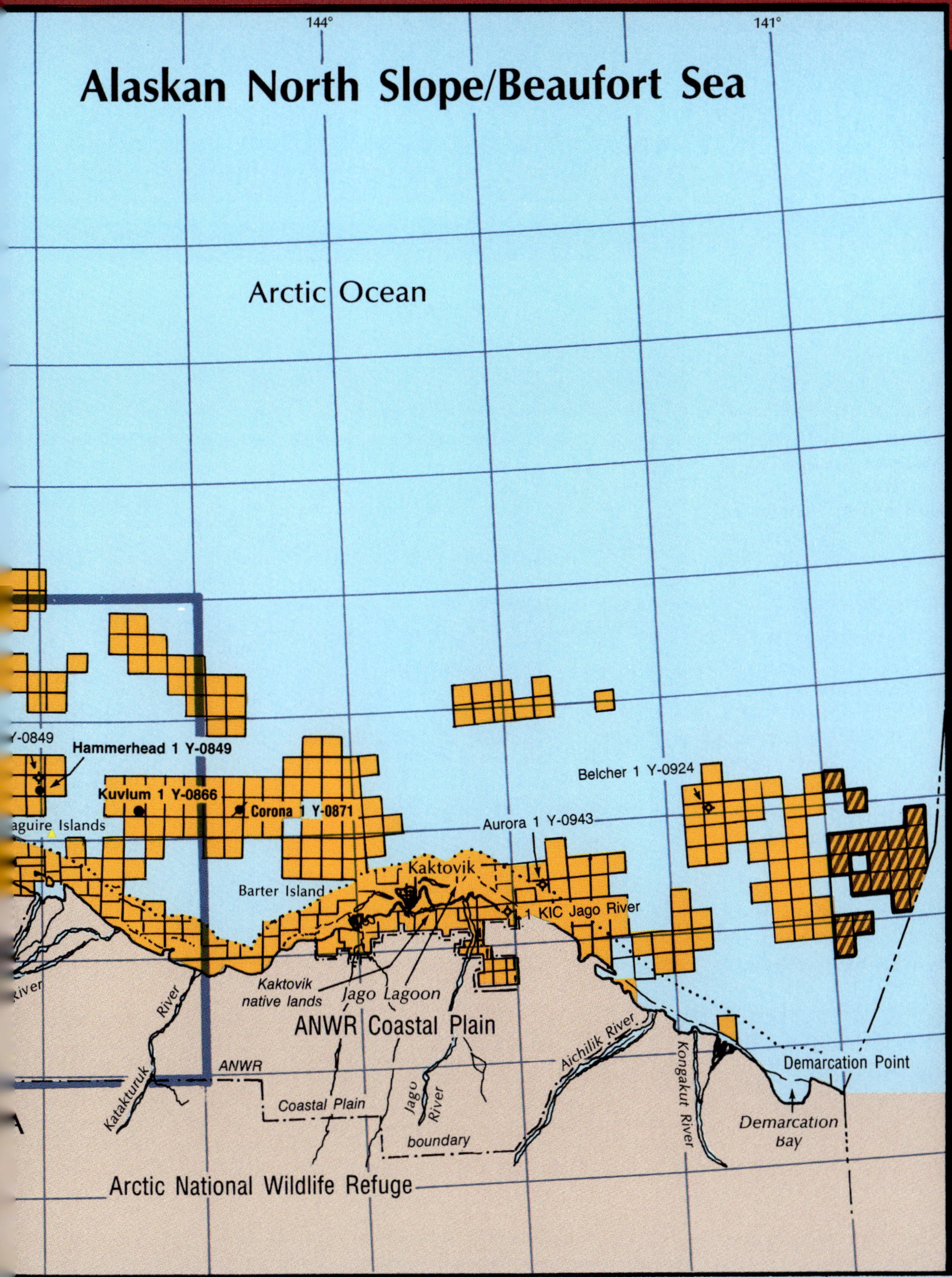

Alaskan North Slope/Beaufort Sea
144°
141°
Arctic Ocean
Hammerhead 1 Y-0849
Kuvlum 1 Y-0866
Corona 1 Y-0871
Belcher 1 Y-0924
Aurora 1 Y-0943
Barter Island
Kaktovik
1 KIC Jago River
Kaktovik native lands
Jago Lagoon
ANWR Coastal Plain
Katakturuk River
ANWR
Coastal Plain
boundary
Jago River
Aichilik River
Kongakut River
Demarcation Point
Demarcation Bay
Arctic National Wildlife Refuge

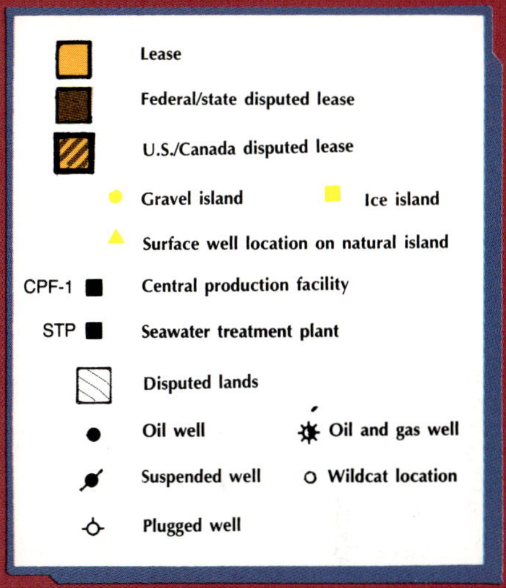

Lease
Federal/state disputed lease
U.S./Canada disputed lease
Gravel island
Ice island
Surface well location on natural island
CPF-1 Central production facility
STP Seawater treatment plant
Disputed lands
Oil well
Oil and gas well
Suspended well
Wildcat location
Plugged well

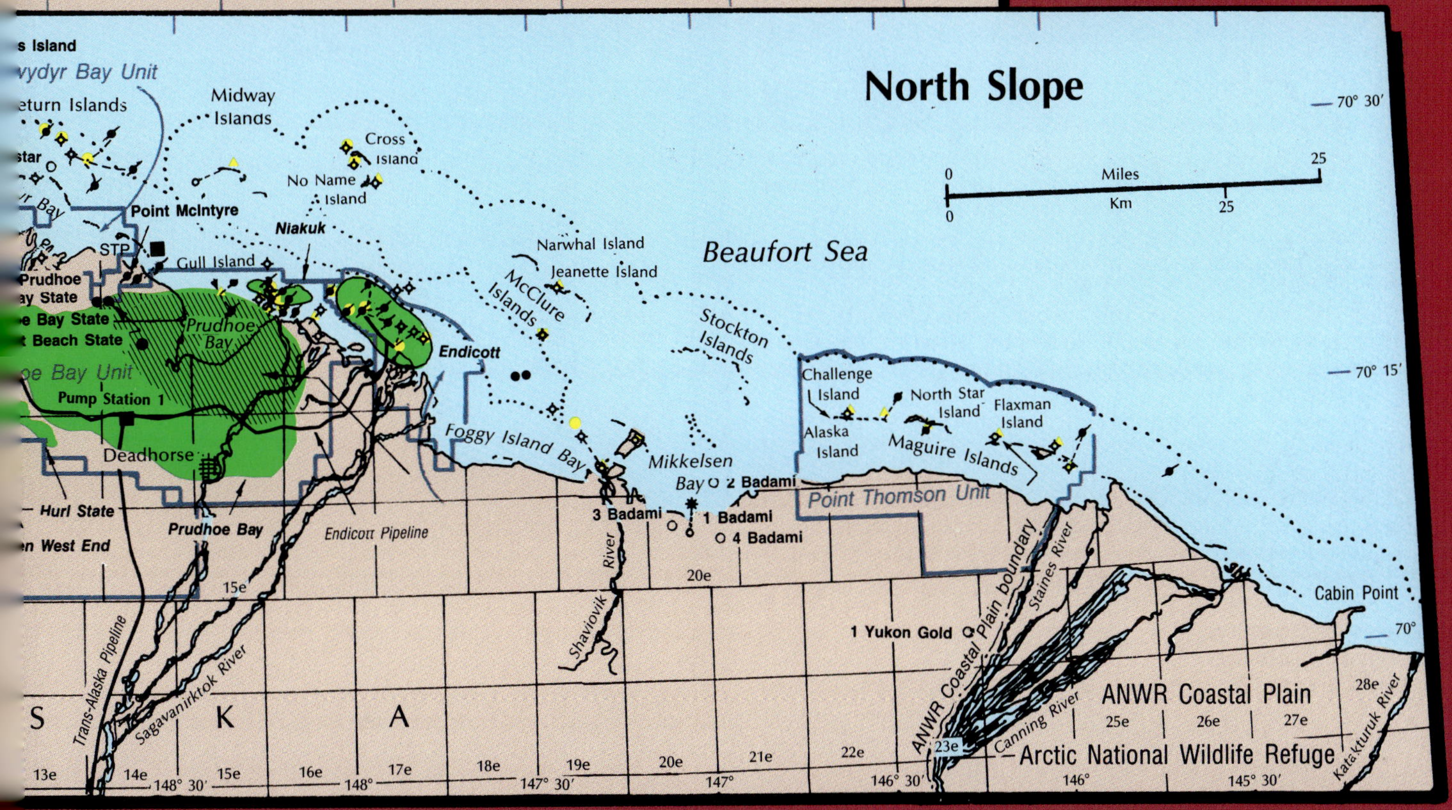

North Slope
Beaufort Sea
Midway Islands
Cross Island
No Name Island
Point McIntyre
Niakuk
Gull Island
STP
Prudhoe Bay
Pump Station 1
Deadhorse
Endicott
Endicott Pipeline
Narwhal Island
Jeanette Island
McClure Islands
Stockton Islands
Foggy Island Bay
Mikkelsen Bay
2 Badami
3 Badami
1 Badami
4 Badami
Challenge Island
North Star Island
Flaxman Island
Alaska Island
Maguire Islands
Point Thomson Unit
Shaviovik River
Staines River
Canning River
1 Yukon Gold
ANWR Coastal Plain boundary
ANWR Coastal Plain
Arctic National Wildlife Refuge
Cabin Point
Katakturuk River
Sagavanirktok River
Trans-Alaska Pipeline
Hurl State
West End
Miles
Km
0
25
70° 30′
70° 15′
70°
148° 30′
148°
147° 30′
147°
146° 30′
146°
145° 30′
S
K
A

Six more stratigraphic columns detailed

Here are stratigraphic columns from six more areas. Areas covered in this installment of an International Petroleum Encyclopedia series range from the superproductive Niger delta of Nigeria and Cameroon to the lighly explored Manitoba part of the Williston basin.

Other areas summarized are the Jeanne d'Arc basin on the Grand Banks off Newfoundland, eastern Uzbekistan, the wide ranging carbonates of Southeast Asia, and the Tarim basin of Northwest China.

Niger delta highly productive

The Tertiary Niger delta of Nigeria and Cameroon is the habitat of 20-30 billion bbl of oil and more than four times the equivalent in gas.

The delta contains more than 12,000 m of miogeosynclinal clastic wedge, which is partitioned into depobelts by east-west major structure building faults.

Nearly all the known hydrocarbons lie in the Agbada formation, sandwiched between the Benin surface sands and the subcompacted clay known as the Akata.

One author points out that the erratic distribution of oil and gas within a given field remains one of the enigmas of the Niger delta. He notes that instances are common where in a single field traps with good closures are wet even though similar ones elsewhere are petroliferous, and deep traps are not filled to spill point whereas shallower ones have larger fills.

Various operators are making heavy use of 3D seismic surveys to explore blocks in 200-1,500 m of water in the Niger delta.

Nigeria is Africa's largest oil producing country. It also rates high in terms of natural gas. Gas reserves at yearend 1992 were an estimated 120 tcf, in Africa second only to Algeria's 128 tcf. Those figures place the countries in the top 10 worldwide in terms of gas reserves.

Jeanne d'Arc basin off Newfoundland

About 113 wells have been drilled in rift basins on the Grand Banks off Newfoundland, resulting in 15 hydrocarbon discoveries in and around the Jeanne d'Arc basin.

The fields include Hibernia oil field, due on production in 1997, and three other oil fields: Terra Nova, Hebron, and Whiterose.

The Grand Banks form a broad, shallow promontory on the passive eastern margin of the North American continent. However, this area was the site of multiple phases of intense tectonism leading to the rifting apart of the Africa, European, and North American continental plates during Mesozoic time.

The effects of alternating tectonically active and quiet periods on the Grand Banks resulted in the creation of distinct sedimentary basins.

Prospective targets in the Jeanne d'Arc basin include the Ben Nevis and older Cretaceous and Jurassic sandstones in a large elongate dome and a lower Tertiary South Mara fan at the mouth of a feeder canyon.

Uzbekistan's geology has good potential

Uzbekistan imports some of its oil, but with production

Lithostratigraphy of Niger delta, Nigeria

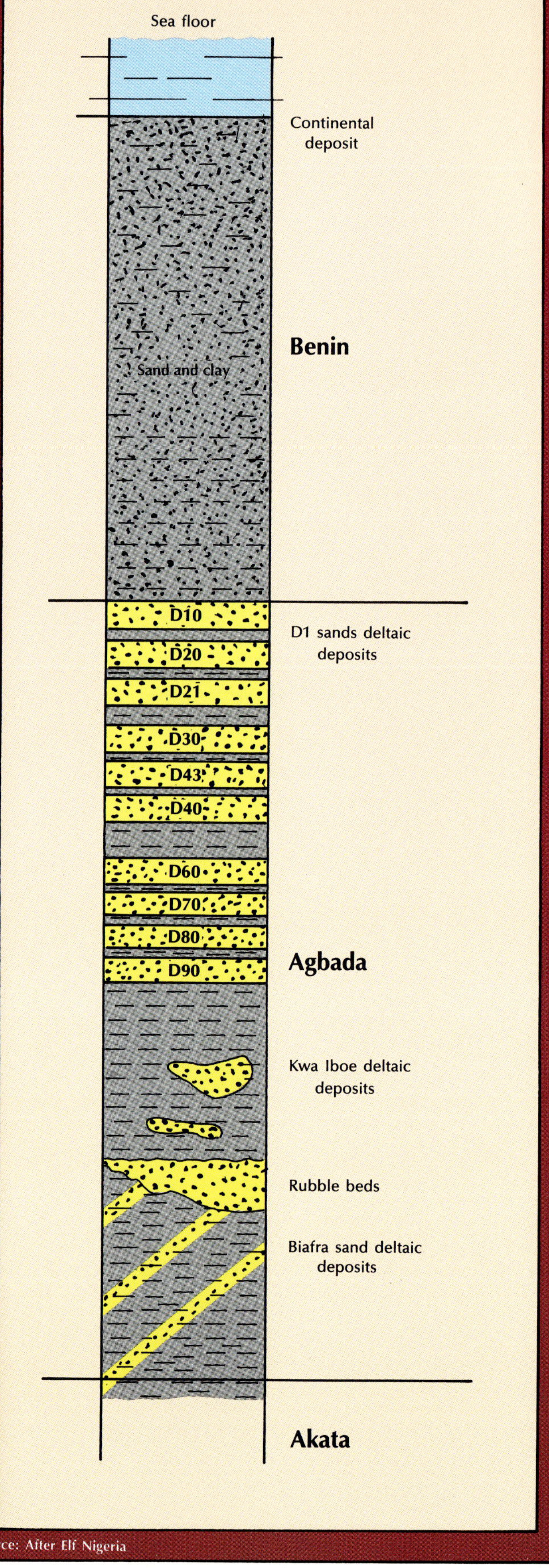

Source: After Elf Nigeria

Grand Banks off Newfoundland

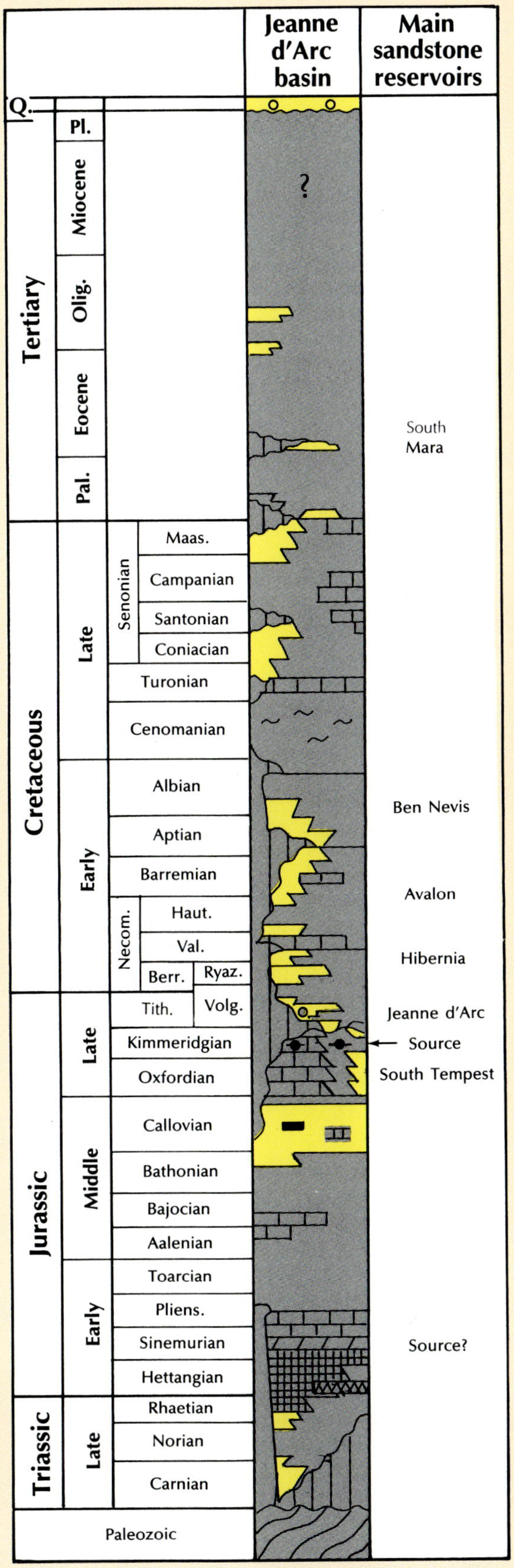

Source: Canada-Newfoundland Offshore Petroleum Board

of 50,000 b/d of oil and 110 MMcfd of gas it is the third largest gas producer among the former Soviet republics known as the Commonwealth of Independent States.

Uzbekistan's large hydrocarbon fields are distributed in a large west-northwest trending belt parallel to the Central Ustyurt system of dislocations. There is also a trend from Dauletabad-Donmez field northward through the Murgab graben to the Zaunguz graben. This trend suggests high potential for the little explored Aral Sea rift, where the Berdakh well flowed 150,000 cu m/day of gas from Jurassic deposits in 1990.

West of the rift on the Ustyurt plateau, Kuanysh and Akchalak gas fields and West Barsakelmes oil field await development of Jurassic reservoirs. The Karachalak reef has yielded 500,000 cu m/day of gas from Paleozoic limestones at 3,521 m—a gas pipeline passes through this field—and oil has been tested from the Karakuduk reef, suggesting significant potential in Paleozoic strata. Shakhpakhty field contains 1.5 tcf of gas in Jurassic clastics.

In the Surkhandarya basin, Gadzhak field contains about 500 bcf of gas, and Kognysai field contains 200 bcf, mostly in Jurassic carbonates.

Oil fields in the Fergana basin are mostly distributed along the northern and southern margins of the basin. This is largely because reservoirs in the deeper, central part of the basin have not been adequately explored and few deep wells have been drilled.

The Fergana basin contains the largest number of pools, mostly oil in Paleogene strata. Although the deeper part of the basin is the main prospective area, the large thickness of Neogene sediment means that hydrocarbons will continue to be found in Cenozoic sediments.

The Bukhara and Chardzhou steps are also prolific and contain mostly gas, predominantly below the late Jurassic salt cap. This distribution is based on discoveries to date. Lack of deep drilling capability has skewed this distribution to the shallower horizons. Several small fields and shows have been found in the Devonian-Triassic section.

More than 30 pay zones, mostly sandstones, have been found in the Fergana basin, although many are not regional in extent. Porosity of the Jurassic reservoir XXIII in South Alamyshik field is 18%. Cretaceous porosities are 10-20%. Paleogene porosities are typically 20% and reach 30%.

In the Surkhandarya basin there are three Jurassic reservoir horizons with porosities of about 10% in the carbonates. Permeabilities exceed 200 md. Flow rates from the Jurassic carbonates of Gadzhak gas field reach 3 million cu m/day.

Porosities in Cretaceous sandstones vary between 10-24%. The best reservoirs are in Paleocene-Eocene carbonates, where porosities commonly are about 20%.

Porosities in the Bukhara and Chardzhou steps are typically 5-15% in the Jurassic limestones, but permeabilities may reach more than 100 md in some of the reefs. Cretaceous sandstones have much better reservoir qualities with porosities of 15-25% and permeabilities of hundreds to thousands of millidarcies.

Early to middle Jurassic fluvial sandstones of the Ustyurt plateau have open porosities of 4-16%. Paleozoic limestone samples have porosities of only 3%, but some are cavernous and it is likely that cores cut to date do not represent permeable intervals. In Shakhpakhty field the Callovian reservoir has 12-16% porosity and 100-1,500 md permeability. Gas flow rates of as much as 1.5 million cu m/day have been recorded.

Southeast Asian carbonates prolific

Tertiary carbonate buildups of all ages are believed to serve as reservoirs for about 40 billion bbl of recoverable hydrocarbons, of which about 40% occurs in Tertiary buildups in Southeast Asia.

Tarim basin, Northwest China

Era	Age		Thickness, m	Lithology	Pays	Fields
Cenozoic	Quaternary		650			
	Neogene	Pliocene	3,600			
		Miocene	4,032		● ●	Kelatuo Kekeya
	Paleocene		1,057			
Mesozoic	Cretaceous		553			
			2,217		●	Yakela
	Jurassic		498			
			1,032		●	Yiqikelike
			973		●	Yakela
	Triassic		660			
			592		●	Lunnan
Paleozoic	Permian		972			
	Carboniferous		395			
			764		●	Akekule
			1,030			
	Devonian		900			
	Silurian		418			
	Ordovician		792		●	Lunnan
			808		● ●	Yakela
	Cambrian		585		●	Yakela
			366			
Pre-cambrian	Sinian		910		●	Yakela

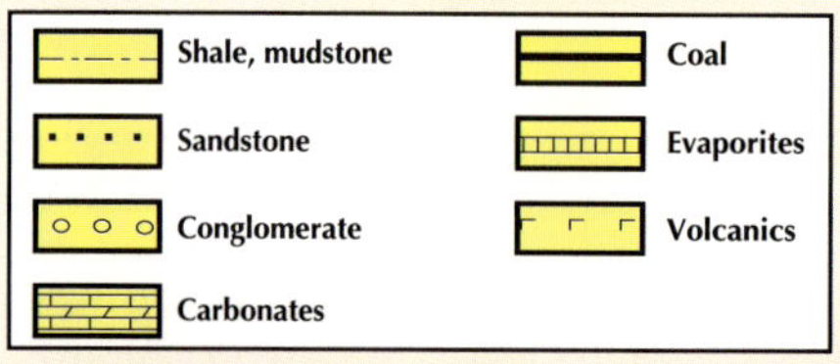

Source: Gustavson Associates Inc. after Wang Qinmin et al.

This volume of reserves makes these the world's single most important set of carbonate buildup reservoirs.

Estimates of reserves in these Miocene buildups continue to grow, as evidenced by large discoveries by Occidental Petroleum Corp. in the Sarawak basin of eastern Malaysia and Vaalco Energy Inc. and Shell International Petroleum Co. Ltd. in the Palawan basin off Philippines.

Also, major development programs are being planned in carbonate reservoirs at the Natuna platform on the L structure in Indonesian waters of the South China Sea, 3DA field off southern Myanmar, and Liuhua field in the Pearl River Mouth basin off southern China.

The level of exploration activity in the area remains high. Productive Tertiary buildups vary widely in terms of their distribution, overall size, facies geometry, and hydrocarbon reserves.

Some of Southeast Asia's major oil producing areas such as the South Sumatra and Northeast Java basins were explored for many decades before large carbonate buildup reservoirs were discovered. This was due in part to the widespread misconception that sandstones offered much more attractive exploration targets.

Many of the rift margin basins around parts of the South China Sea remain underexplored, particularly along Viet Nam's lengthy coast. The potential for large Miocene walled reef complexes in this area is high. Other more stable basin types such as the back-arc and foreland basins of eastern Indonesia also contain good potential as sites for finding low relief carbonate buildup reservoirs.

Williston basin, Manitoba

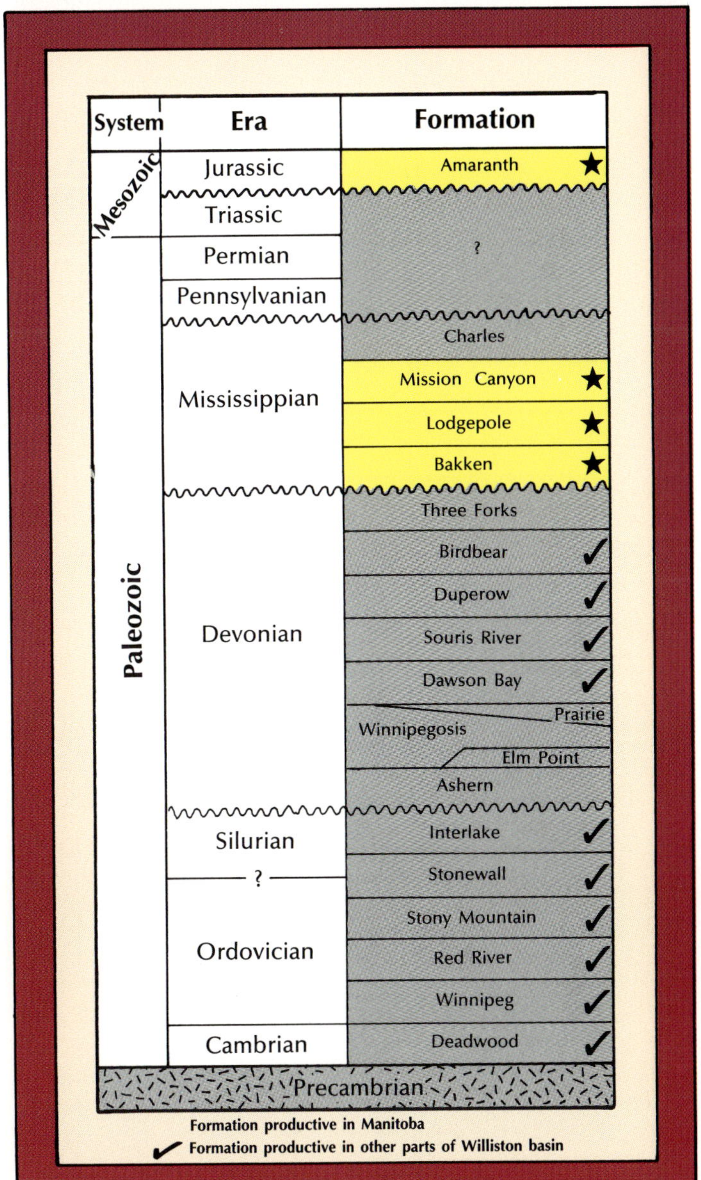

Source: Manitoba Department of Energy and Mines

China's hopes high for Tarim exploration

Intense industry interest homed in on the Tarim basin of far Northwest China when the country opened several blocks to competitive bidding in 1993.

China National Petroleum Corp. hopes Tarim basin production will climb to 100,000 b/d of oil and 57.5 MMcfd of gas by 1995.

The basin's production in 1993 was estimated at 24,000 b/d. Production was to have reached 33,000 b/d by yearend 1993, and the 1994 average was predicted to be 48,000 b/d.

Cumulative production from Tarim is about 19.4 million bbl of oil.

Eleven Tarim basin oil and gas fields produce from 14 pay zones. Five fields on stream in the northern and central parts of the basin—Lunnan, Donghetang, Sangtamu, Jilak, and Jiefangqudong—produced a combined average of 17,780 b/d of oil in 1992.

Tarim's geological surveys started in 1952, and a major exploration and development effort did not begin there until 1985.

Most fields are in the northern part of the basin, but southeastern exploratory drilling near the offered blocks encountered oil in Jurassic sands and as many as 10 pay zones with combined gross thickness of 150 m, the Ministry of Geology and Mineral Resources reported.

Kenneth J. Hsu of Tarim Associates, Zurich, worked the basin 7 years and developed a buried euxenic basin model for generation of Tarim hydrocarbons.

Hsu contended the model explains the genesis of source beds containing postulated hydrocarbons totaling more than 350 million bbl of oil equivalent, the origin of structural and stratigraphic traps, including an oil bearing anticline 230 km long and 120 km wide with 2.4 km of closure, and the history of hydrocarbon maturation and migration.

He cited evidence that large volumes of trapped hydrocarbons may have been prevented from becoming overly mature because of their migration into old traps prior to burial under Cenozoic sedimentary load.

Tarim oil is being produced from Mesozoic and Paleozoic reservoirs at depths of more than 6,000 m. Paradoxically, Hsu contended, gas and condensate are produced from 3,000 m deep Miocene resrvoirs in the basin because of their late migration.

Contrary to current assumptions that the source beds of Tarim oil are Paleozoic platform carbonates, he concluded the Paleozoic source beds are euxenic deposits of a deep marine basin. Oil migrated from lower Paleozoic source beds during the late Paleozoic or Mesozoic into karsts and other porous reservoirs on anticlinal structures and under seal beds above regional unconformities.

Hsu authored a report that also presented a new interpretation of the pre-Mesozoic structures on the Tanan uplift, the main structural elevation of first round blocks.

Gustavson Associates Inc., Boulder, Colo., counted three main types of known or expected reservoir rocks in the basin:

- Clastic sandstone and conglomerate reservoirs of early Silurian, Devonian, Carboniferous, Permian, early and middle Jurassic, Cretaceous, and Tertiary ages.
- Carbonate, mostly marine sedimentary sequences of the upper Silurian, Cambrian, Ordovician, Carboniferous, lower Permian, Cretaceous, and Paleogene.
- Weathered layer reservoirs resulting from frequent uplifts and long exposures of thick zones, especially in the

Eastern Uzbekistan

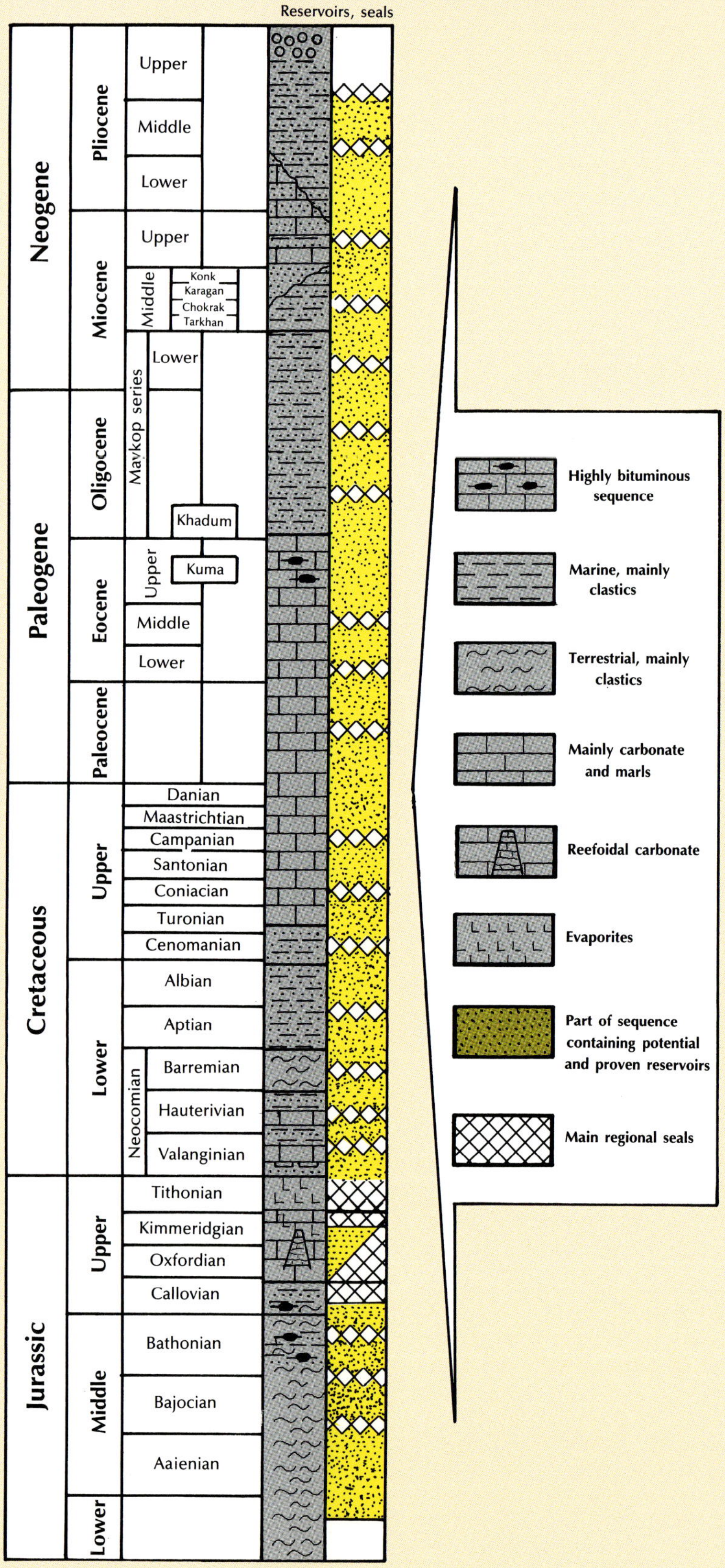

Source: Jebco Seismic Ltd.

Generalized stratigraphic chart of Indonesian petroliferous basins

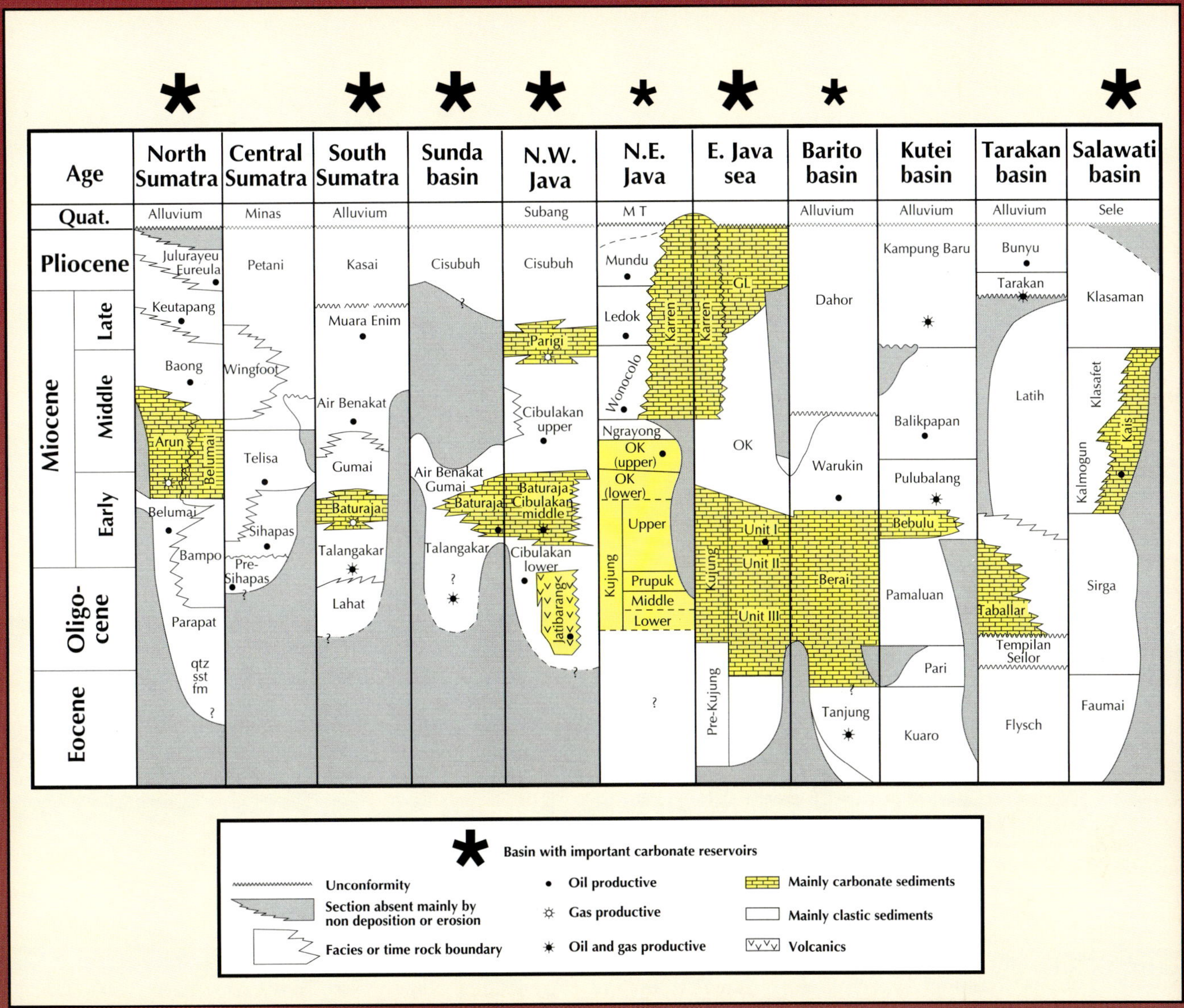

Source: From Nayoan et al., 1981

North Tarim and Central Tarim uplifts where extensive weathering and leaching occurred to carbonates now lying beneath an unconformity.

Gustavson said carbonate reservoirs, generally fractured, fossiliferous, and locally cavernous, karstified, or microcavernous, partially contain skeletal grains in algal mats and bioherm reef complexes. The most important weathered layer reservoirs associated with unconformities are beneath Devonian, Carboniferous, and Triassic sections and between Mesozoic and Cenozoic strata.

Manitoba's deeper section little explored

Paleozoic and Mesozoic age rocks occur in Southwest Manitoba on the Williston basin's northeast flank.

Sediments in Manitoba are 7,500 ft thick in the province's southwest corner. Thickness grades to zero at the margin of the Precambrian shield in eastern Manitoba.

Mississippian and Jurassic formations produce in Manitoba, but many more horizons that produce elsewhere in the Williston basin have not been found productive in the province.

Limestone and dolomitic and argillaceous limestones are dominant in Paleozoic beds. Evaporites, sandstones, and shales are common, mainly toward the base of the stratigraphic sequence.

Jurassic Amaranth and the Mission Canyon, Lodgepole, and Bakken members of Mississippian have produced in the province, where production totaled about 11,000 b/d of oil in 1993. The lower half of the stratigraphic column remains practically unexplored.

Manitoba began producing in 1951. Discovery in 1980 of Waskada field along the border with North Dakota opened oil production from Jurassic lower Amaranth.

Most of the province's main fields are under waterflood. These include Canada's first waterflood, Daly field, where injection began in 1953.

Formations that have produced in the province lie at 2,000-3,500 ft. Most of the oil is a 34° gravity light sour blend, but light sweet and medium variety crudes have also been produced.

The province holds 11 designated oil fields and several other producing areas.

Horizontal drilling is increasing in Manitoba since the province granted exceptions for such wells from production restrictions, royalties, and some taxes. IPE

Salomon Bros. survey shows 1993 spending hike

Salomon Bros. Inc.'s traditional midyear survey of oil and gas exploration and production spending showed significant gains in the U.S. and Canada during 1993.

They were prompted mainly by higher wellhead gas prices, the New York investment firm said. Outside North America, spending levels were slated to dip slightly from 1992 levels.

Overall, Salomon Bros.' survey suggested a moderate year to year increase in 1993 worldwide spending vs. roughly flat spending levels indicated at yearend 1992.

In the firm's largest survey at midyear 1993, 277 companies registered a combined 4.9% advance in that year's world oil and gas E&P spending to $54.3 billion.

In its yearend 1992 survey, 260 companies planned a 0.3% advance in world spending year to year.

Spending hikes in North America accounted for the entire 1993 increase, posting a 14.9% advance in the midyear study vs. only a 2% rise indicated 6 months earlier.

U.S. E&P spending at midyear 1993 was scheduled to climb by 12%—up 22.4% to $4.702 billion by independents and 7.9% to $10.508 billion by majors—compared with a 2.6% increase—6.5% by independents and 1.1% by majors—in Salomon Bros.' 1992 yearend survey.

Spending in Canada was to surge 25.8% vs. flat projections in December 1992. Spending outside North America was slated to dip a mere 0.1%, compared with the 0.6% decline suggested in December 1992.

In total, the 156 independent oil and gas companies Salomon Bros. surveyed at midyear 1993 had raised their 1993 exploration and production budgets 22.4% year to year, more than three times the 6.5% increase suggested in the December 1992 survey of 147 independents.

Overall, 37% of the participating companies had increased U.S. E&P budgets by midyear, while 20% had cut their budgets.

Meantime, a survey by Oil & Gas Journal (OGJ) found that companies of all sizes had earmarked a total of $36.3 billion for U.S. capital and exploration items during 1993, up 6% from an estimated $34.2 billion spent in 1992. The items ran the gamut from drilling and exploration, production, Outer Continental Shelf lease bonuses, processing, transportation, and other activities such as coal mining.

How total U.S. spending has varied*

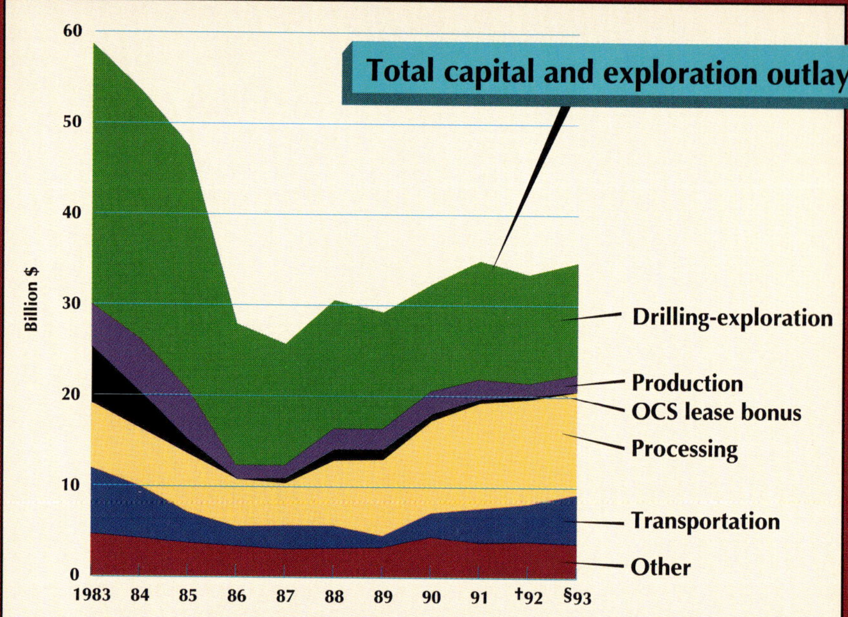

*Domestic projects only. †Estimated. §Budgeted. Source: Oil & Gas Journal

Independents, majors

Many large independents had boosted their 1993 U.S. E&P spending plans by midyear.

Anadarko attributed its increase to Midcontinent development and a pickup in South Texas. Enserch and Maxus cited higher natural gas prices. Gerrity Oil & Gas raised additional funds through the capital markets. Louisiana Land & Exploration shifted funds back to the U.S., citing higher natural gas prices. National Fuel noted a horizontal drilling program and offshore drilling opportunities. Sonat pointed to greater Austin chalk activity.

By contrast, Columbia Gas System reduced its 1993 planned spending from yearend 1992, citing financial constraints, and Nerco Oil & Gas cut its expenditures because the company was up for sale.

U.S. 1993 spending by major oil companies was destined to rise 7.9% from the 1992 level. That was a gain from the 1.1% increase suggested by the same companies in Salomon Bros.' yearend 1992 survey. Amerada Hess, British Petroleum, Conoco, Texaco, and Total Minatome were among companies that increased their budgets, while Chevron, Exxon, and Mobil cut theirs.

Exxon cited a continued shift toward spending outside North America.

Canadian outlays

Canadian exploration and development spending plans, covered by responses from 109 companies, will surge 25.8% year to year, Salomon Bros.' midyear 1993 survey showed.

A resounding 52% of the companies boosted spending estimates from those indicated 6 months earlier, far outnumbering the 15% that reduced them.

Higher than expected natural gas prices was the common theme for the increases.

Anderson Exploration, Talisman Energy (formerly BP Canada), North Canadian Oils, and Saskatchewan Oil & Gas were among companies that said better higher wellhead gas prices fueled their spending hikes.

Canadian Natural Resources cited higher cash flow and production levels as well as a more optimistic outlook. Noranda (Canadian Hunter) noted project success and additional cash flow, and Sceptre Resources and Talisman said additional spending stemmed from asset acquisitions.

Outside North America

Spending outside North America, covered in responses from 92 companies at midyear 1993, will slip 0.1% year to year, showing an improvement from the 0.6% decline indicated by a group of 88 companies in December 1992. A moderately larger percentage of companies (31%) trimmed budgets than raised them (28%).

Among companies that reduced their 1993 budgets from

Breakout of exploration/production spending*

MAJORS, U.S.	1993 (Million U.S. $)	1992 (Million U.S. $)
AGIP	126	112
Amerada Hess	190	175
Amoco	575	550
ARCO	1,190	1,160
British Petroleum	800	900
Chevron	850	861
Du Pont	360	360
Elf Aquitaine	20	30
Exxon	1,850	1,850
Mobil	400	400
Occidental	150	200
Oryx	162	110
Pennzoil	228	95
Phillips	320	256
Royal Dutch/Shell	1,350	1,350
Texaco	625	700
Total Minatome	29	31
Union Pacific	440	530
Unocal	600	475
USX	300	300
Total	**10,565**	**10,445**
Change	**+1.1%**	

SELECTED INDEPENDENTS, U.S.†	1993 (Million U.S. $)	1992 (Million U.S. $)
Alta Energy	11	3
AMAX Oil & Gas	38	17
American Exploration	23	20
Ampol Exploration	9	11
Anadarko	180	136
Apache	150	120
Arkla Exploration	0	27
Barrett Resources	19	7
Basin Exploration	30	32
Beard	12	10
Berry	10	6
BHP	115	59
Bridge Oil	19	12
Brooklyn Union	12	22
Burlington Resources	220	140
Cabot	44	42
Cairn Energy	10	4
Chieftain	21	14
Coastal	90	145
Coho Resources	18	26
Columbia Gas System	100	90
Consolidated Natural Gas	109	88
Devon	27	17
Enron	276	315
Enserch	86	61
Equitable Resources	74	72
Forest	28	14
FM Properties	15	15
Freeport-McMoRan	20	30
General Atlantic	13	11
Gerrity	60	48
Global Natural Resources	14	6
Grace Energy	30	27
Hallwood Energy Partners	15	18
Hardy Oil & Gas	33	29
Helmerich & Payne	33	21
Howell Petroleum	14	10
HS Resources	46	45
J.M. Huber	27	34
Kelley Oil et al.	76	63
Kerr-McGee	105	115
Lasmo	0	13
Louisiana Land	70	75
Maxus	60	60
Mesa	32	71
Mitchell	146	115
Murphy	116	84
National Fuel	40	21
Nerco	90	135
Neste Oy	4	20
Nicor	15	22
Noble Affiliates	106	64
Nomeco	25	30
Norcen	27	36
North Coast Energy	13	10
Nuevo Energy	28	48
Oneok	10	11
Parker & Parsley	100	70
Patrick Petroleum	10	10
Petrofina Group	43	103
Petroleum Development	8	16
PG&E Resources	104	115
Plains Petroleum	23	18
Pogo Producing	51	42
Presidio	15	19
Public Service Enterprise	95	38
Quaker State	10	12
Questar	49	62
Sage Energy	16	15
Santa Fe Energy Resources	85	65
Santos	20	12
Seagull	105	55
Snyder	46	58
Sonat	92	155
Southwestern Energy	37	30
Tesoro	18	9
Texas Meridian	15	5
Transco Coalbed Methane	0	15
Union Texas	18	15
Washington Energy	13	20
Weber Energy	21	15
Whiting Petroleum	10	6
Wiser	13	10
Zilkha Energy	30	20

CANADA†	1993 (Million U.S. $)	1992 (Million U.S. $)
Amerada Hess	30	20
Amoco	200	200
Anadarko	10	8
Anderson Exploration	33	10
Atcor Resources	19	23
Bow Valley Industries	31	29
BP Canada	55	47
Canadaian Natural Resources	47	31
Chauvco Resources	61	31
Chevron	250	256
Conwest Exploration	31	23
Co-Enerco	39	39
CS Resources	14	12
Dekalb Energy	18	17
Discovery West	12	11
Dorset Exploration	19	16
DuPont	30	30
Encor Energy	22	17
Enron	23	15
Gulf Canada	127	204
Hillcrest Resources	9	11
Home	79	59
Imperial	470	430
Intensity Resources	11	7
International Colin	19	15
Inverness	24	19
Kerr-McGee	25	15
Lasmo	11	46
Mark Resources	18	19
Mobil	300	300
Morgan Hydrocarbons	27	23
Morrison Petroleum	24	20
Murphy	32	25
Noble Affiliates	12	7
Noranda	53	64
Norcen	104	81
North Canadian	14	40
Northstar Energy	12	9
Ocelot Industries	26	12
Pacific Enterprises	0	20
PanCanadian	325	280
Paramount Resources	24	10
Petrofina	2	14
Petro-Canada	245	164
Phillips	10	8
Pinnacle Resources	47	24
Poco Petroleum	31	30
Ranchmen's Resources	15	13
Ranger	20	15
Renaissance Energy	171	155
Saskatchewan Oil & Gas	43	43
Sceptre Resources	28	24
Shell Canada	200	466
Suncor	29	13
Tarragon	30	19
Texaco	10	10
Total Canada	23	20
Ulster Petroleum	16	14
Union Pacific	3	43
Unocal	30	30
Westcoast Petroleum	39	45

OUTSIDE NORTH AMERICA†	1993 (Million U.S. $)	1992 (Million U.S. $)
AGIP	2,333	2,729
Amerada Hess	500	712
Amoco	1,075	1,100
Ampol Exploration	177	126
Anadarko	30	21
Apache	15	12
Aran Energy	88	52
ARCO	700	775
BHP	1,000	587
Bow Valley Industries	171	140
Bridge Oil	16	13
British Petroleum	2,200	2,800
Chevron	1,300	1,315
Clyde Petroleum	161	102
DuPont	1,240	1,080
Elf Aquitaine	3,500	3,500
Enron	44	10
Enterprise	550	550
Exxon	3,325	3,325
Garnet Resources	10	8
Goal Petroleum	26	28
Gulf Canada	51	51
Hadson Energy Resources	8	23
Hardy Oil & Gas	50	53
Kerr-McGee	110	110
Lasmo	383	493
Louisiana Land	78	75
Maxus	240	240
Mobil	1,400	1,400
Murphy	95	88
Neste Oy	223	203
Noble Affiliates	12	8
Nomeco	41	30
Norcen	38	23
Norsk Hydro	873	827
Occidental	350	285
Oryx Energy	173	224
PanCanadian	12	12
Petrofina Group	340	555
Petro-Canada	13	16
Phillips	445	355
Premier Consolidated	28	24
Public Service Enterprise	13	7
Ranger	108	105
Royal Dutch/Shell	4,230	3,980
Saga	250	290
Santa Fe Energy Resources	25	15
Santos	130	105
Statoil	1,525	1,370
Sun	40	91
Svenska Petroleum Exploration	35	33
Texaco	1,165	1,080
Total CSB	1,400	1,300
Triton	110	100
Unimar	43	36
Union Pacific	10	10
Union Texas	215	327
Unocal	355	355
USX	500	500

*Estimates for both years. †Companies spending at least $10 million in either year.

Source: Salomon Bros. Inc.

Midyear 1993 E&P budget changes*

	Greater (%)	Same (%)	Reduced (%)
U.S.			
Independents	36.9	43.1	20.0
Majors	45.0	40.0	15.0
CANADA	51.6	33.3	15.1
OUTSIDE NORTH AMERICA	27.7	41.0	31.3
Total	**39.3**	**39.6**	**21.2**

*December 1992 vs. June 1993.

Source: Salomon Bros. Inc.

Where funds were to go for 1993 U.S. projects

	1993 (Million $)	Change 1993-92 (%)	1992 (Million $)	Change 1992-91 (%)	1991 (Million $)
Exploration-production					
Drilling-exploration	12,847	+5.7	12,154	−14.8	14,269
Production	2,589	+6.5	2,431	−14.8	2,854
OCS lease bonus	100	+4.2	96	−71.7	339
Subtotal	**15,536**	**+5.8**	**14,681**	**−15.9**	**17,462**
Other					
Refining	7,277	+7.1	6,795	+ 0.8	6,741
Petrochemicals	2,326	−6.5	2,487	− 1.7	2,530
Marketing	2,528	−0.6	2,544	− 8.1	2,767
Crude and products pipelines	1,865	+139.5	779	+104.4	381
Natural gas pipelines	3,041	−7.6	3,291	+18.8	2,770
Other transporation	728	+11.7	652	−4.8	685
Mining, other energy	764	−28.1	1,062	+16.1	915
Miscellaneous	2,237	+14.2	1,958	−21.0	2,477
Subtotal	**20,766**	**+6.1**	**19,567**	**+1.6**	**19,266**
Total	**36,202**	**+6.0**	**34,248**	**−6.8**	**36,728**

Source: Oil & Gas Journal

the amount suggested at yearend 1992 were Bow Valley Industries, which cited unexpected U.K. petroleum revenue tax changes. Conoco cited a stronger U.S. dollar as well as a moderate shift in spending to the U.S. Elf attributed its reduction to lower revenue expectations and delayed projects.

Largely offsetting those reductions were increases by companies such as Chevron, which raised spending in Indonesia, West Africa, and Russia. Exxon cited a continued shift internationally, driven by its higher return on capital outside North America.

In addition, Maxus pointed to better success in Bolivia, Murphy acquired an interest in the U.K. North Sea, Occidental noted an incremental investment in Congo, Pennzoil allocated funds for development in Azerbaijan, and Saga boosted spending as a result of acquisitions.

U.S. and Canadian independents were largely on target with their budgets, but a significant number of U.S. majors and international companies were underspending theirs.

Worldwide, 26% of respondents was underspending budgets vs. only 10% that was overspending them. The underspending was concentrated outside North America, where 32% of respondents underspent budgets at midyear vs. only 7% that overspent them.

Price outlook

At midyear, independent operators raised their 1993 U.S. natural gas price forecasts by 16.7% to an average $2.03/Mcf, while majors lifted their forecasts 10.3% to $1.92/Mcf from $1.74/Mcf 6 months earlier.

Canadian companies raised their 1993 natural gas price forecasts 16.5% to an average $1.55 (Canadian)/Mcf from $1.33.

Overall, 1993 world oil price expectations were trimmed 3.2% to $19.75/bbl from $20.40.

Preliminary figures suggest higher 1994 world E&P expenditures.

Overall, 47% of Salomon Bros.' respondents indicated higher spending for 1994 vs. only 11% that indicated lower outlays. Anticipated increases once again were concentrated among U.S. independents and companies in Canada. Canadian spending expectations were the most optimistic, with 55% of respondents indicating spending increases, compared with less than 7% suggesting declines.

In the U.S., nearly a five-to-one ratio of independents responding at midyear 1993 expected an increase in 1994 spending. Spending in the U.S. by majors was expected to be relatively flat in 1994, with no companies in this category anticipating lower outlays.

Outside North America, 33% of respondents pointed to higher spending levels in 1994, compared with 19% that suggested lower levels.

Total spending in the U.S.

OGJ's survey of the petroleum industry's plan for 1993 total spending for U.S. capital and exploration items amounted to a rebound from outlays during 1992, when spending was down 6.8% from the $36.7 billion spent in 1991.

Earlier, U.S. capital and exploration outlays advanced 15.7% in 1989 and 8% in 1990. The recent low for spending was $25.2 billion in 1987 in the wake of the disastrous 1986 plunge in oil prices. Planned outlays for 1993 were up a resounding 44.1% from 1987.

Adjusted for inflation, 1993 spending was to be the second highest since 1986, topped only by 1991. But inflation adjusted outlays during 1986-93 have been at the lowest levels since 1973. This was reflected most vividly in the depressed level of upstream activity.

Downstream spending in the U.S. during 1993 was scheduled to exceed upstream spending for the fourth straight year.

Downstream spending topped upstream outlays in 1990 for the first time since 1971.

Sharp drops in drilling cut upstream outlays. At the same time, investments in upgrading, renovation, environmental compliance, marketing, and transportation swelled downstream budgets.

Industry's U.S. exploration and production spending in 1993 was scheduled to rise 5.8% from the 1992 level of $15.5 billion, OGJ figures showed. IPE

N. American storage projects proliferating

North American gas supply reliability during peak winter demand periods likely will depend more than ever on efficient use of available storage capacity.

Precisely how storage utilization will affect North American gas markets is uncertain because many new players are offering storage services through mostly untested contractual arrangements.

But a positive development is that available gas storage capacity in the U.S. is increasing. And that is due in large part to storage's relative value in markets taking on added luster as a result of Federal Energy Regulatory Commission Order 636, which took effect Nov. 1, 1993.

Order 636 in most cases ends interstate pipeline companies' merchant functions, unbundles pipeline interstate gas transportation services and fees, and opens interstate transmission capacity to access by any qualified shipper on firm or interruptible basis. Interstate pipeline gas storage capacity is among the transportation services affected.

As markets set values on controlling or aggregating gas supplies at given points on the U.S. interstate pipeline grid and on transporting those volumes to end use customers, storage will be valued according to its contribution in each supply chain. And because Order 636 allows storage to play a greater role in the supply chain, its value to producers, shippers, and consumers will grow as well.

A valuable commodity

In a 1992 study of U.S. and Canadian gas storage for Gas Research Institute (GRI), Chicago, Pace Consultants Inc., Houston, counted 427 storage facilities with a combined 2.76 tcf of working gas capacity installed before FERC implemented open access transportation in 1986 under Order 436. Industry and government sources now put U.S. working gas storage capacity levels at about 3.9 tcf.

Most U.S. gas storage facilities were built in market areas to serve seasonal and peak demand. Because interstate pipelines managed supplies for most markets, producers had little need for gas storage capacity.

Order 636 allows interstate pipelines to retain only enough storage and transportation capacity to maintain operational control of their interstate transmission systems and to provide no-notice services. The obligation to assure reliability of supplies falls mainly on end users and indirectly on suppliers and shippers serving them.

Many in the U.S. gas industry are concerned that storage facilities designed to serve pipelines' operational objectives will not be used efficiently by the many producers, marketers, and other suppliers that have firm and interruptible storage capacity.

Gas storage can help assure the reliability, deliverability, and flexibility of supplies fragmented beyond the wellhead by Order 636. If efficiently used and marketed under Order 636 by shippers or suppliers, storage capacity could become a commodity as valuable as gas.

That prospect is prompting many interstate pipelines and independent gas marketers to install or expand seasonal and peaking gas storage facilities in supply and market areas.

Gas storage expansions

Companies all over the U.S. are positioning themselves to take advantage of gas storage opportunities derived from Order 636.

Among developments in late 1993:

- Tejas Power Corp. (TPC), Houston, said in September it is ready a year ahead of schedule to provide peaking and swing gas delivery services to a large midwestern utility.
- Mitchell Energy & Development Corp.'s Winnie Pipeline Co. commissioned natural gas storage operations in September at its partially complete Spindletop salt dome storage facility near Beaumont, Tex.

Spindletop gas storage facility developed near Beaumont, Tex., by Mitchell Energy & Development Corp. typifies rapid expansion of U.S. gas storage in recent years. Photo courtesy of Mitchell.

- J. Makowski Co., Boston, started state and federal permitting for a 5 bcf storage facility in Steuben County, N.Y.
- MCN Investment Corp., Detroit, filed an application with the Michigan Public Service Commission for approval to develop and operate a natural gas storage field in Washington Township, Mich.
- Coastal Corp. unit ANR Storage Co. and NGE Enterprises Inc., a unit of New York State Electric & Gas Corp. (Nyseg), disclosed plans to develop a 2 bcf storage facility near Watkins Glen in South Central New York.
- Nyseg also disclosed plans to develop the Seneca Lake storage project, also near Watkins Glen, which will have 800 MMcfd of storage capacity.
- Capacity for gas storage at the proposed Young storage field near Fort Morgan, Colo. was fully subscribed under 30 year contracts, Young Gas Storage Co. Ltd., Colorado Springs, reported.
- Koch Gateway Pipeline Co. held an open season for interruptible storage service at its Bistineau site in Louisiana and a corresponding sale of gas in place. The open season started Oct. 15, and Koch took requests for as much as 60 bcf of interruptible capacity and 11 bcf of storage capacity until Oct. 22. The open season was subject to FERC approval to sell the gas in place at Bistineau. Koch acquired Bistineau with its purchase of United Gas Pipe Line Co. last November.
- GPM Gas Corp., Houston, signed a 5 year lease for as much as 1 bcf/year of natural gas storage at Richfield Gas Storage System's facility in Morton, Stevens, and Grant counties, Southwest Kansas. GPM also reached a long term sales agreement to supply 45 MMcfd to Western Resources Inc., Topeka, and to supply an added 16 MMcfd during peak winter months. GPM laid a 7 mile, 45 MMcfd pipeline from the storage site to connect with Western Resources' intrastate system in Grant County. Gas will be delivered to Western Resources from GPM processing plants in Oklahoma, New Mexico, and West Texas, as well as from GPM leased storage in Kansas.

Supply area storage

The role of supply area storage facilities is growing as suppliers prepare to operate under 636.

Order 636 allows a gas end user with firm transportation capacity to add a relatively secure source of gas to his supply portfolio by acquiring storage capacity from a

How North American working gas storage will grow

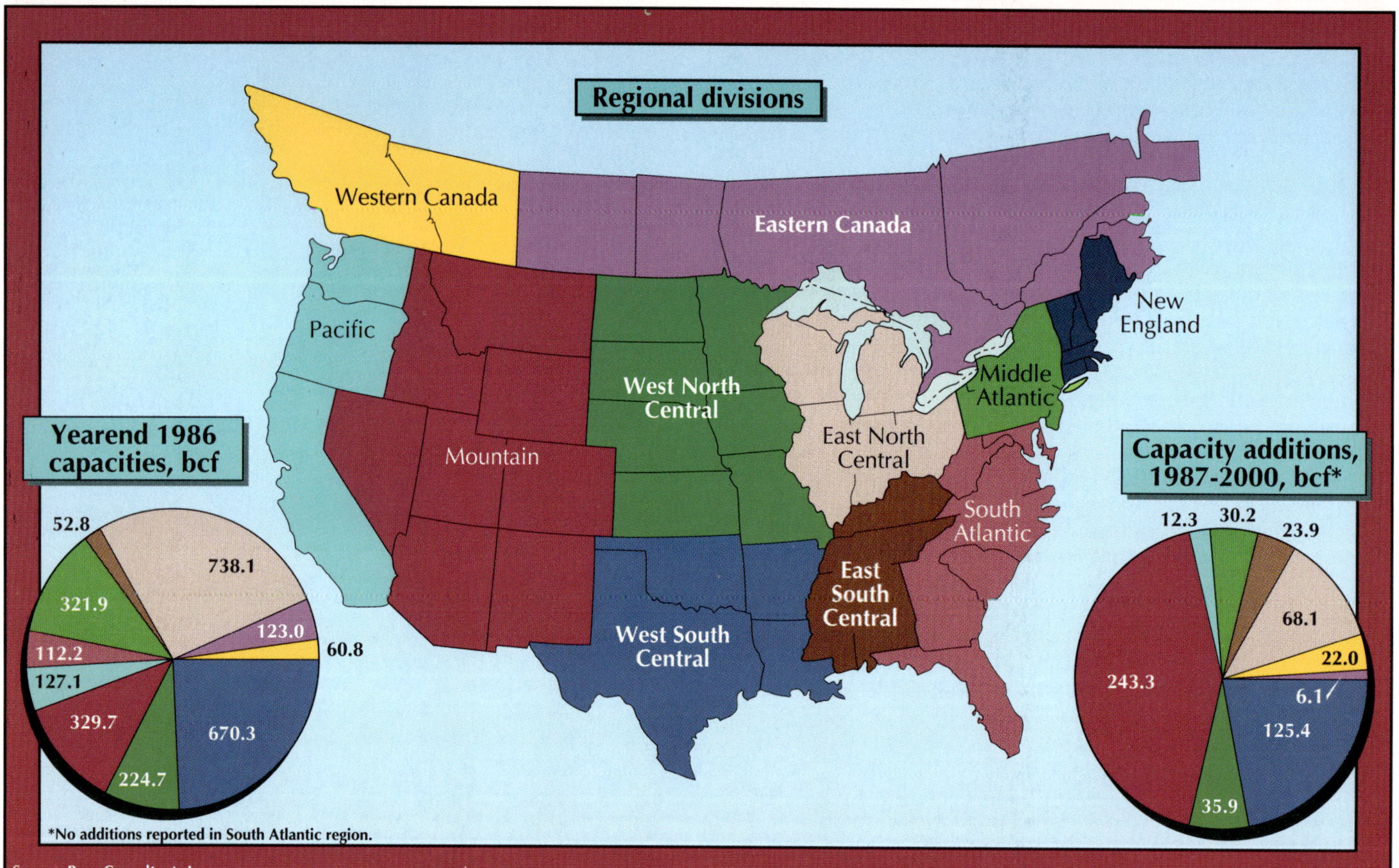

Source: Pace Consultants Inc.

supply area facility and gain reliability not always possible on the spot market. The strategic value of using storage to increase gas supply reliability is prompting storage operators to develop more capacity in supply areas.

AGA in a 1988 survey counted 429 gas storage facilities in the U.S. and Canada, the newest completed in 1984. Of that number, only 88 installations were in net gas supply states or provinces, and most supply area installations were built to serve local or intrastate markets.

In GRI's 1992 study, Pace reported North American supply area working gas capacity accounted for less than 25% of storage capacity installed before Order 436. Since FERC implemented Order 436 in 1986, Pace found 57 storage projects—new facilities, expansion, or retrofits—undertaken in the U.S. and Canada. Thirty-two of the projects are in supply areas, including 17 in Texas.

In addition, Pace concluded that many storage projects reported in market areas are being developed for the dual purposes of serving U.S. Gulf Coast markets as well as for supply area storage.

"While storage proximity to the market area is critical for some applications, the availability of open access transportation allows supply area storage to serve some market area storage needs," Pace said.

Before open access transportation, market area storage especially contributed to supply reliability by providing seasonal and peak day gas. Under 636, new supply area storage capacity also can augment supply reliability for seasonal and peak day demand.

Seasonal vs. peak storage

The growing role of supply area storage expected under Order 636 is increasing the strategic importance of supply area facilities' deliverability and flexibility.

Both attributes are attainable with salt cavern gas storage installations.

Gas storage facilities—whether in market or supply areas—fall generally into two groups:

- Seasonal or reservoir storage for baseload gas supplies during extended periods of high demand.
- High deliverability storage used to handle short peak demand swings.

In a seasonal storage facility, gas injected into a depleted reservoir or aquifer during the summer becomes part of an LDC or other market area end user's winter heating baseload supply when withdrawn during the following winter. Because of their generally larger storage capacities and slow injection and withdrawal capacities, reservoir gas storage facilities release gas over a period of months. Even at peak flow, the rates of gas withdrawal during the winter often are not much different from the injection rates during the summer.

By contrast, high deliverability gas salt caverns or salt dome storage facilities are designed to supply high volumes of gas for very short periods of time. Because the working gas capacities of such facilities are relatively small and rates of injection or withdrawal relatively high, salt caverns can't long sustain peak deliverability.

Salt cavern storage

Because high deliverability facilities in a fully open access gas transportation system can provide supply backup and reliability over several days, more storage operators are developing salt cavern storage projects.

Of the 249 storage projects developed in North America before open access transportation was implemented in the U.S., AGA's 1988 survey counted only 12 in salt caverns.

Pace's 1992 study put those 12 salt caverns' capacity at 29 bcf prior to Order 636 implementation.

Pace found 13 of the 27 gas storage projects developed

North American working gas capacity by facility type

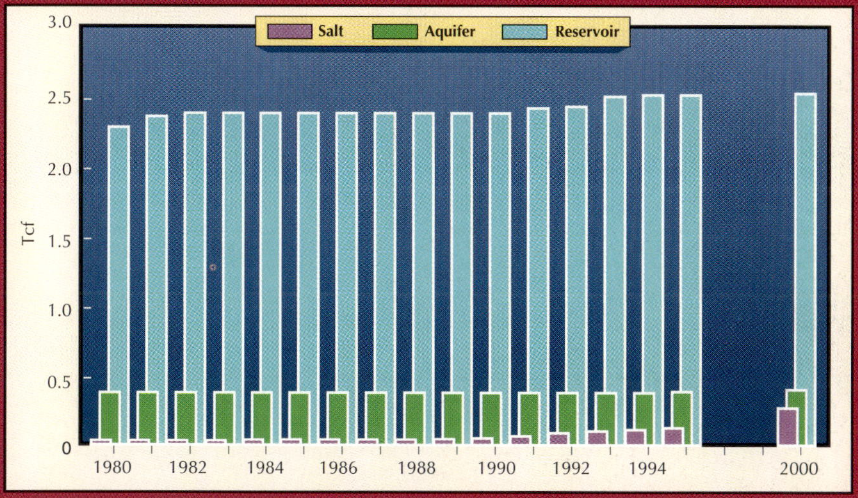

Source: Pace Consultants Inc.

U.S. working gas storage capacity*

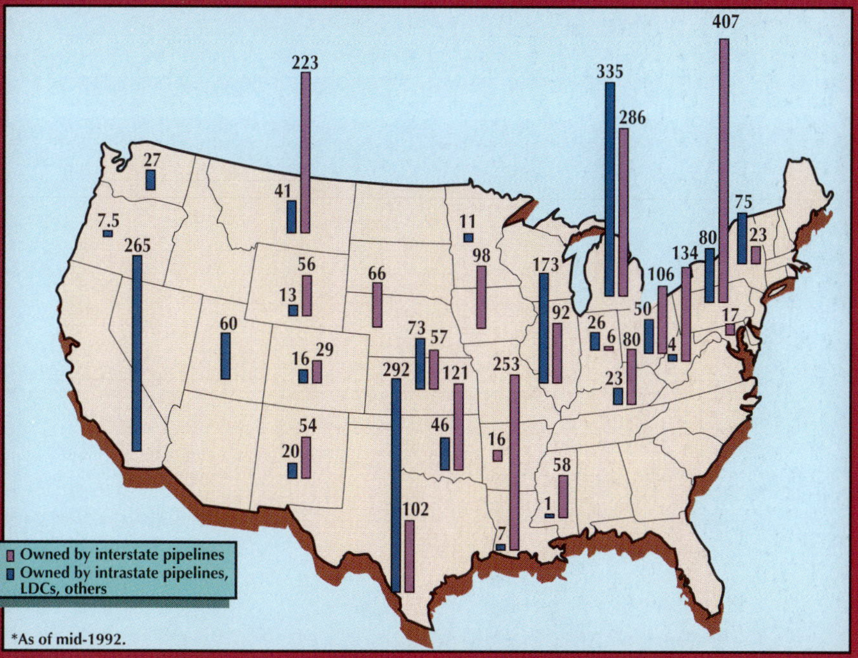

*As of mid-1992.

Source: Pace Consultants Inc.

since Order 636 implementation involved salt caverns. Salt caverns have accounted for 77 bcf of the 252 bcf of gas storage capacity added in North America since 1986.

In addition, Pace concluded six of 16 gas storage projects scheduled to go into service from implementation of Order 636 through 2000 will be in salt caverns. Of the 302 bcf of combined gas storage capacity to come on line during that time, salt cavern facilities will account for about 162 bcf.

Pace said the importance of assuring that baseload gas will be available to end users under Order 636 is underscored by the development of salt cavern gas storage projects in supply regions. Salt caverns have the high deliverability and quick cycling needed to back up baseload gas.

"While certain reservoirs can also provide high deliverability, a salt cavern facility provides the same withdrawal capability with less investment in working gas capacity," Pace said.

Storage service flexibility

Pace's GRI study also lauded operational flexibility salt caverns can provide those trying to balance a portfolio of gas services.

"Salt caverns allow high rates of injection as well as withdrawal, and operators can shift from one mode to another within hours," Pace said. "The flip side of maintaining high deliverability is the ability to move large volumes into storage as needed. The ability to cycle quickly between injections and withdrawals is becoming increasingly important and will become more so after Order 636."

Few projects are being built to provide only daily or hourly storage services. The need for daily or hourly storage in some markets is undercut by the availability of pipeline capacity.

But Pace found many storage developers are building storage facilities based on economics of serving seasonal customers, while retaining flexibility sufficient to provide peak demand storage services. In such cases, a storage developer can negotiate contracts or equity participation with seasonal storage customers to assure an adequate return on his project, then augment that income with storage fees for peak daily or hourly service.

Uses of storage facilities can shift as storage operators optimize capacity developed for other applications. For example, gas stored in a facility built to serve intrastate customers might be diverted later for interstate service if the installation can be tied into interstate pipeline systems. Also, surface facilities can be modified to fit new deliverability needs.

Benefits assured

William W. Elting, TPC chief financial officer, said gas storage is at the center of strategies being devised to manage open access, nondiscriminatory transportation on U.S. interstate pipelines, as shippers try to maximize the loads on gas transmission facilities they're paying to use.

"Every major user of interstate gas pipelines in the U.S. presently is analyzing his gas portfolio to determine what type of services he will need to fulfill his contracts," Elting said. "Each portfolio is going to have to include storage and will require a shipper to use different types of storage in different ways."

Elting said shippers in some cases will offer storage as a separate service and in other cases rebundled with other transportation services.

"Many marketers, gatherers, aggregators, and producers will be including storage as part of a package of gas supplies and services they will sell to end users," he said. "Conversely, at the other end of the pipeline, LDCs and in some cases large industrial customers, particularly electric generating customers, will buy their own gas storage capacity to rebundle gas they buy to provide the type of service they need to manage their portfolio of supplies."

TPC began offering peaking and swing gas delivery services in the 1993-94 winter through Moss Bluff gas storage facility and market center. The company in fall 1993 finished leaching a high deliverability gas salt dome storage cavern—the facility's second—about 9 months sooner than expected.

The addition increased Moss Bluff's working gas capacity to 5.5 bcf from 1.7 bcf. Peak gas injection rate at Moss Bluff is 150 MMcfd and peak withdrawal 450 MMcfd. TPC in October 1993 closed purchase of a 50% interest in Moss Bluff from Phibro Energy U.S.A. Inc., Houston. With the closing, TPC became 100% owner of the facility. Moss Bluff is expected to play a central role in TPC's plan to set up integrated gas marketing hubs throughout the U.S. Gulf Coast.

Control room at Mitchell Energy & Development Corp.'s Spindletop gas storage facility near Beaumont, Tex., typifies the advanced computer technology needed to monitor gas storage and withdrawal operations. Photo courtesy of Mitchell.

Winnie's Spindletop facility

Spindletop gas storage facility being developed in East Texas by Mitchell Energy unit Winnie Pipeline is an example of a storage project where Order 636 is providing an opportunity to provide multiple storage services.

Mitchell by fall 1993 had spent more than $50 million to leach one salt cavern with working gas capacity of 1.7 bcf.

This is the entrance to Spindletop gas storage facility near Beaumont, Tex., which is operated by Mitchell Energy & Development Corp.'s Winnie Pipeline Unit. Photo courtesy of Mitchell.

Leaching plant at Mitchell Energy & Development Corp.'s Spindletop gas storage facility near Beaumont, Tex., is leaching a second cavern that will bring working gas storage capacity at the facility to 12 bcf by 1997. Photo courtesy of Mitchell.

Working gas capacity is to expand to 6 bcf when complete.

Mitchell is leaching a second cavern. That storage cavern will be operable with 6 bcf of capacity by 1995. By 1997, with both caverns complete, combined working gas capacity will be 12 bcf, and Mitchell could add four more 6 bcf caverns for total combined capacity of 36 bcf.

Allen J. Tarbutton Jr., Mitchell Energy's president of gas transmission and processing, said Mitchell intended Spindletop to provide a secure gas supply to East Texas intrastate refining, petrochemical, and industrial customers plagued in the past by weather related supply disruptions.

However, Order 636's no-notice transportation requirements allow any interstate customer in need of no-notice gas supplies to get gas from Spindletop, if the customer has enough transportation space reserved on connecting pipelines. Although supply region storage represents a small part of Mitchell's gas storage business, 636 has increased chances of a facility such as Spindletop serving interstate markets and helped change Mitchell's original plans.

Mitchell has leased storage capacity at Spindletop to three companies, including 1 bcf of capacity under a 10 year agreement with Natural Gas Clearinghouse (NGC), Houston. NGC already is using 500 MMcf of the capacity and will have the other 500 MMcf available when Mitchell completes Spindletop's next storage cavern. Mitchell would have built Spindletop facility with or without Order 636.

"But the opportunities provided by Order 636 are real, and we will pursue that and capitalize on it," Tarbutton said.

MCN Washington 10 project

MCN's Washington 10 project northeast of Detroit will involve converting to storage a nearly depleted natural gas field discovered in 1969.

The facility will have storage capacity of 42 bcf and sustainable delivery of 400-500 MMcfd. MCN said the field should be in operation for the 1995-96 winter heating season. Total cost of the project is about $120 million.

Partners with MCN 40% are TransCanada PipeLines Ltd. 40%, Union Gas Ltd 10%, and Panhandle Eastern Corp. unit Panhandle Storage Co. 10%. Each partner will be storing gas at the site under long term contracts.

The storage unit will be used to serve markets across the North Central U.S. and in Central Canada. The Michigan reservoir is the first storage venture for TransCanada and for Panhandle Storage. MCN is a partner with operator ANR Storage Co. in a $130 million, 42 bcf storage project in northern Michigan's Kalkaska County. The Blue Lake project is in service and has 8 bcf of base gas with 29.5 bcf of working gas as of the end of September 1993.

South Central New York project

ANR and NGE have a project in South Central New York to develop a salt cavern storage facility with initial storage capacity of 2 bcf of gas and a peak withdrawal rate of about 200 MMcfd.

The complex will be connected to the pipeline systems of CNG Transmission Corp. and Columbia Gas Transmission Corp.

Coastal Executive Vice President James F. Cordes said since the project will be developed from existing caverns, it will be possible to complete the project by start of the 1995-96 heating season.

Partners filed an application in 1993 with FERC for construction and an open access blanket certificate to provide storage service. The companies planned an open season for late in 1993 for those interested in contracting for peaking storage services at the site.

Nyseg's South Central New York project calls for developing salt cavern storage that will have storage capacity of about 800 MMcf with a peak withdrawal rate of about 80 MMcfd. The project is to be complete in late 1995.

A 4 mile pipeline will connect the storage area to a Columbia Gas Transmission pipeline.

Young storage field

Young Gas Storage held an open season that ended Aug. 30, 1993, for contract space at its proposed 5.3 bcf storage site in Colorado. The company said it received firm requests for 350 MMcfd of deliverability, 150 MMcfd greater than the field's capacity.

Public Service Co. of Colorado signed a contract for 180 MMcfd of delivery and 4.8 bcf of storage capacity, and Colorado Springs signed for delivery of 20 MMcfd and storage capacity of 500 MMcf.

General partners in the Young Gas Storage venture are two units of Colorado Interstate Gas Co. (CIG). Colorado Springs is a limited partner.

Young said the field will increase the natural gas storage capacity in Colorado by more than 11%. CIG operates about 57% of the natural gas storage capacity in the state.

Young in July 1993 filed an application with FERC to build and operate the storage site. Once approval is received the company will drill about 37 wells and install a 6,000 hp compressor station. About 4 miles of gathering pipelines will connect wells in the field, while 10.8 miles of 20 in. line will connect the field to CIG's pipeline system. The company estimated services will be available for the 1995-96 heating season and development will be complete for the 1998-99 season. IPE

SynSat process scheme

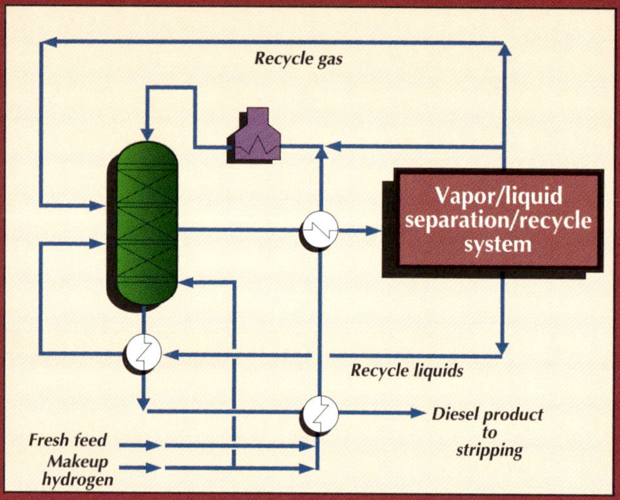

Table 2

Typical blending values for ethers

	MTBE	DIPE	TAME
Octane blending numbers @ 2.0-2.7 wt % fuel oxygen			
Typical regular unleaded			
RON	116	110	114
MON	96	97	99
(R+M)/2	106	103	106
Typical premium unleaded			
RON	117	110	110
MON	98	100	99
(R+M)/2	108	105	104
Rvp, psi	8	5	3
Butane utilization @ 2.0 wt % fuel oxygen			
in 7.8 Rvp blend, vol %	−0.1	+0.7	+1.2
in 9.0 Rvp blend, vol %	+0.1	+1.0	+1.5
in 11.5 Rvp blend, vol %	+0.6	+1.7	+2.1
in 13.5 Rvp blend, vol %	+1.1	+2.2	+2.6
Butane utilization @ 2.7 wt % fuel oxygen			
in 7.8 Rvp blend, vol %	−0.2	+1.0	+1.6
in 9.0 Rvp blend, vol %	+0.1	+1.4	+2.0
in 11.5 Rvp blend, vol %	+0.9	+2.2	+2.8
in 13.5 Rvp blend, vol %	+1.5	+2.9	+3.4

Source: Mobil Research & Development Corp.

Energy Biosystems Corp., The Woodlands, Tex., is developing a bacteriological approach to middle distillate desulfurization. Although the process, called biocatalytic desulfurization (BDS), is not yet commercial, it is one of the more unusual schemes in development.

The BDS process is illustrated in Fig. 4. Possible uses for the process in the refinery flow scheme are desulfurizing crude oil, naphtha, middle distillates, FCC feed, FCC gasoline, or residual fuel oil. The company said capital outlay and operating costs are significantly less than conventional hydrodesulfurization (Fig. 5).

Benzene reduction

Mobil and Badger Co. Inc. have developed several alternatives to reduce benzene in gasoline. All of the processes use reformate benzene and low value offgas from the FCC unit.

Mobil and Badger offer three new benzene alkylation processes:

- A cumene process that uses chemical grade benzene and FCC derived propylene to selectively produce isopropylbenzene or cumene.
- An ethylbenzene process that uses chemical grade

Biocatalytic desulfurization

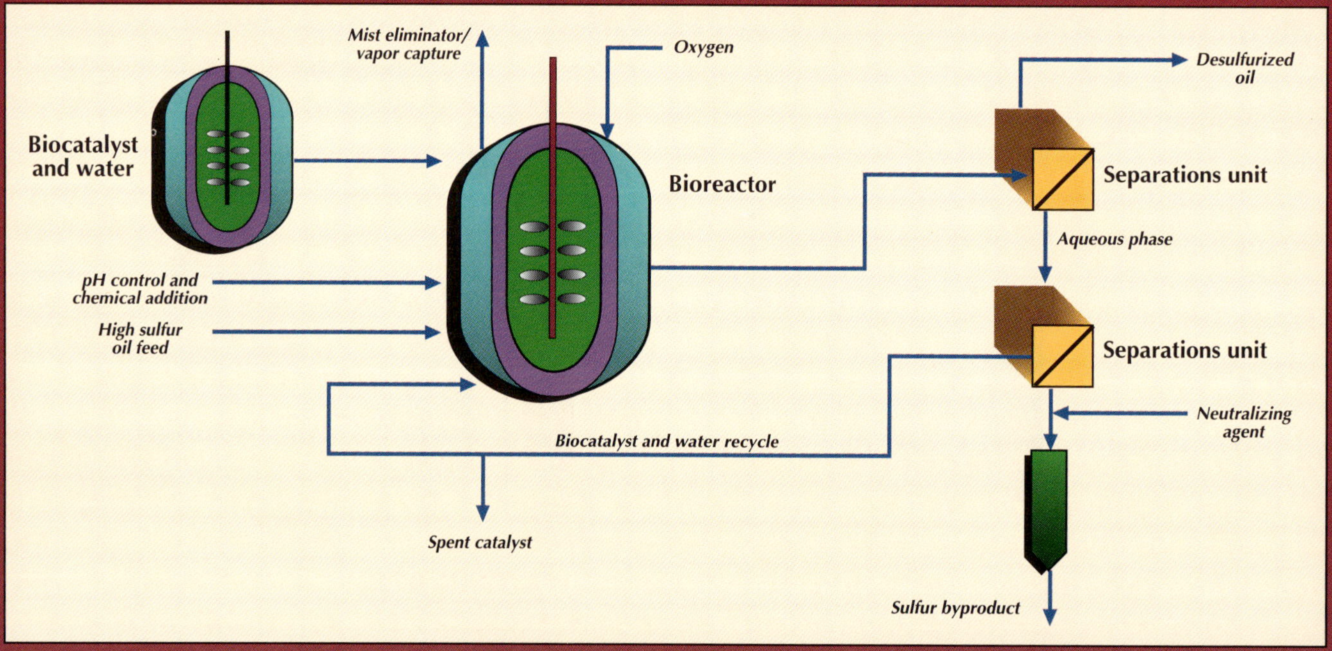

Source: Energy Biosystems Corp.

Processing costs – BDS vs. HDS

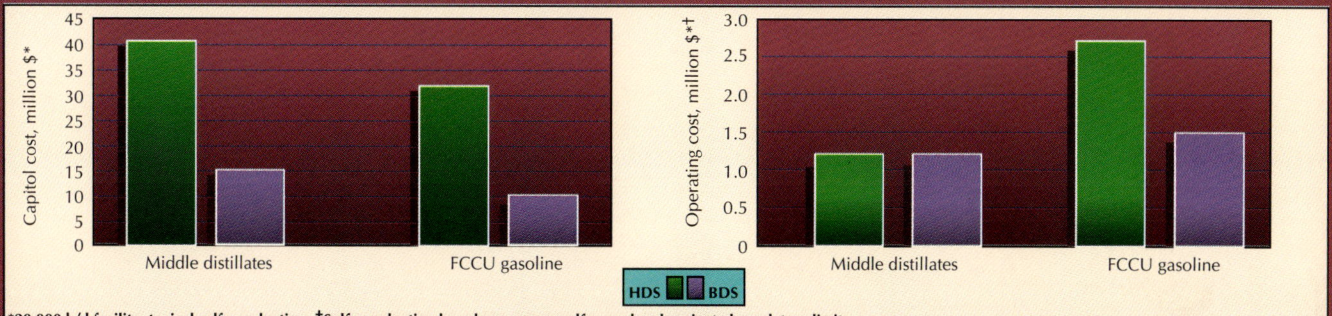

*20,000 b/d facility, typical sulfur reduction. †Sulfur reduction based on average sulfur pool and projected regulatory limits.

Source: Energy Biosystems Corp.

Isodewaxing processing

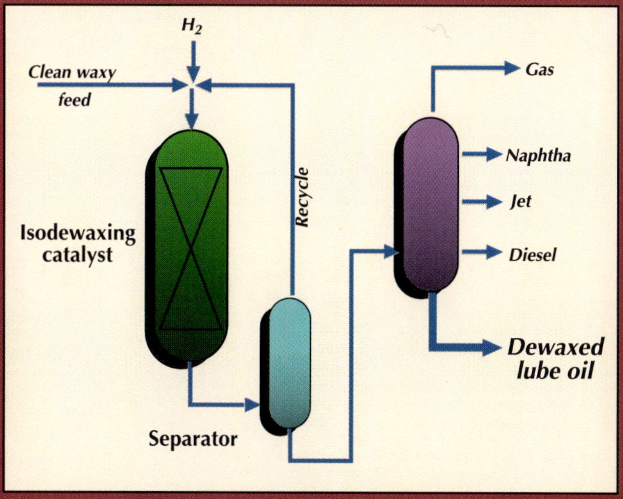

Source: Chevron International Oil Co.

Refinery offgas treatment

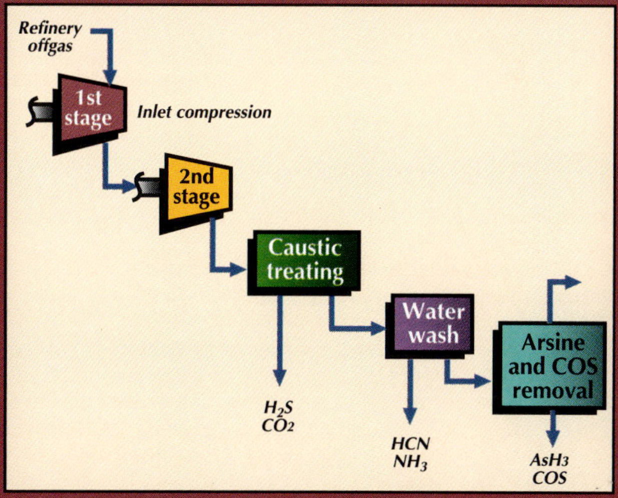

benzene and FCC derived ethylene.

- Mobil's benzene reduction (MBR) process, which mixes dilute light olefins from FCC offgas and coker gas with benzene.

Typical MBR performance characteristics are shown in Table 3.

Other processes

A number of improvements have been made in refining processes not directly related to fuel specifications.

Chevron International Oil Co. achieved a breakthrough in lube base oils manufacturing. Chevron's Isodewaxing process isomerizes waxy paraffin molecules into the isoparaffins required for chemically stable, high viscosity index (VI), low pour point base oils (Fig. 6). The process fits new and existing plants.

Isodewaxing reduces the pour point of even the most difficult feedstocks by converting wax directly to lube base oil rather than removing the wax. The product pour point can be as low as −50° C., Chevron says.

In addition, lower VI feeds can be used, providing greater flexibility in choosing feedstocks.

Table 4 gives a comparison of Isodewaxing and conventional solvent and catalytic dewaxing processes. If hydrocracker feeds are used, Isodewaxing's increased VI, compared with conventional dewaxing, means hydrocracker severity can be reduced. This extends catalyst life, reduces hydrogen consumption, and increases yields of waxy hydrocarbons, Chevron says.

Elsewhere among processes, Uhde GmbH developed and commercialized its new Combined Autothermal Reforming (CAR) technology for hydrogen production. The CAR process combines steam reforming and partial oxidation in a pressure vessel.

The heat contained in the reaction gases is used efficiently as the heat source for the endothermic reaction in the steam reforming section. This means that firing is not required, and CO_2 emissions are reduced considerably, Uhde says.

The process was installed at Chemko Strazske Sp in the Czech Republic. The partial oxidation unit was replaced, resulting in reduced oxygen consumption. CAR also can be used to produce ammonia and methanol.

Stone & Webster Engineering Corp. developed an advanced recovery system (ARS) that can recover high percentages of olefinic material from FCC offgases. This recovery process consists of several stages:

- Gas compression.
- Bulk acid gas removal.
- Contaminant removal or conversion.
- Cryogenic recovery.
- Optional fractionation.

A simplified flow diagram of these stages is shown in Figs. 7 and 8. This technology makes specific use of self-

Product recovery

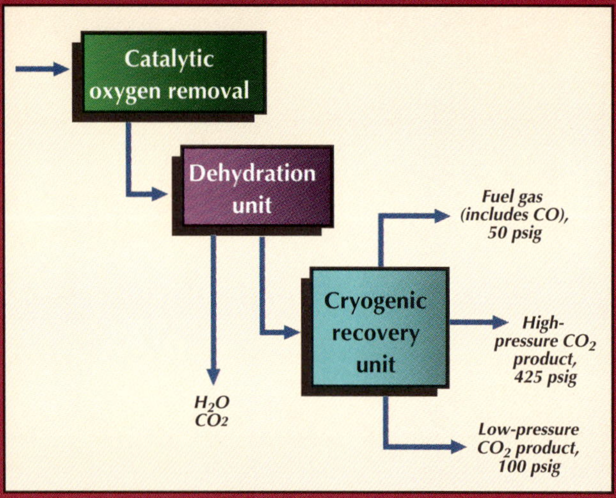

Table 3

Typical MBR performance

	Light reformate C_5 to 215° F. TBP	C_{5+} MBR gasoline Low severity	Mod. severity
Delta uplifts			
RONC	79	Δ = +9	Δ = +13
MONC	74	Δ = +7	Δ = +9
Rvp, psi	5.2	Δ = −0.4	Δ = −0.5
Yield, vol %	100%	Δ = +10%	Δ = +5%
PONA change			
Paraffins, vol %	57%	51%	47%
Olefins, vol %	3%	4%	4%
Naphthenes, vol %	3%	3%	3%
Aromatics, vol %	37%	42%	46%
Benzene, vol %	32%	14%	13%

Table 4

Comparison of dewaxing processes

	Solvent dewaxing	Catalytic dewaxing	Chevron Isodewaxing
Description:	Physical wax crystallization and filtration	Chemical wax cracking	Chemical wax isomerization
Pour point, °C.	−10 to −15	−10 to −50	−10 to −50°C
Byproducts	*Slack wax	Gas, naphtha	Diesel, jet, naphtha
Capital expense	Base (100%)	60-80%	65-85%
Operating expense	Base (100%)	50-60%	55-65%

Source: Chevron International Oil Co.

refluxing plate fin exchangers, called dephlegmators.

Stone & Webster estimates an ARS project can achieve payback in about 15 months.

IFP developed simple, low investment, feedstock treatment processes to eliminate arsenic (As) and mercury (Hg) impurities in raw condensates or crude oils. Left unchecked, these contaminants can cause equipment corrosion and catalyst poisoning.

IFP's processes treat olefinic and nonolefinic feeds and can remove contaminants from either gas or liquid phases. Equipment to achieve contaminant removal includes simple guard beds, trapping masses, demisters, and simple hydrogenolysis reactors.

Capital investments range from about $1.39/ton of treated diesel oil to $1.79/ton of treated condensate. IFP estimates most projects can pay out in about 18 months.

Texaco Development Corp. is licensing Texaco Strategic Total Activity Retention (T-STAR), an ebullated bed hydrotreating/hydrocracking process. The process can be applied to diesel aromatics reduction, FCCU feed pretreating, gas oil hydrocracking, and in line hydrotreating/hydrocracking within an H-Oil unit.

T-STAR combines HRI's well-known H-Oil resid process with Texaco's fixed bed hydrotreating and hydrocracking technologies (Fig. 9). The process uses the same type of reactor as H-Oil but processes lighter feedstocks and uses lower pressures (1,200 psig vs. 1,500 psig).

T-STAR has been commercially demonstrated at Husky Oil's Lloydminster, Alta., refinery.

Kinetics Technology International BV (KTI) developed a process for upgrading refinery butanes to provide useful reactants for alkylation and MTBE production. The Butacracking process is a unique combination of process steps centered around a thermal cracker.

The process converts isomerized butanes to isobutylene, which is then used to produce MTBE. Cracked by-products such as propylene can be used as feed to the alkylation unit or sold as chemical grade product.

Fig. 10 shows a typical refinery flow scheme utilizing the Butacracking process.

Catalysts

As profit margins and regulatory requirements tighten, refiners increasingly look to catalysts as low investment options for changing product yields and impurity levels.

Engelhard Corp., for example, has developed a line of IsoPlus catalysts for increased FCC iso-olefin yields and, therefore, MTBE production. Pilot plant testing has shown that IsoPlus 1000 can increase FCC C_4/C_5 iso-olefin production by 30-70% over traditional catalysts.

IsoPlus 2000 is capable of increasing iso-olefin yields by 100-150%, while increasing production of LPG olefins for alkylation by 40-60%, Engelhard says. These changes potentially can make a difference in a refiner's profits.

In 1993, UOP announced commercialization of its R-132 high performance continuous catalytic reforming (CCR) catalyst.

"Data now confirm that R-132 catalyst offers higher activity, tolerates more regenerations, reduces chloride consumption, and yet maintains its surface area longer than any CCR reforming catalyst on the market," said UOP Pres. Mike Winfield.

At its Sines, Portugal, refinery, Petrogal Refinaria de Sines reported an activity advantage of 20° F. (11° C.) using R-132, compared with UOP's R-32 or R-34. This allowed Petrogal to increase throughput while remaining within heater firing limitations.

In another catalyst development, Intercat Corp. reported enhanced bottoms cracking in a commercial FCC unit using its BCA-105 additive. Coastal Eagle Point Oil Co. reduced heavy fuel yields by 8.5 vol % of fresh feed in a 120 day trial of the additive. Coke and dry gas production

T-STAR process

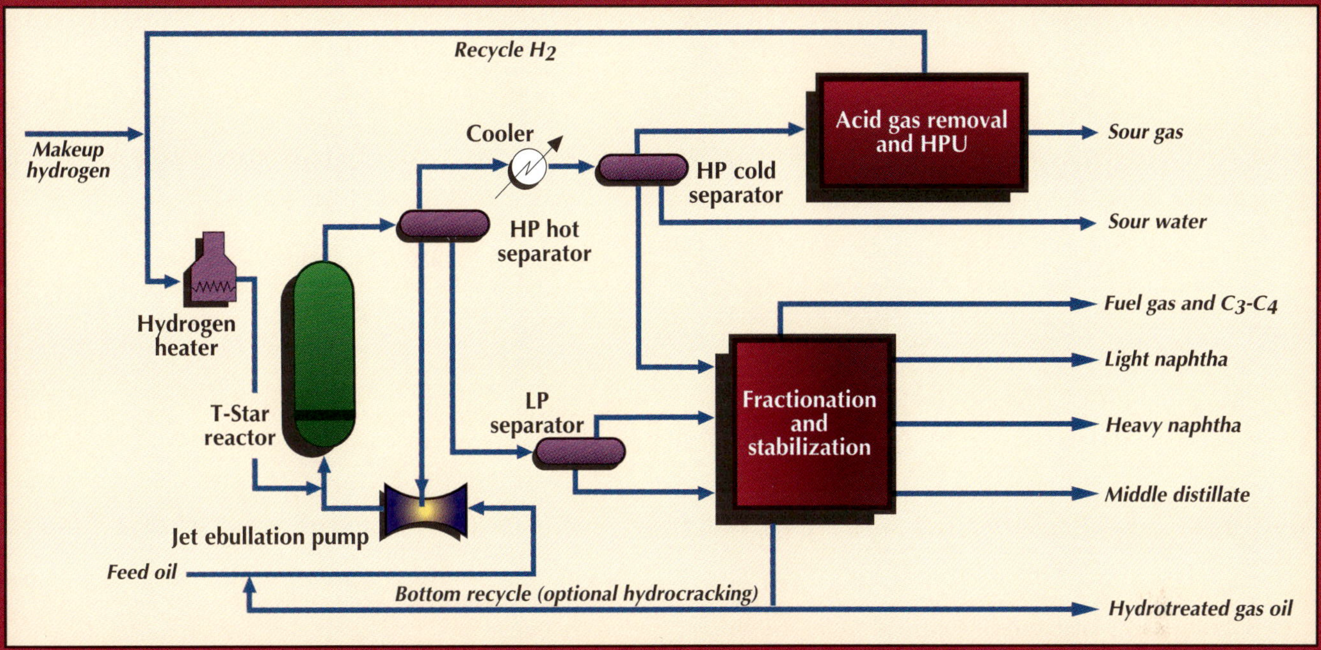

Source: Texaco Development Corp.

Butane upgrading refinery with Butacracking

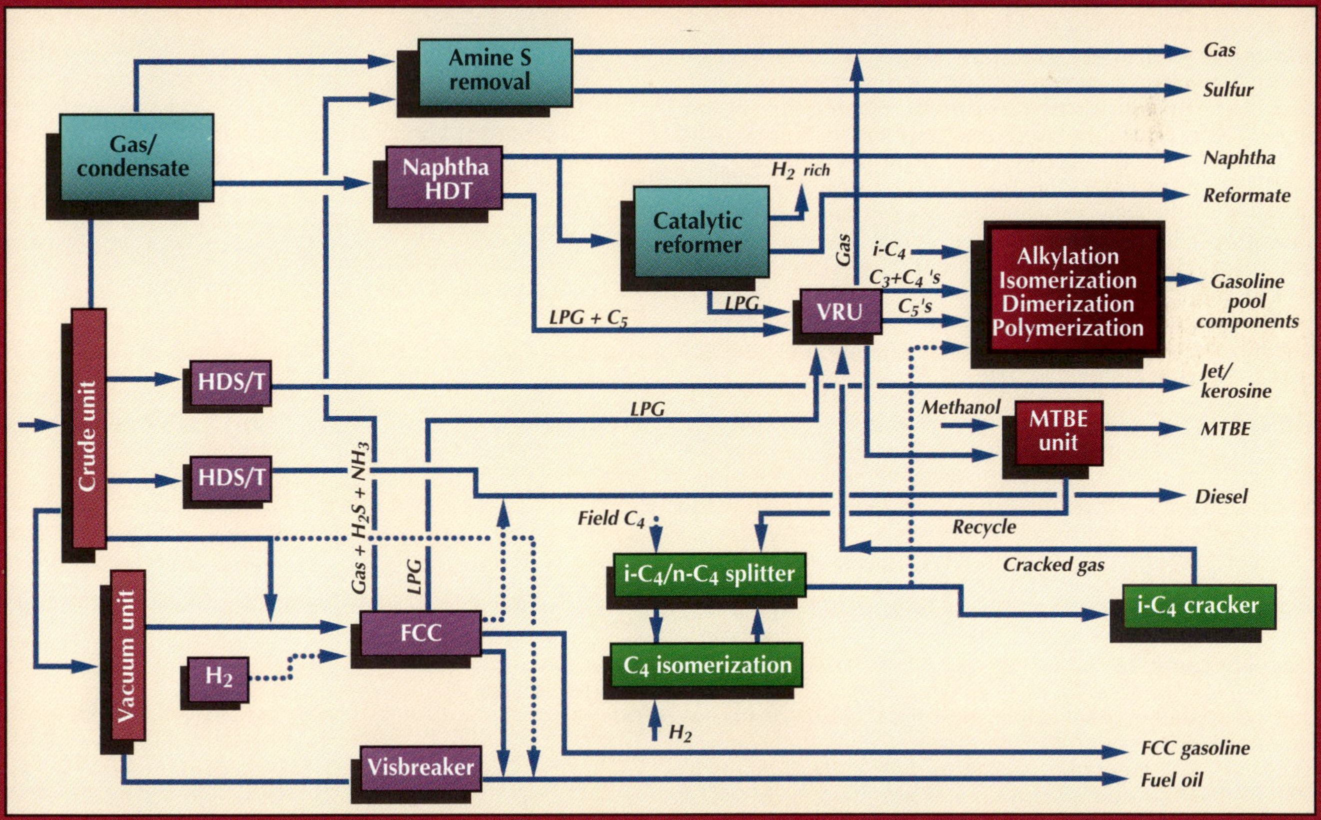

Souce: KTI

remained within normal operating ranges. An increase of 0.1 wt % in delta coke was experienced.

Corresponding increases in transportation fuels and LPG of more than 7 vol % and 3 vol %, respectively, also resulted. These incremental yield shifts were achieved with about 9 wt % additive in the total FCC catalyst inventory. IPE

Assays of world crudes provide valuable data

Crude oil assays are an indispensable tool for refiners and marketers worldwide.

The complex world crude system contains thousands of crudes of varying qualities. Refiners must know, in great detail, how a crude will perform in a particular refinery before purchase or plant design decisions can be made. Costly crude oil assays are essential for this job.

Assay information also determines where a crude lies in the hierarchy of world crudes. The higher its ranking, the higher its market value. This ranking is based on, among other characteristics, the crude's sulfur content and API gravity.

Each assay is headed by the official crude name, with the former or secondary crude name, if applicable, in parentheses. The country of origin comes next, followed by the location and type of loading facilities, where known.

Loading systems are abbreviated as follows:

- CALM—Catenary anchor leg mooring
- SBM—Single buoy mooring
- SPM—Single point mooring.

The first crude in this series is West Texas intermediate (WTI). WTI is not exported. However, it is a world marker crude and, as such, is used as a benchmark for crudes around the world.

FSU crudes

Two low sulfur, high gravity crudes from the former Soviet Union have been assayed. The first is Kumkol from Kazakhstan and the second Siberian light from western Siberia (see map). Both assays come from the Chimkent refinery in Chimkent, Kazakhstan, and both were performed in December 1992.

Kumkol crude is from a field in Southwest Kazakhstan. It is shipped via a leg of the Siberian pipeline system to the Chimkent refinery. Current production is about 10,000 b/d, all of which is shipped to Chimkent. The crude comprises only about 15% of the Chimkent refinery's crude slate.

The field has the potential of producing at a greater rate, which has prompted the Kazakhstan government to consider adding a pipeline leg to the Caspian Sea. Sources said the crude could be exported in the future.

Siberian light is produced in the area near Surgut. It is shipped via the same pipeline system, through the Tyumen area and the city of Omsk, and on to refineries at Pavlodar, Chimkent, and the Fergana Valley in Uzbekistan. Production rates of Siberian light were not available.

In both assays the kerosine range overlaps gasoline and diesel ranges so that, when adding yield percentages, kerosine yields should be omitted.

Australia

High quality Saladin crude oil produced off Western Australia has been assayed. The 48.2° API, 0.02 wt % sulfur crude's characteristics, determined in 1990, are presented here for the first time.

Operator of Saladin field is West Australian Petroleum Pty. Ltd. (Wapet) Group. Partners in the group are Chevron Asiatic Oil Co. (25.71%), Texaco Oil Development Co. (25.71%), Shell Development (Australia) Pty. Ltd. (25.71%), Ampol Exploration Ltd. (12.86%), and Western Mining Corp. (10%).

The estimated 30-40 million bbl field, south of Barrow Island, produces through two platforms in 58 ft of water in block TP 3. Production began in late 1989 from three platforms with three wells each and from two wells drilled directionally from Thevanard Island (Oil & Gas Journal, Dec. 11, 1989, p. 36).

The field went on stream at a rate of 25,000 b/d, peaked at about 60,000 b/d, and is expected to continue producing until 1994-1996. Saladin experienced a natural decline in production in 1991, from about 45,500 b/d to 39,000 b/d. The field, Western Australia's largest producer, contributed 43% of the state's oil production in 1991.

Immediately southwest of Saladin, two small extension fields—Cowle and Yammaderry—went on stream in April 1991 at a combined rate of about 6,500 b/d. Production from these fields, estimated to contain about 4 million bbl each, is expected to continue until about 1995.

Each field produces through from a single well monopad platform.

Yammaderry lies in 30 ft of water, Cowle in 39 ft. Individual flow lines connect the wells to Saladin treating and storage facilities on nearby Thevenard Island. A pipeline from Thevenard moves treated crude to an offshore mooring/tanker loading buoy.

Papua New Guinea

Light, sweet Kutubu crude produced near Lake Kutubu in Papua New Guinea has been assayed. Production of the 44° gravity, 0.04 wt % sulfur, the country's first commercial oil, began in June 1992.

Chevron Niugini Pty. Ltd., acting as operator for a group of six oil companies, discovered oil in the Iagifu-Hedinia and Agogo fields. Other interests are held by the Papua New Guinea government, British Petroleum Co. plc, Ampolex, BHP Petroleum Pty. Ltd., Oil Search Ltd., and Japan PNG Petroleum.

Chevron built production facilities and laid an export pipeline from the central facility to a site on the Kikori River, then to a platform in 66 ft of water in the Gulf of Papua.

Early production was destined for sale to refineries in Australia and Japan. Kutubu production is expected to reach 100,000 b/d—possibly 140,000 b/d.

West Texas intermediate

Midland, Tex., U.S.

Whole crude
Gravity, °API: 40.8
Specific gravity @ 60/60° F.: 0.8212
Sulfur, wt %: 0.34
Vis., cSt @ 60° F. (15.6° C.): 5.24
Neutralization No., mg/g: 0.10
Nitrogen, wt %: 0.08
Pour pt., °F.: −20

Light ends (C_1-C_5)
Yield, wt %: 4.35

Range, °F.: 68-347
Yield, vol %: 32.39
RON, clear: 56.4*
Sulfur, wt %: 0.0314

Range, °F.: 347-563
Yield, vol %: 23.50
Specific gravity @ 60/60° F.: 0.8201
Sulfur, wt %: 0.110
Aniline pt., °F: 150.5
Vis., cSt @ 100° F. (37.8° C.): 1.87
Freeze pt., °F.: −25
Pour pt., °F: −33
Smoke pt., mm: 22.1

Range, °F.: 563-650
Yield, vol %: 8.10
Specific gravity @ 60/60° F.: 0.8529
Sulfur, wt %: 0.289
Aniline pt., °F.: 176.9
Vis., cSt @ 100°F (37.8° C.): 6.21
Freeze pt., °F.: 38*
Pour pt., °F.: 30

Range, °F.: 650-1,049
Yield, vol %: 24.30
Specific gravity @ 60/60° F.: 0.8960
Sulfur, wt %: 0.445
Nitrogen, wt %: 0.105
V/Ni, ppm: 0.04*/0.09*

Range, °F: 650-1,500
Yield, vol %: 33.30
Specific gravity @ 60/60° F.: 0.9153
Sulfur, wt %: 0.72
Vis., cSt @ 140° F. (60.0° C.): 56.8
Pour pt., °F.: 78
V/Ni, ppm: 4.3/4.3
Fe, ppm: 20.2

Range, °F.: 761-1,500
Yield, vol %: 25.30
Specific gravity @ 60/60° F.: 0.9268
Penetration @ 25° C., mm: >2,000†
Vis., cSt @ 212° F. (100.0° C.): 26.70

Range, °F.: 878-1,500
Yield, vol %: 17.95
Specific gravity @ 60/60° F.: 0.9403
Penetration @ 25° C., mm: >2,000†
Vis., cSt @ 212° F. (100.0° C.): 58.7

Range, °F.: 1,049-1,500
Yield, vol %: 9.00
Specific gravity @ 60/60° F.: 0.9672
Penetration @ 25° C., mm: 1,035.5†
Vis., cSt @ 212° F. (100.0° C.): 360.3

*Extrapolated. †Estimated.

FSU crudes Kumkol

Kazakhstan

Whole crude
Gravity, °API: 42.5
Density, g/cu m: 0.813
Total sulfur, wt %: 0.07
Flash pt. in open crucible, °C.: 15
Flash pt. in closed crucible, °C.: −35
Pour pt., °C.: 10
Vis. @ 30° C., cSt: 7.14
Vis. @ 50° C., cSt: 3.71

Coking capacity, %: 1.08
Rel. vis. @ 50° C., Engler degree: 1.28
Corresp. kin. vis., cSt: 3.71
Asphaltenes, %: traces
Silica-gel resins, %: 10
Paraffin melting temp., °C.: 15
V/Ni, mg/g: 0.2/4.6
Nitrogen, %: 0.1
Oxygen, %: 0.18
Acid no., mg/100 cu cm: 0.05

Gas (C_1-C_4)
Yield, wt %: 1.7

Gasoline
Range, °C.: 62-180
Yield, wt %: 24.0
Density, g/cu m: 0.729
Total sulfur, wt %: 0.007
MON: 56
Aromatics, %: 8.1
Paraffins, %: 30.9
Naphthenes, %: 61.0
Acidity, mg/100 cu cm: 0.27

Kerosine
Range, °C.: 120-240
Yield, wt %: 18.5
Density, g/cu m: 0.7742
Total sulfur, wt %: 0.023
Flash pt. in closed crucible, °C.: 28
Freeze pt., °C.: −48
Aromatics, %: 22
Aniline pt., °C.: 65
Vis. @ 20° C., cSt: 1.41
Acidity, mg/100 cu cm: 0.7
Smoke pt., mm: 29.1

Diesel fuel
Range, °C.: 180-350
Yield, wt %: 29.6
Density, g/cu m: 0.8093
Total sulfur, wt %: 0.037
Flash pt. in open crucible, °C.: 64
Pour pt., °C.: −3
Cloud pt., °C.: 0
Cetane no.: 62
Vis. @ 20° C., cSt: 4.16
Acidity, mg/100 cu cm: 1.13

Mazut (resid)
Range, °C.: 350+
Yield, wt %: 44.4
Density, g/cu m: 0.9048
Total sulfur, wt %: 0.16
Flash pt. in open crucible, °C.: 237
Pour pt., °C.: 40
Coking capacity, %: 2.32
Rel. vis. @ 80° C., Engler degree: 3.34

Siberian light

Russia
Whole crude
Gravity, °API: 37.8
Density, g/cu m: 0.8358
Total sulfur, wt %: 0.42
Pour pt., °C.: −17
Vis. @ 20° C., cSt: 6.73
Vis. @ 50° C., cSt: 3.67
Acidity, mg/100 cu cm: 0.78
Coking capacity, %: 2.87

Gas (C_1-C_4)
Yield, %: 1.4

Gasoline
Range, °C.: 62-180
Yield, wt %: 22.44
Density, g/cu m: 0.728
Total sulfur, wt %: 0.01
MON: 54
Aromatics, %: 8.5
Paraffins, %: 54
Naphthenes, %: 37.5

Kerosine
Range, °C.: 120-240
Yield, wt %: 25.42
Density, g/cu m: 0.787
Total sulfur, wt %: 0.02
Freeze pt., °C.: −55
Aromatics, %: 16
Aniline pt., °C.: 63.9
Vis. @ 20° C., cSt: 1.38
Acidity, mg/100 cu cm: 0.49
Smoke pt., mm: 25.5

Diesel fuel
Range, °C.: 180-350
Yield, wt %: 32.06
Density, g/cu m: 0.839
Total sulfur, wt %: 0.2
Pour pt., °C.: −14
Cloud pt., °C.: −3
Color: 0.5
Resins, mg/100 cu cm: 5.0
Diesel index: 59
Vis. @ 20° C., cSt: 4.94
Acidity, mg/100 cu cm: 0.17

Mazut (resid)
Range, °C.: 350+
Yield, wt %: 43.75
Density, g/cu m: 0.934
Total sulfur, wt %: 0.9
Pour pt., °C.: 19
Coking capacity, %: 7.34
Rel. vis. @ 80° C., Engler degree: 5.0
Corresp. kin. vis., cSt: 36.0

Australia
Saladin

Offshore, Northwest Australia
Whole crude
Gravity, °API: 48.2
Sp gr @ 60/60° F.: 0.7873
Density @ 15° C., g/ml: 0.7869
Total sulfur, wt %: 0.02
Mercaptan S, ppm: 4
Sulfur as H_2S, ppm: nil
Rvp, psi (kPa): 4.70 (32.50)
Basic N, ppm: 5
Total N, ppm: 56
Sediment, wt %: <0.01
Water, vol %: <0.025
Pour pt., °C.: −30
Flash pt. (Abel), °C.: −32.0
Wax, wt %: 0.7
Visc. @ 20° C., cSt: 1.732
Visc. @ 40° C., cSt: 1.277
Total acid no., mg KOH/g: 0.01
Concarbon residue, wt %: 0.08
UOP K factor: 12.0
Salt, lb/1,000 bbl (ppm[wt]): <1.0 (<4)
Heat of comb. (gross), MJ/kg: 46.46
Heat of comb. (net), MJ/kg: 43.46
Ash, wt %: <0.01
Asphaltenes, wt %: 0.01
V/Ni, ppm: <0.5/1.5
Cu/Na/Fe, ppm: <0.5/<1.0/<1.0

Naphtha, LSR
Range, °C (°F.): 20.7-70 (69.3-158)
Yield range, vol % (wt %): 2.1-7.9 (1.5-6.3)
Yield, vol % (wt %): 5.8 (4.8)
Gravity, °API: 85.3
Sp gr @ 60/60° F.: 0.6527
Density @ 15° C., g/ml: 0.6526
Rvp, psi (kPa): 15.4 (106.0)
Mole wt.: 77
n-Paraffins, vol %: 47.6
iso-Paraffins, vol %: 41.7
Naphthenes, vol %: 10.5
Aromatics, vol %: 0.1
Total S, ppm: 157
Mercaptan S, ppm: <1
RON (calculated): 71
MON (calculated): 69

Naphtha, HSR
Range, °C (°F.): 70-140 (158-284)
Yield range, vol % (wt %): 7.9-30.7 (6.3-27.8)
Yield, vol % (wt %): 22.8 (21.5)
Gravity, °API: 61.0
Sp gr @ 60/60° F.: 0.7350
Density @ 15° C., g/ml: 0.7347
Rvp, psi (kPa): 1.90 (13.00)
Mole wt.: 103
n-Paraffins, vol %: 26.2
iso-Paraffins, vol %: 26.9
Naphthenes, vol %: 43.4
Aromatics, vol %: 2.7
Unidentified, vol %: 0.8
Total S, ppm: 18
Mercaptan S, ppm: 2
RON (calculated): 50
RON, clear: 53.8

Locale of F.S.U. crude assays

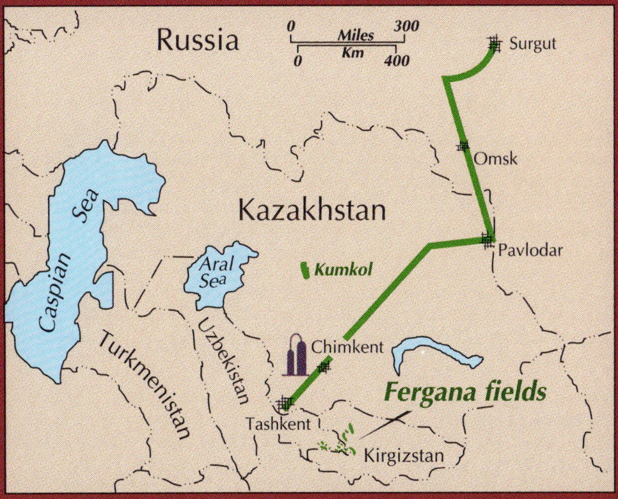

MON, clear: 52.0

Naphtha, HSR
Range, °C (°F.): 140-190 (284-374)
Yield range, vol % (wt %): 30.7-49.6 (27.8-46.2)
Yield, vol % (wt %): 18.9 (18.4)
Gravity, °API: 52.1
Sp gr @ 60/60° F.: 0.7706
Density @ 15° C., g/ml: 0.7703
Rvp, psi (kPa): 0.40 (2.75)
Mole wt.: 129
n-Paraffins, vol %: 30.5
Naphthenes, vol %: 61.5*
Aromatics, vol %: 8.0
Total S, ppm: 42
Mercaptan S, ppm: 2
Aromatics (FIA), vol %: 9.0
Smoke pt., mm: 36
RON (calculated): 40
MON (calculated): 40

*Calculated by difference.

Naphtha, HSR
Range, °C (°F.): 150-200 (302-392)
Yield range, vol % (wt %): 34.7-52.8 (31.7-49.4)
Yield, vol % (wt %): 18.1 (17.7)
Gravity, °API: 50.5
Sp gr @ 60/60° F.: 0.7774
Density @ 15° C., g/ml: 0.7771
Rvp, psi (kPa): 0.00 (0.00)
Aromatics, vol %: 11.0
Total S, ppm: 42
Mercaptan S, ppm: 2
Aromatics (FIA), vol %: 9.9
Smoke pt., mm: 33
Freeze pt., °C.: −70.0
Aniline pt., °C.: 60.40
Visc. @ −20° C., cSt: 2.66
Visc. @ 20° C., cSt: 1.264
Visc. @ 40° C., cSt: 0.965
RONC: 18.0
MONC: 17.0

Kerosine
Range, °C (°F.): 140-260 (284-500)
Yield range, vol % (wt %): 30.7-73.1 (27.8-70.3)
Yield, vol % (wt %): 42.4 (42.5)
Gravity, °API: 46.9
Sp gr @ 60/60° F.: 0.7931
Density @ 15° C., g/ml: 0.7927
Total S, ppm: 60
Mercaptan S, ppm: <1
Freeze pt., °C.: −49.5
Aromatics (FIA), vol %: 10.5
Smoke pt., mm: 28
Flash pt. (Abel), °C.: 45.0
Naphthalenes, vol %: 1.25
Hydrogen, wt %: 14.24
Visc. @ −20° C., cSt: 3.79
Visc. @ 20° C., cSt: 1.655
Visc. @ 40° C., cSt: 1.237
Saybolt color: +30
Total acid no., mg KOH/g: 0.00
Cu strip corrosion (2 hr @ 100° C.): 1A
Aniline pt., °C. (°F.): 63.3 (145.9)
Aniline gravity product: 6,845

Kerosine
Range, °C (°F.): 190-230 (374-446)
Yield range, vol % (wt %): 49.6-63.8 (46.2-60.6)
Yield, vol % (wt %): 14.2 (14.4)
Gravity, °API: 45.0
Sp gr @ 60/60° F.: 0.8018
Density @ 15° C., g/ml: 0.8014
Total S, ppm: 54
Mercaptan S, ppm: <1
Freeze pt., °C.: −47.5
Aromatics (FIA), vol %: 10.2
Smoke pt., mm: 28
Naphthalenes, vol %: 1.71
Visc. @ −20° C., cSt: 4.96
Visc. @ 20° C., cSt: 1.975
Visc. @ 40° C., cSt: 1.439
Saybolt color: +30
Total acid no., mg KOH/g: 0.01
Cu strip corrosion (2 hr @ 100° C.): 1A
Aniline pt., °C. (°F.): 64.6 (148.3)
Aniline gravity product: 6,673

Kerosine
Range, °C (°F.): 200-250 (392-482)
Yield range, vol % (wt %): 52.8-70.5 (49.4-67.5)
Yield, vol % (wt %): 17.7 (18.1)
Gravity, °API: 43.2
Sp gr @ 60/60° F.: 0.8099
Density @ 15° C., g/ml: 0.8095
Total S, ppm: 64
Mercaptan S, ppm: <1
Freeze pt., °C.: −39.0
Aromatics (FIA), vol %: 12.9
Smoke pt., mm: 26
Visc. @ −20° C., cSt: 7.75
Visc. @ 20° C., cSt: 2.417
Visc. @ 40° C., cSt: 1.652
Cloud pt., °C.: −39.0
Cetane index (calculated): 49.1
Total acid no., mg KOH/g: 0.00
Aniline pt., °C. (°F.): 66.7 (152.1)
Aniline gravity product: 6,569

Gas oil
Range, °C (°F.): 230-360 (446-680)
Yield range, vol % (wt %): 63.8-91.7 (60.6-90.6)
Yield, vol % (wt %): 27.9 (30.0)
Gravity, °API: 36.0
Sp gr @ 60/60° F.: 0.8446
Density @ 15° C., g/ml: 0.8442
Total S, ppm: 0.02
Cloud pt., °C.: −9
Pour pt., °C.: −12

Gas oil
Range, °C (°F.): 230-400 (446-752)
Yield range, vol % (wt %): 63.8-94.6 (60.6-93.8)
Yield, vol % (wt %): 30.8 (33.2)
Gravity, °API: 36.1
Sp gr @ 60/60° F.: 0.8442
Density @ 15° C., g/ml: 0.8438

Total S, wt %: 0.02
Total N, ppm: 26
Cloud pt., °C.: 0
Pour pt., °C.: −3
Cetane index: 51.7
Total acid no., mg KOH/g: 0.02
ASTM color: 0.5
Visc. @ 40° C., cSt: 3.668
Concarbon residue, wt %: 0.02†
Aniline pt., °C. (°F.): 74.8 (166.6)
Aniline gravity product: 6,014
Refractive index @ 20° C.: 1.4676
Cu strip corrosion (3 hr @ 50° C.): 1A

†On 10% bottoms.

Gas oil
Range, °C (°F.): 250-300 (482-572)
Yield range, vol % (wt %): 70.5-83.2 (67.5-81.2)
Yield, vol % (wt %): 12.7 (13.7)
Gravity, °API: 37.3
Sp gr @ 60/60° F.: 0.8382
Density @ 15° C., g/ml: 0.8378
Total S, wt %: 0.02
Total N, ppm: 34
Cloud pt., °C.: −15
Pour pt., °C.: −18
Cetane index: 51.5
Total acid no., mg KOH/g: 0.00
ASTM color: L0.5
Visc. @ 40° C., cSt: 2.995
Aniline pt., °C. (°F.): 71.7 (161.1)
Aniline gravity product: 6,009
Br no., g Br/100g: 0.2
Refractive index @ 20° C.: 1.4647

Gas oil
Range, °C (°F.): 300-350 (572-662)
Yield range, vol % (wt %): 83.2-91.2 (81.2-90.1)
Yield, vol % (wt %): 8.0 (8.9)
Gravity, °API: 33.1
Sp gr @ 60/60° F.: 0.8595
Density @ 15° C., g/ml: 0.8591
Total S, wt %: 0.03
Total N, ppm: 32
Cloud pt., °C.: 6
Pour pt., °C.: 6
Cetane index: 52.5
Total acid no., mg KOH/g: 0.03
ASTM color: L0.5
Visc. @ 40° C., cSt: 5.976
Aniline pt., °C. (°F.): 79.6 (175.0)
Aniline gravity product: 5,802
Br no., g Br/100 g: 0.7
Refractive index @ 20° C.: 1.4762
Wax, wt %: <0.1

Vacuum gas oil
Range, °C (°F.): 350-450 (662-842)
Yield range, vol % (wt %): 91.2-97.8 (90.1-97.4)
Yield, vol % (wt %): 6.6 (7.3)
Gravity, °API: 29.6
Sp gr @ 60/60° F.: 0.8782
Density @ 15° C., g/ml: 0.8777
Total S, wt %: 0.06
Total N, ppm: 226
Basic N, ppm: 84
Aniline pt., °C. (°F.): 91.6 (196.9)
Aniline gravity product: 5,828
Pour pt., °C.: 27
Total acid no., mg KOH/g: 0.12
Visc. @ 50° C., cSt: 12.02
Visc. @ 100° C., cSt: 3.644

Residue
Range, °C (°F.): 350+ (662+)
Yield range, vol % (wt %): 91.2-100.0 (90.1-100.0)
Yield, vol % (wt %): 8.8 (9.9)
Gravity, °API: 27.1
Sp gr @ 60/60° F.: 0.8922
Density @ 15° C., g/ml: 0.8917
Total S, wt %: 0.10
Total N, wt %: 0.054
Basic N, wt %: 0.02
Carbon, wt %: 86
Hydrogen, wt %: 13.8
Aniline pt., °C. (°F.): 100.00 (212.0)
Aniline gravity product: 5,745
Pour pt., °C.: 30
Total acid no., mg KOH/g: 0.21
Visc. @ 50° C., cSt: 22.21
Visc. @ 100° C., cSt: 5.357
Concarbon residue, wt %: 1.0
Heat of comb. (gross), MJ/kg: 44.88
Heat of comb. (net), MJ/kg: 42.21
Ash, wt %: 0.03
Asphaltenes, wt %: 0.64
V/Ni, ppm: <0.5/7.8
Cu/Na/Fe, ppm: <0.5/5.0/<1.0
Wax, wt %: 10.9
Wax softening pt., °C.: 50-51
UOP K factor: 12.0

Residue
Range, °C (°F.): 450+ (842+)
Yield range, vol % (wt %): 97.8-100.0 (97.4-100.0)
Yield, vol % (wt %): 2.2 (2.6)
Gravity, °API: 20.9
Sp gr @ 60/60° F.: 0.9286
Density @ 15° C., g/ml: 0.9281
Total S, wt %: 0.13
Total N, wt %: 0.13
Basic N, wt %: 0.04
Pour pt., °C.: 45
Total acid no., mg KOH/g: 0.21
Visc. @ 50° C., cSt: 223.2
Visc. @ 100° C., cSt: 23.53
V/Ni, ppm: 0.6/24.0
Cu/Na/Fe, ppm: <0.5/13.0/4.0

Papua New Guinea

Kutubu

SPM, Gulf of Papua
Whole crude
Gravity, °API: 44.0
Specific gravity, 60/60° F.: 0.8063
Sulfur, wt %: 0.04
Hydrogen, wt %: 13.4*
Total nitrogen, ppm: 340*
Pour pt., °F.: 35
Vis., cSt @ 25° C.: 1.94
Vis., cSt @ 40° C.: 1.56
Vis., cSt @ 50° C.: 1.37
Acid No., mg KOH/g: <0.06
Vis. charac. factor: 11.9*
Ramsbottom C, wt %: 2.2†
Asphaltenes, wt %: 1.4§
Calcium, ppm: 0.3§
V/Ni, ppm: 0.4§/11.4§
Cu/Na/Fe, ppm: 0.8§/10.0§/47.0§¶
Magnesium, ppm: 2.4§
Chromium, ppm: 0.2§
Cobalt, ppm: 0.8§

Range, °C. (°F.): 21-70 (70-158)
Yield range, vol %: 3.7-11.8
Yield range, wt %: 2.6-9.3
Yield, vol %: 8.1
Yield, wt %: 6.7
Gravity, °API: 80.0
Specific gravity, 60/60° F.: 0.6692
Density @ 15° C., g/ml: 0.6691
Molecular wt: 96
P/N//A, vol %: 80.5/18.8/0.7
RON, clear: 71.9
RON, + 3.17 g TEL/U.S. gal: 97.7
MON, clear: 69.6
Rvp, psi (kpa): 11.1 (76.5)
Sulfur, ppm: 7
Mercaptan S, ppm: 2
Total nitrogen, ppm: <1
Acid No., mg KOH/g: 0.05

Range, °C. (°F.): 70-140 (158-284)
Yield range, vol %: 11.8-34.7
Yield range, wt %: 9.3-30.7
Yield, vol %: 22.9
Yield, wt %: 21.4
Gravity, °API: 56.1
Specific gravity, 60/60° F.: 0.7542
Density @ 15° C., g/ml: 0.7539
Molecular wt: 111
P/N//A, vol %: 40.6/49.2/10.2
RON, clear: 62.2
RON, + 3.17 g TEL/U.S. gal: 90.5
MON, clear: 59.4
Rvp, psi (kpa): 2.1 (14.5)
Sulfur, ppm: 8
Mercaptan S, ppm: 2
Total nitrogen, ppm: <1
Acid No., mg KOH/g: 0.05

Range, °C. (°F.): 140-190 (284-374)
Yield range, vol %: 34.7-48.4
Yield range, wt %: 30.7-44.1
Yield, vol %: 13.7
Yield, wt %: 13.4
Gravity, °API: 47.9
Specific gravity, 60/60° F.: 0.7886
Density @ 15° C., g/ml: 0.7882
Molecular wt: 140
P/N//A, vol %: 40.7/40.8/18.5
RON, clear: 58.1
RON, + 3.17 g TEL/U.S. gal: 80.8
Rvp, psi (kpa): 0.2 (1.5)
Sulfur, ppm: 11
Mercaptan S, ppm: 2
Total nitrogen, ppm: <1
Hydrogen, wt %: 13.7
Acid No., mg KOH/g: 0.05
Aniline pt., °C.: 43.3
Aniline gravity product: 5,269
Freeze pt., °C.: −79
Smoke pt., m: 24

Range, °C. (°F.): 140-230 (284-446)
Yield range, vol %: 34.7-57.3
Yield range, wt %: 80.7-53.1
Yield, vol %: 22.5
Yield, wt %: 22.4
Gravity, °API: 45.2
Specific gravity, 60/60° F.: 0.8009
Density @ 15° C., g/ml: 0.8005
Molecular wt: 150
Sulfur, wt %: 0.0014
Mercaptan S, ppm: 2
Total nitrogen, ppm: <1
Hydrogen, wt %: 13.5
Aromatics FIA, vol %: 19.8
Naphthalenes, vol %: 1.89*
Freeze pt., °C.: −60
Smoke pt., m: 22
Pour pt., °C.: −59
Acid No., mg KOH/g: 0.05
Vis., cSt @ −20° C.: 2.80
Vis., cSt @ 40° C.: 1.03
Vis., cSt @ 100° C.: 0.57
Aniline pt., °C.: 47.2
Aniline gravity product: 5,288
UOP K factor: 11.7

Range, °C. (°F.): 190-230 (374-446)
Yield range, vol %: 48.3-57.3
Yield range, wt %: 44.1-53.1
Yield, vol %: 8.8
Yield, wt %: 9.0
Gravity, °API: 41.1
Specific gravity, 60/60° F.: 0.8199
Density @ 15° C., g/ml: 0.8195
Molecular wt: 168
Sulfur, wt %: 0.0018
Mercaptan S, ppm: 2
Total nitrogen, ppm: <1
Hydrogen, wt %: 13.3
Aromatics FIA, vol %: 21.8
Naphthalenes, vol %: 4.83
Freeze pt., °C.: −48
Smoke pt., m: 20
Pour pt., °C.: −48
Acid No., mg KOH/g: 0.05
Vis., cSt @ −20° C.: 4.56
Vis., cSt @ 40° C.: 1.33
Vis., cSt @ 100° C.: 0.68
Aniline pt., °C.: 52.8
Aniline gravity product: 5,220
UOP K factor: 11.7

Range, °C. (°F.): 230-360 (446-680)
Yield range, vol %: 57.3-81.4
Yield range, wt %: 53.1-78.8
Yield, vol %: 24.1
Yield, wt %: 25.7
Gravity, °API: 33.3
Specific gravity, 60/60° F.: 0.8586
Density @ 15° C., g/ml: 0.8582
Molecular wt: 227
Sulfur, wt %: 0.03
Total nitrogen, ppm: 55
Hydrogen, wt %: 12.9
Cloud pt., °C.: −2
Pour pt., °C.: −3
Cetane index: 47.3
Acid No., mg KOH/g: 0.05
Vis., cSt @ 40° C.: 3.16
Vis., cSt @ 100° C.: 1.29
Aniline pt., °C.: 66.1
Aniline gravity product: 5,028
UOP K factor: 11.7
Wax content, wt %: 6.3
Wax softening pt., °C.: 26

Range, °C. (°F.): 360-540 (680-1,004)
Yield range, vol %: 81.4-97.4
Yield range, wt %: 78.8-96.8
Yield, vol %: 16.1
Yield, wt %: 18.0
Gravity, °API: 24.5
Specific gravity, 60/60° F.: 0.9069
Density @ 15° C., g/ml: 0.9064
Molecular wt: 385
Sulfur, wt %: 0.13
Total nitrogen, ppm: 0.10
Basic nitrogen, ppm: 350
Hydrogen, wt %: 12.3
Heat of combustion (gross), MJ/kg: 44.65
Heat of combustion (net), MJ/kg: 42.03
Pour pt., °C.: 39
Acid No., mg KOH/g: 0.05
Vis., cSt @ 50° C.: 24.5
Vis., cSt @ 100° C.: 6.4
Aniline pt., °C.: 87.8
Aniline gravity product: 4,655
UOP K factor: 11.9
Wax content, wt %: 34

Range, °C. (°F.): 360+ (680+)
Yield range, vol %: 81.4-100
Yield range, wt %: 78.8-100
Yield, vol %: 18.7
Yield, wt %: 21.2
Gravity, °API: 22.1
Specific gravity, 60/60° F.: 0.9215
Density @ 15° C., g/ml: 0.9210
Molecular wt: 403
Sulfur, wt %: 0.15
Total nitrogen, ppm: 0.15
Hydrogen, wt %: 11.9
Heat of combustion (gross), MJ/kg: 44.40
Heat of combustion (net), MJ/kg: 41.83
Pour pt., °C.: 42
Acid No., mg KOH/g: 0.05
Vis., cSt @ 50° C.: 43.5
Vis., cSt @ 100° C.: 8.0
Aniline pt., °C.: 87.2
Aniline gravity product: 4,177
UOP K factor: 11.8
Wax content, wt %: 31
Wax softening pt., °C.: 48
Asphaltenes, wt %: 0.2
Ramsbottom C, wt %: 2.8
Con. carbon, wt %: 3.7
V/Ni/Fe, ppm: <0.5/2/7

Range, °C. (°F.): 540+ (1,004+)
Yield range, vol %: 97.4-100
Yield range, wt %: 96.8-100
Yield, vol %: 2.6
Yield, wt %: 3.2
Gravity, °API: 8.3
Specific gravity, 60/60° F.: 1.0125
Density @ 15° C., g/ml: 1.0119
Sulfur, wt %: 0.27
Total nitrogen, ppm: 0.39
Hydrogen, wt %: 10.2
Carbon, wt %: 88.0
Heat of combustion (gross), MJ/kg: 42.79
Heat of combustion (net), MJ/kg: 40.52
Pour pt., °C.: 75
Acid No., mg KOH/g: 0.05
Vis., cSt @ 100° C.: 273
UOP K factor: 11.5
Wax content, wt %: 14.5
Asphaltenes, wt %: 1.5
Ramsbottom C, wt %: 18.7
Con. carbon, wt %: 18.7
V/Ni/Fe, ppm: 0.5/12/49

*Calculated. †Performed on 650° F.+ cut. §Performed on 1,000° F.+ cut. ¶Fe may be high; sample was not washed to

Much new refinery equipment relates to environmental requirements. This is a steam stripper, which removes benzene from waste water streams. Photo courtesy of Nalco Chemical Co.

Refining survey

YOUR USE OF THE INTERNATIONAL REFINing survey may be aided by the following information:

All country-by-country volumes are presented in calendar-day figures. When an asterisk (*) appears beside a company, this indicates that the figure was reported as a stream-day figure and has been converted to a calendar-day figure using the conversion factor of 0.95 for crude oil and distillation units, and 0.90 for cat cracking and cat reforming units.

Barrels per calendar day (b/cd) is the maximum number of barrels of input that can be expressed during a 24-hr period, after making allowances for the following:

- Types and grades of inputs to be processed
- Types and grades of products to be manufactured
- Environmental constraints associated with refining operations
- Scheduled downtime such as technical problems, repairs, and slowdowns.

Barrels per stream day (b/sd) is the amount a unit can process when running at full capacity under optimal feedstock and product slate conditions. IPE

LEGEND

CAT CRACKING
1. Fluid
2. Other

CAT REFORMING
Semiregenerative:
1. Conventional catalyst
2. Bimetallic catalyst

Cyclic:
3. Conventional catalyst
4. Bimetallic catalyst

Continuous regen:
5. Conventional catalyst
6. Bimetallic catalyst

Other:
7. Conventional catalyst
8. Bimetallic catalyst

Cat reforming definitions are:

- **Semiregenerative.** Characterized by shutdown to the reforming unit at specified intervals or at the operator's convenience for in situ catalyst regeneration.
- **Cyclic.** Characterized by continuous or continual regeneration of catalyst in situ in any one of several reactors that can be isolated from and returned to the reforming operation. This is accomplished without changing feed rate or octane.
- **Continuous regeneration.** Characterized by the continuous regeneration of part of the catalyst in a special regenerator, followed by continuous addition of this regenerated catalyst to the reactor.
- **Other.** Includes nonregenerative reforming (catalyst is replaced by fresh catalyst) and moving bed catalyst systems.

Survey of operating refineries worldwide

(capacities as of January 1, 1994)

Country	No. plants	Crude Capacity (b/cd)	Catalytic cracking (b/cd)	Catalytic reforming (b/cd)
Abu Dhabi	2	192,500	....	30,100
Albania	3	40,000	....	....
Algeria	6	530,000	....	74,000
Angola	1	32,100	....	1,900
Argentina	12	709,485	180,100	41,300
Australia	9	705,000	214,500	171,100
Austria	1	210,000	25,100	33,700
Bahrain	1	250,000	39,000	18,000
Bangladesh	1	31,200	....	1,650
Barbados	1	3,000	....	....
Belgium	4	607,000	110,000	95,800
Bolivia	3	45,250	....	14,560
Brazil	13	1,252,860	333,104	....
Brunei	1	8,500	....	5,600
Bulgaria	3	300,000	....	....
Cameroon	1	42,000	....	7,000
Canada	25	1,880,100	391,450	359,450
Chile	3	164,615	38,290	10,100
China	23	2,200,000	....	....
Colombia	5	248,850	90,000	....
C.I.S.	47	9,845,939	433,684	54,312
Congo	1	21,000	....	2,000
Costa Rica	1	15,000	....	1,200
Croatia	2	293,472	32,000	37,400
Cuba	4	280,000	....	....
Cyprus	1	22,000	....	5,100
Czech Republic	6	307,250	....	8,570
Denmark	3	183,500	....	31,600
Dominican Republic	2	48,000	....	9,700
Ecuador	4	148,000	16,000	2,300
Egypt	8	532,153	....	33,540
El Salvador	1	18,000	....	3,200
Ethiopia	1	18,000	....	2,400
Finland	2	200,000	45,200	42,800
France	14	1,861,430	340,060	264,580
Gabon	1	17,300	....	1,400
Germany	21	2,265,060	281,650	409,503
Ghana	1	26,600	....	4,950
Greece	4	395,500	63,700	50,800
Guatemala	2	20,000	....	3,000
Honduras	1	14,000	....	1,800
Hungary	3	241,500	24,000	29,600
India	12	1,086,371	130,505	28,874
Indonesia	6	860,200	13,000	63,600
Iran	8	1,161,700	40,000	105,400
Iraq	8	347,500	....	43,500
Ireland	1	52,500	....	12,000
Israel	2	208,000	49,500	26,000
Italy	17	2,262,000	295,000	270,086
Ivory Coast	2	62,800	....	10,200
Jamaica	1	35,500	....	3,540
Japan	41	4,810,150	707,750	669,550
Jordan	1	100,000	4,250	8,640
Kenya	1	90,000	....	9,600
Korea, N.	1	42,000	....	....
Korea, S.	6	1,147,100	....	122,700
Kuwait	3	624,000	28,000	47,000
Lebanon	2	37,500	7,250	7,442
Liberia	1	15,000	....	2,000
Libya	3	348,400	....	14,000
Lithuania	1	263,420	43,692	51,482
Macedonia	1	51,180	....	10,780
Madagascar	1	16,350	....	2,600
Malaysia	4	269,800	....	30,100
Martinique	1	16,000	....	3,000
Mexico	7	1,524,000	243,000	170,800
Morocco	2	154,600	5,600	26,800
Myanmar	2	32,000	....	....
Netherlands	6	1,183,500	136,000	185,100
Netherlands Antilles	2	477,000	49,000	20,000
New Zealand	1	92,500	....	22,000
Nicaragua	1	16,000	....	3,000
Nigeria	4	433,250	82,700	70,070
Norway	3	287,250	41,000	47,100
Oman	1	76,000	....	15,200
Pakistan	3	132,900	....	4,910
Panama	1	100,000	....	8,000
Paraguay	1	7,500	....	....
Peru	5	184,250	23,300	....
Philippines	4	278,750	22,700	29,300
Poland	9	333,000	32,000	35,000
Portugal	3	294,000	10,100	51,000
Puerto Rico	2	127,000	14,200	26,500
Qatar	1	57,500	....	11,500
Romania	10	680,884	113,295	89,752
Saudi Arabia	8	1,614,000	87,541	184,822
Senegal	1	22,600	....	2,500
Sierra Leone	1	10,000	....	....
Singapore	5	1,055,000	28,000	62,700
Slovakia	1	193,400	....	22,270
Slovenia	1	14,700	....	....
Somalia	1	10,000	....	....
South Africa	4	364,038	85,113	61,490
Spain	10	1,283,000	177,600	197,500
Sri Lanka	1	50,000	....	3,750
Sudan	1	21,700	....	1,900
Sweden	5	427,500	28,000	73,500
Switzerland	2	132,000	....	26,000
Syria	2	242,140	....	29,239
Taiwan	2	542,500	45,200	56,500
Tanzania	1	17,000	....	2,500
Thailand	3	313,750	10,200	45,450
Trinidad	2	305,000	28,000	31,000
Tunisia	1	34,000	....	3,300
Turkey	5	713,000	39,320	66,210
United Kingdom	15	1,866,640	474,900	392,200
United States	178	15,141,556	5,221,475	3,656,608
Uruguay	1	35,000	5,000	3,000
Venezuela	6	1,167,000	233,900	8,870
Virgin Islands	1	545,000	75,000	125,000
Yemen	2	120,000	....	8,500
Yugoslavia	4	281,781	20,750	42,145
Zaire	1	17,000	....	3,500
Zambia	1	23,750	....	5,320
Total	**706**	**73,168,574**	**11,309,679**	**9,337,915**

WORLDWIDE REFINING	Crude capacity (b/cd)	Catalytic cracking (b/cd)	Catalytic reforming (b/cd)
ABU DHABI			
Abu Dhabi National Oil Co.–Ruwais	120,000	....	[2]19,200
Umm Al-Nar 2	72,500	....	[4]10,900
Total	**192,500**	**....**	**30,100**
ALBANIA			
Government-owned refineries at Ballsh, Cerrik, Stalin.			
Total	**40,000**		
ALGERIA			
Sonatrach: Ain Amenas	7,000	....	....
Algiers	65,000	....	15,000
Arzew	60,000	....	11,600
Hassi Messaoud	30,000	....	2,400
Maison Carree	58,000	....	15,000
Skikda	310,000	....	30,000
Total	**530,000**	**....**	**74,000**
ANGOLA			
Fina Petroleos De Angola–Luanda	32,100	....	[1]1,900
Total	**32,100**	**....**	**1,900**
ARGENTINA			
Destileria Argentina de Petróleo SA–Lomas de Zamora	3,585	....	....
Esso SAPA–Campana	93,000	[1]29,600	[2]8,300
Galvan	18,400	....	....
Isaura SA–Bahia Blanca	25,200	[1]8,800	....
Shell Cia. Argentina de Petróleo SA–Buenos Aires	121,700	[1]29,200	[1]12,000
Sol Petroleo SA—San Francisco Solana, Quilmes	6,000	....	....
Yacimientos Petroliferos Fiscales–Campo Duran	32,000	....	....
Dock Sud	4,000	....	....
La Plata	216,000	[1]71,900	[1]9,000
Lujan de Cuyo	129,000	[1]40,600	[1]9,000
Plaza Huincul	23,000	....	[2]3,000
San Lorenzo	37,600	....	....
Total	**709,485**	**180,100**	**41,300**
AUSTRALIA			
Ampol Refineries Ltd.–Lytton	74,000	[1]30,000	[6]20,000
Australian Lubricating Oil Refinery Ltd.–Kurnell	....	....	....
BP Australia–Brisbane	52,000	[1]16,300	[2]11,300
Kwinana	117,000	[1]26,300	[2]13,800
Caltex Refining CPL–Kurnell	107,000	[1]44,000	[2]27,000
Mobil Oil Australia, Ltd.–Adelaide	67,000	....	[2]20,000
Altona	103,000	§[2]25,000	[2]28,000
Shell Refining (Australia) PL.–Clyde	75,000	[1]35,000	[8]20,000
Geelong	110,000	[1]37,900	[2]11,000 [8]20,000
Total	**705,000**	**214,500**	**171,100**

§TCC.

WORLDWIDE REFINING	Crude capacity (b/cd)	Catalytic cracking (b/cd)	Catalytic reforming (b/cd)
AUSTRIA			
OMV–Schwechat	210,000	[1]25,100	[2]15,600 [6]18,100
Total	**210,000**	**25,100**	**33,700**
BAHRAIN			
Bahrain Petroleum Co. BSC–Sitra	250,000	[1]39,000	[2]18,000
Total	**250,000**	**39,000**	**18,000**
BANGLADESH			
Eastern Refinery Ltd.–Chittagong	**31,200**	**....**	**[1]1,650**
BARBADOS			
Mobil Oil Barbados Ltd.–Bridgetown	**3,000**	**....**	**....**
BELGIUM			
Belgian Refining Corp. NV–Antwerp	85,000	....	[2]11,600
Esso Belgium–Antwerp	239,000	[1]32,400	[4]39,200
Fina Raffinaderij Antwerpen—Antwerp	268,000	[1]77,600	[2]45,000
Nynas Petroleum NV—Antwerp	15,000	....	....
Total	**607,000**	**110,000**	**95,800**
BOLIVIA			
Yacimientos Petroliferos Fiscales Bolivianos–Cochabamba	27,250	....	[2]8,160
Santa Cruz	15,000	....	[2]6,400
Sucre	3,000	....	....
Total	**45,250**	**....**	**14,560**
BRAZIL			
Petróleo Brasileiro SA: Araucária, Paraná	159,210	[1]45,700	...
Betim, Minas Gerais	136,470	[1]33,080	....
Canoas, Rio Grande do Sul	68,230	[1]14,040	....
Capuava, São Paulo	38,100	[2]11,580	....
Cubatão, São Paulo	....	[1]49,620	....
Duque de Caxias, Rio de Janeiro	225,730	[1]41,360	....
Fortaleza, Ceará	....	....	....
Manaus, Amazonas	10,230	[1]2,200	....
Mataripe, Bahia	117,960	[1]24,810	....
Paulínia, São Paulo	272,930	[1]50,314	....
São José dos Campos, São Paulo	204,700	[1]57,900	....
Refinaria de Petróleo Ipiranga SA–Rio Grande do Sul	9,300	[1]2,500	....
Refinaría de Petroleos de Manguinhos SA–Rio de Janeiro	10,000	....	....
Total	**1,252,860**	**333,104**	**....**
BRUNEI			
Brunei Shell Petroleum CL–Seria	**8,500**	**....**	**[2]5,600**
BULGARIA			
Government-owned refineries at Burgas, Pleven, Ruse.			
Total	**300,000**		
CAMEROON			
Sonara–National Refining CL–Cape Limboh, Limbe	**42,000**	**....**	**[2]7,000**
CANADA			
Alberta			
Husky Oil Operations Ltd.–Lloydminster	23,500	....	....

WORLDWIDE REFINING	Crude capacity (b/cd)	Catalytic cracking (b/cd)	Catalytic reforming (b/cd)
Imperial Oil Ltd.–Edmonton	176,000	[1]47,500	[4]20,900
Parkland Refining Ltd.–Bowden	6,200	....	[1]3,000
Petro-Canada Products Inc.–Edmonton	115,500	[1]33,300	[2]14,800
Shell Canada Ltd.–Scotford	68,600	....	[6]21,500
Total	**389,800**	**80,800**	**60,200**
British Columbia			
Chevron Canada Ltd.–North Burnaby	45,000	[1]15,000	[2]9,500
Husky Oil Operations Ltd.–Prince George	9,500	[1]3,300	[2]1,400
Imperial Oil Ltd.–Ioco	44,000	[1]13,100	[3]6,900
Total	**98,500**	**31,400**	**17,800**
New Brunswick			
Irving Oil Ltd.–St. John	237,500	[1]19,200	[2]34,650
Total	**237,500**	**19,200**	**34,650**
Newfoundland			
Newfoundland Processing Ltd.–Come By Chance	105,000	....	[2]26,000
Total	**105,000**	**....**	**26,000**
Northwest Territories			
Imperial Oil Ltd.–Norman Wells	**3,600**	....	....
Nova Scotia			
Imperial Oil Ltd.–Dartmouth	82,400	[1]25,700	[4]9,800
Ultramar Canada Inc.–Halifax	22,000	[1]9,600	[2]4,800
Total	**104,400**	**35,300**	**14,600**
Ontario			
Imperial Oil Ltd.–Nanticoke	112,000	[1]40,850	[5]27,400
Sarnia	119,000	[1]25,700	[2]15,200
			[4]15,200
Novacor Chemicals Canada–Sarnia	70,000	....	....
Petro-Canada Products Inc.–Mississauga	....	....	[2]9,800
Oakville	80,000	[1]26,000	[2]13,200
Shell Canada Ltd.–Sarnia	71,000	[1]14,400	[2]21,000
Sunoco Inc.–Sarnia	83,400	[2]15,500	[2]24,700
Total	**535,400**	**122,450**	**126,500**
Quebec			
Petro-Canada Products Inc.–Montreal	87,400	[1]17,200	[2]31,300
Shell Canada Ltd.–Montreal	120,000	[1]24,000	[2]20,700
Ultramar Canada Inc.–St. Romuald	140,000	[1]43,500	[2]18,700
Total	**347,400**	**84,700**	**70,700**
Saskatchewan			
Consumers' Cooperative Refineries Ltd.–Regina	45,200	[1]17,600	[2]9,000
Producers Pipeline–Moose Jaw	13,300	....	....
Total	**58,500**	**17,600**	**9,000**
Total Canada	**1,880,100**	**391,450**	**359,450**

CHILE

ENAP–Gregorio-Magallanes	9,000	....	....
Petrox SA–Talcahuano	89,570	[1]20,790	[2]4,500
Refinería de Concón–Concón	66,045	[1]17,500	[1]5,600
Total	**164,615**	**38,290**	**10,100**

CHINA

Government-owned refineries at Anshan, Beijing, Hangzhou, Karamai-Dushanzi, Lanchow, Lenghu, Nanchong, Nanjing, Shanghai, Shengli, Tianjin, Yumen. Crude Capacity	872,000		
Qilu	160,000	....	....
Yanshan	140,000	....	....
Maoming	170,000	....	....
Daqing	120,000	....	....
Jinlin	140,000	....	....
Gaoqiao	104,000	....	....
Fushun #2	90,000	....	....
Jinxi	100,000	....	....
Guangzhou	104,000	....	....
Balin	100,000	....	....
Dalian	100,000	....	....
Total	**2,200,000**	**....**	**....**

COLOMBIA

Empresa Colombiana de Petróleos			
Apiay	2,250	....	....
Barrancabermeja-Santander	173,000	[1]64,000	....
Cartagena, Bolivar	70,000	[1]26,000	....
Orito, Putumayo	1,800	....	....
Tibu, N. de Santander	1,800	....	....
Total	**248,850**	**90,000**	**....**

C.I.S.

(and former Soviet Union)

Azerbaijan			
Baku	238,340	....	....
Novo-Baku	168,000	[2]34,400	....
Total	**406,340**	**34,400**	**....**
Belarus			
Mozyr	321,280	....	....
Novopolotsk	510,260	....	....
Total	**831,540**	**....**	**....**
Georgia			
Batumi	106,340	....	....
Total	**106,340**	**....**	**....**
Kazakhstan			
Atyrau	103,800	....	....
Chimkent	125,360	....	....
Pavlodar	161,220	[2]40,000	....
Total	**390,380**	**40,000**	**....**
Russia			
Achinsk	134,040	....	....
Afipsky	29,000	....	....
Angarsk	462,960	[2]35,280	....
Groznyi	387,720	[2]40,000	....
Groznyi-Sheripov	40,000	....	....
Khabarovsk	90,000	....	....
Kirishi	386,000	....	....
Komsomolsk	116,140	....	....
Krasnodar	33,980	....	....
Moscow	243,000	[2]40,000	....
Nizhniy	435,060	....	....
Nizhnrkamsk	120,000	....	....
Novo-Groznyi	....	....	....
Novo-Kuibystev	307,420	[2]14,200	....
Novo-Ufa	377,640	[2]23,200	....
Omsk	564,000	[2]43,580	....
Orsk	144,200	....	....
Perm	278,400	....	....
Ryazan	360,760	[2]19,060	....
Salavat	246,160	[2]21,980	....
Samara	119,580	....	....
Saratov	176,200	[1]17,860	....
Syzran	210,360	[2]19,700	....
Tuapse	45,000	....	....
Ufa	233,520	....	....
Ufa	250,220	[1]18,060	....
Ukhta	125,700	....	....
Volgograd	188,920	....	....
Yaroslav	357,340	22,260	....
Total	**6,463,320**	**315,160**	**....**

WORLDWIDE REFINING	Crude capacity (b/cd)	Catalytic cracking (b/cd)	Catalytic reforming (b/cd)
Turkmenistan			
Chardzhou	120,000	....	....
Krasnovodsk	118,000	[2]15,720	....
Total	**238,000**	**15,720**	**....**
Ukraine			
Drogobych	77,600	....	....
Kherson	172,860	....	....
Kremenchug	361,479	[1]28,384	[1]46,972
			[2]7,340
Lisichansk	469,220	....	....
Nadvornaya	73,200	....	....
Odessa	82,340	....	....
Total	**1,236,699**	**28,384**	**54,312**
Uzbekistan			
Fergana	107,120	....	....
Amtyari	66,200	....	....
Total	**173,320**	**....**	**....**
Total C.I.S.	**9,845,939**	**433,684**	**54,312**

Note: Totals may not add due to rounding.
Source: Beicip-Franlab

CONGO

Coraf–Pointe-Noire	**21,000**	**....**	**[2]2,000**

COSTA RICA

Refinadora Costarricense de Petroleo SA–Limón	**15,000**	....	**[2]1,200**

CROATIA

d.d. Zagreb Rafinerija Rijeka–Rijeka	150,000	[1]22,000	[2]18,400
Ina-Industrija Nafte p.o. Zagreb–Sisak	143,472	[1]10,000	[2]19,000
Total	**293,472**	**32,000**	**37,400**

CUBA

Government-owned refineries at Cabaiguan, Cienfuegos, Havana, Santiago de Cuba.			
Total	**280,000**		

CYPRUS

Cyprus Petroleum Refinery Ltd.–Larnaca	**22,000**	**....**	**[2]5,100**

CZECH REPUBLIC

Government owned refineries at Strazke, Zaluzi, Zyolen.			
Crude Capacity	216,810		
Kaucuk s.p.–Kralupy	70,000	....	[2]8,570
Koramo Kolin—Kolin	....	....	....
Paramo Pardubice, s.p.–Pardubice	20,440	....	....
Total	**307,250**	**....**	**8,570**

DENMARK

AS Dansk Shell–Fredericia	62,000	....	[2]13,000
Kuwait Petroleum Refining (Danmark) A/S–Skaelskoer	56,500	....	[2]10,000
Statoil AS–Kalundborg	65,000	....	[2]8,600
Total	**183,500**	**....**	**31,600**

DOMINICAN REPUBLIC

Falconbridge Dominicana C por A–La Bonao	16,000	....	....
Refineria Dominicana de Petróleo SA–Haina	32,000	....	[1]9,700
Total	**48,000**	**....**	**9,700**

WORLDWIDE REFINING	Crude capacity (b/cd)	Catalytic cracking (b/cd)	Catalytic reforming (b/cd)

ECUADOR

Petroecuador–Esmeraldas	90,000	[1]16,000	[5]2,300
La Libertad	47,000	....	....
Amazonas	10,000	....	....
Lago-Agrio	1,000	....	....
Total	**148,000**	**16,000**	**2,300**

EGYPT

Alexandria Petroleum Co.–Alexandria–(El-Mex)	97,850	....	....
Amerya Oil Refining Co.–Alexandria	73,000	....	[2]12,000
Cairo Oil Refining Co.–Mostorod	156,750	....	[2]9,000
Tanta	21,800	....	....
El-Nasr Petroleum Co.–El-Suez	58,400	....	....
Wadi-Feran	9,680	....	....
Suez Oil Processing Co.–El-Suez	62,530	....	[2]12,540
Assiout	52,143	....	....
Total	**532,153**	**....**	**33,540**

El SALVADOR

Refinería Petrólera Acajutla SA–Acajutla	18,000	....	[1]3,200
Total	**18,000**	**....**	**3,200**

ETHIOPIA

Ethiopian Petroleum Corp.–Assab	**18,000**	**....**	**[3]2,400**

FINLAND

Neste Oy–Naantali	40,000	[2]14,200	[5]6,800
Porvoo	160,000	[1]31,000	[5]36,000
Total	**200,000**	**45,200**	**42,800**

FRANCE

CRD—Total France–Gonfreville L'Orcher	309,000	[1]31,500	[2]45,400
La Mède	136,000	[1]29,700	[4]23,400
Mardyck	122,000	[1]29,800	[4]20,500
Cie. Rhénane de Raffinage–Reichstett-Vendenheim	80,000	[1]14,000	[1]16,000
Elf France–Donges	214,300	[1]49,400	[2]29,280
Feyzin	128,130	[1]24,940	[2]9,200
Grandpuits	98,000	[1]30,720	[2]14,800
Esso SAF–Fos sur Mer	100,000	[1]24,000	[1]18,000
Port Jerome	144,000	[1]24,000	[1]18,000
Mobil Oil Francaise–Notre Dame de Gravenchon	65,000	....	[2]13,000
Shell Francaise–Berre l'Etang	127,000	[1]38,000	[2]19,000
Petit Couronne	158,000	[1]20,000	[2]28,000
Ste. Francaise des Petroles BP–Dunkirk	....	....	....
Lavera	180,000	[1]24,000	[2]10,000
Total	**1,861,430**	**340,060**	**264,580**

GABON

Ste. Gabonaise de Raffinage–Port Gentil	**17,300**	**....**	**[3]1,400**

GERMANY

Addinol Mineralol GmbH Lutzkendorf—Krumpa, Kreis Merseburg	11,800	....	....
Beta Raffineriegesellschaft Wilhelmshaven—Wilhelmshaven	160,000	....	[2]39,200
BP Oiltech Gmbh—Hamburg	....	....	....
DEA Mineraloel AG–Heide	80,000	†[2]9,000	[4]20,000
UK-Wesseling	100,000	....	[4]19,000
Deutsche Shell AG–Godorf	170,000	....	[2]17,000
			[8]21,000
Harburg-Grasbrook	98,000	[1]16,000	[8]17,000
Erdoel Raffinerie Neustadt GmbH–Neustadt-Donau	144,000	[1]26,000	[1]17,700
Esso AG–Ingolstadt	100,000	[1]27,700	[2]15,600
Karlsruhe	150,000	....	[2]32,000
Holborn Europa Raffinerie GmbH—Harburg	78,000	[1]18,750	[2]14,700
Hydrierwerk Zeitz GmbH–Zeitz	73,700	....	....
Leune Raffinernegesellschaft—Leuna	100,000	....	[2]15,000

WORLDWIDE REFINING	Crude capacity (b/cd)	Catalytic cracking (b/cd)	Catalytic reforming (b/cd)
Mobil Oil AG–Woerth	100,000	[1]24,000	[2]18,000
Oberrheinische Mineralolwerke GmbH–Karlsruhe	174,000	[1]66,000	[2]11,400
			[6]21,300
OMV Aktiengesellschaft–Burghausen	72,000	....	....
PCK Schwedt AG–Schwedt	230,000	[1]38,000	[2]16,000
			[6]18,000
Raffineriegesellschaft–Vohburg/Ingolstadt	114,000	[1]35,200	[2]16,600
			[8]18,000
Ruhr Oel GmbH—Gelsenkirchen	227,000	[1]21,000	[8]27,000
Wintershall AG–Lingen	79,460	....	[1]8,214
			[2]5,629
			[5]21,160
Salzbergen	3,100	....	....
Total	**2,265,060**	**281,650**	**409,503**

Note: Some of the smaller east german refineries are not accounted for.
†TCC.

GHANA

WORLDWIDE REFINING	Crude capacity	Catalytic cracking	Catalytic reforming
*Ghanaian Italian Petroleum CL–Tema	**26,600**	**....**	**[2]4,950**

GREECE

WORLDWIDE REFINING	Crude capacity	Catalytic cracking	Catalytic reforming
Eko-Hellenic Refineries and Chemicals of Macedonia Industrial & Commerical Co. SA—Thessaloniki	66,500	....	[2]8,100
Hellenic Aspropyrgos Refinery SA–Aspropyrgos	121,000	[1]34,000	[2]9,800
			[6]20,600
Motor Oil (Hellas) Corinth Refineries SA–Aghii Theodori	100,000	[1]29,700	[2]12,300
Petrola Hellas SA–Elefsis	108,000	....	....
Total	**395,500**	**63,700**	**50,800**

GUATEMALA

WORLDWIDE REFINING	Crude capacity	Catalytic cracking	Catalytic reforming
Basic Resources International–Peten	4,000	....	....
Texas Petroleum Co.–Escuintla	16,000	....	[2]3,000
Total	**20,000**	**....**	**3,000**

HONDURAS

WORLDWIDE REFINING	Crude capacity	Catalytic cracking	Catalytic reforming
Refinería Texas de Honduras SA–Puerto Cortes	14,000	....	[2]1,800
Total	**14,000**	**....**	**1,800**

HUNGARY

WORLDWIDE REFINING	Crude capacity	Catalytic cracking	Catalytic reforming
MOL Rt.–Százhalombatta	165,000	[1]24,000	[2]16,000
			[6]13,600
Tiszaojvaros	66,500	....	....
Zalaegerszeg	10,000	....	....
Total	**241,500**	**24,000**	**29,600**

INDIA

WORLDWIDE REFINING	Crude capacity	Catalytic cracking	Catalytic reforming
Bharat Petroleum CL–Mahul, Bombay	134,860	[1]28,755	[1]5,503
Bongaigaon Refinery & Petrochemicals Ltd.–Bongaigaon, Assam	27,110	....	[2]1,833
Cochin Refineries Ltd.–Ambalamugal	93,400	[1]19,300	[1]5,800
Hindustan Petroleum CL–Mahul, Bombay	110,452	[1]12,000	....
Visakhapatnam	90,369	[1]19,200	....
Indian Oil CL–Barauni, Bihar	65,800	....	....
Digboi, Assam	11,700	....	....
Guwahati, Assam	19,920	....	....
Koyali, Gujarat	185,100	[1]20,000	[2]8,300
Haldia, West Bengal	61,000	....	[2]5,300
Mathura, Uttar Pradesh	156,000	[1]20,000	....
Madras Refineries Ltd.–Madras	130,660	[1]11,250	[2]2,138
Total	**1,086,371**	**130,505**	**28,874**

INDONESIA

WORLDWIDE REFINING	Crude capacity	Catalytic cracking	Catalytic reforming
Pertamina–Balikpapan, Kalimantan	257,500	....	[8]20,000
Cilacap, Central Java	302,400	....	[1]12,800
			[8]17,300
Dumai, Central Sumatra	122,500	....	[1]5,400
			[8]8,100
Musi, South Sumatra	126,900	[1]13,000	....
Pangakalan Brandan, North Sumatra	4,800	....	....
Sungai Pakning, Central Sumatra	46,100	....	....
Total	**860,200**	**13,000**	**63,600**

IRAN

WORLDWIDE REFINING	Crude capacity	Catalytic cracking	Catalytic reforming
National Iranian Oil Co.–Abadan	320,000	[1]40,000	[8]41,555
Arak	150,000	....	....
Esfahan	281,000	....	[8]29,600
Kermanshah	35,000	....	....
Lavan	23,000	....	....
Shiraz	38,000	....	[7]6,245
Tabriz	90,000	....	....
Tehran	224,700	....	[8]28,000
Total	**1,161,700**	**40,000**	**105,400**

IRAQ

WORLDWIDE REFINING	Crude capacity	Catalytic cracking	Catalytic reforming
Oil Refineries Administration–Baiji	150,000	....	22,000
			16,500
Basra	70,000	....	....
Doura	100,000	....	5,000
K3-Haditha	7,000	....	....
Khanaqin	12,000	....	....
Mufthia	4,500	....	....
Qaiyarah, Mosul	2,000	....	....
Iraqi Company for Oil Operations–Kirkuk	2,000	....	....
Total	**347,500**	**....**	**43,500**

Note: Present status uncertain due to lack of information.

IRELAND

WORLDWIDE REFINING	Crude capacity	Catalytic cracking	Catalytic reforming
Irish National Refining plc–Whitegate	52,500	....	[2]12,000
Total	**52,500**	**....**	**12,000**

ISRAEL

WORLDWIDE REFINING	Crude capacity	Catalytic cracking	Catalytic reforming
Oil Refineries Ltd.–Ashdod	78,000	[1]27,500	[2]11,000
Haifa	130,000	[1]22,000	[2]15,000
Total	**208,000**	**49,500**	**26,000**

ITALY

WORLDWIDE REFINING	Crude capacity	Catalytic cracking	Catalytic reforming
Agip Plas SpA–Livorno	84,000	....	[6]12,000
Agip Raffinazione SpA–Porto Marghera	80,000	....	[6]15,000
Sannazzaro, Pavia	200,000	[1]34,000	[2]12,000
			[6]16,700
Taranto	84,000	....	[2]17,000
Anonima Petroli Italiana–Falconara, Marittima	78,000	....	[2]14,000
Arcola Petrolifera SpA–La Spezia	33,000	....	....
Cameli Petroli & Co. S.p.A.–Frassino, Mantova	50,000	....	[2]7,500
Esso Italiana SpA–Augusta, Siracusa	180,000	[1]48,000	[2]22,000
Iplom SpA–Busalla	46,500	....	....
Isab–Priolo Gargallo	232,000	....	[4]37,736
Praoil–Gela	105,000	[1]35,000	[2]13,850
Priolo	220,000	[1]32,000	[2]9,000
Raffineria di Roma SpA–Rome	81,500	....	[2]14,300
Raffineria Mediterranea Spa–Milazzo	160,000	[1]45,000	[2]12,000
Saras SpA–Sarroch	285,000	[1]75,000	[5]27,000
Sarpom–S. Martino Di Trecate	248,000	[1]26,000	[1]18,500
			[4]8,100
Tamoil Italia SpA–Cremona	95,000	....	[2]13,400
Total	**2,262,000**	**295,000**	**270,086**

WORLDWIDE REFINING	b/cd Crude capacity	Catalytic cracking	Catalytic reforming
IVORY COAST			
Ste. Ivoirienne de Raffinage–Abidjan	52,800	....	[1]10,200
Ste. Multinationale de Bitumes–Abidjan	10,000	....	....
Total	**62,800**	**....**	**10,200**
JAMAICA			
Petrojam Ltd.–Kingston	35,500	....	[1]3,540
Total	**35,500**	**....**	**3,540**
JAPAN			
*Cosmo Oil CL–Chiba	228,000	[1]31,500	[2]32,850
Sakaide	133,000	[1]17,100	[2]12,600
Sakai	104,500	[1]19,800	[2]7,200
Yokkaichi City	147,250	[1]22,500	[2]5,400 [6]9,900
Fuji Oil CL–Sodegaura,Chiba	145,000	[1]14,000	[5]21,500
General Sekiyu Seisei KK–Sakai	140,400	[1]36,000	[2]30,600
Idemitsu Kosan CL–Chita, Aichi	150,000	[1]33,000	[3]18,000
Himeji, Hyogo	120,000	....	[1]17,000
Ichihara, Chiba	274,000	[1]42,000	[1]17,000
Tokuyama, Yamaguchi	120,000	[1]24,000	....
Tomakomai, Hokkaido	114,000	....	[1]18,000
Kashima Oil CL–Kashima, Ibaragi	165,000	[1]22,000	[5]14,000 [7]7,500
Koa Oil CL–Marifu	110,000	[1]22,000	[5]22,000
Osaka	104,000	[1]25,000	[4]11,000
Kyenus Sekiyu Seisei KK–Kawasaki	66,500	....	....
Kyokuto Petroleum Ltd.–Chiba	138,000	....	[2]22,000
Kyushu Oil CL–Oita	130,000	[1]16,500	[2]8,000
Mitsubishi Oil CL–Kawasaki	75,000	....	[2]10,000
Mizushima	225,000	[1]37,000	[2]25,000 [6]18,500
Nansei Sekiyu KK–Nishihara, Okinawa	72,000	....	[2]10,800
Nihonkai Oil CL–Toyama	60,000	....	[2]7,000
Nippon Mining CL–Chita	95,000	[1]15,000	[2]22,000
Funakawa	6,000	....	....
Mizushima	190,200	[1]40,000	[2]15,000 [6]24,000
*Nippon Oil CL–Niigata	24,700	....	[2]4,500
*Nippon Petroleum Refining CL–Muroran	142,500	[1]20,700	....
Nakagusuku, Okinawa	....	....	[2]2,700
Negishi	342,000	[1]63,900	[2]18,000 [8]27,000
Yokohama	....	....	[2]3,600
Okinawa Sekiyu Seisei–Yonashiro, Okinawa	57,000	....	....
Seibu Oil CL–Yamaguchi	110,100	[1]19,900	[2]20,000
Showa Shell Sekiyu KK–Kawasaki	96,400	....	[2]19,100
Niigata	27,500	....	[2]4,700
Showa Yokkaichi Sekiyu CL–Yokkaichi	220,300	[1]22,600	[2]40,600
Taiyo Oil CL–Ehime	65,000	....	[1]6,000 [3]23,000
Toa Oil CL–Kawasaki	59,700	[1]27,100	[2]7,100
Toho Oil CL–Owase	35,000	....	....
Tohoku Oil CL–Sendai	83,900	[1]25,000	[2]12,000
Tonen Corp.–Kawasaki	228,000	[1]76,950	[2]18,600 [4]22,320
Wakayama	157,700	[1]34,200	[2]33,480
*Wakayama Petroleum Refining Co. Ltd.–Kaiwan City, Wakayama	47,500	....	....
Total	**4,810,150**	**707,750**	**669,550**
JORDAN			
Jordan Petroleum Refinery–Zerka	100,000	[1]4,250	[2]8,640
Total	**100,000**	**4,250**	**8,640**
KENYA			
Kenya Petroleum Refineries Ltd.–Mombasa	90,000	....	[1]4,400 [2]5,200
Total	**90,000**	**....**	**9,600**

WORLDWIDE REFINING	b/cd Crude capacity	Catalytic cracking	Catalytic reforming
KOREA, N.			
Government-owned, Unggi	**42,000**		
KOREA, S.			
Honam Oil Refinery CL–Yocheon	342,000	....	[2]15,300
Kukdong Oil CL–Busan	10,000	....	....
Daesan	60,000	....	[2]3,000
Kyung In Energy CL–Inchon	60,000	....	[2]3,600
Ssangyong Oil Refining CL–Onsan	146,100	....	[2]4,800 [8]27,000
Yukong Ltd.–Ulsan	529,000	....	[2]19,000 [4]50,000
Total	**1,147,100**	**....**	**122,700**
KUWAIT			
Kuwait National Petroleum Co.–Mina Abdulla	211,000	....	....
Mina Al-Ahmadi	288,000	[1]28,000	[2]33,000
Shuaiba	125,000	....	[2]14,000
Total	**624,000**	**28,000**	**47,000**
LEBANON			
Tripoli Oil Installations–Tripoli	20,000	[1]7,250	[2]4,392
Zahrani Oil Installations–Sidon	17,500	....	[1]3,050
Total	**37,500**	**7,250**	**7,442**
LIBERIA			
Liberia Petroleum Refining–Monrovia	15,000	....	[1]2,000
Total	**15,000**	**....**	**2,000**
LIBYA			
Azzawiya Oil Refining Co.–Azzawiya	120,000	....	[2]13,000
Ras Lanuf Oil & Gas Processing Co.–Ras Lanuf	220,000	....	....
Sirte Oil Co.–Brega	8,400	....	[1]1,000
Total	**348,400**	**....**	**14,000**
LITHUANIA			
Mazeikiai	263,420	[1]43,692	[1]25,741 [2]25,741
Total	**263,420**	**43,692**	**51,482**
MACEDONIA			
Rafinerija Skopje–Skopje	51,180	....	[2]10,780
Total	**51,180**	**....**	**10,780**
MADAGASCAR			
Solima–Managareza, Tamatave	**16,350**	**....**	**[1]2,600**
MALAYSIA			
Esso Malaysia Berhad–Port Dickson	59,800	....	[2]13,100
Petronas Penapisan (Melaka) Sdn. Berhad–Kerteh, Kemaman, Terengganu	75,000	....	....
Sarawak Shell Berhad–Luton	45,000	....	....
Shell Refining Co. Berhad–Port Dickson	90,000	....	[1]4,000 [8]13,000
Total	**269,800**	**....**	**30,100**
MARTINIQUE			
Ste. Anonyme de la Raffinerie des Antilles–Fort-de-France	16,000	....	[2]3,000
Total	**16,000**	**....**	**3,000**
MEXICO			
Petroleos Mexicanos—Cadereyta	235,000	[1]40,000	[2]20,000

WORLDWIDE REFINING	b/cd Crude capacity	Catalytic cracking	Catalytic reforming
Ciudad Madero	195,000	[1]43,000	[2]15,000
Minatitlan	200,000	[1]24,000 [2]26,000	[2]31,000
Reynosa	9,000	....	....
Salamanca	235,000	[1]40,000	[2]24,800
Salina Cruz	330,000	[1]40,000	[6]50,000
Tula, Hidalgo	320,000	[1]40,000	[2]30,000
Total	**1,524,000**	**243,000**	**170,800**

MOROCCO

Samir–Mohammedia	129,000	....	[2]24,000
Ste. Cherifienne des Petroles–Sidi Kacem	25,600	[2]5,600	[1]2,800
Total	**154,600**	**5,600**	**26,800**

MYANMAR

Myanma Petrochemical Enterprise–Chauk	6,000	....	....
Thanlyin	26,000	....	....
Total	**32,000**	**....**	**....**

NETHERLANDS

Esso Nederland BV–Rotterdam	175,000	....	[4]26,600
Kuwait Petroleum Europoort BV–Rotterdam	75,500	....	[6]21,000
Netherlands Refining Co.–Europoort & Pernis	399,000	[1]49,000	[2]54,000
Shell Nederland Raffinaderij BV–Pernis	374,000	[1]87,000	[2]17,000 [8]42,000
Smid & Hollander Raffinaderij BV–Amsterdam	10,000	....	....
Total Raffinaderij Nederland NV–Vlissingen	150,000	....	[5]24,500
Total	**1,183,500**	**136,000**	**185,100**

NETHERLANDS ANTILLES

Coastal Aruba Refining Co. N.V.–San Nicolas	157,000	....	....
Refineria Isla Curazao SA–Emmastad	320,000	[1]49,000	[2]20,000
Total	**477,000**	**49,000**	**20,000**

NEW ZEALAND

New Zealand Refining–Whangarei	92,500	....	[2]22,000
Total	**92,500**	**....**	**22,000**

NICARAGUA

Esso Standard Oil SA Ltd.–Managua	16,000	....	[1]3,000
Total	**16,000**	**....**	**3,000**

NIGERIA

Kaduna Refinery & Petrochemical Co. (NNPC)–Kaduna	104,500	[1]18,000	[2]15,300
Port Harcourt Refining Co. (NNPC)–Alesa Eleme	60,000	....	[1]6,000
Rivers State	150,000	[1]40,000	[8]33,000
Warri Refinery & Petrochemical Co. (NNPC)–Warri	118,750	[1]24,700	[2]15,770
Total	**433,250**	**82,700**	**70,070**

NORWAY

Esso Norge AS–Slagen-Toensberg	90,000	....	[2]10,600
Norske Shell AS–Sola	53,000	....	[2]11,000
Statoil Division Mongstad–Mongstad	144,250	§[2]41,000	[1]9,800 [5]15,700
Total	**287,250**	**41,000**	**47,100**

§RCC.

WORLDWIDE REFINING	b/cd Crude capacity	Catalytic cracking	Catalytic reforming

OMAN

Oman Refinery Co.–Mina Al Fahal	**76,000**	**....**	**[2]15,200**

PAKISTAN

Attock Refinery Ltd.–Rawalpindi	30,500	....	....
National Refinery Ltd.–Korangi, Karachi	56,100	....	[2]2,210
Pakistan Refinery Ltd.–Karachi	46,300	....	[4]2,700
Total	**132,900**	**....**	**4,910**

PANAMA

Refinería Panama SA–Las Minas	100,000	....	[2]8,000
Total	**100,000**	**....**	**8,000**

PARAGUAY

Petroleos Paraguayos–Villa Elisa	**7,500**	**....**	**....**

PERU

Petróleos del Peru:			
Conchan	6,500	....	....
Iquitos, Loreto	10,500	....	....
La Pampilla, Lima	102,000	[1]6,700	....
Pucallpa	3,250	....	....
Talara	62,000	[1]16,600	....
Total	**184,250**	**23,300**	**....**

PHILIPPINES

Caltex (Philippines) Inc.–Batangas	65,700	[1]11,900	[2]6,000
*Petron Corp.–Limay, Bataan	147,250	[2]10,800	[2]15,300
Philippine Petroleum Corp.–Pililla	....	....	....
Pilipinas Shell Petroleum–Tabangao	65,800	....	[1]8,000
Total	**278,750**	**22,700**	**29,300**

POLAND

Government-owned refineries at Czechowice, Glinik Mariampolski, Jasto, Jealicze, Kralaty, L. Warynski, Trzebinia. Crude capacity	73,000		
Mazovian Refinery & Petrochemical Works–Plock	200,000	[1]32,000	[2]24,000
Petroleum Refinery Gdansk—Gdansk	60,000	....	[3]11,000
Total	**333,000**	**32,000**	**35,000**

PORTUGAL

Petrogal–Lisboa	....	†[2]10,100	....
Leça da Palmeira, Porto	88,300	....	[2]15,600 [4]10,500
Sines	205,700	....	[4]24,900
Total	**294,000**	**10,100**	**51,000**

†TCC.

PUERTO RICO

Caribbean Petroleum Corp.–Bayamón	42,000	[1]14,200	[1]6,500
Puerto Rico Sun Oil Co.–Yabucoa	85,000	....	[2]20,000
Total	**127,000**	**14,200**	**26,500**

QATAR

National Oil Distribution Co.–Umm Said	57,500	....	[6]9,400 †[8]2,100
Total	**57,500**	**....**	**11,500**

†Nonregenerative.

ROMANIA

Arpechim SA—Pitesti	77,316	[1]23,013	[1]12,233
Astra SA—Pitesti	56,000	....	....

WORLDWIDE REFINING	Crude capacity (b/cd)	Catalytic cracking (b/cd)	Catalytic reforming (b/cd)
Darmanesti Refinery Darmanesti	30,000	....	....
Petrobrazi SA—Ploiesti	159,000	[1]23,000	[1]20,000
			[2]10,000
Petrolsub SA—Barcau	8,000	....	....
Petromidia SA—Midia	110,000	[1]20,000	[2]11,490
Petrotel SA—Ploiesti	104,000	[1]20,900	[2]11,490
Rafo SA—Onesti	108,780	[1]26,382	[1]12,327
			[2]12,212
Steaua SA—Cimpina	9,272	....	....
Vega SA—Ploiesti	18,516	....	....
Total	**680,884**	**113,295**	**89,752**

SAUDI ARABIA

WORLDWIDE REFINING	Crude capacity	Catalytic cracking	Catalytic reforming
Saudi Arabian Oil Co. (Saudi Aramco)–Jeddah	95,000	[1]9,041	[2]2,622
Ras Tanura	290,000	....	[2]49,200
Riyadh	120,000	....	[2]28,600
			[5]5,400
Yanbu	170,000	....	[2]35,000
Arabian Oil CL–Ras Al Khafji	30,000	....	....
Petromin-Mobil-Yanbu	300,000	[1]78,500	[6]45,000
Petromin Petrola Rabigh Refinery Co.—Rabigh	325,000	....	....
Petromin-Shell-Al-Jubail	284,000	....	[8]19,000
Total	**1,614,000**	**87,541**	**184,822**

SENEGAL

WORLDWIDE REFINING	Crude capacity	Catalytic cracking	Catalytic reforming
Ste. Africaine de Raffinage–M'Bao (Dakar)	**22,600**	**....**	**[4]2,500**

SIERRA LEONE

WORLDWIDE REFINING	Crude capacity	Catalytic cracking	Catalytic reforming
Sierra Leone Petroleum Refining Co.–Freetown	**10,000**	**....**	**....**

SINGAPORE

WORLDWIDE REFINING	Crude capacity	Catalytic cracking	Catalytic reforming
BP Refinery Singapore PL–Pasir Panjang	29,000	....	....
Esso Singapore PL–Pulau Ayer Chawan	220,000	....	[2]11,400
Mobil Oil Singapore PL–Jurong	235,000	....	[2]18,000
Shell Eastern Petroleum Ltd.–Pulau Bukom	358,000	[1]28,000	[2]6,900
			[6]12,900
Singapore Refining Co.–Pulau Merlimau	213,000	....	[4]13,500
Total	**1,055,000**	**28,000**	**62,700**

SLOVAKIA

WORLDWIDE REFINING	Crude capacity	Catalytic cracking	Catalytic reforming
Slovnaft–Bratislava	193,400	....	[2]22,270
Total	**193,400**	**....**	**22,270**

SLOVENIA

WORLDWIDE REFINING	Crude capacity	Catalytic cracking	Catalytic reforming
Nafte Lendava–Lendava	**14,700**	....	....

SOMALIA

WORLDWIDE REFINING	Crude capacity	Catalytic cracking	Catalytic reforming
Iraqsoma Ref. Co.–Mogadishu	**10,000**	**....**	**....**

SOUTH AFRICA

WORLDWIDE REFINING	Crude capacity	Catalytic cracking	Catalytic reforming
Caltex Oil SA PL–Cape Town	90,000	[1]25,000	[1]11,000
Gencor–Durban	65,000	[1]15,000	[2]15,000
National Petroleum Refiners of South Africa PL–Sasolburg OFS	86,038	[1]18,113	[1]14,490
Shell and BP South Africa Petroleum Refineries PL–Durban	123,000	[1]27,000	[2]21,000
Total	**364,038**	**85,113**	**61,490**

SPAIN

WORLDWIDE REFINING	Crude capacity	Catalytic cracking	Catalytic reforming
Asfaltos Españoles SA–Tarragona	21,000	....	....
Cia. Española de Petróles–San Roque (Cádiz)	180,000	[1]40,000	[1]35,000
Tenerife	90,000	....	[1]16,000
Ertoil, SA–La Rábida, Huelva	80,000	[1]18,000	[2]17,000
Petrómed—Castellón de la Plana	102,000	[1]21,600	[2]15,000
Petronor SA–Somorrostro, Vizcaya	240,000	[1]40,000	[2]33,500
Repsol Petroleo SA–Cartagena, Murcia	120,000	....	[2]25,000
La Coruña	135,000	[1]28,000	[2]22,000
Puertollano, Ciudad Real	135,000	[1]30,000	[2]18,000
Tarragona	180,000	....	[2]16,000
Total	**1,283,000**	**177,600**	**197,500**

SRI LANKA

WORLDWIDE REFINING	Crude capacity	Catalytic cracking	Catalytic reforming
Ceylon Petroleum Corp.–Sapugaskanda	50,000	....	[2]3,750
Total	**50,000**	**....**	**3,750**

SUDAN

WORLDWIDE REFINING	Crude capacity	Catalytic cracking	Catalytic reforming
Port Sudan Refinery Ltd.–Port Sudan	**21,700**	**....**	**[1]1,900**

SWEDEN

WORLDWIDE REFINING	Crude capacity	Catalytic cracking	Catalytic reforming
AB Nynas Petroleum–Gothenburg	12,500	....	....
Nynäshamn	28,000	....	....
OK Raffinaderi–Gothenburg	106,000	....	[2]20,500
Shell Raffinaderi BV–Gothenburg	81,000	....	[2]17,000
Skandinaviska Raffinaderi AB–Brofjorden-Lysekil	200,000	[1]28,000	[6]36,000
Total	**427,500**	**28,000**	**73,500**

SWITZERLAND

WORLDWIDE REFINING	Crude capacity	Catalytic cracking	Catalytic reforming
Raffinerie de Cressier SA–Cressier	60,000	....	[2]16,000
Raffinerie du Sud-Ouest SA–Collombey	72,000	....	[2]10,000
Total	**132,000**	**....**	**26,000**

SYRIA

WORLDWIDE REFINING	Crude capacity	Catalytic cracking	Catalytic reforming
Banias Refinery Co.–Banias	135,000	....	[6]16,000
Homs Refinery Co.–Homs	107,140	....	[2]3,427
			[6]9,812
Total	**242,140**	**....**	**29,239**

TAIWAN

WORLDWIDE REFINING	Crude capacity	Catalytic cracking	Catalytic reforming
Chinese Petroleum Corp.–Kaohsiung	425,000	[1]45,200	[2]36,160
Tao-Yuan	117,500	....	[2]20,340
Total	**542,500**	**45,200**	**56,500**

TANZANIA

WORLDWIDE REFINING	Crude capacity	Catalytic cracking	Catalytic reforming
Tanzanian & Italian Petroleum Refining CL–Kigamboni, Dar es Salaam	17,000	....	[2]2,500
Total	**17,000**	**....**	**2,500**

THAILAND

WORLDWIDE REFINING	Crude capacity	Catalytic cracking	Catalytic reforming
Esso Standard Thailand Ltd.–Sriracha	75,000	....	[2]9,000
			[4]16,500
Petroleum Authority of Thailand–Bangchak, Bangkok	61,750	....	[2]9,450
Thai Oil CL–Sriracha	177,000	[1]10,200	[1]2,700
			[2]7,800
Total	**313,750**	**10,200**	**45,450**

TRINIDAD

WORLDWIDE REFINING	Crude capacity	Catalytic cracking	Catalytic reforming
Trinidad and Tobago Oil CL–Pointe-a-Pierre	225,000	[1]28,000	[2]25,000
Point Fortin	80,000	....	[2]6,000
Total	**305,000**	**28,000**	**31,000**

TUNISIA

WORLDWIDE REFINING	Crude capacity	Catalytic cracking	Catalytic reforming
Ste. Tunisienne Industries des Raffinage–Bizerte	**34,000**	**....**	**3,300**

TURKEY

WORLDWIDE REFINING	Crude capacity (b/cd)	Catalytic cracking (b/cd)	Catalytic reforming (b/cd)
Anadolu Tasfiyehanesi AS–Mersin	100,000	....	[2]14,000
Turkish Petroleum Refineries Corp.–Aliaga, Izmir	226,400	[1]16,670	[2]10,060
Batman, Siirt	22,000	....	[2]1,300
Izmit	251,600	[1]22,650	[2]20,750
Kirikkale	113,000	....	[2]20,100
Total	**713,000**	**39,320**	**66,210**

UNITED KINGDOM

WORLDWIDE REFINING	Crude capacity (b/cd)	Catalytic cracking (b/cd)	Catalytic reforming (b/cd)
England			
Conoco Ltd.–South Killingholme	130,000	[1]47,000	[2]46,000
Eastham Refinery Ltd.–Eastham, Ellesmere Port	22,000	....	....
Esso Petroleum CL–Fawley	302,400	[1]76,500	[2]35,300 [4]22,500
Lindsey Oil Refinery Ltd.–Killingholm, South Humberside	182,000	[1]48,400	[2]32,600
Mobil Oil CL–Coryton, Essex	173,000	[1]60,000	[6]37,000
Phillips Imperial Petroleum Ltd.–Port Clarence	100,000	....	....
Shell U.K. Ltd.–Shell Haven	92,000	....	[8]39,000
Stanlow	262,000	[1]69,000	[2]29,000 [8]30,000
Total	**1,263,400**	**300,900**	**271,400**
Scotland			
BP Refinery Grangemouth Ltd.–Grangemouth	190,000	[1]20,000	[2]40,000
Nynas UK AB—Dundee	10,240	....	....
Total	**200,240**	**20,000**	**40,000**
Wales			
BP Refinery Llandarcy Ltd.–Llandarcy, Neath	....	....	....
Elf Oil (G.B.) Ltd.–Milford Haven	108,000	[1]32,500	[2]17,800
Gulf Oil-GB–Milford Haven	115,000	[1]31,500	[2]24,000
Pembroke Cracking Co. (65% Tex aco, 35% Gulf Oil-GB)—Milford Haven	....	[1]90,000	....
Texaco Ltd.–Pembroke, Dyfed	180,000	....	[8]39,000
Total	**403,000**	**154,000**	**80,800**
Total United Kingdom	**1,866,640**	**474,900**	**392,200**

UNITED STATES

WORLDWIDE REFINING	Crude capacity (b/cd)	Catalytic cracking (b/cd)	Catalytic reforming (b/cd)
Alabama			
Coastal Mobile Refining Co.—Mobile Bay	14,250	......	......
Hunt Refining Co.—Tuscaloosa	40,000	......	[2]7,200
Louisiana Land & Exploration Co.—Saraland	71,200	......	[2]20,000
Total	**125,450**	**......**	**27,200**
Alaska			
ARCO Alaska Inc.—Kuparuk	14,000	......	......
Prudhoe Bay	17,500	......	......
Mapco Alaska Petroleum—North Pole	122,000	......	......
*Petro Star Inc.—North Pole	8,000	......	......
Valdez	26,300	......	......
Tesoro Alaska Petroleum Co.—Kenai	72,000	......	[2]12,000
Total	**259,800**	**......**	**12,000**
Arizona			
Intermountain Refining Cl—Fredonia	5,710	......	......
Total	**5,710**	**......**	**......**
Arkansas			
Berry Petroleum Co.—Stephens	6,700	......	......
Cross Oil & Refining Co. Inc.—Smackover	6,200	......	......
Lion Oil Co.—El Dorado	48,000	[1]18,500	[6]12,000
Total	**60,900**	**18,500**	**12,000**
California			
Atlantic Richfield Co.—Carson	237,000	[1]90,000	[2]50,000
Chemoil Refining Corp.—Signal Hill	21,700	......	......
Chevron U.S.A. Inc.—El Segundo	226,000	[1]62,000	[2]46,000
Richmond	235,000	[1]62,500	[2]58,000
*Conoco Inc.—Santa Maria	9,500	......	......
Exxon Co.—Benicia	128,000	[1]64,000	[3]31,000
Huntway Refining Co.—Benicia	8,400	......	......
Wilmington	5,500	......	......
Kern Oil & Refining Co.—Bakersfield	21,400	......	[2]2,600
*Lunday-Thagard Co.—South Gate	7,000	......	......
Mobil Oil Corp.—Torrance	127,000	[1]79,000	[2]38,800
Pacific Refining Co.—Hercules	52,250	......	[2]13,500
Paramount Petroleum Corp.—Paramount	39,500	......	[2]7,500
Powerine Oil Co.—Santa Fe Springs	46,500	[1]12,000	[2]9,600
San Joaquin Refining Cl—Bakersfield	18,000	......	......
Shell Oil Co.—Martinez	147,100	[1]65,000 [2]1,000	[3]27,800
*Sunland Refining Corp.—Bakersfield	15,000	......	[2]1,350
*Ten By, Inc.—Oxnard	4,000	......	......
Texaco Refining & Marketing Inc.—Bakersfield	53,000	......	[2]21,000
Wilmington	95,000	[1]30,000	[2]36,000
Tosco Corp.—Martinez	148,000	[1]65,000	[2]18,500 [6]21,500
Ultramar, Inc.—Wilmington	68,000	[2]38,000	[6]14,500
*Unocal Corp.—Los Angeles/Wilmington	118,750	[1]42,300	[1]48,600
San Francisco/Santa Maria	103,645	......	[2]30,600
Witco Corp., Golden Bear Divi sion—Oildale	10,300	......	......
Total	**1,945,545**	**610,800**	**476,850**
Colorado			
Conoco Inc.—Denver	51,500	[1]19,000	[2]10,000
Landmark Petroleum Inc.—Fruita	16,000	....	[2]3,500
Total Petroleum—Commerce City	28,000	[1]8,000	[2]9,000
Total	**95,500**	**27,000**	**22,500**
Delaware			
*Star Enterprise—Delaware City	140,000	[1]63,000	[2]11,700 [5]36,900
Total	**140,000**	**63,000**	**48,600**
Georgia			
*Citgo Petroleum Corp—Savannah	28,000	......	......
Young Refining Corp.—Douglasville	7,500	......	......
Total	**35,500**	**......**	**......**
Hawaii			
BHP Petroleum Americas Refining Inc.—Ewa Beach	92,000	......	[2]12,400
Chevron U.S.A. Inc.—Barber's Point	52,800	[1]20,000	......
Total	**144,800**	**20,000**	**12,400**
Illinois			
Clark Oil & Refining Corp.—Blue Island	66,500	[1]23,800	[2]29,000
Hartford	57,000	[1]24,700	......
*Indian Refining Co.—Lawrence ville	54,000	[1]30,600	[2]13,500
Marathon Oil Co.—Robinson	166,000	[1]43,000	[4]35,000 [5]38,500
Mobil Oil Corp.—Joliet	180,000	[1]93,100	[1]43,000
Shell Oil Co.—Wood River	252,000	[1]77,500	[2]15,000 [3]38,000 [4]31,000
*Uno-Ven Co.—Lemont	145,350	[1]52,200	[2]26,820
Total	**920,850**	**344,900**	**269,820**

WORLDWIDE REFINING	b/cd Crude capacity	Catalytic cracking	Catalytic reforming
Indiana			
Amoco Oil Co.—Whiting	400,000	[1]149,200	[6]85,500
Countrymark Cooperative, Inc. Mt. Vernon	22,000	[1]8,000	[6]6,500
Laketon Refining Corp.—Laketon	8,300	......	......
Marathon Oil Co.—Indianapolis	48,000	[1]19,000	[2]10,300
Total	**478,300**	**176,200**	**102,300**
Kansas			
Farmland Industries Inc.—Coffeyville	62,000	[1]25,000	[2]16,000
National Cooperative Refinery Association—McPherson	70,900	[1]18,900	[5]18,900
Texaco Refining & Marketing Inc.—El Dorado	80,000	[1]29,000	[2]23,000
Total Petroleum Inc.—Arkansas City	56,000	[1]18,000	[2]16,600
Total	**268,900**	**90,900**	**74,500**
Kentucky			
Ashland Petroleum Co.—Catlettsburg	213,400	[1]58,200 §[2]38,800	[7]42,195
Somerset Refinery Inc.—Somerset	5,500	......	[2]1,000
Total	**218,900**	**97,000**	**43,195**

§Reduced crude converter.

WORLDWIDE REFINING	b/cd Crude capacity	Catalytic cracking	Catalytic reforming
Louisiana			
American International Refinery, Inc.—Lake Charles	30,000	.....	[2]8,500
*Atlas Processing Co., Division of Pennzoil—Shreveport	46,200	......	[2]4,500
BP Oil Co.—Belle Chasse	221,000	[1]91,000	[2]37,000
Calcasieu Refining Co.—Lake Charles	13,500	......	......
*Calumet Lubricants Co.—Princeton	4,376	......	......
Canal Refining Co.—Church Point	10,000	......	......
*Citgo Petroleum Corp.—Lake Charles	304,000	[1]135,000	[2]46,800 [4]48,600
Conoco Inc.—Lake Charles	182,000	[1]50,000	[2]17,000 [6]13,000
Exxon Co.—Baton Rouge	424,000	[1]180,000	[4]87,000
Kerr-McGee Refining Corp.—Cotton Valley	7,800	......	......
Marathon Oil Co.—Garyville	225,000	[1]90,000	[5]42,500
Mobil Oil Corp.—Chalmette	170,000	[1]58,000	[2]26,165 [4]17,500
Murphy Oil USA Inc.—Meraux	95,000	[1]36,000	[2]21,500
*Phibro Refining Inc.—Krotz Springs	56,700	[1]25,200	[2]10,800
Saint Rose	28,300	......	......
Placid Refining Co.—Port Allen	48,000	[1]18,915	[2]9,700
Shell Oil Co.—Norco	215,000	[1]85,000	[4]36,000
*Star Enterprise—Convent	225,000	[1]76,500	[2]36,000
Total	**2,305,876**	**845,615**	**462,565**
Michigan			
Crystal Refining Co.—Carson City	4,000	......	......
*Lakeside Refining Co.—Kalamazoo	5,600	......	[1]900
Marathon Oil Co.—Detroit	70,000	[1]27,000	[2]15,000
Total Petroleum Inc.—Alma	45,600	[1]18,500	[2]12,000
Total	**125,200**	**45,500**	**27,900**
Minnesota			
Ashland Petroleum Co.—St. Paul Park	67,100	[1]22,310	[2]22,795
*Koch Refining Co.—Rosemount	218,500	[1]49,500	[2]5,400 [8]23,400
Total	**285,600**	**71,810**	**51,595**
Mississippi			
*Amerada-Hess Corp.—Purvis	30,000	[2]14,400	[2]5,220
Chevron U.S.A. Inc.—Pascagoula	295,000	[1]63,000	[2]71,000

WORLDWIDE REFINING	b/cd Crude capacity	Catalytic cracking	Catalytic reforming
Ergon Refining Inc.—Vicksburg	25,000	......	......
Southland Oil Co.—Lumberton	5,800	......	......
Sandersville	11,000	......	......
Total	**366,800**	**77,400**	**76,220**
Montana			
Cenex—Laurel	41,450	[1]13,500	[2]12,000
Conoco Inc.—Billings	49,500	[1]19,000	[2]12,000
Exxon Co.—Billings	42,000	[1]20,000	[1]9,500
Montana Refining Co.—Great Falls	7,000	[1]2,400	[2]1,000
Total	**139,950**	**54,900**	**34,500**
Nevada			
Petro Source Refining Partners—Eagle Springs	**7,000**	**......**	**......**
New Jersey			
*Amerada-Hess Corp.—Port Reading	......	[1]45,000	......
Bayway Refining Co.—Linden	200,000	[1]125,000	[4]26,000
*Chevron U.S.A. Inc.—Perth Amboy	80,000	......	......
Citgo Asphalt Refining Co.—Thorofare	80,000	.....	.....
Coastal Eagle Point Oil Co.—Westville	110,000	[1]50,000	[2]27,000
Mobil Oil Corp.—Paulsboro	113,000	[1]41,300	[2]22,560
Total	**583,000**	**261,300**	**75,560**
New Mexico			
Bloomfield Refining Co.—Bloomfield	18,000	[1]6,200	[2]4,000
Giant Industries Inc.—Gallup	20,800	[1]8,500	[2]6,800
Navajo Refining Co.—Artesia	57,000	[1]20,000	[2]5,000 [5]12,500
Total	**95,800**	**34,700**	**28,300**
New York			
Cibro Petroleum Products Co.—Albany	**41,850**	**......**	**......**
North Dakota			
Amoco Oil Co.—Mandan	58,000	[1]24,700	[4]11,500
Total	**58,000**	**24,700**	**11,500**
Ohio			
Ashland Petroleum Co.—Canton	66,000	[1]24,250	[7]19,400
BP Oil Co.—Lima	155,000	[1]30,000	[4]51,000
Toledo	121,000	[1]52,000	[4]40,000
*Sun Refining & Maketing Co.—Toledo	125,000	[1]54,000	[2]41,040
Total	**467,000**	**160,250**	**151,440**
Oklahoma			
Barrett Refining Corp.—Thomas	10,500	......	......
Conoco Inc.—Ponca City	140,000	[1]53,000	[2]36,000
Cyril Petrochemical Corp.—Cyril	12,500	......	......
Kerr-McGee Refining Corp.—Wynnewood	43,000	[1]20,000	[2]12,500
*Sinclair Oil Corp.—Tulsa	50,000	[1]16,200	[1]10,800
*Sun Refining & Marketing Co.—Tulsa†	85,000	[1]27,000	[2]21,600
Total Petroleum Inc.—Ardmore	68,000	[1]23,000	[6]15,000
Total	**409,000**	**139,200**	**95,900**

†Lube production only.

WORLDWIDE REFINING	b/cd Crude capacity	Catalytic cracking	Catalytic reforming
Oregon			
Chevron U.S.A. Inc.—Portland	.....	.....	.....
Pennsylvania			
BP Oil Co.—Marcus Hook	185,000	[1]48,000	[6]47,500
Chevron U.S.A. Inc.—Philadelphia	172,000	[1]59,500	[2]31,600
*Pennzoil Products Co.—Rouseville	15,700	......	[2]5,238
Sun Refining & Marketing Co.—Marcus Hook	175,000	[1]87,000	[2]24,000
Philadelphia	130,000	[1]36,000	[2]45,000

WORLDWIDE REFINING	Crude capacity (b/cd)	Catalytic cracking (b/cd)	Catalytic reforming (b/cd)
United Refining Co.—Warren	66,700	......	[1]22,000 [4]16,000
Witco Corp.—Bradford	10,000	......	[2]1,650
Total	**754,400**	**230,500**	**192,988**
Tennessee			
Mapco Petroleum Inc.—Memphis	75,000	[1]38,000	[2]14,000
Total	**75,000**	**38,000**	**14,000**
Texas			
AGE Refining & Manufacturing—San Antonio	5,000	......	.....
Amoco Oil Co.—Texas City	433,000	[1]193,800	[4]131,000
Chevron U.S.A. Inc.—El Paso∞	81,900	[1]23,500	[2]23,000
Port Arthur	185,000	[1]63,000	[4]45,000
*Citgo—Corpus Christi	132,500	[1]68,850	[7]48,600
Coastal Refining & Marketing Inc.—Corpus Christi	90,250	[1]20,000	[4]11,000 [6]17,000
*Crown Central Petroleum Corp.—Houston	100,000	[1]50,400	[2]12,600 [5]19,800
Diamond Shamrock Corp.—Sunray	120,000	[1]45,000	[2]15,000 [5]25,000
Three Rivers	70,000	[1]20,000	[2]10,000 [3]20,000
Exxon Co. U.S.A.—Baytown	396,000	[1]190,000	[3]58,000 [4]61,000
Fina Oil & Chemical Co.—Big Spring	55,000	[1]22,000	[2]21,000
Port Arthur	145,000	[1]57,500	[5]35,000
Howell Hydrocarbons Inc.—Channelview	1,300	......	.....
*Koch Refining Co.—Corpus Christi	125,000	[1]36,000	[3]13,500 [8]30,150
LaGloria Oil & Gas Co.—Tyler	52,000	[1]17,000	[2]4,500 [5]12,000
Lyondell-Citgo Refining Co.—Houston	265,000	[1]92,000	[2]108,000
Marathon Oil Co.—Texas City	70,000	[1]32,000	[1]10,000
Mobil Oil Corp.—Beaumont	310,000	[1]97,900	[2]27,800 [5]97,400
*Phibro Refining Inc.—Houston	67,000	[1]54,900	[2]12,150
Texas City	123,500	[1]45,000	[2]9,900 [6]10,800
Phillips Petroleum Co.—Borger	105,000	[1]57,000	[2]24,700
Sweeny	175,000	[1]80,500 [2]11,100	[2]34,000
Pride Refining Inc.—Abilene	44,100	......	[2]6,750
Shell Oil Co.—Deer Park	215,900	[1]61,100 [2]4,700	[2]20,200 [4]40,900
Odessa	28,600	[1]9,500	[2]9,000
Southwestern Refining Co., Inc.—Corpus Christi	104,000	[1]50,000	[2]30,000
*Star Enterprise—Port Arthur	235,000	[1]72,000	[5]45,000
Trifinery—Corpus Christi	30,000	......	......
Valero Refining Co.—Corpus Christi	29,900	†[2]62,900	[6]29,200
Total	**3,794,950**	**1,537,650**	**1,128,950**

†Heavy oil cracker.
∞Chevron—El Paso capacities includes portions of the former El Paso Refining Co.

WORLDWIDE REFINING	Crude capacity (b/cd)	Catalytic cracking (b/cd)	Catalytic reforming (b/cd)
Utah			
Amoco Oil Co.—Salt Lake City	40,000	[1]17,100	[4]7,200
Big West Oil Co.—Salt Lake City	24,000	†[2]5,000	[2]5,000
Chevron U.S.A.—Salt Lake City	45,000	[1]11,500	[2]7,000
Crysen Refining Inc.—Woods Cross	12,500	......	[2]3,000
*Pennzoil Products Co.—Roosevelt	8,000	[1]5,400	[2]1,800
Phillips Petroleum Co.—Woods Cross	25,000	†[2]8,000	[2]5,700
Total	**154,500**	**47,000**	**29,700**

†TCC.

WORLDWIDE REFINING	Crude capacity (b/cd)	Catalytic cracking (b/cd)	Catalytic reforming (b/cd)
Virginia			
Amoco Oil Co.—Yorktown	53,000	[1]26,100	[4]9,700
Total	**53,000**	**26,100**	**9,700**
Washington			
Atlantic Richfield Co.—Ferndale	185,000	......	[2]52,000
*Chevron U.S.A. Inc.—Seattle	.....	......	......
Shell Oil Co.—Anacortes	96,600	[1]42,000 [2]2,800	[3]24,300
Sound Refining Inc.—Tacoma	11,900	......	......
Texaco Refining & Marketing Inc.—Anacortes	132,000	[1]48,000	[2]27,000
Tosco—Ferndale	88,500	[2]25,000	[4]14,500
*U.S. Oil & Refining Co.—Tacoma	32,775	......	[1]5,400
Total	**546,775**	**117,800**	**123,200**
West Virginia			
Quaker State Oil Refining Corp.—Newell	10,500	......	[2]3,300
St. Mary's Refining Co.—St. Mary's	5,000	......	[1]1,350
Total	**15,500**	**......**	**4,650**
Wisconsin			
Murphy Oil USA Inc.—Superior	**33,200**	**[1]10,400**	**[2]7,600**
Wyoming			
Frontier Oil & Refining Co.—Cheyenne	41,000	[1]11,500	[2]6,600
Little America Refining Co.—Casper	22,000	[1]13,300	[4]5,700
Sinclair Oil Corp.—Sinclair	54,000	[1]20,600	[2]14,200
*Wyoming Refining Co.—Newcastle	12,000	[2]4,950	[1]2,475
Total	**129,000**	**50,350**	**28,975**
Total U.S.	**15,141,556**	**5,221,475**	**3,656,608**

URUGUAY

Ancap-La Teja, Montevideo	35,000	[1]5,000	[3]3,000
Total	**35,000**	**5,000**	**3,000**

VENEZUELA

Corpoven–El Palito, Carabobo	105,000	[1]50,400	[4]8,870
El Toreño, Barinas	4,800	....	....
Puerto La Cruz, Anzoategui	195,000	[1]13,600	....
San Roque, Anzoategui	5,200	....	....
Lagoven–Judibana, Falcon	571,000	[1]97,200	....
Maraven–Punta Cardon, Falcon	286,000	[1]72,700	....
Total	**1,167,000**	**233,900**	**8,870**

VIRGIN ISLANDS

Hess Oil Virgin Islands Corp.–St. Croix	545,000	[1]75,000	[2]125,000
Total	**545,000**	**75,000**	**125,000**

YEMEN

Aden Refinery Co.–Little Aden	110,000	....	[1]6,000
Yemen Hunt Oil Co.–Marib	10,000	....	[1]2,500
Total	**120,000**	**....**	**8,500**

YUGOSLAVIA

Bosanski Brod	112,835	....	[2]7,612 [8]13,500
Novi Sad	60,246	....	[2]12,233
Rafinerija Nafte Pancevo–Pancevo	108,000	[1]20,750	[2]8,800
Zagreb	700	....	....
Total	**281,781**	**20,750**	**42,145**

ZAIRE

Sozir–Muanda	**17,000**	**....**	**[2]3,500**

ZAMBIA

Indeni Petroleum Refinery CL–Bwana Nkubwa Area, Ndola	**23,750**	**....**	**[2]5,320**

This HF alkylation unit horizontal reactor/settler is part of a $50 million expansion commissioned at Navajo Refining Co.'s Artesia, N.M., refinery.

UNITED STATES

Survey of operating refineries in the U.S. (state capacities as of January 1, 1994)

		b/cd		
Country	No. plants	Crude Capacity	Catalytic cracking	Catalytic reforming
Alabama	3	125,450	......	27,200
Alaska	6	259,800	......	12,000
Arizona	1	5,710	......	.....
Arkansas	3	60,900	18,500	12,000
California	25	1,945,545	610,800	476,850
Colorado	3	95,500	27,000	22,500
Delaware	1	140,000	63,000	48,600
Georgia	2	35,500	......	......
Hawaii	2	144,800	20,000	12,400
Illinois	7	920,850	344,900	269,820
Indiana	4	478,300	176,200	102,300
Kansas	4	268,900	90,900	74,500
Kentucky	2	218,900	97,000	43,195
Louisiana	18	2,305,876	845,615	462,565
Michigan	4	125,200	45,500	27,900
Minnesota	2	285,600	71,810	51,595
Mississippi	5	366,800	77,400	76,220
Montana	4	139,950	54,900	34,500

		b/cd		
Country	No. plants	Crude Capacity	Catalytic cracking	Catalytic reforming
Nevada	1	7,000	......	......
New Jersey	6	583,000	261,300	75,560
New Mexico	3	95,800	34,700	28,300
New York	1	41,850	......	......
North Dakota	1	58,000	24,700	11,500
Ohio	4	467,000	160,250	151,440
Oklahoma	7	409,000	139,200	95,900
Oregon	1	.....	......	......
Pennsylvania	7	754,400	230,500	192,988
Tennessee	1	75,000	38,000	14,000
Texas	29	3,794,950	1,537,650	1,128,950
Utah	6	154,500	47,000	29,700
Virginia	1	53,000	26,100	9,700
Washington	7	546,775	117,800	123,200
West Virginia	2	15,500	......	4,650
Wisconsin	1	33,200	10,400	7,600
Wyoming	4	129,000	50,350	28,975
Total	**178**	**15,141,556**	**5,221,475**	**3,656,608**

World 1992: Oil production, refining capacity, oil movements, reserves

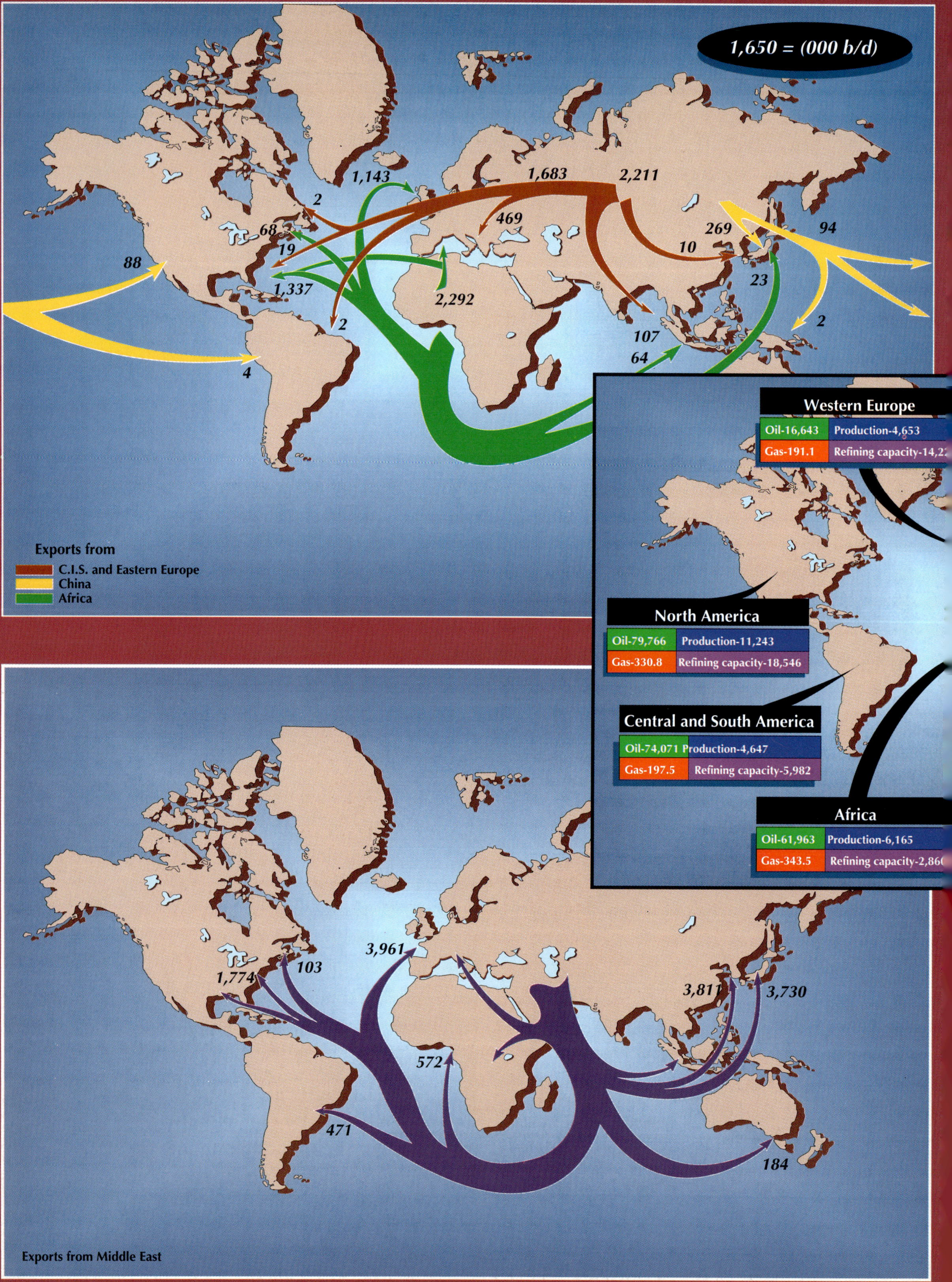

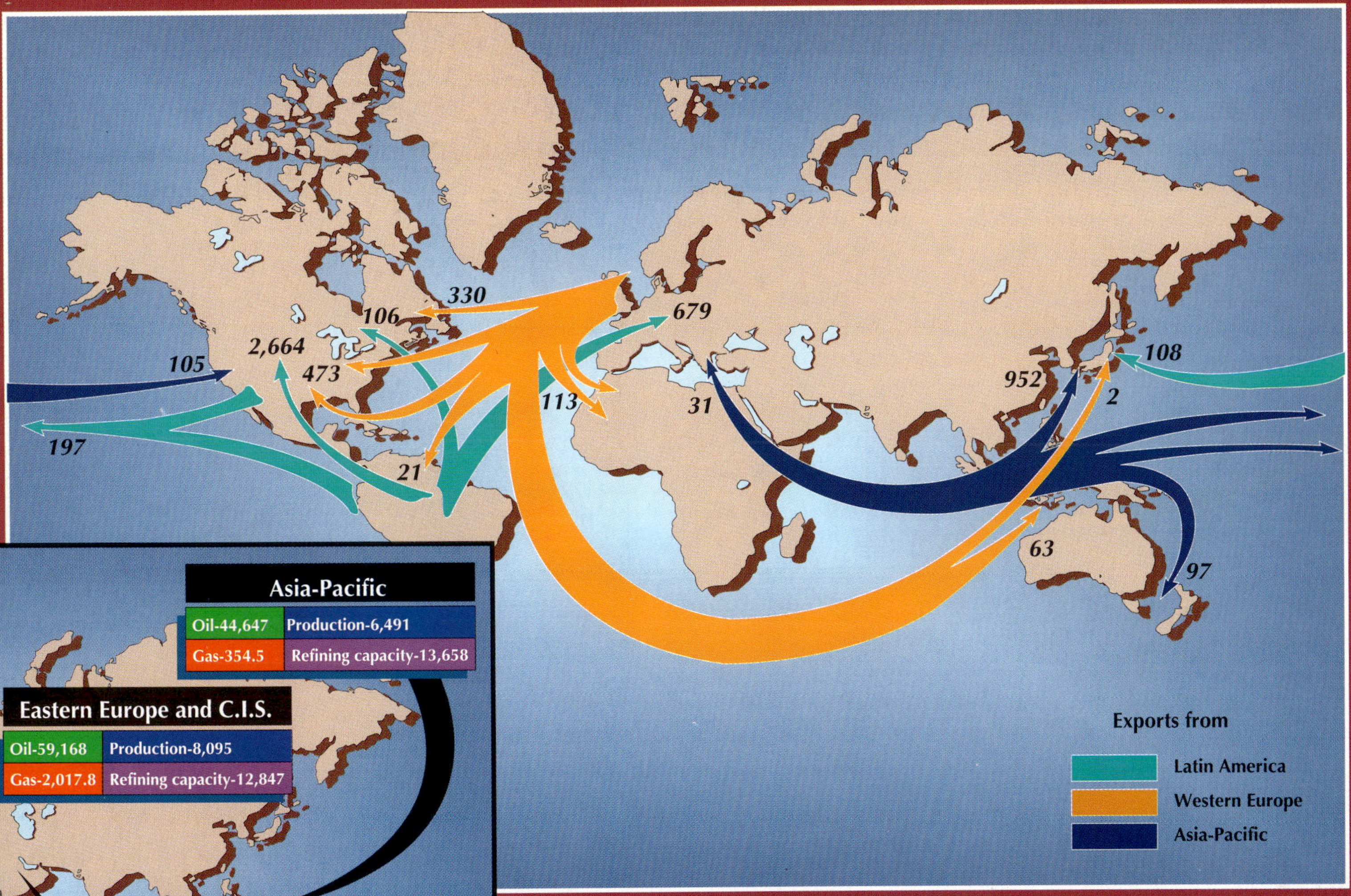
330
106
679
2,664
473
105
108
952
2
113
31
197
21
63
97
Exports from
Latin America
Western Europe
Asia-Pacific

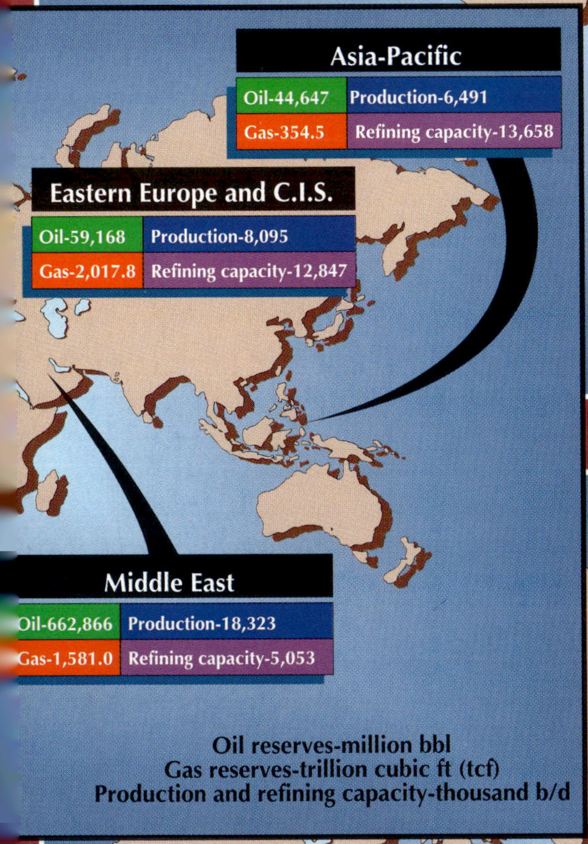
Asia-Pacific
Oil-44,647
Production-6,491
Gas-354.5
Refining capacity-13,658
Eastern Europe and C.I.S.
Oil-59,168
Production-8,095
Gas-2,017.8
Refining capacity-12,847
Middle East
Oil-662,866
Production-18,323
Gas-1,581.0
Refining capacity-5,053
Oil reserves-million bbl
Gas reserves-trillion cubic ft (tcf)
Production and refining capacity-thousand b/d

1,050
71
15
16
8
237
16
8
104
298
298
8
171
135
23
Exports from
U.S.
Canada

Refinery throughput

	1983	1984	1985	1986	1987	1988	1989	1990	1991	1992
	(Thousand barrels per calendar day)									
U.S.	11,685	12,045	12,000	12,715	12,855	13,245	13,400	13,410	13,300	13,410
Canada	1,455	1,460	1,430	1,395	1,460	1,535	1,555	1,590	1,490	1,450
Latin America	5,955	5,845	5,585	5,615	5,765	6,000	6,165	6,290	6,175	6,260
OECD Europe	10,165	10,255	9,950	10,510	10,380	10,725	11,600	11,895	12,525	12,895
Non-OECD Europe	12,255	12,005	11,725	11,990	12,030	12,070	11,950	11,235	10,065	8,460
Middle East	3,080	3,320	3,665	3,765	3,935	4,265	4,350	4,340	3,800	4,150
Africa	1,700	1,795	1,850	1,920	2,000	2,100	2,185	2,355	2,360	2,375
China	1,600	1,765	1,785	1,865	1,950	2,035	2,110	2,155	2,280	2,430
Japan	3,255	3,355	3,120	2,990	2,910	2,990	3,175	3,435	3,655	3,880
South Asia	880	885	1,020	1,120	1,150	1,150	1,230	1,235	1,215	1,280
Other Asia	2,410	2,485	2,685	2,800	2,785	3,155	3,370	3,580	4,100	4,535
Australasia	605	615	580	580	645	680	700	715	745	745
Total world	**55,045**	**55,830**	**55,395**	**57,265**	**57,865**	**59,950**	**61,790**	**62,235**	**61,710**	**61,870**
Of which OECD*	27,165	27,730	27,080	28,190	28,250	29,175	30,430	31,045	31,715	32,380
Of which LDCs†	15,625	16,095	16,590	17,085	17,585	18,705	19,410	19,955	19,930	21,030

*Organisation for Economic Cooperation and Development.
†Lesser developed countries.
Source: British Petroleum

Historical reserve data (billion bbl oil)*

	1960	1970	1975	1980	1985	1988	1989	1990	1991	1992	1993	1994
Asia-Pacific												
Australia	—	2.0	1.7	2.4	1.4	1.7	1.7	1.7	1.6	1.5	1.8	1.6
Brunei	—	1.0	2.0	1.7	1.4	1.4	1.4	1.4	1.4	1.4	1.4	1.4
India	—	—	—	2.5	3.5	4.3	6.4	7.5	7.9	6.1	6.0	5.9
Indonesia	9.5	10.0	14.0	9.5	8.6	8.4	8.2	8.2	11.0	6.6	5.8	5.8
Malaysia	—	—	2.5	3.0	3.0	2.9	2.9	3.0	2.9	3.0	3.7	4.3
Europe												
Norway	—	1.0	7.0	5.5	8.3	14.8	10.4	11.5	7.6	7.6	8.8	9.3
United Kingdom	—	1.0	16.0	14.8	13.6	5.2	5.2	4.2	3.8	3.9	4.1	4.6
Middle East												
Abu Dhabi	—	11.8	29.5	29.0	30.5	92.2	92.2	92.2	92.2	92.2	92.2	92.2
Divided Zone	6.0	25.7	6.4	6.0	5.4	5.2	5.2	5.2	5.0	5.0	5.0	5.0
Dubai	—	—	1.0	1.4	1.4	4.0	4.0	4.0	4.0	4.0	4.0	4.0
Iran	35.0	70.0	64.5	57.5	48.5	92.8	92.8	92.8	92.9	92.9	92.9	92.9
Iraq	27.0	32.0	34.3	30.0	44.5	100.0	100.0	100.0	100.0	100.0	100.0	100.0
Kuwait	62.0	67.1	68.0	64.9	90.0	91.9	91.9	94.5	94.5	94.0	94.0	94.0
Oman	—	—	5.0	2.3	3.5	4.0	4.0	4.2	4.3	4.3	4.5	4.7
Qatar	2.5	3.5	5.8	3.6	3.4	3.2	3.1	4.5	4.5	3.7	3.7	3.7
Saudi Arabia	50.0	128.5	148.6	165.0	169.0	166.9	169.9	254.9	257.5	258.7	257.8	258.7
Syria	1.0	1.0	1.0	1.3	1.4	1.7	1.7	1.7	1.7	1.7	1.7	1.7
Yemen	—	—	—	—	—	—	—	3.0	4.0	4.0	4.0	4.0
Africa												
Algeria	5.2	7.0	7.4	8.2	9.0	8.5	8.4	9.2	9.2	9.2	9.2	9.2
Angola-Cabinda	—	—	1.3	1.2	1.8	1.1	2.0	2.0	2.1	1.8	1.5	1.5
Egypt	—	4.5	3.9	2.9	3.2	4.3	4.3	4.5	4.5	4.5	6.2	6.3
Gabon	—	—	—	—	—	0.6	0.7	0.7	0.7	0.7	0.7	0.7
Libya	2.0	29.2	26.0	23.0	21.1	21.0	22.0	22.8	22.8	22.8	22.8	22.8
Nigeria	—	9.3	20.2	16.7	16.7	16.0	16.0	16.0	17.1	17.9	17.9	17.9
Tunisia	—	—	1.0	1.7	1.5	1.8	1.8	1.8	1.7	1.7	1.7	1.7
Western Hemisphere												
Argentina	2.2	4.5	2.5	2.5	2.3	2.3	2.3	2.3	2.3	1.6	1.6	1.6
Brazil	—	—	—	1.3	1.9	2.3	2.5	2.8	2.8	2.8	3.0	3.6
Canada	—	10.7	7.1	6.4	7.1	6.8	6.8	6.1	5.8	5.6	5.3	5.1
Colombia	—	1.7	1.7	1.0	0.624	1.6	2.0	2.0	2.0	1.9	1.9	1.9
Ecuador	—	—	2.5	1.1	1.4	1.6	1.4	1.5	1.4	1.6	1.6	2.0
Mexico	2.3	3.2	9.5	44.0	48.6	48.6	54.1	56.4	51.9	51.3	51.3	50.9
Venezuela	18.5	14.0	17.7	17.9	25.8	56.3	58.0	58.5	59.0	59.1	62.7	63.3
U.S.	33.5	37.0	33.0	26.4	27.3	25.3	26.5	25.9	26.2	26.3	24.7	23.7
Subtotal	**268.5**	**511.4**	**555.7**	**562.2**	**614.6**	**808.1**	**823.6**	**917.5**	**916.3**	**932.3**	**937.9**	**939.8**
U.S.S.R.†	33.5	77.0	80.4	63.0	63.0	59.0	58.5	24.0	57.0	57.0	57.0	57.0
China	—	—	—	—	19.1	18.4	23.6	24.0	24.0	24.0	24.0	24.0
Other	—	—	—	—	—	1.8	1.7	1.7	1.8	1.8	2.2	2.2
Total world	**302.0**	**611.4**	**658.7**	**648.5**	**698.7**	**887.3**	**907.4**	**1,001.6**	**999.1**	**991.0**	**997.0**	**999.1**

*Columns do not add to totals because countries with lesser reserves are omitted. †C.I.S. after 1991.

Libertad 13, an old well in northern Cebu, Philippines, flowed gas on retest in 1993 at rates calculated at more than 2 MMcfd through a 2 in. choke. Cophil Inc., a Manila venture of Coplex Resources NL and E.F. Durkee & Associates Inc., planned to drill in the area in 1994. Photo by E.F. Durkee.

World gas production (bcf)

Country	1993*
Argentina	610.3
Barbados	0.7
Bolivia	108.3
Brazil	117.6
Canada	5,474.3
Chile	40.7
Colombia	143.6
Ecuador	3.6
Guatemala	0.3
Mexico	1,307.6
Peru	46.0
Trinidad	180.5
United States	19,209.0
Venezuela	786.7
Western Hemisphere	**28,029.2**
Austria	51.2
Denmark	147.9
France	125.4
Germany	624.5
Greece	2.6
Ireland	88.3
Italy	677.8
Netherlands	3,019.9
Norway	872.5
Spain	24.8
Turkey	7.1
United Kingdom	2,226.6
Western Europe	**7,868.6**
Albania	4.8
Bulgaria	0.4
C.I.S.	26,680.7
Croatia	67.1
Czechoslovakia	24.7
Hungary	191.1
Poland	180.1
Romania	778.4
Serbia	31.7
Eastern Europe & C.I.S.	**27,959.0**

Country	1993*
Algeria	1,786.6
Angola	19.8
Egypt	306.9
Gabon	3.6
Libya	224.3
Morocco	0.9
Nigeria	166.0
Tunisia	8.3
Africa	**2,516.4**
Bahrain	184.4
Iran	1,027.0
Iraq	66.8
Israel	0.8
Jordan	7.3
Kuwait	158.8
Oman	125.3
Qatar	284.9
Saudi Arabia	1,134.7
Syria	130.8
United Arab Emirates	859.4
Middle East	**3,980.2**
Afghanistan	10.4
Australia	849.4
Bangladesh	213.5
Brunei	299.6
China	586.3
India	577.3
Indonesia	1,881.2
Japan	76.0
Malaysia	732.4
Myanmar (Burma)	29.4
New Zealand	168.2
Pakistan	521.3
Papua New Guinea	2.9
Taiwan-China	25.6
Thailand	313.9
Viet Nam	24.6
Asia-Pacific	**6,312.0**
TOTAL WORLD	**76,665.4**

*Estimated
Source: Oil & Gas Journal

Crude oil prices

	$/bbl												
	U.S. refiner acquisition cost			U.S. landed cost of imports							Average export prices		
					Selected countries								
Year	Domestic	Imported	Composite	Total	Canada	Mexico	United Kingdom	Nigeria	Saudi Arabia	Venezuela	Worldwide	OPEC	Non-OPEC
1976	8.84	13.48	10.89	13.32	13.36	12.64	NA	13.81	13.06	11.89	NA	NA	NA
1977	9.55	14.53	11.96	14.36	14.13	13.82	14.83	15.29	13.69	13.11	NA	NA	NA
1978	10.61	14.57	12.46	14.35	14.41	13.56	14.53	14.88	13.94	12.84	NA	NA	NA
1979	14.27	21.67	17.72	21.45	20.22	20.77	22.97	22.97	18.95	17.65	NA	NA	NA
1980	24.29	33.89	28.07	33.67	30.11	31.77	35.68	37.15	29.80	25.92	NA	NA	NA
1981	34.93	37.05	35.24	36.47	32.32	33.70	37.29	39.66	34.20	29.91	34.81	34.42	36.16
1982	31.22	33.55	31.87	33.18	27.15	28.63	34.25	36.16	34.99	24.93	33.33	33.71	32.30
1983	28.87	29.30	28.99	28.93	25.63	25.78	30.87	30.85	29.27	22.94	29.66	29.89	29.13
1984	28.53	28.88	28.63	28.54	26.56	26.85	29.45	30.36	29.20	25.19	28.60	28.61	28.60
1985	26.66	26.99	25.75	26.67	25.71	25.63	28.36	28.96	24.72	24.43	27.60	28.00	26.98
1986	14.82	14.00	14.55	13.49	13.43	12.17	14.63	15.29	12.84	11.52	14.98	15.45	14.21
1987	17.76	18.13	17.90	17.65	17.04	16.69	18.78	19.32	16.81	15.76	17.41	17.31	17.57
1988	14.74	14.56	14.67	14.08	13.50	12.58	15.82	15.88	13.37	13.66	13.79	13.67	14.04
1989	17.87	18.08	17.97	17.68	16.81	16.35	18.74	19.19	17.34	16.78	16.65	16.40	17.11
1990	22.59	21.76	22.22	21.13	20.48	19.64	22.65	23.33	21.82	20.31	21.35	20.98	22.07
1991	19.33	18.70	19.06	18.02	17.16	15.89	21.37	21.39	17.22	15.92	17.82	17.36	18.52
1992	18.63	18.20	18.43	17.75	17.04	15.60	20.63	20.78	17.48	15.13	17.95	17.65	18.51

From: Oil & Gas Journal Energy Database
Source: U.S. Department of Energy

Petroleum product prices

	(¢/gallon)												
	Motor gasoline-pump price					Refiner resale (wholesale) product prices							
Year	Leaded Regular	Unleaded Regular	Unleaded Premium	Average All-Types	Heating Oil Residential	Motor Gasoline	Aviation Gasoline	Kerosene Jet Fuel	Kerosene	No. 2 Fuel Oil	No. 2 Diesel Oil	Propane	Resid
1978	62.6	67.0	NA	65.2	49.0	43.4	53.7	38.6	40.4	36.9	36.5	23.7	26.3
1979	85.7	90.3	NA	88.2	70.4	63.7	72.1	66.0	62.4	56.9	57.4	29.1	39.9
1980	119.1	124.5	NA	122.1	97.4	94.1	112.8	86.8	86.4	80.3	80.1	41.5	52.8
1981	131.1	137.8	147.0	135.3	119.4	106.4	125.0	101.2	106.6	97.6	97.2	46.6	66.3
1982	122.2	129.6	141.5	128.1	116.0	97.3	122.8	95.3	101.8	91.4	91.4	42.7	61.2
1983	115.7	124.1	138.3	122.5	107.8	88.2	117.8	85.4	89.2	81.5	80.8	48.4	60.9
1984	112.9	121.2	136.6	119.8	109.1	83.2	116.5	83.0	91.6	82.1	80.3	45.0	65.4
1985	111.5	120.2	134.0	119.6	105.3	83.5	113.0	79.4	87.4	77.6	77.2	39.8	57.7
1986	85.7	92.7	108.5	93.1	83.6	53.1	91.2	49.5	60.6	48.6	45.2	29.0	30.5
1987	89.7	94.8	109.3	95.7	80.3	58.9	85.9	53.8	59.2	52.7	53.4	25.2	38.5
1988	89.9	94.6	110.7	96.3	81.3	57.7	85.0	49.5	54.9	47.3	47.3	24.0	30.0
1989	99.8	102.1	119.7	106.0	90.0	65.4	95.0	58.3	66.9	56.5	56.7	24.7	36.0
1990	114.9	116.4	134.9	121.7	106.3	78.6	106.3	77.3	83.9	69.7	69.4	38.6	41.3
1991	112.5	114.0	132.1	119.6	101.9	69.9	100.1	65.0	72.2	62.2	61.5	34.9	31.4
1992	111.2	112.7	131.6	119.0	93.4	67.7	99.1	60.4	63.2	57.9	59.0	32.8	30.7

From: Oil & Gas Journal Energy Database
Source: U.S. Department of Energy

Comparative energy prices

	Natural gas delivered to consumers ($/MCF)					Cost of fossil fuels delivered to utilities (¢/Million BTU)				Electricity retail prices (¢/Kilowatt Hour)			
Year	Residential	Commercial	Industrial	Electric Utilities	Consumer Average	Coal	Heavy oil	Natural gas	All fuels	Residential	Commercial	Industrial	Total
1973	1.29	0.94	0.50	0.38	0.68	40.5	78.5	33.8	47.6	2.54	2.41	1.25	1.96
1974	1.46	1.07	0.67	0.51	0.84	70.9	189.0	48.2	91.4	3.10	3.04	1.69	2.49
1975	1.71	1.35	0.96	0.77	1.12	81.4	200.5	75.2	104.4	3.51	3.45	2.07	2.92
1976	1.98	1.64	1.24	0.98	1.37	84.8	195.2	103.4	111.9	3.73	3.69	2.21	3.09
1977	2.35	2.04	1.50	1.32	1.66	94.7	219.8	129.1	129.7	4.05	4.09	2.50	3.42
1978	2.56	2.23	1.70	1.48	1.85	111.6	212.5	142.2	141.1	4.31	4.36	2.79	3.69
1979	2.98	2.73	1.99	1.81	2.21	122.4	298.8	174.9	163.9	4.64	4.68	3.05	3.99
1980	3.68	3.39	2.56	2.27	2.80	135.1	426.7	219.9	192.8	5.36	5.48	3.69	4.73
1981	4.29	4.00	3.14	2.89	3.39	153.2	533.4	280.5	225.6	6.20	6.29	4.29	5.46
1982	5.17	4.82	3.87	3.48	4.15	164.7	483.0	337.6	224.9	6.86	6.86	4.95	6.13
1983	6.06	5.59	4.18	3.58	4.82	165.6	457.0	347.4	220.6	7.18	7.01	4.97	6.29
1984	6.12	5.55	4.22	3.70	4.85	166.4	481.0	360.3	219.1	7.54	7.33	5.04	6.52
1985	6.12	5.50	3.95	3.55	4.72	164.8	424.4	344.4	209.4	7.79	7.47	5.16	6.71
1986	5.83	5.08	3.23	2.43	4.13	157.9	240.1	235.1	175.0	7.41	7.13	4.90	6.42
1987	5.54	4.77	2.94	2.32	4.05	150.6	297.6	224.0	170.6	7.41	7.01	4.72	6.32
1988	5.47	4.63	2.95	2.33	4.09	146.6	240.5	226.3	164.3	7.49	7.07	4.62	6.31
1989	5.64	4.74	2.96	2.43	4.22	144.5	284.6	235.5	167.5	7.64	7.23	4.69	6.43
1990	5.80	4.83	2.93	2.39	4.19	145.5	331.9	232.1	168.9	7.85	7.33	4.74	6.56
1991	5.82	4.81	2.69	2.18	4.07	144.7	246.5	215.3	160.3	8.10	7.50	4.80	6.80
1992	5.89	4.88	2.84	2.36	4.17	141.2	247.5	232.8	159.0	8.20	7.60	4.80	6.80

From Oil & Gas Journal Energy Database
Source: U.S. Department of Energy

Price history, crude oil, natural gas and motor gasoline

	Actual current prices						Inflation adjusted real prices				
	Crude oil			Natural gas			Crude oil			Natural gas	
Year	U.S. average wellhead $/bbl	West Texas Inter-mediate posted $/bbl	Motor* Gasoline pump price $/gal	U.S. average wellhead $/Mcf	Consumer average $/Mcf	Producer Price index 1982-100	U.S. average wellhead $/bbl	West Texas Inter-mediate posted $/bbl	Motor Gasoline pump price $/gal	U.S. average wellhead $/Mcf	Consumer average $/Mcf
1925	1.68	1.79	0.222	0.094	0.223	17.8	9.438	10.056	1.247	0.528	1.253
1926	1.88	1.90	0.234	0.095	0.229	17.2	10.930	11.047	1.359	0.552	1.311
1927	1.30	1.28	0.211	0.088	0.220	16.5	7.879	7.758	1.278	0.533	1.333
1928	1.17	1.36	0.209	0.089	0.232	16.7	7.006	8.144	1.254	0.533	1.389
1929	1.27	1.45	0.214	0.082	0.216	16.4	7.744	8.841	1.306	0.500	1.317
1930	1.19	0.95	0.200	0.076	0.214	14.9	7.987	6.376	1.339	0.510	1.436
1931	0.65	0.77	0.170	0.070	0.233	12.6	5.159	6.111	1.348	0.556	1.849
1932	0.87	0.69	0.179	0.064	0.247	11.2	7.768	6.161	1.601	0.571	2.205
1933	0.67	1.00	0.178	0.062	0.237	11.4	5.877	8.772	1.563	0.544	2.079
1934	1.00	1.00	0.189	0.060	0.223	12.9	7.752	7.752	1.461	0.465	1.729
1935	0.97	1.00	0.188	0.058	0.224	13.8	7.029	7.246	1.365	0.420	1.623
1936	1.09	1.10	0.195	0.055	0.220	13.9	7.842	7.914	1.399	0.396	1.583
1937	1.18	1.22	0.200	0.051	0.220	14.9	7.919	8.188	1.342	0.342	1.477
1938	1.13	1.02	0.195	0.049	0.218	13.5	8.370	7.556	1.445	0.363	1.615
1939	1.02	1.02	0.188	0.049	0.216	13.3	7.669	7.699	1.410	0.368	1.624
1940	1.02	1.02	0.184	0.045	0.217	13.5	7.556	7.556	1.364	0.333	1.607
1941	1.14	1.17	0.192	0.049	0.221	15.1	7.550	7.748	1.274	0.325	1.464
1942	1.19	1.17	0.204	0.051	0.227	17.0	7.000	6.882	1.202	0.300	1.335
1943	1.20	1.17	0.205	0.052	0.223	17.8	6.742	6.573	1.153	0.292	1.253
1944	1.21	1.17	0.206	0.051	0.215	17.9	6.760	6.536	1.150	0.285	1.201
1945	1.22	1.17	0.205	0.049	0.214	18.2	6.703	6.429	1.126	0.269	1.176
1946	1.41	1.36	0.208	0.053	0.220	20.8	6.779	6.524	0.999	0.255	1.058
1947	1.93	1.84	0.231	0.060	0.232	25.6	7.539	7.188	0.903	0.234	0.906
1948	2.60	2.57	0.259	0.065	0.241	27.7	9.386	9.278	0.934	0.235	0.870
1949	2.54	2.57	0.268	0.063	0.254	26.3	9.658	9.772	1.019	0.240	0.966
1950	2.51	2.57	0.268	0.065	0.266	27.3	9.194	9.414	0.980	0.238	0.974
1951	2.53	2.57	0.272	0.073	0.298	30.4	8.322	8.454	0.893	0.240	0.980
1952	2.53	2.57	0.276	0.078	0.332	29.6	8.547	8.682	0.931	0.264	1.122
1953	2.68	2.72	0.287	0.092	0.355	29.2	9.178	9.298	0.983	0.315	1.216
1954	2.78	2.82	0.290	0.101	0.381	29.3	9.488	9.625	0.991	0.345	1.300
1955	2.77	2.82	0.291	0.104	0.400	29.3	9.454	9.625	0.992	0.355	1.365
1956	2.79	2.82	0.299	0.108	0.415	30.3	9.208	9.307	0.988	0.356	1.370
1957	3.09	3.04	0.310	0.113	0.431	31.2	9.904	9.753	0.992	0.362	1.381
1958	3.01	3.06	0.304	0.119	0.462	31.6	9.525	9.677	0.961	0.377	1.462
1959	2.90	2.98	0.305	0.129	0.477	31.7	9.148	9.385	0.962	0.407	1.505
1960	2.88	2.97	0.311	0.140	0.500	31.7	9.085	9.369	0.982	0.442	1.577
1961	2.89	2.97	0.308	0.151	0.510	31.6	9.146	9.399	0.973	0.478	1.614
1962	2.90	2.97	0.306	0.155	0.514	31.7	9.148	9.369	0.967	0.489	1.621
1963	2.89	2.97	0.304	0.158	0.512	31.6	9.146	9.399	0.963	0.500	1.620
1964	2.88	2.95	0.304	0.154	0.519	31.6	9.114	9.320	0.962	0.487	1.642
1965	2.86	2.92	0.312	0.156	0.522	32.3	8.854	9.040	0.966	0.483	1.616
1966	2.88	2.94	0.321	0.157	0.523	33.3	8.649	8.817	0.964	0.471	1.571
1967	2.92	3.03	0.332	0.160	0.520	33.4	8.743	9.060	0.994	0.479	1.557
1968	2.94	3.07	0.337	0.164	0.504	34.2	8.596	8.977	0.985	0.480	1.474
1969	3.09	3.30	0.348	0.167	0.515	35.6	8.680	9.256	0.978	0.469	1.447
1970	3.18	3.35	0.357	0.171	0.550	36.9	8.618	9.079	0.967	0.463	1.491
1971	3.39	3.56	0.364	0.182	0.590	38.0	8.921	9.368	0.958	0.479	1.553
1972	3.39	3.56	0.361	0.186	0.630	39.8	8.518	8.945	0.907	0.467	1.583
1973	3.89	3.87	0.388	0.216	0.680	45.0	8.644	8.604	0.862	0.480	1.511
1974	6.74	10.37	0.532	0.304	0.840	53.5	12.598	19.383	0.994	0.568	1.570
1975	7.56	11.16	0.567	0.445	1.120	58.4	12.945	19.110	0.971	0.762	1.918
1976	8.14	12.65	0.590	0.580	1.370	61.1	13.322	20.704	0.966	0.949	2.242
1977	8.57	14.30	0.622	0.790	1.660	64.9	13.205	22.034	0.958	1.217	2.558
1978	8.96	14.85	0.626	0.905	1.850	69.9	12.818	21.245	0.896	1.295	2.647
1979	12.51	22.40	0.857	1.178	2.210	78.7	15.896	28.463	1.089	1.497	2.808
1980	21.59	37.37	1.191	1.590	2.800	89.7	24.069	41.661	1.328	1.773	3.122
1981	31.77	36.67	1.311	1.980	3.390	98.0	32.418	37.418	1.338	2.020	3.459
1982	28.52	32.75	1.296	2.460	4.150	100.0	28.520	32.750	1.296	2.460	4.150
1983	26.19	30.25	1.241	2.590	4.820	101.2	25.879	29.891	1.226	2.559	4.763
1984	25.88	29.83	1.212	2.660	4.850	103.6	24.981	28.793	1.170	2.568	4.681
1985	24.09	28.08	1.202	2.510	4.720	103.1	23.366	27.236	1.166	2.435	4.578
1986	12.51	16.44	0.927	1.940	4.130	100.1	12.498	16.421	0.926	1.938	4.046
1987	15.40	18.21	0.948	1.670	4.050	102.8	14.981	17.714	0.922	1.625	3.940
1988	12.58	15.52	0.946	1.690	4.090	106.9	11.768	14.518	0.885	1.581	3.826
1989	15.86	18.29	1.021	1.690	4.220	112.2	14.135	16.301	0.910	1.506	3.761
1990	20.03	23.17	1.164	1.710	4.190	116.3	17.223	19.923	1.001	1.470	3.603
1991	16.54	20.42	1.140	1.640	4.071	116.5	14.197	17.528	0.979	1.408	3.494
1992	15.99	19.67	1.127	1.740	4.173	117.2	13.643	16.783	0.962	1.485	3.561

* Leaded gasoline price prior to 1982, unleaded gasoline price beginning in 1982.
From: Oil & Gas Journal Energy Database
Source: U.S. Department of Energy

International active rig count

						1991			1992			1993		
Region	1986	1987	1988	1989	1990	Land	Off.	Total	Land	Off.	Total	Land	Off.	Total
NORTH AMERICA														
Canada	178	181	196	130	138	120	1	121	96	0	96	183	1	184
United States*	964	936	936	869	1,010	779	81	860	669	52	721	673	82	755
Subtotal	**1,142**	**1,117**	**1,132**	**999**	**1,148**	**899**	**82**	**981**	**765**	**52**	**817**	**856**	**83**	**939**
LATIN AMERICA														
Argentina	47	61	63	55	64	60	0	60	49	0	49	42	0	42
Bolivia	4	6	5	6	6	10	0	10	16	0	16	11	0	11
Brazil	77	60	39	28	23	14	7	21	14	12	26	13	12	25
Chile	6	7	5	5	6	3	3	6	3	1	4	2	2	4
Colombia	17	14	19	15	11	9	0	9	12	0	12	13	0	13
Costa Rica	0	1	0	0	0	0	0	0	0	0	0	0	0	0
Ecuador	3	3	6	5	2	7	0	7	4	0	4	6	0	6
Mexico	163	143	155	103	90	76	33	109	51	29	80	12	24	36
Peru	13	15	11	10	6	2	1	3	3	3	6	3	3	6
Trinidad	11	9	7	6	8	2	4	6	2	1	3	0	2	2
Venezuela	29	18	25	28	32	28	17	45	34	16	50	34	24	58
Other	1	1	0	0	0	0	0	0	0	0	0	0	0	0
Subtotal	**371**	**338**	**335**	**261**	**248**	**211**	**65**	**276**	**188**	**62**	**250**	**136**	**67**	**203**
FAR EAST														
Australia	15	16	19	14	14	7	5	12	6	4	10	6	5	11
Bangladesh	5	5	2	5	4	0	0	0	0	0	0	0	0	0
Brunei	5	3	4	4	5	1	3	4	0	2	2	0	1	1
China-offshore	7	4	8	4	3	0	2	2	0	3	3	15	6	21
India	70	116	131	135	137	114	22	136	106	18	124	84	21	105
Indonesia	62	37	44	46	52	39	14	53	37	13	50	27	13	40
Japan	15	10	11	6	7	5	1	6	6	1	7	6	2	8
Malaysia	8	8	9	12	13	0	13	13	0	10	10	4	6	10
Myanmar (Burma)	33	29	26	23	19	22	0	22	18	0	18	14	1	15
New Zealand	4	3	2	2	1	1	0	1	1	0	1	0	2	2
Papua New Guinea	1	2	3	3	3	2	0	2	2	0	2	1	0	1
Philippines	1	2	4	7	6	4	1	5	3	4	7	8	1	9
Taiwan	7	4	4	5	3	2	1	3	3	1	4	2	1	3
Thailand	4	6	6	6	5	1	3	4	2	3	5	1	4	5
Other	0	0	1	1	1	0	0	0	0	0	0	0	0	0
Subtotal	**237**	**245**	**274**	**273**	**273**	**198**	**65**	**263**	**184**	**59**	**243**	**168**	**63**	**231**
AFRICA														
Algeria	41	40	32	24	35	31	0	31	30	0	30	29	0	29
Angola	7	10	10	8	8	0	6	6	0	4	4	0	4	4
Congo	2	3	2	1	2	1	2	3	1	3	4	0	2	2
Gabon	3	3	5	6	5	3	2	5	2	1	3	2	1	3
Kenya	0	1	1	3	1	0	0	0	1	0	1	0	0	0
Libya	20	12	16	17	15	14	1	15	15	0	15	15	0	15
Nigeria	10	11	13	14	18	8	13	21	7	9	16	2	8	10
South Africa	3	3	4	3	3	1	1	2	0	1	1	0	2	2
Tunisia	3	4	3	3	3	4	2	6	2	1	3	1	2	3
Other	7	4	5	4	2	0	0	0	1	0	1	1	1	2
Subtotal	**96**	**91**	**91**	**83**	**92**	**62**	**27**	**89**	**59**	**19**	**78**	**50**	**20**	**70**
EUROPE														
Austria	6	4	4	2	1	2	0	2	2	0	2	1	0	1
Denmark	3	3	2	2	2	0	2	2	0	3	3	1	2	3
France	15	8	8	7	6	4	0	4	3	0	3	3	0	3
Germany	19	12	9	8	9	7	0	7	6	0	6	7	0	7
Greece	0	2	2	1	0	0	0	0	0	0	0	0	0	0
Hungary	N/A	N/A	N/A	N/A	N/A	4	0	4	2	0	2	2	0	2
Italy	33	26	24	22	21	19	6	25	17	6	23	11	3	14
Netherlands	15	10	14	12	14	3	8	11	3	7	10	3	4	7

International active rig count—continued

Region	1986	1987	1988	1989	1990	1991			1992			1993		
						Land	Off.	Total	Land	Off.	Total	Land	Off.	Total
Norway	12	12	15	12	12	0	16	16	0	15	15	0	13	13
Poland	N/A	N/A	N/A	N/A	N/A	22	0	22	18	0	18	18	0	18
Romania	N/A	N/A	N/A	N/A	N/A	N/A	N/A	N/A	1	0	1	4	0	4
Spain	7	2	2	3	1	0	1	1	0	0	0	0	0	0
Turkey	27	26	21	20	17	19	0	19	15	0	15	16	0	16
United Kingdom	43	44	57	46	48	1	38	39	0	33	33	0	24	24
Yugoslavia	30	31	31	32	31	15	0	15	0	1	1	0	0	0
Other	1	1	1	0	0	1	0	1	1	0	1	1	0	1
Subtotal	**211**	**181**	**190**	**167**	**162**	**97**	**71**	**168**	**68**	**65**	**133**	**67**	**46**	**113**
MIDDLE EAST														
Abu Dhabi	15	9	7	6	9	6	7	13	8	7	15	7	8	15
Dubai	3	2	3	3	3	1	2	3	1	2	3	1	2	3
Egypt	33	23	21	18	15	9	8	17	7	9	16	5	8	13
Iran	18	18	18	20	19	20	1	21	36	3	39	42	3	45
Iraq†	21	10	23	26	23	N/A	N/A	N/A	0	0	0	0	0	0
Jordan	2	2	2	3	2	2	0	2	2	0	2	0	0	0
Kuwait†	6	6	6	4	2	1	0	1	4	0	4	9	0	9
Oman	13	10	9	13	17	14	0	14	12	0	12	16	0	16
Pakistan	17	14	13	13	11	10	0	10	11	0	11	9	0	9
Qatar	1	1	3	3	2	2	1	3	2	2	4	2	2	4
Saudi Arabia	6	5	4	5	10	15	1	16	16	6	22	15	6	21
Syria	27	22	24	24	17	14	0	14	17	0	17	15	0	15
Yemen	4	4	3	2	3	5	0	5	8	0	8	6	0	6
Other	3	1	0	1	1	0	0	0	2	0	2	1	0	1
Subtotal	**169**	**127**	**136**	**141**	**134**	**99**	**20**	**119**	**126**	**29**	**155**	**128**	**29**	**157**
Total	**2,226**	**2,099**	**2,158**	**1,924**	**2,057**	**1,566**	**330**	**1,896**	**1,390**	**286**	**1,676**	**1,405**	**308**	**1,713**

*U.S. land includes inland water rigs: 17 in 1991, 14 in 1992, 14 in 1993. †1990 data for Iraq and Kuwait is average rigs January through August.

Source: Baker Hughes Inc.

International seismic crew count

	1988			1989			1990		
	Land	Marine	Total	Land	Marine	Total	Land	Marine	Total
	(Average crews operating)								
United States	147	29	176	109	22	131	102	23	125
Canada	69	1	70	57	1	58	51	1	52
Mexico	21	0	21	17	0	17	18	0	18
Central & South America	53	4	57	53	5	58	48	5	53
Europe	49	26	75	42	25	67	38	25	63
Middle East	42	2	44	41	2	43	37	2	39
Africa	36	6	42	46	8	54	50	7	57
Far East	75	11	86	58	14	72	60	16	76
Total	**492**	**79**	**571**	**423**	**77**	**500**	**404**	**79**	**483**

	1991			1992			1993*		
	Land	Marine	Total	Land	Marine	Total	Land	Marine	Total
	(Average crews operating)								
United States	81	20	101	64	12	76	64	16	80
Canada	37	1	38	23	0	23	19	0	19
Mexico	22	1	23	21	1	22	18	0	18
Central & South America	53	4	57	40	6	46	30	5	35
Europe	37	38	75	28	42	70	24	35	59
Middle East	35	1	36	35	3	38	32	3	35
Africa	41	10	51	40	11	51	42	10	52
Far East	53	16	69	39	18	57	24	14	38
C.I.S.	1	0	1	1	0	1	5	0	5
Total	**360**	**91**	**451**	**291**	**93**	**384**	**258**	**83**	**341**

*First half.
Source: Society of Exploration Geophysicists

Worldwide production statistics

NORTH AMERICA

Upper figures = 1,000 b/d
Lower figures = million t/y

	1940	1950	1960	1970	1980	1985	1986	1987	1988	1989	1990	1991	1992	1993
Canada	23.5	79.6	525.6	1,263.6	1,412.0	1,453.0	1,475.3	1,508.3	1,604.6	1,588.0	1,508.0	1,548.4	1,603.7	1,684.1
	1.2	4.0	26.2	62.9	70.3	72.4	73.5	75.1	79.9	79.1	75.1	77.1	79.9	83.9
U.S.	3,707.4	5,407.1	7,054.6	9,630.0	8,569.0	8,919.0	8,790.0	8,276.7	8,165.9	7,675.6	7,220.0	7,416.5	7,171.1	6,895.9
	184.6	269.3	351.3	479.6	426.7	444.2	437.7	412.2	406.7	382.2	359.6	369.3	357.1	343.4
Total	**3,730.9**	**5,486.7**	**7,580.2**	**10,893.6**	**9,981.0**	**10,372.0**	**10,265.3**	**9,785.0**	**9,770.5**	**9,263.6**	**8,728.0**	**8,964.9**	**8,774.8**	**8,580.0**
	185.8	**273.3**	**377.5**	**542.5**	**497.0**	**516.6**	**511.2**	**487.3**	**486.6**	**461.3**	**434.7**	**446.4**	**437.0**	**427.3**

MIDDLE EAST

Upper figures = 1,000 b/d
Lower figures = million t/y

	1940	1950	1960	1970	1980	1985	1986	1987	1988	1989	1990	1991	1992	1993
Abu Dhabi	...	...	...	693.8	1,350.0	705.1	949.0	969.4	1,012.6	1,359.0	1,587.0	1,947.0	1,842.0	1,784.4
	...	...	...	34.6	67.2	35.1	47.3	48.3	50.4	67.7	79.0	97.0	91.7	88.9
Bahrain	19.3	30.2	45.1	76.6	48.0	41.9	44.0	43.2	42.3	42.2	42.0	38.1	36.7	36.0
	1.0	1.5	2.2	3.8	2.4	2.1	2.2	2.2	2.1	2.1	2.1	1.9	1.8	1.8
Dubai	...	...	...	...	349.0	348.0	350.0	381.2	355.3	410.6	469.0	434.0	402.0	363.8
	...	...	...	...	17.4	17.3	17.4	19.0	17.7	20.4	23.4	21.6	20.0	18.1
Iran	181.2	664.3	1,067.6	3,328.8	1,467.0	2,279.0	1,806.3	2,341.7	2,207.5	2,934.3	3,120.0	3,358.0	3,455.0	3,640.0
	9.0	33.1	53.2	165.8	73.1	113.5	90.0	116.6	109.9	146.1	155.4	167.2	172.1	181.3
Iraq	60.7	136.2	1,004.2	1,548.6	2,638.0	1,396.8	1,787.7	2,095.8	2,679.2	2,830.0	2,083.0	283.0	425.0	435.8
	3.0	6.8	50.0	77.1	131.4	69.6	89.0	104.4	133.4	140.9	103.7	14.1	21.2	21.7
Israel	...	...	2.5	70.2	1.0	0.1	0.1	1.0	0.3	0.3	0.3	0.2	0.2	0.2
	...	...	0.1	3.5	...		...	...	...	...	...	...	...	...
Jordan	...	...	...	...	...	...	...	4.0	0.5	0.3	0.4	0.1	0.1	0.1
	...	...	...	...	...	...	...	0.2	...	...	...	...	...	...
Kuwait	...	344.4	1,628.2	2,734.5	1,382.0	823.3	1,202.0	1,095.8	1,254.2	1,542.6	1,080.0	136.0	880.0	1,692.1
	...	17.2	81.1	136.2	68.8	41.0	59.9	54.6	62.5	76.8	53.8	6.8	43.8	84.3
Neutral Zone	...	...	136.2	500.6	540.0	357.8	325.7	399.2	316.3	396.6	315.0	129.0	341.0	361.7
	...	...	6.8	24.9	26.9	17.8	16.2	19.9	15.8	19.8	15.7	6.4	17.0	18.0
Oman	...	...	...	332.4	283.0	484.3	540.8	565.0	596.7	622.7	658.0	700.7	742.0	774.8
	...	...	...	16.6	14.1	24.1	26.9	28.1	29.7	31.0	32.8	34.9	37.0	38.6
Qatar	...	33.6	175.1	362.4	472.0	297.0	332.3	284.4	349.2	394.6	387.0	386.0	425.0	427.5
	...	1.7	8.7	18.0	23.5	14.8	16.5	14.2	17.4	19.7	19.3	19.2	21.2	21.3
Ras Al Khaimah	...	...	...	...	...	9.1	11.0	10.0	10.0	10.0	10.0	0.8	1.0	1.0
	...	...	...	...	...	0.5	0.5	0.5	0.5	0.5	0.5	...	...	...
Saudi Arabia	13.9	546.7	1,247.1	3,548.9	9,630.0	3,295.3	4,719.7	4,054.2	4,708.3	4,935.6	6,215.0	8,158.0	8,137.0	7,981.3
	0.7	27.2	62.1	176.7	479.6	164.1	235.0	201.9	234.5	245.8	309.5	406.3	405.2	397.5
Sharjah	...	...	...	...	10.0	66.1	65.0	66.9	65.0	65.0	35.0	34.6	40.0	40.0
	...	...	...	...	0.5	3.3	3.2	3.3	3.2	3.2	1.7	1.7	2.0	2.0
Syria	...	...	...	83.1	165.0	161.7	185.0	231.7	273.3	300.6	385.0	478.0	518.0	576.3
	...	...	...	4.1	8.2	8.1	9.2	11.5	13.6	15.0	19.2	23.8	25.8	28.7
Yemen	...	...	...	...	...	...	10.0	10.0	172.4	186.6	179.0	197.8	182.0	208.4
	...	...	...	...	...	...	0.5	0.5	8.6	9.3	8.9	9.9	9.1	10.4
Total	**275.1**	**1,755.4**	**5,306.0**	**13,279.9**	**18,335.0**	**10,265.5**	**12,328.6**	**12,553.5**	**14,043.1**	**16,031.0**	**16,565.7**	**16,281.3**	**17,427.0**	**18,323.4**
	13.7	**87.5**	**264.2**	**661.3**	**913.1**	**511.3**	**613.8**	**625.2**	**699.3**	**798.3**	**825.0**	**810.8**	**867.9**	**912.6**

WESTERN EUROPE

Upper figures = 1,000 b/d
Lower figures = million t/y

	1940	1950	1960	1970	1980	1985	1986	1987	1988	1989	1990	1991	1992	1993
Austria	7.7 0.4	27.9 1.4	45.4 2.3	53.3 2.7	29.0 1.4	21.7 1.1	22.2 1.1	21.6 1.2	23.2 1.2	23.3 1.2	24.0 1.2	24.3 1.2	23.1 1.2	24.5 1.2
Denmark	... ...	... ...	... ...	... ...	6.0 0.3	58.0 2.9	74.7 3.7	94.0 4.7	95.5 4.8	111.0 5.5	118.0 5.9	142.3 7.1	156.9 7.8	162.5 8.1
France	1.4 0.1	2.5 0.1	38.6 1.9	45.8 2.3	26.0 1.3	49.0 2.4	59.0 2.9	63.5 3.2	68.7 3.4	65.3 3.3	62.0 3.1	59.7 3.0	58.0 2.9	54.4 2.7
Germany	20.1 1.0	23.2 1.2	108.3 5.4	148.4 7.4	100.0 5.0	90.5 4.5	90.0 4.5	83.8 4.2	85.9 4.3	81.7 4.1	78.0 3.9	68.8 3.4	65.0 3.2	61.2 3.0
Greece	... ...	... ...	... ...	... ...	... ...	26.4 1.3	26.4 1.3	24.4 1.2	21.8 1.1	17.0 0.8	15.0 0.7	14.0 0.7	13.7 0.7	11.3 0.6
Italy	0.2 ...	0.2 ...	37.5 1.9	26.3 1.3	39.0 1.9	41.7 2.1	49.0 2.4	50.0 2.5	92.9 4.6	94.7 4.7	97.0 4.8	87.6 4.4	83.0 4.1	86.0 4.3
Netherlands	... ...	13.4 0.7	36.0 1.8	35.8 1.8	25.0 1.2	69.0 3.4	79.0 3.9	91.3 4.5	83.7 4.2	87.0 4.3	71.0 3.5	65.0 3.2	57.6 2.9	45.7 2.3
Norway	... ...	... ...	... ...	... ...	528.0 26.3	771.0 38.4	823.3 41.0	973.3 48.5	1,069.1 53.2	1,469.8 73.2	1,574.0 78.4	1,876.1 93.4	2,144.3 106.8	2,247.3 111.9
Spain	... ...	... ...	... ...	3.1 0.2	32.0 1.6	45.3 2.3	37.2 1.9	32.8 1.6	30.4 1.5	18.0 0.9	14.3 0.7	20.3 1.0	20.7 1.0	18.4 0.9
Turkey	... ...	0.3 ...	7.0 0.3	67.9 3.4	44.0 2.2	41.5 2.1	46.8 2.3	52.0 2.6	50.0 2.5	55.5 2.8	70.0 3.5	85.2 4.2	82.0 4.1	75.8 3.8
United Kingdom	0.3 ...	0.9 ...	1.7 0.1	1.7 0.1	1,619.0 80.6	2,519.8 125.5	2,602.0 129.6	2,445.9 121.8	2,376.3 118.3	1,743.0 86.8	1,860.0 92.6	1,823.2 90.8	1,863.8 92.8	1,866.2 92.9
Total	**29.7** **1.5**	**68.4** **3.4**	**274.5** **13.7**	**382.3** **19.2**	**2,448.0** **121.8**	**3,733.9** **186.0**	**3,909.6** **194.6**	**3,932.6** **195.9**	**3,997.5** **199.1**	**3,766.3** **187.6**	**3,983.3** **198.3**	**4,266.5** **212.4**	**4,568.1** **227.5**	**4,653.3** **231.7**

AFRICA

Upper figures = 1,000 b/d
Lower figures = million t/y

	1940	1950	1960	1970	1980	1985	1986	1987	1988	1989	1990	1991	1992	1993
Algeria	... ...	... ...	182.7 9.1	1,029.1 51.2	1,016.0 50.6	643.3 32.0	600.7 29.9	647.9 32.3	666.8 33.2	698.0 34.8	797.0 39.7	803.0 40.0	772.0 38.4	750.0 37.4
Angola (inc. Cabinda)	... ...	... ...	1.3 0.1	13.7 0.7	150.0 7.5	228.5 11.4	280.7 14.0	341.3 17.0	449.3 22.4	454.7 22.6	480.0 23.9	497.0 24.8	531.0 26.4	498.4 24.8
Cameroon	... ...	... ...	... ...	... ...	58.0 2.9	133.5 6.6	180.0 9.0	170.0 8.5	170.0 8.5	172.7 8.6	164.0 8.2	152.0 7.6	137.0 6.8	118.8 5.9
Congo	... ...	... ...	... ...	... ...	56.0 2.8	115.1 5.7	115.0 5.7	118.0 5.9	134.8 6.7	150.0 7.5	161.0 8.0	159.2 7.9	172.0 8.6	200.0 10.0
Egypt	18.1 0.9	44.9 2.2	61.6 3.1	327.3 16.3	596.0 29.7	885.8 44.1	773.7 38.5	899.3 44.8	851.3 42.4	854.3 42.5	873.0 43.5	878.0 43.7	873.0 43.5	894.3 44.5
Gabon	... ...	... ...	16.1 0.8	108.8 5.4	180.0 9.0	150.0 7.5	146.2 7.3	155.8 7.8	175.0 8.7	197.3 9.8	269.0 13.4	294.0 14.6	300.0 14.9	296.7 14.8
Ivory Coast	... ...	... ...	... ...	... ...	... ...	28.0 1.4	20.0 1.0	16.0 0.8	12.9 0.6	2.0 0.1	2.1 0.1	1.3 0.1	1.0 ...	0.5 ...
Libya	... ...	... ...	... ...	3,318.0 165.2	1,785.0 88.9	1,045.3 52.1	1,030.7 51.3	1,019.7 50.8	1,012.5 50.4	1,100.6 54.8	1,369.0 68.2	1,509.0 75.1	1,493.0 74.4	1,367.1 68.1
Nigeria	... ...	... ...	17.3 0.9	1,083.3 53.9	2,057.0 102.4	1,445.5 72.0	1,464.0 72.9	1,238.6 61.7	1,358.3 67.6	1,605.3 79.9	1,808.0 90.0	1,867.0 93.0	1,902.0 94.7	1,895.8 94.4
Tunisia	... ...	... ...	3.0 0.1	87.7 4.4	110.0 5.5	125.0 6.2	106.0 5.3	104.3 5.2	102.8 5.1	102.0 5.1	93.0 4.6	107.6 5.4	106.2 5.3	100.1 5.0
Zaire	... ...	... ...	... ...	... ...	20.0 1.0	30.5 1.5	32.1 1.6	31.8 1.6	29.5 1.5	27.5 1.4	28.0 1.4	29.1 1.4	23.8 1.2	24.0 1.2
Others	... ...	0.8 ...	3.0 0.1	1.5 0.1	4.0 0.2	8.9 0.4	7.6 0.4	12.3 0.6	5.0 0.2	4.8 0.2	4.3 0.2	4.2 0.2	10.4 0.5	19.1 1.0
Total	**18.1** **0.9**	**45.7** **2.2**	**285.0** **14.2**	**5,969.4** **297.2**	**6,032.0** **300.5**	**4,839.4** **240.9**	**4,756.7** **236.9**	**4,755.0** **237.0**	**4,968.2** **247.3**	**5,369.2** **267.3**	**6,048.4** **301.2**	**6,301.4** **313.8**	**6,321.4** **314.7**	**6,164.8** **307.1**

Drilling rig works for Occidental Petroleum Corp. on its Block 15 in the Oriente jungle region of Ecuador, which is opening more of its petroleum sector to private participation. Photo courtesy Oxy.

ASIA-PACIFIC

Upper figures = 1,000 b/d
Lower figures = million t/y

	1940	1950	1960	1970	1980	1985	1986	1987	1988	1989	1990	1991	1992	1993
Australia	...	...	...	175.6	379.0	556.0	472.0	566.1	552.3	488.8	582.0	544.0	533.0	512.8
	...	...	...	8.7	18.9	27.7	23.5	28.2	27.5	24.3	29.0	27.1	26.5	25.5
Bangladesh	...	...	...	...	...	...	...	0.5	0.9	0.9	0.9	0.5	1.1	1.0
	...	...	...	...	...	...	...	...	...	...	...	...	0.1	...
Brunei	19.3	84.8	92.9	148.0	230.0	149.5	170.0	133.7	138.3	143.3	143.0	144.7	165.3	158.8
	1.0	4.2	4.6	7.4	11.5	7.4	8.5	6.7	6.9	7.1	7.1	7.2	8.2	7.9
China	...	4.0	100.0	500.0	2,119.0	2,496.0	2,620.0	2,700.0	2,733.0	2,770.0	2,755.0	2,800.0	2,835.0	2,895.4
	...	0.2	5.0	24.9	105.5	124.3	130.5	134.5	136.1	137.9	137.2	139.4	141.2	144.2
China, Taiwan	...	...	0.4	1.8	5.0	2.7	1.9	2.6	2.6	2.6	2.5	1.9	1.2	1.1
	...	...	...	0.1	0.2	0.1	0.1	0.1	0.1	0.1	0.1	0.1	0.1	0.1
India	3.9	5.1	9.0	138.6	182.0	614.0	622.7	608.7	631.8	673.3	679.0	625.0	547.0	524.3
	0.2	0.3	0.4	6.9	9.1	30.6	31.0	30.3	31.5	33.5	33.8	31.1	27.2	26.1
Indonesia	169.4	132.6	411.2	853.6	1,576.0	1,219.5	1,243.8	1,185.8	1,137.5	1,208.6	1,274.0	1,410.7	1,346.0	1,322.5
	8.4	6.6	2.05	42.5	78.5	60.7	61.9	59.1	56.6	60.2	63.4	70.3	67.0	65.9
Japan	7.2	5.6	10.2	16.3	10.0	10.0	12.8	12.1	12.3	11.3	10.5	15.1	17.2	15.5
	0.4	0.3	0.5	0.8	0.5	0.5	0.6	0.6	0.6	0.6	0.5	0.8	0.9	0.8
Malaysia	...	...	...	...	288.0	432.8	503.3	467.4	540.0	557.3	605.0	652.0	661.0	640.0
	...	...	...	...	14.3	21.6	25.1	23.3	26.9	27.8	30.1	32.5	32.9	31.9
Myanmar (Burma)	21.1	1.5	11.1	16.4	30.0	30.0	30.0	24.0	15.0	12.0	13.0	12.0	13.0	14.0
	1.1	0.1	0.6	0.8	1.5	1.5	1.5	1.2	0.7	0.6	0.6	0.6	0.6	0.7
New Zealand	...	...	...	...	7.0	18.0	29.3	28.3	28.0	28.0	39.0	40.0	36.8	34.9
	...	...	...	...	0.3	0.9	1.5	1.4	1.4	1.4	1.9	2.0	1.8	1.7
Pakistan	2.4	3.5	7.2	9.9	10.0	34.3	41.0	41.6	47.0	45.0	60.0	72.0	75.0	62.5
	0.1	0.2	0.4	0.5	0.5	1.7	2.0	2.1	2.3	2.2	3.0	3.6	3.7	3.1
Papua New Guinea	...	4.8	4.2	...	...	...	...	...	...	...	...	0.2	52.4	124.6
	...	0.2	0.2	...	...	...	...	...	...	...	...	...	2.6	6.2
Philippines	...	...	...	...	15.0	9.3	6.3	5.1	9.2	4.0	5.0	3.0	8.9	10.1
	...	...	...	...	0.7	0.5	0.3	0.3	0.5	0.2	0.2	0.1	0.4	0.5
Thailand	...	...	...	0.4	...	34.3	35.6	31.3	38.4	39.0	41.0	43.4	51.4	51.4
	...	...	...	...	...	1.7	1.8	1.6	1.9	1.9	2.0	2.2	2.6	2.6
Viet Nam*	...	...	...	...	...	...	...	...	...	...	40.0	80.0	105.0	122.5
	...	...	...	...	...	...	...	...	...	...	2.0	4.0	5.2	6.1
Total	**223.3**	**241.9**	**646.2**	**1,860.6**	**4,851.0**	**5,606.4**	**5,788.7**	**5,807.2**	**5,886.3**	**5,984.1**	**6,249.9**	**6,444.5**	**6,449.3**	**6,491.4**
	11.2	**12.1**	**32.2**	**92.6**	**241.5**	**279.2**	**288.3**	**289.4**	**293.0**	**297.8**	**310.9**	**321.0**	**321.0**	**323.3**

*Prior to 1990, included in other Eastern Europe - C.I.S.

LATIN AMERICA

Upper figures = 1,000 b/d
Lower figures = million t/y

	1940	1950	1960	1970	1980	1985	1986	1987	1988	1989	1990	1991	1992	1993
Argentina	56.3 2.8	64.0 3.2	171.9 8.6	382.9 19.1	487.0 24.3	447.5 22.3	430.2 21.4	419.1 20.9	449.6 22.4	459.0 22.9	473.0 23.6	489.2 24.4	554.0 27.6	574.8 28.6
Bolivia	0.8 ...	1.7 0.1	9.8 0.5	16.3 0.8	30.0 1.5	20.0 1.0	18.7 0.9	18.5 0.9	18.8 0.9	20.3 1.0	19.2 1.0	22.2 1.1	21.2 1.1	20.3 1.0
Brazil	... ...	0.9 ...	80.9 4.0	160.5 8.0	182.0 9.1	540.5 26.9	578.0 28.8	562.8 28.0	555.6 27.7	597.1 29.7	633.0 31.5	623.6 31.1	625.7 31.2	631.3 31.4
Chile	... ...	1.7 0.1	19.8 1.0	34.6 1.7	29.0 1.4	32.5 1.6	34.4 1.7	28.3 1.4	24.8 1.2	23.7 1.2	20.3 1.0	17.8 0.9	15.8 0.8	14.3 0.7
Colombia	69.9 3.5	93.3 4.6	152.7 7.6	214.0 10.7	125.0 6.2	178.8 8.9	324.2 16.1	387.8 19.3	346.7 17.3	405.0 20.2	445.0 22.2	429.0 21.4	458.3 22.8	449.8 22.4
Cuba*	... ...	... ...	... ...	... ...	... ...	... ...	... ...	... ...	... ...	... ...	15.0 0.7	15.0 0.7	20.0 1.0	25.0 1.2
Ecuador	6.4 0.3	7.2 0.4	7.7 0.4	4.1 0.2	222.0 11.1	273.3 13.6	270.1 13.5	157.2 7.8	310.1 15.4	286.6 14.3	287.0 14.3	300.2 14.9	321.0 16.0	340.8 17.0
Guatemala	... ...	... ...	... ...	... ...	5.0 0.2	3.1 0.2	4.9 0.2	3.9 0.2	3.7 0.2	3.7 0.2	4.0 0.2	4.1 0.2	5.6 0.3	6.7 0.3
Mexico	120.3 6.0	198.5 9.9	270.6 13.5	430.2 21.4	1,936.0 96.4	2,797.0 139.3	2,468.0 122.9	2,537.8 126.4	2,527.3 125.9	2,618.0 130.4	2,633.0 131.1	2,774.0 138.1	2,667.7 132.9	2,663.4 132.6
Peru	33.2 1.7	41.1 2.0	52.6 2.6	72.2 3.6	191.0 9.5	189.0 9.4	179.0 8.9	166.7 8.3	141.7 7.1	129.0 6.4	132.0 6.6	114.9 5.7	117.0 5.8	126.3 6.3
Trinidad & Tobogo	60.9 3.0	56.5 2.8	115.7 5.8	139.8 7.0	211.0 10.5	179.0 8.9	167.9 8.4	163.1 8.1	149.3 7.4	153.7 7.7	151.0 7.5	144.0 7.2	135.4 6.7	119.5 6.0
Venezuela	507.0 25.2	1,498.0 74.6	2,846.1 141.7	3,708.0 184.7	2,167.0 107.9	1,669.0 83.1	1,664.9 82.9	1,591.9 79.3	1,658.0 82.6	1,731.6 86.2	2,118.0 105.5	2,349.6 117.0	2,314.0 115.2	2,331.7 116.1
Others	... ...	... ...	... ...	... ...	... ...	2.9 0.1	3.6 0.2	3.6 0.2	3.7 0.2	4.9 0.2	5.2 0.3	5.4 0.3	5.8 0.3	6.5 0.3
Total	**854.8 42.5**	**1,962.9 97.7**	**3,727.8 185.7**	**5,162.6 257.2**	**5,585.0 278.1**	**6,332.6 315.3**	**6,143.9 305.9**	**6,040.7 300.8**	**6,189.3 308.3**	**6,432.6 320.4**	**6,935.7 345.5**	**7,289.0 363.0**	**7,261.5 361.7**	**7,310.4 363.9**
World Total, excl. E. Europe-C.I.S.	**5,131.9 255.6**	**9,561.0 476.2**	**17,819.7 887.5**	**37,048.4 1,870.0**	**47,232.0 2,352.0**	**41,149.8 2,049.3**	**43,192.8 2,150.7**	**42,874.0 2,135.6**	**44,854.9 2,233.6**	**46,846.8 2,332.7**	**48,511.0 2,415.6**	**49,547.6 2,467.4**	**50,802.1 2,529.8**	**51,523.3 2,565.9**

*Prior to 1990, included in other Eastern Europe - C.I.S.

EASTERN EUROPE — C.I.S.

Upper figures = 1,000 b/d
Lower figures = million t/y

	1940	1950	1960	1970	1980	1985	1986	1987	1988	1989	1990	1991	1992	1993
Bulgaria	... ...	... ...	4.0 0.2	6.7 0.3	6.0 0.3	3.0 0.1	3.0 0.1	6.0 0.3	5.0 0.2	5.0 0.2	5.0 0.2	4.0 0.2	1.0 ...	1.0 ...
C.I.S.*	624.9 31.1	760.6 37.9	2,969.1 147.9	7,089.2 353.0	12,031.0 599.1	11,900.0 592.6	12,300.0 612.5	12,480.0 621.5	12,452.0 620.1	12,140.0 604.6	11,500.0 572.7	10,296.0 512.7	8,949.0 445.7	7,846.5 390.8
Croatia	... ...	... ...	... ...	... ...	... ...	... ...	... ...	... ...	... ...	... ...	... ...	44.5 2.2	43.0 2.1	40.0 2.0
Czechoslovakia	... ...	0.8 ...	2.7 0.1	4.1 0.2	1.9 0.1	2.0 0.1	2.0 0.1	2.0 0.1	3.0 0.1	3.0 0.1	3.0 0.1	3.0 0.1	2.0 0.1	2.0 0.1
Hungary	... ...	10.1 0.5	24.3 1.2	38.7 1.9	40.6 2.0	40.0 2.0	40.0 2.0	40.0 2.0	40.0 2.0	39.0 1.9	40.0 2.0	38.6 1.9	33.4 1.7	31.7 1.6
Poland	... ...	3.3 0.2	3.9 0.2	8.5 0.4	6.6 0.3	5.0 0.2	4.0 0.2	4.0 0.2	4.0 0.2	3.0 0.1	3.0 0.1	3.3 0.2	3.6 0.2	3.4 0.2
Romania	... ...	87.7 4.4	231.0 11.5	268.6 13.4	230.9 11.5	226.0 11.3	220.0 11.0	215.0 10.7	188.0 9.4	180.0 9.0	160.0 8.0	140.0 7.0	136.0 6.8	139.0 6.9
Yugoslavia	0.3 ...	2.1 0.1	18.8 0.9	57.8 2.9	84.5 4.2	81.0 4.0	83.0 4.1	84.0 4.2	73.0 3.6	70.0 3.5	65.0 3.2	22.2 1.1	22.0 1.1	23.6 1.2
Others	... ...	2.7 0.1	14.9 0.7	37.1 1.8	50.0 2.5	86.0 4.3	77.0 3.8	80.0 4.0	85.0 4.2	86.0 4.3	30.0 1.5	30.0 1.5	11.0 0.5	8.0 0.4
Total	**625.2 31.1**	**867.3 43.2**	**3,268.7 162.7**	**7,510.7 373.9**	**12,451.5 620.0**	**12,343.0 614.6**	**12,729.0 633.8**	**12,911.0 643.0**	**12,850.0 639.8**	**12,526.0 623.7**	**11,806.0 587.8**	**10,581.6 526.9**	**9,201.0 458.2**	**8,095.2 403.2**
*Includes gas liquids														
World Total	**5,757.1 286.7**	**10,428.3 519.4**	**21,088.4 1,050.2**	**45,059.1 2,243.9**	**59,683.5 2,972.0**	**53,492.8 2,663.9**	**55,921.8 2,784.5**	**55,785.0 2,778.6**	**57,704.9 2,873.4**	**59,372.8 2,956.4**	**60,317.0 3,003.4**	**60,129.2 2,994.3**	**60,003.1 2,988.0**	**59,618.5 2,969.1**

Worldwide oil consumption

Upper figures = 1,000 b/d
Lower figures = million t/y

	1990	1991	1992
U.S.	16,305 781.8	16,000 765.6	16,240 781.0
Canada	1,690 77.7	1,630 74.8	1,665 76.4
Total North America	**17,995 859.5**	**17,630 840.4**	**17,905 857.4**
Argentina	390 18.5	410 19.2	420 19.6
Brazil	1,275 58.4	1,290 59.0	1,305 59.7
Mexico	1,510 69.6	1,575 72.1	1,605 73.3
Venezuela	395 18.3	405 18.5	430 19.6
Other	1,420 68.6	1,445 69.9	1,455 69.8
Total Latin America	**4,990 233.4**	**5,125 238.7**	**5,215 242.0**
Austria	225 10.8	240 11.6	235 11.3
Belgium and Luxembourg	510 24.8	545 26.5	555 26.9
Denmark	185 9.0	185 9.1	215 10.5
Finland	230 11.0	220 10.6	215 10.4
France	1,910 89.4	2,020 94.6	2,010 94.4
Germany	2,705 127.2	2,835 133.1	2,850 134.3
Greece	320 15.7	325 15.8	330 16.0
Iceland	15 0.6	15 0.6	15 0.7
Ireland	90 4.4	95 4.6	95 4.6
Italy	1,930 93.6	1,920 92.4	1,945 94.2
Netherlands	765 35.0	765 35.8	785 36.1
Norway	205 9.2	195 8.7	190 8.7
Portugal	230 11.1	240 11.5	265 12.9
Spain	1,040 48.7	1,055 49.4	1,065 50.0
Sweden	340 16.4	330 15.7	345 16.5
Switzerland	275 12.8	275 13.0	280 13.1
Turkey	470 22.1	470 22.1	500 23.5
United Kingdom	1,750 82.3	1,755 82.5	1,745 82.4
Total OECD Europe	**13,195 624.1**	**13,485 637.6**	**13,640 646.5**

	1990	1991	1992
C.I.S.	8,160 406.3	7,845 390.6	6,675 333.4
Czechoslovakia	300 15.4	260 13.2	225 11.5
Hungary	180 8.4	170 8.0	170 8.3
Poland	295 14.1	285 13.5	300 14.1
Other	1,015 49.5	660 32.7	490 24.2
Total Non-OECD Europe	**9,950 493.7**	**9,220 458.0**	**7,860 391.5**
Middle East	**3,400 162.8**	**3,405 162.5**	**3,525 168.4**
Africa	**1,990 95.1**	**2,015 96.3**	**2,025 97.0**
China	2,255 110.3	2,410 117.9	2,615 128.1
India	1,200 57.9	1,220 58.9	1,285 62.0
Indonesia	645 31.0	675 32.4	730 35.1
Japan	4,305 247.7	5,410 252.1	5,545 258.8
Malaysia	270 13.0	290 13.9	295 14.0
Philippines	235 11.4	240 11.7	280 13.7
Singapore	370 19.5	380 19.9	400 21.0
South Korea	1,025 48.5	1,235 58.7	1,500 71.2
Taiwan	540 26.2	570 27.9	590 28.6
Thailand	405 19.3	440 21.1	495 23.7
Other Asia	645 31.4	635 30.7	680 33.1
Total Asia	**11,895 616.2**	**13,505 645.2**	**14,415 689.3**
Australia	700 31.6	685 30.8	685 31.1
New Zealand	105 4.8	110 5.1	115 5.2
Total Australasia	**805 36.4**	**795 35.9**	**800 36.3**
Total World	**64,220 3,121.2**	**65,180 3,114.6**	**65,385 3,128.4**
Of which OECD*	37,300 1,767.7	37,320 1,766.0	37,890 1,799.0
Of which LDCs†	17,970 859.8	18,640 890.6	19,635 937.9

*Organisation for Economic Cooperation and Development
†Lesser developed countries.
Source: British Petroleum

Worldwide refining capacity

NORTH AMERICA

Upper figures = 1,000 b/d
Lower figures = million t/y

	1940	1950	1960	1970	1980	1986	1987	1988	1989	1990	1991	1992	1993	1994
Canada	203	329	920	1,400	2,222	1,856	1,760	1,869	1,856	1,852	1,882	1,905	1,872	1,880
	10.1	16.4	45.8	69.7	110.7	92.4	87.6	93.1	92.4	92.2	93.7	94.9	93.2	93.6
U.S.	4,460	6,696	10,250	12,600	17,720	15,182	15,258	15,288	15,557	16,244	15,559	15,327	15,210	15,142
	222.1	333.5	510.5	627.5	882.5	756.1	759.8	761.3	774.7	809.0	774.8	763.3	757.5	754.1
Total	**4,663**	**7,025**	**11,170**	**14,000**	**19,942**	**17,038**	**17,018**	**17,157**	**17,413**	**18,096**	**17,441**	**17,232**	**17,082**	**17,022**
	232.2	**349.9**	**556.3**	**697.2**	**993.2**	**848.5**	**847.4**	**854.4**	**867.1**	**901.2**	**868.5**	**858.2**	**850.7**	**847.7**

WESTERN EUROPE

Upper figures = 1,000 b/d
Lower figures = million t/y

	1940	1950	1960	1970	1980	1986	1987	1988	1989	1990	1991	1992	1993	1994
Austria	...	26	47	94	280	204	204	204	204	204	204	210	210	210
	...	1.3	2.3	4.7	13.9	10.2	10.2	10.2	10.2	10.2	10.2	10.5	10.5	10.5
Belgium	12	16	171	669	1,064	652	648	631	631	614	602	602	607	607
	0.6	0.8	8.5	33.3	53.0	32.5	32.3	31.4	31.4	30.6	30.0	30.0	30.2	30.2
Cyprus	...	...	...	...	16	16	16	16	17	18	19	19	19	22
	...	...	...	...	0.8	0.8	0.8	0.8	0.8	0.9	0.9	0.9	0.9	1.1
Denmark	1	1	1	186	214	166	166	177	177	187	186	186	184	184
	...	...	...	9.3	10.7	8.3	8.3	8.8	8.8	9.3	9.3	9.3	9.2	9.2
Finland	...	...	25	174	336	241	241	241	241	241	241	241	200	200
	...	...	1.2	8.7	16.7	12.0	12.0	12.0	12.0	12.0	12.0	12.0	10.0	10.0
France	152	306	769	2,317	3,385	1,947	1,834	1,941	1,876	1,820	1,816	1,816	1,851	1,861
	7.6	15.2	38.3	115.4	168.6	97.0	91.3	96.7	93.4	90.6	90.4	90.4	92.2	92.7
Germany	48	81	573	2,359	2,986	1,933	1,720	1,648	1,518	1,507	2,065	2,062	2,233	2,265
	2.4	4.0	28.5	117.5	148.7	96.3	85.7	82.1	75.6	75.0	102.8	102.7	111.2	112.8
Greece	...	...	30	102	431	390	385	385	385	385	395	396	396	396
	...	...	1.5	5.1	21.5	19.4	19.2	19.2	19.2	19.2	19.7	19.7	19.7	19.7
Ireland	11	...	40	55	56	56	56	56	56	56	56	56	56	53
	0.5	...	2.0	2.7	2.8	2.8	2.8	2.8	2.8	2.8	2.8	2.8	2.8	2.6
Italy	41	107	773	2,963	4,131	2,738	2,679	2,563	2,450	2,804	2,385	2,386	2,420	2,262
	2.0	5.3	38.5	147.6	205.7	136.4	133.4	127.6	122.0	139.6	118.8	118.8	120.5	112.6
Netherlands	15	77	359	1,362	1,828	1,468	1,401	1,381	1,381	1,381	1,197	1,219	1,231	1,184
	0.7	3.8	17.9	67.8	91.0	73.1	69.8	68.8	68.8	68.8	59.6	60.7	61.3	59.0
Norway	1	1	2	123	264	240	240	240	239	295	288	288	285	287
	...	...	0.1	6.1	13.1	12.0	12.0	12.0	11.9	14.7	14.3	14.3	14.2	14.3
Portugal	4	8	23	37	378	296	294	294	313	313	294	294	294	294
	0.2	0.4	1.1	1.8	18.8	14.7	14.6	14.6	15.6	15.6	14.6	14.6	14.6	14.6
Spain	11	17	146	685	1,455	1,367	1,305	1,305	1,285	1,293	1,321	1,321	1,301	1,283
	0.5	0.8	7.3	34.1	72.5	68.1	65.0	65.0	64.0	64.4	65.8	65.8	64.8	63.9
Sweden	4	26	48	230	458	429	437	437	427	428	428	435	428	428
	0.2	1.3	2.4	11.5	22.8	21.4	21.8	21.8	21.3	21.3	21.3	21.7	21.3	21.3
Switzerland	...	...	...	106	137	137	65	132	132	132	132	132	132	132
	...	...	...	5.3	6.8	6.8	3.2	6.6	6.6	6.6	6.6	6.6	6.6	6.6
Turkey	...	1	7	149	356	460	460	676	725	725	729	704	714	713
	...	...	0.3	7.4	17.7	22.9	22.9	33.7	36.1	36.1	36.3	35.1	35.6	35.5
United Kingdom	133	197	963	2,301	2,527	1,792	1,780	1,803	1,803	1,831	1,867	1,856	1,843	1,867
	6.6	9.8	48.0	114.6	125.8	89.2	88.6	89.8	89.8	91.2	93.0	92.4	91.8	93.0
Total	**433**	**864**	**3,977**	**13,912**	**20,302**	**14,532**	**13,931**	**14,130**	**13,860**	**14,234**	**14,225**	**14,223**	**14,404**	**14,248**
	21.3	**42.7**	**197.9**	**692.9**	**1,010.9**	**723.9**	**693.9**	**703.9**	**690.3**	**708.9**	**708.4**	**708.3**	**717.4**	**709.6**

LATIN AMERICA

Upper figures = 1,000 b/d
Lower figures = million t/y

	1940	1950	1960	1970	1980	1986	1987	1988	1989	1990	1991	1992	1993	1994
Argentina	103	152	238	457	676	667	670	690	690	689	696	696	709	709
	5.1	7.6	11.9	22.8	33.7	33.2	33.4	34.4	34.4	34.3	34.7	34.7	35.3	35.3
Bahamas	...	...	...	...	500	...	...	...	...	...	...	...	...	...
	...	...	...	...	24.9	...	...	...	...	...	...	...	...	...
Barbados	...	...	...	3	3	3	3	3	3	3	3	3	3	3
	...	...	...	0.1	0.1	0.1	0.1	0.1	0.1	0.1	0.1	0.1	0.1	0.1
Bolivia	1	7	12	12	74	47	47	47	58	58	45	45	45	45
	...	0.3	0.6	0.6	3.7	2.3	2.3	2.3	2.9	2.9	2.2	2.2	2.2	2.2
Brazil	4	12	156	502	1,205	1,305	1,321	1,407	1,407	1,397	1,412	1,406	1,403	1,253
	0.2	0.6	7.8	25.0	60.0	65.0	65.8	70.1	70.1	69.6	70.3	70.0	69.9	62.4
Chile	1	4	24	91	139	141	150	141	147	147	146	143	144	165
	...	0.2	1.2	4.5	6.9	7.0	7.5	7.0	7.3	7.3	7.3	7.1	7.2	8.2
Colombia	15	24	51	137	193	211	226	226	227	227	247	274	264	249
	0.7	1.2	2.5	6.8	9.6	10.5	11.3	11.3	11.3	11.3	12.3	13.6	13.1	12.4
Costa Rica	...	...	...	8	11	15	16	16	16	16	15	15	15	15
	...	...	...	0.4	0.5	0.7	0.8	0.8	0.8	0.8	0.7	0.7	0.7	0.7
Cuba	4	7	87	93	160	160	160	160	160	160	280	280	280	280
	0.2	0.3	4.3	4.6	8.0	8.0	8.0	8.0	8.0	8.0	13.9	13.9	13.9	13.9
Ecuador	2	5	7	33	86	88	88	122	123	145	142	142	147	148
	0.1	0.2	0.3	1.6	4.3	4.4	4.4	6.1	6.1	7.2	7.1	7.1	7.3	7.4
El Salvador	...	...	...	14	16	16	17	17	17	17	16	16	15	18
	...	...	...	0.7	0.8	0.8	0.8	0.8	0.8	0.8	0.8	0.8	0.7	0.9
Guatemala	...	...	...	21	16	16	16	16	16	16	16	16	16	20
	...	...	...	1.0	0.8	0.8	0.8	0.8	0.8	0.8	0.8	0.8	0.8	1.0
Honduras	...	...	...	10	14	14	14	14	14	14	14	14	14	14
	...	...	...	0.5	0.7	0.7	0.7	0.7	0.7	0.7	0.7	0.7	0.7	0.7
Jamaica	...	...	...	32	34	36	36	36	34	34	34	34	32	36
	...	...	...	1.6	1.7	1.8	1.8	1.8	1.7	1.7	1.7	1.7	1.6	1.8
Mexico	98	160	357	495	1,394	1,269	1,349	1,354	1,354	1,514	1,679	1,574	1,524	1,524
	4.9	8.0	17.8	24.7	69.4	63.2	67.2	67.4	67.4	75.4	83.6	78.4	75.9	75.9
Netherlands Antilles	495	617	680	840	792	320	320	320	320	320	320	470	470	477
	24.7	30.7	33.9	41.8	39.4	15.9	15.9	15.9	15.9	15.9	15.9	23.4	23.4	23.8
Nicaragua	...	...	...	22	14	15	15	15	15	15	16	16	16	16
	...	...	...	1.1	0.7	0.7	0.7	0.7	0.7	0.7	0.8	0.8	0.8	0.8
Panama	...	...	...	78	100	100	100	100	100	100	100	40	100	100
	...	...	...	3.9	5.0	5.0	5.0	5.0	5.0	5.0	5.0	2.0	5.0	5.0
Paraguay	...	...	...	5	7	8	8	8	8	8	8	8	8	8
	...	...	...	0.2	0.3	0.4	0.4	0.4	0.4	0.4	0.4	0.4	0.4	0.4
Peru	22	35	49	92	170	176	176	182	172	172	189	189	189	184
	1.1	1.7	2.4	4.6	8.5	8.8	8.8	9.1	8.6	8.6	9.4	9.4	9.4	9.2
Puerto Rico	...	...	84	155	284	121	121	123	123	123	125	123	127	127
	...	...	4.2	7.7	14.1	6.0	6.0	6.1	6.1	6.1	6.2	6.1	6.3	6.3
Trinidad & Tobago	49	104	182	430	456	260	300	300	300	300	246	246	246	305
	2.4	5.2	9.1	21.4	22.7	12.9	14.9	14.9	14.9	14.9	12.3	12.3	12.3	15.2
Uruguay	5	16	28	40	40	45	45	45	33	33	33	29	29	35
	0.2	0.8	1.4	2.0	2.0	2.2	2.2	2.2	1.6	1.6	1.6	1.4	1.4	1.7
Venezuela	55	254	886	1,313	1,446	1,230	1,225	1,201	1,201	1,201	1,167	1,171	1,167	1,167
	2.7	12.6	44.1	65.4	72.0	61.3	61.0	59.8	59.8	59.8	58.1	58.3	58.1	58.1
Virgin Islands	...	...	...	220	728	600	600	545	545	545	545	545	545	545
	...	...	...	11.0	36.3	29.9	29.9	27.1	27.1	27.1	27.1	27.1	27.1	27.1
Others	...	...	...	11	62	57	57	61	60	61	60	60	64	64
	...	...	...	0.5	3.1	2.8	2.8	3.0	3.0	3.0	3.0	3.0	3.2	3.2
Total	**854**	**1,397**	**2,841**	**5,114**	**8,620**	**6,920**	**7,080**	**7,149**	**7,143**	**7,315**	**7,554**	**7,555**	**7,572**	**7,507**
	42.3	**69.4**	**141.5**	**254.5**	**429.2**	**344.4**	**352.5**	**355.8**	**355.5**	**364.0**	**376.0**	**376.0**	**376.8**	**373.7**

Saudi Arabian Oil Co.'s Ras Tanura refinery is being upgraded and rebuilt to the 530,000 b/d capacity it possessed prior to a fire in 1990. Photo courtesy of Saudi Aramco.

ASIA-PACIFIC

Upper figures = 1,000 b/d
Lower figures = million t/y

	1940	1950	1960	1970	1980	1986	1987	1988	1989	1990	1991	1992	1993	1994
Australia	4	14	237	594	725	623	626	637	644	675	706	703	698	705
	0.2	0.7	11.8	29.6	36.1	31.0	31.2	31.7	32.1	33.6	35.2	35.0	34.8	35.1
Bangladesh	...	...	...	...	31	31	31	31	31	31	31	31	31	31
	...	...	...	...	1.5	1.5	1.5	1.5	1.5	1.5	1.5	1.5	1.5	1.5
Brunei-Malaysia-Singapore	10	10	46	320	1,093	1,240	1,183	1,080	1,071	1,049	1,098	1,113	1,301	1,333
	0.5	0.5	2.3	15.9	54.4	61.8	58.9	53.8	53.3	52.2	54.7	55.4	64.8	66.4
China	...	8	98	300	1,600	2,150	2,200	2,200	2,200	2,200	2,200	2,200	2,200	2,200
	...	0.4	4.9	14.9	79.7	107.1	109.6	109.6	109.6	109.6	109.6	109.6	109.6	109.6
China, Taiwan	...	...	28	119	425	543	543	600	570	570	543	543	543	543
	...	...	1.4	5.9	21.2	27.0	27.0	29.9	28.4	28.4	27.0	27.0	27.0	27.0
Guam*	...	...	...	...	44	...	...	...	...	...	...	...	...	...
	...	...	...	...	2.2	...	...	...	...	...	...	...	...	...
India	10	6	112	446	557	867	991	1,059	1,051	1,080	1,122	1,122	1,047	1,086
	0.5	0.3	5.6	22.2	27.7	43.2	49.4	52.7	52.3	53.8	55.9	55.9	52.1	54.1
Indonesia	170	203	274	268	528	636	636	714	714	714	814	860	860	860
	8.5	10.1	13.6	13.3	26.3	31.7	31.7	35.6	35.6	35.6	40.5	42.8	42.8	42.8
Japan	53	34	640	3,140	5,509	4,613	4,790	4,567	4,363	4,198	4,383	4,612	4,741	4,810
	2.6	1.7	31.9	156.4	274.3	229.7	238.5	227.4	217.3	209.1	218.3	229.7	236.1	239.5
Korea, North†	...	...	...	...	...	...	...	...	...	...	42	42	42	42
	...	...	...	...	...	...	...	...	...	...	2.1	2.1	2.1	2.1
Korea, South	...	...	...	180	601	782	862	820	880	867	867	1,163	1,147	1,147
	...	...	...	9.0	29.9	38.9	42.9	40.8	43.8	43.2	43.2	57.9	57.1	57.1
Myanmar (Burma)	...	...	9	26	26	26	26	26	26	32	32	32	32	32
	...	...	0.4	1.3	1.3	1.3	1.3	1.3	1.3	1.6	1.6	1.6	1.6	1.6
New Zealand	...	...	...	66	74	53	54	82	88	95	95	95	84	93
	...	...	...	3.3	3.7	2.6	2.7	4.1	4.4	4.7	4.7	4.7	4.2	4.6
Okinawa*	...	...	...	...	198	110	...	...	...	...	...	...	...	...
	...	...	...	...	9.9	5.5	...	...	...	...	...	...	...	...
Pakistan	...	5	6	107	98	130	130	130	130	121	121	121	121	133
	...	0.2	0.3	5.3	4.9	6.5	6.5	6.5	6.5	6.0	6.0	6.0	6.0	6.6
Philippines	...	...	22	182	253	216	286	286	254	284	279	287	278	279
	...	...	1.1	9.1	12.6	10.8	14.2	14.2	12.6	14.1	13.9	14.3	13.8	13.9
Sri Lanka	...	...	...	44	38	50	50	50	50	50	50	50	50	50
	...	...	...	2.2	1.9	2.5	2.5	2.5	2.5	2.5	2.5	2.5	2.5	2.5
Thailand	...	...	1	62	186	192	192	192	191	215	221	221	221	314
	...	...	...	3.1	9.3	9.6	9.6	9.6	9.5	10.7	11.0	11.0	11.0	15.6
Others	27	4	...	12	...	...	...	...	...	...	...	...	...	...
	1.3	0.2	...	0.6	...	...	...	...	...	...	...	...	...	...
Total	**274**	**284**	**1,472**	**5,866**	**11,986**	**12,262**	**12,600**	**12,474**	**12,263**	**12,181**	**12,604**	**13,195**	**13,396**	**13,658**
	13.6	**14.1**	**73.3**	**292.1**	**596.9**	**610.7**	**627.5**	**621.2**	**610.7**	**606.6**	**627.7**	**657.0**	**667.0**	**680.0**

***Prior to 1980, included in others**

†Prior to 1991, included in other C.I.S.—Eastern Europe

AFRICA

Upper figures = 1,000 b/d
Lower figures = million t/y

	1940	1950	1960	1970	1980	1986	1987	1988	1989	1990	1991	1992	1993	1994
Algeria	...	...	...	48	122	465	465	465	465	465	465	465	530	530
	...	...	...	2.4	6.1	23.2	23.2	23.2	23.2	23.2	23.2	23.2	26.4	26.4
Angola-Cabinda	...	...	2	14	36	32	32	32	32	32	38	32	32	32
	...	...	0.1	0.7	1.8	1.6	1.6	1.6	1.6	1.6	1.9	1.6	1.6	1.6
Cameroon	...	...	...	...	...	43	43	43	42	43	42	42	42	42
	...	...	...	...	...	2.1	2.1	2.1	2.1	2.1	2.1	2.1	2.1	2.1
Congo	...	...	...	14	...	21	21	21	21	21	21	21	21	21
	...	...	...	0.7	...	1.0	1.0	1.0	1.0	1.0	1.0	1.0	1.0	1.0
Egypt	17	39	87	175	234	434	452	452	489	489	523	523	532	532
	0.8	1.9	4.3	8.7	11.7	21.6	22.5	22.5	24.4	24.4	26.0	26.0	26.5	26.5
Ethiopia	...	...	...	11	14	18	18	18	18	18	18	18	18	18
	...	...	...	0.5	0.7	0.9	0.9	0.9	0.9	0.9	0.9	0.9	0.9	0.9
Gabon	...	...	...	13	20	23	16	23	24	24	24	24	24	17
	...	...	...	0.6	1.0	1.1	0.8	1.1	1.2	1.2	1.2	1.2	1.2	0.8
Ghana	...	...	...	29	26	28	28	28	27	27	27	27	27	27
	...	...	...	1.4	1.3	1.4	1.4	1.4	1.3	1.3	1.3	1.3	1.3	1.3
Ivory Coast	...	...	...	19	50	90	73	60	60	69	69	69	69	63
	...	...	...	0.9	2.5	4.5	3.6	3.0	3.0	3.4	3.4	3.4	3.4	3.1
Kenya	...	...	...	44	95	95	95	90	90	95	90	90	90	90
	...	...	...	2.2	4.7	4.7	4.7	4.5	4.5	4.7	4.5	4.5	4.5	4.5
Liberia	...	...	...	10	15	15	15	15	15	15	15	15	15	15
	...	...	...	0.5	0.7	0.7	0.7	0.7	0.7	0.7	0.7	0.7	0.7	0.7
Libya	...	...	...	10	138	330	329	329	329	329	348	348	348	348
	...	...	...	0.5	6.9	16.4	16.4	16.4	16.4	16.4	17.3	17.3	17.3	17.3
Madagascar	...	...	...	13	11	16	16	16	16	16	16	16	16	16
	...	...	...	0.6	0.5	0.8	0.8	0.8	0.8	0.8	0.8	0.8	0.8	0.8
Morocco	...	...	2	35	72	81	81	155	155	155	155	155	155	155
	...	...	0.1	1.7	3.6	4.0	4.0	7.7	7.7	7.7	7.7	7.7	7.7	7.7
Mozambique	...	...	...	20	16	17	17	...	...	...	...	...	...	...
	...	...	...	1.0	0.8	0.8	0.8	...	...	...	...	...	...	...
Nigeria	...	...	...	46	160	250	250	270	415	433	433	433	433	433
	...	...	...	2.3	8.0	12.4	12.4	13.4	20.7	21.6	21.6	21.6	21.6	21.6
Senegal	...	...	...	13	20	30	30	30	30	24	24	24	23	23
	...	...	...	0.6	1.0	1.5	1.5	1.5	1.5	1.2	1.2	1.2	1.1	1.1
Sierra Leone	...	...	...	10	10	10	10	10	10	10	10	10	10	10
	...	...	...	0.5	0.5	0.5	0.5	0.5	0.5	0.5	0.5	0.5	0.5	0.5
South Africa	1	...	25	184	478	389	389	434	434	434	431	431	431	364
	...	...	1.2	9.2	23.8	19.4	19.4	21.6	21.6	21.6	21.5	21.5	21.5	18.1
Sudan	...	...	...	21	26	24	24	21	21	25	22	22	22	22
	...	...	...	1.0	1.3	1.2	1.2	1.0	1.0	1.2	1.1	1.1	1.1	1.1
Tanzania	...	...	...	14	17	14	14	14	14	14	17	17	17	17
	...	...	...	0.7	0.8	0.7	0.7	0.7	0.7	0.7	0.8	0.8	0.8	0.8
Togo	...	...	...	...	20	20	20	20	...	...	...	...	...	...
	...	...	...	...	1.0	1.0	1.0	1.0	...	...	...	...	...	...
Tunisia	...	...	...	23	34	34	34	34	34	34	34	34	34	34
	...	...	...	1.1	1.7	1.7	1.7	1.7	1.7	1.7	1.7	1.7	1.7	1.7
Zaire	...	...	...	21	16	17	17	17	17	17	17	17	17	17
	...	...	...	1.0	0.8	0.8	0.8	0.8	0.8	0.8	0.8	0.8	0.8	0.8
Others	...	...	...	...	35	35	35	34	31	35	35	34	34	34
	...	...	...	...	1.7	1.7	1.7	1.7	1.5	1.7	1.7	1.7	1.7	1.7
Total	**18**	**39**	**116**	**787**	**1,665**	**2,531**	**2,524**	**2,631**	**2,789**	**2,824**	**2,874**	**2,867**	**2,940**	**2,860**
	0.8	**1.9**	**5.7**	**38.8**	**82.9**	**125.7**	**125.4**	**130.8**	**138.8**	**140.4**	**142.9**	**142.6**	**146.2**	**142.1**

MIDDLE EAST

Upper figures = 1,000 b/d Lower figures = million t/y

	1940	1950	1960	1970	1980	1986	1987	1988	1989	1990	1991	1992	1993	1994
Abu Dhabi	...	...	...	...	14	185	185	180	180	180	193	193	193	193
	...	...	...	...	0.7	9.2	9.2	9.0	9.0	9.0	9.6	9.6	9.6	9.6
Bahrain	33	150	187	205	250	250	250	250	243	243	243	243	243	250
	1.6	7.5	9.3	10.2	12.4	12.4	12.4	12.4	12.1	12.1	12.1	12.1	12.1	12.4
Iran	333	502	495	645	921	530	530	530	530	530	720	720	1,089	1,162
	16.6	25.0	24.7	32.1	45.9	26.4	26.4	26.4	26.4	26.4	35.9	35.9	54.2	57.9
Iraq	4	9	56	104	168	319	319	319	319	319	319	319	319	348
	0.2	0.4	2.8	5.2	8.4	15.9	15.9	15.9	15.9	15.9	15.9	15.9	15.9	17.3
Israel	...	83	87	107	195	170	180	180	180	180	180	208	221	208
	...	4.1	4.3	5.3	9.7	8.5	9.0	9.0	9.0	9.0	9.0	10.4	11.0	10.4
Jordan	...	...	...	7	21	100	100	100	100	100	100	100	100	100
	...	...	...	0.3	1.0	5.0	5.0	5.0	5.0	5.0	5.0	5.0	5.0	5.0
Kuwait*	...	25	270	569	645	634	618	628	817	819	819	819	368	624
	...	1.2	13.5	28.4	32.1	31.6	30.8	31.3	40.7	40.8	40.8	40.8	18.3	31.1
Lebanon	...	6	24	38	53	17	37	37	37	37	38	38	38	38
	...	0.3	1.2	1.9	2.6	0.8	1.8	1.8	1.8	1.8	1.9	1.9	1.9	1.9
Qatar	...	...	1	1	11	56	62	62	62	62	62	60	60	58
	...	...	...	...	0.5	2.8	3.1	3.1	3.1	3.1	3.1	3.0	3.0	2.9
Saudi Arabia*	...	140	189	377	487	1,115	1,125	1,375	1,375	1,484	1,863	1,863	1,863	1,614
	...	7.0	9.4	18.8	24.3	55.5	56.0	68.5	68.5	73.9	92.8	92.8	92.8	80.4
Syria	...	...	...	59	223	229	229	229	244	244	237	237	237	242
	...	...	...	2.9	11.1	11.4	11.4	11.4	12.2	12.2	11.8	11.8	11.8	12.1
Yemen	...	...	120	178	175	162	170	170	162	162	172	115	115	120
	...	...	6.0	8.9	8.7	8.1	8.5	8.5	8.1	8.1	8.6	5.7	5.7	6.0
Others	...	...	...	...	...	48	58	87	87	87	76	80	80	76
	...	...	...	...	...	2.4	2.9	4.3	4.3	4.3	3.8	4.0	4.0	3.8
Total	**370**	**916**	**1,429**	**2,290**	**3,163**	**3,815**	**3,863**	**4,147**	**4,336**	**4,447**	**5,022**	**4,995**	**4,926**	**5,033**
	18.4	**45.5**	**71.2**	**114.0**	**151.4**	**190.0**	**192.4**	**206.6**	**216.1**	**221.6**	**250.3**	**248.9**	**245.3**	**250.8**
Total, exclu. C.I.S.-E. Europe	**6,612**	**10,524**	**21,005**	**41,969**	**65,678**	**57,098**	**57,016**	**57,688**	**57,804**	**59,097**	**59,720**	**60,067**	**60,320**	**60,328**
	328.6	**523.5**	**1,045.9**	**2,089.5**	**3,270.5**	**2,843.2**	**2,839.1**	**2,872.7**	**2,878.5**	**2,942.7**	**2,973.8**	**2,991.0**	**3,003.4**	**3,003.9**

***Includes Neutral Zone.**

C.I.S.-EASTERN EUROPE

Upper figures = 1,000 b/d Lower figures = million t/y

	1940	1950	1960	1970	1980	1986	1987	1988	1989	1990	1991	1992	1993	1994
Albania*	...	...	...	...	...	...	...	...	...	...	40	40	40	40
	...	...	...	...	...	...	...	...	...	...	2.0	2.0	2.0	2.0
Bulgaria*	...	...	...	...	...	...	...	...	...	...	300	300	300	300
	...	...	...	...	...	...	...	...	...	...	14.9	14.9	14.9	14.9
C.I.S.	...	900	2,800	5,640	10,950	12,260	12,260	12,260	12,300	12,300	12,300	12,300	9,851	9,846
	...	44.8	139.4	280.9	545.3	610.5	610.5	610.5	612.5	612.5	612.5	612.5	490.6	490.3
Croatia	...	...	...	...	...	...	...	...	...	...	...	...	143	293
	...	...	...	...	...	...	...	...	...	...	...	...	7.1	14.6
Czechoslovakia†	...	16	46	220	455	455	455	455	455	455	455	455	457	307
	...	0.8	2.3	11.0	22.7	22.7	22.7	22.7	22.7	22.7	22.7	22.7	22.8	15.3
Hungary	...	20	58	130	290	242	242	220	220	220	220	220	235	242
	...	1.0	2.9	6.5	14.4	12.1	12.1	11.0	11.0	11.0	11.0	11.0	11.7	12.1
Lithuania	...	...	...	...	...	...	...	...	...	...	...	...	267	263
	...	...	...	...	...	...	...	...	...	...	...	...	13.3	13.1
Macedonia	...	...	...	...	...	...	...	...	...	...	...	...	51	51
	...	...	...	...	...	...	...	...	...	...	...	...	2.5	2.5
Poland	...	4	20	156	385	385	385	385	385	385	385	333	333	333
	...	0.2	1.0	7.8	19.2	19.2	19.2	19.2	19.2	19.2	19.2	16.6	16.6	16.6
Romania	...	182	260	356	608	617	617	617	617	617	617	657	737	681
	...	9.1	12.9	17.7	30.3	30.7	30.7	30.7	30.7	30.7	30.7	32.7	36.7	33.9
Slovakia	...	...	...	...	...	...	...	...	...	...	...	...	...	193
	...	...	...	...	...	...	...	...	...	...	...	...	...	9.6
Slovenia	...	...	...	...	...	...	...	...	...	...	...	...	15	15
	...	...	...	...	...	...	...	...	...	...	...	...	0.7	0.7
Yugoslavia	5	8	10	179	296	302	483	608	609	609	609	609	444	282
	0.2	0.4	0.5	8.9	14.7	15.0	24.1	30.3	30.3	30.3	30.3	30.3	22.1	14.0
Others	...	4	40	332	852	894	894	852	910	852	...	...	...	...
	...	0.2	2.0	16.5	42.4	44.5	44.5	42.4	45.3	42.4	...	...	...	...
Total	**5**	**1,134**	**3,234**	**7,013**	**13,836**	**15,155**	**15,336**	**15,397**	**15,496**	**15,438**	**14,926**	**14,914**	**12,873**	**12,846**
	0.2	**56.5**	**161.0**	**349.3**	**689.0**	**754.7**	**763.8**	**766.8**	**771.7**	**768.8**	**743.3**	**742.7**	**641.0**	**639.6**
World Total	**6,617**	**11,658**	**24,239**	**48,982**	**79,514**	**72,253**	**72,352**	**73,085**	**73,300**	**74,535**	**74,646**	**74,981**	**73,193**	**73,174**
	328.8	**580.0**	**1,206.9**	**2,438.8**	**3,959.5**	**3,597.9**	**3,602.9**	**3,639.5**	**3,650.2**	**3,711.5**	**3,717.1**	**3,733.7**	**3,644.4**	**3,643.5**

***Prior to 1991, included in others. †Beginning 1994, Czech Republic.**

WORLDWIDE LOOK AT RESERVES AND PRODUCTION

Country	ESTIMATED PROVED RESERVES Jan. 1, 1994 Oil (1,000 bbl)	Jan. 1, 1994 Gas (bcf)	Jan. 1, 1993 Oil (1,000 bbl)	Jan. 1, 1993 Gas (bcf)	OIL PRODUCTION Producing oil wells* Dec 31, 1992	Estimated 1993 (1,000 b/d)	Change from 1992 (%)	Actual 1992 (1,000 b/d)
ASIA-PACIFIC								
Afghanistan	...	3,500	...	3,500	...	...	...	...
Australia	1,614,594	19,607	1,767,900	18,254	1,081	512.8	−3.8	533.0
Bangladesh	4,000	25,200	1,500	25,400	25	1.0	−9.1	1.1
Brunei	1,350,000	14,000	1,350,000	14,000	738	158.8	−3.9	165.3
China	24,000,000	59,000	24,000,000	49,400	49,700	2,895.4	2.1	2,835.0
China, Taiwan	4,000	2,400	4,000	2,400	84	1.1	−8.3	1.2
India	5,920,856	25,354	6,049,068	25,954	2,681	524.3	−4.1	547.0
Indonesia	5,779,000	64,388	5,779,000	64,388	8,015	1,322.5	−1.7	1,346.0
Japan	54,243	1,046	59,850	968	274	15.5	−9.9	17.2
Malaysia	4,300,000	76,700	3,700,000	67,800	561	640.0	−3.2	661.0
Myanmar (Burma)	50,000	9,800	50,000	9,800	450	14.0	7.7	13.0
New Zealand	156,000	3,155	169,570	3,386	52	34.9	−5.2	36.8
Pakistan	203,284	22,940	412,000	31,000	115	62.5	−16.7	75.0
Papua New Guinea	253,400	15,000	340,000	14,200	23	124.6	137.8	52.4
Philippines	279,000	2,980	147,540	1,600	4	10.1	13.5	8.9
Thailand	178,565	5,740	241,900	8,463	314	51.4	0.0	51.4
Viet Nam	500,000	3,700	500,000	500	NA	122.5	16.7	105.0
Total Asia-Pacific	**44,646,942**	**354,510**	**44,572,328**	**341,013**	**64,117**	**6,491.4**	**0.7**	**6,449.3**
WESTERN EUROPE								
Austria	94,730	669	93,200	569	1,149	24.5	6.1	23.1
Denmark	779,936	4,273	729,618	3,985	120	162.5	3.6	156.9
France	177,434	1,240	177,434	1,240	609	54.4	−6.2	58.0
Germany	449,314	12,113	449,314	12,113	2,308	61.2	−5.8	65.0
Greece	41,000	300	41,000	300	13	11.3	−17.5	13.7
Ireland	...	613	...	700	...	...	...	...
Italy	620,500	10,665	746,977	12,996	234	86.0	3.6	83.0
Netherlands	132,086	68,157	144,650	68,860	200	45.7	−20.7	57.6
Norway	9,283,760	70,488	8,805,734	70,629	386	2,247.3	4.8	2,144.3
Spain	22,518	700	22,518	700	45	18.4	−11.1	20.7
Turkey	487,670	371	474,761	608	732	75.8	−7.6	82.0
United Kingdom	4,554,260	21,542	4,143,630	19,070	735	1,866.2	0.1	1,863.8
Total Western Europe	**16,643,208**	**191,131**	**15,828,836**	**191,770**	**6,531**	**4,653.3**	**1.9**	**4,568.1**
EASTERN EUROPE and C.I.S.								
Albania	165,000	350	165,000	700	NA	8.0	−27.3	11.0
Bulgaria	15,000	250	15,000	250	100	1.0	0.0	1.0
C.I.S.	57,000,000	1,997,000	57,000,000	1,942,300	148,990	7,846.5	−12.3	8,949.0
Croatia	150,925	1,252	162,462	1,316	919	40.0	−7.0	43.0
Czechoslovakia	15,000	460	15,000	470	400	2.0	0.0	2.0
Hungary	139,031	3,412	146,956	3,758	1,776	31.7	−5.1	33.4
Poland	36,793	5,477	42,208	5,597	2,302	3.4	−5.6	3.6
Romania	1,568,754	8,000	1,568,754	7,335	250	139.0	2.2	136.0
Serbia	77,500	1,580	77,500	1,580	646	23.6	7.3	22.0
Total Eastern Europe and C.I.S.	**59,168,003**	**2,017,781**	**59,192,880**	**1,963,306**	**155,383**	**8,095.2**	**−12.0**	**9,201.0**
MIDDLE EAST								
Abu Dhabi	92,200,000	188,400	92,200,000	188,400	993	1,784.4	−3.1	1,842.0
Bahrain	69,584	5,900	69,584	5,810	352	36.0	−1.9	36.7
Dubai	4,000,000	4,400	4,000,000	4,500	151	363.8	−9.5	402.0
Iran	92,860,000	730,000	92,860,000	699,200	730	3,640.0	5.4	3,455.0
Iraq	100,000,000	109,500	100,000,000	109,500	58	435.8	2.5	425.0
Israel	3,351	14	3,418	15	11	0.2	0.0	0.2
Jordan	1,000	200	4,000	215	4	0.1	0.0	0.1
Kuwait	94,000,000	52,400	94,000,000	52,400	350	1,692.1	92.3	880.0
Neutral Zone	5,000,000	1,000	5,000,000	1,000	231	361.7	6.1	341.0
Oman	4,700,000	20,000	4,483,000	16,900	1,243	774.8	4.4	742.0
Qatar	3,729,000	250,000	3,729,000	227,000	268	427.5	0.6	425.0
Ras al Khaimah	400,000	1,100	400,000	1,200	7	1.0	0.0	1.0
Saudi Arabia	258,703,000	185,360	257,842,000	182,600	1,400	7,981.3	−1.9	8,137.0
Sharjah	1,500,000	10,700	1,500,000	10,500	30	40.0	0.0	40.0
Syria	1,700,000	7,000	1,700,000	7,000	963	576.3	11.3	518.0
Yemen	4,000,000	15,000	4,000,000	13,900	179	208.4	14.5	182.0
Total Middle East	**662,865,935**	**1,580,974**	**661,791,002**	**1,520,140**	**6,970**	**18,323.4**	**5.1**	**17,427.0**

	ESTIMATED PROVED RESERVES				OIL PRODUCTION			
	Jan. 1, 1994		Jan. 1, 1993					
Country	Oil (1,000 bbl)	Gas (bcf)	Oil (1,000 bbl)	Gas (bcf)	Producing oil wells* Dec 31, 1992	Estimated 1993 (1,000 b/d)	Change from 1992 (%)	Actual 1992 (1,000 b/d)
AFRICA								
Algeria	9,200,000	128,000	9,200,000	128,000	1,446	750.0	−2.8	772.0
Angola	1,500,000	1,800	1,500,000	1,800	467	498.4	−6.1	531.0
Benin	19,900	...	19,900	...	8	2.8	−6.7	3.0
Cameroon	400,000	3,900	400,000	3,900	193	118.8	−13.3	137.0
Congo	830,000	2,700	830,000	2,700	396	200.0	16.3	172.0
Egypt	6,300,000	15,400	6,200,000	15,400	1,014	894.3	2.4	873.0
Equatorial Guinea	3,600	1,300	3,600	1,300	NA	4.0	300.0	1.0
Ethiopia	...	800	...	800	...	...	...	...
Gabon	730,000	500	730,000	400	329	296.7	−1.1	300.0
Ghana	500	800	500	...	NA	6.0	233.3	1.8
Ivory Coast	50,000	500	100,000	3,500	12	0.5	−50.0	1.0
Libya	22,800,000	45,800	22,800,000	46,200	1,092	1,367.1	−8.4	1,493.0
Madagascar	...	70	...	70	...	...	...	...
Morocco	1,063	32	1,604	37	9	0.2	0.0	0.2
Mozambique	...	2,300	...	2,300	...	...	...	...
Namibia	...	5,200	...	5,200	...	...	...	...
Nigeria	17,899,820	120,000	17,899,820	120,000	1,799	1,895.8	−0.3	1,902.0
Rwanda	...	2,000	...	2,000	...	...	...	...
Somalia	...	200	...	200	...	...	...	...
South Africa	41,000	943	...	1,900	5	4.1	0.0	4.1
Sudan	300,000	3,000	300,000	3,000	...	2.0	566.7	0.3
Tanzania	...	4,100	...	4,100	...	...	...	...
Tunisia	1,700,000	3,200	1,700,000	3,000	181	100.1	−5.7	106.2
Zaire	187,000	1,000	187,000	1,058	60	24.0	0.8	23.8
Total Africa	61,962,883	343,545	61,872,424	346,865	7,011	6,164.8	−2.5	6,321.4
WESTERN HEMISPHERE								
Argentina	1,569,987	26,500	1,569,987	22,700	8,402	574.8	3.8	554.0
Barbados	3,501	7	5,892	7	102	1.2	−7.7	1.3
Bolivia	108,392	3,931	112,136	4,108	332	20.3	−4.2	21.2
Brazil	3,600,000	4,828	3,030,000	4,400	6,249	631.3	0.9	625.7
Canada	5,095,551	94,823	5,291,630	95,734	40,667	1,684.1	5.0	1,603.7
Chile	300,000	3,900	300,000	3,900	355	14.3	−9.5	15.8
Colombia	1,935,200	10,000	1,935,200	7,200	3,079	449.8	−1.9	458.3
Cuba	100,000	100	100,000	100	200	25.0	25.0	20.0
Ecuador	2,014,000	3,800	1,599,793	3,899	999	340.8	6.2	321.0
Guatemala	207,000	10	207,000	10	16	6.7	19.6	5.6
Mexico	50,925,000	70,954	51,298,000	70,900	4,740	2,663.4	−0.2	2,667.7
Peru	380,900	7,039	380,866	7,054	3,157	126.3	7.9	117.0
Suriname	56,200	...	49,200	...	223	5.3	17.8	4.5
Trinidad & Tobago	466,000	8,465	572,600	8,702	3,262	119.5	−11.7	135.4
United States	23,745,000	165,015	24,682,000	167,062	602,197	6,895.9	−3.8	7,171.1
Venezuela	63,330,000	128,900	62,650,000	126,492	12,140	2,331.7	0.8	2,314.0
Total Western Hemisphere	153,836,731	528,272	153,784,304	522,268	686,120	15,890.4	−0.9	16,036.3
TOTAL WORLD	999,123,702	5,016,213	997,041,774	4,885,362	926,132	59,618.5	−0.6	60,003.1
Total OPEC	772,130,820	2,020,448	770,589,820	1,961,780	29,039	24,691.0	2.6	24,075.0

EDITOR'S NOTE: All reserves figures except those for the C.I.S. (and gas for Canada) are reported as proved reserves recoverable with present technology and prices. C.I.S. figures are "explored reserves" which include proved, probable, and some possible. Canadian gas figure, under criteria adopted by Canadian Petroleum Association in 1980, includes proved and some probable. *Does not include shut in, injection, or service wells.

Worldwide oil production

• Offshore (e) Estimated (c) Condensate (NA) Not available

		Name of field, discovery date	Depth, ft	No. of wells — Producing	No. of wells — Total	Production — 1992 average b/d	Production — Cumulative to Dec. 31, 1992 bbl	°API gravity
ABU DHABI(e)	ADNOC, OTHERS	Asab, 1965	8,000-8,700	106	248			40.8
		Bab, 1958	8,600-9,300	100	232			40.6
		Bu Hasa, 1962	8,300-9,200	201	371			39.0
		Sahil, 1967	9,500-9,800	22	52			39.7
		• Umm Shaif, 1958	9,800-12,000	142	203			37.0
		• Zakum Lower, 1963	8,000-9,000	131	254			39.0
		• Zakum Upper, 1963	7,000-8,000	142	262			35.0
		• Umm Addalkh, 1969	7,700-8,500	32	54	1,842,000	11,384,588,000	32.5
		• Satah, 1975	8,550-9,500	10	21			39.8
		• Arzanah, 1973	11,000-11,300	14	34			44.6
		• Abu Al Bukhoosh, 1969	8,900-10,900	42	48			32.0
		• Mubarraz, 1969	8,300-10,000	28	28			37.0
		• Bunduq, 1964	9,400-10,900	23	40			40.0
		Total Abu Dhabi		**993**	**1,847**	**1,842,000**	**11,384,588,000**	
ALGERIA*	SONATRACH	Acheb W. & Kreb, 1963	7,360	6	14	2,374		45.6
		Ait Kheir, 1971	8,860	9	16	22,157		53.2
		Alrar(c), 1961	8,202	...	...	29,868		...
		Amassak, 1970	6,565	42	50	9,723		41.1
		Askarene, 1962	8,200	2	9	519		41.7
		Assakaifaf N. & S., 1958	2,800	2	5	354		42.7
		Bab El Hattabet, 1982	11,880	1	1	959		43.4
		Djebel Bissa, 1977	...	...	...	2,327		37.0
		Djebel Onk, 1960	3,940	6	14	817		37.0
		Djoua West, 1967	5,580-6,240	4	6	810		41.1
		Draaa Ettamra, 1971	11,600	2	2	102		43.5
		Edeyen, 1964	5,580	2	7	558		41.5
		Edjeleh, 1956	1,500-2,800	142	243	8,882		37.2
		El Adeb Lareche, 1958	4,100	7	29	668		43.3
		El Borma, 1967	8,200	27	29	9,605		42.9
		El Gassi-El Agreb, 1959	10,170	26	66	23,328		47.9
		Gara, 1962	9,420	2	2	825		42.9
		Gassi-Touil + Est, 1961	6,890	47	63	16,632		50.6-50.8
		Guellala + Guellala N.E., 1969	11,320	9	18	12,042		42.2-42.3
		Guelta, 1962	8,960	2	9	440		41.7
		Haniet El Beida, 1978	12,045	1	1	385		43.2
		Haniet El Mokta, 1972	12,375	1	1	8		43.5
		Hassi Chergui, 1963	8,200	7	9	2,319		50.6
		Hassi Mazoula B., 1965	4,920	2	3	110		42.0
		Hassi Mazoula S., 1963	4,430	5	10	1,187		40.2
		Hassi Messaoud N., 1956	10,500	82	263	146,701		45.9
		Hassi Messaoud S., 1956	10,500	369	575	217,116		46.1
		Hassi R'Mel, 1979	7,760	22	41	9,220	8,165,871,000	42.7
		Hassi R'Mel(c), 1956	7,218	...	...	322,152		...
		H. Berkaoui + Ben Kahla, 1965	11,000	44	60	32,470		43.6
		La Reculeé, 1957	5,414-5,742	...	...	472		43.4
		Makouda, 1975	6,000	1	1	141		53.4
		Mereksen, 1974	9,420	15	18	11,837		45.8
		Mesdar E. & W., 1972	11,155	19	27	10,540		40.3
		Mokh El Kebch, 1984	12,870	2	2	951		43.5
		Nezla Nord, 1965	8,530	5	6	...		51.5
		Nord In Aménas(c), 1959	...	...	...	462		...
		N'Goussa, 1974	11,715	2	3	55		42.2
		Ohanet N. & S., 1960	7,878	21	46	4,873		43.8
		Oued Guétérini, 1946	6,600	13	180	47		38.4
		Oued Noumer, 1969	8,860	3	8	5,368		48.3
		Ras Toumb, 1976	9,190	...	8	55		35.7
		Rhourde Chegga, 1959	10,170	2	2	2,562		44.0
		Rhourde El Baguel, 1962	9,850	24	32	24,240		40.2
		Rhourde Nouss, 1972	8,530	15	16	3,113		39.7
		Rhourde Nouss(c), 1962	8,202-8,859	...	...	54,037		...
		Sahane, 1972	12,649	1	1	354		43.5
		Sidi Mezghiche, 1976	8,085	1	1	1,006		53.3
		Stah, 1971	9,515	41	45	23,273		45.8
		Tamendjelt, 1970	5,585-6,890	8	8	2,727		41.1
		Tiguentourine, 1966	2,160-4,265	31	49	2,727		44.4
		Timedratine + Est, 1964	8,155	20	33	6,822		42.7
		Tin Fouye F6 + TF Nord, 1962	4,265	48	50	10,627		39.6
		Tin Fouye Tabankort, 1966	4,600-6,560	184	206	82,067		41.1-41.3

WORLDWIDE PRODUCTION

*Table shows 1991 field production of crude oil condensate and NGL.
1992 crude production estimated at 772,000 B/D.
Cumulative crude oil production at end 1992 8,897,651,000.

	Name of field, discovery date	Depth, ft	No. of wells Producing	No. of wells Total	Production 1992 average b/d	Production Cumulative to Dec. 31, 1992 bbl	°API gravity
	Zarzaitine, 1957	4,700	117	194	36,911		42.2
	Zemlet Nouss, 1982	8,745	2	3	1,234		42.0
	Total Algeria		**1,446**	**2,485**	**1,161,159**	**8,165,871,000**	
ANGOLA CHEVRON	• Kali, 1969	11,400-11,660	1	1	7	4,358,577	29.0
	• Kambala, 1971	10,500	4	7	5,057	27,681,216	32.8
	• Kungulo, 1975	4,000	17	17	5,982	60,656,521	35.0
	• Lifuma, 1984	7,800	...	1	...	291,421	31.4
	• Limba, 1969	2,660-7,350	13	14	9,323	106,447,808	34.3
	• Livuite, 1979	8,000	...	1	1,050	4,361,819	31.7
	• Malongo North, 1966	1,170-9,250	52	63	16,882	240,632,291	30.0
	• Malongo South, 1966	1,150-1,440	26	31	6,168	85,592,168	25.0
	• Malongo West, 1969	1,680-8,800	21	33	20,332	272,269,785	22.5
	• Numbi, 1980	3,446	43	61	84,683	109,738,772	34.0
	• Takula, 1971	3,070-9,090	71	86	110,813	345,870,469	31.5
	• Vuko, 1983	3,700	11	11	11,686	20,872,091	35.0
	• Wamba, 1982	2,700	12	13	33,074	62,966,485	29.6
ELF ANGOLA	Bisfals, 1982	9,760	8	14	22,110	29,066,300	39.0
	Impula, 1983	8,850	1	1	1,430	525,000	40.0
	Impula South East, 1985	7,880	7	12	23,180	27,579,000	36.0
	• Pacassa, 1982	10,660	14	25	64,690	135,534,500	38.0
	• Palanca, 1981	8,530	15	23	46,850	125,427,000	40.0
TEXACO-SONANGOL-TOTAL-BRASPETRO	• Cuntala, 1978	7,346	...	1	...	2,265,104	30.5
	• Essungo, 1975	6,532	10	17	2,874	25,773,298	36.5
	• Lombo Este, 1984	7,845	11	12	29,063	43,433,898	38.3
	• Raia, 1988	8,800	2	2	1,146	419,596	34.5
	• Sulele, 1982	6,891	1	1	107	39,031	39.8
	• Sulele West, 1985	7,800	6	6	3,421	6,173,927	40.2
	• Sulele South 1, 1986	7,290	13	13	8,302	4,916,937	39.2
	• Tubarao, 1984	7,845	7	8	12,107	19,878,433	38.3
FINA ANGOLA[(e)]	Benfica, 1955	8,300	1	5			30.0
	Bento, 1972	7,350	1	3			26.0
	C. Cobra, 1969	4,000	7	11			41.0
	Cacuaco, 1958	7,870	1	4			30.0
	Galinda, 1959	7,220	...	8			30.0
	Ganda, 1975	5,000	10	12			36.0
	Legua, 1972	6,300	1	1			31.0
	Luango, 1977	3,800	3	3			38.0
	Lumueno, 1977	4,600	9	12			37.0
	Mulenvos, 1966	5,900-6,600	5	12	25,397	262,946,000	25.0
	N'Zombo, 1973	5,000-6,300	11	14			33.0
	N'Z Pangala, 1973	5,000-5,100	8	13			33.1
	N'Z Kitona D & I, 1973	5,200-5,670	7	8			33.0
	Pambo, 1982	4,600	5	5			39.0
	Pinda, 1966	6,710	...	1			26.0
	Quenguela Norte, 1968	6,000	20	32			31.0
	Quinfuquena, 1975	6,800-7,100	11	20			32.0
	Quinguila, 1972	3,400	16	28			36.0
	Sereia, 1974	6,200	3	4			27.0
	Tobias, 1961	1,970	1	10			31.0
	Total Angola		**486**	**680**	**545,734**	**2,025,717,447**	
ARGENTINA YPF	Chubut, 1907	1,908-9,168	656	3,323	18,316	638,694,029	21.1-26.0
	Formosa, 1984	11,466-11,490	...	...	7,045	18,666,443	42.0-42.9
	Jujuy, 1969	2,461-16,732	...	...	478	62,416,318	46.5-47.0
	La Pampa, 1968	3,609-4,921	...	...	57	36,703,873	31.0
	Mendoza, 1932	2,178-16,778	478	1,211	28,123	787,721,306	23.7-32.0
	Neuquen, 1918	1,312-12,498	803	1,739	99,898	568,255,130	24.0-41.0
	Rio Negro, 1960	2,953-7,546	491	994	15,863	339,635,111	22.5-41.0
	Salta, 1928	6,551-15,092	...	...	4,818	124,398,293	29.0-59.0
	Santa Cruz, 1946	1,468-17,060	1,995	5,598	47,244	985,275,193	21.1-30.0
	Tierra del Fuego, 1959	5,249-7,824	...	...	484	149,018,201	43.4-46.0
AMOCO[(e)]	Cerro Dragon, 1932	7,218	1,120	1,600			24.0
ASTRA C.A.P.S.A.[(e)]	Kilometro 20, 1912	5,249	78	78			23.0
	Cañason Seco/Meseta Espinosa, 1978	7,546	137	137			24.0
	Rio Tunoyan, 1978	7,874	22	22			28.0
	Manantiales Behr, 1980	5,577	146	146			23.0
	Restinga Ali, 1990	5,577	17	17			23.0
	Flanco Sur, 1990	4,921	55	55			24.0
	Vizcacheras, 1991	5,906	169	169			28.0
BRIDAS[(e)]	Al Sur de la Dorsal, 1941	2,850-4,265	53	160			24.6
	El Cordón, 1955	3,600-5,900	99	460			27.5
	Est. Fernández Oro, 1969	8,530	18	60			38.6
	Lindaro Atravesado[(c)], 1961	6,550-9,850	46	62			63.3
	Piedra Clavada, 1964	4,900-6,230	164	257			21.0

WORLDWIDE PRODUCTION

	Name of field, discovery date	Depth, ft	No. of wells Producing	No. of wells Total	Production 1992 average b/d	Production Cumulative to Dec. 31, 1992 bbl	°API gravity
COMPANIA NAVIERA[e] PEREZ COMPANQ	Bajada del Palo, 1978	7,874	11	38			30.0
	Cacheuta, 1933	8,202	22	75			31.0
	Cañadon Amarillo, 1969	6,234	13	21			32.5
	Caranda[c], 1970	6,562	34	103			40.0
	Catrieloeste, 1959	2,953	136	208			22.5
	Colpa[c], 1961	9,186	16	53			50.0
	Dorsal, 1934	4,593	60	147			32.0
	El Manzano, 1978	5,906	3	11			25.0
	Entre Lomas, 1966	4,757	232	349			33.0
	Jaguel de Los Machos, 1974	4,757	12	22	331,674	2,430,260,000	30.0
	Koluel Kaike, 1957	5,413	166	290			24.0
	Medanito, 1962	4,757	293	460			30.0
	Pampa del Castillo, 1935	5,906	163	302			22.5
	Paro Virgenes, 1977	5,249	8	10			57.0
	Piedras Coloradas, 1953	6,070	93	140			31.5
	Puesto Hernandez, 1969	3,937	359	553			33.5
	Puntilla Huincan, 1979	5,249	2	8			25.0
	Santa Cruz II, 1968	6,070	68	77			46.8
PLUSPETROL S.A.[e]	Aguada Baguales, 1963	3,450	23	60			38.0-39.0
	Agua Blanca, 1926	2,460	...	8			22.5
	Centenario, 1961	8,200	63	118			39.2
	Cuchuma, 1990	1,080	4	5			31.3
	El Porvenir, 1941	3,600	36	192			24.0
	Ramos[c], 1976	11,975	6	6			61.0
TOTAL AUSTRAL[e]	Ara-Alfa, 1981	5,250	11	11			42.0
	El Huenul	...	...	400			...
	Hidra, 1982	5,410	21	25			38.0
OTHER OPERATORS[e]	All fields	...	...	...			...
	Total Argentina		8,402	19,780	554,000	6,141,043,897	
AUSTRALIA BHP PETROLEUM PTY. LTD.	• Challis/Cassini, 1984	5,000	6	7	20,672	31,000,000	39.5
	• Jabiru, 1983	4,786	4	4	25,524	74,000,000	42.3
	• Skua, 1986	7,545	3	3	21,206	8,000,000	42.5
BRIDGE OIL LTD	Taylor, 1986	5,538	4	7	510	1,100,000	47.0
MARATHON AUSTRALIA	• Talisman, 1984	9,600	...	...	2,900	7,700,000	42.0
MINORA ENERGY (Australasia) PTY. LTD.	Blina, 1981	4,300	5	6	210	1,508,800	37.0
	Boundary, 1990	3,940	1	1	43	34,300	37.0
	Lloyd, 1987	4,806	1	2	15	158,900	39.0
	Sundown, 1983	3,600	2	4	36	219,800	33.0
	West Terrace, 1985	3,800	1	2	14	158,160	32.8
WESTERN MINING CORP.	Chervil, 1983	3,450	2	2	1,162	2,214,264	44.0
	North Herald, 1983	4,000	1	1	360	3,776,268	42.0
	South Pepper, 1982	4,000	9	10	7,961	10,868,062	42.0
DISCOVERY[e] PETROLEUM	Mount Horner, 1965	1,200	7	13			35.5
HADSON ENERGY LTD.[e]	• Harriet, 1983	6,230	12	12			38.2
	• Tanami, 1991	6,010	1	1			43.0
AGL PETROLEUM[e]	Roma	...	...	...			...
	Moonie, 1961	5,800	19	38			44.5
	Mereenie, 1964	4,000	24	37			48.9
PETROLEUM SECURITIES[e] ENERGY LIMITED	Blina, 1981	4,300	5	8			37.0
	Boundary, 1990	3,940	1	1			37.0
	Lloyd, 1987	4,806	1	2			39.0
	Sundown, 1983	3,600	2	4			33.0
	West Terrace, 1985	3,800	2	2			32.8
WOODSIDE[e]	• North Rankin, 1972	410	18	23			54.5
ESSO-BHP[e]	• Barracouta, 1965	4,550	7	10			62.8
	• Bream, 1969	6,200	16	24			45.4
	• Cobia, 1972	7,700	6	12			44.6
	• Dolphin, 1967	4,000	1	1			45.0
	• Flounder, 1968	8,200	10	22			46.7
	• Fortescue, 1978	7,700	10	28			43.0
	• Halibut, 1967	7,700	10	16			43.3
	• Kingfish, 1967	7,500	14	30			46.9
	• Mackerel, 1969	7,700	13	18			45.6
	• Marlin, 1966	5,100	12	22			50.0
	• Perch, 1968	3,700	1	1			41.0
	• Seahorse, 1978	4,700	1	1			53.0
	• Snapper, 1968	4,350	11	27			47.0
	• Tarwhine, 1981	4,500	1	1			...
	• Tuna, 1968	6,500	9	18			40.5
	• West Kingfish, 1968	7,500	16	23			46.9
	• Whiting, 1983	4,800	4	5			...
AMPOLEX (PPL) PTY. LTD.[e]	Cranstoun, 1987	5,141	1	1			43.0
	Endeavour, 1989	5,085	2	2			48.0

WORLDWIDE PRODUCTION

	Name of field, discovery date	Depth, ft	No. of wells Producing	No. of wells Total	Production 1992 average b/d	Production Cumulative to Dec. 31, 1992 bbl	°API gravity
	Ipundu, 1985	4,376	2	2			49.0
	Ipundu North, 1989	4,353	2	2			48.0
	Monler, 1987	5,063	2	2			49.0
	Talgeberry, 1985	4,458	2	3			46.0
	Takyah, 1986	4,301	...	1			49.0
	Tarbat, 1988	3,117	1	1			47.0
	Tintaburra, 1983	4,055	5	7			43.6
	Toobunyah, 1985	3,989	4	4			47.0
SANTOS LTD.[e]	Alwyn, 1985	5,200					41.0
	Big Lake, 1984	6,600			452,387	3,375,789,000	45.0-48.0
	Bogala, 1984	6,900					40.0
	Bookabourdie, 1985	7,000					46.0
	Brolga, 1983	9,547					53.0
	Challum, 1983	...					...
	Chookoo, 1985	6,000					45.0
	Cook, 1985	...					...
	Corella, 1989	...					...
	Cooroo/Cooroo North, 1986	...					...
	Dingera, 1987	...					...
	Dirkala, 1986	7,000					56.0
	Dullingari, 1979	5,200					54.0
	Fly Lake, 1971	9,300					53.0
	Gidgealpa, 1984	6,150					45.0-50.0
	Gidgee, 1989	...					...
	Gunna, 1983	6,000					40.0
	James, 1989	...					...
	Jackson, 1981	4,750					40.0-48.0
	Jackson South, 1982	4,800					40.0
	Jena, 1985	5,200					41.0
	Kercummura, 1985	...					50.8
	Kerinna, 1984	4,300					41.0
	Limestone Creek/Biala, 1984	5,100					41.0
	Mawson, 1987	...					...
	McKinlay, 1985	5,100					41.0
	Meranji, 1985	5,660					1.0
	Merrimelia, 1983	5,250-7,070					5.5
	Mooliampah, 1985	6,000					...
	Moorari, 1971	7,705-9,400					6.9
	Munro, 1988	...	346	346			...
	Muteroo, 1985	6,000					5.0
	Naccowlah, 1989	...					...
	Naccowlah East, 1988	...					...
	Naccowlah South, 1983	6,000					...
	Naccowlah West, 1983	5,700					...
	Narcoonowie, 1983	6,000					...
	Natan	...					...
	Nungaroo, 1985	4,150					...
	Pinaroo, 1889	...					...
	Pintari North, 1988	...					...
	Pitchery, 1988	...					...
	Sigma, 1983	5,700					...
	Spencer, 1986	6,500					7.0
	Strzelecki, 1978	5,800					...
	Sturt, 1988	...					...
	Sturt East	...					...
	Taloola, 1988	...					...
	Tantanna, 1988	...					...
	Tickalara, 1984	5,600					...
	Tinpilla, 1983	6,000					...
	Tirrawarra, 1970	9,500					2.0
	Toby, 1987	...					...
	Ulandi, 1985	5,200					0.4
	Wandilo, 1989	...					...
	Wancoocha, 1984	6,000					54.0
	Watson, 1985	...					...
	Watson South, 1985	5,500					40.0
	Wilson, 1983	5,000					...
	Woolkina, 1982	9,700					...
	Yanda, 1984	8,000					...
WAPET GROUP[e]	Barrow Island, 1964	1,200-6,700	405	569			37.7
	Dongara, 1969	5,250-5,577	2	9			35.5
	Dongara[c], 1969	5,250-5,577	1	5			50.0
	Thevenard Island, 1985	4,600-5,750	9	12			48.2
OIL COMPANY OF AUSTRALIA[e]	Black Stump, 1986	3,815	1	1			45.2
	Bodalla South, 1984	4,789-5,215	6	7			47.8
	Bora Creek, 1982	4,900	1	1			34.0
	Cogoon River West, 1990	4,660	...	1			...
	Kenmore, 1985	4,593-4,986	9	11			48.2
	Kincora, 1980	4,700	1	1			34.0
	Riverslea, 1981	4,950	2	2			37.0
	Sandy Creek, 1982	4,900	...	1			34.0

WORLDWIDE PRODUCTION

Country	Operator	Name of field, discovery date	Depth, ft	No. of wells — Producing	No. of wells — Total	Production — 1992 average b/d	Production — Cumulative to Dec. 31, 1992 bbl	°API gravity
		Waratah, 1982	5,350	3	4			34.0
		Washpool, 1985	5,250	1	1			35.0
		Total Australia		**1,081**	**1,445**	**533,000**	**3,516,527,554**	
AUSTRIA	MOBIL/SHELL/EVN	Gaiselberg, 1938	5,200	27	59	696	33,406,352	25.0
		Other	...	69	164	2,058	59,094,672	34.0
	OMV	Hochleiten, 1974	3,100-5,000	36	45	1,459	3,373,300	21.3
		Matzen, 1949	6,100	742	877	10,644	402,712,750	25.7
		Muehlberg, 1942	3,900-5,500	28	45	276	40,931,303	35.0
		Piberbach, 1968	3,500	7	7	146	4,215,784	39.0
		Pirawarth, 1957	1,600-6,500	59	61	939	14,021,977	28.4
		Schoenkirchen-Prottes Tief, 1962	7,800-10,100	19	21	4,547	65,532,575	31.1
		St. Ulrich, 1938	1,600-4,300	72	146	662	44,125,431	31.1
		Other	...	90	118	1,664	41,532,557	27.0
		Total Austria		**1,149**	**1,543**	**23,091**	**708,946,701**	
BAHRAIN	BANOCO	Awali, 1932	1,850-4,600	352	352	36,747	854,926,642	33.0
		Total Bahrain		**352**	**352**	**36,747**	**854,926,642**	
BANGLADESH	PETRO-BANGLA	Bakhrabad, 1969	9,309	6	8	251	489,501	40.2
		Feni, 1981	10,496	1	1	44	24,915	44.5
		Habiganj, 1963	11,506	5	5	5	16,753	30.3
		Kailastila, 1962	13,576	1	3	209	630,968	45.0
		Sylhet, 1955	7,803	1	2	26	512,950	46.0
		Syl-7, 1986	6,773	1	1	170	492,838	29.3
		Titas, 1962	12,326	10	11	359	1,373,541	39.1
		Total Bangladesh		**25**	**31**	**1,064**	**3,541,466**	
BARBADOS	BARB. NOC	Woodbourne, 1966	1,665-6,701	102	180	1,308	6,794,344	26.1
		Total Barbados		**102**	**180**	**1,308**	**6,794,344**	
BENIN(e)	PROJET PÉTROLIER DE SÈMÈ	Seme, 1968	6,232	8	8	3,000	17,742,000	22.6
		Total Benin		**8**	**8**	**3,000**	**17,742,000**	
BOLIVIA	YPFB	Buena Vista, 1956	2,085	3	8	19	6,856	45.8
		Caigua, 1973	3,280	2	5	18	6,618	42.4
		Camatindi, 1929	4,110	5	6	80	29,366	42.4
		Cambeiti, 1976	5,787	6	6	90	1,100,935	48.0
		Camiri, 1927	3,936	41	59	327	48,950,180	55.0
		Caranda, 1960	13,032	...	...	...	54,359,920	60.0
		Carrasco, 1991	15,650	1	1	797	293,076	56.1
		Cascabel, 1985	11,319	6	7	682	824,936	45.6
		Colpa, 1962	9,184	...	...	...	18,284,329	65.0
		H. Suarez R., 1982	7,820	5	7	283	934,053	35.0
		Katari, 1992	16,210	1	1	201	73,410	57.5
		La Pena, 1965	8,856	19	49	2,342	32,465,742	43.0
		Monteagudo, 1967	4,592	22	29	766	33,574,236	41.0
		Naranjillos, 1964	4,806	13	26	77	193,257	65.3
		Palmar, 1964	10,758	2	5	25	1,805,700	65.0
		Rio Grande, 1962	9,512	33	48	1,213	62,978,101	62.0
		Santa Cruz, 1965	9,350	5	11	94	1,953,697	65.0
		San Roque, 1981	7,170	16	18	2,846	4,556,624	71.0
		Sirari, 1985	13,999	7	7	1,353	1,663,048	65.0
		Tatarenda, 1964	2,296	8	12	122	7,607,845	65.3
		Tita, 1976	12,949	1	15	4	4,441,439	51.0
		Tundy, 1992	10,597	1	1	99	36,094	48.6
		Vivora, 1988	12,949	10	10	1,826	1,982,566	67.9
		Villamontes, 1987	9,770	1	2	42	685,398	57.7
		Vuelta Grande, 1978	8,220	24	24	2,728	6,190,189	57.0
		Yapacani, 1968	8,400	7	15	775	566,579	34.1
		Other	...	8	14	185	23,931,205	...
	OCCIDENTAL	Porvenir, 1978	9,600	7	7	957	13,332,112	60.5
	TESORO BOLIVIA	Escondido, 1980	11,293	2	2	50	23,934	60.0
		La Vertiente, 1977	15,230	10	15	1,107	5,889,001	50.0
		Taiguati, 1981	10,174	2	2	359	655,090	63.0
	PEREZ	Caranda, 1960	13,032	29	56	389	436,800	60.0
		Colpa, 1962	9,184	17	36	387	423,792	65.0
	SOL PETROL	Buena Vista, 1956	2,085	...	...	1	20,093	45.8
		Camatindi, 1929	4,110	...	...	8	77,171	42.4
		Caigua, 1973	3,280	...	...	2	10,819	42.4
	SOPETROL	Palmer, 1964	10,758	...	...	24	8,861	65.0

WORLDWIDE PRODUCTION

		Name of field, discovery date	Depth, ft	No. of wells — Producing	No. of wells — Total	Production — 1992 average b/d	Production — Cumulative to Dec. 31, 1992 bbl	°API gravity
	PLUSPETROL	Bermejo, 1924	2,085	15	30	438	331,000	42.0
		Toro, 1954	2,982	2	2	248	137,693	27.0
	MAXUS ENERGY	Surubi, 1992	14,078	1	2	216	79,061	44.0
		Total Bolivia		**332**	**538**	**21,180**	**330,920,826**	
BRAZIL	PETROBRAS/SDJCL	Agua Grande, 1951	820-4,265	112	129	5,423	282,412,127	40.0
		• Agulha, 1975	5,412	4	4	1,060	8,737,444	33.0
		• Albacora, 1984	8,430	15	22	31,705	44,370,890	26.0
		Alto Rodrigues, 1981	990-1,120	326	328	4,612	12,808,429	18.0
		Anequim, 1981	8,500	2	2	2,241	7,712,081	27.0
		Aracas, 1965	3,280-8,856	143	154	7,020	121,110,942	41.0
		• Area RJS-387, 1987	11,300	1	1	734	3,560,341	34.0
		Atalaia Sul., 1961	1,450	...	1	...	2,688,650	24.0
		• Atum, 1983	2,300-3,200	14	20	4,670	15,266,653	31.0
		• Badejo, 1981	8,500-10,500	6	7	3,926	23,253,931	28.0
		• Bagre, 1983	8,700-9,000	5	5	2,979	11,399,822	25.0
		• Bicudo, 1982	8,000-11,000	12	13	17,310	63,883,378	22.0
		• Bonito, 1977	7,000-8,000	14	19	15,816	53,415,253	27.0
		Brejo Grande, 1967	1,670	8	10	43	2,175,493	26.0
		Buracica, 1959	1,476-1,840	169	192	5,495	145,530,918	35.0
		• Cacao, 1978	9,184	6	6	2,213	13,672,622	32.0
		• Caioba, 1969	6,232-6,890	10	10	805	32,282,372	36.0
		• Camurim, 1970	6,232	40	55	3,833	24,725,085	25.0
		• Candeias Mar, 1941	3,300-7,400	2	3	512	7,859,756	30.0
		Candeias Terra, 1941	3,280-7,415	41	45	2,837	82,081,498	30.0
		Canto Do Amaro, 1986	2,500-3,600	678	708	36,723	62,001,884	25.0
		Carapeba, 1982	10,100	32	35	28,854	41,189,195	23.0
		Carmopolis, 1963	2,460	867	918	20,252	206,486,756	24.0
		Cassarongongo, 1959	1,475-4,755	116	133	1,854	19,246,744	32.0
		Cexis, 1983	1,670-5,230	40	46	3,161	8,429,816	30.0
		Cherne, 1982	9,900-10,200	31	33	37,480	107,089,191	27.0
		• Corvina, 1983	9,000-11,000	6	6	11,487	49,594,318	25.0
		• Curiman, 1978	6,888	10	15	3,781	23,738,240	32.0
		• Dom Joao Mar, 1947	920-3,000	138	167	1,813	86,026,871	38.0
		Dom Joao Terra, 1947	920-1,350	34	62	456	21,529,426	38.0
		• Dourado, 1970	3,620-3,635	1	4	300	4,996,702	36.0
		• Enchova, 1976	6,973-7,885	16	16	18,082	95,332,289	23.0
		• Espada, 1983	5,750-6,900	5	6	1,756	6,159,067	26.0
		Estreito, 1983	650-820	469	470	7,443	19,609,735	16.0
		Faz Alvorada, 1983	2,100-4,900	65	75	3,099	12,131,741	31.0
		Faz Balsamo, 1983	2,500-4,000	118	125	4,393	16,344,177	31.0
		Faz Belem, 1980	1,150-2,500	314	322	4,368	11,699,015	13.0
		Faz Cedro, 1962	5,248	14	14	569	10,934,509	28.0
		Faz Imbe, 1964	2,130-5,575	44	55	973	17,311,599	40.0
		Furado, 1969	4,440-5,500	36	39	1,325	17,348,325	35.0
		• Garoupa, 1974	11,382	15	19	11,435	93,691,156	30.0
		• Garoupinha, 1974	12,693	2	3	2,338	15,210,987	30.0
		Guamare, 1983	1,400-1,600	25	27	461	2,297,677	25.0
		• Guaricema, 1968	3,335-4,430	8	9	3,105	42,225,189	41.0
		Lagoa Suruaca, 1982	6,400	14	16	1,191	6,944,215	30.0
		• Linguado, 1981	9,000-11,000	16	16	22,322	99,101,403	30.0
		Livramento, 1986	2,500-3,300	27	27	1,076	6,560,041	41.0
		Lorena, 1984	1,200-1,700	30	30	640	1,432,565	28.0
		• Malhado, 1990	10,824	1	1	1,865	2,108,026	22.0
		Malombe, 1966	3,280	9	11	273	7,343,625	41.0
		• Marimba, 1983	8,900-9,200	10	10	28,294	46,433,630	28.0
		• Marlim, 1991	11,808	7	7	21,003	9,707,614	15.0
		Mata/Remanso, 1971	4,445-4,560	29	57	1,205	22,461,033	35.0
		Miranga, 1965	3,280-3,936	245	267	6,136	192,982,450	40.0
		• Moreia, 1984	8,200-9,500	4	5	2,638	7,816,019	22.0
		Mossoro, 1980	3,280	24	26	1,335	1,430,432	28.0
		• Namorado, 1975	10,083	31	32	46,205	195,024,010	28.0
		• Pampo, 1977	6,215	31	32	25,242	149,842,334	21.0
		• Parati, 1982	8,900-9,200	1	1	946	6,529,409	35.0
		• Pargo, 1975	11,840	16	21	11,402	21,370,413	20.0
		Pilar, 1981	4,100-5,100	48	55	3,193	18,683,700	40.0
		• Pirauna, 1982	8,000-10,500	7	8	10,268	44,430,769	27.0
		Riacho Da Barra, 1982	2,910-3,998	34	43	2,030	15,165,136	39.0
		Riachuelo, 1961	1,475	196	210	2,302	26,232,480	28.0
		Rio Do Bu, 1984	2,900-3,000	51	56	3,057	7,808,082	33.0
		Rio Dos Ovos, 1979	2,410-3,750	9	10	357	2,476,854	35.0
		Rio Itaunas, 1977	5,400	27	36	682	4,839,589	28.0
		Rio Pojuca, 1982	5,600-6,200	28	29	1,659	7,113,971	35.0
		Rio Preto, 1976	1,460-2,600	15	18	205	2,017,761	...
		Santana, 1962	2,400-4,000	4	6	67	5,750,494	37.0
		Sao Mateus, 1969	5,750	34	37	952	6,052,316	27.0
		Serraria, 1982	3,200-3,900	34	37	1,010	4,499,025	27.0
		Sesmaria, 1963	3,100-4,600	26	31	862	4,553,601	40.0
		Siririzinho, 1967	1,640	160	172	2,956	42,283,927	30.0
		Tab Do Martins, 1957	3,900-4,950	22	29	281	4,556,532	27.0
		Taquipe, 1958	3,115-3,935	92	103	2,962	90,491,554	40.0

WORLDWIDE PRODUCTION

Country	Operator	Name of field, discovery date	Depth, ft	No. of wells: Producing	No. of wells: Total	Production: 1992 average b/d	Production: Cumulative to Dec. 31, 1992 bbl	°API gravity
		• Trilha, 1983	8,800-11,000	3	3	1,394	14,111,669	28.0
		• Ubarana, 1973	7,216-8,365	71	78	11,452	70,482,989	40.0
		Upanema, 1985	1,110-1,800	14	16	1,036	3,233,259	28.0
		• Vermelho, 1983	11,200	55	59	34,411	47,913,935	24.0
		• Viola, 1981	9,200-9,900	7	9	5,467	20,877,041	13.0
		• Xareu, 1977	5,717-5,766	16	28	3,964	19,360,640	39.0
		Other fields	...	787	866	44,590	153,226,795	....
		Total Brazil		**6,249**	**6,831**	**625,672**	**3,342,821,632**	
BRUNEI	SHELL	• Ampa SW, 1963	7,068-8,155	160	222	59,555	686,791,544	41.0
		• Champion, 1970	4,300	162	249	58,590	367,471,705	23.0
		• Fairley, 1969	10,740	42	52	11,257	132,842,876	40.0
		• Fairley-Baram, 1973	9,500	4	5	3,124	20,383,083	40.0
		Gannet	...	6	6	2,794	6,298,665	40.0
		• Magpie, 1975	...	13	21	10,755	93,230,352	31.0
		Rasau, 1929	...	15	19	11,021	12,519,065	30.0
		Seria-Total, 1929	...	336	950	14,770	1,278,332,682	34.0
		Total Brunei		**738**	**1,524**	**171,866**	**2,597,869,972**	
CAMEROON	ELF	• Asoma Centre, 1977	4,000	14	18			26.1
		• Asoma Marine, 1973	3,600	7	10			19.0
		• Bakassi Marine, 1976	6,500	1	1			37.1
		• Barombi Marine, 1975	7,400	1	1			34.1
		• Bavo, 1976	5,600	31	37			29.3
		• Betika, 1972	5,300	7	12			42.5
		• Betika S. Marine, 1979	6,300	2	5			30.0
		• Biboundi, 1974	6,400	2	7			34.6
		• Boa, 1976	5,900	7	12			35.0
		• Ekoundou Horst plus Center, 1972	7,200	19	28			29.0
		• Ekoundou S., 1975	6,200	13	32			24.0
		• Ekoundou N. Marine, 1977	5,250	8	24			36.0
		• Inoua Marine, 1976	6,200	7	10	102,000	540,444,805	32.8
		• Itindi, 1977	5,600	9	9			26.1
		• Kita E. Marine, 1978	6,900	9	12			40.8
		• Kole, 1974	5,500	15	17			31.0
		• Kole Superieur, 1978	4,100	7	8			23.0
		• Kombo Center, 1976	5,900	19	27			37.0
		• Kombo Nord, 1978	5,700	3	8			36.0
		• Nyangassa, 1977	5,100	12	14			29.8
	PECTEN INTERNATIONAL[e]	• Makoko NE plus Abana, 1977	4,600	...	...			29.0
		• Makoko S. Marine, 1977	3,300	...	...			17.0
	OTHER OPERATIONS[e]	Other fields		...	...	35,000	125,446,000	...
	TOTAL[e]	• Moudi, 1979	5250-5,900	...	...			40.0
		• Moudi D., 1980	3,300	...	...			21.3
		Total Cameroon		**193**	**292**	**137,000**	**665,890,805**	
CANADA	ALBERTA	Acheson, 1950	3,390	56	127	3,268	136,509,345	38.0
		Bantry, 1940	2,574-3,300	199	259	10,281	64,450,947	...
		Bellshill Lake, 1955	2,950	390	478	9,443	78,085,709	...
		Bonnie Glen, 1952	3,959-6,779	91	165	4,374	513,877,667	34.0-44.0
		Caroline	...	192	301	6,388	76,316,783	...
		Carson Creek N., 1958	8,787-8,935	37	58	6,082	150,548,109	44.0
		Cessford, 1950	2,800-3,028	209	341	3,150	35,022,969	...
		Clive, 1951	6,074-6,208	122	190	5,431	58,599,207	...
		Countess, 1951	2,858-4,272	135	209	6,941	73,421,434	...
		Fenn-Big Valley, 1950	3,900-5,400	103	218	3,166	324,983,953	32.0
		Gilby, 1962	4,201-7,000	86	168	3,369	51,667,611	...
		Golden Spike, 1949	3,570-5,961	56	79	1,848	184,657,983	36.0
		Grand Forks	...	361	505	20,647	122,834,938	...
		Harmattan East, 1954	6,358-8,766	89	142	3,201	78,232,333	...
		Harmatton Elkton, 1955	8,796-10,757	59	83	3,525	67,985,652	...
		Hayter, 1968	2,030-2,815	301	482	9,379	24,566,736	...
		Innisfail, 1956	6,730-8,440	24	69	2,656	80,056,007	...
		Joarcam, 1949	3,179	172	243	5,090	110,929,457	...
		Joffre, 1953	4,983-6,779	87	180	2,572	88,270,719	...
		Judy Creek, 1951	8,307-8,701	134	269	15,504	421,330,313	43.0
		Kaybob, 1957	4,870-9,577	75	102	6,719	112,464,288	...
		Kaybob S., 1958	5,600-10,042	60	109	13,649	93,661,192	...
		Leduc-Woodbend, 1947	3,062-5,380	103	610	1,473	380,209,553	40.0
		Lloydminster, 1933	1,690-1,945	303	803	3,559	55,997,105	...
		Medicine River, 1954	5,435-7,600	248	381	10,132	86,114,782	...
		Mitsue, 1964	5,908	283	436	29,327	326,382,858	43.0
		Nipisi, 1965	5,648-5,726	281	383	22,882	312,201,875	41.0
		Pembina, 1953	3,000-6,133	3,335	4,285	77,412	1,370,203,260	32.0-37.0
		Provost, 1946	2,760-2,898	2,491	3,272	82,979	171,965,893	...
		Rainbow, 1965	4,994-6,160	281	405	36,945	569,192,625	38.0-42.0
		Rainbow South, 1965	6,102-6,400	51	82	5,451	72,963,313	...
		Red Earth	4,184-4,878	187	268	4,290	44,837,959	...

WORLDWIDE PRODUCTION

Country	Region	Name of field, discovery date	Depth, ft	No. of wells Producing	No. of wells Total	Production 1992 average b/d	Production Cumulative to Dec. 31, 1992 bbl	°API gravity
		Redwater, 1956	2,012-3,200	578	874	9,795	801,170,474	35.0
		Simonette, 1958	10,500	22	46	2,310	41,573,218	...
		Snipe Lake, 1962	8,500	35	93	2,526	58,579,699	...
		Sturgeon Lake S., 1953	4,912-8,471	74	147	9,104	160,329,117	...
		Swan Hills, 1957	8,100	538	817	38,047	736,207,916	40.0
		Swan Hills S., 1959	8,400	146	200	8,766	360,656,976	...
		Taber N., 1966	3,110-3,270	96	138	8,913	25,514,442	...
		Turner Valley, 1913	3,100-9,150	111	151	2,469	143,661,821	39.0
		Utikuma Lake, 1963	5,624	110	148	10,308	66,930,967	...
		Valhalla, 1973	1,815-7,060	312	370	10,268	24,629,035	...
		Virginia Hills, 1957	9,210	124	191	17,352	172,975,898	34.0
		Wainwright, 1925	1,903-2,200	601	750	8,310	81,739,979	...
		Westerose, 1952	6,818	31	43	4,680	140,360,581	...
		Westpem	...	19	29	7,003	41,726,764	...
		Willisden Green, 1956	5,157	497	826	6,363	127,680,448	...
		Wizard Lake, 1951	4,044-5,973	34	49	4,484	332,213,202	...
		Zama, 1966	3,702	106	248	5,435	84,707,696	33.0-37.0
		Other	...	8,992	13,869	336,998	1,907,006,482	...
	BRITISH COLUMBIA	Boundary Lake, 1957	3,418-4,575	251	355	8,906	185,079,605	40.0
		Eagle West, 1976	2,170-3,940	46	95	3,972	34,075,892	...
		Other	...	395	850	22,083	263,936,574	...
	MANITOBA	All fields	...	1,387	1,547	11,286	183,927,626	...
	SASKATCHEWAN	Steelman, 1950	...	614	703	7,025	266,758,517	...
		Weyburn, 1955	4,600	870	944	17,067	312,891,668	...
		Other	...	12,948	17,281	205,579	1,989,960,575	...
	ONTARIO	All fields	...	970	1,180	3,857	68,229,740	...
	NEW BRUNSWICK	All fields	...	...	...	...	805,261	...
	EASTCOAST OFFSHORE	All fields	...	4	4	18,604	3,627,840	...
	NORTHWEST TERRITORIES	All fields	...	155	165	32,651	112,927,783	
		Total Canada*		**40,667**	**57,815**	**1,245,294**	**15,068,428,371**	
CHILE	ENAP-ISLA	Calafate, 1956	6,045	11	61	227	26,524,538	49.3
		Catalina, 1956	5,756	7	26	149	6,773,484	38.7
		Catalina Sur, 1962	5,779	6	43	68	6,414,713	44.0
		Cullen, 1954	5,730	9	106	251	41,521,294	41.8
		Flamenco, 1954	...	6	6	163	3,453,389	40.2
		Gaviota Norte, 1990	7,000	6	6	833	1,493,792	38.0
		Gorrion, 1986	6,025	4	10	370	1,442,631	39.0
		Tres Lagos, 1957	5,724	15	84	311	16,528,973	40.1
		Other	...	30	264	1,342	57,555,454	...
	ENAP-CONTINENTE	Canadon, 1962	6,090	8	62	79	14,720,514	38.4
		Daniel, 1960	5,806	8	121	131	34,756,660	25.4
		Daniel Este, 1961	5,523	9	66	359	28,892,576	25.0
		Delgada Este, 1958	7,694	9	31	147	5,100,382	35.0
		Dungenes, 1962	5,251	5	19	98	4,206,254	28.6
		Faro Este, 1959	7,381	5	32	131	3,502,191	35.0
		Posesion, 1961	5,622	38	76	585	22,390,371	63.6
		Other	...	16	118	1,532	17,403,036	...
	ENAP-COSTA AFUERA	Daniel Este-Dungeness, 1984	6,126	19	36	835	5,743,584	35.9
		Ostion, 1979	7,156	9	13	284	8,770,970	33.0
		Pejerrey, 1984	7,229	22	67	532	10,614,212	39.0
		Posesion, 1980	6,727	19	39	292	5,986,836	34.5
		Skua, 1988	5,932	23	39	1,421	5,424,571	32.0
		Spiteful, 1979	6,298	26	82	945	45,257,114	34.5
		Spiteful Norte, 1980	7,150	3	32	192	9,960,323	39.0
		Other	...	42	42	4,501	4,158,124	...
		Total Chile		**355**	**1,481**	**15,778**	**388,595,986**	
CHINA(e)	CNPC	• Cnooc	...			79,400		...
		Changqing	...			32,900		...
		Dagang, 1964	...			78,500		...
		Daqing, 1959	...			1,116,400		...
		Henan	...			47,300		...
		Huabei	...			98,100		...
		Jianghan	...			15,900		...
		Jiangsu	...			17,000		...
		Jidong	...			7,800		...
		Jilin, 1978	...	49,700	49,700	68,600	17,757,340,000	...
		Liaohe, 1964	...			276,400		...
		Qinghai	...			20,400		...
		Sichuan	...			2,800		...
		Shengli, 1962	...			673,600		...
		Tarim	...			16,400		...
		Xinjiang	...			147,400		...

*Excludes oil sands bitumen production and wells.

WORLDWIDE PRODUCTION

Country	Company	Name of field, discovery date	Depth, ft	No. of wells: Producing	No. of wells: Total	Production: 1992 average b/d	Production: Cumulative to Dec. 31, 1992 bbl	°API gravity
		Yumen	...			18,700		...
		Zhongyuan, 1981	...			117,400		...
		Total China		**49,700**	**49,700**	**2,835,000**	**17,757,340,000**	
CHINA, TAIWAN	CHINESE PETROLEUM	Chingtsohu & CBK, 1967	...	7	24			50.0
		Chinshui & Yunghoshan, 1971	...	20	35			45.0
		Chuhuangkeng, 1904	...	31	34	1,160	27,066,862	50.0
		Tiechengshan, 1962	...	19	31			51.0
		Other	...	7	10			...
		Total China, Taiwan		**84**	**134**	**1,160**	**27,066,862**	
COLOMBIA	ELF AQUITAINE	Barquerena, 1982	9,300	3	3	726	4,277,000	32.3
		Caño Garza, 1979	7,900	6	7	3,145	6,128,300	40.2
		Caño Garza Norte, 1983	7,900	1	2	375	2,598,400	40.2
		Chaparrito, 1988	5,400	2	2	411	150,000	34.4
		Cravo Este, 1987	11,376	1	1	170	187,000	42.0
		Cravo Sur, 1982	10,600	2	2	997	4,270,700	40.2
		Gloria Norte, 1983	12,100	3	3	4,945	7,077,300	17.2
		La Flora, 1985	8,600	2	2	2,140	1,331,700	34.2
		La Gloria, 1990	14,000	5	6	7,192	5,690,500	17.8
		Las Abejas, 1990	5,600	1	1	112	41,000	33.0
		Los Toros, 1989	5,300	1	1	658	240,000	34.2
		Morichal, 1984	15,500	1	2	627	1,186,900	30.0
		Palmarlto, 1988	5,200	1	1	822	300,000	41.1
		Tierra Blancha, 1987	14,493	1	1	359	708,600	32.0
		Tocaria, 1980	13,500	2	3	992	2,308,700	32.1
		Trinidad, 1974	9,000	6	7	7,216	12,297,800	32.4
	OCCIDENTAL	Caño Limón, 1984	7,600	10	10	23,690	69,705,004	29.0
		Cano Yarumal, 1987	7,600	4	4	37,134	39,129,862	29.0
		La Yuca, 1984	7,600	42	45	113,380	273,603,081	29.4
		Matanegra, 1984	7,600	12	14	21,562	65,353,990	29.7
		Redondo, 1984	7,600	4	4	5,966	14,369,702	30.2
	ESSO COLOMBIANA LTD.	Arauca, 1980	19,500	4	4	...	6,500,000	40.5
		Rio Ceibas, 1988	4,000	2	2	44	16,000	21.5
		Yaguará, 1987	3,000	25	27	3,019	1,150,000	22.6
	CHEVRON PETROLEUM COMPANY	Castilla, 1969	6,500	19	21	14,063	44,065,217	13.5
		Chichimene NE, 1969	6,500	4	4	1,422	2,095,738	21.3
	TEXACO	Cocorná, 1963	1,600-2,300	31	50	1,237	16,290,552	12.8
		Nare Sur, 1963	1,600-2,300	48	51	2,988	4,378,405	12.8
		Teca, 1963	1,600-2,300	135	270	10,400	44,020,523	12.8
		Velazquez, 1946	5,600-7,400	82	104	3,196	170,596,268	19.7-28.5
	HOCOL SA (SHELL)	Neiva, 1961	...	...	...	19,500	122,890,000	21.0
		San Francisco, 1985	2,800	41	41	25,400	53,150,000	27.0
	ARGOSY ENERGY INTERNATIONAL[(e)]	Burdine, 1974	11,100	2	5			...
		Maxine, 1976	11,200	...	1			...
		Nancy, 1974	11,300	1	1			...
	HOCOL SA (SHELL)[(e)]	Balcon, 1988	...	...	...			21.0
		Tello, 1972	7,500	30	34			22.0
	LASMO OIL (COLOMBIA) LTD.[(e)]	Santiago, 1985	10,000	6	8			...
	CONSOLIDATED EUROCAN VENTURES[(e)]	Aguas Claras	...	1	1			...
		Curazon	...	1	1			...
		Payda/La Salina	55	102	102			...
	PETROLEOS COLOMBIANOS[(e)]	Andalucia Sur, 1980	2,200	16	31			...
	PETROMINEROS[(e)]	Totare, 1987	1,200	5	8			...
	AMERICAN INTERNATIONAL PETROLEUM CORPORATION[(e)]	Toqui Toqui, 1985	3,000	8	9			...
	ECOPETROL[(e)]	Infantas, 1917	2,000	236	568			23.0
		La Cira, 1926	3,300	612	1,125			23.0
		Colorado, 1923	4,000	17	82			30.0
		Aguas Blancas, 1962	7,000	2	7			30.0
		San Luis, 1926	5,000	2	4			35.0
		Tenerife, 1968	7,500	1	3			19.0
		Lisama, 1965	8,000	93	144			30.0
		Nutria, 1981	8,000	18	26			30.0
		Tesoro, 1979	8,000	24	29			30.0
		Peroles, 1977	7,000	6	12			30.0
		Llanito, 1961	7,000	45	75			20.0
		Gala, 1985	7,500	11	12			20.0
		Galan, 1945	5,000	43	122			19.0
		San Silvestre, 1958	3,500	3	17			19.0
		Casabe, 1941	4,000	361	361			19.0
		Peeas Blancas, 1958	4,000	4	7			22.0
		Yarigui, 1953	8,500	55	69			16.0
		Cantagallo, 1942	7,500	10	23			20.0

WORLDWIDE PRODUCTION

		Name of field, discovery date	Depth, ft	No. of wells Producing	No. of wells Total	Production 1992 average b/d	Production Cumulative to Dec. 31, 1992 bbl	°API gravity
		Cristalina, 1959	9,500	1	8			29.0
		Garzas, 1957	6,000	1	2			31.0
		Pavas, 1983	11,000	1	1			21.0
		Totumal, 1955	8,000	2	4			24.0
		Bca. Lebrija, 1955	8,000	2	11			24.0
		Sogamoso, 1973	10,000	2	4			29.0
		Bajo Rio, 1973	8,000	1	2			23.0
		Tibu, 1940	4,500	384	554			31.0
		Sardinata, 1941	7,000	11	45			45.0
		Petrolea, 1933	600	7	207			47.0
		Rio de Oro, 1920	8,000	14	65			45.0
		Yuca, 1957	5,500	1	5			...
		Carbonera, 1939	3,000	3	7			...
		Pto. Barco	7,000	...	4			...
		Orito, 1963	6,800	46	97			30.0
		Caribe, 1959	7,900	3	3			30.0
		Sucio, 1970	7,900	1	1			30.0
		Hormiga, 1970	11,100	1	3			30.0
		Loro, 1968	10,600	3	7			30.0
		Acae, 1965	10,800	8	9			30.0
		Temblom, 1964	10,600	...	2	124,112	3,010,409,914	32.0
		San Antonio, 1969	9,600	3	4			30.0
		San Miguel, 1965	10,800	2	4			30.0
		Churuyaco, 1968	9,200	4	4			31.0
		Sucombios, 1959	9,000	3	3			31.0
		Quilili, 1958	9,600	2	2			25.0
		Sibumdoy, 1958	9,600	1	1			32.0
		Mansoya, 1929	9,600	1	1			33.0
		Pieuea, 1991	9,550	...	1			35.0
		Cencella, 1991	9,620	...	1			32.0
		Alborada, 1991	9,580	...	1			32.0
		Quillacinga, 1988	9,500	...	1			30.0
		Quinde, 1983	10,000	...	1			17.0
		Curiquinga, 1983	9,600	...	1			35.0
		Alea, 1983	8,300	...	1			31.0
		Cedral, 1967	10,500	...	1			14.0
		Cohembi, 1969	9,580	...	1			15.0
		Mecaya, 1989	7,360	...	1			27.0
		Yurilla, 1991	9,700	...	1			31.0
		Apiay, 1982	10,650	13	15			21.0
		Guatiquia, 1985	10,750	2	2			21.0
		Apiay Este, 1987	10,650	1	1			19.0
		Suria, 1985	10,350	4	4			33.0
		Suria Sur, 1985	10,350	3	4			33.0
		Pompeya, 1982	10,400	1	1			33.0
		Austral, 1987	10,550	1	1			33.0
		Castilla Est., 1989	7,050	1	1			13.0
		Castilla Norte, 1988	7,050	2	2			13.0
		Guayuriba, 1984	10,600	...	2			...
		Camoa, 1989	6,950	...	1			19.0
		Libertad, 1986	10,850	...	3			33.0
		Reforma, 1985	11,050	...	2			26.0
		Libertad Norte, 1987	11,050	...	4			33.0
		Ocoa, 1987	10,600	...	1			26.0
		Togo, 1987	10,950	...	1			17.0
		Tanane, 1986	10,600	...	3			33.0
		Entrerios, 1985	8,950	...	1			17.0
		La Punta, 1986	8,300	...	1			23.0
		Rancho Hermoso, 1984	9,050	...	2			17.0
		Saurio, 1982	10,650	...	1			33.0
		Vanguardia, 1990	14,500	...	1			15.0
		Palagua, 1954	5,000	118	188			25.0
		Caipal-Erm. 1954	6,000	3	19			17.0
		Ortega, 1955	6,000	9	15			27.0
		Toldado, 1987	6,000	6	7			20.0
		Toy, 1987	900	3	3			19.0
		Pacande, 1989	6,000	1	1			28.0
		Quimbaya, 1983	6,000	1	2			19.0
		Cicuco, 1957	8,000	5	28			42.0
		Boquete, 1961	8,000	7	24			42.0
		Total Colombia		**2,905**	**4,994**	**438,000**	**3,986,518,156**	
COMMONWEALTH OF INDEPENDENT STATES(e)	REPUBLICS	Azerbaijan*	...					...
		Belarus	...					...
		Georgia*	...					...
		Kazakhstan	...	148,990	169,440	8,949,000	114,383,361,000	...
		Kirgizstan	...					...
		Russia	...					...
		Tajikistan	...					...
		Turkmenistan	...					...

WORLDWIDE PRODUCTION

Country	Company	Name of field, discovery date	Depth, ft	No. of wells: Producing	No. of wells: Total	Production: 1992 average b/d	Production: Cumulative to Dec. 31, 1992 bbl	°API gravity
		Ukraine	...					...
		Uzbekistan	...					...
*Have not joined C.I.S.		**Total Commonwealth of Independent States**		**148,990**	**169,440**	**8,949,000**	**114,383,361,000**	
CONGO	AGIP	• Loango, 1972	3,000	63	63	22,074	119,101,000	25.5
		• Zatchi, 1980	2,000	35	36	13,360	13,903,000	24.7
	ELF CONGO	Bindi, 1982	4,400	1	1	27	175,009	34.2
		• Emeraude, 1969	820	103	103	13,834	192,988,130	22.3
		Kunji, 1980	4,000	2	2	110	678,617	35.8
		• Likouala, 1979	4,270	28	41	11,358	77,309,195	32.3
		Mengo, 1979	6,500	3	3	106	827,185	34.0
		Pointe Indienne, 1951	5,000	5	5	127	7,302,466	36.9
		• Sendji, 1973	3,940	54	80	20,376	92,283,430	29.5
		• Tchendo, 1979	1,600	18	18	15,904	5,857,291	32.6
		• Tchibouela, 1983	2,110	30	30	42,928	81,449,237	27.0
		• Yanga, 1979	2,950	54	69	12,637	90,947,846	29.4
		Total Congo		**396**	**451**	**152,841**	**682,822,406**	
CROATIA	INA-NAFTAPLIN	Benicanci, 1966	6,752-5,938	21	24	1,870	108,674,078	30.2
		Bilogora, 1966	3,346-5,052	30	50	554	9,257,533	40.8
		Bizovac, 1989	5,483-5,499	7	7	403	319,056	33.8
		Boksic-Klokocevci, 1975	7,218-10,171	7	9	51	292,109	31.0
		Bunjani, 1951	1,800-2,550	23	27	126	2,638,905	30.2
		Crnac, 1977	6,814-8,061	2	3	76	230,938	34.6
		Deletovci, 1982	3,143-3,422	...	48	...	8,396,728	28.9
		Dugo Selo, 1953	3,445-4,183	13	16	365	10,351,064	31.1
		Ferdinandovac, 1959	5,625-5,915	6	7	120	1,305,538	36.0
		Ilaca, 1982	3,445-3,855	...	12	...	561,511	31.1
		Ivanic Grad, 1959	5,282-5,709	47	47	2,688	48,620,649	33.4
		Jagnjedovac, 1961	2,864-3,570	24	35	403	7,019,346	33.0
		Jamarice, 1965	3,281-5,512	86	106	1,693	16,747,691	37.6
		Jezevo, 1963	6,234-7,382	10	11	284	1,044,192	36.7
		Kalinovac(c), 1978	11,841-14,682	14	15	7,024	14,645,897	51.5
		Klostar, 1952	2,297-4,593	60	82	1,139	41,811,593	36.3
		Kozarice, 1972	3,281-3,609	18	19	221	2,014,135	23.6
		Kriz, 1948	1,600-2,250	32	50	101	6,942,417	17.4
		Kucanci-Kapelna, 1973	6,562-7,218	13	14	693	2,247,546	25.2
		Lacici, 1979	7,605-8,012	7	9	1,102	5,919,333	35.6
		Lepavina, 1969	1,780-2,200	8	9	57	1,066,388	22.8
		Leticani, 1989	4,584-4,817	4	5	76	152,373	35.0
		Lipovljani, 1960	3,937-5,709	73	103	1,549	24,635,318	36.7
		Lupoglav, 1971	3,860-5,600	5	5	76	241,611	33.4-37.0
		Mihovljani, 1973	5,150-5,760	4	4	89	144,576	32.5
		Molve(c), 1974	10,263-11,345	17	22	1,467	495,463	57.2
		Mramor Brdo, 1949	2,625-4,101	4	4	32	4,802,530	31.1
		Obod, 1967	6,378-7,550	3	5	151	2,349,371	32.5
		Okoli, 1962	7,382-8,366	4	4	158	834,072	35.5
		Okoli(c), 1962	7,218-7,874	14	18	629	4,705,475	80.6
		Privlaka, 1983	2,933-3,041	...	17	...	2,016,129	25.9
		Sandrovac, 1962	2,250-3,700	107	150	3,241	42,701,882	37.8
		Stari Gradac(c), 1980	11,720-13,630	4	7	1,183	1,592,473	51.4
		Stevkovica, 1978	7,710-9,843	13	17	579	4,627,948	31.9
		Struzec, 1956	2,362-3,773	60	81	4,638	108,214,480	38.8
		Vezisce, 1965	6,135-6,562	...	5	...	59,484	31.6
		Zutica, 1963	5,742-6,923	165	174	5,947	95,783,820	33.8
		Zutica(c), 1963	...	7	15	44	2,093,077	87.9
		Other(c)	5,250-11,800	7	8	31	310,730	87.9
		Total Croatia		**919**	**1,244**	**38,860**	**585,867,459**	
DENMARK	DUC	• Dagmar, 1993	4,200	2	2	5,252	4,910,605	33.8
		• Dan, 1971	6,000-6,400	57	65	46,410	95,049,820	29.7
		• Gorm, 1971	6,500-7,000	23	33	28,579	113,007,461	33.9
		• Kraka, 1966	6,000	3	3	3,527	2,197,035	31.5
		• Rolf, 1981	5,900	2	3	5,225	17,370,845	35.3
		• Skjold, 1977	5,000	6	11	39,195	97,162,969	30.3
		• Tyra, 1968	6,500	27	46	28,701	50,149,967	50.4
		Total Denmark		**120**	**163**	**156,889**	**379,848,702**	
DUBAI	DUPETCO(e)	• Fateh, 1966	7,900-9,000					31.8
		• S.W. Fateh, 1970	7,500-9,000					30.3
		• Falah, 1972	8,100	141	156	390,429	2,614,251,014	25.5
		• Rashid, 1973	9,400-11,500					38.0
	ARCO	Margham, 1981	11,200	10	16	11,571	49,347,000	43.5
		Total Dubai		**151**	**172**	**402,000**	**2,663,598,014**	

WORLDWIDE PRODUCTION

		Name of field, discovery date	Depth, ft	No. of wells — Producing	No. of wells — Total	Production — 1992 average b/d	Production — Cumulative to Dec. 31, 1992 bbl	°API gravity
ECUADOR	**PETROECUADOR ORIENTE**	Anaconda, 1991	10,200	1	1			25.0
		Atacapi, 1968	9,855	4	5			31.0
		Auca, 1970	10,578	28	28			28.0
		Auca Sur, 1980	9,100-10,300	1	2			25.6
		Cononaco, 1972	11,233	12	12			33.3
		Culebra, 1973	10,626	2	2			17.5
		Dureno/Guanta, 1968	9,575-10,290	9	10			31.0
		Lago Agrio, 1967	10,175	20	33			28.5
		Palanda, 1991	10,700	1	1			29.0
		Parahuacu, 1968	10,173	2	5			30.0
		Pindo, 1991	10,800	1	1			30.0
		Rumiyacu, 1982	10,191	...	1			21.0
		Sacha, 1969	10,160	101	105			28.0
		Shushufindi/Aguarico, 1959	9,772	57	70			30.0
		Yuca, 1970	10,423	6	7			28.0
		Yuca Sur, 1979	9,600-10,000	1	1	321,000	1,742,190,000	14.5
		Yulebra, 1980	10,345	3	3			21.4
		Bermejo Sur + Norte, 1967	5,000	21	26			32.5
		Chanangue, 1990	...	1	1			32.0
		Charapa, 1967	10,000	1	4			29.5
		Coca, 1991	8,500	...	...			24.0
		Cuyabeno, 1972	8,050	14	19			27.4
		Frontera, 1991	9,200	2	2			29.0
		Libertador*	9,800	49	50			29.7
		Paraiso, 1991	9,700	2	2			28.0
		Parayacu, 1991	9,500	1	1			...
		Payamino, 1991	8,500	...	...			24.0
		Pucuna, 1990	9,800	5	5			30.0
		Sansahuari, 1984	8,200	5	5			26.0
		Singue, 1991	8,100	1	1			28.0
		Tetete-Tapi, 1980/1984	9,300	8	11			29.1
		Tiguino, 1990	10,700	4	4			27.0
		V.H.R., 1991	8,300	5	5			31.0
	SANTA ELENA PENINSULA	Ancon, 1921	4,000	439	439			35.7
		Carpet Zona Norte, 1921	4,000	99	99			36.4
		Cautivo, 1921	4,000	79	79			36.4
	CITY IVESTING	Fanny-18B, 1972	7,700	8	8			22.5
		Joan	...	...	...			20.0
		Mariann, 1971	7,700	5	5			20.0
		Tarapoa, 1973	8,300	1	2			20.0
		Total Ecuador		**999**	**1,055**	**321,000**	**1,742,190,000**	
EGYPT	**AGIP**	• Ashrafi, 1987	6,240-8,500	5	5	3,952	1,520,000	...
		Belayim Blend, 1953	7,000-8,500	121	129	40,887	485,889,000	21.5
		• Belayim Marine, 1961	8,500	84	94	165,292	842,161,000	29.0
		Meleiha, 1972	5,950	65	85	21,077	36,793,000	41.5
		Ras Garra, 1964	10,725	3	3	11,275	6,288,000	35.2
		West Razzak, 1985	6,240	7	7	3,889	1,350,000	31.7
	BADR PETROLEUM/SHELL	Badr El Din, 1982	10,800	10	10	30,100	58,400,000	38.0
	WEPCO/PHILLIPS PETROLEUM	El Alamein, 1966	8,700	16	20	1,992	70,200,507	34.0
		Umbarka, 1976	10,700	4	7	3,273	11,050,248	44.0
		Yidma, 1971	9,000	4	4	780	22,843,761	43.6
	SUESSO (ESSO)	• East Zeit, 1982	11,000	7	11	19,038	45,666,000	39.9
	MARATHON	Gazwarina, 1987	6,700	1	1	2,000	986,300	42.7
	GUPCO/AMOCO	Abu Gharadig, 1972	9,500	24	31			35.0-55.0
		• Badri (315), 1981	6,400	23	27			27.0-28.0
		• East Badri, 1991	7,100	1	1			29.0
		• El Morgan, 1965	5,100-6,400	100	122			27.0-32.0
		• El Morgan Rudeis, 1988	6,515	1	1			27.0-32.0
		• GH (376), 1981	8,200-9,750	3	7			24.0-28.0
		• GS (277) 1980	8,610	1	1			34.0
		• GS (345/346), 1982	7,500	1	2			35.0-38.0
		• GS (365), 1980	10,140	...	1			42.0
		• GS (381), 1985	8,700	...	1			37.0
		• GS (327), 1983	5,600	2	2			20.0
		• GS (172) ASL, 1990	9,500	2	2			30.0
		• GS (172) Nubia, 1990	11,250	2	2			28.0
		• Hilal (404), 1982	9,200-10,000	8	10			32.0
		• July, 1973	8,100-10,000	35	48			33.0
		• July Lower Rudeis, 1989	9,820	1	1			36.0
		NAG-9/9-1, 1985	10,000	2	2	409,500	3,419,548,000	35.0
		NAG-3, 1991	11,900	1	1			33.0
		NEAG-9/15, 1984	10,635-12,128	2	2			35.0
		• Nessim (336), 1981	6,700	2	4			29.0
		• October, 1978	11,000	35	45			27.0-31.0
		• N. October (Asi & Nub), 1989	9,500-11,250	2	2			28.0-35.0
		• Ramadan, 1974	11,400	29	36			30.5

WORLDWIDE PRODUCTION

	Name of field, discovery date	Depth, ft	No. of wells Producing	No. of wells Total	Production 1992 average b/d	Production Cumulative to Dec. 31, 1992 bbl	°API gravity
	• Ramadan U. Rudeis, 1990	9,750	1	1			35.0
	Razzak and E. Razzak, 1972	5,730-7,710	11	16			38.0
	• SB (305), 1981	8,100	...	1			27.0
	• SB (339), 1981	4,600	1	1			29.5
	• SB (367), 1984	7,500	...	1			39.0
	• SB (294), 1986	11,600	2	2			30.5
	• SB (374), 1981	10,200	...	2			30.0
	• SG-300, 1976	5,876-6,200	3	5			24.0
	• Shoab Ali, 1978	4,700-5,500	18	31			28.0-33.0
	• Sidki (382), 1977	10,500	...	6			34.0
	• Waly (356), 1982	6,900	1	2			36.0
	WD-33, 1972	9,700	...	2			35.0
	WD-33/15, 1985	5,800	2	2			30.0
	• Younis (347), 1981	4,300	2	5			32.0
BADR PETROLEUM/SHELL[(e)]	• Gemsa Offsh, 1989	...	...	...			29.6
	Sitra, 1982	...	...	...			40.0
REPSOL[(e)]	Hayat, 1986	9,000	10	13			40.0
	Khalda, 1983	7,000	11	21			35.0
	Safir, 1986	12,000	4	4			36.0
	Safir North, 1986	11,600	1	1			36.0
	Salam, 1985	8,000	25	42			35.0
	Tut, 1987	9,000	10	10			35.0
	Tut West, 1988	9,000	1	1			35.0
	Yasser, 1988	8,000	1	1			40.0
SUCO[(e)]	• Ras Budran, 1978	12,500	17	20			23.0-24.0
	• LL 87-2, 1975	12,000	...	1			34.0
	• Ras Fanar, 1978	3,000	5	6			30.0
	• Zeit Bay, 1981	6,000	22	32			34.5
GEYSUM[(e)]	• Geysum, 1980	5,500	5	5			17.0
EGPC[(e)]	Gharib, 1938	2,200-4,500	81	215			24.3
	Bakr/W. Bakr, 1958/1978	3,200-1,500	48	83			18.0-19.0
	Umel Yusr, 1968	3,000-3,300	19	37			21.5
	Amer, 1965	3,500	21	39			21.5
	El Ayoun, 1968	3,300	5	12	159,945	897,081,900	22.0
	Kareem, 1958	2,300	6	35			16.6
	Kheir, 1973	1,050	1	4			28.5
	Shukheir, 1966	2,500	5	11			34.0
	Ras El Behar, 1983	4,500	6	11			32.5
	Esh Elmalaha	3,700	2	2			44.0
	Sudr, 1946	2,600	12	20			21.6
	Asl, 1947	3,000	3	6			21.6
	Matarma, 1948	2,000	2	2			14.6
	El Khaligue, 1980	1,600	...	1			29.4
	Abu Senan T&Y, 1981	6,000	12	20			...
	Harghada, 1913	...	...	...			...
EPEDECO[(e)]	West Bakr, 1978	1,300-1,500	20	25			18.0
EGYPTCO[(e)]	Meleha, 1972	6,000	12	12			40.0
SHUKAIR MARINE CO.[(e)]	O. Suco	...	1	1			...
EL ALAMEIN[(e)]	Mors	...	1	1			...
OSOCO/TOTAL[(e)]	Amal	...	..	...			...
	Sukmeir Bay, 1978	6,300	1	1			34.5
	Sukmeir Gamma	...	...	...			...
	Total Egypt		1,015	1,499	873,000	5,899,777,716	

FRANCE

	Name of field, discovery date	Depth, ft	No. of wells Producing	No. of wells Total	Production 1992 average b/d	Production Cumulative to Dec. 31, 1992 bbl	°API gravity
ALSACE	All fields	...	12	20	212	3,362,068	...
	Abandoned	...	...	...	...	24,388,362	...
AQUITAINE BASIN	Castéra-Lou, 1976	9,600	6	11	288	2,974,371	21.7
	Cazaux, 1959	7,550	39	74	3,860	75,736,330	34.2-37.3
	Lacq Supérieur, 1949	2,170	29	54	582	27,087,136	22.3
	Lagrave, 1984	6,500	4	5	4,425	13,408,607	42.5
	Lavergne, 1962	11,480	7	8	343	12,765,964	41.7
	Les Arbousiere, 1991	10,180	2	2	744	272,810	34.4
	Lugos, 1958	5,250	16	25	404	10,767,379	20.8
	Parentis, 1954	7,710	49	104	4,266	205,174,145	33.4
	Pécorade, 1974	10,660	8	20	1,511	12,393,942	29.3
	Vic Bilh, 1979	7,100	24	45	3,421	22,079,246	26.6
	Other	...	7	14	138	10,568,081	...
	Abandoned	...	...	...	...	8,849,099	...
PARIS BASIN	Champotran, 1985	8,400	8	14	584	1,596,226	31.1
	Châteaurenard, 1958	1,800	19	41	257	5,360,678	27.5
	Chaunoy, 1983	7,350	55	90	15,277	43,820,122	37.1
	Chuelles, 1961	1,800	28	48	370	7,564,436	27.5
	Coulommes, 1958	6,230	22	50	307	13,827,703	32.6
	Courtenay, 1964	1,800	21	40	601	3,349,752	27.5
	Dommartin-Lettrée, 1985	7,020	3	4	542	1,328,001	35.0
	Donnemarie, 1979	8,040	1	8	180	2,440,268	36.2

WORLDWIDE PRODUCTION

Country	Operator	Name of field, discovery date	Depth, ft	No. of wells — Producing	No. of wells — Total	Production — 1992 average b/d	Production — Cumulative to Dec. 31, 1992 bbl	°API gravity
		Fontaine-au-Bron, 1986	6,140	10	13	954	2,703,901	35.4
		Itteville, 1990	5,020	5	6	980	403,542	32.3
		L'Ile-du-Gord, 1986	7,260	8	11	1,009	1,237,507	35.0
		Soudron, 1976	4,600	17	28	331	2,604,664	34.4
		St. Firmin, 1960	1,800	30	58	471	10,667,192	27.5
		St. Germain, 1984	7,200	...	15	417	2,839,188	36.0
		St. Martin-de-Bossenay, 1959	4,590	4	19	206	8,820,756	33.0
		Vert-le-Grand, 1986	6,230	7	8	3,633	5,758,464	34.8
		Villeperdue, 1959	6,100	111	162	7,823	31,099,345	34.4
		Other	...	57	148	3,895	17,845,865	...
		Abandoned	...	...	...	...	11,364,099	...
	JURA & SOUTHEAST	All fields abandoned	...	...	...	...	223,062	...
		Total France		**609**	**1,145**	**58,031**	**604,682,311**	
GABON	BRITISH GAS	• Obando, 1984	8,650	3	3	7,002	6,974,301	32.0
		• Octopus, 1984	10,300	3	3	788	2,127,029	33.0
		• Pelican, 1985	9,300	4	4	4,112	10,710,310	31.0
	ELF	• Anguille, 1962	9,200					31.0
		• Anguille NE, 1968	7,500					32.3
		• Anguille SE, 1962	8,200					33.0
		• Ayol Marine, 1978	7,700					19.8
		• Baliste, 1975	4,900					27.5
		• Barbier, 1972	6,000					32.6
		Batanga, 1960	5,900					39.2
		• Baudroie Marine, 1968	7,500					29.7
		• Baudroie N., 1980	7,500					31.5
		• Brème, 1973	5,000					37.0
		Clairette, 1956	4,400					32.5
		Coucal, 1987	4,100					33.2
		• Dorée, 1972	7,300	233	267	108,140	1,021,400,000	30.4
		• Girelle, 1973	9,400					31.0
		• Gonelle, 1972	5,800					25.7
		• Grand-Anguille, 1983	8,400					31.1
		• Grondin Marine, 1971	7,500					26.6
		• Konzi, 1962	3,600					39.0
		M'Bya Sud, 1969	6,700					20.5
		M'Boumba, 1978	2,800					32.6
		• Mandaros, 1972	4,900					36.2
		• Mérou Sardine, 1984	9,900					30.2
		M'Polunié, 1980	6,600					37.0
		M'Polunié W, 1982	1,700					18.0
		N'Tchengué, 1957	4,100					31.0
		N'Tchengué Océan, 1963	6,600					30.2
		Olendé, 1976	6,700					34.0
		• Pageau, 1972	11,500					19.8
		• Pélicon Est, 1985	9,600					31.9
		• Port-Gentil Sud Marine, 1975	7,500					32.2
		Port-Gentil Océan, 1964	6,500					32.1
		• Torpille, 1968	8,200					34.0
		• Torpille NE, 1982	8,200					32.3
	SHELL GABON	Rabi, 1985	...	...	...	145,400	177,800,000	34.3
	SHELL GABON(e)	Gamba/Ivinga, 1963	3,000-3,100	51	84			31.6
		Lucina, 1971	5,900	15	22			38.4
		Lucina W., 1982	...	...	...	34,558	388,766,579	38.4
	AMOCO	Gombe Maria Beta, 1962	3,400	2	2			24.0
		Oguendjo B and C, 1981	5,000-6,000	16	16			36.0
		Oguendjo Z, 1988	5,000-6,000	2	2			36.0
		Total Gabon		**329**	**403**	**300,000**	**1,607,778,219**	
GERMANY	BEB	Annaveen, 1963	4,428	...	4	...	499,164	32.3
		Barenburg, 1953	2,300-3,300	40	72	1,649	42,812,916	31.1
		Barver, 1963	3,640	2	5	22	522,201	35.4
		Bodenteich, 1960	4,641	1	2	15	388,811	31.3
		Eich, 1983	5,995	7	8	1,072	3,743,347	39.4
		Eilte-West, 1956	6,051	12	16	57	3,716,859	31.5
		Eldingen, 1949	5,028	29	41	537	21,871,893	36.2
		Elsfleth, 1956	5,175	...	...	...	2,940,459	26.2
		Eystrup 2, 1960	3,526	...	1	...	439,043	17.1
		Eystrup 3-6, 1959	1,075	4	4	20	811,484	17.1
		Eystrup-Verden, 1977	3,526	...	4	...	177,756	17.1
		Georgsdorf, 1944	1,800-2,600	185	368	3,786	111,590,195	24.5
		Gross Lessen, 1969	3,300	6	10	571	22,209,595	31.3
		Hankensbüttel-Mitte, 1954	5,182	5	9	346	7,537,591	31.3
		Hankensbüttel-Nord, 1955	5,182	4	4	260	6,850,390	31.3
		Hankensbüttel-Ost, 1954	5,182	1	1	24	1,340,453	31.3
		Hebelermeer, 1955	3,899	5	15	61	4,636,371	26.6
		Hillerse-Nord, 1958	2,683	1	1	10	767,149	31.5
		Hillerse-West, 1957	6,077	1	1	2	31,068	39.6

WORLDWIDE PRODUCTION

	Name of field, discovery date	Depth, ft	— No. of wells — Producing	Total	— Production — 1992 average b/d	Cumulative to Dec. 31, 1992 bbl	°API gravity
	Hohenassel, 1943	1,705	...	1	...	3,555,748	30.4
	Hohnebostel, 1969	8,341	2	6	3	1,196,556	41.1
	Königsgarten, 1984	6,235	3	3	91	428,015	38.2
	Lingen-Dalum (incl. Wd), 1942	3,129	34	126	83	17,199,348	33.0
	Lüben-West, 1958	4,690	5	6	87	1,684,120	36.2
	Lüben, 1955	3,950	8	11	421	11,557,526	33.6
	Meerdorf, 1954	6,077	9	25	30	1,666,794	40.0
	Meppen-Schwefingen, 1959	4,257	21	24	764	13,725,195	29.5
	Mölme-Feldbergen, 1934	1,410	2	2	6	170,640	30.8
	Mölme-Wachtel, 1934	3,886	2	8	3	913,821	36.6
	Nienhagen, 1909	3,079	11	26	143	5,226,438	30.2
	Nienhagen-Dachtmissen, 1959	3,427	5	8	60	1,133,427	27.8
	Pfullendorf-Ostrach, 1962	3,558	13	23	113	2,744,011	37.8
	Pötrau, 1959	5,248	...	1	...	31,136	14.8
	Rühlermoor-Malm, 1962	4,519	7	10	87	6,761,836	30.4
	Rühlermoor-Valendis, 1949	2,381	226	444	8,473	151,471,806	24.5
	Rühme, 1954	1,950	37	47	485	11,431,540	33.4
	Steimbke-Alt+Nord, 1935	1,246	13	19	20	934,535	18.9
	Steimbke-Lichtenmoor, 1943	2,722	16	28	113	43,309	16.0
	Steimbke-Ost, 1959	4,428	...	7	...	3,852,375	34.6
	Stockstadt, 1952	5,402	8	21	96	8,022,875	39.6
	Suderbruch, 1949	5,175	42	61	353	24,247,230	30.6
	Sulingen, 1973	3,345	9	9	563	5,876,609	30.0
	Varel, 1957	5,084	5	15	123	5,948,371	25.4
	Varloh, 1984	4,129	6	6	63	444,327	32.7
	Volkensen, 1960	7,540	4	4	57	873,552	21.6
	Wehrbleck, 1947	3,400	46	61	435	17,447,047	32.1
	Wietingsmoor, 1954	2,820	22	29	316	13,822,157	30.6
MOBIL	Bavaria	4,660-6,190	...	20	...	7,526,952	20.0-25.0
	Bavaria[C]	...	2	2	1	337,472	...
	Hannover	980-6,630	52	65	759	31,891,244	25.0-35.0
	Hofolding, 1980	11,100	2	3	185	1,271,299	28.0-30.0
	Süd-Oldenburg, 1950	2,630-6,800	72	128	1,017	41,094,243	31.0-39.0
	Voigtei, 1953	980-1,900	80	103	629	25,870,196	32.0-34.0
	Weser Ems	2,000-2,450	18	24	310	6,426,452	24.0
RWE-DEA A.G.	Boostedt, 1952	5,900	21	43	396	19,065,675	32.0
	Hankensbuettel-Sued, 1954	4,900	32	77	1,503	85,007,026	27.0-35.0
	Hardesse, 1957	8,200	16	46	380	9,474,761	40.0
	Hohne, 1951	5,250	23	52	557	41,881,235	33.0
	Leiferde, 1956	2,600	15	21	284	20,902,992	33.0
	• Mittelplate, 1987	7,500-9,350	5	6	5,528	9,413,700	19.3
	Oerrel-Sued, 1954	4,600	20	46	497	12,550,639	26.0
	Other, 1935	2,300-8,200	7	10	371	5,838,000	19.0-37.0
	Ploen-Ost, 1958	8,850	19	82	599	52,172,921	33.0
	Preetz, 1962	8,200	16	33	373	12,130,885	26.0
	• Schwedeneck-See, 1978	4,900	15	18	4,681	18,992,104	28.0
	Wesendorf, 1943	4,650	14	45	350	18,037,395	25.0-40.0
WINTERSHALL A.G.	Aitingen, 1977	6,232	4	4	746	5,710,785	31.1
	Aldorf, 1952	4,100	20	103	183	17,538,496	33.0
	Arlesried, 1964	4,920-8,550	12	18	358	14,653,570	38.5
	Bockstedt, 1954	4,264	25	54	719	22,085,043	27.4
	Dickel, 1953	3,936	33	68	213	7,097,610	28.3
	Dueste, 1954	3,936	33	78	263	12,533,301	33.0
	Eicklingen, 1937	2,460	7	12	40	3,727,317	30.0
	Emlichheim, 1944	3,116	78	128	2,526	48,619,399	24.5
	Hauerz, 1985	8,639	5	5	98	391,224	40.4
	Kirchdorf, 1982	5,737	1	1	15	76,468	38.5
	Landau, 1955	3,936	74	78	884	28,592,415	36.9
	Maximiliansau, 1959	2,100-5,363	...	...	...	14,387	41.0
	Meckelfeld, 1954	6,396	13	25	182	14,347,543	27.5
	Moenchsrot, 1958	5,248	13	21	245	10,956,537	38.1
	Nienhagen, 1925	3,729	10	12	52	11,007,147	30.0
	Oberschwarzach, 1978	4,920	4	4	56	787,050	37.7
	Pattensen, 1954	2,165	5	7	39	1,129,113	29.3
	Rheinzabern, 1959	3,750-5,576	1	1	117	124,312	37.9
	Ruehlertwist, 1949	2,350-2,900	52	113	1,363	33,725,597	25.0
	Ruelzheim, 1984	7,544	1	2	98	92,690	37.7
	Schwabmünchen, 1976	4,428	...	1	...	39,491	25.7
	Schwedeneck, 1956	5,182	1	3	108	5,361,624	27.5
	Wathlingen, 1939	2,624	3	9	13	1,441,391	30.0
DEILMANN ERDÖL ERDGAS GMBH	Adorf, 1948	5,000	24	38	825	10,138,707	28.0
	Allermöhe, 1979	6,700	1	3	32	503,980	28.0
	Assling, 1981	11,400	2	4	171	8,570,960	39.0
	Bramberge, 1958	2,580-3,300	59	83	4,780	111,300,266	29.0
	Brolstedt, 1937	2,800	3	15	23	1,371,879	33.0
	Darching, 1967	12,200	2	5	228	8,270,583	39.0
	Eddesse-Nord, 1950	460	23	87	91	604,148	28.0
	Fronhofen, 1964	8,800	6	16	103	2,651,843	37.0
	Höhenrain, 1962	12,900	1	4	26	1,999,285	29.0
	Höver, 1957	2,200	9	16	49	2,248,954	25.0
	Hummling, 1953	5,000-8,200	29	33	205	4,350,103	25.0-30.0

WORLDWIDE PRODUCTION

Country	Company	Name of field, discovery date	Depth, ft	No. of wells — Producing	No. of wells — Total	Production — 1992 average b/d	Production — Cumulative to Dec. 31, 1992 bbl	°API gravity
		Illmensee, 1967	6,500	2	5	24	1,083,992	37.0
		Kronsberg, 1953	2,000	14	20	124	2,543,627	25.0
		Lehrte, 1952	4,000	11	23	49	3,266,484	31.0
		Oelhelm-Büd/Rhät, 1988	2,000	21	25	486	5,663,738	33.0
		Reitbrook-Kreide, 1937	2,300	7	180	386	16,057,601	21.0
		Reitbrook-West, 1960	6,100	15	23	324	22,894,450	30.0
		Soheemom, 1949	3,550-3,750	88	154	2,209	57,176,338	30.0
		Sinstorf, 1960	7,100	8	23	357	19,980,143	30.0
		Vorhop, 1952	5,000	21	38	875	18,736,549	39.0
		Vorhop-Knesebeck, 1958	4,100	23	35	1,095	21,235,779	35.0
	DST[e]	Bramberge, 1958	2,560-3,300	64	82			29.0
		Fronhofen, 1964	3,800	8	13			37.0
		Illmensee, 1967	3,800	2	5			37.0
		Markdorf, 1960	7,700	...	2			37.0
		Memmingen-M, 1958	5,000	...	3			37.0
	PREUSSAG[e]	Allermöhe, 1979	6,700	1	4			28.0
		Assling, 1961	11,400	2	4			39.0
		Broistedt, 1937	2,600	3	17			33.0
		Darching, 1967	12,200	2	5			39.0
		Eddesse-Nord, 1950	460	31	87			26.0
		Höhenrain, 1962	12,900	1	4	4,380	124,106,150	39.0
		Höver, 1957	2,200	10	16			25.0
		Kronsberg, 1953	2,000	14	20			25.0
		Lehrte, 1952	4,000	10	23			31.0
		Oelheim-Süd/Rhät, 1968	2,000	23	25			33.0
		Reitbrook-Kreide, 1937	2,300	13	159			21.0
		Reitbrook-West, 1960	6,100	16	21			30.0
		Sinstorf, 1960	7,100	12	23			30.0
		Vorhop, 1952	5,000	21	38			39.0
		Vorhop-Knesebeck, 1958	4,100	23	35			35.0
		Total Germany		**2,308**	**4,584**	**65,000**	**1,649,990,349**	
GHANA[e]	AGRI-PETCO	• Saltpond, 1977	8,500	...	3	1,800	4,596,000	36.0-39.0
		Total Ghana		...	**3**	**1,800**	**4,596,000**	
GREECE	N. AEGEAN PET. CORP.	• Prinos, 1974	8,000-10,000	13	15	13,721	86,505,000	28.5
		Total Greece		**13**	**15**	**13,721**	**86,505,000**	
GUATEMALA	MINISTERIO DE ENERGIA Y MINAS	Caribe, 1981	6,830-7,280	3	3	633	3,807,253	22.0-24.0
		Rubelsanto, 1974	5,300-7,000	5	7	550	6,845,074	26.0-30.0
		San Diego, 1981	17,150	...	1	...	135,974	35.0
		Tierra Blanca, 1983	11,230	2	3	646	1,297,934	19.5-23.0
		Tortugas, 1970	2,300	...	2	...	67,804	25.0
		West Chinaja, 1977	3,200-3,800	3	3	204	7,621,367	31.0-35.0
		Xan, 1985	8,427	3	4	3,583	2,293,716	17.0
		Yalpemach, 1980	9,000	...	1	...	168,102	31.0-35.0
		Total Guatemala		**16**	**24**	**5,616**	**22,237,224**	
HUNGARY	HUNGARIAN OIL & GAS	Algyô, 1965	5,320-8,317	737	1,226	20,305	190,139,121	41.0
		Almosd, 1977	8,143	2	2	1	1,560	33.8
		Asotthalom, 1967	3,199	15	20	614	10,764,476	32.1
		Battonya, 1959	3,041	17	28	72	1,443,645	36.0
		Battonya East, 1970	2,913	114	126	864	4,504,968	46.1
		Belezna, 1963	8,176	5	31	6	289,441	41.1
		Budafa-Kiscsehi, 1937	2,713-4,167	123	514	526	38,911,155	27.5
		Csanádapáca, 1967	6,063	6	6	120	231,857	37.0
		Demjén, 1954	725-3,937	193	224	215	9,469,559	38.2
		Devavanya, 1986	7,930-8,034	2	20	4	1,323	46.7
		Ferencszállás, 1969	7,001-7,674	29	52	281	3,311,624	41.1
		Ferencszallas East-Kiszombor, 1973	7,283	13	16	54	3,219,713	41.1
		Kaba South, 1978	6,548	3	4	162	840,139	34.1
		Kaszaper South, 1971	5,056	14	14	15	231,578	32.8
		Kelebia NE, 1968	3,113-3,258	4	4	20	1,111,194	37.0
		Kelebia South, 1970	2,428	6	8	223	3,405,685	17.4
		Kiskundorozsma, 1964	4,846-9,711	26	36	1,705	5,390,330	41.9
		Kiskunhalas NE, 1974	5,905-6,594	32	108	989	11,886,117	31.2
		Kiskunhalas North, 1983	7,546	4	10	4	3,723	35.0
		Kiskunhalas NW, 1990	5,660	2	2	30	19,049	31.2
		Kiskunmajsa South, 1977	6,712	18	28	4	15,573	28.2
		Kismarja, 1979	2,178-2,536	27	36	74	1,203,159	25.1
		Lovâszi, 1940	3,510-4,301	75	479	345	50,410,112	41.3
		Mezôcsokonya West, 1987	2,216	3	9	150	97,559	27.5
		Mezôhegyes, 1960	3,504-3,589	6	9	63	1,376,325	38.9
		Mezôsas, 1978	8,432	3	3	428	341,435	28.7
		Nagylengyel, 1951	5,545-8,235	72	462	4,132	132,723,610	16.0
		Ortaháza, 1970	4,580-6,089	10	36	118	4,136,476	31.1
		Ortaháza East, 1986	5,413	4	12	191	205,823	31.1

WORLDWIDE PRODUCTION

		Name of field, discovery date	Depth, ft	No. of wells — Producing	No. of wells — Total	Production — 1992 average b/d	Production — Cumulative to Dec. 31, 1992 bbl	°API gravity
		Pitvaros North, 1992	5,945-6,322	1	2	13	4,880	43.8
		Pusztaapáti, 1973	8,300	2	11	32	1,065,600	22.3
		Pusztaföldvár, 1958	5,341-5,538	69	107	450	15,838,145	30.2
		Ruzsa, 1979	6,445-9,403	3	3	100	336,677	45.4
		Szank, 1964	5,827-5,932	22	110	609	17,006,297	38.7
		Szank NW, 1977	5,495	3	8	104	984,323	41.1
		Szeged, 1971	8,629	18	25	914	24,713,371	41.7
		Szeghalom, 1980	6,545-6,608	42	55	1,595	2,414,641	44.5
		Szeghalom North, 1982	7,600	2	12	1	1,567	44.5
		Szeghalom West, 1989	7,463-7,528	1	4	42	17,177	44.6
		Tázlár, 1966	6,365-6,890	2	23	42	421,492	24.0
		Tázlár North, 1986	7,129-7,415	1	7	56	87,293	32.5
		Totkomlos South, 1992	5,272-5,295	1	1	116	42,249	35.6
		Tura, 1991	2,611-2,628	3	4	4	1,301	21.6
		Ullés, 1962	5,848-7,464	35	39	419	262,192	36.0
		Zalakaros-Sávoly, 1979	5,463-5,791	6	34	1,121	3,723,137	20.7
		Other	...	...	61	...	23,016,871	...
		Total Hungary		**1,776**	**4,031**	**37,333**	**565,623,542**	
INDIA	**WESTERN**	Ankleshwar, 1960	1,969-6,564	138	215	16,331		47-60.0
		Balol, 1970	3,280-3,600	101	114	4,070		16.0
		Gandhar, 1983	9,846-12,145	52	72	21,389		49.9
		Jhalora, 1976	3,940-4,270	86	93	18,471		25.7
		Jotana, 1977	3,600-4,930	31	52	9,556		39.0
		Kadi, N., 1967	3,610-7,220	183	221	12,129		22.3
		Kadi, S., 1968	6,892-9,846	10	21	...		41.1
		Kalol, 1961	4,595-10,830	219	251	11,190		37.0
		Lakwa, 1972	4,595-10,830	75	84	9,062		15.0
		Motwan and Sisodara, 1974	1,500-1,700	16	31	6,907		45.4
		Nada, 1988	9,180-11,490	5	8	2,108		47.6
		Nawagam, 1963	4,595-10,174	91	102	2,879		35.0
		Sanand, 1962	4,267-10,174	25	35	...		25.7
		Santhal, 1974	3,281-3,600	99	111	10,954		17.0
		Sobhasan, 1968	4,595-9,846	103	134	16,325		33.0
		Viraj, 1978	4,595-4,923	35	39	2,936	3,467,200,000	22.3
		Others	...	140	228	...		...
	EASTERN	Champang, 1973	8,861-10,502	10	16	1,518		33.0
		Demulgaon, 1976	...	4	7	...		33.0
		Geleki, 1968	9,518-14,769	85	120	...		30.0
		Lakwa-Lakhmani, 1964	6,564-14,113	214	284	...		33.0
		Rudrasagar, 1960	7,877-13,128	46	68	...		33.0
		Others	...	13	22	...		...
	BOMBAY OFFSHORE	• B-38 (Heera), 1977	4,267-6,564	32	32	19,909		37.0
		• Bombay High, 1974	3,282-7,877	271	339	330,433		39.0
		• Neelam, 1987	...	4	20	1,266		38.0
		• Panna (North Bassein), 1976	4,923-7,549	10	38	2,824		38.0
		• Others	...	3	7	463		...
	SOUTHERN	Narimanam, 1985	6,236-7,549	20	20	8,850		43-45.0
		Others	...	15	30	...		48-52.0
	OTHER FIELDS	Bakrol, 1966	2,626-9,846	3	6	...		32.7
		Digboi, 1890	300-7,500	225	1,001	400		...
		Hapjan, 1987	10,000-11,000	10	20	3,600		...
		Jorajan, 1972	7,000-12,000	59	97	5,800		25.0-34.0
		Kathana, 1965	5,250-5,906	...	33	...		25.0
		Kharsang/Ningru/Kumehai, 1976	3,500-5,500	24	54	1,200		...
		Kosamba, 1963	1,313-8,533	...	23	...		...
		Moran, 1956	10,000-12,000	59	97	5,800		...
		Nahorkatiya, 1953	8,500-10,000	132	281	14,000		...
		Sabamati, 1981	...	...	2	...		36.6
		Shalmari, 1986	11,000-12,000	13	18	3,400		...
		Others	...	20	69	3,200		...
		Total India		**2,681**	**4,515**	**546,970**	**3,467,200,000**	
INDONESIA	**TOTAL CFP**	• Bekapai, 1972	4,600-15,600	34	73	9,247	170,980,892	40.0
		• Handil, 1974	2,900-19,200	190	326	47,224	734,123,579	33.0
		Semanlu, 1990	8,778-12,141	1	1	9	193,793	...
		• Tambora, 1974	8,000-13,000	15	36	6,522	14,407,617	33.0
		Tunu(C), 1973	8,001-13,000	10	34	3,877	3,256,037	55.0
	ARCO	• Ardjuna, 1969	2,380-7,250	362	447	110,940	826,397,000	37.0
		• Arimbi, 1972	2,943	32	35	5,306	37,295,000	33.0
		• Bima, 1983	3,000-3,888	54	63	6,074	8,795,000	34.0
		• Northwest Corner, 1974	5,411-5,800	27	44	7,227	10,234,000	31.5
	SANTOS PETROLEUM (SERAM) LTD.	• Bula, 1897	280-1,300	88	377	1,023	17,081,101	23.0
	CHEVRON & TEXACO	Batang, 1975	690	11	13	837	3,193,871	19.0
		Buaya, 1978	2,035	...	2	...	190,489	28.0
		Jingga, 1984	7,000	2	3	246	1,450,967	45.0
		Lindai, 1971	1,100	18	20	1,182	8,436,031	32.0

WORLDWIDE PRODUCTION

	Name of field, discovery date	Depth, ft	No. of wells — Producing	No. of wells — Total	Production — 1992 average b/d	Production — Cumulative to Dec. 31, 1992 bbl	°API gravity
	Menggala S., 1968	3,765	2	2	567	16,620,646	34.0
	Rintis, 1984	6,900	1	1	480	580,609	40.0
	Tanjung Medan, 1976	2,770	2	7	514	2,981,680	34.0
	Ujung Tanjung, 1975	2,554	...	2	...	98,455	30.0
	Benua, 1978	2,400	19	24	4,798	29,539,543	37.0
	Beruk, 1974	1,750	21	28	9,163	63,165,483	38.0
	Beruk N., 1976	1,900	5	7	854	2,149,482	38.0
	Beruk N.E., 1976	1,800	3	6	197	1,877,664	40.0
	Bungsu, 1976	2,050	8	9	925	12,218,475	38.0
	Butun, 1982	4,500	9	9	11,334	18,799,229	44.0
	Damar, 1974	4,650	5	5	231	5,218,979	34.0
	Doral, 1992	4,145	3	3	1,195	437,408	40.0
	Dusun, 1979	2,550	6	7	421	6,453,653	42.0
	Gatam, 1977	2,385	...	3	...	627,756	34.0
	Idris, 1982	2,450	1	3	67	1,081,075	39.0
	Kasikan, 1972	660	24	29	1,414	7,845,597	28.0
	Langgak, 1976	1,380	24	27	1,966	7,877,111	32.0
	Osam, 1978	900	5	7	252	1,215,038	32.0
	Paitan, 1978	2,315	1	1	96	945,336	32.0
	Pak, 1988	3,667	4	4	1,406	636,176	35.0
	Pedada, 1973	950	29	31	10,160	84,927,073	33.0
	Pusaka, 1977	2,250	19	25	4,327	21,422,933	36.0
	Sabak, 1974	2,350	14	23	1,320	20,520,952	36.0
	Terantam, 1973	900	8	10	671	1,261,623	32.0
	Zamrud, 1975	9,600	49	55	16,095	91,217,484	40.0
CALTEX PACIFIC INDONESIA	Aman, 1974	4,700	10	10	1,882	38,774,866	37.0
	Ampuh, 1981	6,000	14	15	2,402	4,999,601	40.0
	Antara, 1978	1,425	5	6	1,026	5,132,689	36.0
	Ayu, 1986	4,410	6	6	737	797,337	34.0
	Balam SE, 1972	4,500	13	13	613	7,565,813	29.0
	Balam S., 1969	1,600	78	80	12,438	113,196,287	33.0
	Bangko, 1970	1,950	104	115	29,764	386,532,372	34.0
	Batang CPI, 1978	690	7	7	442	1,673,096	19.0
	Bekasap, 1955	2,950	68	78	18,169	432,555,494	34.0
	Bekasap S., 1968	3,900	7	14	835	28,457,291	34.0
	Benar, 1973	2,450	28	33	3,470	63,152,748	33.0
	Cebakan, 1974	4,700	5	5	450	5,011,580	30.0
	Cucut, 1981	5,800	3	3	130	453,655	33.0
	Duri, 1941	770	2,016	2,664	197,440	727,171,569	22.0
	Garuk, 1980	4,600	...	2	...	119,164	31.0
	Genting, 1982	3,610	4	4	448	376,047	25.0
	Hitam, 1975	6,690	4	5	927	3,393,263	39.0
	Hiu, 1983	5,900	5	6	873	1,727,948	40.0
	Intan, 1977	3,350	6	7	1,230	8,839,910	33.0
	Jambon, 1992	4,004	1	1	100	36,433	31.0
	Jorang, 1972	5,500	21	24	5,516	48,372,845	38.0
	Kerang, 1977	3,900	11	14	1,120	7,749,491	39.0
	Kopar, 1974	5,300	17	24	2,640	17,460,230	37.0
	Kotabatak, 1952	5,500	90	147	18,917	164,553,375	29.0
	Kulin, 1970	2,050	72	81	5,473	39,303,713	20.0
	Libo, 1968	6,000	3	12	13,265	12,342,876	35.0
	Libo SE, 1973	5,900	26	39	1,047	82,544,105	42.0
	Lincak, 1981	2,570	4	4	188	3,120,133	35.0
	Mangga, 1988	7,800	3	3	61	201,297	34.0
	Menggala North, 1968	3,850	18	21	3,465	69,595,611	33.0
	Menggala South, 1968	3,765	6	7	1,089	33,639,886	34.0
	Minas, 1944	2,600	487	691	211,208	3,680,618,898	35.0
	Mindal, 1971	3,400	2	4	...	3,127,612	37.0
	Mutiara, 1976	3,500	2	2	254	2,094,486	34.0
	Nella, 1977	5,000	3	5	360	1,044,455	35.0
	Obor, 1978	7,370	3	3	948	3,202,085	40.0
	Pager, 1974	4,300	28	29	5,771	84,991,698	38.0
	Pelita, 1977	6,600	6	7	344	5,439,922	39.0
	Pemburu, 1981	2,300	10	11	2,300	12,873,717	35.0
	Permatang, 1959	3,750	45	52	9,350	230,608,151	32.0
	Permatang Bow, 1969	5,300	15	19	2,348	10,382,454	33.0
	Perkebunan, 1977	2,300	1	1	146	506,425	29.0
	Petani, 1964	4,750	41	44	11,040	290,348,142	33.0
	Petapahan, 1971	4,500	22	33	3,082	98,534,738	28.0
	Pinang, 1971	3,950	21	24	3,000	30,957,812	30.0
	Pinggir, 1972	3,300	4	6	209	4,794,300	35.0
	Pinggir S., 1973	3,300	1	4	59	2,111,741	39.0
	Pudu, 1972	5,950	7	13	1,384	30,211,463	35.0
	Pukat, 1983	6,200	3	3	153	1,141,617	40.0
	Puncak, 1979	2,400	16	18	2,207	18,061,974	36.0
	Pungut, 1951	3,400	26	27	3,819	52,417,444	38.0
	Rangau, 1968	6,100	4	13	291	8,751,886	42.0
	Rantaubais, 1972	1,070	14	18	349	1,317,997	18.0
	Rokiri, 1979	6,100	1	4	...	1,048,776	41.0
	Sakti, 1991	8,205	1	1	183	79,777	34.0
	Seruni, 1972	3,000	26	27	4,615	53,697,473	35.0
	Sidingin, 1992	8,161	2	5	989	362,000	38.0

WORLDWIDE PRODUCTION

	Name of field, discovery date	Depth, ft	No. of wells Producing	No. of wells Total	Production 1992 average b/d	Production Cumulative to Dec. 31, 1992 bbl	°API gravity
	Sikladi, 1975	4,340	19	21	2,478	22,989,113	40.0
	Singa, 1976	3,200	1	1	46	460,062	31.0
	Sintong, 1971	3,400	25	26	2,816	58,968,061	32.0
	Sintong SE, 1973	4,500	1	1	34	1,256,075	36.0
	Suram, 1971	1,750	3	7	197	4,003,416	28.0
	Tandun, 1969	3,000	13	13	1,880	15,791,900	33.0
	Tanggul, 1977	3,700	2	2	746	645,410	30.0
	Tegar, 1992	7,900	1	1	251	92,005	32.0
	Telinga, 1975	4,205	8	9	942	4,397,786	33.0
	Titian, 1992	8,185	1	1	363	132,675	35.0
	Topaz, 1977	5,200	7	8	1,755	18,977,331	33.0
	Topi, 1979	3,630	1	1	13	622,351	37.0
	Tunas, 1983	4,819	2	2	231	751,026	35.0
	Ubi, 1976	4,350	24	25	1,989	16,578,861	35.0
	Waduk, 1975	6,100	4	11	362	5,115,944	38.0
ASAMERA (NORTH SUMATRA)	Alur Cimon, 1972	1,697	1	34	33	1,390,943	51.6
	Geudondong, 1967	2,827-2,882	2	31	45	5,217,106	50.0-52.0
	Iee Tabeue, 1971	1,992-2,288	2	67	3	11,917,350	50.3
	Julu Rayeu, 1970	2,389-2,404	7	62	128	12,515,058	52.8-54.0
	Merbau, 1980	1,907	...	5	...	47,596	53.0
	Peudawa, 1978	1,250-1,905	1	21	19	1,574,542	47.7-48.5
	Peutouw, 1980	1,690-2,270	...	9	...	544,288	57.0-57.4
	Tualang, 1973	2,348	11	68	767	31,451,779	50.6
ASAMERA (SOUTH SUMATRA)	Bijak, 1987	1,705	...	1	...	3,188	21.3
	Koban, 1941	1,358	2	9	103	754,000	49.0
	Rawa, 1984	3,661	29	40	3,135	9,437,801	37.5-38.0
	Rawa S., 1985	3,763	2	3	29	101,155	38.3
	Rebonjaro, 1929	858-1,351	6	6	400	569,000	46.7
	Supat, 1984	3,703-4,216	12	13	2,463	2,928,019	31.5
	Tengah, 1984	3,245	...	4	...	16,935	34.7
	Keri, 1985	3,957	2	2	19	128,919	41.5
	Bertak, 1988	4,364	...	4	...	...	27.4
	Bentayan, 1932	4,446	27	32	4,150	5,104,781	22.0
	Kluang, 1900	2,591	25	74	472	34,578,106	42.6
	Mangunjaya, 1934	2,700	3	98	196	23,272,515	32.0
	Panerokan, 1976	4,915	2	5	15	218,197	39.4
	Ramba, 1982	3,130	75	79	9,125	62,635,282	37.0
	Tanjung Laban, 1982	3,590	24	31	3,192	12,530,553	38.1
	Tempino, 1931	4,493	3	181	89	1,518,686	40.9
MAXUS ENERGY	• Cinta, 1970	3,500	44	55	13,300	196,146,000	34.0
	• Duma, 1982	2,300	...	...	...	774,000	16.0
	• Farida, 1981	7,000	18	23	3,140	16,181,000	32.0
	• Gita, 1972	5,000	2	9	380	8,219,000	33.3
	• Indri, 1989	3,500	1	2	1,860	680,760	34.2
	• Intan, 1987	3,200	16	18	22,500	37,109,000	32.0
	• Intan NE, 1987	3,300	6	6	4,830	5,223,000	32.6
	• Karmila, 1983	4,650	7	9	7,040	38,381,000	36.0
	• Kitty, 1971	2,700	4	8	780	15,502,000	18.0
	• Krisna, 1976	4,500	28	42	6,100	69,362,000	37.0
	• Nora, 1973	2,800	3	5	680	9,591,000	29.0
	• Rama, 1974	3,200	36	80	7,590	101,060,000	31.0
	• Selatan, 1978	2,800	6	9	790	22,033,000	20.0
	• Sundari, 1981	4,400	12	15	1,950	13,808,000	25.0
	• Titi, 1981	6,100	2	4	235	1,764,000	33.0
	• Wanda, 1982	4,800	2	5	260	1,746,000	36.0
	• Widuri, 1988	3,500	44	47	93,070	81,472,000	24.0
	• Yvonne, 1980	5,000	7	11	2,000	16,201,000	36.0
	• Zelda, 1971	6,300	22	41	3,525	44,218,000	32.0
SANTA FE ENERGY RESOURCES LTD.	Arar Gas(C), 1977	8,748	...	1	...	288,850	...
	Candrawasih, 1976	3,226	16	26	652	13,808,500	26.7
	Jaya, 1973	2,843	12	19	762	37,141,800	39.7
	Kasim, 1973	3,031	41	54	1,679	50,927,000	36.5
	Kasim Barat, 1975	4,129	1	4	51	3,101,000	34.5
	Kasim Utara, 1973	4,832	3	7	107	8,424,000	25.7
	Klagagi, 1986	4,561	...	...	...	21,400	...
	Klagagi W(C), 1986	5,450	...	...	...	39,200	...
	Klalin	...	4	3	277	71,100	41.1
	Klalin(C), 1986	8,025	4	5	1,182	1,351,800	55.0
	Klaserai, 1981	2,884	1	...	...	25,300	33.0
	Lao-lao Gas(C), 1983	3,896	...	...	...	7,100	...
	Maniwar Gas(C), 1980	3,635	...	...	...	2,500	...
	Moi, 1979	4,373	3	3	33	851,000	16.2
	Walio, 1975	2,750	19	20	340	445,100	34.3
	Walio Kais, 1973	2,337	196	221	9,384	167,133,000	34.3
	Walio Text II, 1973	2,268	17	45	475	1,599,100	34.3
STANVAC (CENTRAL SUMATRA)	Binio, 1972	1,600	16	23	229	12,272,715	34.0
	East Kayuara, 1983	3,500	46	48	2,699	11,671,867	35.0
	Gemuruh, 1983	2,800	18	21	1,127	3,972,520	34.0
	Kayuara, 1982	3,700	8	9	358	2,568,343	35.0
	Kerumutan, 1980	3,000	13	31	427	5,968,663	35.0
	Merbau, 1979	2,700	21	30	1,217	8,309,522	36.0

WORLDWIDE PRODUCTION

	Name of field, discovery date	Depth, ft	No. of wells — Producing	No. of wells — Total	Production — 1992 average b/d	Production — Cumulative to Dec. 31, 1992 bbl	°API gravity
	North Merbau, 1980	2,600	5	11	195	3,234,363	36.0
	Mutiara, 1985	2,700	1	2	26	167,208	34.0
	Panduk, 1981	2,500	6	9	222	1,884,089	46.0
	Pekan, 1976	3,700	13	44	248	8,319,848	35.0
	Parum, 1987	4,200	19	23	1,914	3,905,500	35.0
	Other	4,600	...	...	...	9,175	36.0
STANVAC (SOUTH SUMATRA)	Ibul, 1970	5,700	14	47	392	22,774,859	36.0
	Jene, 1985	6,500	31	42	11,990	36,803,364	35.0
	Kerang (PSC), 1984	2,900	4	10	268	1,588,079	35.0
	Lagan, 1986	2,400	4	5	216	541,487	56.8
	Pian, 1986	6,000	3	10	158	433,878	36.0
	Rambutan, 1972	4,500	...	15	...	3,283,312	42.0
	Rimbabat (PSC), 1984	3,200	2	3	32	233,206	34.0
	Tabuan (PSC), 1986	4,000	4	9	579	2,278,660	34.0
	Teras, 1976	4,500	3	3	99	912,075	42.0
	South Tabuan (PSC), 1981	2,200	1	1	3	50,610	34.0
	Langkap (PSC), 1987	2,900	12	17	2,073	2,027,511	34.0
	Gunung Kembang, 1987	3,600	2	8	726	595,242	36.0
	Other	4,500	...	...	576	2,201,296	36.0
VIRGINIA INDONESIA	Badak, 1972	5,190-12,200	145	175	14,917	105,465,000	34.4
	Mutiara, 1984	900-8,500	23	34	6,177	9,994,000	38.5
	Nilam, 1974	7,500-14,500	102	137	14,622	58,350,000	33.4
	Pamaguan, 1974	900-5,100	15	26	1,370	8,988,000	26.8
	Semberah, 1974	2,900-10,500	32	43	10,376	4,391,000	30.0
	Wailawi[(C)], 1974	5,000-6,000	6	6	...	50,000	32.0
UNOCAL	• Attaka, 1970	1,200-11,000	93	114	42,553	537,157,414	32.0-42.0
	• Kerindingan, 1972	3,100-7,100	5	8	617	5,264,977	20.0-36.0
	• Melahin, 1972	2,900-4,900	6	8	4,340	9,542,462	16.0-29.0
	• Sepinggan, 1973	3,400-11,600	35	40	19,496	73,210,361	25.0-38.0
	• Yakin C, 1976	670-2,235	3	11	834	9,339,610	13.0-22.0
	• Yakin N, 1977	2,180-6,600	9	10	2,163	12,248,753	16.2-28.0
	• Yakin W, 1982	2,050-2,530	4	8	677	7,055,215	18.4-21.0
MOBIL OIL	Arun, 1971	9,367	74	89	112,661	509,656,000	55.0
MARATHON	• Kakap KH, 1980	6,350	9	9	2,500	19,800,000	43.0
	• Kakap KF, 1984	2,875	14	14	18,200	26,900,000	50.5
CONOCO GROUP	• Belida, 1989	5,000	19	20	4,475	1,637,936	47.0
	• Ikan Pari, 1983	4,100-4,500	3	4	2,355	5,226,328	47.4
	• Udang, 1974	5,595	17	28	3,387	63,601,383	39.0
	Wiriagar, 1984	3,500	...	3	600	1,589,143	39.7
PERTAMINA[(e)] UEP I	Besitang, 1977	3,973	3	22			51.5
	Gebang, 1977	2,917	13	50			54.0
	Kuala Dalam, 1983	3,872	13	20			50.8
	K. Simpang Barat, 1971	3,740-5,019	24	43			45.0
	K. Simpang Timur, 1967	3,576	8	30			45.0
	P. Tabuhan Barat, 1937	3,350-5,893	17	44			50.8
	P. Tabuhan Timur, 1968	2,460	22	127			50.8
	Pulau Panjang, 1928	3,300	...	66			...
	Rantau, 1929	2,764	71	542			45.3
	Serang Jaya, 1926	3,701	18	84			47.5
	Securai, 1980	4,100-6,890	7	15			...
	Sungai Buluh, 1984	3,701	4	8			51.0
PERTAMINA UEP II[(e)]	Abab, 1951	6,000	19	137			36.0
	Air Serdang, 1988	4,860	7	7			...
	Andan, 1950	1,969	2	15			29.0
	Bajubang, 1930	3,822-3,225	18	130			25.7
	Belimbing, 1965	5,371	4	29			35.5
	Benakat, 1932	4,167	62	273			35.0
	Benakat East, 1973	4,000-4,600	12	55			...
	Benuang, 1942	6,103	18	72			35.1
	Beringin, 1975	6,365-6,925	...	15			38.8
	Betun, 1935	7,057	1	18			...
	Betung, 1966	4,003	8	25			...
	Betung Jambi, 1951	328	8	204			...
	Blimbing, 1952	4,922	3	9			...
	Candi, 1937	8,012	1	2			...
	Deras, 1937	6,618	2	7			...
	Dewa, 1936	8,337	6	49			...
	Gn. Kemala, 1938	5,548	20	76			38.0
	Jirak, 1930	2,093	30	222			...
	Karangan, 1975	5,250	2	10			23.0
	Kenali Asam, 1931	2,077	58	244			29.0
	Ketaling Timur, 1982	1,837	16	20			...
	Kruh N., 1935	4,652	1	18			...
	Kuang, 1940	5,249	6	17			25.1
	Limau, 1928	3,898-4,626	31	217			...
	Limau Niru, 1954	5,118	19	47			...
	Lirik, 1939	1,600	27	74			34.0
	Molek, 1956	2,600	8	38			34.0
	Musi, 1985	2,717	10	25			...

WORLDWIDE PRODUCTION

	Name of field, discovery date	Depth, ft	No. of wells — Producing	No. of wells — Total	Production — 1992 average b/d	Production — Cumulative to Dec. 31, 1992 bbl	°API gravity
	N. Pulai, 1941	1,800	13	72			34.0
	S. Pulai, 1941	1,800	10	32			34.0
	Ogan Timur, 1943	3,896-4,626	8	28			...
	Prabumulih Barat, 1953	5,315	4	13			...
	Raja, 1950	6,000	21	118			38.0
	Sago, 1940	1,800	66	107			34.0
	Setiti, 1960	1,837	9	32			...
	Sukaraja, 1934	7,379	1	10			...
	Sungai Gelam, 1958	5,479-6,201	2	8			...
	Sungai Lilin, 1964	2,690	23	30			...
	Talang Akar, 1922	6,562-8,681	27	482			35.0
	Talang Jimar, 1937	4,123	35	204			31.0
	Tanjung Tiga, 1938	3,750	30	107			28.0
	Tempino, 1931	1,647	42	179			23.0
	T. Miring, 1935	3,511	26	50			...
	Ukui/Pondok, 1950	1,378-1,641	4	15			...
	Other	...	...	9			...
PERTAMINA UEP III[(e)]	Banyubang, 1990	1,312	1	1	67,547	2,579,397,534	...
	Cemara, 1976	3,964-7,104	18	37			...
	Jatibarang, 1969	7,136	77	159			29.0
	Kandang Haur Timur, 1979	7,546	3	10			...
	Kawengan, 1926	1,600-2,625	39	142			...
	Ledok, 1893	1,380	16	251			...
	MB STR, 1985	10,400	12	14			...
	Nglobo + Semanggi, 1896	1,182	19	131			...
	Pegaden, 1986	6,890	1	6			...
	Randegan, 1973	5,578	4	13			...
	Tugu Barat, 1979	3,014	13	26			...
	Other	...	...	32			...
PERTAMINA UEP IV[(e)]	Bangkudulis, 1960	3,250	1	5			39.5
	Bunyu Nibung, 1923	5,315	31	144			33.2
	Dahor Selatan, 1966	3,610	1	6			...
	Kambitin, 1981	3,491	2	11			...
	Sangkimah	4,265	...	2			...
	Sangatta, 1939	4,265-4,593	...	121			33.0
	Sembakung, 1976	2,600	16	19			36.0
	Tanjung, 1938	7,956	61	103			39.9
	Tanta, 1975	...	1	1			...
	Tapa-Teritip, 1975	6,660	8	22			...
	Tapian Timur, 1967	4,396	10	25			38.8
	Warukin, 1966	2,067	22	48			26.9
PERTAMINA UEP V[(e)]	Klamono, 1936	574	60	97			19.0
	Linda, 1977	3,281	10	28			19.0
	Salawati, 1977	5,500	9	69			...
	Sele, 1951	2,156	12	29			35.0
KODECO ENERGY[(e)]	KE-3/Camar, 1970	4,178-4,260	...	...			36.6-37.0
	Madura KE-2, 1978	3,370-4,080	3	3			45.5-47.2
	Madura KE-23, 1973	6,700-8,350	...	...			35.1-42.0
AMOSEAS[(e)]	• Anoa, 1975	3,415-6,610	10	10			44.5-45.0
LASMO (MALACCA STRAIT)[(e)]	• Kurau, 1986	4,500-9,000	20	23			47.4
	• Lalang, 1980	3,100-3,600	11	18			39.6
	Melibur JS, 1984	900-1,300	17	25			35.6
	• Mengkapan, 1981	3,850-4,450	9	10			42.0
	MSAI S	...	...	...			...
	MSBH S	...	...	...			...
	MSBK S	...	...	...			...
	MSBM	...	...	...	66,726	688,126,048	...
	MSBT S	...	...	...			...
	MSBZ	...	...	...			...
	MSN	...	...	...			...
	• Selatan, 1987	3,900-4,500	12	16			...
TESORO[(e)]	Tarakan, 1906	...	114	1,167			...
	[Juata], 1926	1,278-3,600	...	...			19.0-20.5
	[Mengatai], 1938	1,050-5,250	...	...			19.0-32.0
	[Pamusian], 1906	213-4,249	...	...			18.0-19.0
	[Sesanip], 1938	750-1,050	...	...			19.0-20.0
	Sanga-sanga	...	110	253			...
	[Anggana], 1903	230-2,772	...	...			25.0-29.0
	[Kutai Lama N.], 1938	426-5,637	...	...			27.0-33.0
	[Kutai Lama S.], 1902	1,640-2,953	...	...			25.0-30.0
	[Louise C.], 1898	102-1,898	...	...			25.0-30.0
	[Louise N.], 1902	62-2,264	...	...			25.0-29.0
	[Louise S.], 1897	82-2,379	...	...			25.0-30.0
	[Maura], 1902	869-2,297	...	...			25.0-29.0
	[Nonny], 1975	295-3,248	...	...			27.0-31.0
	[T. Una], 1936	1,263-2,297	...	...			27.0-29.0
	Samboja	...	24	102			...
	[Jambatan B], 1934	951-1,213	...	...			26.0
	[Samboja], 1909	65-4,387	...	...			18.0-27.0
	Mamburungan, 1985	...	2	3			...

WORLDWIDE PRODUCTION

Country	Company	Name of field, discovery date	Depth, ft	No. of wells — Producing	No. of wells — Total	Production — 1992 average b/d	Production — Cumulative to Dec. 31, 1992 bbl	°API gravity
	ENIM OIL[(e)]	Harimau	...	7	7			...
	ENTERPRISE OIL[(e)]	• North Camar, 1970	4,350	1	2			36.6-36.9
		• South Camar, 1970	4,178-4,260	3	4			37.0
		Total Indonesia		**8,047**	**17,126**	**1,504,558**	**15,380,277,486**	
IRAN[(e)]	NIOC	• Abouzar, 1969	2,800					27.0
		Agha Jari, 1936	7,440					34.0
		Ahwaz Asmari, 1958	8,150					32.0
		Ahwaz Bangestan, 1958	12,000					24.3
		• Bahregansar, 1960	9,000					30.0
		Bibi Hakimeh, 1961	4,570					29.9
		Binak, 1959	10,440					29.9
		Chesmeh Khush, 1967	10,700					27.7
		Chillingar, 1974	3,350					39.3
		• Darius, 1961	11,000					34.0
		Dehluran, 1972	11,900					31.0
		• Forozan, 1966	7,000					28.5
		Gachsaran, 1937	3,400					31.0
		Haft Kel, 1927	1,980					37.7
		• Hendijan, 1968	11,000					23.0
		Karanj, 1963	5,360					33.9
		Kharg, 1961	10,900					31.6
		Kupal, 1965	10,500					32.8
		Lab-e Safid, 1968	4,350	688	688	3,455,000	41,669,617,000	35.3
		Mansuri, 1963	7,100					28.4
		Marun, 1963	9,400					32.6
		Masjid-e Suleiman, 1908	1,600					39.3
		Naft Safid, 1938	5,000					36.8
		Naft-Shahre, 1923	2,000-3,000					43.0
		• Nowruz, 1966	8,200					21.0
		Par-e Siah, 1964	4,250					37.7
		Paris, 1964	4,670					38.8
		Pazanan, 1961	7,250					37.2
		Rag-e Safid, 1964	7,230					28.6
		Ramin, 1966	12,200					32.9
		Ramshir, 1962	8,750					27.6
		• Resalat, 1969	7,650					35.8
		• Reshadat, 1966	7,000					34.4
		• Salman, 1965	7,500-8,100					34.0
		Shadegan, 1989	...					...
		Soroush, 1962	7,000					18.3
		Sulabedar, 1971	3,400					30.4
		Total Iran		**688**	**688**	**3,455,000**	**41,669,617,000**	
IRAQ[(e)]	INOC	Abu Ghurab, 1971	9,800	12	16			24.0
		Ain Zalah, 1939	5,200-6,500	15	28			31.0
		Bai Hassan, 1953	4,800-5,400	56	84			34.0
		Balad, 1983	6,000-7,000	4	8			25.0
		Butmah, 1953	3,900	1	17			31.0
		Buzurgan, 1970	12,000	20	22			24.0
		East Baghdad, 1975	6,000-7,000	32	79			23.0
		Hamrin, 1973	2,000	...	14			30.0
		Jabal Fauqui, 1974	10,000	20	23			25.5
		Jambur, 1954	5,500-12,500	27	50	425,000	22,689,258,000	38.0
		Kirkuk, 1927	2,800-4,200	71	269			36.0
		Luhais, 1961	8,200	12	21			33.0
		Naft Khaneh, 1909	3,500	7	11			42.5
		Nahr Umr, 1949	8,200	5	13			42.0
		Qaiyanah, 1929	650	33	50			16.0
		Rumaila North, 1958	10,300	303	443			33.0
		Rumaila South, 1953	10,200	133	208			34.0
		Suffayah, 1978	3,900	11	39			25.0
		West Tikrit, 1985	8,200	2	6			23.0
		Zubair, 1949	10,200	56	110			35.0
*Wells prior to conflict in the Persian Gulf.		**Total Iraq**		***820**	**1,511**	**425,000**	**22,689,258,000**	
ISRAEL	LAPIDOTH	Heletz-Brur, 1955	5,200	3	18	75	12,419,000	29.2
		Kochav, 1962	5,500	6	13	83	4,534,000	27.9
		Zuk Tamrur, 1982	6,107	...	...	...	8,197	27.0
	OEIL	Ashdod, 1976	8,400	2	2	26	198,333	32.5
	EMOG	Gurim, 1984	3,280	...	...	...	6,954	17.0
		Total Israel		**11**	**33**	**184**	**17,166,484**	
ITALY	ELF	• Rospo Mare, 1975	4,600	28	29	16,930	48,120,220	11.0
		• Santa Maria a Mare, 1974	7,900	8	12	540	18,409,900	23.0
		• Sarago Mare, 1979	5,200	4	7	3,260	14,616,350	8.0

WORLDWIDE PRODUCTION

Country	Company	Name of field, discovery date	Depth, ft	No. of wells Producing	No. of wells Total	Production 1992 average b/d	Production Cumulative to Dec. 31, 1992 bbl	°API gravity
	MONTEDISON	Vega, 1980	8,500	20	24	8,845	28,716,496	16.0
	AGIP	Benevento, 1974	10,335	1	1			39.0
		Cammarata/Pozzillo, 1959	10,499	4	4			8.3
		Cavone, 1973	9,350	10	18			21.5
		Cortemaggiore, 1949	6,562	5	8			33.2
		Costamolina, 1983	12,795	1	1			17.8
		• Emilio, 1971	8,530	1	1			10.7
		Gaggiano, 1982	14,764	1	2			37.2
		Gela, 1956	10,827	63	85	53,425	385,781,000	10.0
		Giaurone, 1982	10,170	5	5			10.2
		Malossa, 1973	17,060	1	3			52.9
		Monte Alpi, 1988	11,483	1	2			37.2
		• Perla, 1976	7,874	...	4			13.3
		Pisticci-S. Cataldo, 1960	6,980	5	12			12.0
		Ponte Dirillo, 1959	9,514	8	11			15.6
		• Prezioso, 1982	15,748	8	9			19.7
		Ragusa, 1953	4,593	38	41			19.3
		S. Maria Inbaro, 1963	7,546	3	4			15.9
		Torrente Tona, 1963	4,101	8	14			31.9
		Villafortuna/Trecate, 1984	18,044	11	12			46.3
		Total Italy		**234**	**309**	**83,000**	**495,643,966**	
IVORY COAST(e)	PETROCI	• Belier, 1975	6,500-7,000	12	18	1,000	51,016,000	34.5
		• Espoir, 1980	6,500 11,000	...	...	...		32.3
		Total Ivory Coast		**12**	**18**	**1,000**	**51,016,000**	
JAPAN	JAPEX	• Aga-Oki, 1972	7,590	4	7	446	3,887,581	24.8
		Higashi-Niigata, 1965	4,500-9,200	1	2	583	8,563,403	53.9
		• Iwafune-Oki, 1984	5,050-8,660	8	8	2,543	1,517,647	38.2
		Katakai, 1960	3,310-16,300	...	1	465	1,302,535	50.0
		Sarukawa, 1958	2,000-3,200	82	121	1,215	18,923,864	32.3
		Yoshii, 1968	8,000	...	...	651	9,811,709	61.8
		Yurihara, 1976	1,310-7,610	6	6	298	756,107	43.5
		Other	...	24	28	328	22,715,688	...
	IDEMITSU OIL	• Agaoki-Kita, 1982	4,950	10	10	1,241	7,164,310	37.0
	TEIKOKU OIL CO., LTD.	Higashi-Kashiwazaki, 1970	6,000-8,900	6	7	562	8,911,458	62.0
		Kubiki, 1959	1,100-6,400	68	79	608	14,111,027	29.7
		Minami-Aga, 1964	7,300-9,600	9	16	564	14,609,020	38.5
		Minami-Nagaoka, 1984	13,200-15,100	6	17	669	1,777,595	55.7
		Yabase, 1933	1,100-5,800	44	55	354	33,762,668	33.0
		Other	...	6	7	73	2,808,042	...
		Total Japan		**274**	**364**	**10,600**	**150,622,654**	
JORDAN	NRA	Hamzah, 1984	9,500	4	17	58	688,507	30.0
		Total Jordan		**4**	**17**	**58**	**688,507**	
KUWAIT(e)	KOC	Burgan, 1938	4,800		393			30.8
		Ahmadi, 1952	4,800		84			31.7
		Magwa, 1951	4,800		113			33.0
		Raudhatain, 1955	8,600		53			34.4
		Bahra, 1956	8,500	295	2	880,000	24,485,608,000	...
		Sabriya, 1957	8,300		44			36.3
		Minagish, 1959	10,000		21			34.4
		Umm Gudair, 1962	9,000		33			26.9
		Total Kuwait		**295**	**743**	**880,000**	**24,485,608,000**	
LIBYA	AGIP	• Bouri, 1979	8,300	55	55	80,052	102,727,000	25.6
		Bu Attifel, 1968	14,000	38	50	178,118	1,078,077,000	40.6
		Katib, 1969	12,300	2	2	230	714,000	35.9
		N. Rimal, 1985	11,500	6	6	8,921	3,293,000	36.2
		Rimal, 1965	11,000-14,000	5	5	2,598	11,627,000	33.1
	LNOC(e)	Ali, 1975	3,900					50.0
		Almas, 1975	4,000					49.5
		Amal, 1959	9,900					37.5
		Arshad, 1966	10,100					42.0
		Aswad, 1978	6,250					46.0
		Bahi, 1968	3,500-3,750					43.4
		Balat, 1983	5,100-5,400					32.2
		Belhedan, 1962	7,000					33.6
		Beda, 1959	4,000					36.0
		Bualawn, 1972	6,000					40.0
		Dahra, 1959	3,100-3,700					37.0-41.0
		Defa, 1969	5,400-5,900					35.6
		Dor, 1967	6,400					33.2
		Fidaa, 1979	7,000					41.7

WORLDWIDE PRODUCTION

		Name of field, discovery date	Depth, ft	No. of wells — Producing	No. of wells — Total	Production — 1992 average b/d	Production — Cumulative to Dec. 31, 1992 bbl	°API gravity
		Ghani, 1978	6,000					40.0
		Gialo, 1961	2,200-6,300					35.7
		Hakim, 1978	6,700					48.5
		Harsh, 1978	7,400					34.6
		Intisar A, 1967	9,750					45.0
		Intisar B, 1974	9,600					50.0
		Intisar C, 1967	9,650	986	986	1,223,081	21,390,855,040	37.5
		Intisar D, 1967	9,400					39.0
		Intisar E, 1968	5,300-7,100					34.7
		Jebel, 1962	8,100					38.0
		Khalifa, 1978	5,400-7,000					34.6
		Kotla, 1963	5,500					34.0
		Lehib-Dor Marada Unit, 1965	9,500					50.0
		Majid, 1968	9,000-9,800					...
		Don Mansour Unit	5,600					...
		Masrab, 1978	10,000					39.6
		Nafoora-Augila, 1966	2,200-10,000					35.5
		Nasser, 1959	5,500-7,600					38.0
		Ora, 1962	7,250					40.2
		Raguba, 1961	5,400					43.0
		Sabah, 1977	5,500					42.0
		Sahabi B & D, 1961	9,000-10,000					39.0
		Samah, 1961	9,000					33.4
		Sarir, 1961	9,000					37.2
		Waha, 1960	6,000					36.0
		Zaggut, 1962	6,000					36.4
		Zella, 1977	6,750					48.1
		Other	7,900-8,900					40.4
		Total Libya		**1,092**	**1,104**	**1,493,000**	**22,587,293,040**	
MALAYSIA	**ESSO MALAYSIA**	• Bekok, 1976	5,500	15	32			47.0
		• Guntong, 1978	5,200	73	94			44.0
		• Irong Barat, 1979	3,300	14	18			35.0
		• Kepong, 1979	6,000	5	10			47.0
		• Palas, 1979	6,300	10	18			49.0
		• Pulai, 1973	4,300	16	31	333,000	1,184,000,000	42.3
		• Seligi, 1971	4,300	101	124			50.0
		• Semangkok, 1980	3,700	29	41			42.5
		• Tabu, 1978	5,600	6	24			44.0
		• Tapis, 1975	6,700	44	80			45.0
		• Tinggi, 1980	4,400	18	31			48.0
		• Tiong, 1978	5,800	18	41			47.0
	SABAH SHELL PETROLEUM (SSPC)	• Barton, 1971	2,400	4	4	10,100	24,500,000	37.0
		• Erb W, 1971	6,785	15	19	22,100	60,400,000	30.0
		• Samarang, 1972	5,700	56	80	36,100	321,500,000	37.0
		• St. Joseph, 1975	2,200	11	14	35,200	74,800,000	32.0
	SARAWAK SHELL BERHAD (SSB)	• Bayan, 1976	4,200	7	9	15,800	52,800,000	37.3
		• Temana, 1972	3,100	19	29	19,700	63,800,000	35.0
	SABAH SHELL PETROLEUM (SSPC)(e)	• South Furious, 1974	3,300	7	15			30.0
	SARAWAK SHELL BERHAD (SSB)(e)	• D-18, 1981	4,000	5	5			39.9
	PETRONAS CARIGALI(e)	• Tembungo, 1971	6,000	8	16			40.0
	SARAWAK SHELL/PETRONAS CAG.(e)	• Bakau	...	...	...			...
		• Baram	...	...	...			...
		• Baronia, 1967	7,900	19	40			42.0
		• Betty, 1967	7,900	13	17			38.0
		• Bokor, 1972	2,000	18	19	189,000	778,035,000	35.0
		• Fairley Baram, 1973	8,655	...	2			39.5
		• Siwa, 1973	4,100	4	4			25.0
		• Tukau, 1966	5,000	26	48			29.2
		• West Lutong	...	...	...			39.5
		Total Malaysia		**561**	**865**	**661,000**	**2,559,835,000**	
MEXICO(e)	**NORTH ZONE, FRONTERA NE**	Misión, 1945	5,440					...
		Monterrey, 1950	6,510					42.4
		Tigrillo, 1971	3,342					...
		Cañon	...					...
		Comitas	...					...
	NORTH ZONE, ALTAMIRA	Arenque, 1970	11,362			27,160		36.0
		Constituciones, 1956	6,350					17.0
		Ebano-Pánuco, 1901	1,450					12.0
		Tamaulipas	4,200					19.0
	NORTH ZONE, CERRO AZUL	Chicontepec, 1973	4,264					...
		• Isla de Lobos, 1963	6,875					34.1
		• Marsopa, 1974	10,198					40.0
		Naranjos-C. Azul, 1909	1,800					34.0
		Tres Hermanos, 1959	6,960					27.0

WORLDWIDE PRODUCTION

	Name of field, discovery date	Depth, ft	No. of wells Producing	No. of wells Total	Production 1992 average b/d	Production Cumulative to Dec. 31, 1992 bbl	°API gravity
NORTH ZONE, POZA RICA	• Atún, 1966	9,040					37.0
	• Bagre, 1973	10,919					33.0
	Escolín, 1942	7,216					26.3
	Hallazgo, 1955	10,170					28.7
	Jiliapa, 1958	7,390					35.6
	Mecatepec, 1941	7,544			65,980		29.8
	Miquetla, 1959	6,480					29.8
	Poza Rica, 1930	7,090					35.0
	Pres. Alemán, 1949	8,036					24.0
	Remolino, 1962	10,745					29.5
	San Andrés, 1956	10,410					29.0
NORTH ZONE, FAJA DE ORO	Acuatempa, 1955	4,085					21.0
	Muro, 1965	3,966					17.0
	Ordoñez, 1952	5,220			5,982		21.0
	Ocotepec, 1953	3,737					20.9
	Santa Agueda, 1953	4,780					16.0
NORTH ZONE, VERACRUZ	Matapionche, 1974	11,129					...
	Angostura	...					36.9
	Copite	...					14.0
	Other	...					32.3
SOUTH ZONE, NANCHITAL	Ixhuatlán Ote, 1965	1,960					20.4
	Moloacán, 1962	500					22.0
	Santa Rosa, 1952	1,312					25.4
SOUTH ZONE, AGUA DULCE	Agata, 1966	3,830					34.6
	Bacal, 1976	3,500					35.0
	Blasillo, 1967	7,216					37.0
	Cinco Ptes, 1960	6,862					32.0
	Concepcion, 1974	1,600					32.7
	Cuichapa, 1935	2,200					30.0
	El Plan, 1931	1,700					30.0
	El Burro, 1931	2,200			60,159		31.0
	Lacamango, 1973	1,700					31.6
	Los Soldados, 1953	4,492					37.0
	La Venta, 1954	4,730					33.0
	Ogarrio, 1957	5,790					33.0
	Otates, 1965	7,469					32.0
	Rodador, 1971	11,398					32.0
	S. Magallenes, 1957	4,240					30.0
	San Ramon, 1967	9,515					28.0
	Tonala, 1928	1,770	4,740	4,740		19,424,397,000	32.0
SOUTH ZONE, REFORMA	Agave, 1977	13,450					43.1
	Arroyo Zanapa, 1978	14,599					39.2
	Arteza, 1977	11,800					24.4
	Cacho López, 1977	14,250					29.0
	Cáctus, 1972	14,100					36.0
	Carmito, 1980	10,050					45.6
	Carrizo, 1962	3,500					23.3
	Comoapa, 1979	14,432					33.2
	Copano, 1977	11,890					47.3
	Chiapas, 1979	12,136					47.3
	Giraldas, 1977	15,225			186,967		46.0
	Iride, 1974	13,775					28.5
	Iris, 1979	14,432					38.8
	Muspac, 1982	9,676					58.5
	Mundo Nuevo, 1977	11,800					49.3
	Nispero, 1974	14,100					37.7
	Platanal, 1978	15,900					31.2
	Rio Nuevo, 1975	14,950					35.6
	Samaria (Cret), 1973	14,209					31.0
	Sitio Grande, 1972	13,766					35.0
	Sunuapa, 1978	12,887					31.8
	Tapijulapa, 1982	15,613					39.8
	Topen, 1978	11,172					39.6
	Arteza (Terc), 1984	...					28.0
	Samaria (Terc), 1965	4,920					13.9
SOUTH ZONE, COMALCALCO	Ayapa, 1972	8,200					27.2
	Castarrical, 1967	10,080					30.9
	Cunduacan, 1974	13,775					27.6
	El Golpe, 1963	3,500			165,902		28.5
	Mora, 1981	...					36.1
	Oxiacaque	11,150					27.9
	Santuario, 1966	9,617					37.0
	Tintal, 1968	5,904					23.0
	Tupilco, 1959	9,685					27.0
	Bellota, 1982	17,056					36.8
	Caparroso, 1982	18,368					33.2
SOUTH ZONE, CARDENAS	Cardenas, 1979	17,548					38.0
	Eden, 1983	17,590					37.0
	Fenix, 1979	17,876					40.0

WORLDWIDE PRODUCTION

Country	Operator	Name of field, discovery date	Depth, ft	No. of wells — Producing	No. of wells — Total	Production — 1992 average b/d	Production — Cumulative to Dec. 31, 1992 bbl	°API gravity
		Jolote, 1983	18,119					35.7
		Jujo, 1980	17,548			240,737		34.0
		Paredón, 1978	15,690					38.8
		Tecominoacan, 1983	19,519					33.2
	SOUTH ZONE, OCOSINGO	Fortuna Nacional	...					34.0
		Jose Colomo	...			558		34.0
		Vernet	..					34.0
	MARINE ZONE, CD. DEL CARMEN	• Cantarell, 1976	8,528			1,295,691		21.3
		• Ku, 1979	10,000					22.0
	MARINE ZONE, DOS BOCAS	• Abkatun, 1978	11,800					27.6
		• Pol, 1979	12,600					33.0
		• Chuc, 1982	13,100			618,589		35.0
		• Ixtoc, 1974	...					29.9
		Total Mexico		**4,740**	**4,740**	**2,667,725**	**19,424,397,000**	
MOROCCO		Toukimt & Meskala, 1976	10,500	6	9	80	490,786	65.0
		Sidi Rhalem, 1962	6,722	2	2	99	17,597,910	39.0
		Sidi Fili, 1951	4,590	1	1	35		39.0
		Total Morocco		**9**	**12**	**214**	**18,088,696**	
MYANMAR(e)	MYANMA	Chauk-Lanywa, 1902	1,500-4,000					36.0
		Mann, 1970	1,600-7,000					37.5
		Myanaung, 1964	2,000-4,000	450	450	13,000	513,438,000	40.5
		Prome, 1965	850-5,000					36.0
		Yenangyaung, 1902	100-6,700					36.0
		Total Myanmar		**450**	**450**	**13,000**	**513,438,000**	
NETHERLANDS	UNOCAL	• Haven, 1980	5,300	2	2	1,924	3,279,537	25.5
		• Helder, 1980	4,800	13	16	4,898	27,476,286	21.6
		• Helm, 1979	4,400	7	8	1,514	10,110,533	17.9
		• Hoorn, 1981	5,000	8	9	2,277	24,052,625	25.8
	CONOCO	• Kotter, 1980	5,618	8	14	13,382	56,047,040	32.4
		• Logger, 1982	6,022	5	12	5,549	24,240,049	32.7
	MOBIL	P/6, 1968	8,800	8	8	80	905,480	66.0
		P/12, 1986	9,200	5	5	435	403,192	54.0
	AMOCO	• Rijn, 1982	6,500	10	27	3,240	21,500,000	36.0
	NAM(e)	Berkel, 1954	...	17	17			32.1
		DeLier, 1955	5,200	22	33			33.2
		Ijsselmonde, 1956	...	59	86			26.8
		Pernis West, 1987	...	...	...	24,281	398,972,846	28.2
		Pijnacker, 1955	...	4	5			36.3
		Rotterdam, 1984	...	2	2			30.5
		Schoonebeek, 1944	2,500	...	495			12.3
		Werkendam, 1958	...	...	...			33.0
		Zoetermeer, 1957	3,200	30	30			16.5
		Total Netherlands		**200**	**769**	**57,580**	**566,987,588**	
NEUTRAL ZONE	AOC(e)	• Hout, 1969	4,900-8,800	21	34	283,792	3,066,412,412	33.0
		• Khafji, 1961	4,400-7,300	137	167			28.2
	TEXACO & KOC(e)	S. Fuwaris, 1963	6,300	...	9			25.0
		S. Umm Gudair, 1966	8,900	...	17	57,208	1,833,545,000	24.3
		Wafra, 1953	1,100-7,000	...	427			18.9-23.5
		Total Neutral Zone		**158**	**654**	**341,000**	**4,899,957,412**	
NEW ZEALAND	SHELL-BP-TODD	• Kapuni, 1959	11,700	11	14			54.0
		• Maui, 1969	11,000	11	14			51.0
	PCNZ	Ahuroa, 1987	8,688	...	1			46.1
		Kaimiro, 1982	11,900	3	3	36,815	138,880,210	42.0
		McKee, 1980	7,848-8,251	20	27			38.5
		Tariki, 1986	9,065	...	1			52.7
		Waihapa, 1988	10,125-10,643	5	5			36.6
	WESTERN MINING CORP	Kupe, 1987	9,514-12,467	...	...			...
	NEW ZEALAND OIL & GAS	Ngatoro, 1972	...	2	2			35.0
		Total New Zealand		**52**	**67**	**36,815**	**138,880,210**	
NIGERIA	AGIP	• Agbara, 1981	7,500-12,000	14	14	19,073	28,665,000	36.2
		Agwe, 1975	10,800	2	2	986	2,842,000	44.0
		Akri, 1967	9,600-10,600	10	11	9,775	73,161,000	40.8
		Akri W., 1972	7,500-10,200	2	2	...	482,000	43.0
		Ashaka, 1968	9,700-12,000	1	1	880	4,093,000	37.4

WORLDWIDE PRODUCTION

	Name of field, discovery date	Depth, ft	No. of wells Producing	No. of wells Total	Production 1992 average b/d	Production Cumulative to Dec. 31, 1992 bbl	°API gravity
	Azuzuama, 1975	9,700-14,900	3	3	2,269	2,642,000	34.1
	Beniboye, 1978	9,000-10,000	9	9	5,250	20,706,000	28.6
	Beniku, 1974	12,000	2	2	22	3,474,000	28.0
	Clough Creek, 1976	10,000-12,400	9	9	7,297	35,061,000	25.5
	Ebegoro, 1975	11,500-12,000	10	10	7,182	71,450,000	37.1
	Ebegoro S., 1975	14,100	2	2	576	211,000	36.0
	Ebocha, 1965	8,000-10,900	16	16	8,049	123,087,000	36.3
	Idu, 1967	7,750-10,700	10	10	...	23,863,000	30.0
	Kwale, 1967	10,000-11,500	7	10	5,612	17,792,000	37.1
	Mbede, 1966	7,300-9,400	17	17	15,867	184,109,000	43.1
	Obama, 1973	11,600-14,300	10	10	10,595	115,846,000	30.7
	Obiafu, 1967	9,300-12,200	19	25	17,129	111,778,000	41.0
	Obrikom, 1967	8,000-10,000	12	16	7,969	65,324,000	43.4
	Odugri, 1972	12,500	2	2	2,416	12,543,000	45.9
	Ogbogene, 1972	10,000	2	2	95	3,338,000	33.0
	Ogbogene W., 1982	13,000	1	1	...	353,000	45.0
	Okpai, 1968	10,000-12,000	8	8	4,673	25,453,000	42.8
	Omoku W., 1975	11,500	1	1	985	5,937,000	40.0
	Oshi, 1972	9,600-11,300	11	11	12,895	85,285,000	35.2
	Pirigbene, 1987	13,700	1	1	1,802	660,000	43.0
	Taylor Creek, 1985	9,300-10,400	1	1	2,094	3,235,000	37.3
	Tebidaba, 1972	10,300-11,300	10	10	11,807	139,520,000	30.4
	Umuoru, 1975	14,100	4	4	2,511	8,880,000	40.3
ASHLAND	• Adanga, 1980	8,500	8	8	6,289	29,322,064	31.9
	• Akam, 1980	6,700	8	8	4,307	24,816,557	34.0
	• Bogi, 1989	7,500	3	3	1,154	2,176,274	28.7
	• Ebughu, 1983	7,250	3	3	603	1,441,655	23.8
	Izombe, 1974	8,000	11	11	8,404	63,946,519	34.3
	• Mimbo, 1984	7,500	4	4	2,243	4,609,017	31.1
	Ossu, 1974	8,000	1	1	411	6,059,482	31.6
	• Ukpam, 1989	5,500	1	1	1,262	2,413,000	40.6
ELF	Aghigo, 1972	7,303	29	35	7,340	33,800,000	24.8
	Erema, 1972	11,211	6	7	5,350	19,300,000	24.6
	Obagi, 1964	8,377	77	89	54,700	452,600,000	24.6
	Obodo-Jatumi, 1966	7,656	16	20	15,040	92,600,000	24.8
	Okpoko, 1967	6,745	19	22	7,500	42,500,000	24.8
	Upomami, 1965	6,256	19	23	3,920	22,350,000	24.8
CHEVRON	Abiteye, 1970	5,750-9,400	23	49	10,991	76,386,461	39.7
	• Delta, 1965	5,600-9,500	34	61	19,407	197,944,573	37.3
	• Delta South, 1965	7,100-10,179	22	63	29,100	317,042,657	38.4
	• Isan, 1970	5,900-9,000	7	17	3,516	46,624,023	40.4
	Jisike, 1975	6,300-7,600	6	12	9,780	11,667,095	41.1
	Makaraba, 1973	7,100-12,005	27	46	21,540	125,111,674	27.7
	• Malu, 1969	4,800-6,300	15	35	15,836	125,663,845	40.4
	• Mefa, 1965	8,570-12,030	19	21	15,977	30,855,363	38.1
	• Meji, 1965	5,200-10,900	35	47	21,019	180,227,564	31.9
	• Meren, 1965	5,000-7,500	60	118	64,579	586,092,508	31.9
	• Mina, 1985	6,700-9,300	5	6	3,701	6,495,922	40.3
	• Okan, 1964	5,500-9,245	77	158	45,200	503,549,857	38.1
	• Parabe/Eko, 1968	4,500-8,200	33	59	12,079	118,843,002	40.4
	Robertkiri, 1964	11,484-13,190	20	25	25,735	46,072,683	40.2
	• Tapa, 1978	8,150-10,842	11	21	12,476	42,979,393	39.5
	Utonana, 1971	7,400-9,165	3	9	879	9,173,058	20.4
	• W. Isan, 1971	7,825-10,229	8	23	2,305	38,501,482	40.4
	Yorla South, 1974	11,389-12,635	3	4	2,356	2,017,021	41.0
TEXACO-CHEVRON-NNPC	• Funiwa, 1978	8,000	19	27	21,603	72,410,909	35.5
	• Middleton, 1972	9,000	6	6	5,541	20,941,232	33.3
	• North Apoi, 1973	8,500	26	30	28,835	212,012,810	35.5
	• Pennington, 1965	11,000	2	2	2,173	28,324,458	38.5
	• Sengana, 1969	12,000	...	...	...	666,102	...
SHELL	Adibawa, 1966	11,950	14	14	12,000	96,900,000	26.4
	Afisere, 1966	8,500	24	24	11,700	133,300,000	19.6
	Afremo, 1972	9,263	8	8	21,300	33,400,000	36.8
	Agbada, 1960	10,000	32	32	23,800	213,400,000	23.9
	Akaso, 1979	13,122	6	6	22,400	17,000,000	37.1
	Benisede, 1973	8,941	13	13	31,400	149,600,000	21.9
	Cawthorne Chan, 1963	11,000	21	21	47,500	351,100,000	36.9
	Egwa, 1967	9,350	17	17	21,600	162,500,000	34.1
	Ekulama, 1958	10,483	23	23	38,300	261,500,000	31.6
	Escravos Beach, 1969	8,176	7	7	10,100	73,800,000	31.2
	Etelebou, 1971	12,000	8	8	14,500	142,800,000	31.3
	Forcados Yorki, 1968	10,859	55	55	62,300	633,900,000	24.4
	Imo River, 1959	7,900	37	37	30,300	524,100,000	30.3
	Jones Creek, 1967	8,000	29	29	50,800	463,300,000	29.7
	Kalaekule, 1972	9,236	7	7	13,100	23,100,000	40.5
	Kokori, 1961	8,900	22	22	16,900	334,500,000	39.9
	Kolo Creek, 1971	12,000	18	18	19,700	159,400,000	39.1
	Nembe Creek, 1973	12,213	39	39	94,500	354,300,000	31.0
	Nun-River, 1960	14,200	5	5	11,700	40,300,000	32.7
	Oben, 1972	12,036	18	18	18,900	147,500,000	37.6

WORLDWIDE PRODUCTION

	Name of field, discovery date	Depth, ft	No. of wells Producing	No. of wells Total	Production 1992 average b/d	Production Cumulative to Dec. 31, 1992 bbl	°API gravity
	Obigbo North, 1963	8,250	30	30	16,900	196,300,000	23.6
	Odeama Creek, 1981	12,897	5	5	21,900	51,400,000	35.5
	Odidi, 1967	10,980	30	30	33,300	265,600,000	36.2
	Oguta, 1965	10,300	13	13	15,000	153,400,000	43.2
	Olomoro, 1963	8,500	24	24	20,200	305,500,000	21.5
	Opukushi, 1962	7,823	11	11	21,600	149,600,000	28.3
	Otumara, 1969	8,176	23	23	38,000	216,600,000	24.9
	Soku, 1958	11,500	15	15	12,600	127,000,000	27.7
SHELL(e)	Adibawa N.E., 1972	10,663	3	3			25.3
	Afam, 1956	6,000	8	8			45.6
	Afam Umu, 1957	10,024	...	4			...
	Ahia, 1965	11,561	4	4			38.2
	Ajokpori, 1967	11,139	1	1			39.3
	Ajuju, 1970	13,324	1	1			34.7
	Akpor, 1965	10,149	...	3			...
	Akuba, 1967	10,434	1	1			22.7
	Alakiri, 1959	10,169	13	13			44.5
	Amukpe, 1970	11,459	2	2			41.7
	Apara, 1960	9,000	3	3			38.4
	Assa, 1961	9,738	2	2			20.6
	Bodo East, 1962	9,700	9	9			27.1
	Batan, 1968	11,000	8	8			21.7
	Bomu, 1958	7,000	7	7			37.3
	Bonny, 1959	12,254	7	7			32.6
	Buguma Creek, 1960	12,496	2	2			37.6
	Diebu Creek, 1966	13,119	7	7			40.3
	Ebubu, 1958	9,411	3	3			21.2
	Egbema, 1960	8,731	6	6			33.1
	Egbema W., 1965	9,096	5	5			41.2
	Elelenwa, 1960	11,000	6	6			35.8
	Enwhe, 1965	10,170	...	6			...
	Eriemu, 1961	12,500	14	14			20.6
	Evrweni, 1967	10,900	8	8			25.6
	Gbaran	...	3	3			...
	Ibigwe, 1965	...	3	3			...
	Isimiri, 1964	8,450	10	10			29.3
	Isoko, 1960	12,479	4	4			29.8
	Korokoro, 1959	10,222	5	5			35.8
	Krakama, 1958	9,890	6	6			22.9
	Nkali, 1963	12,000	5	5			39.2
	Nun-River, 1960	14,200	5	5			32.7
	Obeakpu, 1975	8,274	2	2			26.3
	Obele, 1964	10,040	...	5			...
	Ogini, 1964	5,860	7	7			18.2
	Onne, 1965	10,384	3	3			35.3
	Opuama, 1972	9,214	3	3			43.7
	Opobo South, 1974	9,849	4	4			43.7
	Opukushi North, 1977	11,100	1	1			35.6
	Oloibiri, 1956	10,367	...	18	498,740	4,555,321,500	...
	Oroni, 1964	12,000	6	6			22.6
	Orubiri, 1971	9,700	1	1			38.2
	Osioka, 1967	11,721	2	2			26.2
	Otamini, 1973	9,440	3	3			21.0
	Oweh, 1964	12,300	9	9			25.8
	Rapele, 1965	11,627	1	1			44.0
	Rumuekpe, 1961	11,016	1	1			29.3
	Saghara, 1970	8,590	5	5			31.8
	Sapele, 1969	12,788	8	8			42.9
	Tai, 1965	10,367	1	1			38.3
	Ubie, 1961	14,380	5	5			27.6
	Ughada, 1966	8,618	1	1			37.7
	Ughelli East, 1959	11,800	7	7			38.9
	Ughelli West, 1960	8,800	9	9			20.5
	Umuchem, 1959	8,250	8	8			37.2
	Urhure	...	1	1			...
	Utapate South, 1974	10,497	8	8			38.5
	Utorogu, 1964	9,000	18	18			26.8
	Uzere East, 1960	8,500	10	10			28.6
	Uzere West, 1963	8,500	10	10			25.4
	Warri River, 1961	12,264	2	2			29.8
	Yorla, 1970	11,917	10	10			39.2
	Yorla South, 1979	12,190	...	2			...
MOBIL(e)	• Adua, 1967	6,970	4	4			34.8
	• Asabo, 1966	5,600	5	5			34.6
	• Asabo "D", 1979	7,621	...	2			34.6
	• Ata, 1964	8,823	...	5			...
	Edop, 1988	9,215	8	8			37.4
	• Ekpe, 1966	8,200	7	7			34.9
	• Ekpe-WW, 1977	6,810	6	6			31.9
	• Eku, 1966	5,420	2	2			30.4
	• Enang, 1968	6,600	6	6			37.4
	• Etim, 1968	6,200	5	5			37.1

WORLDWIDE PRODUCTION

Country	Company	Name of field, discovery date	Depth, ft	No. of wells — Producing	No. of wells — Total	Production — 1992 average b/d	Production — Cumulative to Dec. 31, 1992 bbl	°API gravity
		● Idoho, 1966	9,020	1	1			30.5
		● Inim, 1966	5,850	5	5			37.5
		● Isobo, 1968	7,345	2	2			30.4
		● Iyak	...	1	1			...
		● Iyak S.E., 1979	7,441	10	10			38.6
		● Mfem, 1968	5,200	2	2			36.1
		● Ubit, 1968	5,400	18	18			36.1
		● Unam, 1967	5,180	4	4			33.3
		● Usari, 1965	8,431	...	4			...
		● Utue, 1966	5,700	5	5			38.6
	PAN OCEAN(e)	Ogharefe, 1973	9,900	5	5			46.6
	DUBRIL(e)	Gilli-Gilli, 1967	9,357	2	2			47.1
	TENNECO(e)	Abura, 1979	13,471	3	3			44.7
		Total Nigeria		1,824	2,295	1,902,000	15,103,749,760	
NORWAY	AMOCO	● Hod, 1974	8,860	5	5	22,127	19,098,375	34.0
		● Valhall, 1975	8,200	23	26	69,536	216,450,260	35.0
	BP	● Gyda, 1981	13,123	11	11	68,000	57,003,000	48.0
		● Ula, 1976	11,800	9	9	133,000	240,606,000	44.6
	CONOCO	● Murchison*, 1975	9,900	21	34	8,054	72,947,650	38.0
	NORSK HYDRO	● Oseberg B & C, 1979	8,200	26	54	450,825	494,344,810	34.0
		● Mime, 1982	13,450	1	1	2,049	1,907,685	30.0
	PHILLIPS	● Albuskjell, 1972	10,400	7	19	1,496	44,556,866	45.0
		● Cod, 1968	9,500	3	6	843	16,582,061	55.0
		● Edda, 1972	10,480-10,600	6	7	2,912	26,188,127	41.5
		● Ekofisk, 1968	10,000	56	60	155,103	1,098,515,442	41.1
		● Eldfisk, 1970	9,350-10,000	42	144	48,062	320,207,296	41.1
		● Tor, 1970	10,000	13	15	6,247	117,522,934	38.0
		● West Ekofisk, 1970	10,300	9	12	1,705	73,615,401	43.9
	STATOIL	● Gullfaks, 1978	9,500-9,700	53	73	431,025	551,000,000	29.3
		● Statfjord, 1974	9,500	63	92	635,672	2,123,328,400	38.2
		● Tommeliten, 1976	10,400	6	6	7,535	16,300,000	47.0
		● Veslefrikk, 1981	10,100	9	14	67,240	65,000,000	39.9
	ELF NORGE	● Heimdal(c), 1972	...	7	7	65,310	19,291,849	...
		Frigg	...	11	11	59	992,937	...
		East Frigg	...	3	3	59	99,006	...
		North East Frigg	...	2	2	29	233,849	...
*Norway share (22.2%) of total production.		**Total Norway**		386	611	2,176,888	5,575,791,948	
OMAN	ELF GROUP	Ramlat, 1979	12,000	1	1	26	191,713	46.0
		Sahmah, 1978	12,400	21	21	12,862	36,626,794	46.0
	OCCIDENTAL	Al Barakah, 1992	5,000	2	2	832	304,485	42.0
		Safah, 1983	6,500	94	104	26,088	43,686,970	42.0
	SHELL	Al Huwaisah, 1969	5,320	43	47	16,700	197,100,000	39.0
		Amal, 1977	...	...	...	12,000	22,900,000	19.0
		Anzauz, 1974	7,098	4	4	10,400	17,400,000	46.5
		Fahud NW, 1963	2,499	153	235	36,600	689,500,000	32.0
		Fahud SE, 1963	...	...	...	11,500	162,600,000	32.0
		Karim W., 1982	3,546	26	27	18,900	37,200,000	23.0
		Lekhwair, 1974	4,352	47	63	58,500	115,200,000	37.5
		Marmul, 1977	5,087	169	185	31,100	171,900,000	21.5
		Natih, 1963	4,788	36	68	21,600	385,700,000	33.4
		Nimr, 1980	3,296	117	119	58,200	92,000,000	21.0
		Rima, 1979	3,247	47	48	79,200	234,600,000	33.0
		Saih Nihayda, 1972	8,700	21	28	17,200	191,600,000	39.6
		Sayyala, 1982	3,690	14	14	36,600	74,200,000	50.1
		Yibal, 1963	8,465	194	206	165,300	903,500,000	38.5
		Zareef, 1985	3,264	10	11	12,000	25,000,000	47.0
	SHELL	Al Burj, 1987	6,890	6	7			28.0
		Al Dhabi, 1985	...	...	...			19.0
		Al Ghubar, 1978	1,916	3	6			29.0
		Alam, 1987	...	...	...			33.0
		Amal S, 1982	...	34	39			30.0
		Amin, 1975	...	...	...			20.0
		Bahja, 1984	5,697	10	12			47.2
		Barik, 1978	8,856	3	3			47.4
		Dhiab, 1985	...	...	...			26.5
		Fahud West, 1965	...	...	...			34.0
		Fayyadh, 1983	...	...	...			27.0
		Ghaba N., 1972	2,227	14	25			33.5
		Habur, 1975	5,800	1	2			15.0
		Haima, 1984	...	...	...			23.3
		Hasirah, 1984	8,708	10	12			48.8
		Ihsan, 1984	5,578	5	6			23.5

WORLDWIDE PRODUCTION

		Name of field, discovery date	Depth, ft	No. of wells Producing	No. of wells Total	Production 1992 average b/d	Production Cumulative to Dec. 31, 1992 bbl	°API gravity
		Jalmud, 1979	4,236	3	3			24.0
		Jalmud North, 1980	...	...	...			19.0
		Jawdah, 1984	3,570	6	7			28.5
		Kaukab, 1983	8,282	1	1			31.0
		Maqtaa, 1991	...	...	...			44.0
		Musallim, 1971	...	...	...	106,550	436,638,300	27.0
		Natih West	...	...	...			33.8
		Qaharir, 1978	4,536	27	32			30.0
		Quharir East, 1978	...	2	4			25.5
		Qarat Al Milh, 1973	...	...	...			29.0
		Qarn Alam, 1972	...	...	...			17.0
		Qata, 1979	4,707	11	16			28.0
		Rahab, 1978	3,237	21	28			25.0
		Rajaa, 1988	...	...	...			50.0
		Ramlat Rawl, 1987	2,720	1	1			37.0
		Rasha, 1982	...	...	...			24.0
		Reihan, 1986	...	...	...			21.3
		Runib, 1979	3,215	10	10			21.1
		Sadad, 1986	...	...	...			25.0
		Saih Nihayda SE, 1975	...	...	...			39.0
		Saih Rawl, 1973	9,826	14	24			39.0
		Simsim, 1985	5,250	11	12			30.6
		Suwaihat, 1982	6,452	10	10			37.5
		Thafut, 1981	...	...	...			19.0
		Thamoud, 1987	3,773	...	...			23.0
		Thuleilat, 1980	...	14	15			25.5
		Wadi Haka, 1989	...	...	...			32.0
		Wafra, 1983	5,510	7	8			38.8
		Warad, 1985	4,922	5	5			22.0
		Zahra, 1986	6,562	5	5			32.0
		Zauliyah, 1981	8,177	10	14			44.3
		Zumurrud, 1988	...	...	...			29.0
		Total Oman		**1,243**	**1,490**	**732,158**	**3,837,848,262**	
PAKISTAN	**OCCIDENTAL OF PAKISTAN, INC.**	Balkassar, 1946	8,300	4	17	442	32,415,117	27.0
		Bhangali, 1989	12,650	1	2	703	1,588,179	32.5
		Dhurnal, 1984	13,100	3	7	10,012	42,614,545	38.0
	OGDC	Bhal Syedan, 1989	13,120	1	1	189	69,124	40.6
		Bobi, 1988	7,897-9,515	1	5	397	665,794	52.0
		Chak-Naurang, 1986	9,187	2	4	1,135	1,978,294	12.1
		Dakhni, 1983	14,080	2	5	893	1,081,280	48.5
		Fimkassar, 1982	11,465	2	2	3,660	4,345,673	29.4
		Ghotana, 1986	7,300	...	1	...	165,548	41.1
		Kunnar, 1988	7,700	1	4	1,232	475,736	40.2
		Lashari Centre, 1988	6,916	1	1	2,034	2,906,323	44.2
		Lashari East, 1988	8,377	1	1	84	15,208	42.0
		Messa-Keswal, 1991	1,500	2	2	2,806	1,188,398	41.2
		Neyun Ismeal, 1991	7,869	1	1	...	56,167	42.0
		Pasahki, 1989	7,153	2	2	3,133	2,994,680	43.0
		Pasaki North, 1989	7,206	1	1	689	1,188,147	44.0
		Sadkal, 1992	14,925	1	1	49	17,854	41.5
		Sono, 1988	7,618	3	5	2,741	3,370,811	40.5
		Tando Alam, 1984	6,890-7,448	4	10	2,180	8,687,671	41.7
		Thora, 1987	7,546	4	4	3,532	6,509,034	42.2
		Thora East, 1987	7,546	1	1	247	859,067	...
		Toot, 1968	15,135	3	11	669	11,237,386	39.0
	PAKISTAN OILFIELD LTD.	Dhulian, 1936	8,899	11	49	20	41,356,761	48.0
		Joya Mair, 1944	6,900	1	4	571	6,698,828	15.0
		Khaur, 1915	9,306	1	397	5	4,183,903	33.0
		Meyal, 1968	12,513	7	15	2,551	34,897,348	41.3
	UNION TEXAS	Bari, 1992	2,600	1	1	47	16,988	42.0
		Bhatti, 1989	5,800	3	3	92	33,686	44.0
		Bukhari, 1985	5,800	3	3	603	220,233	53.0
		Dabhi, 1984	5,975	2	4	730	3,102,714	42.0
		Golarchi, 1984	6,200	3	3	64	132,944	51.1
		Halipota, 1986	6,000	2	2	387	238,380	44.4
		Khaskeli, 1981	3,600	3	10	320	7,970,083	39.0
		Laghari, 1983	2,500	4	7	2,342	18,081,367	42.0
		Liari, 1988	6,600	2	2	4,299	3,166,138	42.5
		Makhdumpur, 1986	7,000	2	2	2	770	54.7
		Matli, 1986	5,400	4	4	81	235,493	43.7
		Mazari, 1984	3,800	6	6	4,892	12,136,530	43.0
		Nari, 1985	5,900	2	2	67	24,490	47.0
		Sonro, 1985	3,960	2	2	109	355,791	43.7
		South Mazari, 1985	3,920	4	5	3,707	4,658,424	43.0
		Turk, 1985	6,150	6	6	459	772,090	37.8
	PAKISTAN PETROLEUM LTD.	Adhi Gas Condensate Field, 1978	8,360-8,439	5	5	2,506	3,741,172	51.0
		Total Pakistan		**115**	**620**	**60,681**	**266,454,169**	

WORLDWIDE PRODUCTION

Country	Operator	Name of field, discovery date	Depth, ft	No. of wells — Producing	No. of wells — Total	Production — 1992 average b/d	Production — Cumulative to Dec. 31, 1992 bbl	°API gravity
PAPUA NEW GUINEA	CHEVRON	Agogo, 1989	4,200-5,400	4	4	1,335	488,517	44.0
		Iagifu/Hedinia, 1986	7,400-8,500	19	23	51,045	18,716,023	45.0
		Total Papua New Guinea		23	27	52,380	19,204,540	
PERU	PETROMAR S.A.	• Litoral, 1955	1,500-6,000	76	76	1,445	39,984,726	36.0-40.0
		• Lobitos, 1960	3,500-7,500	146	146	5,687	62,819,876	38.0
		• Organos	...	3	3	6	316,420	38.0
		• Pena Negra, 1960	3,000-8,500	185	185	7,143	120,515,169	38.0
		• Providencia, 1967	3,000-7,000	47	47	1,204	23,477,126	38.0-40.0
	OCCIDENTAL	Bartra, 1979	8,000-8,930	13	16	4,718	24,744,420	11.6
		Capahuari Sur, 1973	11,800-12,900	18	22	6,698	142,235,272	33.7
		Dorissa, 1978	9,800-10,600	9	15	7,634	46,897,894	32.0
		Forestal, 1973	9,000-9,900	8	12	4,085	41,728,836	18.5
		Huayuri Sur, 1977	10,000-10,800	10	13	3,166	23,136,639	28.5
		Jibarito, 1981	9,000-9,650	6	8	9,030	19,704,702	10.5
		Jibaro, 1974	10,110-10,155	6	7	2,844	11,556,940	10.6
		San Jacinto, 1978	7,200-9,760	14	22	6,337	37,516,777	14.9
		Shiviyacu, 1973	9,100-10,100	20	26	13,807	88,867,216	20.3
		Other	...	2	12	708	3,564,382	21.3
	OCCIDENTAL/BRIDAS	Talara area, 1978	1,343-6,080	715	1,309	5,324	54,811,634	34.8
	PETROPERU	Agua Caliente, 1939	1,000-15,000	9	11	133	14,501,517	42.0
		Capirona, 1972	9,800-12,000	4	5	353	4,093,956	25.0
		Chambira, 1991	12,600-14,600	...	2	...	...	26.5
		Corrientes, 1971	10,000-12,600	31	37	15,219	121,657,263	24.9
		Maquia, 1957	2,100-2,400	21	26	681	14,293,145	37.5
		N. Esperanza, 1980	10,200-10,700	1	3	110	6,010,880	42.5
		Pacaya, 1983	...	...	3	...	52,397	...
		Pavayacu, 1972	9,600-11,300	8	14	2,226	20,333,554	29.5
		Talara-Lima area, 1869	2,000-9,000	1,800	5,249	17,579	786,651,025	32.3
		Valencia, 1975	10,500-11,200	...	2	...	572,138	43.0
		Yanayacu, 1974	11,400-13,600	5	9	855	5,426,600	18.8
		Total Peru		3,157	7,280	116,992	1,715,470,504	
PHILIPPINES	ALCORN PHILIPPINES	• Galoc, 1981	7,260-7,280	...	1	...	383,460	37.2
		• Matinloc, 1978	6,656-6,750	...	3	...	10,388,488	43.8
		• Nido, 1977	6,576-6,885	1	5	287	16,523,000	27.1
		• North Matinloc, 1987	6,830-7,021	...	3	...	2,063,964	43.2
		• West Linapacan, 1990	7,025-8,140	3	3	8,655	3,159,000	33.5
	ALCORN PALAWAN	• Cadlao, 1977	5,734-5,881	...	2	...	11,235,334	45.6
		• South Tara, 1987	4,335-4,436	...	1	...	221,408	40.4
		Total Philippines		4	18	8,942	43,974,654	
POLAND	PGNIG	Buk, 1974	8,540	1	2	213	959,000	40.6
		Daszewo, 1980	9,100	2	3	98	1,410,000	34.6
		Daszewo N, 1987	9,320	1	1	121	323,000	35.4
		Dominikowice-Kryg, 1860	200-2,250	510	721	99	12,070,000	30.2
		Grabownica, 1896	980-3,840	138	161	98	7,134,000	45.4
		Grobla, 1962	2,460	36	53	175	19,705,000	44.3
		Jeniniec, 1986	9,580	1	1	182	346,000	36.2
		Kamień Pomorski, 1972	7,360	4	10	278	13,323,000	34.0
		Kije, 1986	6,540	5	5	104	219,000	39.2
		Nosówka, 1988	10,870	2	3	334	282,000	29.3
		Osobnica, 1953	1,310-2,230	83	94	89	3,858,000	36.2
		Plawowice, 1962	1,970	14	54	85	3,926,000	35.9
		Wańkowa, 1848	490-2,030	315	348	107	10,428,000	36.9
		Weglówka, 1890	980-3,770	82	87	96	6,760,000	34.6
		Wysoka Kamieńska, 1979	9,960	4	8	292	2,133,000	30.7
		Other	...	1,104	1,597	1,260	53,728,000	...
		Total Poland		2,302	3,148	3,631	136,604,000	
QATAR(e)	QGPC	• Bul Hanine, 1970	6,000	20	35			35.0
		Dukhan, 1940	6,500	182	237			41.1
		• Idd El Shargi, 1960	4,500-8,250	17	42	425,000	4,762,334,000	35.0
		• Maydan-Mahzan, 1963	7,000	19	41			38.0
		Total Qatar		238	355	425,000	4,762,334,000	
RAS al KHAIMAH(e)	RAS AL KHAIMAH GAS COMMISSION	• Saleh, 1983	14,500	7	7	1,000	13,506,000	50.0
		Total Ras al Khaimah		7	7	1,000	13,506,000	
SAUDI ARABIA(e)	ARAMCO	Abqaiq, 1940	6,690					37.0
		Abu-Hadriya, 1940	9,340					35.0
		Abu Jifan	...					...
		• Abu-Safah, 1963	6,600					30.0
		Bakr	...					...

WORLDWIDE PRODUCTION

Country	Company	Name of field, discovery date	Depth, ft	No. of wells — Producing	No. of wells — Total	Production — 1992 average b/d	Production — Cumulative to Dec. 31, 1992 bbl	°API gravity
	TRINTOPEC	Central Los Bajos, 1973	...	168	278	2,083	12,767,000	...
		Coora/Quarry, 1936	14,000	144	744	1,476	94,284,000	...
		Fyzabad/Apex/Quarry, 1920-38	11,000	282	1,049	2,091	175,634,000	...
		Galeota, 1963	6,304	37	105	1,832	18,988,000	...
		Palo Seco/Erin/McKenzie, 1926	12,718	484	1,657	6,769	131,167,000	...
		Other	...	252	1,291	3,102	74,785,000	...
	PCOL	All fields	...	113	524	800	22,093,000	...
	TRINTOMAR	Pelican	...	2	9	2,560	3,180,000	...
		Total Trinidad and Tobago		**3,262**	**14,060**	**135,415**	**2,655,137,000**	
TUNISIA	ELF-ETAP	• Ashtart, 1971	10,500	22	27	22,285	215,222,000	29.8
	ELF-SEREPT	Douleb/Semmama, 1966	3,937	12	13	490	18,831,000	40.0
		Tamesmida, 1967	3,937	1	2	140	2,311,000	36.9
	ETAP-AGIP-FINA (SODEPS)	Debech, 1980	7,500	1	1	6	272,254	42.8
		Larich, 1979	7,550	1	2	116	433,959	41.0
		Makrouga, 1980	6,230	4	6	5,767	13,754,761	45.6
	AGIP	Couech es Saida, 1971	12,600	3	3	120	3,036,000	43.2
		Ech Chouech, 1971	11,800	1	1	124	90,000	39.0
	AGIP-Tunisia (SITEP)	El Borma, 1964	8,250-8,900	122	145	45,709	593,051,000	41.3
	AGIP-SAMEDAN-EAT	Zinnia, 1989	9,850	2	2	207	72,000	39.0
	BRITISH GAS	El Ain/Gremda, 1980	8,700	3	5	2,520	2,340,800	42.0
		El Hajeb, 1982	7,050	1	1	156	1,581,000	29.0
		Guebiba, 1981	9,500	1	1	247	241,000	35.0
	MARATHON	Belli, 1991	4,800	1	1	12,000	4,400,000	38.0
		Ezzaouia, 1986	6,800	6	8	13,800	13,900,000	40.0
	CFPT	Mahares	...	...	...	70	118,000	...
		Sidi El Itayem, 1971	7,500-7,800	...	30	2,400	25,736,000	44.0
		Total Tunisia		**181**	**248**	**106,157**	**895,390,774**	
TURKEY	ALADDIN MIDDLE EAST LTD.	Kahta, 1958	3,500	12	26	166	4,363,802	11.0
		Yasince, 1974	6,400	2	4	3	19,602	28.0
		Zeynel, 1990	4,700	1	2	3	6,133	23.0
	MOBIL	Bulgurdağ, 1961	4,500	2	2	35	2,535,587	38.0
		Selmo, 1964	5,800	23	28	3,473	72,109,021	34.0
	TPAO	Adiyaman, 1971	5,800	9	30	847	9,193,920	27.6
		Akpinar, 1984	10,500	...	4	...	157,937	31.6
		Alcik, 1983	6,500	...	2	...	96,864	32.0
		B. Firat, 1986	8,150	10	16	565	2,335,195	34.8
		B. Haznemir, 1992	7,200	1	1	10	206	22.5
		B. Kozluca, 1985	5,000	16	27	438	2,203,265	12.0
		B. Migo, 1992	11,500	1	1	253	45,540	36.8
		B. Selmo, 1981	5,500	3	5	75	870,007	34.0
		Bati Raman, 1961	4,200	172	201	12,044	49,946,597	13.3
		Besikli, 1989	6,234	9	11	2,038	1,936,970	25.6
		Beycayir, 1976	7,700	1	3	6	227,053	26.4
		Bolukyayla, 1977	10,300	...	3	...	167,663	35.0
		Camurlu, 1976	5,600	13	13	197	1,458,374	12.2
		Celikli, 1964	10,500	10	18	238	4,285,713	35.2
		Cemberlitas, 1983	6,500	10	29	843	9,511,825	30.5
		Cendere (J.V.), 1989	8,856	10	11	3,741	2,500,205	29.4
		Cukurtas, 1985	10,000	...	2	...	219,021	36.0
		D. Besikli, 1991	6,100	4	4	473	284,245	19.4
		Devecatak, 1973	4,800	1	4	23	317,476	37.0
		G. Adiyaman, 1977	5,000	...	2	...	115,836	20.4
		G. Dincer, 1981	5,350	10	29	504	5,585,372	16.7
		G. Karakus, 1989	7,900	4	6	859	1,202,913	26.5
		G. Kayakoy, 1976	8,600	2	8	326	2,542,975	30.4
		G. Sahaban, 1978	5,450	8	15	457	3,690,050	33.2
		G. Saricak, 1973	5,250	9	19	268	4,790,087	31.5
		Garzan, 1956	4,700	30	91	1,007	34,520,896	26.2
		Germik, 1968	6,500	7	14	125	3,682,756	18.3
		Ikizce, 1991	7,500	1	1	136	60,628	26.4
		Ikiztepe, 1976	4,900	...	18	...	88,051	11.3
		K. Karakus, 1989	8,489	10	14	12,271	10,851,059	29.0
		K. Osmancik, 1971	3,800	2	4	188	2,398,813	37.6
		Karaali (J.V.), 1989	9,941	1	2	85	123,308	24.6
		Karadut, 1992	9,000	1	1	96	17,303	32.0
		Karakus, 1988	8,850	20	30	15,401	23,063,955	30.0
		Karteltepe, 1982	6,550	3	4	214	1,297,781	32.0
		Kastel (J.V.), 1989	7,625	3	6	752	713,806	34.7
		Kurtalan, 1961	5,750	...	1	...	260,949	31.9
		Kuzey Adiyaman, 1977	9,500	...	1	...	59,375	32.0
		Magrip, 1961	5,700	12	48	475	15,686,057	18.4
		Mehmetdere, 1982	6,700	...	6	...	157,439	31.0

WORLDWIDE PRODUCTION

Country	Company	Name of field, discovery date	Depth, ft	No. of wells: Producing	No. of wells: Total	Production: 1992 average b/d	Production: Cumulative to Dec. 31, 1992 bbl	°API gravity
		O. Sungurlu, 1991	9,200	2	3	732	276,736	37.2
		Oyuktas, 1972	7,600	...	5	...	592,005	31.0
		Raman, 1940	4,500	80	226	3,058	55,839,327	18.7
		S. Sinan, 1987	3,900	1	3	40	163,703	13.0
		Saricak, 1973	5,250	4	7	203	2,951,972	31.5
		Sezgin, 1970	5,600	...	1	...	87,559	17.0
		Silivanka, 1962	8,200	11	19	723	9,002,072	23.0
		Sivritepe, 1977	8,200	...	2	...	591,789	33.4
		Tokaris, 1991	7,900	2	4	284	187,024	24.1
		Yanarsu, 1990	4,725	1	1	41	29,532	14.7
		Yeniköy, 1973	6,400	19	30	2,444	12,149,919	31.4
	ROYAL DUTCH/SHELL[e]	Barbes, 1972	...	18	22			37.2
		Baysu, 1985	...	6	7			34.5
		Bektas, 1985	...	1	1			34.7
		Beykan, 1964	...	38	48			33.2
		Karaali, 1988	...	...	...			24.6
		Kastel, 1990	...	...	...			34.7
		Katin, 1971	...	3	8			29.7
		Kay West (deep), 1970	...	...	...			34.1
		Kayako W (OF), 1964	...	19	29			34.1
		Kayakoy, 1960	...	14	27			38.2
		Kopru, 1990	...	...	...	15,840	244,997,000	35.2
		Kurkan, 1963	...	29	35			31.4
		Kurkan S, 1965	...	5	10			34.5
		Malatepe, 1970	...	14	15			32.7
		Northern Trend, 1972	...	...	...			30.0
		Piyanko, 1968	...	1	3			35.3
		Sahaban, 1965	...	7	13			34.5
		Sahaban SE, 1983	...	2	2			34.5
		Sincan, 1979	...	8	9			31.4
		Yatir E, 1973	...	9	11			30.9
		Yenikoy E, 1974	...	2	2			31.3
		Yesildere, 1985	...	3	3			34.5
		Total Turkey		**732**	**1,313**	**82,000**	**602,570,268**	
UNITED KINGDOM	MARATHON	• Central Brae, 1976	14,100-15,000	4	7	17,600	18,500,000	33.0
		• North Brae, 1975	11,900-12,475	11	18	49,000	108,000,000	45.0
		• South Brae, 1977	12,700-13,437	13	29	12,800	239,000,000	35.0
	MOBIL	Beryl, 1972	9,900-11,500	47	62	107,557	536,142,530	37.0
		Linnhe, 1988	11,500-12,065	...	...	437	160,000	37.8
		Ness, 1986	9,900-10,112	3	5	7,985	23,164,989	37.0
		Camelot Complex[C]	...	...	...	136	177,000	...
		Tristan[C]	...	...	...	5	2,000	...
	SHELL	• Auk, 1971	...	9	11	7,230	91,646,000	38.0
		• Brent, 1971	...	53	130	219,400	1,449,600,000	40.0
		• Cormorant, 1972	...	17	37	50,500	391,000,000	36.0
		• Dunlin, 1973	...	26	40	28,400	313,000,000	36.0
		• Eider, 1976	...	...	...	32,000	52,000,000	34.0
		• Fulmar, 1975	...	13	28	84,200	482,000,000	40.0
		• Kittiwake, 1981	...	...	...	26,800	21,400,000	30.0
		• Osprey, 1974	...	...	...	27,600	16,500,000	31.0
		• Tern, 1975	...	...	...	73,500	64,000,000	34.0
	PHILLIPS	• Maureen, 1973	8,000-8,700	7	7	20,997	185,846,145	35.9
		• Moira, 1988	8,800	1	1	1,924	2,668,000	42.0
	TEXACO NORTH SEA	• Highlander, 1976	8,759-9,372	5	7	12,000	58,525,207	34.0
		• Petronella, 1975	8,110	1	1	13,500	27,334,789	40.2
		• Tartan, 1974	10,217-12,614	9	21	11,000	90,288,589	38.0
	TOTAL OIL MARINE	• Alwyn North, 1975	10,170-11,155	10	30	77,984	157,442,000	39.0
	UNOCAL	• Heather, 1973	10,800	26	41	9,645	95,321,859	34.3
	ELF ENTERPRISE	• Claymore, 1974	8,100	29	46	45,678	374,249,311	30.1
		• Piper, 1973	8,100	...	...	...	834,416,000	37.0
		• Scapa, 1975	8,500	7	12	28,128	50,206,874	32.0
	CHEVRON	• Ninian, 1974	10,000	57	106	68,011	982,761,761	35.6
	CONOCO	• Hutton, 1973	10,000	13	15	22,205	142,550,000	33.0
		• Murchison*, 1975	9,900	16	34	27,685	244,115,000	38.0
	AMOCO	• Arbroath, 1969	8,100	12	15	35,478	33,702,100	38.0
		• Montrose, 1969	8,150	1	22	2,503	81,752,100	37.5
		• Hutton North West, 1975	11,500	26	37	9,533	114,470,000	35.0
	STATOIL	• Statfjord, 1975	9,500	10	16	103,961	367,671,600	38.2
	BP	• Beatrice (A & B), 1976	6,750	20	41	14,800	131,914,000	38.4
		• Buchan, 1974	8,700	9	14	13,200	88,626,000	33.0
		• Clyde, 1978	12,400	14	20	26,400	83,266,000	38.1
		• Cyrus, 1979	8,265	...	2	...	4,500,000	35.5
		• Deveron, 1972	9,000	3	3	1,400	14,010,000	38.2
		• Don, 1976	11,200	3	4	6,200	6,996,000	42.0

	Name of field, discovery date	Depth, ft	— No. of wells — Producing	Total	Production 1992 average b/d	Cumulative to Dec. 31, 1992 bbl	°API gravity
	• Donan, 1987	6,480	2	2	6,500	2,386,000	39.0
	• Forties and Echo, 1970	7,000	73	103	160,050	2,248,500,000	37.0
	Kimmeridge, 1959	1,705	1	1	100	3,000,000	40.0
	• Leven, 1981	12,400	1	1	3,200	1,183,000	38.0
	• Magnus, 1974	9,500	10	8	155,400	486,046,000	39.0
	• Miller, 1982	13,400	5	7	59,600	21,820,000	39.0
	• Thistle, 1973	9,000	35	35	21,500	379,875,000	38.0
	Warcham, 1984	...	1	1	1,921	1,376,000	34.0
	Wytch Farm, 1974	3,031-5,198	54	81	77,860	85,870,000	37.0
IPC EXPRO LTD.	Beckering, 1990	5,324	1	1	...	3,429	34.8
	Nettleham, 1983	3,870	1	2	293	1,156,070	35.1
	Scampton N, 1985	4,850	6	7	314	587,071	35.6
	Stainton, 1984	4,364	1	3	33	101,070	28.1
	Welton, 1981	4,450	28	37	3,093	6,518,586	34.2
HAMILTON BROS. OIL & GAS LTD.	• Argyll, 1971	9,000-9,400	7	7	3,840	72,959,603	38.0
	• Duncan, 1980	9,600-10,000	2	2	740	18,961,896	36.0
	• Innes, 1983	12,300-12,600	...	...	...	5,753,622	45.0
SUN	• Balmoral, 1975	...	1	1	28,050	72,220,000	39.3
	• Glamis, 1982	...	1	1	3,480	11,284,000	42.0
AMERADA HESS	• Angus, 1987	...	2	2	24,500	8,967,000	36.0
	• Hamish, 1988	...	1	1	2,280	2,767,000	39.0
	• Ivanhoe, 1975	...	4	7	25,030	29,904,000	30.0
	• Rob Roy, 1984	...	5	8	36,550	46,326,000	39.0-41.0
LASMO	• Staffa, 1985	...	2	2	5,981	2,189,000	39.6
EDINBURGH OIL & GAS	• Crosby Warren, 1986	...	1	1	126	414,000	40.8
MIDLAND & SCOTTISH ENERGY	• Emerald, 1981	...	1	1	6,948	2,543,000	23.0
TULLOW	West Finsby, 1988	...	2	4	199	113,550	38.0
VELT	Horndean, 1982	...	1	1	350	579,760	37.0
	Humbly Grove, 1980	...	1	1	959	3,121,660	39.0
	Singleton, 1989	...	1	1	445	302,710	39.0
EAST MIDLAND	Farley's Wood, 1983	...	2	2	49	185,000	39.7
	Whisby, 1985	...	1	1	71	78,000	32.7
PENTEX	Beckingham West, 1985	...	1	1	11	59,200	35.6
	Kirklington, 1985	...	1	1	5	21,900	29.6
	Long Clawson, 1986	...	1	1	199	204,400	35.7
	Rempstone, 1985	...	1	1	52	29,200	34.2
	Stockbridge, 1984	...	1	1	740	956,000	36.8
CAIRN	Palmers Wood, 1983	...	1	1	880	1,150,000	36.4
	Total United Kingdom		**735**	**1,200**	**1,928,731**	**11,498,438,581**	

•U.K. share (77.8%) of total production.

UNITED STATES*

Name of field, discovery date	Depth, ft	Producing	Total	1992 average b/d	Cumulative to Dec. 31, 1992 bbl	°API gravity
Alabama, 1944	...	876	...	51,981	359,199,900	...
Alaska, 1905	...	1,573	...	1,713,995	10,405,980,840	...
Arizona, 1958	...	22	...	257	19,932,960	...
Arkansas, 1921	...	8,393	...	28,033	1,664,946,935	...
California, 1861	...	43,014	...	950,934	23,400,958,120	...
Colorado, 1887	...	6,334	...	81,385	1,645,621,145	...
Florida, 1943	...	92	...	14,822	613,826,115	...
Illinois, 1889	...	32,369	...	52,743	3,368,325,965	...
Indiana, 1889	...	7,484	...	8,240	521,029,075	...
Kansas, 1889	...	47,027	...	146,484	5,663,969,050	...
Kentucky, 1860	...	22,838	...	14,967	718,373,855	...
Louisiana, 1902	...	27,104	...	1,148,915	25,488,897,180	...
Michigan, 1900	...	4,440	...	42,566	1,121,772,175	...
Mississippi, 1889	...	1,885	...	68,803	2,273,269,895	...
Missouri, 1889	...	616	...	388	4,451,015	...
Montana, 1916	...	4,068	...	50,495	1,348,869,965	...
Nebraska, 1939	...	1,941	...	14,956	465,195,970	...
Nevada, 1954	...	55	...	10,169	37,718,925	...
New Mexico, 1911	...	17,737	...	191,183	5,655,636,165	...
New York, 1865	...	3,741	...	1,101	80,976,050	...
North Dakota, 1951	...	3,521	...	89,877	1,085,502,180	...
Ohio, 1860	...	29,968	...	25,128	1,000,362,850	...
Oklahoma, 1891	...	97,752	...	278,164	13,482,177,115	...
Pennsylvania, 1859	...	21,228	...	5,839	1,330,313,100	...
South Dakota, 1954	...	153	...	4,254	24,756,940	...
Tennessee, 1860	...	793	...	1,369	15,533,085	...
Texas, 1889	...	187,610	...	1,841,661	62,409,949,470	...
Utah, 1907	...	2,058	...	62,077	965,792,140	...
Virginia, 1943	...	24	...	25	589,950	...
West Virginia, 1860	...	15,920	...	5,650	562,173,970	...
Wyoming, 1894	...	11,561	...	264,508	5,858,592,875	...
Total United States		**602,197**	**...**	**7,170,969**	**171,594,694,975**	

*Discovery date indicates first year of production in state. There are 33,719 oil fields in the U.S.

WORLDWIDE PRODUCTION

		Name of field, discovery date	Depth, ft	No. of wells Producing	No. of wells Total	Production 1992 average b/d	Production Cumulative to Dec. 31, 1992 bbl	°API gravity
VENEZUELA	PDVSA/ AMACURO	Pedernales	...	...	...			...
		Tucupita, 1945	...	...	32			27.4
	PDVSA/ ANZOATEGUI	Adas, 1957	5,000	27	59			13.9
		Aguasay	...	...	16			...
		Anaco	...	...	2			...
		Araibel	...	2	26			...
		Araibel-R	...	4	20			...
		Arecuna	1,300	58	198			11.2
		Bare	1,100	191	279			10.2
		Boca, 1971	9,500	7	59			32.2
		Boca-R	...	1	21			...
		Bote	...	...	1			...
		Bucaral	...	2	8			...
		Cachama-R	...	...	11			...
		Cachicamo	...	...	9			...
		Caico Este	...	3	34			...
		Caico Seco, 1946	6,500-7,300	1	16			31.8
		Cantaura	...	...	5			...
		Capacho	...	...	2			...
		Caracoles	...	7	61			...
		Caricari	...	...	10			...
		Carina	...	4	10			...
		Cascaroncito	...	...	1			...
		Caz-501x	...	...	12			...
		Cerro Pelado	...	...	...			...
		Chimire, 1948	7,000-7,200	4	79			31.1
		Chimire-Chv.	...	2	33			...
		Chimire-Norte	...	5	27			...
		Chimire-R	...	10	74			...
		Coporo	...	...	...			...
		Dacion, 1957	6,700	82	143			23.0
		El Robie, 1939	3,500-11,500	18	70			49.3
		El Toco, 1948	2,300	30	45			23.7
		Elotes, 1954	5,000-6,470	29	117			...
		Elotes Norte	...	3	18			...
		Esquina-R	...	1	9			...
		Esquina Sur	...	7	30			...
		Finca	...	7	38			...
		Freites	...	2	22			...
		Ganso	...	19	52			...
		GG-401	...	13	41			...
		GM-2	...	6	19			...
		GM-4	...	21	86			...
		Guamache	...	...	1			...
		Guara Central, 1946	5,000-10,000	5	19			27.0
		Guara Este	...	49	132			...
		Guara Oeste	...	34	117			...
		Guario, 1940	3,400	14	63			43.6
		Guere	...	...	59			...
		Guico, 1944	4,500-7,000	15	69			25.4
		Guico Oeste	...	5	10			...
		Guico Sur	...	1	11			...
		Inca-Isia, 1948	7,250	13	82			28.2
		Ipire	...	...	...			...
		Ira	...	12	23			...
		Juanita	...	...	...			...
		Junta	...	3	21			...
		Kilo	...	7	25			...
		La Ceiba, 1946	9,450	14	41			41.9
		La Ceibita, 1963	9,878	19	57			42.0
		La Freitera	...	...	1			...
		La Vieja	...	...	15			...
		Largo	...	1	1			...
		Lejos	...	51	59			...
		Leona, 1938	2,200-12,800	15	80			26.7
		Lestes	...	...	3			...
		Levas	...	10	29			...
		Libro	...	1	10			...
		Lido	...	26	56			...
		Limon	...	29	146			...
		Lobo	...	...	16			...
		Lustro	...	...	1			...
		Mapiri	...	10	84			...
		Mapiri C.	...	2	24			...
		Mapiri E.	...	1	14			...
		Mapvey	...	...	5			...
		Mata, 1954	8,970-10,516	41	410			35.9

WORLDWIDE PRODUCTION

	Name of field, discovery date	Depth, ft	No. of wells Producing	No. of wells Total	Production 1992 average b/d	Production Cumulative to Dec. 31, 1992 bbl	°API gravity
	Mata-R	...	30	69			...
	Maulpa	...	...	1			...
	Melones	...	132	238			...
	Melones C.	...	92	126			...
	Merey, 1937	5,400-5,700	29	46			12.5
	Miga	...	72	199			...
	Nardo	...	22	155			...
	Nieblas	...	5	35			...
	Nigua	...	10	48			...
	Nipa, 1945	6,000-8,500	18	134			...
	Nipa 100	...	9	19			...
	Nipa C.	...	4	30			...
	Nipa Oeste	...	4	18			...
	Oca	...	4	7			...
	Oficina Central, 1937	5,900	34	127			26.1
	Oficina Norte	...	6	36			...
	Oleos	...	37	61			...
	Oscurote	...	1	8			...
	Oscurote Norte, 1952	9,513	17	83			24.2
	Ostra	...	41	102			...
	Oueja	...	75	175			...
	Pato Este, 1988	...	1	1			41.9
	Pelayo	...	1	1			...
	Pradera-R	...	1	13			...
	Quiamare	...	5	28			...
	Rincon Largo	...	2	4			...
	Rosa R.V.	...	1	3			...
	San Joaquin, 1939	6,550	61	143			51.3
	Santa Ana, 1936	8,500	41	116			45.6
	Santa Fe	...	1	5			...
	Santa Rosa, 1941	8,500	160	253			45.7
	Sanvi	...	...	...			...
	Socororo	...	4	36			...
	Soto Este, 1950	9,500	5	72			...
	Soto Norte	...	6	31			...
	Tascabana	...	5	22			...
	Torta	...	1	1			...
	Trico	...	26	128			...
	Yopales Central, 1937	4,600	38	137			22.6
	Yopales Norte	...	6	35			...
	Yopales Sur	...	42	73			...
	Zacarias	...	2	22			...
	Zanjas, 1958	13,270	6	20			38.1
	Zapatos, 1955	11,500	12	71	2,314,000	44,725,391,475	32.4
	Zarza	...	2	10			...
	Zorro, 1953	11,100	7	34			29.3
	Zulus, 1957	12,710	9	28			41.2
	Zumo, 1954	9,200	11	39			25.3
	Zumo Norte	...	2	11			...
PDVSA/BARINAS	Caipe	...	3	5			24.7
	Estero	...	...	...			...
	Guafita, 1984	8,500	28	32			29.1
	Hato Viejo, 1965	9,550	16	18			22.7
	La Victoria, 1987	...	15	15			34.5
	Maporal, 1957	10,950	16	16			25.7
	Nuirias	...	...	...			...
	Paez (Mingo), 1963	...	43	67			19.9
	Palmita	...	1	2			27.8
	Palmita Sur	...	...	...			...
	Sinco, 1953	8,500-9,100	37	86			21.5
	Silvan, 1949	10,860	12	23			26.8
	Silvestre, 1948	8,860	22	56			22.1
PDVSA/FALCON	All fields, 1920	...	...	...			...
PDVSA/GUARICO	Anibal	...	...	1			...
	Barso, 1959	...	3	6			24.0
	Belen	...	...	4			...
	Budare	...	22	50			...
	Cocomon	...	...	1			...
	Copa	...	1	2			...
	Dakoa	...	1	10			...
	GC	...	6	6			...
	Gala	...	...	...			...
	Galan	...	...	...			...
	Gamo	...	...	...			...
	Grico	...	...	3			...
	Guavinita	...	...	13			...

WORLDWIDE PRODUCTION

	Name of field, discovery date	Depth, ft	No. of wells: Producing	No. of wells: Total	Production: 1992 average b/d	Production: Cumulative to Dec. 31, 1992 bbl	°API gravity
	Ipire	...	1	9			...
	Jobal, 1960	...	8	14			33.6
	Las Mercedes, 1942	4,500	...	279			32.0
	Macoya	...	3	8			...
	Palacio	...	...	86			...
	Piragua	...	...	3			...
	Punzon	...	...	6			...
	Retumbo	...	...	1			...
	Ruiz, 1949	4,500	...	67			32.0
	Saban, 1947	...	14	95			43.9
	Taman	...	...	15			...
	Tucupido	...	1	44			...
	Valle 13	...	...	3			...
	Valle 3	...	...	5			...
PDVSA/MONAGAS	Acema 100	...	3	80			...
	Acema 200	...	23	57			...
	Acema 300	...	2	5			...
	Acema 400	...	...	2			...
	Acema 500	...	1	3			...
	Acema A, 1960	...	21	56			26.4
	Acema B	...	...	4			...
	Acema Casma	4,023	36	60			24.3
	Adas	...	5	7			...
	Aguasay 3	...	5	16			...
	Aguasay C, 1955	8,100-13,400	8	50			41.3
	Aguasay E	...	...	10			...
	Aguasay N, 1955	4,849	3	11			42.4
	Amarilis	...	...	...			...
	Boqueron	...	3	3			...
	Cachipo	...	...	...			...
	Carisito, 1984	4,881	7	8			41.7
	Carito, 1988	...	10	10			31.4
	Cerro Negro	1,200	102	102			9.0
	El Corozo	...	...	...			...
	El Furrial, 1986	...	28	28			26.3
	El Salto	...	...	...			...
	Florida	...	1	1			...
	Isleños	...	...	...			...
	Jobo I, 1956	3,600-4,000	177	177			10.3
	Jobo II, 1956	3,600-4,000	39	39			10.3
	Jusepin, 1938	4,253	5	5			26.5
	Manresa, 1954	...	...	...			16.7
	Mata Grande	...	...	72			...
	Morichal, 1958	3,312	40	209			9.2
	Mulata	16,500	19	21			32.0
	Onado, 1971	4,791	5	26			28.0
	Oritupano Norte	...	22	54			...
	Oritupano Sur	...	15	123			...
	Oritupano, 1950	7,657	37	69			18.4
	Orocual, 1953	2,959	12	61			25.1
	Pilon, 1937	...	59	141			11.6
	Pirital, 1958	450-1,100	...	93			...
	Quiriquire, 1928	7,000-7,200	3	28			27.7
	Santa Barbara, 1941	5,020-6,500	4	165			31.4
	S.B. Meyer	...	...	...			...
	Tacat, 1953	1,820-3,670	...	...			30.0
	Temblador, 1936	3,500-4,500	...	60			18.7
	Travieso	...	...	9			...
	Uracoa	...	...	...			...
	Uracoa-Bombal	...	...	...			...
PDVSA/ ZULIA	Alpuf	...	2	2			...
	Alturitas	...	6	16			32.2
	Amana	...	...	...			...
	Ambrosio	...	...	...			...
	Aricuaisa	...	...	...			...
	Bachaquero, 1930	10,720	1,276	1,276			22.9
	Barua, 1928	...	15	15			20.8
	Boscan, 1946	9,708	213	501			10.2
	Cabimas, 1917	2,200	188	499			22.4
	Centro, 1959	9,600-11,000	297	365			33.6
	Ceuta	16,380	119	207			30.2
	Colon Casigua	...	62	62			...
	Garcia	...	...	...			...

WORLDWIDE PRODUCTION

		Name of field, discovery date	Depth, ft	No. of wells Producing	No. of wells Total	Production 1992 average b/d	Production Cumulative to Dec. 31, 1992 bbl	°API gravity
		La Concepcion, 1953	3,148-8,000	59	196			...
		La Ensenada	...	1	1			...
		Lago, 1959	11,450	28	28			...
		Lagunillas, 1926	3,000	3,215	3,215			22.0
		Lama, 1957	8,320	219	219			32.2
		Lamar, 1958	13,000	121	224			33.5
		Las Palmas	...	...	...			...
		La Paz, 1925	4,268-8,000	54	180			30.1
		Las Moroches	...	...	...			...
		Macoa	...	...	...			...
		Machiques	...	1	1			...
		Mamon	...	...	...			...
		Mara, 1945	5,240	49	118			26.6
		Mene Grande, 1914	4,130	211	595			19.9
		Motatan, 1952	...	21	31			22.0
		Netick	...	...	...			...
		Punta Benitez	...	347	347			...
		Punta Icotea La Rosa	...	78	78			...
		San José	...	2	2			...
		San Julian	...	...	...			...
		Sibucara, 1948	3,000	1	3			30.8
		Sur Lago	17,830	10	10			33.6
		Tia Juana, 1928	3,000	2,339	2,339			17.9
		Totumos	...	1	1			...
		Urdaneta, 1970	10,000	213	311			22.0
		Total Venezuela		**12,140**	**20,431**	**2,314,000**	**44,725,391,475**	
VIET NAM	PETROVIETNAM	White Tiger	...	...	...	105,000	105,298,000	...
		Total Viet nam		**...**	**...**	**105,000**	**105,298,000**	**...**
YEMEN	HUNT OIL	Al-Shura	7,600	3	4	1,800	2,032,000	40.0
		Alif, 1984	5,750	111	130	118,500	273,475,000	40.0
		Al-Wihdah, 1989	7,500	1	1	200	174,000	40.0
		Asa'ad Al-Kamil, 1987	6,100	34	46	33,200	25,889,000	40.0
		Azal, 1987	6,850	20	23	19,500	35,416,000	40.0
		Jabal Nuqum, 1988	3,780	1	1	700	1,483,000	44.0
		Jabal Samadan, 1990	3,950	2	2	300	135,000	39.0
		Raydan, 1986	2,875	3	7	400	448,000	26.0
		Saif Ben The-Yazen, 1988	7,600	1	2	1,000	1,626,000	40.0
		Shaharah	...	1	1	200	56,000	...
		Wadi Bana	...	2	2	200	79,000	...
		Total Yemen		**179**	**219**	**176,000**	**340,813,000**	
ZAIRE	CHEVRON	• GCO, 1970	7,000-9,000	6	7	203	6,612,358	38.7
		• Libwa, 1981	5,500	3	4	1,362	1,637,443	33.0
		• Lukami, 1983	12,400	4	6	711	8,516,759	33.2
		• Mibale, 1973	5,000-6,000	12	14	11,058	99,883,838	30.0
		• Moko, 1972	12,310	1	1	192	311,276	32.0
		• Motoba, 1976	11,098	8	8	5,754	12,815,848	31.5
		• Mwambe, 1979	6,150	1	2	83	869,981	29.7
	ZAIREP (FINA)	East Mibale, 1978	6,000	...	1	7	418,502	34.0
		Kifuku, 1983	3,400	...	1	...	59,241	31.0
		Kinkasi, 1972	3,700	14	42	524	9,771,499	31.0
		Liawenda, 1972	3,900	6	33	235	5,118,321	31.0
		Makelekese, 1983	3,600	...	8	...	1,269,404	31.0
		Muanda, 1972	3,600	...	6	...	1,044,604	31.0
		Nsiamfumu, 1977	8,300	...	...	...	5,347	34.0
		Tshiende, 1978	6,000	5	14	3,634	12,100,144	34.0
		Total Zaire		**60**	**147**	**23,763**	**160,434,565**	

U.S. fields with ultimate oil recovery exceeding 100 million bbl

State Field Disc. year	Est. 1993 prod. (1,000 bbl)	Cum prod. 1-1-94 (1,000 bbl)	Est. rem. reserves (1,000 bbl)	Est. No. prod. wells
ALABAMA				
Citronelle, 1955	1,445	155,740	5,572	177
ALASKA				
•Endicott, 1978	39,100	242,100	237,000	56
•Granite Point, 1965	2,500	120,400	46,000	29
Kuparuk River, 1969	114,800	1,070,200	1,308,000	379
Lisburne, 1967	8,400	99,200	104,000	48
•McArthur River, 1965	6,613	561,311	58,387	80
•Middle Ground Shoal, 1962	2,134	170,698	37,000	47
Point McIntyre, 1988*	7,200	7,200	333,800	15
Prudhoe Bay, 1968	†385,000	§8,308,300	3,740,000	869
Swanson River, 1957	1,583	218,531	12,000	28

*Started production on Oct. 14, 1993. †Includes 52.27 million bbl of condensate. §Includes 343.24 million bbl of condensate.

State Field Disc. year	Est. 1993 prod. (1,000 bbl)	Cum prod. 1-1-94 (1,000 bbl)	Est. rem. reserves (1,000 bbl)	Est. No. prod. wells
ARKANSAS				
Magnolia, 1938	485	164,208	4,210	55
Smackover, 1922	2,040	568,108	7,004	1,704
CALIFORNIA				
San Joaquin Valley				
Belridge South, 1911	46,297	944,871	261,699	4,140
Buena Vista, 1909	1,502	655,045	19,131	503
Coalinga, 1890	10,316	810,058	95,613	1,818
Coalinga Nose, 1938	474	502,441	5,673	52
Coles Levee North, 1938	250	162,166	2,507	55
Cuyama South, 1949	429	220,254	4,876	112
Cymric, 1909	9,749	249,300	45,660	1,013
Edison, 1928	662	139,047	15,075	553
Elk Hills, 1911	23,624	1,034,112	467,980	945
Fruitvale, 1928	560	118,146	11,116	219
Greeley, 1936	205	113,472	938	18
Kern Front, 1912	2,311	184,361	43,935	779
Kern River, 1899	44,909	1,428,542	518,579	7,018
Kettleman North Dome, 1928	158	457,949	452	40
Lost Hills, 1910	9,975	216,819	99,011	1,484
McKittrick, 1896	1,149	274,837	83,354	531
Midway-Sunset, 1894	60,827	2,178,243	574,115	9,618
Mount Poso, 1926	2,584	286,578	55,952	457
Rio Bravo, 1937	83	116,434	293	11
Yowlumne, 1974	2,322	101,715	7,596	54
Coastal Area				
•Carpinteria, 1966	1,034	94,813	22,914	49
Cat Canyon E. & W., 1908	1,199	295,169	25,756	311
•Dos Cuadras, 1969	3,436	230,269	36,675	139
•Elwood, 1928	70	106,550	1,438	3
•Hondo, 1969	7,473	130,835	69,766	23
Orcutt, 1901	744	169,845	9,194	120
•Pescado, 1969	200	200	*--	20
•Point Arguello, 1981	26,086	47,201	252,799	24
•Point Pedernales, 1982	4,657	38,504	†33,979	13
•Rincon, 1927	885	153,546	11,383	148
San Ardo, 1947	4,602	428,402	102,852	346
Santa Maria Valley, 1934	662	202,943	15,801	111
South Mountain, 1916	690	149,490	8,529	319
Ventura, 1919	5,379	925,765	75,494	409
Los Angeles basin				
•Beta, 1976	3,792	63,510	67,447	71
Beverly Hills, 1900	1,080	130,577	34,395	60
Brea-Olinda, 1880	1,483	392,911	45,770	602
Coyote East, 1909	487	111,360	10,515	86
Coyote West, 1909	382	252,554	4,896	48
Dominguez, 1923	476	271,984	4,730	90
•Huntington Beach, 1920	3,746	1,086,593	51,819	431
Inglewood, 1924	1,923	356,628	42,974	225
•Long Beach, 1921	1,629	919,716	7,257	290
Montebello, 1917	415	194,671	7,953	104
Richfield, 1919	768	194,432	22,145	174
Santa Fe Springs, 1919	655	617,033	16,406	116
•Seal Beach, 1924	636	206,656	10,580	142
•Torrance, 1922	926	217,779	30,151	220
•Wilmington, 1932	21,959	2,415,307	372,555	1,403

*Started production on Dec. 4, 1993. Hondo, Pescado, and Sacate fields are in Exxon Corp.'s Santa Ynez Unit. Combined oil reserves of the three fields are estimated at 500 million bbl, although Exxon has not disclosed individual field reserves estimates for Sacate and Pescado. Sacate development has been deferred indefinitely because of permit considerations. †Revised to reflect only developed portion of field. Original estimate of total field reserves is 350 million bbl.

State Field Disc. year	Est. 1993 prod. (1,000 bbl)	Cum prod. 1-1-94 (1,000 bbl)	Est. rem. reserves (1,000 bbl)	Est. No. prod. wells
COLORADO				
Rangely, 1933	10,517	799,333	104,667	465
FLORIDA				
Jay, 1970	3,695	379,482	26,850	50
ILLINOIS				
Clay City, 1938	2,011	404,574	7,418	674
Dale Consolidated, 1940	135	105,186	1,007	97
Lawrence, 1906	2,359	407,134	7,578	249
Louden, 1938	555	392,611	2,450	156
Main, 1906	1,610	242,179	8,387	535
New Harmony, 1938	755	157,752	4,173	191
Salem, 1938	910	231,090	1,303	30
KANSAS				
Bemis-Shutts, 1928	1,102	249,968	5,591	688
Chase-Silica, 1931	830	306,555	2,333	702
El Dorado, 1917	801	302,817	3,476	685
Hall-Gurney, 1931	892	150,931	3,607	771
Kraft-Prusa, 1937	419	130,272	2,556	364
Trapp, 1937	911	231,463	5,560	599
LOUISIANA				
Offshore				
•Bay Marchand Block 2, 1949	12,595	655,249	47,000	189
•Breton Sound Block 20, 1953	2,870	103,132	18,881	84
•Eugene Island Block 126, 1950	1,312	118,816	5,882	21
•Eugene Island Block 276, 1964	3,450	80,626	24,104	45
•Eugene Island Block 330, 1971	6,350	301,723	26,275	99
•Grand Isle Block 16, 1948	1,898	276,533	12,841	52
•Grand Isle Block 43, 1956	2,900	289,193	68,751	70
•Grand Isle Block 47, 1955	1,837	87,461	28,980	41
•Green Canyon Block 65, 1984	16,915	48,292	91,708	17
•Main Pass Block 35, 1951	575	97,412	5,861	47
•Main Pass Block 41, 1957	1,979	248,949	12,289	72
•Main Pass Block 69, 1948	2,512	251,564	10,883	89
•Main Pass Block 144, 1967	3,111	104,418	21,655	56
•Main Pass Block 299, 1962	11,635	81,522	72,850	56
•Main Pass Block 306, 1969	945	207,657	5,183	64
•Mississippi Canyon Block 194 (Cognac), 1980	7,040	145,511	84,489	40
•Ship Shoal Block 113, 1955	3,815	90,308	26,125	32
•Ship Shoal Block 207, 1967	698	91,698	9,118	25
•Ship Shoal Block 208, 1962	2,645	174,577	14,218	42
•South Marsh Island Block 128, 1974	2,401	106,257	12,585	39
•South Marsh Island Block 130, 1975	4,474	155,142	21,417	75
•South Pass Block 24, 1950	2,960	268,781	14,855	225
•South Pass Block 27, 1954	2,486	141,702	12,407	213
•South Pass Block 61, 1968	2,115	177,558	14,945	53
•South Pass Block 62, 1965	2,065	117,763	20,195	65
•South Pass Block 65, 1965	1,938	113,916	14,265	40
•South Pass Block 89, 1969	7,814	122,626	54,176	57
•South Timbalier Block 21, 1939	1,864	226,573	10,519	39
•South Timbalier Block 135, 1956	939	145,468	7,707	26
•West Delta Block 30, 1949	5,011	474,303	22,046	103
•West Delta Block 73, 1962	6,050	213,431	66,560	77
•West Delta Block 79, 1966	2,217	142,986	17,949	57
Onshore South				
Avery Island, 1942	925	96,928	5,215	14
Bay de Chene, 1941	165	97,694	5,255	13
Bay St. Elaine, 1928	267	166,567	1,902	9
Bayou Sale, 1941	425	163,873	1,111	8
Black Bay West, 1953	1,465	153,835	5,618	62
Caillou Island, 1930	2,514	613,905	5,319	114
Cote Blanche Bay West, 1940	819	186,407	4,613	91
Cote Blanche Island, 1948	325	185,148	1,418	13
Delta Farms, 1944	683	118,855	3,515	34
Garden Island Bay, 1934	1,130	226,762	5,675	98
Golden Meadow, 1938	425	137,879	2,100	114
Grand Bay, 1938	960	174,452	2,262	67
Hackberry, East, 1927	359	111,366	5,324	37
Hackberry, West, 1928	714	146,852	2,557	75
Iberia, 1917	705	98,529	2,182	45
Iowa, 1931	58	99,744	356	20
Jennings, 1901	187	117,533	975	131
Lafitte, 1935	1,143	261,435	3,965	72
Lake Barre, 1929	659	208,165	5,919	24
Lake Washington, 1931	1,907	253,117	8,853	78
Leeville, 1931	571	144,616	4,251	31
Quarantine Bay, 1937	852	177,052	11,376	30
Romere Pass, 1950	430	99,222	2,014	20
Timbalier Bay, 1938 onshore	1,708	301,990	24,146	90
Venice, 1937	530	185,064	4,404	33
Vinton, 1910	289	162,579	2,070	84
Weeks Island, 1945	1,370	231,703	15,592	34
West Bay, 1940	993	233,268	11,333	44
White Castle, 1929	581	92,439	7,687	38
North				
Caddo-Pine Island, 1905	2,555	375,508	6,199	8,975
Delhi, 1944	185	213,332	3,026	21
Haynesville, 1921	1,130	173,990	6,750	90
Homer, 1919	309	100,210	1,303	146
Nebo-Hemphill, 1940	2,211	94,560	13,003	247
Rodessa, 1935	360	108,000	1,529	72

U.S. fields with ultimate oil recovery exceeding 100 million bbl (continued)

State Field Disc. year	Est. 1993 prod. (1,000 bbl)	Cum prod. 1-1-94 (1,000 bbl)	Est. rem. reserves (1,000 bbl)	Est. No. prod. wells
MISSISSIPPI				
Baxterville, 1944	1,395	248,321	9,108	134
Heidelberg, 1944	2,286	186,801	7,000	203
Tinsley, 1939	569	222,883	2,620	73
MONTANA				
Bell Creek, 1967	320	131,817	1,889	39
Cabin Creek, 1953	1,235	96,556	5,081	93
Cut Bank, 1926	672	166,220	3,437	512
Elk Basin, 1915	540	98,020	2,524	47
Pennel, 1955	2,570	86,952	17,311	167
Pine, 1951	1,055	111,185	4,214	102
NEW MEXICO				
Denton, 1949	707	142,646	3,061	79
Empire-Abo, 1957	950	225,842	4,739	255
Eunice-Monument, 1929	2,175	137,298	11,137	454
Grayburg-Jackson, 1929	1,766	103,853	6,543	880
Hobbs, 1928	4,892	326,511	20,421	320
Langlie-Mattix, 1935	1,157	131,131	5,771	645
Maljamar, 1926	1,414	152,887	9,226	518
Vacuum, 1929	7,717	476,184	27,810	690
NORTH DAKOTA				
Beaver Lodge, 1951	1,260	118,772	9,384	53
OKLAHOMA				
Bowlegs, 1927	118	163,268	1,539	89
Burbank, 1920	776	541,073	5,421	902
Cement, 1917	1,613	138,266	5,528	722
Cushing, 1912	1,425	504,143	9,038	1,249
Earlsboro, 1926	185	203,017	1,307	81
Eola-Robberson, 1921	803	137,521	3,813	437
Fitts, 1933	1,950	211,520	7,410	498
Glenn Pool, 1905	622	331,682	6,829	557
Golden Trend, 1945	2,999	491,284	26,678	1,216
Healdton, 1913	1,449	343,003	7,870	915
Hewitt, 1919	1,975	278,592	11,359	731
Little River, 1927	189	141,874	2,354	88
Oklahoma City, 1928	851	821,306	7,829	216
Postle, 1960	1,009	111,809	9,505	213
Seminole, 1926	240	194,282	1,741	128
Sho-Vel-Tum, 1919	14,601	1,247,121	59,879	6,459
Sooner Trend, 1965	3,649	313,513	30,923	2,901
St. Louis, 1925	788	178,699	3,237	390
TEXAS				
Dist. 1—Southwest				
Darst Creek, 1929	465	156,701	2,478	817
Luling-Branyon, 1922	974	152,408	3,967	1,622
Pearsall, 1936	9,212	141,468	81,706	944
Dist. 2—Central Gulf Coast				
Greta, 1933	337	149,709	1,773	65
Lake Pasture, 1953	1,406	95,840	5,588	168
Tom O'Connor, 1934	4,343	774,486	28,244	637
West Ranch, 1938	1,264	388,321	8,794	194
Dist. 3—Upper Gulf Coast				
Anahuac, 1935	339	287,404	2,051	55
Barbers Hill, 1916	241	112,422	1,865	85
Conroe, 1931	1,501	737,391	7,972	194
Giddings, 1971	27,913	362,455	99,783	2,498
Goose Creek, 1906	595	120,832	2,509	195
Hastings East, 1934	1,449	701,699	7,092	150
High Island, 1922	305	136,778	1,237	83
Hull, 1918	398	169,728	3,190	215
Humble, 1905	143	99,012	2,501	186
Magnet Withers, 1936	1,079	117,404	9,339	134
Manvel, 1931	384	102,128	1,226	65
Oyster Bayou, 1941	413	163,039	1,958	24
Raccoon Bend, 1927	559	103,150	2,213	104
Sour Lake, 1902	853	100,294	9,439	336
Thompson, 1921	1,425	483,151	9,896	139
Tomball, 1933	222	122,361	1,216	55
Webster, 1937	2,363	588,700	16,695	169
West Columbia, 1917	211	101,721	2,754	79
Dist. 4—South				
Agua Dulce-Stratton, 1928	111	147,509	1,991	56
Borregos, 1945	78	114,653	826	20
Kelsey, 1938	115	115,674	965	39
Plymouth, 1925	156	123,800	2,323	44
Seeligson, 1925	136	272,330	869	39
TCB, 1944	74	113,287	554	11
White Point, E., 1938	72	104,494	726	24
Dist. 5—East				
Van, 1928	2,270	535,107	11,263	486
Dist. 6—East				
East Texas, 1930	30,050	5,176,112	183,497	7,579
Fairway, 1960	1,612	201,729	8,739	96
Hawkins, 1940	4,701	850,381	22,306	368
Neches, 1953	676	108,146	1,982	118
Quitman, 1948	889	126,783	3,003	138
Talco, 1931	1,615	281,779	12,142	413
Dist. 7B—West Central				
Stephens County Reg., 1916	4,253	158,237	24,130	1,034
Dist. 7C—West Central				
Big Lake, 1923	397	132,217	1,309	74
McCamey, 1925	240	125,630	1,061	560
Pegasus, 1949	921	139,615	2,713	171
Dist. 8—West				
Andector, 1946	453	184,529	3,859	43
Block 31, 1945	2,112	232,686	11,733	195
Cowden, N., 1930	13,038	555,543	47,635	1,370
Cowden, S., Foster, Johnson 1930	6,501	539,232	28,066	1,457
Dollarhide, 1945	2,855	206,965	23,042	252
Dune, 1938	1,819	193,730	10,266	649
Fuhrman-Mascho, 1930	1,268	112,612	5,955	700
Fullerton, 1942	6,943	384,929	37,150	914
Goldsmith, 1934	4,511	782,270	24,813	1,239
Hendrick, 1926	449	254,705	1,441	182
Howard-Glasscock, 1925	4,808	430,087	26,512	1,907
Iatan, E., 1926	2,373	156,314	10,497	1,205
Jordan, 1937	684	132,827	2,805	206
Kermit, 1928	330	115,236	1,811	392
Keystone, 1930	1,278	320,218	5,210	627
Mabee, 1943	2,095	98,161	11,470	398
McElroy, 1926	8,825	506,341	45,310	1,458
Means, 1934	5,375	257,122	36,773	714
Midland Farms, 1944	1,954	254,536	9,746	245
Penwell, 1926	1,868	117,438	10,673	397
Sand Hills, 1931	1,903	258,495	11,312	1,054
Shafter Lake, 1947	1,105	101,055	6,990	448
Spraberry Trend, 1951	18,875	728,675	96,382	7,587
TXL, 1944	1,373	270,835	6,174	502
Waddell, 1927	425	103,222	3,376	166
Ward, S., 1938	473	107,653	2,807	592
Ward Estes, N., 1929	3,206	382,125	58,488	970
Westbrook, 1920	1,676	97,528	7,005	496
Yates, 1926	17,249	1,276,507	677,998	1,164
Dist. 8A—West				
Adair, 1947	928	112,119	5,956	131
Anton-Irish, 1944	2,229	189,761	16,754	234
Cedar Lake, 1939	2,207	86,519	16,211	174
Cogdell, 1949	820	297,369	6,053	173
Diamond M, 1948	1,247	246,776	9,115	276
Garza, 1926	2,402	118,179	14,954	531
Jo-Mill, 1953	2,250	138,065	13,199	287
Kelly-Snyder, 1948	5,635	1,281,513	68,487	539
Levelland, 1938	14,710	563,224	64,255	3,045
Prentice, 1951	3,795	185,375	27,750	382
Robertson, N., 1956	5,285	92,867	35,555	536
Salt Creek, 1950	8,005	293,655	60,000	179
Seminole, 1936	14,755	604,096	64,196	446
Sharon Ridge, 1923	1,805	94,280	16,973	2,053
Slaughter, 1936	16,173	1,115,577	96,642	2,466
Wasson, 1936	28,400	1,853,968	134,696	2,220
Welch, 1942	2,559	158,284	29,500	767
Dist. 9—North				
Archer County Reg., 1911	1,315	185,361	6,407	4,035
Cooke County Reg., 1926	851	104,673	4,810	1,451
K-M-A, 1931	861	144,180	2,909	688
Walnut Bend, 1938	444	125,521	2,475	151
Wichita County Reg., 1910	2,474	345,338	12,711	7,175
Wilbarger County Reg., 1915	569	148,418	2,429	963
Young County Reg., 1917	1,063	103,543	5,187	2,809
Dist. 10—Panhandle				
Panhandle, 1921	5,741	1,454,533	18,467	10,934
UTAH				
Altamont, 1955	2,898	103,892	24,238	219
Bluebell, 1949	3,477	128,491	25,233	217
Greater Aneth, 1956	5,475	381,889	27,469	467
WYOMING				
Byron, 1918	704	124,633	3,652	58
Elk Basin, 1915	2,159	444,358	17,699	169
Frannie, 1928	427	119,080	1,861	51
Garland, 1906	2,267	174,985	7,145	237
Grass Creek, 1914	1,790	195,341	7,839	238
Hamilton Dome, 1918	2,095	241,900	9,179	252
Hartzog Draw, 1976	3,201	87,677	19,947	154
Lance Creek, 1918	161	108,604	1,384	29
Little Buffalo Basin, 1914	1,480	128,015	5,518	158
Lost Soldier, 1916	3,615	226,584	34,733	91
Oregon Basin, 1912	6,110	422,721	29,706	530
Salt Creek, 1889	3,019	648,735	27,412	992
Steamboat Butte, 1943	419	92,587	9,193	31
Wertz, 1921	1,444	110,733	20,412	51

Sources: OGJ estimates from operator, Petroleum Information Corp., state geological survey, and oil and gas agency data
•Field partially or wholly offshore.

Drilling/completion depths and costs per foot in selected U.S. areas

Depth category, area	1989	1990	1991	1992
	Cost per foot, $			
0-1,249 ft				
California-East Central	$103.46	$ 84.87	$ 97.30	89.61
Illinois	31.58	29.83	26.52	28.14
Indiana	19.15	16.88	18.82	22.81
Kansas	13.19	20.09	17.88	15.32
Kentucky	14.19	21.09	24.48	21.18
Louisiana-North	23.58	27.68	24.97	24.54
Wyoming	43.19	65.26	122.35	68.16
1,250-2,499 ft				
Alabama	66.07	68.88	66.98	81.68
Appalachian	28.62	30.05	34.17	37.21
California-East Central	76.07	68.13	77.45	72.36
California-South	71.42	71.93	76.18	116.60
Indiana	15.99	17.54	17.85	20.03
Kentucky	18.32	26.52	22.81	28.37
Michigan	49.43	50.74	49.85	71.03
Montana	35.26	45.95	44.93	35.49
Oklahoma	34.88	30.68	35.86	29.78
Texas-Dist. 1 South	25.87	22.24	24.79	27.82
Texas-Dist. 7B West Central	17.66	16.35	19.76	19.44
Texas-Dist. 7C West Central	34.02	39.56	33.59	31.34
Texas-Dist. 9 North	14.82	16.25	16.95	18.94
2,500-3,749 ft				
Alabama	49.93	100.42	78.97	82.57
Appalachian	29.37	31.49	30.39	31.62
California-Central Coastal	91.73	105.77	119.06	121.51
Illinois	23.29	35.57	31.16	28.76
Kansas	33.22	32.97	34.72	27.00
New Mexico-West	86.66	87.47	91.37	85.63
Texas-Dist. 2 Central Gulf Coast	19.34	22.68	26.09	26.24
Texas-Dist. 5 East	45.89	69.69	48.91	35.93
Texas-Dist. 10 Panhandle	48.44	49.09	53.19	46.83
3,750-4,999 ft				
Appalachian	30.66	31.29	32.25	30.99
California-Offshore	158.58	202.39	192.50	163.15
Kansas	18.49	19.51	21.01	22.88
Nebraska	12.69	14.89	15.97	12.45
New Mexico-East	62.54	56.54	64.53	59.01
Texas-Dist. 7B West Central	23.87	26.94	32.04	25.11
Texas-Dist. 8 West	40.12	46.36	51.73	44.96
5,000-7,499 ft				
Appalachian	31.64	30.15	28.59	31.65
Arkansas	46.73	56.83	46.76	48.63
California-North	55.99	55.28	54.82	39.61
Colorado	34.61	36.89	39.79	27.31
Louisiana-State Offshore	224.42	312.57	213.00	233.34
Michigan	62.78	57.51	49.22	38.92
Mississippi	21.81	38.34	27.79	23.99
Nebraska	20.09	22.79	18.01	16.23
Oklahoma	42.05	46.11	44.16	40.59
Texas-Dist. 2 Central Gulf Coast	26.52	30.58	34.94	32.59
Texas-Dist. 3 Upper Gulf Coast	46.06	46.10	49.56	42.56
Texas-Dist. 4 Lower Gulf Coast	34.93	38.34	37.63	36.20
Texas-Dist. 6 East	45.16	47.53	49.68	42.72
Texas-Dist. 7C West Central	41.62	44.76	45.50	37.07
Texas-Dist. 8A West	35.29	46.31	47.48	42.97
Texas-Dist. 9 North	35.42	38.15	38.25	40.01
Texas-Dist. 10 Panhandle	41.02	38.62	45.85	40.04
Utah	52.64	80.69	65.08	66.91
7,500-9,999 ft				
Arkansas	45.41	49.23	54.87	53.49
California-North	59.83	49.94	56.25	46.68
Colorado	42.12	42.46	52.50	32.78
Gulf of Mexico-North Federal	421.96	282.95	484.78	445.73
Montana	46.09	65.40	59.41	55.34
New Mexico-East	59.02	62.30	67.78	64.45
North Dakota	45.21	46.36	48.33	51.41
Texas-Dist. 1 South	50.53	78.89	87.12	68.19
Texas-Dist. 4 Lower Gulf Coast	70.72	76.22	78.01	63.86
Texas-Dist. 6 East	90.49	93.27	92.39	86.07
Texas-Dist. 8 West	54.01	50.06	45.04	46.16
Texas-Dist. 8A West	44.22	51.73	49.14	48.33
Texas-Federal Offshore	NA	NA	223.35	218.26
Texas-State Offshore	227.41	257.72	196.65	197.05
Wyoming	49.81	62.84	51.75	55.56
10,000-12,499 ft				
Alaska Land	288.43	249.82	257.20	235.40
Alaska Offshore	610.40	597.81	270.85	420.70
Louisiana-Federal Offshore	NA	NA	235.14	233.91
Louisiana-South	81.81	88.91	99.15	82.22
Louisiana-State Offshore	219.72	225.63	249.94	232.94
Texas-Dist. 1 South	72.99	92.75	101.11	83.35
Texas-Dist. 3 Upper Gulf Coast	83.63	95.41	96.64	85.57
Texas-Dist. 5 East	84.27	101.58	97.47	78.35
Texas-Federal Offshore	NA	NA	298.21	228.49
Wyoming	66.08	67.31	77.20	71.04
12,500-14,999 ft				
Gulf of Mexico-North Federal	419.53	529.86	548.78	546.92
Louisiana-Federal Offshore	NA	NA	288.78	222.75
Louisiana-South	122.89	128.84	155.24	135.69
Oklahoma	116.16	128.31	126.74	86.99
Utah	154.14	166.60	121.70	131.03
15,000-17,499 ft				
Florida	89.56	47.48	43.68	46.22
Louisiana-South	222.38	215.72	239.96	176.13
Mississippi	94.90	110.41	99.01	88.51
New Mexico-East	130.96	101.36	103.68	94.44
Texas-Dist. 3 Upper Gulf Coast	167.25	196.02	196.42	114.06
17,500-19,999 ft				
Louisiana-Federal Offshore	NA	NA	300.31	277.76
Louisiana-South	371.39	287.15	282.47	356.82
Mississippi	214.63	265.39	143.15	134.57
Oklahoma	254.82	209.55	174.95	143.44
Wyoming	364.73	NA	300.28	217.01
20,000 ft and deeper				
Alabama	NA	855.14	720.99	1,091.13
Gulf of Mexico-North Federal	804.61	NA	955.49	914.25
Louisiana-South	530.88	503.58	366.91	351.89
Oklahoma	829.14	NA	544.87	313.12
Texas-Dist. 8 West	374.75	NA	149.35	NA

Data for most depth categories in all states during 1954-1992 available from American Petroleum Institute. Figures represent the cost per foot as determined by dividing the total expenditure for oil wells, gas wells, and dry holes, if applicable, in each area by the total footage drilled in that depth category in that area. Average depths and the relative distribution of oil wells, gas wells, and dry holes vary by state and within each depth category. The cost per foot represents the average dollar amount spent from the time locations were made until the wells were completed as productive or abandoned. Appalachian includes New York, Ohio, Pennsylvania, and West Virginia. Alabama figures may include offshore wells. NA Not available or not broken out.

Source: Joint Association Survey on Drilling Costs, sponsored by API, Independent Petroleum Association of America, and Mid-Continent Oil & Gas Association

U.S. drilling and production depth records

Shaky energy markets trim pipeline construction plans

Pipeline construction, 1994 and beyond*

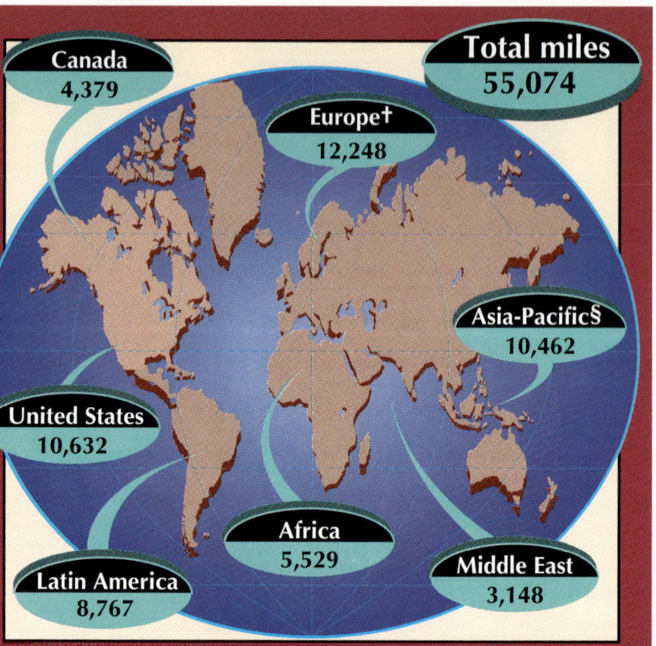

*Projects planned to start in 1994 and be completed in that year or later. Includes some probable major projects whose installation will begin in1995 or later. †Including western Russia and former Soviet republics west of the Ural Mountains. §Including eastern Russia and former Soviet republics east of the Ural Mountains

Plans for pipeline construction worldwide in the mid-1990s reflect the uncertainties in energy markets.

Total mileage planned for the long term has shrunk compared with what companies said earlier in the decade they planned to build.

The slide in crude oil prices in 1993 has even cast a shadow over the bright prospects for natural gas.

Nevertheless, as 1994 began, significant expansions were under way or planned for Latin America, Asia and the Pacific regions, and Europe.

Companies will lay more than 55,000 miles of crude oil, product, and natural gas pipeline in 1994 and beyond. The figure includes projections for pipeline construction in Russia and former republics of the Soviet Union.

Western Russia and all countries west of the Ural Mountains are included under totals for Europe, eastern Russia and countries east of the Urals under totals for the Asia-Pacific region.

Companies will spend nearly $47 billion in the mid to late 1990s.

These figures are based on the latest survey of worldwide pipeline operators, industry sources, and published information conducted early in 1994 by Oil & Gas Journal.

South American construction plans

Industry construction plans in South America include large diameter crude oil pipelines in Colombia and Ecuador and gas pipelines in Bolvia, Brazil, and Chile.

Earlier in the 1990s, BP Exploration Co. (Colombia) Ltd. said it planned to expand oil pipelines in Colombia to handle increasing production from its Cusiana and Cupiagua fields.

Through the middle of 1994, about 10,000 b/d of Cusiana oil was to have moved through a 12 in. Empresa Colombiana de Petroleos (Ecopetrol) pipeline from Araguanay to the Caribbean Sea coastal oil terminal at Covenas.

By 1995, BP wants to lay a 20 in. oil pipeline from Cusiana field to the 75,000 b/d Central Llanos pipeline at El Porvenir and a 30 in. loop to upgrade the Central Llanos section of the 150,000 b/d Oleoducto de Colombia pipeline shipping oil to Covenas.

State owned Petroleos del Ecuador (Petroecuador) early in 1994 disclosed plans to begin expanding throughput of the 312 mile trans-Ecuadorian pipeline to 450,000 b/d from 325,000 b/d.

In Chile, feasibility studies were almost complete for a proposed expansion of gas transportation and distribution pipelines. The proposal included linking gas supplies in Argentina's Neuquen basin to an expanded Chilean gas pipeline system.

Plans called for construction to get under way in 1995. Parts of the expanded pipeline system could begin moving gas in early 1997. Work on the network's transmission lines would be completed first, while construction of the distribution lines could span 10-15 years.

Preliminary plans called for what would be the largest pipeline project in Chile's history: 750 miles of 24-28 in. mainline and laterals in Chile and about 5,000 miles of distribution lines as large as 28 in. to Santiago and other cities.

Estimated costs included $600 million for the transmission portion of the system and as much as $1.5 billion for the entire network.

Argentina wanted to be included in plans to expand and integrate the international gas pipeline grid to serve emerging markets in Brazil.

At the time, plans called for the Bolivia-Brazil gas pipeline's first segment to be laid beginning in second half 1995, with service to begin by the end of 1997.

With volumes in the first year reaching as much as 280 MMcfd and doubling by the end of the third year of operations, gas on the first pipeline would move from Rio Grande, Bolivia, to Porto Alegre, Brazil.

First phase construction costs were estimated at $1.5 billion.

As envisioned, the entire Southern Cone gas pipeline grid would stretch 1,200 miles, with construction costs estimated at $5-6 billion. Earlier plans called for a 1,900 mile system extending north from Campo Duran, Argentina, to Santa Cruz, Bolivia, then east to Sao Paulo, Brazil, with laterals to large markets in areas traversed by the pipeline.

European gas pipelines

By 1995, efforts will peak on expansions of pipeline capacity to deliver Alergian gas to customers in Europe.

Expansion of the trans-Mediterranean pipeline will more than double capacity to 2.5 bcfd. Adding compression after 1995 could boost capacity to 2.9 bcfd.

Enough of the trans-Mediterranean expansion had been completed by yearend 1993 to allow incremental gas throughput of 671 MMcfd.

Farther west, contractors were marshalling equipment to start construction on the Maghreb-Europe gas pipeline system that will by 1996 move gas to markets in Spain and, later, in Portugal, France, and Germany.

Plans call for the 856 mile Maghreb-Europe pipeline to be laid in four sections, including 330 miles of 48 in. line in Algeria, 335 miles of 48 in. in Morocco, 30 miles of twin 22 in. subsea lines across the Strait of Gibralter, and 170 miles of 36 in. and 48 in. in Spain.

First stage construction will create capacity of about 700 MMcfd at a cost of $1.5 billion. More compression could expand the system's capacity to 1.8 bcfd.

In the region of the former Soviet Union, members of the Caspian Pipeline Consortium (Kazakhstan, Russia, Azerbaijan, and Oman) had not set a starting date for construction of the proposed oil pipeline from Kazakhstan's Tengiz field to Novorossysk on the Black Sea.

The group settled on a 1,300 mile route around the northern shore of the Caspian Sea.

Latest plans called for a 500 mile, 42 in. pipeline to be laid from Novorossysk to Komsomolsk, Russia, on the Caspian Sea. From there, an existing 40 in. pipeline would be upgraded to cover the last 800 miles to a point near Tengiz. The project is estimated to cost more than $850 million.

Capacity at first would be about 300,000 b/d, expandable to 1.5 million b/d in phases as more capacity is needed. Eventually, Russian and Azeri crude oil also could reach world markets across the system.

Plans also were advancing for construction of a trans-Urals pipeline system to move NGLs from western Siberia to the Urals-Volga region.

The system, initially envisioned as three 16 in. lines covering more than 1,000 miles, will move a mostly C_3-C_5 stream from South Balyk in western Siberia to petrochemical plants near Minnibaevo, Tatarstan, via Tchaikovsky.

Much pipeline construction activity in Europe revolved around plans to create infrastructure to serve growing gas demand in the western part of the region, where gas consumption could increase to 420-480 billion cu m/year (bcmy) from about 315 bcmy at present.

Wingas, a joint venture of Germany's Wintershall AG and Russia's Gazprom, late in 1993 finished integrating the 200 mile Stegal pipeline in eastern Germany and the 397 mile Midal pipeline in the western part of the country.

Russian and Polish officials agreed to construct a $10 billion gas pipeline to transport gas from the Yamal Peninsula by way of Poland to western Europe.

Construction of the 670 km Polish segment of the pipeline is to start this year. It will run from Kondratki on Poland's eastern border to Odra on the western border.

Early in 1994, crews began excavating a 1.6 mile tunnel beneath protected wetlands in a German coastal national park near Dornumersiel, Germany.

The 385 mile, 40 in. Europipe will pass through the tunnel en route to Emden.

By October 1995, Norway's Den norske stats oljeselskap expects to begin transporting about 12 bcmy of gas from the North Sea through Europipe to markets in Germany.

Meantime, Norwegian officials late in 1993 approved Statoil's new plan for transporting West Troll field oil by pipeline to Mongstad terminal. The 53 mile, 16 in. Troll pipeline is to be laid in 1995, with first oil to flow from West Troll field Jan. 1, 1996. Troll line's capacity is to be 157,000 b/d.

Pipeline construction in 1994 and beyond*

	4-10 in.	12-20 in.	22-30 in.	32+ in.	Total
	Miles				
GAS PIPELINES					
U.S.	794	2,000	2,266	2,709	**7,769**
Canada	577	724	282	1,842	**3,425**
Latin America	47	1,226	3,555	0	**4,828**
Asia-Pacific†	330	655	1,406	3,891	**6,282**
Europe§	2,151	1,998	989	3,674	**8,812**
Middle East	0	52	25	974	**1,051**
Africa	0	394	144	2,400	**2,938**
Total	**3,899**	**7,049**	**8,667**	**15,490**	**35,105**
CRUDE OIL PIPELINES					
U.S.	27	534	518	270	**1,349**
Canada	0	934	0	0	**934**
Latin America	0	128	39	105	**272**
Asia-Pacific†	18	542	117	237	**914**
Europe§	31	218	1,475	789	**2,513**
Middle East	211	519	54	60	**844**
Africa	908	515	0	0	**1,423**
Total	**1,195**	**3,390**	**2,203**	**1,461**	**8,249**
PRODUCT PIPELINES					
U.S.	700	814	0	0	**1,514**
Canada	0	20	0	0	**20**
Latin America	1,140	2,436	0	0	**3,576**
Asia-Pacific†	1,016	1,410	840	0	**3,266**
Europe§	502	323	5	93	**923**
Middle East	6	1,147	0	100	**1,253**
Africa	535	633	0	0	**1,168**
Total	**3,899**	**6,783**	**845**	**193**	**11,720**
WORLD TOTALS					
Gas	**3,899**	**7,049**	**8,667**	**15,490**	**35,105**
Crude oil	**1,195**	**3,390**	**2,203**	**1,461**	**8,249**
Product	**3,899**	**6,783**	**845**	**193**	**11,720**
Total	**8,993**	**17,222**	**11,715**	**17,144**	**55,074**

*Projects planned to begin in 1994 and be completed this year or later. Includes some probable major projects whose installation will begin in 1995 or later. †Including eastern Russia and former Soviet republics east of the Ural Mountains. §Including western Russia and former Soviet republics west of the Ural Mountains.

Asia and Pacific regions

Early in 1994, Gulf of Thailand gas flow began increasing to fuel an Electricity Generating Authority of Thailand power project near Khanom, in southern Thailand on the Malay Peninsula. The gas was flowing through a recently completed Petroleum Authority of Thailand pipeline from Unocal Thailand Ltd.'s offshore Erawan field processing center.

Flow of gas from the gulf to markets in eastern Thailand could double early in 1996 when a second large diameter subsea gas line will start moving gas from Erawan north to Rayong, Thailand.

Unocal is to produce some of the gas bound for Khanom through the 100 mile, 24 in. line from Erawan, with the balance coming from Total Exploration & Production Thailand's Bongkot field, also in the gulf. Bongkot gas is flowing through a 107 mile, 32 in. pipeline that began moving gas to Erawan late in 1993.

With Thailand's existing 270 mile, 34 in. subsea trunkline from Erawan to Rayong in eastern Thailand, the new Bongkot-Erawan-Khanom system increased gas pipeline capacity out of the gulf to more than 1.1 bcfd.

That capacity wil nearly double by March 1996, when the Petroleum Authority completes a 264 mile, 36 in. gas pipeline paralleling the Erawan-Rayong line. The $700 million expansion is also to include a gas receiving terminal ashore, 68 miles of 28 in. pipeline from Rayong to Bang Pakong near Bangkok, and onshore and offshore compression.

Pipelaying began late in 1993 on the subsea gas trunkline

Natural gas pipeline transmission expenses*

	Expenses, $ 1990-majors†	Expenses, $ 1991-majors§	Cost/mile, $ 1990-majors†	Cost/mile, $ 1991-majors§
Operation expenses				
Supervision and engineering	197,306,089	460,111,770	775.19	2,275.89
System control and load dispatching	35,277,653	37,007,971	138.60	183.06
Communications-system expenses	34,509,554	39,141,120	135.58	193.61
Compressor-station labor and expenses	341,491,401	330,501,629	1,341.67	1,634.79
Gas for compressor fuel	345,640,346	306,957,716	1,357.97	1,518.33
Other fuel and power for compressor stations	64,069,550	60,700,554	251.72	300.25
Mains	197,727,858	190,148,842	776.84	940.55
Measuring and regulating-station expenses	71,104,907	71,904,598	279.36	355.67
Transmission and compression of gas by others	861,285,682	813,756,736	3,383.87	4,025.15
Other transmission expenses	61,517,640	70,706,490	241.69	349.74
Rents	22,900,198	20,516,109	89.97	101.48
Total operation expenses	**2,232,830,878**	**2,401,453,535**	**8,772**	**11,878**
Maintenance expenses				
Supervision and engineering	34,053,355	35,139,856	133.79	173.82
Structures and improvements	32,216,414	29,799,377	126.57	147.40
Mains	112,179,368	98,740,962	440.74	488.41
Compressor-station equipment	270,315,337	279,135,266	1,062.03	1,380.71
Measuring and regulating-station equipment	16,698,512	17,106,276	65.61	84.61
Communications equipment	12,153,982	11,389,816	47.75	56.34
Other equipment	9,088,418	7,268,786	35.71	35.95
Total maintenance expenses	**486,705,386**	**478,580,339**	**1,912.20**	**2,367.24**
Total transmission expenses, $	**2,719,536,264**	**2,880,033,874**	**10,684.67**	**14,245.75**
Total transmission pipeline, miles	**254,527**	**202,168**		

Source: Statistics of Interstate Natural Gas Pipeline Companies—1990 & 1991 (final report in series), U.S. Department of Energy, Washington, D.C.

*U.S. interstate pipelines for calendar years 1990 & 1991. †44 of 140 companies filing annual reports with the U.S. Federal Energy Regulatory Commission for 1990; majors are companies whose combined gas sold for resale and gas transported or stored for a fee exceeded 50 bcf (14.7 psi; 60° F.) in each of the 3 previous calendar years. §44 of 125 companies filing annual reports with the U.S. FERC for 1991.

to link ARCO China's $1.2 billion Yacheng 13-1 field development project in the South China Sea to markets in Hong Kong.

Pipelay contractor SEJV, a joint venture of Saipem and European Marine Contractors, expected to complete the 440 mile, 30 in. line by early summer 1994.

Earlier in 1994, the $150 million (Australian) Southwest Queensland pipeline began moving the first shipment of Queensland gas to markets in South Australia.

The 118 mile line traverses Cooper Creek floodplain en route to gas treatment and distribution at Moomba, in the northeast corner of South Australia.

By fourth quarter 1994, Northwest Shelf gas project partners were to begin moving gas and gas liquids from Goodwyn A and North Rankin A offshore platforms through a two phase, 97½ mile pipeline to a processing and LNG facility at Dampier on the northwest coast of Western Australia.

Gas processed at Dampier is to move through a proposed $400 million (Australian), 95 MMcfd pipeline to gold and nickle mining operations near Kalgoorlie and Kambalda in south central Western Australia. Total length of the pipeline is estimated at 850 miles.

Construction of the Dampier-Kalgoorlie pipeline could begin in 1995, pending completion of studies by a joint venture formed in late 1993 by companies expected to take as much as 90% of initial deliveries.

The system would have a 16 in. mainline as far as Newman on the eastern slope of the Hammerly Range mountains and 14 in. to Kalgoorlie.

Southeast U.S. projects

Sponsors of large interstate pipelines competing for new gas customers in the U.S. Southeast were advancing their projects as 1994 began.

Meanwhile, Gulf of Mexico producers in 1993 began expanding offshore gas gathering systems to provide transportation for growing supplies off Louisiana, Mississippi, and Alabama.

Florida Gas Transmission Co. was on schedule to have its Phase III expansion in service by yearend 1994.

The project is to increase FGT system capacity by 530 MMcfd to about 1.4 bcfd.

Early in 1994, the company began building three compressor stations and increasing compression at eight other stations as part of its $900 million Phase III expansion.

FGT also began laying 815 miles of 30 and 36 in. line in four southeastern U.S. states.

Competition began heating up in 1993 among Gulf of Mexico gathering pipelines to tie in supplies of gas to serve new markets in the U.S. Southeast.

In 1993, Dauphin Island Gathering Partners, a 50-50 venture of Offshore Energy Development Corp., The Woodlands, Tex., and the Enron Gas Gathering Inc. unit of Enron Gas Services, Houston, completed construction of a 20 in., 12 mile extension of Dauphin Island Gathering System from north of Dauphin Island to shore.

The project created a 70 mile, 400 MMcfd pipeline corridor from Mobile Block 821 in federal water off Alabama.

Dauphin Island Gathering Partners in January 1994 began laying a 40 mile, 20 in. extension of the gas gathering system across federal waters south of Mobile Bay.

That span in third quarter 1994 is to begin gathering low sulfur gas from wells in the Mobile federal planning area and in the northern tier of Viosca Knoll area tracts.

Growing gas production in and around Mobile Bay prompted Tenneco Gas Gathering Co. late in 1993 to unveil plans to lay a 200 MMcfd gathering pipeline in the area.

The 20 in., 30 mile line at first will connect Mobile Block 916 wells to the interstate gas pipeline systems of Koch Gateway and Transcontinental Gas Pipe Line Corp. in southwest Alabama.

The line could be in service by October or November 1994.

Meantime, Tenneco and Leviathan Gas Pipeline Partners L.P., Houston, early in 1994 agreed to lay 95 miles of 20 in. subsea gas pipeline off Louisiana from a Shell Offshore Inc. platform in Main Pass Block 252 to a connection with a Tennessee Gas Pipeline Co. transmission line in South Pass Block 55.

Plans included laying a 6 mile, 16 in. lateral from Viosca Knoll Block 817 to an interconnect with a Southern Natural Gas Co. line in Main Pass Block 289.

Efforts to integrate large North American gas transmission pipelines slowed somewhat in 1993, but a handful of big projects went on line.

Northwest Pipeline Corp., Salt Lake City, in mid-1993 began moving gas through its $432 million, 443 MMcfd expansion. The project included laying 362 miles of pipeline loops, adding 113,572 hp of compression, and

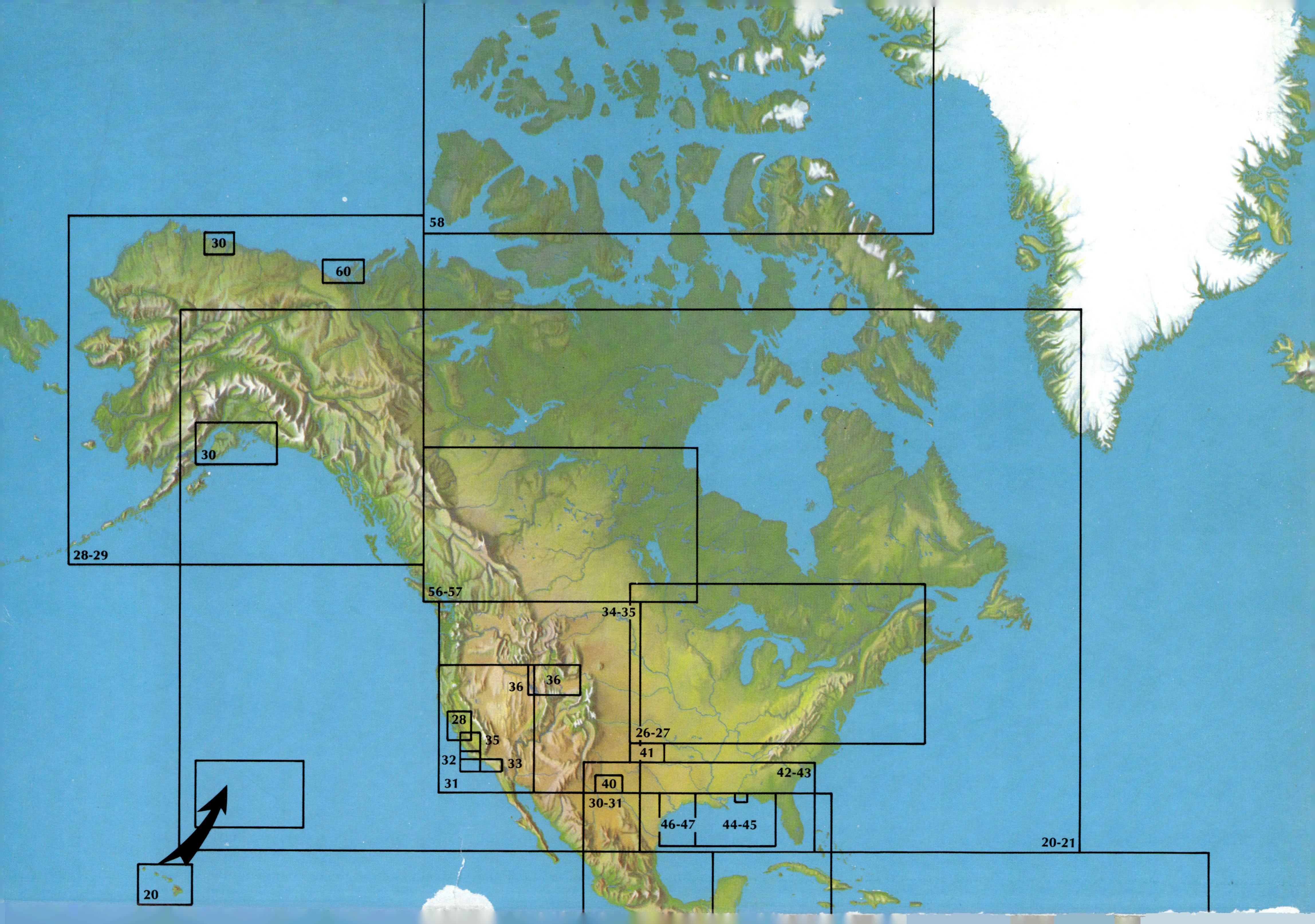
58
30
60
28-29
30
56-57
34-35
36
36
28
35
32
33
31
26-27
41
42-43
40
30-31
46-47
44-45
20-21
20

U.S. pipeline costs: estimated vs. actual, 1992-1993*

Size, in.	Location	Length, miles	Materials	Labor	Misc.†	R.O.W. & damages	Total	$/mile
			Costs, $					
	Land pipelines							
4	Ohio (R)	0.2						
	Estimated		35,631	54,410	29,713	3,338	123,092	615,459
	Actual		40,844	79,813	21,174	4,988	146,819	734,094
4	Ohio (R)	0.4						
	Estimated		71,529	109,230	59,648	6,702	247,109	617,772
	Actual		81,996	160,227	42,506	10,013	294,742	736,854
8	North Dakota	49.30						
	Estimated		1,754,286	1,685,200	691,940	—	4,131,426	83,801
	Actual		1,145,082	1,393,479	451,218	—	2,989,779	60,644
12	New Hampshire (R)	2.50						
	Estimated		247,000	425,000	305,500	—	977,500	391,000
	Actual		214,906	346,766	329,074	—	890,746	356,298
16	Ohio (R)	2						
	Estimated		222,800	351,900	54,253	9,800	638,753	319,376
	Actual		208,219	642,300	148,600	14,602	1,013,721	506,860
16	Delaware	4.3						
	Estimated		594,700	3,135,300	1,263,900	957,100	5,951,000	1,383,953
	Actual		874,731	3,811,453	1,298,000	1,471,640	7,455,824	1,733,912
20	Ohio (R)	0.9						
	Estimated		160,740	245,460	134,040	15,060	555,300	616,999
	Actual		184,260	360,060	95,520	22,500	662,340	735,933
20	Ohio (R)	1.40						
	Estimated		193,000	258,200	42,253	8,100	501,553	358,252
	Actual		164,560	292,597	51,193	13,950	522,300	373,071
20	West Virginia (R)	4.50						
	Estimated		706,100	1,033,300	583,754	93,000	2,416,154	536,923
	Actual		788,600	1,515,500	542,700	44,700	2,891,500	642,555
20	New Jersey (L)	4.56						
	Estimated		897,325	2,031,700	881,713	251,150	4,061,888	890,764
	Actual		1,059,939	414,7294	1,978,632	1,364,701	8,550,566	1,875,124
24	Pennsylvania	3.70						
	Estimated		745,500	1,546,000	554,200	232,800	3,078,500	832,027
	Actual		800,026	1,919,098	423,339	99,115	3,241,578	876,102
24	Pennsylvania	7.20						
	Estimated		1,470,300	2,720,200	1,023,800	456,200	5,670,500	787,569
	Actual		1,452,639	3,426,939	385,006	255,416	5,520,000	766,666
24	Pennsylvania	7.30						
	Estimated		1,856,300	2,925,900	1,194,300	464,000	6,440,500	882,260
	Actual		1,807,277	4,003,036	337,890	248,797	6,397,000	876,301
24	Pennsylvania	8.40						
	Estimated		1,677,700	3,018,200	1,123,500	534,100	6,353,500	756,369
	Actual		1,408,307	3,864,442	764,595	51,577	6,088,921	724,871
24	Pennsylvania	25.19						
	Estimated		5,711,000	17,790,000	2,998,990	1,462,000	27,961,990	1,110,043
	Actual		5,329,022	17,754,687	894,952	361,291	24,339,952	966,254
26	Pennsylvania	11.20						
	Estimated		3,073,000	4,643,000	2,278,500	177,000	10,171,500	908,169
	Actual		2,869,015	7,230,769	1,152,820	119,264	11,371,868	1,015,345
30	Pennsylvania (L)	3.43						
	Estimated		1,473,094	2,111,658	1,045,449	84,911	4,715,112	1,374,668
	Actual		1,865,072	2,997,097	1,351,916	364,339	6,578,424	1,917,907
30	Texas							
	Estimated	8.3	2,125,000	1,637,000	695,000	187,000	4,644,000	559,518
	Actual	8.38	2,196,750	1,982,228	381,360	176,357	4,736,695	565,238
36	Pennsylvania (L)	1.61						
	Estimated		915,000	898,482	517,492	75,650	2,406,624	1,494,797
	Actual		1,112,962	1,360,076	666,612	240,099	3,379,749	2,099,222
36	Pennsylvania	3.1						
	Estimated		1,732,000	2,166,000	1,080,400	210,100	5,188,500	1,673,709
	Actual		1,486,074	2,610,461	422,681	25,107	4,544,323	1,465,910
36	Pennsylvania (L)	3.39						
	Estimated		1,607,334	1,810,489	1,017,921	262,368	4,698,112	1,385,873
	Actual		1,651,541	2,449,156	1,054,527	239,393	5,394,617	1,591,332
36	Pennsylvania (R)	4.25						
	Estimated		2,098,000	4,285,000	931,292	500,000	7,814,292	1,838,656
	Actual		2,160,077	4,992,747	173,314	342,316	7,668,454	1,804,342
36	Pennsylvania (R)	5.25						
	Estimated		3,674,000	8,225,000	1,548,780	644,000	14,091,780	2,684,148
	Actual		3,174,000	8,164,000	305,856	226,000	11,869,856	2,260,924
36	Pennsylvania	7.7						
	Estimated		3,597,000	4,568,000	2,219,500	91,000	10,475,500	1,360,454
	Actual		2,859,809	6,049,251	891,646	64,897	9,865,603	1,281,247
36	Pennsylvania (L)	8.5						
	Estimated		3,723,000	4,862,000	2,387,240	124,000	11,096,240	1,305,440
	Actual		3,218,575	5,428,657	925,195	134,913	9,707,340	1,142,040
36	Pennsylvania	8.9						
	Estimated		4,199,000	5,411,000	2,644,240	380,000	12,634,240	1,419,577
	Actual		2,961,313	7,321,664	929,488	83,890	11,296,355	1,269,253
36	Pennsylvania (L)	10.72						
	Estimated		5,966,669	5,933,026	3,434,241	503,000	15,836,936	1,477,326
	Actual		5,739,331	6,317,947	3,267,515	1,087,466	16,412,259	1,530,994
42	New Jersey	3.86						
	Estimated		2,786,770	5,159,357	2,371,789	645,300	10,963,216	2,840,211
	Actual		3,355,505	9,412,408	4,136,568	1,978,220	18,882,701	4,891,891
Total, miles		**191.76**						
	Total estimated		**53,090,978**	**88,688,112**	**33,059,095**	**8,367,879**	**183,206,064**	**955,392**
	Total actual		**50,002,213**	**109,391,852**	**23,275,297**	**9,030,949**	**191,700,311**	**999,688**

Source: U.S. Federal Energy Regulatory Commission completed-project costs filings, July 1, 1992, to June 30, 1993. No offshore projects were reported for the period. *Actual cost data must be filed within 6 months following final hydrostatic test of pipeline. Not all projects proposed (estimated costs) are built (actual costs). L = loop; R = replacement. †Generally includes engineering, supvervision, interest, administration and overheads, contingencies, allowance for funds used during construction, and FERC fees.

requalifying 89 miles of existing pipeline at a higher line pressure.

Northwest's interstate pipeline system stretches from the U.S.-Canadian border near Sumas, Wash., to Blanco hub in northwest New Mexico. The expansion increased system capacity to 2.4 bcfd.

U.S. compressor-station costs: estimated vs. actual, 1992-1993*

Location		Size, hp	Materials	Labor	Costs, $ Misc. †	Land	Total	$/hp
North Dakota		1,100						
	Estimated	1,705,870	976,490	556,206	15,000	3,253,566	2,957	
	Actual	1,716,691	768,599	455,884	15,464	2,956,638	2,687	
Pennsylvania †		1,350						
	Estimated	1,045,000	420,000	203,500	—	1,668,500	1,235	
	Actual	1,172,248	894,234	152,518	—	2,219,000	1,643	
Pennsylvania †		4,000						
	Estimated	3,812,000	1,800,000	886,500	—	6,498,500	1,624	
	Actual	4,147,035	3,708,808	679,367	—	8,535,210	2,133	
Pennsylvania †		4,000						
	Estimated	3,850,000	1,734,000	912,240	—	6,496,240	1,624	
	Actual	4,816,225	3,741,666	1,232,109	—	9,790,000	2,447	
Pennsylvania		6,500						
	Estimated	6,633,000	3,046,000	1,945,220	20,000	11,644,220	1,791	
	Actual	6,451,000	5,384,000	1,442,144	—	13,277,144	2,042	
Pennsylvania		6,500						
	Estimated	6,396,000	2,971,000	1,912,549	20,000	11,299,549	1,738	
	Actual	6,438,000	3,670,000	2,991,736	—	13,099,736	2,015	
Virginia		7,660						
	Estimated	4,562,000	1,806,000	1,046,988	—	7,414,988	968	
	Actual	5,497,887	2,987,533	3,994,849	—	12,480,269	1,629	
Louisiana		8,100						
	Estimated	9,127,534	2,360,000	1,822,398	30,000	13,339,932	1,646	
	Actual	6,417,584	3,384,201	2,282,332	10,000	12,094,117	1,493	
Texas		9,000						
	Estimated	7,576,000	2,651,000	1,861,000	—	12,088,000	1,343	
	Actual	6,617,000	2,400,000	941,000	—	9,958,000	1,106	
Pennsylvania		9,600						
	Estimated	9,600,000	3,893,000	1,905,500	—	15,398,500	1,604	
	Actual	10,443,544	5,397,592	1,928,864	110,000	17,880,000	1,862	
Pennsylvania		11,000						
	Estimated	5,843,000	3,777,000	2,806,459	30,000	12,456,459	1,132	
	Actual	6,021,000	2,392,000	2,902,117	—	11,315,117	1,028	
New Jersey		12,000						
	Estimated	8,977,500	4,242,800	4,701,196	375,000	18,296,496	1,524	
	Actual	9,655,209	9,111,739	5,313,182	591,850	24,671,980	2,055	
Pennsylvania		12,600						
	Estimated	5,376,000	1,423,500	2,051,636	—	8,851,136	702	
	Actual	5,806,577	2,388,091	2,556,953	—	10,751,621	853	
Texas		16,560						
	Estimated	16,015,000	4,210,000	2,455,000	57,000	22,737,000	1,373	
	Actual	15,245,917	5,698,266	4,740,394	224	25,684,801	1,551	
Total hp		**109,970**						
Total estimated			**90,518,904**	**35,310,790**	**25,066,392**	**547,000**	**151,443,086**	**1,377**
Total actual			**90,445,917**	**51,926,729**	**31,613,449**	**727,538**	**174,713,633**	**1,588**

Source: U.S. Federal Energy Regulatory Commission completed-project costs filings, July 1, 1992, to June 30, 1993. No offshore projects were reported for the period.
*Actual cost data must be filed within 6 months following commissioning of installed compression equipment. Not all projects proposed (estimated costs) are built (actual costs). †Generally includes engineering, supvervision, interest, administration and overheads, contingencies, allowance for funds used during construction, and FERC fees.

Northwest planned to expand capacity again near yearend 1995 by adding gas turbine compressor stations at Sumas and Tumwater, Wash., and Lava Hot Springs and Owyhee, Ida.

Pacific Gas Transmission Co. (PGT), San Francisco, on Nov. 1, 1993, began flowing gas through facilities in a $1.6 billion, 900 MMcfd expansion of its Alberta-California mainline.

About 800 miles of mainline were included in the project.

By yearend 1995, PGT expects to have added another 300 MMcfd of capacity to its system. Included in the $200 million expansion were two 12 in. laterals, a 75 mile, 40 MMcfd line to Boardman, Ore., and an 18 mile, 30 MMcfd line to Medford, Ore.

Part of PGT's new capacity in 1995 is to transport gas for customers in northeast California, northern Nevada, and Reno served by Tuscarora Gas Transmission Co., a unit of Sierra Pacific Resources Co., Reno.

The gas is to flow through a 200 mile, 120 MMcfd pipeline Tuscarora and TransCanada PipeLines Ltd., Calgary, plan to lay between Reno and PGT.

The $130 million Tuscarora-TransCanada line also is to be complete in 1995.

Construction costs

World pipeline construction costs have risen marginally in recent years despite relatively strong levels of construction activity.

Costs of natural gas pipelines in the U.S. reflect those in the rest of the world, with allowances for remote locations and rough terrains.

These costs, both estimated and actual, are collected by the Federal Energy Regulatory Commission throughout the year.

For proposed U.S. gas pipeline projects during 1992-93, the period most recently surveyed by Oil & Gas Journal, the average onshore cost was more than $860,000/mile.

The only offshore pipeline project proposed during 1992-93 showed an estimated cost of almost $550,000/mile.

The four major categories of pipeline construction costs are material, labor, right-of-way (R.O.W.), and miscellaneous.

Material costs include those for line pipe, pipe coating, and cathodic protection. R.O.W. cost figures include surveying, obtaining rights-of-way, and allowances for damages.

Miscellaneous costs generally cover engineering, supervision, contingencies, allowances for funds used during construction, administration and overhead, and FERC filing fees.

For 1992-93, cost per mile data for the four categories are material $275,399/mile, labor $338,190/mile, R.O.W. and damages $46,844/mile, and miscellaneous $200,082/mile.

Average costs per mile for projects show few clear trends related to either length or geographic area. In general, however, costs per mile within a given diameter indicate that the longer the pipeline, the lower the incremental cost for construction.

More deepwater pipelining in the U.S. Gulf of Mexico is likely before the end of the 1990s. Here, a semisubmersible derrick/pipelay barge lays 20 in. pipeline for Southern Natural Gas Co., Birmingham, Ala., en route to Exxon Corp.'s Alabaster platform in Mississippi Canyon Block 397. Pipelay reached a water depth of 1,229 ft.

Northwest Pipeline Corp., Salt Lake City, major 1993 expansion of its system was one of two major projects to serve more U.S. western and Pacific Northwest markets with natural gas. In the other, Pacific Gas Transmission completed a major expansion of its system to bring gas from Canada to California markets.

And, broadly, lines built nearer populated areas tend to have higher per mile costs.

Road, highway, river or channel crossings, and marshy or rocky terrain each strongly affects pipeline construction costs, as well. Material and labor make up more than 71% of the cost of laying onshore pipelines.

Operating costs

Once a natural gas pipeline is laid, compressor stations built, and meters installed, operating costs become the next consideration.

As an aid in estimating this element, transmission expenses for U.S. interstate gas pipelines for 1990 and 1991, the most recent years for which data are available from the U.S. Department of Energy, are shown in an accompanying table.

The highest component of transmission-operating expenses is transmission and compression of gas by others. For majors in 1991, this component was $4,025/mile. For majors in 1990, this component was $3,384/mile.

The highest cost component of maintenance expense covers compressor station equipment.

For majors in 1991, this was $1,381/mile and for majors in 1990 $1,062/mile.

During 1990, major U.S. natural gas pipeline companies spent slightly more than $2.2 billion for operation costs and $486 million on maintenance. Total spending by majors for operation and maintenance costs reached more than $2.7 billion in 1990.

Total operation expenses for majors in 1991 increased to more than $2.4 billion, the second year of increase following a drop of almost $31 million from 1988 to 1989.

Total maintenance expenses in 1991 fell to more than $478 million from a 1990 cost of $486.7 million. These expenses had shown a rise of more than 17% in 2 years beginning in 1989.

Total transmission expenses showed a rise in 1990-91. In the latter year, they reached nearly $2.9 billion. IPE

National oil companies, ministries listed

Drilling crew makes a connection for Petroleos de Venezuela unit Maraven SA on Lake Maracaibo.

A

Abu Dhabi

Abu Dhabi National Oil Co. (ADNOC)
Box 898
Abu Dhabi, UNITED ARAB EMIRATES
Ph: 971 2 666-000
Fax: 971 2 602-3389
Telex: 949 22215, 23688
Cable: ADNOC

Ministry of Petroleum & Mineral Resources
Box 59
Abu Dhabi, UNITED ARAB EMIRATES
Ph: 971 2 62810
Telex: 2544 MPMR AH

Afghanistan

Afghan National Petroleum Co. (ANPC)
Shari Nau
Kabul, AFGHANISTAN

Albania

Albpetrol (formerly DPNG)
c/o Geological Institute for Oil & Gas
Fieri, ALBANIA
Ph: 355 423-4542
Fax: 355 423-4204
Telex: 4204 MIN EN

Ministry of Mines & Energy
Tirana, ALBANIA
Telex: 4296

Algeria

Sonatrach Inc.
Immeuble Mauretania Pl. Perou
80 Ave. Ahmed Ghermoul
Algiers, ALGERIA
Ph: 213 2 60-80-11
Fax: 213 2 60-37-91
Telex: 62115 DZ
Cable: SONATRACH ALGIERS

Ministry of Energy & Petrochemical Industries
80 Rue Ahmed Ghermoul
Algiers, ALGERIA
Ph: 213 2 66-33-00
Telex: 52790, 52916

Angola

Sociedad Nacional de Combustiveis de Angola (SONANGOL)
Box 1316
Luanda, ANGOLA
Ph: 244 2 334143
Telex: 3148

Argentina

Gas del Estado SE
Alsina 1169
C. Postal 1311
Buenos Aires, ARGENTINA
Ph: 54 1 383-2091
Fax: 54 1 381-4462
Telex: 33 19021 GASES AR

Yacimientos Petroliferos Fiscales SA (YPF)
C.P. 1364
Buenos Aires, ARGENTINA
Ph: 54 1 326-7271, 326-2545
Fax: 54 1 326-0641
Telex: 21999 21792

Yacimientos Petroliferos Fiscales SA (YPF)
5 Greenway Plaza Suite 210
Houston, Texas 77046
Ph: 713 621-4850
Telex: 0762441

Australia

Australian Geological Survey Organisation (AGSO)
P.O. Box 378
Canberra City, A.C.T. 2601
AUSTRALIA
Ph: 61 62 72-5120
Fax: 61 62 48-8178

Austria

Federal Ministry of Economic Affairs
Department VII/3
Landstrasser Hauptstrasse 55-57
A-1030 Vienna, AUSTRIA
Ph: 43 222 711-02-270, 711-02-271
Fax: 43 222 711-02-352
Telex: 131300

OMV Aktiengesellschaft
Otto Wagner-Platz 5
A-1090 Vienna, AUSTRIA
Ph: 43 222 404-4090
Fax: 43 222 404-4091
Telex: 114801
Cable: ERDOEL-WIEN

B

Bahrain

Bahrain National Oil Co.
P.O. Box 25504
Awali, BAHRAIN
Ph: 973 754666
Fax: 973 753203
Telex: 8670 BANOCO BN

Bangladesh

Bangladesh Oil, Gas & Mineral Corp. (PETROBANGLA)
Chamber Bldg.
122-124, Motijheel Comm'l Area
Dhaka 1000, BANGLADESH
Ph: 880 2 53131, 53133
Telex: 642765
Cable: BOGMC, DHAKA

Minister of Energy & Mineral Resources
Dhaka, BANGLADESH

Benin

Ste. Nat. de Commercialisation des Produits Petrolieres (SONACOP)

B.P. 245
Cotonou, BENIN
Ph: 229 30-03-75
Fax: 229 30-11-38
Telex: 5173

Bolivia

Yacimientos Petroliferos Fiscales Bolivianos (YPFB)
Box 401
La Paz, BOLIVIA
Ph: 591 2 56541, 56540, 25289
Telex: YACIBOL BX2567, 3560006
Cable: YACIBOL

Brazil

Petroleo Brasileiro SA (PETROBRAS)
Caixa Postal 809
Avenida Republica do Chile 65
20031.170 Rio de Janeiro, RJ, BRAZIL
Ph: 55 21 534-1731, 534-1745, 534-4477
Fax: 55 21 262-2627, 262-0145, 262-0271
Telex: 038 212-3335 PETRA BR
Cable: PETROBRASRIO

Brunei

Director of Petroleum Unit
Petroleum Unit
Jabatan Perdana Menteri
Bandar Seri Begawan 2007
Negara BRUNEI Darussalam
Ph: 673 2 28932
Fax: 673 2 40270
Telex: BRUPET BU 2209

Burkina Faso

Burkina & Shell
(formerly Petrovolta)
Box 601
Ouagadougou, BURKINA FASO
Ph: 226 310071
Fax: 226 333936
Telex: 5929

C

Cameroon

Societe Nationale des Hydrocarbures (SNH)
Box 955
Yaounde, CAMEROON
Ph: 237 22-19-10
Cable: HYDROCAM 8514 KN

Chile

Empresa Nacional del Petroleo (ENAP)
Ahumada 341, 8 Piso
Box 3556
Santiago, CHILE
Ph: 56 2 638-1845
Fax: 56 2 638-0164
Telex: 034 240427, 240447 ENAP CL
Cable: ENAP 40447; 40427

China

Ministry of Petroleum Industry
P.O. Box 1411
Beijing
PEOPLE'S REPUBLIC OF CHINA

China National Offshore Oil Corp. (CNOOC)
Jia 1 Sidaokou Rd., Haidan
Box 9623
Beijing
PEOPLE'S REPUBLIC OF CHINA
Ph: 86 1 2014906
Telex: 22611; 222301 CNOOC CN
Cable: 3113

China National Petroleum Corp.
P.O. Box 766
Beijing
PEOPLE's REPUBLIC OF CHINA
Ph: 86 1 201-5544
Telex: 22312 PCPRC CN

China National Petroleum Corp.
One Westchase Center Suite 840
10777 Westheimer Rd.
Houston, Tex. 77042
Ph: 713 784-8598
Fax: 713 784-9197
Telex: 378-7801

Colombia

Empresa Colombiana del Petroles (ECOPETROL)
Carrera 13, No. 3624
Apartado Aereo 5938
Bogota, COLOMBIA
Ph: 57 1 285-6400
Fax: 57 1 285-1531
Telex: 035 44464, 42237, 44707 ECP CO
Cable: ECOPETROL

Ministerio de Minas y Energia de Colombia
Centro Administrative Nacional
Avenida El Dorado
Bogota, COLOMBIA
Ph: 57 1 44520, 44525
Telex: MINENERGIA

Congo

General Direction of Hydrocarbons
Ministry of Mines & Energy - S.P. 2120
Brazzaville, CONGO
Ph: 242 831281; 835873; 835974
Fax: 242 836243
Telex: 55 47 KG

Hydro Congo
B.P. 2008
Pointe Noire, Brazzaville, CONGO
Ph: 242 813560; 814023
Telex: 5215; 5220

Costa Rica

Refinadora Costarricense de Petroleo SA (RECOPE)
P.O. Box 4351-1000
San Jose, COSTA RICA
Ph: 506 33-96-11
Fax: 506 53-19-17
Telex: 0376 3381 RECOPE CR

Cote d'Ivoire

Director of the Office of the Minister—Ministere des Mines
Box V 50
Abidjan, COTE D'IVOIRE
Ph: 225 21 50 03
Fax: 225 21 53 20

Societe Nationale d'Operations Petrolieres de la Cote d'Ivoire (PETROCI)
Box V 194
Abidjan, COTE D'IVOIRE
Ph: 225 20-25-82
Fax: 225 21-68-24
Telex: 969 22135

Croatia

INA-NAFTAPLIN
Subiceva ulica 29
Zagreb, CROATIA
Ph: 385 41 418 011; 440 011; 440 022; 418 443
Telex: 22428

Cuba

Ministerio de la Industria Basica
Empedarado 113
Ciudad de la Habana, CUBA

Comercial Cupet SA (CUPET)
Calle 11 No. 511 E/D Yevedado
Ciudad Habana, CUBA
Ph: 53 7 62-40-11
Fax: 53 7 33-34-60, 33-80-27, 33-30-72
Telex: 511452 MMG CU

Czech Republic

Ministry of Economic Policy & Development
tr. SNG 65 - 101 60 Prague 10
CZECH REPUBLIC
Ph: 42 2 712-1111
Fax: 42 2 731-357
Telex: 122298

D

Denmark

Danish Energy Agency
11 Landemaerket
K-1119 Copenhagen, DENMARK
Ph: 45 3 392-6700
Fax: 45 3 311-4743
Telex: 22450

Dubai

Dubai Petroleum Co. (DUPETCO)
Box 2222
Dubai, UNITED ARAB EMIRATES
Ph: 971 4 442990

Telex: 45423
Cable: DUPETCO DUBAI

E

Ecuador

Petroleos del Ecuador (Petroecuador)
Av. Amazonas y La Nina
Edifico Torrealba 10° Piso
Box 5007 y 5008
Quito, ECUADOR
Ph: 593 2 55-06-11, 52-14-36
Fax: 593 2 56-97-38
Telex: 0308 22213, 22297 PETROEC ED
Cable: CEPE QUITO

Ministerio de Recursos Naturales y Energeticos
Direccion General de Geologia y Miras
Carrion 1016 y Paez
Casilla 23-A
Quito, ECUADOR
Ph: 593 2 231-204
Telex: 2271

Egypt

Egyptian General Petroleum Corp. (EGPC)
Box 2130
Cairo, EGYPT
Ph: 20 2 603899; 603968; 603087
Telex: 92049
Cable: PETMISR CAIRO

Ministry of Petroleum
2 Latin America St.
Garden City
Cairo, EGYPT
Ph: 20 2 25022, 25023
Telex: 92049 PETMISR UM

Ethiopia

Ethiopian Petroleum Corp.
Box 3375
Addis Ababa, ETHIOPIA
Ph: 251 1 51-32-88
Fax: 251 1 51-29-38
Telex: 21054
Cable: REFINERY ADDIS

Ministry of Mines, Energy & Water Resources
Box 486
Addis Ababa, ETHIOPIA
Ph: 251 1 44-82-50

F

Falkland Islands

British Geological Survey
19 Grange Terrace
Edinburgh EH9 2LF
UNITED KINGDOM
Ph: 44 31 667-1000
Fax: 44 31 668-4930

Fiji

Mineral Resources Department
Ministry of Lands, Energy & Mineral Resources
Private Mail Bag
Suva, FIJI
Ph: 679 381-611
Fax: 679 386-684
Telex: 2330
Cable: GEOLOGY

Finland

Ministry of Trade & Industry
00170 Helsinki 17, FINLAND
Ph: 358 0 1601
Fax: 358 0 160-3666
Telex: 124645

France

Service de Conservation des Gisements d'Hydrocarbures
Ministry of Industry
366 Avenue Napoleon Bonaparte
Rueil-Malmaison, FRANCE

G

Gabon

Petrogab
Box 564
Libreville, GABON

Ghana

Ghana National Petroleum Corp.
Private Mail Bag
Accra-North
Republic of GHANA
Ph: 233 21 318
Fax: 233 21 712916
Telex: 974 2188

Greece

Ministry for Industry, Energy & Technology
80 Michalakopoulos St.
Athens 611, GREECE
Ph: 30 1 770-5111
Telex: 21 5089, 21 6620

Guatemala

Ministerio de Energia y Minas
Directorate of Hydrocarbons
Apartado Postal 1421
01901 Guatemala City, GUATEMALA
Ph: 502 2 770901, 770903
Fax: 502 2 762044, 763175

Guinea

Office National des Hydrocarbures
Societe Guineane des Petroles (SOGUIP)
Conakry, GUINEA
Ph: 224 4 461256
Telex: 776

Guinea-Bissau

Empresa Nacional de Pesquisa e Exploracao Petroliferas e Mineras E P (PETROMINAS)
Caixa Postal 387
Bissau, GUINEA-BISSAU
Ph: 245 212389

H

Honduras

Direccion General de Minas e Hidrocarburos
Apartado Postal 981
Ministerio de Recursos Naturales
Boulevard a la Colonia Kennedy
Tegucigalpa, DC, HONDURAS
Ph: 504 32-78-48, 32-67-21, 32-84-55

Hungary

Hungarian National Oil & Gas Corp.
H-1502, P.O. Box 22
Xi, Schonherz Z. U. 18
1117 Budapest
HUNGARY
Ph: 36 1 166-2646
Fax: 36 1 186-8281
Telex: 22 4762 OKGT H

I

India

Oil & Natural Gas Commission (ONGC)
7th Floor, Bank of Baroda Bldg.
Parliament Street
New Delhi 110 001, INDIA
Ph: 91 11 331-7205, 371-5291
Fax: 91 11 331-6413
Telex: 031-65184, 031-66262

Director General of Hydrocarbons
Ministry of Petroleum & Natural Gas
2nd Floor, Shastri Bhavan
New Delhi 110 001, INDIA

Indonesia

Perusahaan Pertambangan Minyak Dan Gas Bumi Negara (PERTAMINA)
Jalan Merdeka Timur
Box 12
Jakarta 10110, INDONESIA
Ph: 62 21 3031
Fax: 62 21 343882; 363585
Telex: 796 44152, 44302, 46554, 46549

Iran

National Iranian Oil Co. (NIOC)
Ave. Takhte Jamshid

Box 1863
Tehran, IRAN
Ph: 98 21 6151
Telex: 2114
Cable: NAFTMELLI-TEHRAN

Petroleum Ministry of Islamic Republic of Iran
Box 1863
Tehran, IRAN
Ph: 98 21 661120
Fax: 98 21 666293
Telex: 215582 DVNI
Cable: NAFTMELLI

Iraq

Iraq National Oil Co. (INOC)
Al-Khullani Square
Box 476
Baghdad, IRAQ
Ph: 964 1 80066, 80069
Telex: 943 212204
Cable: PETROLNA

Ministry of Oil
Box 6178
Al-Mansour, Baghdad, IRAQ
Ph: 964 1 541-0031
Telex: 943 212216
Cable: NAFTIYAH BAGHDAD

Ireland

Department of Energy
Petroleum Affairs Division
Beggars Bush, Haddington Rd.
Dublin 4, IRELAND
Ph: 353 1 715233
Fax: 353 1 604462, 681782
Telex: 90870

Israel

Israel National Oil Co. Ltd.
Beit Gibor, 6 Kaufman St.
Tel-Aviv 68012, ISRAEL
P.O. Box 50199
Tel-Aviv 61500, ISRAEL
Ph: 972 3 514-2020
Fax: 972 3 514-2061
Telex: 341751
Cable: ISNEFT TEL AVIV

Ministry of Energy & Infrastructure
234 Yaffo St.
Box 1442/91013
Jerusalem, ISRAEL
Ph: 972 2 551211

Italy

Ufficio Nazionale Idrocarburi
Ministero dell Industria del Commercio e dell Artigianato
Via Veneto
33 Roma, ITALY

Ente Nazionale Idrocarburi (ENI)
1, Piazzale Enrico Mattei
00144 Rome, ITALY
Ph: 39 6 59002
Telex: 610082; 610086
Cable: ENIDRO-ROMA

J

Jamaica

Petroleum Corp. of Jamaica (PCJ)
PCJ Resource Center 5th Floor
36 Trafalgas Rd.
P.O. Box 579
Kingston 10, JAMAICA W.I.
Ph: 809 929-5380
Fax: 809 929-2400
Telex: 291 2356 PETCORP J

Japan

Japan National Oil Corp. (JNOC)
Fukoku-Seimei Bldg.
2-2 Uchisaiwaicho, 2-chome
Chiyoda-ku, Tokyo 100, JAPAN
Ph: 81 3 3580 5411
Fax: 81 3 3591 0172
Telex: 22166 J
Cable: SEKIYUKODAN TOKYO

Jordan

Natural Resources Authority
Government of Jordan
Box 7
Amman, JORDAN
Ph: 962 6 811 300, 811 306
Fax: 962 6 811 866
Telex: 925 21415
Cable: NRA JO

K

Kazakhstan

Ministry of Fuel & Energy
Bakenbay-Batyj St. 142
480091 Alma-Ata
KAZAKHSTAN
Ph: 327 2 62-64-10
Fax: 327 2 62-66-30

Kenya

Ministry of Energy & Regional Development
Nyayo House
Box 30582
Nairobi, KENYA
Ph: 254 2 33-88-18
Telex: 23094
Cable: MINIPOWER NAIROBI

National Oil Corp. of Kenya
National Bank Bldg.
Box 58567
Nairobi, KENYA
Ph: 254 2 340030
Telex: 23094
Cable: MINIPOWER NAIROBI

Kuwait

Kuwait Petroleum Corp. (KPC)
Box 26565
13126 Safat, KUWAIT
Ph: 965 245-5455
Fax: 965 246-7159
Telex: 44874/5/6/7/8
Cable: PETCORP KUWAIT

Ministry of Oil
P.O. Box 5077
Safat, KUWAIT
Telex: PETROL 959 22363KT

L

Latvia

Department of Geology
11. Novembra bulv. 31
226285 Riga, LATVIA
Ph: 0132 222-390

Liberia

Ministry of Lands, Mines & Energy
Box 9024
Monrovia, LIBERIA
Ph: 231 221488
Telex: 4532

Libya

General Department of Technical Affairs
Ministry of Petroleum
P.O. Box 256
Tripoli, LIBYA

National Oil Corp. (NOC)
Box 2655
Tripoli, LIBYA
Ph: 218 21 46181
Telex: 20270 LINAFT LY
Cable: LINAFT

M

Madagascar

Office Militaire National pour les Industries Strategiques (OMNIS)
21 Lalana Razanakombana
Box 1 Bis
Antananarivo, MADAGASCAR
Ph: 261 2 24283
Fax: 261 2 22985
Telex: 22370

Malaysia

Petroliam Nasional Berhad (PETRONAS)
Menara Dayabumi, Komplex Dayabumi
Jalan Sultan Hishamuddin
Box 12444
50778 Kuala Lumpur, MALAYSIA

Statoil towed out the Sleipner A platform from Stavanger in early June 1993 in preparation for startup of gas production as required by the Troll gas sales contract.

Ph: 60 3 274-3833; 274-8011; 274-8022
Fax: 60 3 274-0217
Telex: 31123 PETRON MA
Cable: PETRONAS

Malta

Government of Malta
Oil Exploration Division
Office of the Prime Minister
Castille
Valletta, MALTA
Ph: 356 237921
Fax: 356 248015
Telex: 1458 OED OPM

Mauritania

Direction des Mines et de la Geologie
Box 199
Nouakchott, MAURITANIA
Ph: 222 2 53225
Fax: 222 2 53225
Telex: 531 BMTN

Mexico

Petroleos Mexicanos (PEMEX)
Avenida Marina Nacional 329
Mexico 17, D.F., MEXICO
Ph: 52 5 250-2611, 254-2044
Fax: 52 5 625-4385, 533-0418
Telex: PMX GRP MEX 017-7-912
Cable: PEMEX

Morocco

Bureau de Recherches et de Participations Minieres Morocco (BRPM)
5-7 Carli Moulay Hassan
Box 99
Rabat, MOROCCO
Ph: 212 7 630-35
Fax: 212 7 76-24-10
Telex: 310-66

Office National de Reserches & Exploitations Petrolieres (ONAREP)
17 Av. Michlifen-Agdal
Rabat, MOROCCO
Ph: 212 7 77 57 12, 77 05 05
Fax: 212 7 77 58 66
Telex: OFREP 317-15M

Mozambique

Empresa Nacional de Hidrocarbonetos de Mocambique
Avenida Fernao de Magalhaes 34
P.O. Box 4787
Maputo, MOZAMBIQUE
Ph: 258 1 427191
Fax: 258 1 430181
Telex: 6-382
Cable: PETRO MO

Republic Popular de Mocambique
Ph: 258 1 429456, 429467
Fax: 258 1 30931
Telex: 6-478 ENH MO

Myanmar

Burma Ministry of Mines
Minister's Office
Yangon, MYANMAR
Cable: BURMIN

Myanma Oil & Gas Enterprise
Box 1049
Yangon, MYANMAR
Ph: 95 1 82266
Fax: 95 1 22964
Telex: 713 21307

N

Namibia

Ministry of Mines & Energy
Private Bag 13297
Windhoek, NAMIBIA
Ph: 264 61 226571
Fax: 264 61 38643
Telex: 50 908 487 WK

National Petroleum Corp. of Namibia Pty. Ltd. (NAMCOR)
Box 13196
Windhoek, NAMIBIA
Ph: 264 61 221699, 225723
Fax: 264 61 221785

Nepal

Petroleum Exploration Project (former Department of Mines & Geology)
Lainchaur
Kathmandu, NEPAL
Ph: 977 413541; 414740
Fax: 977 411783
Telex: 2320

Nepal Oil Corp. Ltd.
Babar Mahal
Box 1140
Kathmandu, NEPAL
Ph: 977 214600
Fax: 977 226799
Telex: 2246 NEANI NP
Cable: NEANI

Netherlands

Ministry of Economic Affairs
General Energy Policy Division
Postbus 20101
2500 The Hague
NETHERLANDS
Ph: 31 70 79 89 11
Fax: 31 70 79 63 58

New Zealand

Ministry of Energy
Petroleum Unit
Box 2337
Wellington, NEW ZEALAND
Ph: 64 4 733-805
Fax: 64 4 739-930
Telex: ENMIN NZ 31488

Nicaragua

Direccion General de Hydrocarburos
P.O. Box 3104
Managua, NICARAGUA
Ph: 505 2 27817
Fax: 505 2 27323

Empresa Nicaraguense del Petroleo (PETRONIC)
Iglesia Santo Domingo 2 Cuadras Abajo
Escombros de Managua
P.O. Box 3104
Managua, NICARAGUA
Ph: 505 2 27-140, 27-156, 27-168, 27-513
Fax: 505 2 62-10-84, 62-10-49, 27-244
Telex: 0375 1278 CABLE PETRONIC

Niger

Ministere des Mines et de l'Energie
Direction de l'Energie
Naimey, NIGER
Ph: 227 73-55-21
Telex: 5418 Minimi

Ste. Nigerienne de Produits Petroliers (SONIDEP)
Box 1735 & 11702
Naimey, NIGER
Ph: 227 733334
Telex: 5343

Nigeria

Ministry of Petroleum Resources
44, Eric Moore, Surulere
P.M.B. 12701
Lagos, NIGERIA
Ph: 234 1 603100
Telex: 961 21126

Nigerian National Petroleum Corp. (NNPC)
Falomo Office Complex
Private Mail Bag 12701 (Ikoyi)
Lagos, NIGERIA
Ph: 234 1 603100/28
Telex: 21610
Cable: NAPETCO

Norway

Den Norske Stats Oljeselskap AS (STATOIL)
Box 300
N-4001 Stavanger, NORWAY
Ph: 47 4 80-80-80
Fax: 47 4 80-70-42
Telex: 73600 STAST N
Cable: STATOIL

Royal Ministry of Petroleum & Energy
Ploensgt 8
Box 8148 Dep. 0033
Oslo 1, NORWAY
Ph: 47 2 34-90-90
Fax: 47 2 34-95-65
Telex: 21486 OEDEP N

O

Oman

Ministry of Petroleum & Minerals
Sultanate of Oman
Box 551
Muscat, OMAN
Ph: 968 603333
Telex: 5280 ON

Petroleum Development Oman (PDO)
Box 81
Mina Al-Fahal, OMAN
Ph: 968 678111
Fax: 968 677106
Telex: 5212
Cable: PETRO MUSCAT

P

Pakistan

Ministry of Petroleum & Natural Resources
19-A, Pak Plaza, Jinnah Ave.
Islamabad, PAKISTAN
Ph: 92 51 824 993, 818 352
Fax: 92 51 213 245
Telex: 54089 TWPET

Oil & Gas Development Corp.
H-14 Almarkaz F-8
Islamabad, PAKISTAN
Ph: 92 51 851945
Fax: 92 51 858939
Telex: 5692 OGDC PK
Cable: OILGASCORP

Papua New Guinea

Papua New Guinea Department of Minerals & Energy (Mines Division)
Box 352
Konedobu, Port Moresby
PAPUA NEW GUINEA
Ph: 675 214011
Telex: 22211

Paraguay

Petroleos Paraguayos (PETROPAR)
Oliva 299 Casi Chile
Edificio Finamericana 3RO y 4TO Piso
Casilla Correo 571
Asuncion, PARAGUAY
Ph: 595 21 44-85-03
Fax: 595 21 49-62-32
Telex: 0305 25025 PY

Peru

Ministerio de Energia & Minas y Direccion General de Hidrocarburos
Esq. Avenida Javier Prado Este y Avenida Aviacion
Edifico Sol Gas, 5 Piso
Lima 34, PERU
Ph: 51 14 35-2996, 36-6236
Telex: 25731 PU MEM

Petroleos del Peru SA (PETROPERU)
Paseo de la Republica 3361, San Isidro
Box 3126 & 1081
Lima, PERU
Ph: 51 14 41-50-00, 41-44-02, 42-50-00
Fax: 51 14 40-07-47, 41-14-24
Telex: 036 20303, 25592 PETPERU
Cable: PETROPERU

Philippines

Ministry of Energy
7901 Makati Ave.
Box 1031 MCC
Makati, Metro Manila, PHILIPPINES
Ph: 63 2 85-88-56/9
Telex: 22259; 63667; 5270
Cable: PHILNOCO MANILA

Philippine National Oil Co.
PNOC Bldg.
7901 Makati Ave.
Makati, Metro Manila 1200, PHILIPPINES
Ph: 63 2 85-90-61; 85-88-56; 88-75-51

Fax: 63 2 815-3094; 810-6728
Telex: 22259 PNO PH
Cable: PHILNOCO MANILA

Poland

Bureau of Geological Concessions
Ministry of Environmental Protection, Natural Resources, and Forestry
Wawelska 52/54
00-922 Warsaw, POLAND
Ph: 48 22 254393
Fax: 48 22 253972
Telex: 817857 PL

Polish Oil & Gas Co. (PGNiG)
ul. Krucza 36
00-921 Warsaw, POLAND
Ph: 48 22 281642
Fax: 48 22 290856
Telex: 813466 PL

Portugal

Petroleos de Portugal EP (PETROGAL)
Rua das Flores No. 7
Box 2539
1200 Lisboa, Codex, PORTUGAL
Ph: 351 1 328035; 370371
Telex: 12521
Cable: PETROGAL, LISBON

Gabinete Para a Pesquisa e Exploracao de Petroleo (GPEP)
Rua Vale do Pereiro, 4
1200 Lisbon, PORTUGAL
Ph: 351 1 388 3541
Fax: 351 1 388 2775

Q

Qatar

Ministry of Finance & Petroleum
Department of Petroleum Affairs
Box 2233
Doha, QATAR
Ph: 974 414123; 413571
Telex: 957 4891
Cable: PETROFAIRS DOHA

Qatar General Petroleum Corp. (QGPC)
Box 3212
Doha, QATAR
Ph: 974 491491
Fax: 974 831125
Telex: 4343
Cable: PETROL DOHA

R

Romania

Rompetrol
109 Calea Victoriei
70176 Bucharest, ROMANIA
Ph: 40 0 59 43 25
Fax: 40 0 12 06 87
Telex: 10155

Russia

Ministry of the Gas Industry
Stroitelei St. 8
Moscow 117311
RUSSIA

Ministry of the Petroleum Industry
Naberezhnaya Morisa
Toreza 26/1
Moscow 113035
RUSSIA

S

Saudi Arabia

Saudi Arabian Oil Co. (SAUDI ARAMCO)
Administration Bldg. Room 2220
Box 5000
Dhahran 31311, SAUDI ARABIA
Ph: 966 3875-6110
Fax: 966 3873-8190
Telex: 801220 A SAO SJ
Cable: ARAMCO DAMMAM

Senegal

Ministere du Developpement Industriel et de L'artisanat Societe des Petroles de Senegal (PETROSEN)
B.P. 2076, 56/58 Ave. Faidherbe
Dakar, SENEGAL
Ph: 221 22 04 44, 22 05 95
Fax: 221 22 83 40
Telex: (0906) 61 249 PETRO SG

Seychelles

Seychelles National Oil Co. Ltd. (SNOC)
Box 230
Victoria, Mahe, SEYCHELLES
Ph: 248 222282, 222283
Fax: 248 225177
Telex: 2342

Sharjah

The Director
Petroleum & Mineral Department
P.O. Box 188
Sharjah, UNITED ARAB EMIRATES

Somalia

Ministry of Mineral Resources & Water (SOMALIA)
Government of Somalia
Box 774
Mogadishu, Southern Region, SOMALIA
Telex: 639

National Petroleum Agency
Box 573
Mogadishu, Southern Region, SOMALIA
Ph: 252 20747, 20561
Telex: 605

South Africa

Soekor (Pty.) Ltd.
P.O. Box 307
Parow 7500
Cape Province
SOUTH AFRICA
Ph: 27 21 938 3911
Fax: 27 21 938 3144
Telex: 555307
Cable: SOEKOR

South Korea

Korea Petroleum Development Corp.
Songam Bldg.
84 Nonhyun-Dong
Gangnam-Gu
Seoul
KOREA
Ph: 82 2 547-5321/9
Fax: 82 2 540-0880, 546-4776
Telex: PEDCO K24593

Spain

Jefe del Servicio de Hidrocarburos de la Direccion General de la Energia del Ministerio de Industria y Energia
Madrid 16, SPAIN

Hispanica de Petroleos SA (HISPANOIL)
Box 14355
Madrid 30, SPAIN
Ph: 34 1 274-72-00; 274-97-00
Telex: 22114

Sri Lanka

Ceylon Petroleum Corp.
113 Galle Rd.
Colombo 3, SRI LANKA
Ph: 94 1 25231
Telex: 21167, 21235
Cable: CEYPETCO

Sudan

General Petroleum Corp. of Sudan
Khartoum, SUDAN

Suriname

Staatsolie Maatschappij Suriname NV
Industrieterrein 21, Flora
P.O. Box 4069
Paramaribo, SURINAME
Ph: 597 49-96-49
Fax: 597 49-11-05, 48-08-11
Telex: 217 SUROIL SN
Cable: 597 91105

Syria

Ministry of Petroleum & Mineral Resources
Government of Syria
Jadet Al Katib
Damascus, SYRIAN ARAB REPUBLIC

Syrian Petroleum Co.
Box 2849
Damascus, SYRIAN ARAB REPUBLIC
Ph: 963 11 228298; 228271
Telex: 924 411031 SY; 411726
Cable: PETSYRIA

T

Taiwan

Chinese Petroleum Corp.
83 Chung Hwa Road
Taipei 10031, TAIWAN
Ph: (02) 3610221; 3717121
Fax: 886 2 3319645
Telex: TP 11215
Cable: CHINOL TAIPEI

Thailand

Department of Mineral Resources
Ministry of Industry
Rama VI Rd.
Bangkok 10400, THAILAND
Ph: 66 2 511-0215; 511-0749
Fax; 66 2 245-9855
Telex: 84763
Cable: MINDEPT BANGKOK THAILAND

Petroleum Authority of Thailand (PTT)
14 Vighavadi Rangsit Rd.
Bangkok 10900, THAILAND
Ph: 66 2 279-5010/9; 279-5130/9
Telex: 87940
Cable: PTT BANGKOK

Trinidad & Tobago

Ministry of Energy & State Enterprises
4th Floor, Salvatori Bldg.
Box 96
Port of Spain, TRINIDAD
Ph: 809 62-38841-6

Trinidad & Tobago Oil Co. Ltd. (TRINTOC)
Box 601
5th Floor, Salvatori Bldg.
Port of Spain, TRINIDAD
Ph: 809 658-4200, 648-4354
Fax: 809 658-1213, 658-1315
Telex: 0294 39366, 39367 TTOCPAP WG

Tunisia

Entreprise Tunisienne d'Activites Petrolieres (ETAP)
27 Bis, Ave. Kherredine Pacha
Tunis 1002 de Belvedere, TUNISIA
Ph: 782288; 783233
Fax: 216 1 784092
Telex: 15128; 13877
Cable: ETAPETROLE

Turkey

General Directorate of Mineral Research & Exploration (MTA)
MTA, Ankara, TURKEY
Ph: 90 41 287-34-30
Fax: 90 41 222-28-78
Telex: 42-721
Cable: METEA ANKARA

Turkiye Petrolleri Anonim Ortakligi (TPAO)
Mustafa Kemal Mahallesi
2. Cadde No. 86
06520 Ankara, TURKEY
Ph: 90 41 286 9100
Fax: 90 41 286 9000, 9001
Telex: 42426 TPAO TR

U

Uganda

Uganda Petroleum Co. Ltd.
Box 2133
Kampala, UGANDA
Ph: 256 41 258516, 257066, 241849, 241807
Telex: 61299

Ukraine

Counsel of Ministries
Vladimirskaya St. 34
252003 Kiev, UKRAINE
Ph: 044 226-2007
Telex: 131419 TOPAZ

United Kingdom

U.K. Department of Energy
1 Palace St.
London SW1E 5HE
UNITED KINGDOM
Ph: 44 1 238-3000
Fax: 44 1 834-03771
Telex: 918777

Uruguay

Administracion Nacional de Combustibles, Alcohol y Portland (ANCAP)
Casilla 1090
Montevideo, URUGUAY
Ph: 598 2 92-06-08, 98-95-04, 98-93-55
Fax: 598 2 92-11-36
Telex: ANCAP UY 23168

Asistencia Reciproca Petrolera Empresa Latinoamericana (ARPEL)
Javier de Viana 2345
Casilla de Correo 1006
11200 Montevideo, URUGUAY
Ph: 598 2 40-69-93, 40-74-54
Fax: 598 2 23-70-23
Telex: ARPEL UY 22560

Uzbekistan

Ministry of Foreign Economic Relations
Buyuk Ipak Iuli 75
700077 Tashkent, UZBEKISTAN
Ph: 371 2 68-76-31
Fax: 371 2 68-72-31, 68-74-77

Uzneftegazdobycha
Amir Temir Ulitsa 66
700084 Tashkent, UZBEKISTAN
Ph: 371 2 35-83-23
Fax: 371 2 35-83-37

V

Venezuela

Ministerio de Energia y Minas
Edificio Parque Central
Caracas, VENEZUELA
Ph: 58 2 483-2033, 483-4033, 483-2317

Petroleos de Venezuela SA (PDVSA)
Avenida Libertador, Torre Este, La Campina
Apartado 169
Caracas 1010-A, VENEZUELA
Ph: 58 2 708-4111
Fax: 58 2 708-4461, 708-4420
Telex: 031 23489, 21890 PDVSA VE

Viet Nam

Ministry of Petroleum & Gas
193/6A Nam Ky Khoi Nghia
Ho Chi Minh City 3, VIET NAM

Vietnam Oil & Gas Co. (PETROVIETNAM)
Oil & Gas General Dept.
80 Nguyen Du Ha Noi
Hanoi, VIET NAM
Ph: 84 4 252526, 252357, 252910
Fax: 84 4 265942
Telex: 412280 PVN-VT

Y

Yemen

Ministry of Oil & Mineral Resources
REPUBLIC OF YEMEN
Sanaa, YEMEN
Ph: 967 2 52 268
Fax: 967 2 52 253

Yemen Petroleum Co.
YPC Bldg.
Box 3360
Al Hudaydah
YEMEN

Z

Zaire

Enterprise Petroliere du Zaire
B.P. 7617
Kinshasa 1, ZAIRE
Ph: 243 12 25 356; 24 416; 23 445
Telex: 21066

INDEX

ADVERTISERS' INDEX INTERNATIONAL PETROLEUM ENCYCLOPEDIA 1994